PHYSICAL CHEMISTRY CALCULATIONS

with

Excel
Visual Basic
Visual Basic for Applications
Mathcad
Mathmatica

RODNEY J. SIME

San Francisco Boston New York
Cape Town Hong Kong London Madrid Mexico City
Montreal Munich Paris Singapore Sydney Tokyo Toronto

Acquisitions Editor: Jim Smith
Editorial Assistant: Cinnamon Hearst
Managing Editor, Production: Erin Gregg
Production Supervisor: Jane Brundage
Marketing Manager: Scott Dustan
Manufacturing Buyer: Mike Early

ISBN 0-8053-3089-5

2 3 4 5 6 7 8—CRK—08 07 06 05

www.aw-bc.com

For my wife, Ruth

Table of Contents

Part I. Spreadsheets

Chapter 1. Introduction to Spreadsheets **1**
History **1**
The Microsoft Excel Workspace Layout **2**
Functions **5**
Excel Example 1: X-Ray Powder Diffraction:
Tabular Calculations and Statistics **6**
Short Review of the Science **7**
Preparing the Worksheet **8**
Cell References **9**
Using Some of Excel's Statistical Functions **10**
Defining and Using the Name of a Range **11**
The Geometry of Crystals and Diffraction **12**
The Diffraction of X-Rays **16**
Excel Example 2: Student Grade Sheet and Histogram **17**
Preparing the Worksheet **17**
Sorting **19**
Creating the Histogram **20**

Chapter 2. Thermodynamics **23**

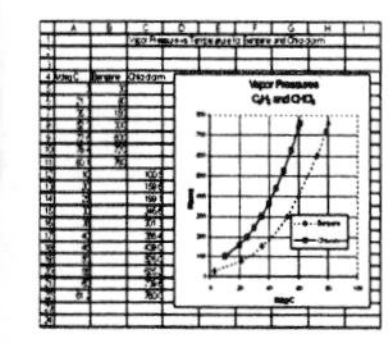

Excel Example 3: A Single Point-Plot: Heat Capacity of Silver **23**
Review: The First Law of Thermodynamics **23**
Internal Energy **24**
Enthalpy **24**
Heat Capacity **25**
Preparing the Worksheet **25**
Creating the Chart **26**
Chart Wizard – Step 1 of 4 – Chart Type **27**
Chart Wizard – Step 1 of 4 – Chart Source Data **27**
Chart Wizard – Step 1 of 4 – Chart Options **27**
Chart Wizard – Step 1 of 4 – Chart Location **27**
Exploring the Chart **27**
Size and Position of the Chart on the Worksheet **28**
Editing the Title **29**
Improving the Plot Area **30**
Customizing the X-Axis **30**
Customizing the Y-Axis **31**
Relocating the Chart **31**
Excel Example 4: Calculation of Vapor Pressures with the
Antoine Equation: Plotting a Function **32**
Review of Vapor Pressures **32**
Excel Calculation of the Vapor Pressure of Acetone **34**
Creating the Vapor-Pressure Chart **35**
Excel Example 5: Multiple Point-Plots with Nonidentical *x* Values:
Vapor Pressure of Benzene and Chloroform **36**
Preparing the Worksheet Data **37**
Creating the Chart **37**
Editing the Chart **39**

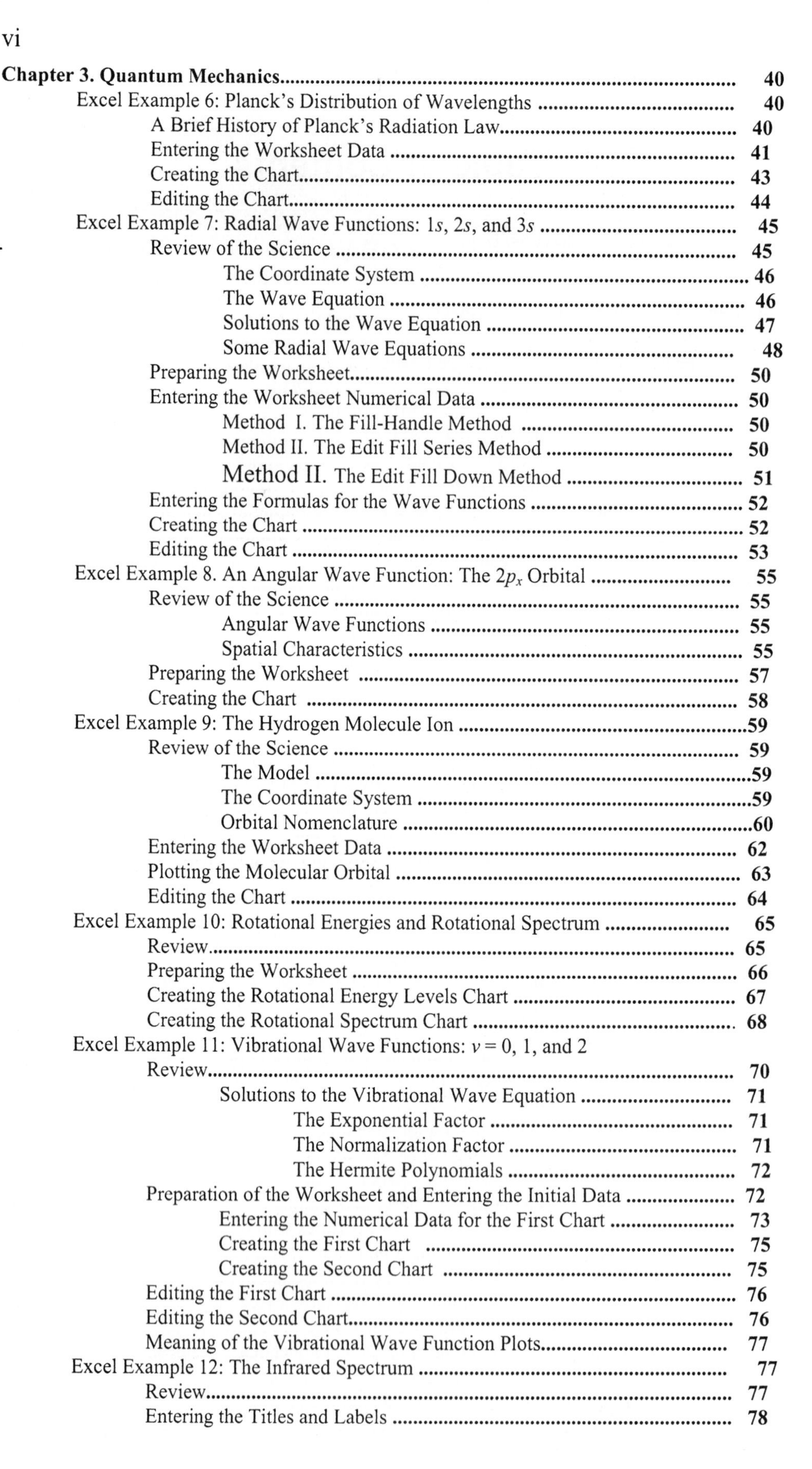

Chapter 3. Quantum Mechanics **40**
Excel Example 6: Planck's Distribution of Wavelengths **40**
A Brief History of Planck's Radiation Law **40**
Entering the Worksheet Data **41**
Creating the Chart **43**
Editing the Chart **44**
Excel Example 7: Radial Wave Functions: 1*s*, 2*s*, and 3*s* **45**
Review of the Science **45**
The Coordinate System **46**
The Wave Equation **46**
Solutions to the Wave Equation **47**
Some Radial Wave Equations **48**
Preparing the Worksheet **50**
Entering the Worksheet Numerical Data **50**
Method I. The Fill-Handle Method **50**
Method II. The Edit Fill Series Method **50**
Method II. The Edit Fill Down Method **51**
Entering the Formulas for the Wave Functions **52**
Creating the Chart **52**
Editing the Chart **53**
Excel Example 8. An Angular Wave Function: The $2p_x$ Orbital **55**
Review of the Science **55**
Angular Wave Functions **55**
Spatial Characteristics **55**
Preparing the Worksheet **57**
Creating the Chart **58**
Excel Example 9: The Hydrogen Molecule Ion **59**
Review of the Science **59**
The Model **59**
The Coordinate System **59**
Orbital Nomenclature **60**
Entering the Worksheet Data **62**
Plotting the Molecular Orbital **63**
Editing the Chart **64**
Excel Example 10: Rotational Energies and Rotational Spectrum **65**
Review **65**
Preparing the Worksheet **66**
Creating the Rotational Energy Levels Chart **67**
Creating the Rotational Spectrum Chart **68**
Excel Example 11: Vibrational Wave Functions: $v = 0$, 1, and 2
Review **70**
Solutions to the Vibrational Wave Equation **71**
The Exponential Factor **71**
The Normalization Factor **71**
The Hermite Polynomials **72**
Preparation of the Worksheet and Entering the Initial Data **72**
Entering the Numerical Data for the First Chart **73**
Creating the First Chart **75**
Creating the Second Chart **75**
Editing the First Chart **76**
Editing the Second Chart **76**
Meaning of the Vibrational Wave Function Plots **77**
Excel Example 12: The Infrared Spectrum **77**
Review **77**
Entering the Titles and Labels **78**

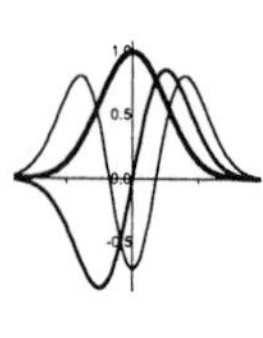

Creating the Infrared Spectrum of HCl Chart.......... 80
Editing the Infrared Spectrum of HCl Chart.......... 81
Excel Example 13. The Emission Spectrum and DesLandres Table.......... 81
Vibrational Levels in Electronic States 81
The DesLandres Table.......... 82
Preparing the Worksheet for the DesLandres Table.......... 83
Discussion.......... 84
Progressions 84
Sequences 84

Chapter 4. Statistical Thermodynamics **86**

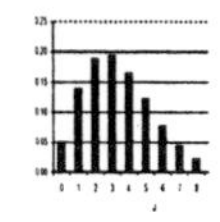

Excel Example 14: Calculation of the Vibrational Heat Capacities of I_2, CL_2, and N_2 86
Review of the Science.......... 86
Molecular Energies.......... 86
Energy and the Partition Functions.......... 88
Translational and Rotational Contributions to the Energy.......... 91
The Vibrational Contribution to the Energy.......... 92
The Vibrational Heat Capacity of a Gas.......... 92
Preparation of the Worksheet 93
Creating the Chart.......... 95
Editing the Chart 96
Meaning of the Chart of Heat Capacities 97
Excel Example 15: Population of Rotational Energy Levels 97
Review 97
The Rotational Energy.......... 97
The Rotational Partition Function 98
The Boltzmann Distribution 98
Creating the Worksheet 99
Calculating the Rotational Partition Function 99
Entering the Data for the Chart 99
Plotting the Population Distribution as a Bar Graph 99
Editing the Plot Area and Chart Area 100
The Total Population 100
Discussion 100

Chapter 5. Gases **102**

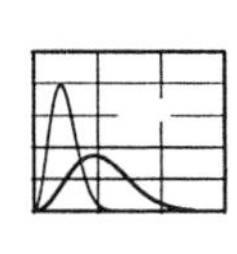

Excel Example 16: The Lennard-Jones Potential 102
Review 102
Creating the Worksheet 102
Creating the Chart 103
Editing the Chart 104
Excel Example 17: The van der Waals Equation of State 105
Review 105
Entering the Titles, Constants, and Their Labels 106
Creating the Chart 107
Editing the Chart 108
Excel Example 18: Velocity and Speed Distribution Functons 109
Review 109
One-Dimensional Velocity 109
Entering the Constants and Labels for the Velocity Distribution 112
Creating the Chart 113
Editing the Chart 113
Three-Dimensional Speed Distribution 114

Entering the Constants and Labels for the Speed Distribution 116
Creating the Chart 117
Editing the Chart 118
The Most Probable Speed 119
Average Quantities 120
Average Speed 120
The Root-Mean-Square (RMS) Velocity 121

Chapter 6. Kinetics **122**
Excel Example 19: Kinetics: Consecutive Rate Processes: A δ B δ C **122**

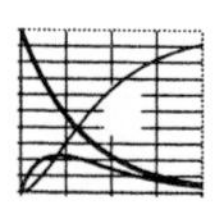

Review of the Science **122**
Entering the Title, Labels, and Constants **123**
Creating the Chart **124**
Editing the Chart **125**
Results **126**
Excel Example 20: Linear Regression and the Trendline: Enzyme Kinetics **126**
Review of the Science **126**
The Michaelis and Menten Mechanism **126**
Determination of V_{max} and K_m **130**
The Lineweaver-Burk Plot **130**
The Eadie-Hofstee Plot **130**
Entering the Worksheet Data **130**
Creating the Chart and Adding the Trendline **131**
Editing the Chart **133**
Using the Linear Regression Data **133**

Chapter 7. Statistics **134**
Excel Example 21: Linear Regression with Data Analysis:
The Hydrogen Sulfide Equilibrium **134**
Review **134**
Entering the Data **135**
Creating the Chart **137**
Adding the Trendline **138**
Editing the Chart **139**
Using the Linear Regression Data **139**
Related Excel Regression Functions **140**
The Slope Function **140**
The Intercept Function **140**
The Linest Function **140**
Excel Example 22: Multiple Linear Regression: *emf* versus *T* **143**
Review of the Science **143**
Preparing the Worksheet **144**
Creating the Chart **144**
Editing the Chart **144**
Plotting the Linear Regression with Trendlines **146**
Calculating the Thermodynamic Functions **146**
Excel Example 23: Nonlinear Regression: A Calibration Curve **148**
Entering the Worksheet Title and Data **148**
Calculating a Linear Regression **149**
Calculating a Nonlinear Regression **149**
Adding a Polynomial (Nonlinear) Trendline **150**
Editing the Chart **150**
Discussion **152**
Excel Example 24: Statistics: Errors of Measurement **153**
Review **153**

The Normal Error Distribution 153
The Standard Deviation 154
Confidence Levels and Confidence Factors 155
The Standard Deviation of the Mean 156
Excel's TINV() Statistical Function: Student's *t*-Distribution 157
Creating a Table of Critical Values of *t* 159
Filling the Table with Degrees of Freedom 159
Filling in the Student's *t* Values 159

Chapter 8. Three-Dimensional Plots **161**

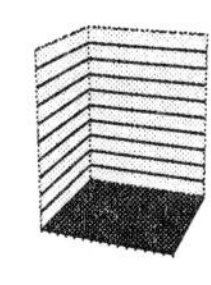

Excel Example 25. Connected Points in Space 161
Entering the Worksheet Title and Data 161
Creating the 3-D Chart 162
Excel Example 26. Three-Dimensional Charts: Wave Functions 164
Review of Quantum Mechanics 164
Operatior Equations 164
The Postulates of Quantum Mechanics 165
The Particle in a Two-dimensional Box: ψ_{22} and ψ_{22}^2 166
The System 166
Constructing the Hamiltonian Operators 166
Solving the Schrödinger Wave Equation 168
Normalizing the Wave Function 169
Entering the Worksheet Title and Data 169
Creating the 3-D Chart 171
Discussion 174

Chapter 9. Lotus 1-2-3 and Quattro Pro **175**
Lotus 1-2-3 Introduction 175
Lotus 1-2-3 Example 1: Linear Regression and Enzyme Kinetics 177
Entering the Experimental Data 179
Creating the Chart 179
Editing the Chart 179
Lineweaver-Burk Plot 182
Creating the Linear Regression Line with Lotus 1-2-3 183
Eadie-Hofstee Plot 183
Lotus 1-2-3 Example 2: Plotting an Angular Wave Function 184
Entering the Data 184
Calculating the p_x, x, and y Data for the plot 186
Creating the Chart 187
Editing the Chart 187
Lotus 1-2-3 Example 3: The Hydrogen Molecule Ion 188
Entering Data into the Lotus 1-2-3 Spreadsheet 188
Entering the Incremented Value of *r* 189
Method I. Fill Arrows 189
Method II. Range Fill 190
Plotting the Molecular Orbital 190
Editing the Chart 190
Quattro Pro Introduction 191
Quattro Pro Example 1: Linear Regression: The H_2S Equilibrium 192
Entering the Formulas and Numerical Data 193
Creating the Chart for the H_2S Equilibrium Data 194
Expert – Step 1 of 4. Select Data 194
Expert – Step 2 of 4. Select Category and Type 194
Expert – Step 3 of 4. Enter Titles and Labels 195

Expert – Step 4 of 4. Choose a Color Scheme 195
Editing the Chart 196
Options for Editing the Chart 199
Creating the Linear Regression Line 200
Editing the Hydrogen Sulfide Equilibrium Chart 200
Data Analysis 203
Comparison of Linear Regression in Excel, Lotus, and Quattro Pro.......... 205
Quattro Pro Example 2: Vibrational Wave Functions 205
Entering the Spreadsheet Data 205
Entering Values with the Edit Fill Series Method 206
Entering the Formulas with the Edit Copy Cell Method 207
Creating the Chart of Unsquared Wave Functions 207
Editing the Chart 209
Creating the Chart of Squared Wave Functions 210

Part II. Visual Basic

Chapter 10. Visual Basic—Introduction 211
The Nature of a Windows Program 212
What the User Sees 212
What the User Does 212

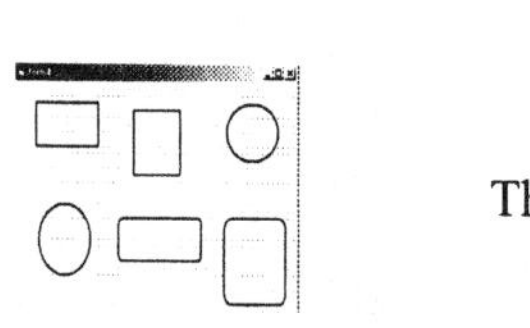

What the Programmer Does 213
The Visual Basic Programming Environment 214
The Design Window 214
The Project Window 214
The Form Window 215
The Tool box 216
The Properties Window 216
The Code Window 216
The Command Button Control.......... 217
The Click() Event 217
The Structure of a Procedure 218
The Shape Control 218
Object Browser 219
The Color Properties Program 220
Starting the Project 220
Creating the User Interface 221
Setting the Properties of the Controls 223
Naming Conventions for Standard Visual Basic Controls 223
Entering the Code for the Color Properties Program 228
Automatic Code Completion 229
The Procedure List 230
The Events List 231
Running the Color Properties Program 232
Creating an Executable File 232
Understanding the Color Properties Program 233
Summary and Review of the Color Properties Program 234

Chapter 11. Visual Basic Controls: Properties, Methods, and Events **236**
The Text Box Control **236**
The Label Control **237**
The Justify Program **237**
Entering the Code of the Justify Program **238**
Executing the Justify Program **240**
Understanding the Code of the Justify Program **240**
The Text Box **241**
The Command Buttons **241**
Set Focus Statement **242**
Tab Index **242**
Variables and Data **243**
Data Types **243**
Choosing Data Types **244**
Choosing Variable Names **244**
Type Prefixes for Names **244**
The Assignment Operator **247**
The Form **248**
Printing to the Form **249**
Printing to the Printer **250**
The Print Demo Program **250**
Understanding the Print Demo Program **253**
Separators for the Print Method **254**
Printing and Formatting Numbers **255**
Printing and Formatting Dates and Time **257**
The Printer, Its Methods and Properties **258**
Option Explicit and Declaring Variables **258**
Programs OptExp1, OptExp2, OptImp1, and OptImp2 .... **258**

Chapter 12. Visual Basic Operators, Control Structures, and Functions **263**
Relational Operators **263**
Conditional Operators **263**
The Boolean Operator Program **265**
Control Structures **268**
The IfThenElse Program **270**
The Random Guess Program **271**
The Rnd Function **271**
The Randomize Statement **271**
The Timer Function **272**
The MsgBox Statement **272**
The Scrollbar Control **278**
The pH Program **279**
The ColorBox Program **283**
The List Box Control **288**
The Select Case Decision Statement **288**
The List Box Program **289**
The List Box Program, Version 2 .... **292**
Arithmetic Operators **292**
The Arithmetic Program **294**
Creating a Menu Control **296**
Getting the Menu Editor **296**
Using the Menu Editor **296**
Conversion Functions **302**
The VarType Function **302**
Handling Errors **303**
The Options Button Control **305**

The Functions Program ... 306
Mathematical Function ... 306
Previously Used Functions ... 310
The MsgBox Function ... 311
Appearance of the MsgBox ... 311
The Values Returned by the Message Box Function ... 312

Chapter 13. Visual Basic Loops without Arrays ... 313
Unconditional Looping ... 313
Conditional Looping ... 314
The Loops Program ... 315
The Nested Loops Program ... 318
The HKL Program ... 320
The ASCII Character Codes ... 322
The ASCII Print Program ... 322
The Keypress Event ... 326
The KeyPress Program ... 327
The Object Browser ... 329
The Timer Control ... 329
The Timer A to Z Program ... 330
The Rotational Energy Program ... 334
The AddItem Method ... 338
User-Defined Functions ... 339
The Vapor Pressure Program ... 340

Chapter 14. Visual Basic Loops with Arrays ... 346
Arrays ... 346
The Array of Pressures Program ... 347
The IR Spectrum Program ... 350
The Heat Capacity Program ... 354
Boxes Again: The Combo Box, the Text Box, and the List Box ... 359
The Text Box to List Box Program ... 360
The Combo Box Control ... 362
Properties and Methods of the Text Box, List Box and Combo Box ... 363
The Combo Box to Array Program ... 363
The Trapezoid Integration Program ... 370
The Two-D Array Program ... 376
The TotoLoto Program ... 380
The Control array ... 381
The Control Array Dialog Box ... 381
The Bubble Sort Program ... 383
The DesLandres Program ... 386

Chapter 15. Visual Basic Files ... 392
Opening, Closing, and Deleting a File ... 392
Opening a File for Output ... 392
Opening a File for Append ... 393
Opening a File for Input ... 393
Closing Files ... 393
Deleting Files ... 394
Writing to a File ... 394
The Write to File Program ... 395
The Write to File 2 Program ... 397

The Write to File 3 Program 398
The Write to a File from a List Program 400
The Write to a File from a Table Program 402
Reading from Files 404
The Read a File of Names Program 404
Creating a File for the Program to Read 404
The Read X_iY_i File Program 407
The Common Dialog Boxes Control 410
The Show the Boxes Program 412
The CD Color Program 414
The CD Printer Program 417
The CD Fonts Program 419
The CD Open File Program 421

Chapter 16. Visual Basic Graphics **426**
Objects and Controls for Drawing and Graphics 426
Graphic Formats Supported by the Form, Picture Box, and Image Controls 427

The Image Control 427
The Image Program 427
The Picture Box Control and the Line Control 429
The Picture Box Program 429
The Control Array 431
Properties for Drawing and Graphics 432
The Coordinate System 433
Scale Mode 434
Properties Relating to the Coordinate System 434
The Coordinate Demo Program 435
Methods for Drawing and Graphics 437
DrawWidth and DrawStyle 438
Settings and Constants for the DrawStyle Property 438
The Line and Box Program 439
The Circle and Scale Methods 441
The Graph Circle Program 442
Units and Scaling 443
FillColor and FillStyle 444
Drawing Arcs 444
The Resize the Form Program 445
Graphing with the Pset Method 445
The $r2R2$ Program 446
The $2p_x$ Orbital Program 448
The Graph Vib Program: Vibrational Wave Functions 450
The MouseMove Event 452
The Sketch Program 453
The Random Points Program 454
Printing Graphic Methods to the Printer 455
The Circle to Printer Program 456
Methods of the Form and of the Printer 456
The Coordinate System 456
The Pset to Printer Program 457
The NoForm Program 458

Part III. Visual Basic for Applications

Chapter 17. Visual Basic for Applications: Microsoft Word **462**
The Relationship Between VB and VBA **462**
Macros **463**
Word VBA Example 1: The Ariel10B Macro **463**
Recording the Arial10B Macro........ **463**
The Record Macro Dialog Box **464**
The Stop Recording Toolbar **465**
Running the Arial10B Macro **465**
The Virus Warning Message Box **465**
The Macros Dialog Box **466**
Enabling and Disabling Macros **466**
The Security Dialog Box **467**
Word VBA Example 2: The Times12 Macro **467**
The Customize Keyboard Dialog Box **468**
The Visual Basic for Applications Design Window **468**
The Relationship between Macros and Visual Basic for Applications **469**
The VBA Editor **469**
The VBA Project Explorer **469**
The VBA Properties Window **470**
Global versus Local Macros **470**
The VBA Code Window **470**
Word VBA Example 3: The Table3_5 Macro **471**
Objects, Properties, Methods, and Events **473**
Examples of Assignments: Object.Property = Value **474**
Object Browser **475**
Writing a VBA Procedure without a Form **475**
Word VBA Example 4: The VBA CountStuff Procedure,
aka the CountStuffMacro **476**
Word VBA Example 5: The DateName Macro **479**
Running a VBA Procedure from the VBA Design Window **482**
Running a VBA Procedure from the Macros Dialog Box **482**

Chapter 18. Visual Basic for Applications: Microsoft Excel **485**
Excel and Word: Macros and VBA **485**
Excel VBA Example 1: The SwimHeader Macro **486**
Observations on the SwimHeader Macro **487**
The VBA Editor's Code, Project, and Properties Windows for the
SwimHeader Macro **488**
Absolute and Relative Cell References **488**
Referencing Objects **489**
Excel VBA Example 2: The AutoGraph Macro **490**
Writing Excel VBA Procedures without a Form **492**
Excel VBA Example 3: The CheckNames Procedure **492**
Organization of the CheckNames Procedure........ **493**
A Compact Version of the CheckNames Macro **494**
Summary of the CheckNames and CheckNamesShort Macros **494**
The Essence of Excel VBA Macros **495**
More on the Ranges **495**
Excel VBA Example 4: The ColorBlock Procedure **496**
Understanding the VBA ColorBlock Procedure **497**
The Cells Property of the ActiveSheet **497**
Excel VBA Example 5: The ChessBoard Procedure **498**
Understanding the Excel VBA ChessBoard Procedure **500**

Excel VBA Example 6: The DesLandresTable Procedure 501
Writing Excel VBA Procedures with a Form 502
Excel VBA Example 7: The First Form Procedure 503
The Visual Basic Editor Design Window for Host for Form.xls 504
Running the VBA Procedure from the VBA Editor 504
Running the VBA Procedure from the Macros Dialog Box 505
The VBA Demonstration Form, frmFirstForm 505
Review and Preview 505
Summary of VBA Procedures with a Form 506
Nomenclature for Excel VBA Projects 506
Excel VBA Example 8: The NumberStats Procedure 506
Data in Sheet1 with the Statistical Functions on the Form 508
The Visual Basic Editor Design Window 509
Excel VBA Example 9: The SilverEntropy Procedure 510
The form of the VBA SilverEntropy procedure 513
The VBA Editor 514
Excel VBA Example 10: TheABCUser Procedure 516
The ABCUser Program Dialog Box (frmABCUser) 516
Sheet1 of the ConsecReacts.xls Workbook 519
Six Different Snapshots of the Concentrations of *A*, *B*, and *C* 520
Excel VBA Example 11: The ABCScroll Procedure 522
The Dialog Box for the VBA ABCScroll Program (frmABCScroll 523
The Project Window for the ConsecReact.xls Project 525
The Design Window for the ConsecReact.xls Project 527
Excel VBA Example 12: The Eutectic Procedure 528
Creating the Host Excel Program, Eutectic Host.xls 528
Writing the VBA Program 528
The Form for the VBA Eutectic Program 530
Running the VBA Eutectic Program from the Macros Dialog Box 532
The Eutectic Freezing Point Diagram 534

Part IV. Mathcad and Mathematica

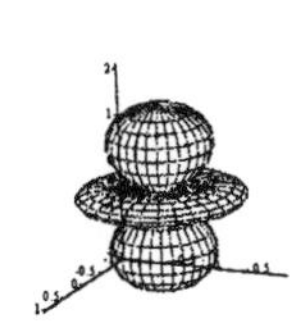

Chapter 19. Mathcad 537
Introduction: The Mathcad Worksheet 537
Cursors and Calculations 539
Assigning a Value to a Variable 539
The Mathcad Editing Lines 539
Positioning Math Regions and Making Assignments to Variables 540
Elementary Calculations 540
Range Variables and Iteration 541
Mathcad's Built-in Functions 541
User-Defined Functions 542
Creating Lists of Raw Data: Subscripted Variable 544
Calculating with Lists 545
Mathcad Worksheet 1: Calculating with Lists:
X-Ray Powder Diffraction 545
Two-Dimensional Plots
Mathcad Worksheet 2: Plotting with Lists:
The Vapor Pressure of Benzene and Chloroform 548
Plotting a Table of *x,y* Points 548
Entering the Data for the Vapor Pressure Plots..... 548

Creating Two Vapor Pressure Plots on the Same Graph........... 548
Editing the Vapor Pressure Graph.. 549
The Formatting Currently Selected X-Y Plot Dialog Box........ 550
Mathcad Worksheet 3: Calculating with Subscripted Variables:
A Linear Regression Line..550
Linear Regression Calculation and Plot with Subscripted Variables 551
Entering the Data 551
Calculating the Slope and Intercept 551
Mathcad Regression Functions .... .551
Plotting the Experimental Data and the Regression Line 552
To Graph the Data Markers 552
To Graph the Linear Regression Line 552
The Linear Regression Calculation and Plot with Vectors 553
Defining Vectors 553
Arithmetic Operators and Vectors 554
The Vectorize Operator 555
Mathcad Worksheet 4: Calculating with Vectors:
A Linear Regression Line.. 555
Assigning the Data to Vectors.. 556
Calculating the Slope and Intercept... 556
Plotting the Experimental Data and the Regression Line........... 557
Graphing the Data Markers... 557
Graphing the Linear Regression Line 557
Plotting Functions 557
Mathcad Worksheet 5: Plotting Two Functions 559
Mathcad Worksheet 6: Plotting a One-Dimensional Velocity
Distribution Function for N_2 560
Mathcad Worksheet 7: Plotting a Three-Dimensional Speed
Distribution Function for N_2 561
Mathcad Worksheet 8: Multiple Plots: Three Atomic Wave Functions
on One Graph 561
Mathcad Worksheet 9: Plotting H Atom Angular Wave Functions
in Two Dimensions.. 565
Three-Dimensional Plots 566
Mathcad Worksheet 10: A Simple 3-D Surface Plot 568
Mathcad Worksheet 11: A 3-D Surface Plot for a Particle in a
Two-Dimensional Box 569
Mathcad Worksheet 12: A 3-D Parametric Surface Plot of a
d_{z^2} Orbital.. 570
Mathcad Worksheet 13: A 3-D Parametric Surface Plot of the
$2p_x$ Orbital 571
Symbolic Calculations 572
Symbolic Operators and Keywords 572
The Symbolic Equal Sign with Keywords 573
Some Keywords and Their Functions 573
Simplify 573
Expand 574
Factor 574
Substitute 575
Calculus 575
Derivatives 575
Integrals 576
Definite Integrals 577
The Keyword *Assume* 578
Integration with a Range Variable 578
Summation and Product Operators 579
Some Derivatives and Integrals from Statistical Thermodynamics 580

Mathcad Worksheet 14: Partition Functions 580
Mathcad Worksheet 15: Translational and Rotational Internal Energy 580
Mathcad Worksheet 16: Vibrational Internal Energy 580
Mathcad Worksheet 17: Vibrational Heat Capacity 581
Mathcad Worksheet 18: The Population of Rotational Energy Levels 585
Mathcad Worksheet 19: The Heat Capacities of Iodine, Chlorine, and Nitrogen 587
Solutions of Algebraic Equations 588
Mathcad Worksheet 20: The Minimum and Maximum in a Curve 589
Mathcad Worksheet 21: Probability Distribution Function for a 1*s* Orbital: r_{mp} and r_{ave} 590
Mathcad Worksheet 22: The Most Probable Speed of a Gas 590
Mathcad Worksheet 23: The Average Speed of a Gas 590
Mathcad Worksheet 24: The Lennard-Jones Potential 592
Mathcad Worksheet 25: The Minimum in the Lennard-Jones Potential.. 594
Solutions of Simultaneous Equations 596
Mathcad Worksheet 26: The Matrix Method for Solving Linear Equations 596
Mathcad Worksheet 27: The *Solve* Keyword Method for Solving Linear Equations 598
Mathcad Worksheet 28: The Solve Block Method for Solving Linear Equations 599

Chapter 20. Mathematica 601
Introduction: The Mathematica Notebook 601
Notebook Regions 601
First Text Region 601
Numerical Regions 602
The Workspace of a New Mathematic Notebook 602
The File Menu 603
Save and Open 603
Palettes 603
Help! 605
Elementary Calculations 606
Arithmetic Operators 606
Predefined Constants 607
User-Defined Constants 608
The Clear Function 609
Functions 610
Built-in Functions 610
Elementary 611
Numerical Functions 612
User-Defined Functions 612
Evaluating and Tabulating Functions 613
Practical Calculations with Lists 614
HCl Infrared Spectral Lines 614
Mathematica Notebook 1: The DesLandres Table for the PN Molecule 615
Mathematica Notebook 2: X-Ray Diffraction 615
Mathematica Packages 617
Directory Names of Mathematica Packages 617

More on Lists 619
Operating on Lists 619
Mathematica Notebook 3: The Heat Capacity of Iodine..... 619
Mathematica Notebook 4: Plotting Functions with Plot 621
Plot Options 621
Plotting Lists 623
Mathematica Notebook 5: Plotting Lists of Numbers with ListPlot 623
Mathematica Notebook 6: The DisplayTogether Function 625
Linear Regression 626
Mathematica Notebook 7: The Fit Function 626
Mathematica Notebook 8: Plotting the Linear Regression Line..... 627
Mathematica Notebook 9: The Linear Regression Line and DisplayTogether 627
Plotting Functions from Quantum Mechanics..... 630
Mathematica Notebook 10: Plotting the Vibrational Wave Function..... 631
Mathematica Notebook 11: Plots of $R_{10}(r)$, $R_{20}(r)$, and $R_{30}(r)$ 634
Three Dimensional Plots 636
Mathematica Notebook 12: Surface Plots with Plot3D..... 636
Mathematica Notebook 13: Spherical Harmonics..... 637
Mathematica Notebook 14: Plotting s, p, and d Angular Wave Functions with SphericalPlot3D..... 638
Mathematica Notebook 15: The Particle in a Box: ψ_{22} and ψ_{22}^2 640
Mathematica Notebook 16: 3-D Atom Radial Wave Functions ψ_{2s} and ψ_{2s}^2 640
Bar Charts 643
Mathematica Notebook 17: Bar Chart of the Population of Rotational Energy Levels..... 643
Symbolic Calculations 645
Expand 645
Factor 645
Simplify 645
The Replacement Operator 645
Mathematica Notebook 18: Multiple Plots with Plot 646
Calculus 647
Derivatives 647
Integrals 648
Equations in Mathematica 649
Solve 650
NSolve 650
FindRoot 650
Mathematica Notebook 19: Solving Simultaneous Equations..... 651
Mathematica Notebook 20: The r_{mp} and r_{ave} for a 1s Orbital 653
Mathematica Notebook 21: The Gas Speed Distribution Function at 298 K, 500 K, and 1000 K 653
Minima and Maxima 653
Mathematica Notebook 22: The u_{mp} and u_{ave} for a Gas..... 653
Mathematica Notebook 23: The Lennard-Jones Plot..... 653
Mathematica Notebook 24: The Lennard-Jones Minimum..... 653
Mathematica Notebook 25: Miscellaneous Information in Mathematica 659

Bibliography **661**

Index **663**

Preface

Who This Book Is For

This book is written for you, the scientists, engineers, and students who do numerical and graphical calculations. It is written for those of you who are open to exploring alternative approaches and widening your computer background. You should already know the basics of computer hardware and software, such as word processors, and a little about Microsoft Windows. You probably have already enjoyed using some kind of spreadsheet.

In any case, this book covers the fundamentals from the beginning. Little previous experience is expected for Part I on spreadsheets, and no previous knowledge is required for the remainder of the book. For example, you probably remember from your elementary chemistry courses that an *s* orbital looks like a circle, a *p* orbital resembles a dumbbell, and a *d* orbital is similar to a flower; in this book you will review the chemistry, physics, and mathematics underlying the particular geometries of these orbitals and learn to calculate their graphs.

How This Book Is Organized

Part I, Spreadsheets, consists of eight chapters that provide examples for doing numerical calculations and graphs with Microsoft Excel, by far the most widely used spreadsheet. These chapters cover thermodynamics, quantum mechanics, statistical thermodynamics, gases, kinetics, statistics, and three-dimensional plots. Part I includes nearly all the physics and physical chemistry used for the application examples in the remainder of the book. The final chapter in Part 1 provides a brief introduction to Lotus 1-2-3 and Quattro Pro.

Part II, Visual Basic, is a complete primer for the Microsoft Visual Basic (VB) language. Its purpose in this book is twofold; the first purpose is to provide a source book and index for the VB language used in Microsoft Visual Basic for Applications (VBA), the subject of Part III of this book. The second purpose is to provide a stand-alone introduction to the VB language, with an emphasis on numerical calculations, something ignored by most books on the VB language. Part II uses the physical chemistry presented in Part I but is otherwise completely independent of other parts of the book. It's not necessary to master VB to use VBA, but it sure is fun.

Part III, Visual Basic for Applications, is an introduction to VBA. Chapter 17 and Chapter 18 introduce VBA for Microsoft Word and VBA for Microsoft Excel, respectively. You might not realize it, but VBA is included in many Microsoft applications you may already use and still more in non-Microsoft applications that you may also be using. VBA is the language of *macros*, those underused utilities that can greatly multiply your application's power and versatility. With few exceptions, Part III uses the physical chemistry background presented in Part I. Part III also uses the VB developed in Part II as a source book and index.

Part IV, Mathcad and Mathematica, covers these applications in Chapters 19 and 20, respectively. Both chapters use the physical chemistry presented in Part I but are otherwise independent of other parts of the book. Mathcad and Mathematica are powerful applications not only for numerical calculating and graphing but also for symbolic calculations.

Rodney J. Sime
Sacramento, California
November, 2004

Part I. Spreadsheets

Chapter 1. Introduction to Spreadsheets
Chapter 2. Thermodynamics
Chapter 3. Quantum Mechanics
Chapter 4. Statistical Thermodynamics
Chapter 5. Gases
Chapter 6. Kinetics
Chapter 7. Statistics
Chapter 8. Three-Dimensional Plots
Chapter 9. Lotus 1-2-3 and Quattro Pro

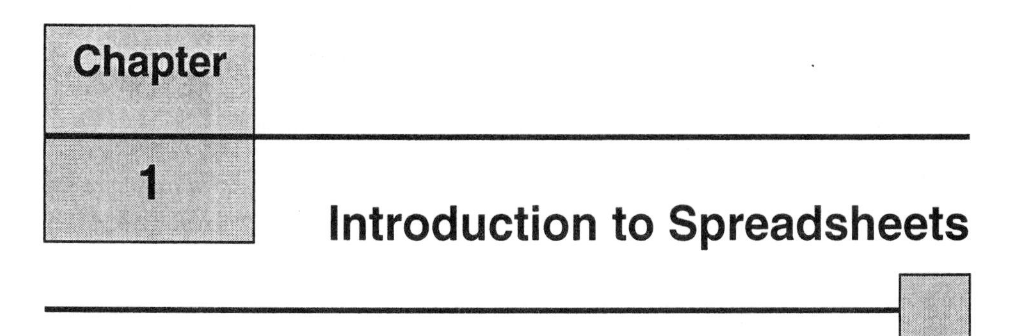

Chapter 1 Introduction to Spreadsheets

History

The personal computer and its software have evolved in great jumps separated by rather short periods of consolidation. The Apple computer was born in a garage (1974) and the first IBM PC in the more sanitary confines of the IBM Corporation (1981). The Apple Macintosh introduced the graphical user interface, or GUI (1981). Microsoft followed in 1985 with Windows 1.0, an operating system that rapidly replaced MS-DOS. The first serious personal computer software was Apple Writer, a popular word processing program, but the rocket that launched the personal computer revolution as we know it today was a spreadsheet, simple by today's standards, known as VisiCalc (1978). With VisiCalc up and running on an Apple computer, you would instantly recognize a rectangular array of columns labeled alphabetically and rows labeled numerically: a spreadsheet. It could not graph data, but it could calculate, and the calculations were instantly visible on the screen. VisiCalc and the Apple computer presented to the business world an irresistible combination of power and economy. And that combination led to today's powerful and affordable personal computers and software.

VisiCalc is not around anymore; the company that made it was long ago sold to another computer industry entrepreneur. The business history of the computer industry presents a fascinating chronology of foresight and its lack, good and bad luck, good and bad judgments. Borland International, Inc., developed the Quattro Pro spreadsheet, which is currently part of Corel Office. The Lotus Corporation developed the Lotus 1-2-3 spreadsheet, which was subsequently absorbed by IBM. Later, we will look briefly at these two excellent spreadsheets, but we will begin with Excel, a product of the Microsoft Corporation. Most of the important features of all three spreadsheets are accessed in virtually identical manners, so shifting from one spreadsheet to another is relatively smooth.

Physical chemistry enjoys a long and distinguished history. The classic period (before quantum mechanics) includes many illustrious names:

Atomic theory: Dalton, Avogadro, Berzelius, Dumas, Bunsen, Lothar-Meyer.
Gases: Gay-Lussac, Charles, Boyle, van der Waals, Lennard-Jones.
Thermodynamics: Carnot, Clausius, Claypeyron, Joule, Thomson, Kelvin, Gibbs, Ostwald, van't Hoff, Rault, Kohlrausch.
Kinetics: Arrhenius, Berthollet, Guldberg, Waage, Bodenstein, Maxwell, Boltzmann.

Physical chemistry was a couple of hundred years old by the time it was recognized as a unique discipline with the founding of its first journal, the *Zeitschrift für physikalische Chemie,* by Ostwald in 1887. The publication of Max Planck's paper on the quantum of radiation in 1901 marks the birth of modern chemistry and physics. Planck's revolutionary paper was published in *Annalen der Physik*, volume 4, 553.

In the next thirty years, Bohr, Boltzmann, de Broglie, Born, Einstein, Dirac, Heisenberg, Planck, Schrödinger, and many others led the development of quantum mechanics, which became as important to chemistry as to physics.

None of the aforementioned scientists had access to computers. But today, the ubiquitous computer, taken for granted, plays a significant roll in advances in modern physics and chemistry. In the first nine chapters, we shall examine how the personal computer and spreadsheets, beginning with Excel, can assist in calculations in chemistry and physics.

The Microsoft Excel Workspace Layout

Figure 1-1 displays the screen layout for Excel. The toolbars at the top are similar to those in other Windows applications, and the worksheet is similar to worksheets in other spreadsheet applications, such as Lotus 1-2-3 and Quattro Pro. The rows and columns of worksheet cells dominate the workspace.

The Worksheet. A worksheet is a rectangular array of cells arranged into columns and rows as seen in Figure 1-1. The columns are labeled A, B, C, D, ... from left to right, and the rows are labeled 1, 2, 3, 4, ... from top to bottom. The intersections of the columns and rows define the cells of a worksheet. The intersecting column and row labels form a unique address or reference for each cell. For example, in Figure 1-1, the address of the active cell is G10, because it is located in column G and row 10. An active cell is one that has been selected by clicking in it. Its heavy black border identifies the selected cell.

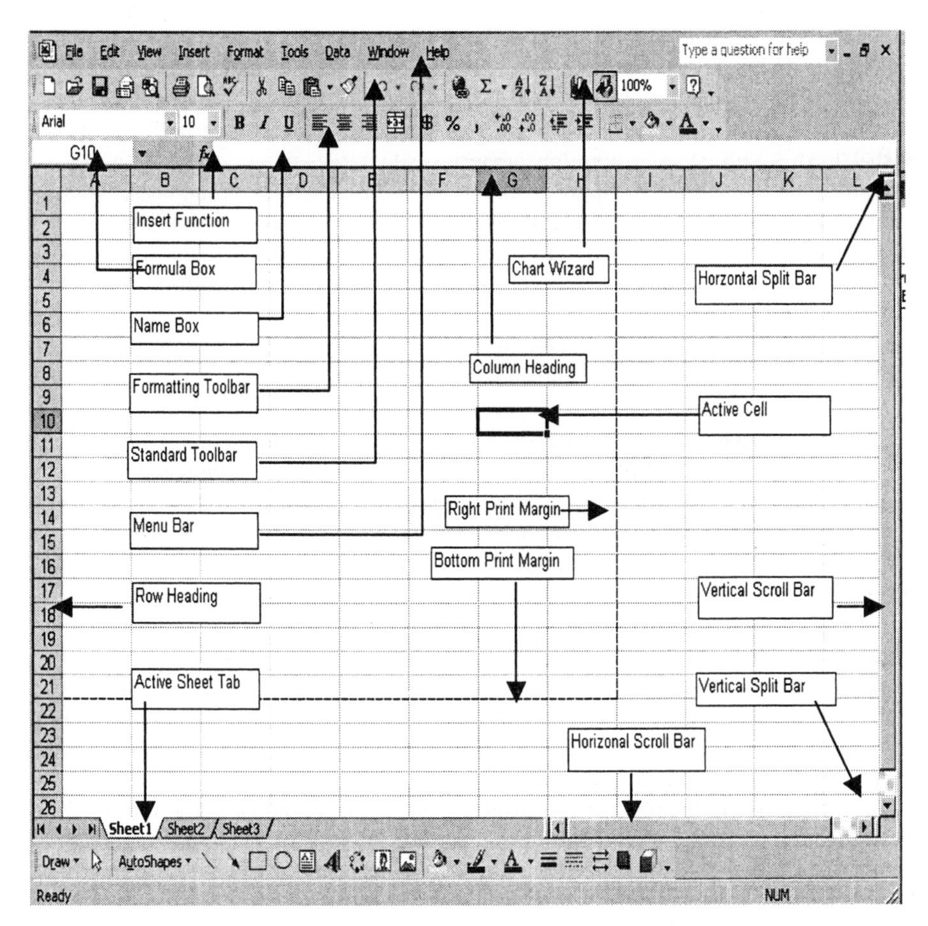

Figure 1-1. The Excel Workspace

The Menu Bar. This is the uppermost bar on the **workspace** (Figure 1-1). The **menu bar** displays the familiar items (left to right) **File, Edit, View, Insert, Format, Tools, Data, Window,** and **Help** menus. The menu bar is essentially the same for most Windows applications. Clicking a menu item precipitates a drop-down menu. The commands in the menu differ somewhat from application to application.

The Standard Toolbar. The **standard toolbar** displays some familiar icons for (from left to right) **New Page, Open Folder, Email, Search, Print, Cut, Copy, Paste,** and other options. Further to the right on the toolbar, you can locate the **Σ** icon for summing a list of numbers. Continuing to the right, we find the $\frac{A}{Z}\downarrow$ icon, which provides quick access to sorting a list. To its right is the **Chart Wizard** icon, which is the key to simple but powerful tools for creating and editing charts, plots, and graphs.

The Formatting Toolbar. The **formatting toolbar** lies just below the standard toolbar. (Figure 1-1). Two drop-down list boxes at the left of the formatting toolbar display and permit selecting the font type (Arial in Figure 1-1) and point size (10 points in the figure). Located further to their right are icons for **Bold, Italic, Underline,** and **Alignment.**

The Formula Toolbar. The lowest toolbar at the top of Figure 1-1, located below the formatting toolbar, is the **formula toolbar.** It displays the formula, if any, that you have entered into a selected cell. The cell address of the selected cell in Figure 1-1 is G10. This cell address appears in the **name box** at the leftend of the formula toolbar. Because cell G10 is blank and contains no text, number, or formula, the **formula box** is also blank.

Other Toolbars. Clicking **Toolbars** on the **View** menu opens a submenu that lists many more toolbars, for example, the **Drawing** toolbar, which is used to create the text box labels on Figure 1-1. The **Drawing** toolbar is located at the bottom of Figure 1-1.

Cell Entries. A cell can contain alphanumeric text or real numbers. Text entries are important to label, describe, explain, remind, and in general communicate what is happening in neighboring cells. The number observed in a cell may be just a number, or it might be there because it is calculated by a formula in the cell. The formula is invisible in the cell, but if the cell containing it is selected, the formula box at the top of the sheet displays the formula, while the selected cell displays the calculated number. Select a cell by clicking in it. Once you select a cell, you can do nearly everything that you can do with a word processor: enter text or numbers, copy, cut, paste, delete, justify, and format.

The Name Box. When a cell is selected, its address is shown in the name box (Figure 1-1). As a number, formula, or function is entered into a cell, the number, formula, or function is displayed in the formula box (Figure 1-1). When the ENTER key is subsequently pressed, the value of the number, formula, or function is displayed in the cell. An equal sign (=) must precede the entry of a formula or function. Functions are discussed in more detail later.

Examples of Entries of Numbers, Formulas, and Functions. The first four examples in Table 1-1 display the entry of a number or an arithmetic calculation. When a calculation is involved in the cell, an equal sign must precede the entry. The last three entries show the use of two of Excel's many functions. The next section describes Excel's functions in more detail.

As the data is entered, the cell display is the same as shown in the formula box. After you press the ENTER key or click in an empty cell, the cell displays the calculation result in the cell. The display of the formula in the formula bar disappears. Editing of an entry is done in the formula box, not in the cell.

In example number four, the value stored in cell B3 is **143** (not shown) and the value stored in cell C3 is **17** (not shown).

Table 1-1

Examples of Cell Entries

Example Number	Cell Entry Formula Box Displays	Press ENTER Cell Displays
1	5	5
2	=3^6	729
3	=4.1*5.6/8.3	2.76626506
4	=B3/C3	8.411765
5	=SIN(3.14*5/180)	0.087111671
6	=SIN(PI()*5/180)	0.087155743
7	=LN(5)	1.609437912

Functions

Excel supports a large number of functions, and so that you do not have to memorize their names, you can click the **Function** button (f_x) (Figure 1-1) or, alternatively, on the **Insert** menu, click **Function**. Both operations cause the **Insert Function** dialog box to appear onscreen (Figure 1-2).

For ease of selection, the functions are categorized. In Figure 1-2, the **Statistical** category is selected, causing a drop-down list of category selections to appear. The **AVERAGE** category is selected in the **Select a function** box. This is how we know that the proper function we want is spelled AVERAGE, not AVGE or AVG or some other spelling.

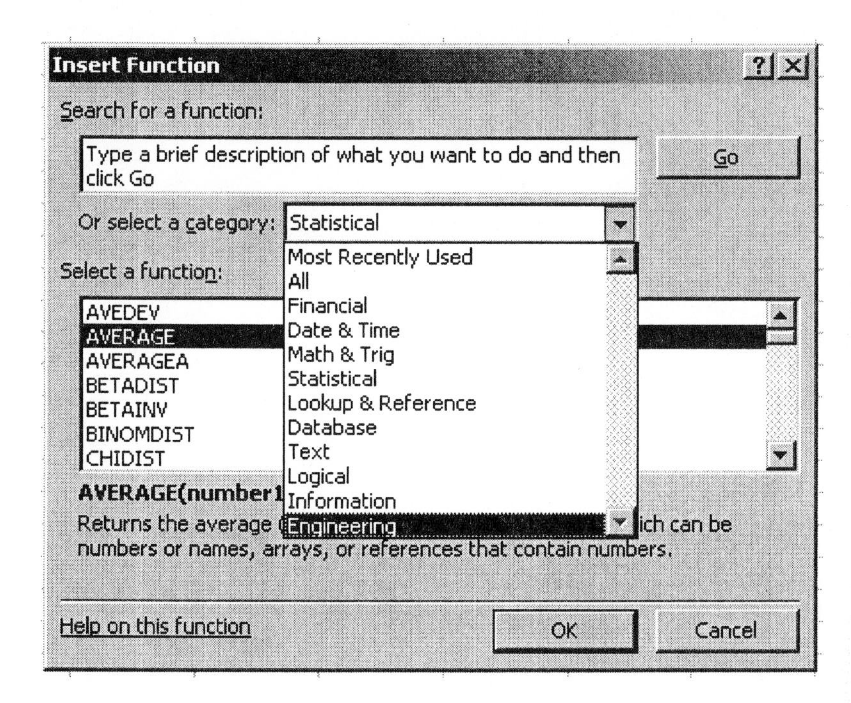

Figure 1-2. The **Insert Function** Dialog Box

The argument of the **AVERAGE** function is the range of numbers to be averaged. If you select the **AVERAGE** function and click **OK,** the **Function Arguments** dialog box appears (Figure 1-3). In the text box for **Number1,** you can enter the range of the numbers to be averaged, for example, F7:F12 (ranges in Excel are indicated by the first cell in the range followed by a colon and the last cell in the range), or you can enter the name of the range if you have defined a name (described in the next section) for the range of interest.

When a function is selected, a brief description of the function and its use is written in the area just below the **Select a function** list box (Figure 1-2). For further information, click **Help on this function,** located in the lower left corner of the dialog box.

If you want the sine of an angle, you use the **SIN()** function. The argument of a trigonometric function is the angle in radians. If you want the sine of an angle in degrees, for example, 5.0 degrees, you would enter into a cell: **=SIN(5*PI()/180)**. Remember to use uppercase letters for the names of all functions. Thus, **SIN(),** not Sin(), and **PI(),** not Pi(). **PI()** is a function that takes no argument but returns a value for π, accurate to 15 digits: 3.14159265358979.

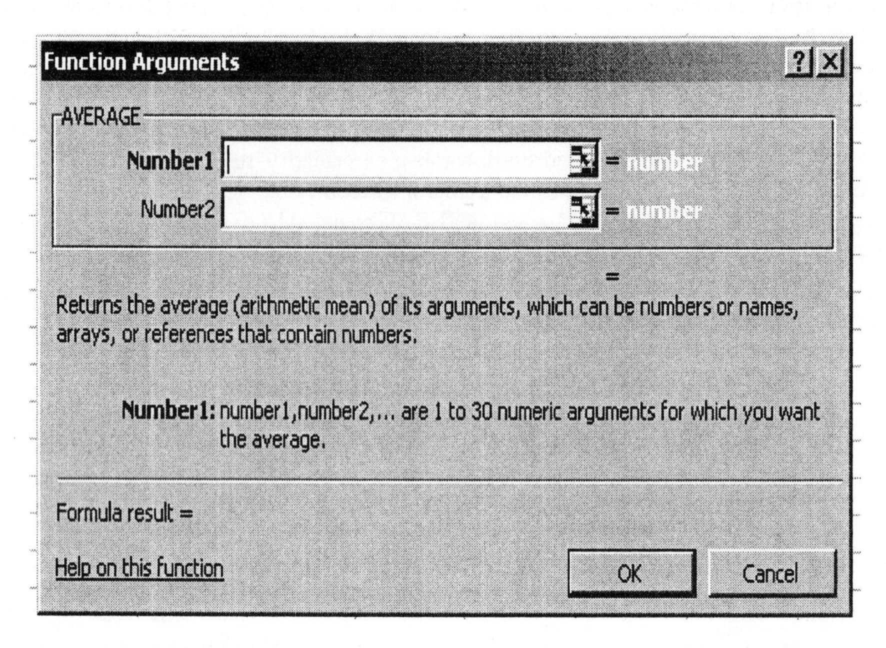

Figure 1-3. The **Function Arguments** Dialog Box

The next 26 Excel Examples serve to demonstrate and clarify many of the common features of Excel with typical problems from chemistry and physics.

Excel Example 1

X-Ray Powder Diffraction: Tabular Calculations and Statistics

The calculation of X-ray powder diffraction data provides an example of typical laboratory calculations. No graphing is involved, just repetitive calculations arranged in tabular form. Even if you are not familiar with the nature of X-ray diffraction calculations, doing Excel Example 1 illustrates some frequently encountered spreadsheet features and Excel functions.

Short Review of the Science

You can think of a cubic crystal as a solid bounded by six orthogonal planes separated from each other by a distance a_0. Further, you can imagine all kinds of planes that could be drawn through the edges and sides of the cube (or unit cell). Some of these planes form families of planes that are parallel to each other and are separated by a distance d. After a little consideration, you might notice that the spacing of the planes must be less than the spacing of the unit cell a_0. The families of parallel planes are characterized by a set of three integers, h, k, and ℓ , called Miller indices.

Measurement of the diffraction angle θ from various families of planes permits calculation of the separation between planes, d_{hkl}, and the size of the unit cell itself, a_0. The length of the edge of a cubic cell is a_0. The wavelength λ is an apparatus constant. The relevant equations are:

$$d_{hkl} = \frac{\lambda}{2\sin\theta} \tag{1-1}$$

and

$$a_0 = d_{hk\ell}\sqrt{h^2 + k^2 + \ell^2} \tag{1-2}$$

The science underlying these equations is discussed in detail at the end of this section. For now, we shall just use these two equations for practicing with Excel.

The experimental determination of the unit cell dimension a_0 consists of three parts:

1. At constant X-ray wavelength λ, the angles θ at which diffraction occurs are measured (Table 1-2). This permits calculation of the d spacings with Equation 1-1.
2. Each reflection is indexed to determine the planes $hk\ell$ from which the reflections originated, so they can now be labeled $d_{hk\ell}$.
3. Redundant values of a_0 can then be calculated with Equation 1-2 and averaged.

Table 1-2

Diffraction Angles for Copper at λ = 1.54180Å

θ/deg	h	k	ℓ
21.811	1	1	1
25.369	2	0	0
37.218	2	2	0
45.132	3	1	1
47.754	2	2	2
58.603	4	0	0

Preparing the Worksheet

Open a new Excel worksheet and enter the labels for this Excel Example on Sheet1. Then enter the numerical data as directed. This example involves only tabular calculations. No chart is created.

Entering the Labels

- In your worksheet, enter the labels shown in Figure 1-4 in the range A1:F5. You'll see that the entry **Wavelength =** does not fit the default width of column A.
- To widen the column, click the boundary between the column A and column B labels at the top of the worksheet, and drag it to the right until **Wavelength =** fits in cell A3.
- Click and drag to select the *h, k,* and *ℓ* in cells C5:E5, and center these labels by clicking the **Center Align** button in the standard toolbar. If you are not sure which icon that is, place the pointer on each icon (without clicking). In a moment, a yellow label box appears informing you of the name or function of the icon.

	A	B	C	D	E	F	G	H
1		Powder Diffraction Data for Copper						
2								
3	Wavelength =	1.5418						
4								
5	Theta/deg	d/angstroms	h	k	l	a/angstroms		
6								
7	21.811	2.075	1	1	1	3.594		
8	25.369	1.799	2	0	0	3.599		
9	37.218	1.275	2	2	0	3.605		
10	45.132	1.088	3	1	1	3.608		
11	47.754	1.041	2	2	2	3.607		
12	58.603	0.903	4	0	0	3.613		
13						3.604	= average	
14								
15						0.0068	= standard deviation	
16								
17						0.0053	= average deviation	
18								
19						3.613	= maximum a	
20								
21						3.594	= minimum a	
22								

Figure 1-4. The Excel Worksheet for Powder Diffraction

Entering the Data

- Enter the numerical data in the range A7:A12, as shown in Figure 1-4.
- Enter the integers in the range C7:E12, and make columns C, D, and E narrower by dragging the lines between column C and column D, between column D and column E, and between column E and column F.

- Select the **1.5418** in cell B3, and then click the **Align Left** icon on the formatting toolbar.
- If the **1.5418** you entered appears as **1.542,** then change its format. Click in cell B3, and then, on the **Format** menu, click **Number.** The **Format Cells** dialog box opens. The default number of decimal places is 2 (Figure 1-5). In the dialog box, change the number in the **Decimal places** box to **4.**

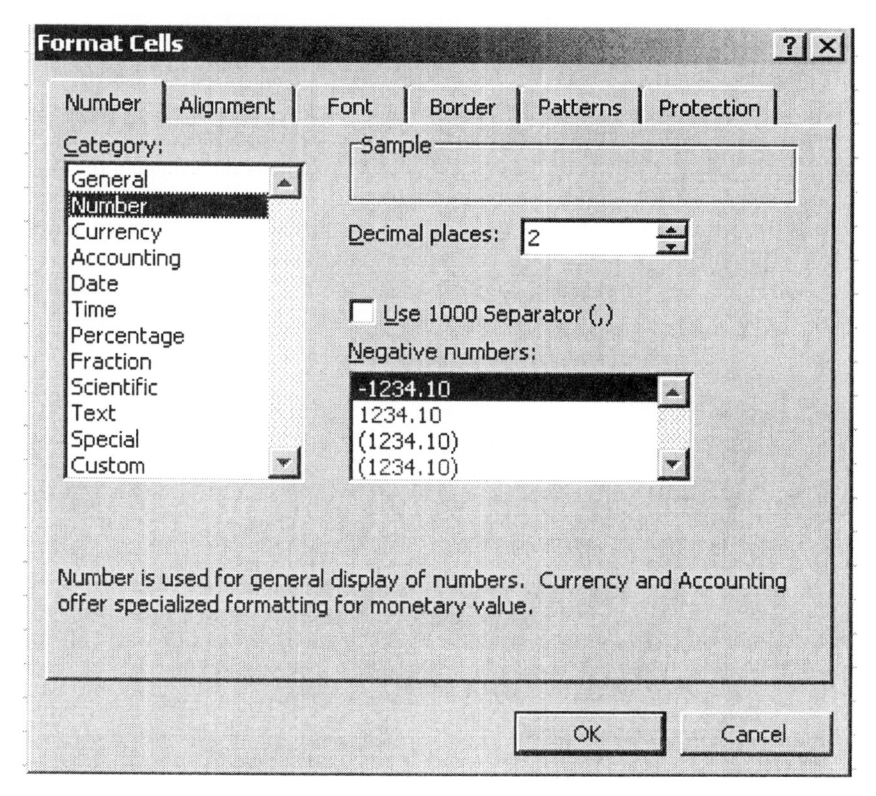

Figure 1-5. The **Format Cells** Dialog Box

- In cell B7, type **=$B$3/(2*SIN(A7/PI()/180))** as the formula for $d_{hk\ell}$.
- Click in cell F7, and enter **=B7*SQRT(C7^2+D7^2+E7^2)** as the formula for a_0.

The first formula reads like this: take the value stored in cell B3 (1.5418) and divide it by 2 times the sine of the value stored in cell A7 (21.811) divided by pi over 180.

The cell reference $B$3 is an absolute cell reference. All other cell references in these two formulas are relative cell references. The differences and uses of absolute cell references and relative cell references are discussed in the next section.

Cell References

The difference between relative and absolute cell references will become clearer as we proceed with this example.

- The first formula appears in cell B7 as you enter it. As soon as you finish and press the ENTER key, the formula disappears and a calculated value (**2.075**) appears in its place in cell B7. The formula itself reappears in the formula box, just above the label for column A (Figure 1-1). The name box displays the name of the selected cell, B7.

- Now calculate the remaining d_{hkl} values. Select cell B7. Note the small black square (■) at the lower right corner of cell B7, the selected cell. Click this fill handle, and drag it down to cell B12. The series B7:B12fills with d_{hkl} values that reflect the underling formulas.
- Select cell F7, and drag the fill handle down to cell F12. The series F7:F12 fills with a_0 values.

Next examine the formula contents of cells in the range B7:B12. Click cell B7, and the formula you entered manually appears in the formula box. Now click cell B8 and notice that the absolute reference, $B$3, remains unchanged, but A7 has changed to A8, its correct address relative to cell B8 where the formula that uses the contents of cell A8 is located. A spreadsheet finds each cell reference by its position in relation to the cell containing the formula, not by its address. When you copy a cell containing an absolute address in a formula, no adjustment to the cell address is made. The absolute cell reference always refers to the original cell, regardless of where you place the copied formula. This is a powerful feature common to all spreadsheets, not only Excel but also Lotus 1-2-3, Quattro Pro, and others. Continue to click each member of the range B8:B12to see how the relative cell reference changes and how the absolute cell reference remains constant.

Using Some of Excel's Statistical Functions

We have already discussed Excel's functions. The current Excel Example offers an opportunity to use some of Excel's statistical functions. A few of them are listed in Table 1-3. It is convenient at this time to use the calculated X-ray diffraction data (Figure 1-4) to try out a few more of Excel's statistical functions. Excel can calculate

Table 1-3

A Few of Excel's Basic Statistical Functions

Function Name	Value Returned
AVERAGE( )	The average of a list of values
COUNT ( )	The number of values in a list
DEVSQ ( )	The sum of the squares of the deviations from the mean
GEOMEAN ( )	The geometric mean of the numbers in a list
HARMEAN ( )	The harmonic mean of the numbers in a list
MAX ( )	The maximum value in a list
MIN ( )	**The minimum value in a list**
STDEVP ( )	The standard deviation of the values in a list

three standard deviations: **STDEV, STDEVA,** and **STDEVP.** To find out exactly what any Excel function calculates, click **Help on this function** in the lower left corner of the **Function Arguments** dialog box (Figure 1-3). For example, to calculate the standard deviation based on the entire population, select **STDEVP.** The standard deviation is a measure of how widely values are dispersed about the average value. The Excel **STDEVP** function uses the formula

$$s = \sqrt{\frac{n\sum x^2 - \left(\sum x\right)^2}{n(n-1)}} \tag{1-3}$$

Excel Example 2 in this chapter contains more information about Excel statistical functions. To calculate the average of the cell lengths in the range F7:F12:

- Select cell F13, and type **=AVERAGE (F7:F12)** and press ENTER. You must use uppercase letters.
- In cell G13, type **'= average** as the label. The apostrophe must precede the equal sign to let Excel know that text, not a formula or a number, is going to be entered.

Alternatively:

- On the **Insert** menu, click **Function**. The **Insert Function** dialog box appears.
- Select **AVERAGE()**. The **Function Arguments** dialog box appears with the range F7:F12 already in the **Number1** box. If this range is not in the box already, then type it yourself, and click **OK.**

Defining and Using the Name of a Range

You can assign a name of your choice to a cell, a range, or even a formula. Let us give a name to the range of a_0 values, F7:F12.

- First, click and drag over the range F7:F12 to select it,
- Click **Insert_Name_Define** to open the **Define Name** dialog box (Figure 1-6).
- For the range containing the list of a_0 values, type **aList** as the name, and click **OK.**

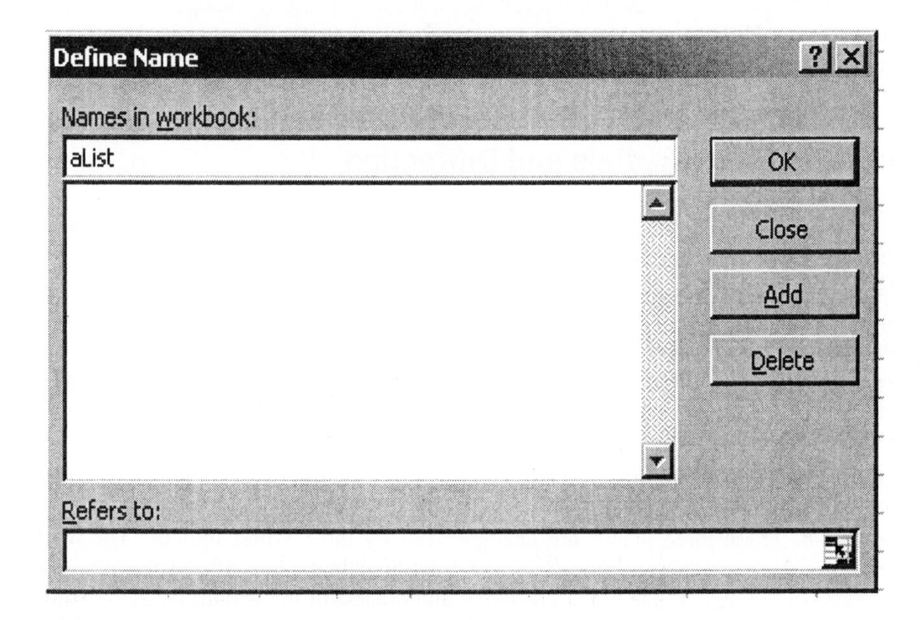

Figure 1-6. The **Define Name** Dialog Box

- Next, select cell F15, and on the **Insert** menu, click **Function** (Figure 1-1) to open the **Insert Function** dialog box again.
- Now select **Statistical** as the category, select **STDEVP** (standard deviation), and click **OK** to open the **Function Arguments** dialog box.
- In the **Number1** box, type **aList** as the name of the range to be averaged. The desired standard deviation appears in cell F15.

Now examine the formula box, which displays **STDEVP(aList)**. Add the label **'= standard deviation** in the neighboring cell (G15). Now that you have named the range, it is especially easy to calculate the average deviation, the maximum, and the minimum. Before doing so, look once more at the **Function Arguments** dialog box in Figure 1-3. In the lower left corner is the statement **Help on this function**. Clicking this leads you to detailed help and information on the function in question.

Select any blank cell, and notice how the name box displays the cell name or address. Now click the down arrow on the name box. A drop-down list shows the name you gave to the range. Click this name to select it, and you will see that the range for which it stands is selected.

To illustrate further how powerful name ranges can be, use the procedure just described (on the **Insert** menu, point to **Name**, and click **Define**) with reference to Figure 1-4 to name the following ranges :

- Cell B3 Name: **Lambda**
- Range A7:A12 Name: **Theta**
- Range B7:B12 Name: **dhkl**
- Range C7:C12 Name: **h**
- Range D7:D12 Name: **k**
- Range E7:E12 Name: **l**
- In cell B7, enter the formula **=Lambda(2*SIN(Theta*PI()/180))** and drag the Fill-Handle down to cell A12.
- In cell F7, enter the formula = **dhkl*SQRT(h^2 + k^2 + l^2)** and drag the Fill-Handle down to cell F12.

These formulas are much more readable than those entered into cells B7 and F7. They appear almost like Equation 1-1 and Equation 1-2. Other than what appears in the formula box, the worksheet appears exactly as it does in Figure 1-4. Click each cell in the range A7:A12 and you will see that the formula does not appear to change. Click on each cell in the range F7:F12, and again the formula does not appear to change. Excel keeps track of the elements in the ranges that you have named.

The Geometry of Crystals and Diffraction

To the casual observer, the beauty of crystals is characterized by their color and the symmetry of their planes. The external symmetry and planes of crystals arise from the regular three-dimensional periodic repetitions (translations) of the smallest unit of the crystal, often an atom or a molecule. The number of different periodic translations is surprisingly small, and these are described by the fourteen Bravais lattices (Figure 1-7). When side and body centering are excluded, the Bravais lattices reduce to the six crystal systems (Table 1-4).

Miller Indices. In addition to the planes defining the unit cell, any number of equally spaced parallel planes may be drawn through the unit cell. Three such planes are shown in Figure 1-8. These three planes are characterized by their Miller indices. The Miller indices of a plane are the reciprocals of the fractional intercepts of a plane along the crystal axis. In Figure 1-8, the plane labeled *I,* and closest to the origin, intercepts the x-axis at (1/3) a_0, (¾) b_0 and (½) c_0. Equally spaced parallel planes have the same Miller indices as shown in Table 1-5. Compare the intercepts of the planes in Figure 1-8 with the intercepts listed in Table 1-5. These planes are referred to as the 946 planes of the crystal, and the spacing between them is d_{946}. In general the spacing between planes is labeled $d_{hk\ell}$.

Table 1-4

The Six Crystal Systems

Crystal System	Lattice Parameters
Cubic	$a = b = c$ $\alpha = \beta = \gamma = 90^\circ$
Hexagonal	$a = b \neq c$ $\alpha = \beta = \gamma = 120^\circ$
Tetragonal	$a = b \neq c$ $\alpha = \beta = \gamma = 90^\circ$
Orthorhombic	$a \neq b \neq c$ $\alpha = \beta = \gamma = 90^\circ$
Monoclinic	$a \neq b \neq c$ $\alpha = \beta = 90^\circ \neq \gamma$
Triclinic	$a \neq b \neq c$ $\alpha \neq \beta \neq 90^\circ; \gamma$

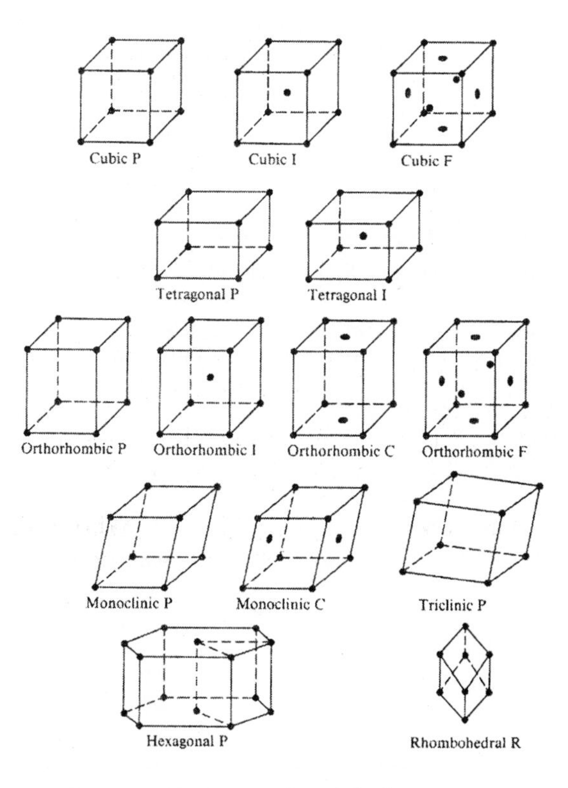

Figure 1-7. The Fourteen Bravais Lattices

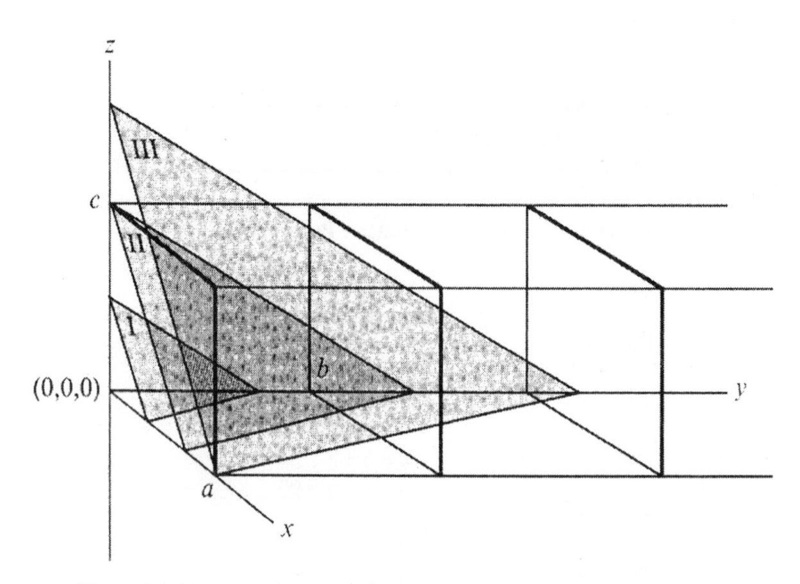

Figure 1-8. Members of a Family of Planes

The smallest unit of a crystal that has all the properties of the bulk crystal is called the unit cell (Figure 1-8). It is useful to visualize the unit cell bounded by planes that are separated by the translational repeat distance a_0, b_0, and c_0. The angle between a_0 and b_0 is γ, the angle between c_0 and a_0 is β, and the angle between b_0 and c_0 is β. The angles are not labeled in Figure 1-8.

Table 1-5

Miller Indices of Planes I, II, and III in Figure 1-8

Intercepts on the *a, b,* and *c* axes	Reciprocal of Intercepts	Cleared Fractions	Miller Indices
Plane I			
1/3	3	9/3	9
3/4	4/3	4/3	4
1/2	2	6/3	6
Plane II			
2/3	3/2	9/6	9
6/4	4/6	4/6	4
1	1	6/6	6
Plane III			
1	1	9/9	9
9/4	4/9	4/9	4
3/2	2/3	6/9	6

Miller Indices and the Spacing between Planes. Figure 1-9 facilitates the calculation of the spacing between planes. In Figure 1-9, the line OP is $d_{hk\ell}$ and is the perpendicular distance from the origin at O to the plane drawn through ABC. The intercepts on the axes are shown in parentheses.

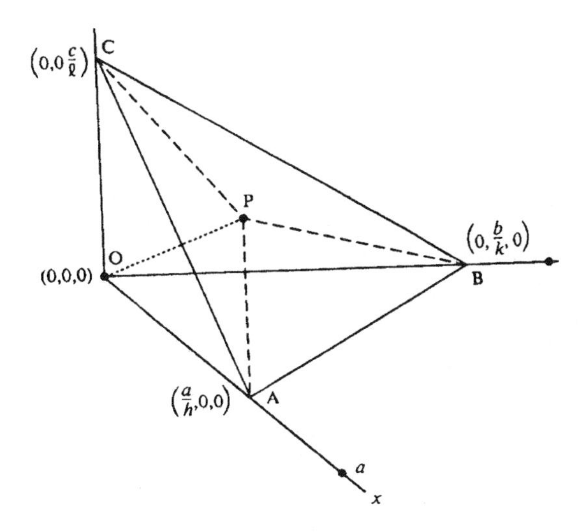

Figure 1-9. The line OP, which equals $d_{hk\ell}$, is the distance between the plane ABC and a plane (not shown) passing through the origin and parallel to plane ABC. The Miller indices of a plane are its fractional intercepts (shown in parentheses) on the axes.

The direction cosines of the line OP are

$$\cos \text{POA} = \frac{d_{hk\ell}}{a/h} \tag{1-4}$$

$$\cos \text{POB} = \frac{d_{hk\ell}}{b/k} \tag{1-5}$$

$$\cos \text{POV} = \frac{d_{hk\ell}}{c/\ell} \tag{1-6}$$

The law of direction cosines for the line PO in Figure 1-9 is

$$\cos^2 \text{POA} + \cos^2 \text{POB} + \cos^2 \text{POC} = 1 \tag{1-7}$$

Combining Equation 1-7 with Equations 1-6, 1-7, and 1- 8 gives

$$\frac{d_{hk\ell}^2 h^2}{a^2} + \frac{d_{hk\ell}^2 k^2}{b^2} + \frac{d_{hk\ell}^2 \ell^2}{c^2} = 1 \tag{1-8}$$

Separating out $d_{hk\ell}$. gives

$$d_{hk\ell} = \frac{1}{\sqrt{\frac{h^2}{a^2} + \frac{k^2}{b^2} + \frac{\ell^2}{c^2}}} \tag{1-9}$$

For a cubic crystal, $a = b = c$, and Equation 1-9 reduces to

$$d_{hk\ell} = \frac{a}{\sqrt{h^2 + k^2 + \ell^2}} \tag{1-10}$$

The Diffraction of X-Rays

In Figure 1-10, a beam of parallel X-rays traveling from left to right is reflected from planes drawn through the atoms in a crystal lattice. The reflected rays on the right are in phase with each other only when the path difference BC + CD between the rays equals an integral number of wavelengths. The angle theta is the diffraction angle.

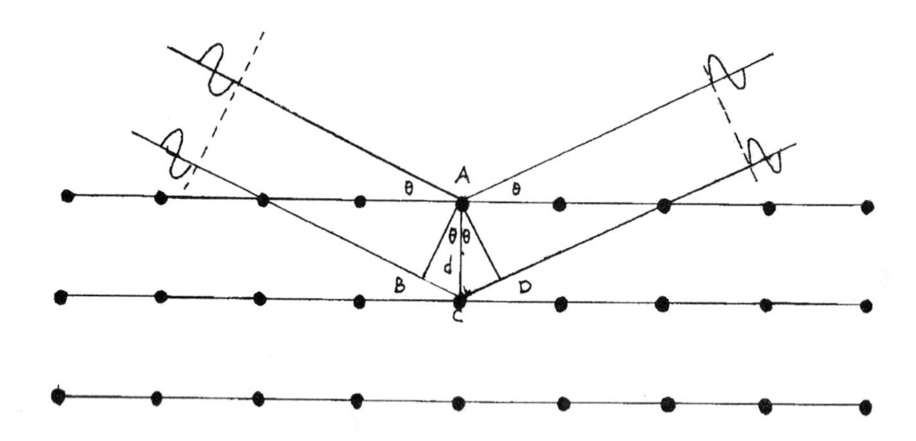

Figure 1-10. The Diffraction of X-Rays by Crystal Planes

When the diffraction angle theta corresponds to constructive interference, the wavelets are in phase so that the distance BC + CD equals an integral number of wavelengths.

$$\mathrm{BC} + \mathrm{CD} = n\lambda \tag{1-11}$$

From the geometry of the reflection

$$BC = d \sin\theta \tag{1-12}$$

and

$$CD = d \sin\theta \tag{1-13}$$

so that

$$n\lambda = 2d \sin\theta \tag{1-14}$$

This equation for the constructive interference of reflected X-rays is known as Bragg's law. Because higher order (n) reflections are equivalent to reflection from higher order planes, Bragg's law is usually written

$$\lambda = 2d_{hk\ell} \sin\theta \tag{1-15}$$

Wilhelm Konrad Röntgen, physics professor at Würzburg University, discovered X-rays in 1895. Reasoning (correctly) that the layers of atoms in a crystal might serve as a diffraction grating for X-rays, Max von Laue and his coworkers were the first to observe the diffraction of X-rays. For this work, Laue was awarded the Nobel Prize for physics in 1914. The quantitative interpretation of diffraction (Bragg's law) was made by Sir William Henry Bragg and his son Sir William Lawrence Bragg. The Braggs were awarded the Nobel Prize for physics in 1915. That the Nobel committee awarded consecutive prizes indicates that the committee recognized the importance of X-ray diffraction. Even today the elucidation of the crystal and molecular structures of drugs, proteins, and other solids depends on Bragg's law.

Excel Example 2

Student Grade Sheet and Histogram

Preparing the Worksheet

The final worksheet is shown in Figure 1-11.The instructor constructs this worksheet over a period of time, a semester. She begins by entering a title beginning in cell B1. The title includes a section number to distinguish it from other sections. She leaves row 2 blank and enters the column labels **Last** in cell A3, **First** in cell B3, and **SSN** in cell C3. Several column labels for quizzes, exams, labs, final exams, and total scores follow these entries.

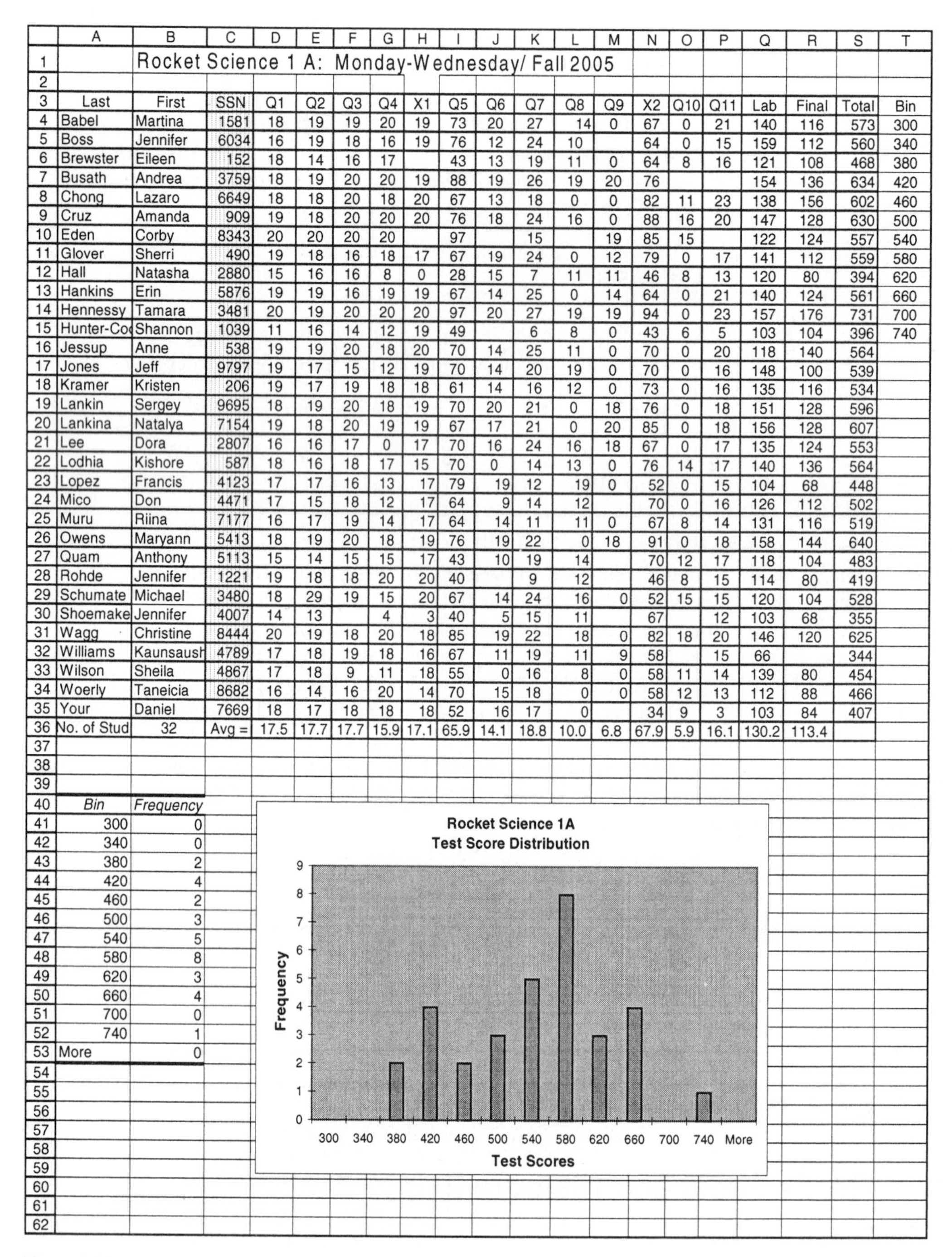

	A	B	C	D	E	F	G	H	I	J	K	L	M	N	O	P	Q	R	S	T
1		Rocket Science 1 A: Monday-Wednesday/ Fall 2005																		
2																				
3	Last	First	SSN	Q1	Q2	Q3	Q4	X1	Q5	Q6	Q7	Q8	Q9	X2	Q10	Q11	Lab	Final	Total	Bin
4	Babel	Martina	1581	18	19	19	20	19	73	20	27	14	0	67	0	21	140	116	573	300
5	Boss	Jennifer	6034	16	19	18	16	19	76	12	24	10		64	0	15	159	112	560	340
6	Brewster	Eileen	152	18	14	16	17		43	13	19	11	0	64	8	16	121	108	468	380
7	Busath	Andrea	3759	18	19	20	20	19	88	19	26	19	20	76			154	136	634	420
8	Chong	Lazaro	6649	18	18	20	18	20	67	13	18	0	0	82	11	23	138	156	602	460
9	Cruz	Amanda	909	19	18	20	20	20	76	18	24	16	0	88	16	20	147	128	630	500
10	Eden	Corby	8343	20	20	20	20		97		15		19	85	15		122	124	557	540
11	Glover	Sherri	490	19	18	16	18	17	67	19	24	0	12	79	0	17	141	112	559	580
12	Hall	Natasha	2880	15	16	16	8	0	28	15	7	11	11	46	8	13	120	80	394	620
13	Hankins	Erin	5876	19	19	16	19	19	67	14	25	0	14	64	0	21	140	124	561	660
14	Hennessy	Tamara	3481	20	19	20	20	20	97	20	27	19	19	94	0	23	157	176	731	700
15	Hunter-Co	Shannon	1039	11	16	14	12	19	49		6	8	0	43	6	5	103	104	396	740
16	Jessup	Anne	538	19	19	20	18	20	70	14	25	11	0	70	0	20	118	140	564	
17	Jones	Jeff	9797	19	17	15	12	19	70	14	20	19	0	70	0	16	148	100	539	
18	Kramer	Kristen	206	19	17	19	18	18	61	14	16	12	0	73	0	16	135	116	534	
19	Lankin	Sergey	9695	18	19	20	18	19	70	20	21	0	18	76	0	18	151	128	596	
20	Lankina	Natalya	7154	19	18	20	19	19	67	17	21	0	20	85	0	18	156	128	607	
21	Lee	Dora	2807	16	16	17	0	17	70	16	24	16	18	67	0	17	135	124	553	
22	Lodhia	Kishore	587	18	16	18	17	15	70	0	14	13	0	76	14	17	140	136	564	
23	Lopez	Francis	4123	17	17	16	13	17	79	19	12	19	0	52	0	15	104	68	448	
24	Mico	Don	4471	17	15	18	12	17	64	9	14	12		70	0	16	126	112	502	
25	Muru	Riina	7177	16	17	19	14	17	64	14	11	11	0	67	8	14	131	116	519	
26	Owens	Maryann	5413	18	19	20	18	19	76	19	22	0	18	91	0	18	158	144	640	
27	Quam	Anthony	5113	15	14	15	15	17	43	10	19	14		70	12	17	118	104	483	
28	Rohde	Jennifer	1221	19	18	18	20	20	40		9	12		46	8	15	114	80	419	
29	Schumate	Michael	3480	18	29	19	15	20	67	14	24	16	0	52	15	15	120	104	528	
30	Shoemake	Jennifer	4007	14	13		4	3	40	5	15	11		67		12	103	68	355	
31	Wagg	Christine	8444	20	19	18	20	18	85	19	22	18	0	82	18	20	146	120	625	
32	Williams	Kaunsaush	4789	17	18	19	18	16	67	11	19	11	9	58		15	66		344	
33	Wilson	Sheila	4867	17	18	9	11	18	55	0	16	8	0	58	11	14	139	80	454	
34	Woerly	Taneicia	8682	16	14	16	20	14	70	15	18	0	0	58	12	13	112	88	466	
35	Your	Daniel	7669	18	17	18	18	18	52	16	17	0		34	9	3	103	84	407	
36	No. of Stud	32	Avg =	17.5	17.7	17.7	15.9	17.1	65.9	14.1	18.8	10.0	6.8	67.9	5.9	16.1	130.2	113.4		

	A	B
40	*Bin*	*Frequency*
41	300	0
42	340	0
43	380	2
44	420	4
45	460	2
46	500	3
47	540	5
48	580	8
49	620	3
50	660	4
51	700	0
52	740	1
53	More	0

Figure 1-11. A Student Grade Worksheet 11

Entering Names and SSN. When she receives the enrollment printout from the registrar's office, she enters the students' names in columns A and B and their Social Security numbers in column C. If some of the names are too long to fit, she drags the line between the column A label and the column B label until the longest name fits. She does the same for the first names if necessary. After she enters the last student's name, she enters **No. of Stud.** in cell A36. She enters **=ROWS(B4:B35)** in cell B36. That formula appears in the formula box, and the number **32** appears in cell B35. If a couple of students enroll late, she inserts new rows by clicking **Rows** on the **Insert** menu and filling in the students' names and Social Security numbers. The enrollment number changes automatically from 32 to the new total enrollment. This is especially useful in courses with enrollments in the hundreds.

The columns for recording the quiz grades are wider than necessary, so the instructor clicks the column D label and drags to column P to select these columns. On the **Format** menu, she points to **Column**, and clicks **Width** and sets the width to **4**. Similarly, she sets the widths for the columns labeled Lab, Final, and Total to **5**.

Entering Grades. She tells the students that they will receive a short quiz at the end of this and every week except exam week. After giving the first quiz, she records the grades in the Q1 column in the range D4:D35. She enters the formula **=AVERAGE(D4:D35)** in cell D36, and the average for the first quiz appears: **17.5**. The average is high, but she intentionally made the first quiz on the easy side in order not to intimidate the new students.

At the end of each week the instructor enters the grades and the command to calculate the average. In cell S4, she enters the Excel command **=SUM(D4:R4)** under the column labeled Total. After two weeks, the value in cell S4 is 37 (the sum of 18 and 19, the values entered in cells D4 and E4); at the end of the semester, the value is 573. She enters this command in all the cells in column S by clicking cell S4 and dragging to cell S35.

Sorting

At the end of the semester it is time to assess the students' totals and assign grades. First, to sort the grades in descending order, the teacher selects the block of data to be sorted (including column labels).

- She selects the range A3:S35 by clicking cell A3 and then dragging to cell S35.
- On the **Sort** menu, she clicks **Data** to open the **Sort** dialog box (Figure 1-12).
- In the **Sort by** box, she scrolls until she reaches **Total** in the drop-down list.
- She selects **Descending** to sort from the highest grade to the lowest grade.

Note: she can get the original alphabetical order by sorting on the column labeled Last.

When she clicks OK on the Sort dialog box, the entire worksheet is sorted with the highest grade at the top. She sees that the minimum total is 344 and the maximum total is 731. In a very high enrollment class she probably would have use Excel's MIN (range) and MAX (range) commands. On this worksheet the range of interest is S4:S35.

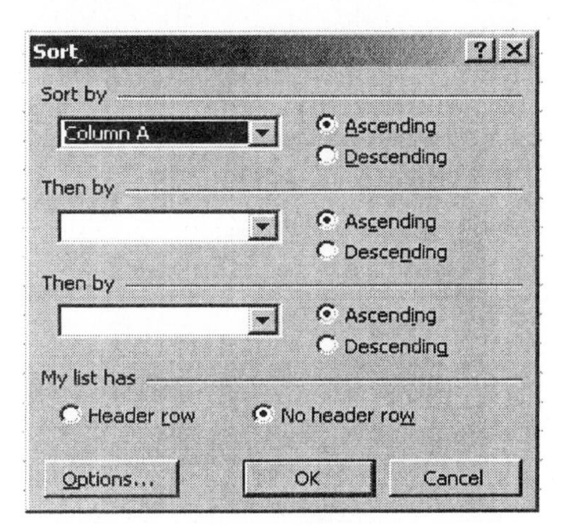

Figure 1-12. The *Sort* Dialog Box

Creating the Histogram

To assist her in assigning grades to the test totals, the instructor prepares a histogram of the test-total frequency distribution.

Based on the minimum and maximum totals, she divides the test-total range into twelve "bins" and enters their values into worksheet column T under the label Bin (Figure 1-11). The bin values range from 300 to 740, which includes the minimum and maximum scores.

- On the **Tools** menu, she clicks **Data Analysis,** which opens the **Data Analysis** dialog box (Figure 1-13).
  Note: if the **Data Analysis** command does not appear on the **Tools** menu, see the "Add-Ins" section at the end of this chapter.
- She scrolls to **Histogram** and clicks **OK**, which opens the **Histogram** dialog box (Figure 1-14).

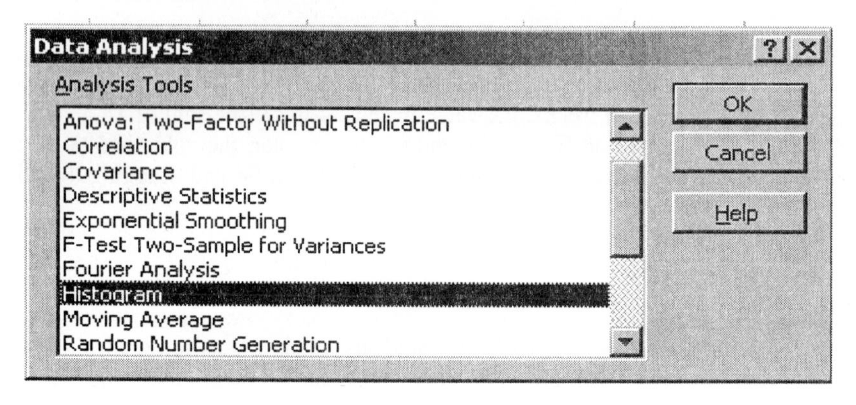

Figure 1-13. The **Data Analysis** Dialog Box

- She types **S4:S35** in the **Input Range** box, and she types **T4:T15** in the **Bin Range** box.
- She then selects the **Output Range** option and types **A40** (a cell in a clear area on the worksheet).

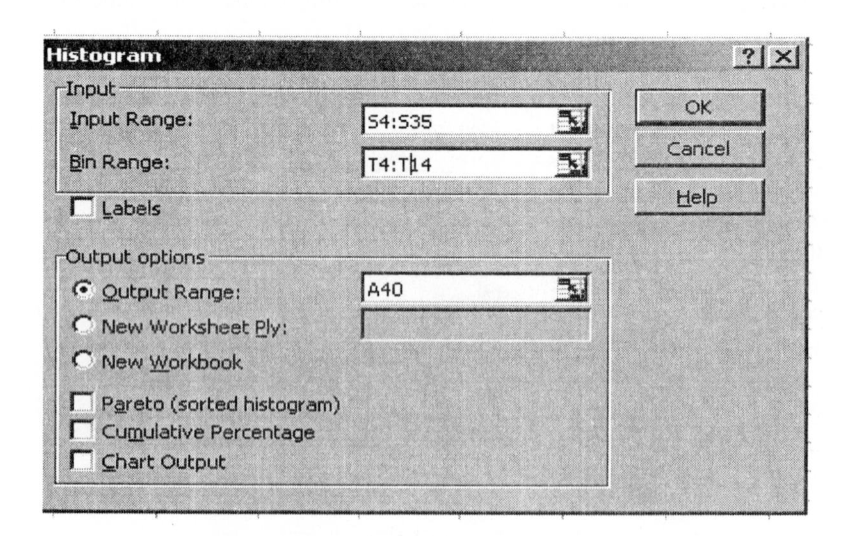

Figure 1-14. The **Histogram** Dialog Box 14

After the teacher clicks OK, a table of bins and frequencies appears in the range A40:B53 (Figure 1-15). Its upper left corner is located at cell A40. A histogram appears to the right of the table. The default histogram (Figure 1-15) needs editing. Figure 1-11 shows the location of the table and the histogram after editing.

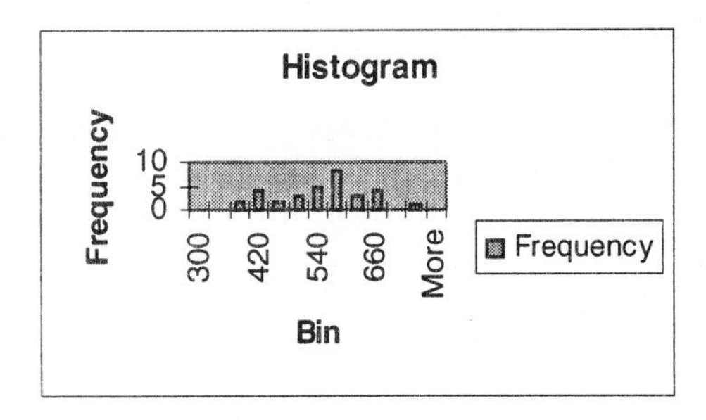

Figure 1-15. The Default Histogram 14

Editing the Histogram

The default histogram is shown in Figure 1-15. It is squashed in appearance, its title is not very informative, it is too small, and the legend box gives redundant information. First format the size and shape of the chart.

- Click in a blank area of the chart so that size handles (■) appear at the chart's corners and sides.
- Place the pointer on the bottom middle size handle, and when it changes to a double arrow, drag the size handle down to about row 20 to double the chart height.
- Right-click the legend box (on the right, with the word *Frequency*); then click **Clear**.

Next, edit the chart title and axis titles.

- Place the pointer in the chart title (the default is **Histogram**).
- Click twice slowly (do not double-click). The pointer remains in the title.
- Edit the title as you would with a word processor to resemble the tilte in Figure 1-11.
- Repeat for the x-axis title (**Bin**). Type **Test Scores** as the new title. The y-axis default title (**Frequency**) does not need editing, but you would edit it the same way.

The chart now appears as in Figure 1-11. We could do some more editing, but we will come back to chart editing later. However, if you wish to experiment with the chart at this time, several right-click operations are useful in editing. We will come back to the details of chart editing later.

- Right-click the chart title or axis titles to change font and colors.
- Right-click the bars in a bar graph to change their color.
- Right-click the chart background to change its colors or borders.
- Right-click the x-axis or y-axis numbers to change:
  1. Patterns (borders and tic marks)
  2. Scale (more important on line graphs)
  3. Font (type, size, italic, bold, color)
  4. Number (general, currency, scientific, or other format)
  5. Alignment (parallel, perpendicular, or slanted)

Add-Ins

Not all of the available tools are listed in the **Tools** menu, but you can add them to the menu if you like. For example, if the **Data Analysis** command is missing from the **Tools** menu, here's how to include it.

- On the **Tools** menu, click **Add-Ins** to open the **Add-Ins** dialog box (Figure 1-16).
- Select the **Analysis ToolPak** check box.
- Click **OK** and proceed. (The **Data Analysis** command now appears in the **Tools** menu.)

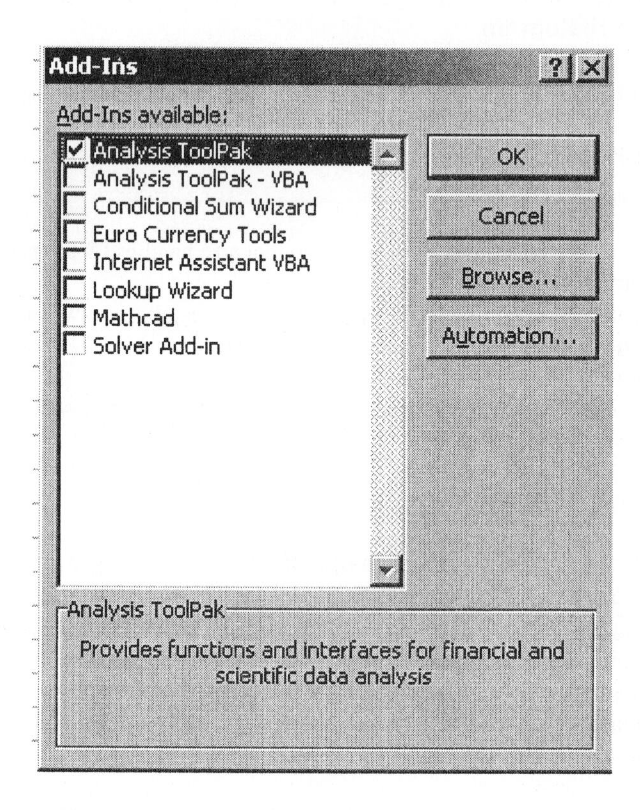

Figure 1-16. The **Add-Ins** Dialog Box

Chapter 2

Thermodynamics

Excel Example 3

A Single Point-Plot: Heat Capacity of Silver

Review: The First Law of Thermodynamics

Before using Excel to plot the heat capacity of silver, it is useful to review some of the relevant equations and definitions from thermodynamics. The first law of thermodynamics, in the form of an equation, is given by

$$\Delta U = q + w \tag{2-1}$$

This equation states that the increase in the internal energy ΔU of the system is equal to the heat q absorbed by the system plus the work w done on the system. When heat is released from the system, q is negative; when work is done by the system, w is negative. Many different kinds of work can be done on a system, for example, electrical, surface tension (creating new surface area), elongation (stretching a rubber band), and pressure-volume (PV) work of expansion (or compression) of a gas. It is PV work that is most common in chemical and engineering thermodynamics:

$$\delta w = -P_{op}dV \tag{2-2}$$

or

$$w = -\int_{V_1}^{V_2} P_{op}dV \tag{2-3}$$

P_{op} is the opposing or external pressure against which an expansion is done, and V_1 and V_2 are the initial and final volumes of the system. The choice of sign in Equation 2-3 is conventional, but arbitrary. Once a plus sign is used in Equation 2-1, then a minus sign must be used in Equations 2-2 and 2-3. Suppose a gas is compressed with a constant opposing (external) pressure. Then upon integration, Equation 2-3 becomes

$$w = -P_{op}(V_2 - V_1) \tag{2-4}$$

In compression, work is done on the system and $V_2 < V_1$, so a negative sign is required to give a positive value to *w*, which is needed when work is done on the system.

Internal Energy. When the first law (Equation 2-1) in differential form is combined with Equation 2-2, the result is

$$dU = \delta q - P_{op} dV \tag{2-5}$$

If the system undergoes a change of state under conditions of constant volume, then *dV* equals zero, δw equals zero, and no *PV* work is done on or by the system. Then

$$dU = \delta q_v \tag{2-6}$$

At constant volume, the increase in the internal energy of a system is equal to the heat absorbed by the system and

$$\Delta U = q_v \tag{2-7}$$

When heat is absorbed at constant pressure, Equation 2-5 may be written

$$U_2 - U_1 = q_P - P_{op}(V_2 - V_1) \tag{2-8}$$

Enthalpy. Experimentally, it is much more common to carry out changes of state under conditions of constant pressure (isobaric) than constant volume (isochoric). For example, most chemical reactions are carried out in vessels open to the constant pressure of the atmosphere. Because constant pressure processes are so common, it is convenient to invent a new thermodynamic function, *H*, the enthalpy, such that

$$H = U + PV \tag{2-9}$$

The change in enthalpy for a change of state taking place at constant pressure is then

$$H_2 - H_1 = U_2 - U_1 + P(V_2 - V_1) \tag{2-10}$$

Comparison of Equation 2-10 with Equation 2-8 shows that

$$H_2 - H_1 = q_P \tag{2-11}$$

or

$$\Delta H = q_P \tag{2-12}$$

The heat absorbed by a system undergoing a change of state at constant pressure is equal to the increase in the enthalpy of the system. By inventing the enthalpy function, the expressions for the heat absorbed at constant volume (Equation 2-7) and for the heat absorbed at constant pressure (Equation 2-12) become equally simple.

The Heat Capacity. The heat capacity is defined historically as the amount of heat required to raise the temperature of a substance by 1 degree C and has the symbol C and unit of J/K.

$$C = \frac{\Delta q}{\Delta t} \tag{2-13}$$

Since the heat absorbed depends on whether it is absorbed at constant pressure or constant volume, the conditions must be specified, giving rise to two kinds of heat capacities, defined by

$$C_V = \frac{\mathrm{d}q_V}{\mathrm{d}T} = \left(\frac{\partial U}{\partial T}\right)_V \tag{2-14}$$

$$C_P = \frac{\mathrm{d}q_P}{\mathrm{d}T} = \left(\frac{\partial H}{\partial T}\right)_P \tag{2-15}$$

Heat capacities are ordinarily measured at constant pressure, meaning in containers open to atmospheric pressure. The slight change in temperature dT is measured upon heating a sample with a carefully measured dq_P supplied by electrical energy for a short time dt. The electrical energy $dH = I^2Rdt$ is measured at constant pressure, that is, in an apparatus open to the (constant) pressure of the atmosphere. Data for elemental silver are tabulated in Table 2-1. The heat capacity of silver decreases as the temperature decreases.

The worksheet created in this Excel Example will be used in Excel VBA Example 9 in Chapter 18. A Visual Basic for Applications program will determine the entropy of silver by calculating the area under the *Cp* versus *lnT* plot.

Preparing the Worksheet

Our goal is to enter the data in Table 2-1 into an Excel worksheet and prepare a plot of *Cp* versus *lnT*. The first part of the project consists of entering the text labels and numerical data into the worksheet. The second part concerns generating the chart and then editing its appearance to our taste.

Entering the Labels and Data

- Beginning in cell D1, enter **The Heat capacity of Silver** as the title.
- In cells A3, B3, and C3, enter **T/K**, **lnT**, and **Cp(J/K/mol)** as the column labels.
- Drag the line between column B and C so the label in cell C3 fits.
- In the range A5:A20, enter the temperature data of Table 2-1.
- In cell B5, enter **=LN(A5)** as the formula.
- Click cell B5, and drag the fill handle down to cell B20.
- Select the range B5:B20, and on the **Format** menu, click **Cell** to open the **Format Cells** dialog box (Figure 1-5).
  1. Select the **Number** category.
  2. In the **Decimal places** box, type **2** and click **OK**.

Creating the Chart

Table 2-1

Heat Capacity of Silver

Cp (J/K/mol)	T/K
0.67	15
4.77	30
11.65	50
16.33	70
19.13	90
20.96	110
22.13	130
22.97	150
23.61	170
24.09	190
24.42	210
24.73	230
25.03	250
25.31	270
25.44	290
25.50	300

Data from Mead, P. F., W. K. Forsythe, and W. F. Giauque. *J. Am. Chem. Soc.* 63: 1902 (1941).

Excel provides a large number of different kinds of charts, many of which are especially important in business applications. The chart that is most important to scientists and engineers is the XY (scatter) chart. When you plot y values (dependent variable) and x values (independent variable) in a series of experiments or tests, you are plotting what Excel categorizes as an XY (scatter) chart. Scientists and engineers usually refer to *plots* or *graphs*, rather than *charts*, a word from the business world.

With its Chart Wizard, Excel makes it easy to create a chart from x,y data. The icon for the Chart Wizard (Figure 2-1) is located on Excel's standard toolbar (Figure 1-1) and appears as a tiny, blue, yellow, and red bar chart. The Chart Wizard leads the user through a four-step process of creating the chart and labeling it. Clicking the **Chart Wizard** icon opens the **Chart Wizard – Step 1 of 4 – Chart Type** dialog box (Figure 2-2). The chart that the Chart Wizard creates is quite good, although the user will almost always find it profitable to edit the resulting chart. Excel provides easily accessible methods for editing the chart, and we will use several of them.

Figure 2-1. *Chart Wizard Icon*

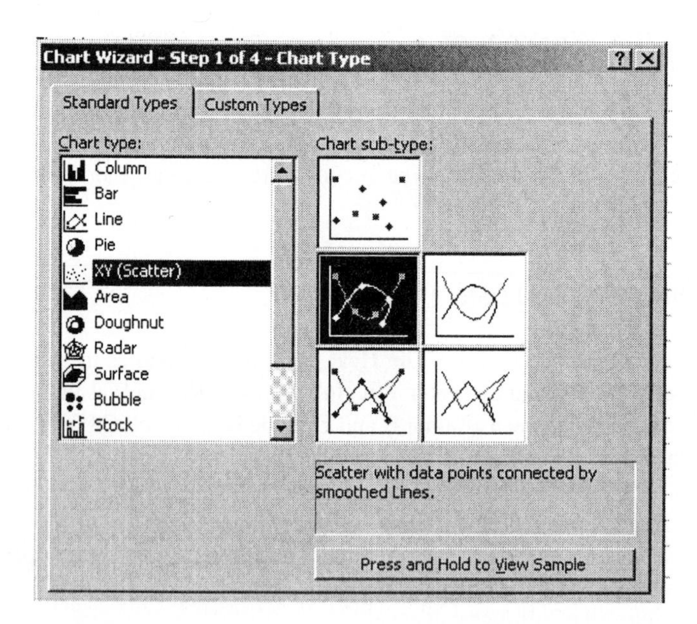

Figure 2-2. The **Chart Wizard – Step 1 of 4 – Chart Type** Dialog Box

To use the Chart Wizard, first select the numerical data to be plotted, and then click the **Chart Wizard** icon.

- Select the *x,y* data to be graphed, the block B5:C20. Excel expects the *x* data to be in a column to the left of the *y* data. This is why we placed the *lnT* data in the column to the left of the *Cp* data.
- Click the **Chart Wizard** icon.
- In the **Chart Wizard – Step 1 of 4 – Chart Type** dialog box (Figure 2-2):
    1. Click the second chart icon (selected in Figure 2-3); its label reads **Scatter with data points connected by smoothed lines.**
    2. Click **Next.**
- In the **Chart Wizard – Step 2 of 4 – Chart Source Data** dialog box (Figure 2-3):
    1. You get a first look at the chart in this step.
    2. The data range you chose is displayed, and in the **Series in** options, **Columns** is selected.
    3. Click **Next** if the chart appears to be what you expected.
- In the **Chart Wizard – Step 3 of 4 – Chart Options** dialog box (Figure 2- 4):
    1. Click the **Gridlines** tab (Figure 2-4), and for both **Value (X) axis** and **Value (Y) axis**, select **Major gridlines.**
    2. Click the **Titles** tab (Figure 2-5):
        a. In the **Chart title** box, type **The Heat Capacity of Silver** as the title.
        b. In the **Value (X) axis** box, type **lnT** as the value.
        c. In the **Value (Y) axis** box, type **C_P (J/K/mol)** as the value.
        d. Click **Next.**
- In the **Chart Wizard – Step 4 of 4 – Chart Location** dialog box (Figure 2-6):
    1. Select **As object in Sheet1**. This selection embeds the chart in the worksheet.
    2. Selecting **As new sheet** covers a whole new worksheet with the chart and names the worksheet Chart1 (unless you eventually rename it). Keep your selection **As object in Sheet1** for now.

The chart produced at this point is probably not in the best position on the worksheet. The chart may be too small or too large. The plot area may be too small or too large. The title may need a second line or change of font size or type. The x-axis and y-axis numbers may need changing, their labels may need editing, and the scale might not be quite right. The line colors may need changing, and the lines themselves may be too thick or too thin. Excel's editing options allow all these deficiencies to be corrected. The approach to solving these problems is nearly the same in each case. Just right-click the object to be edited, and then proceed. In these procedures *click* means to use the button on the left side of the mouse. *Right-click* means to use the button on the right side of the mouse. Figure 2-7 displays the edited chart embedded in the worksheet.

Exploring the Chart

Before systematically editing the chart, explore the chart a little.

- Notice that the menu bar does not contain a **Chart** menu.
- Click anywhere in the chart, and you see the **Tools, Data,** and **Window** menus in the menu bar change to the **Tools, Chart,** and **Window** menus.
- Click the **Chart** menu to see what commands it contains. Click each of these commands, and some of the dialog boxes that appear will be familiar. Cancel each dialog box you open so you don't make any changes yet. This is one approach to editing the chart.

Now right-click each different type of object (area, line, text, or number) that you can see on your chart successively. Each time you right-click an object, a shortcut appears. The first item on this menu is **Format [*the object you just right-clicked*]**. Click the plot area, and it reads **Format Plot Area**. Click the chart area, and it reads **Format Chart Area**. Click a number, and it reads **Format Axis**. This is a second approach to editing the chart and the one that we will use for the most part.

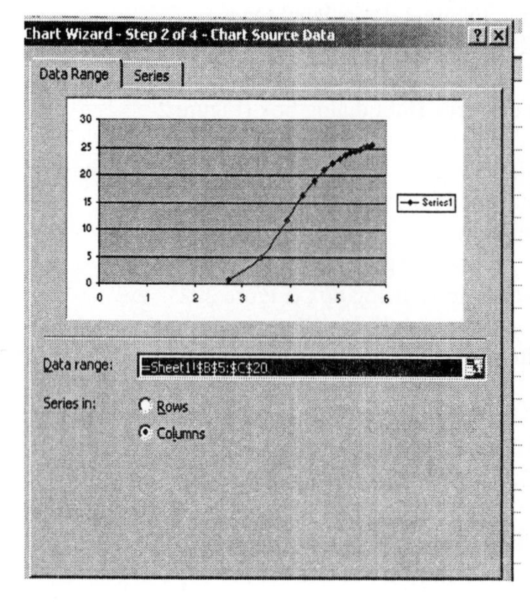

Figure 2-3. The **Chart Wizard – Step 2 of 4 – Chart Source Data** Dialog Box

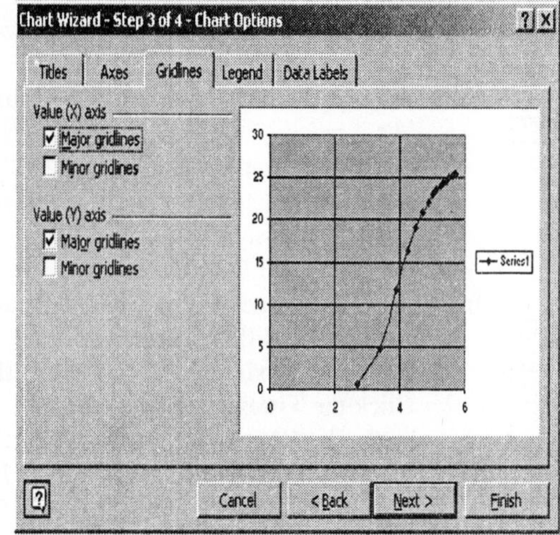

Figure 2-4. The **Chart Wizard – Step 3 of 4 – Chart Options** Dialog Box with the **Gridlines** Tab Selected

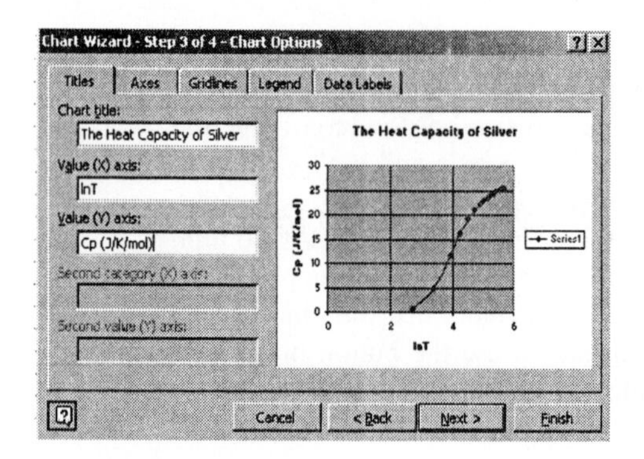

Figure 2-5. The **Chart Wizard – Step 3 of 4 – Chart Options** Dialog Box with the **Titles** Tab Selected

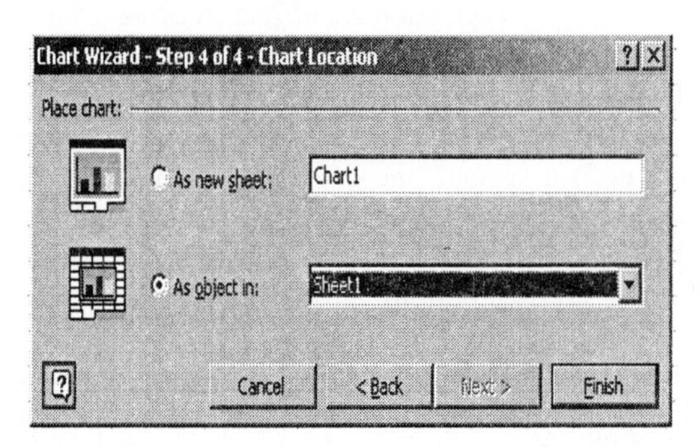

Figure 2-6. The **Chart Wizard – Step 4 of 4 – Chart Location** Dialog Box

Size and Position of the Chart on the Worksheet

- Click the legend box on the chart to select it; then press the DELETE key.
- Click anywhere in a blank area of the chart, and drag the chart to where you would like to position it on the worksheet.
- Scroll the worksheet to reveal a few rows under the chart, and then click the bottom edge size handle (■) and drag it to increase the size of the chart vertically. Experiment with the side size handles.

- Click anywhere in the empty part of the plot area, and then enlarge the plot area by clicking and dragging the side or bottom size handles.

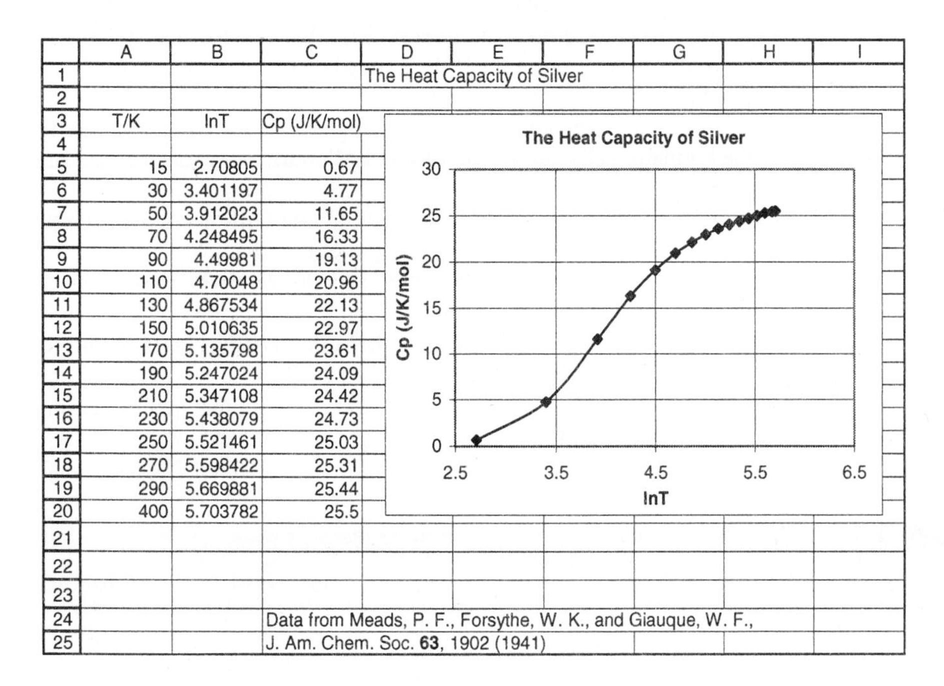

	A	B	C	D	E	F	G	H	I
1				The Heat Capacity of Silver					
2									
3	T/K	lnT	Cp (J/K/mol)						
4									
5	15	2.70805	0.67						
6	30	3.401197	4.77						
7	50	3.912023	11.65						
8	70	4.248495	16.33						
9	90	4.49981	19.13						
10	110	4.70048	20.96						
11	130	4.867534	22.13						
12	150	5.010635	22.97						
13	170	5.135798	23.61						
14	190	5.247024	24.09						
15	210	5.347108	24.42						
16	230	5.438079	24.73						
17	250	5.521461	25.03						
18	270	5.598422	25.31						
19	290	5.669881	25.44						
20	400	5.703782	25.5						
21									
22									
23									
24			Data from Meads, P. F., Forsythe, W. K., and Giauque, W. F.,						
25			J. Am. Chem. Soc. **63**, 1902 (1941)						

Figure 2-7. The Spreadsheet for the Heat Capacity of Silver

Editing the Title

- Click twice anywhere in the chart title (but don't double-click). An insertion point, which looks like a vertical line, appears.
- Move the insertion point to the right end of the title by pressing the END key. You could write more at this point, but don't.
- Instead, press the ENTER key, and the pointer drops to the next line below. Type **15 to 300K** for the second line. You could enter several lines if you want.
- Right-click the chart title, then click **Format Chart Title** to open the **Format Chart Title** dialog box (Figure 2-8); then click the **Font** tab, and change the font to 12-point bold or whatever appears satisfactory to you.

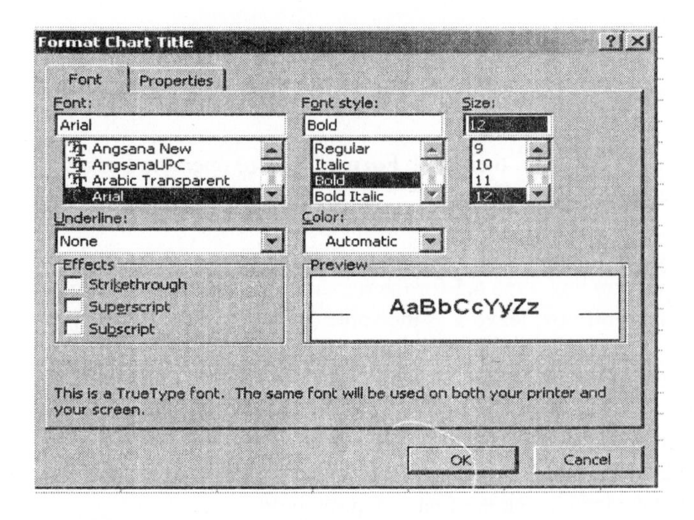

Figure 2-8. The **Format Chart Title** Dialog Box

Improving the Plot Area

- Right-click in the plot area between the horizontal lines to open a shortcut menu.
- Click **Format Plot Area** to open the **Format Plot Area** dialog box (Figure 2-9).
  1. In the **Area** side (the right side of the dialog box), change the gray background to white by clicking a white square in the palette.
  2. In the **Border** side (the left side of the dialog box), click **Color**. Then click the black square on the color palette. Click **Weight**, and choose a medium-heavy line. Click **OK**.

  (If you right-click in the plot area and then click **Chart Options** in the shortcut menu, you get a dialog box that looks like step 3 of the Chart Wizard, with five tabs labeled **Titles, Axes, Gridlines, Legend,** and **Data Labels.** Excel provides many ways of letting you change your mind.)
- Right-click anywhere along the plotted line, then click **Format Data Series** to open the **Format Data Series** dialog box (Figure 2-10).
- In the **Color** box, click the down arrow and select black (in the upper left corner of palette), and then click **OK**.

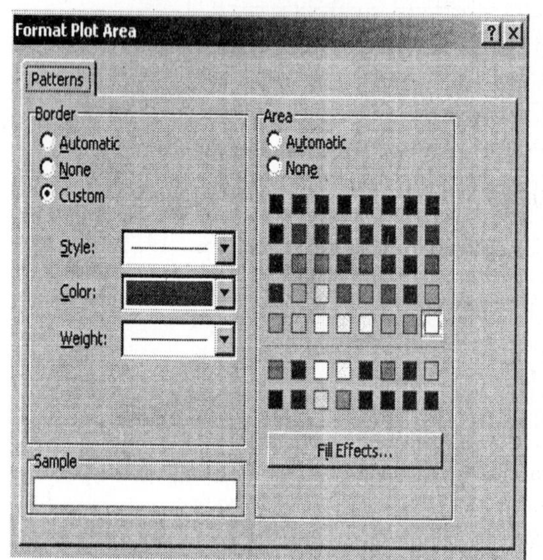

Figure 2-9. The **Format Plot Area** Dialog Box

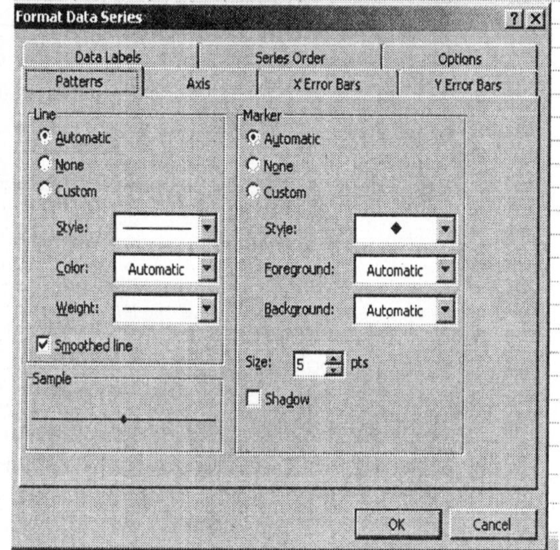

Figure 2-10. The **Format Data Series** Dialog Box

Customizing the X-Axis

- Right-click an x-axis number, then click **Format Axis** to open the **Format Axis** dialog box (Figure 2-11).
- Click the Scale tab.
  1. In the **Minimum** box, type **2.5** for the minimum value.
  2. In the **Maximum** box, type **6.5** for the maximum value.
  3. In the **Major unit** box, leave **2** as the value (it was set automatically by Excel).
  4. In the **Minor Unit** box, leave **0.4** as the value (it was set automatically by Excel).
  5. In the **Value (Y) axis crosses at** box, leave **0** as the value (it was set automatically by Excel).
- Click the **Font** tab, and change the font to 8 point bold.

- Click the **Number** tab; click **Number** in the **Category** list, and change the number of decimal places to **1**.

Customizing the Y-Axis

- Right-click any y-axis number, then click **Format Axis** to open the **Format Axis** dialog box (Figure 2-11).
- Click the **Scale** tab. In the **Maximum** box, type **30** as the value; in the **Minimum** box, type **0** as the value.
- Click the **Font** tab, and change the typeface to 8 point bold.
- Click the **Number** tab, and change the number of decimal places to 0.

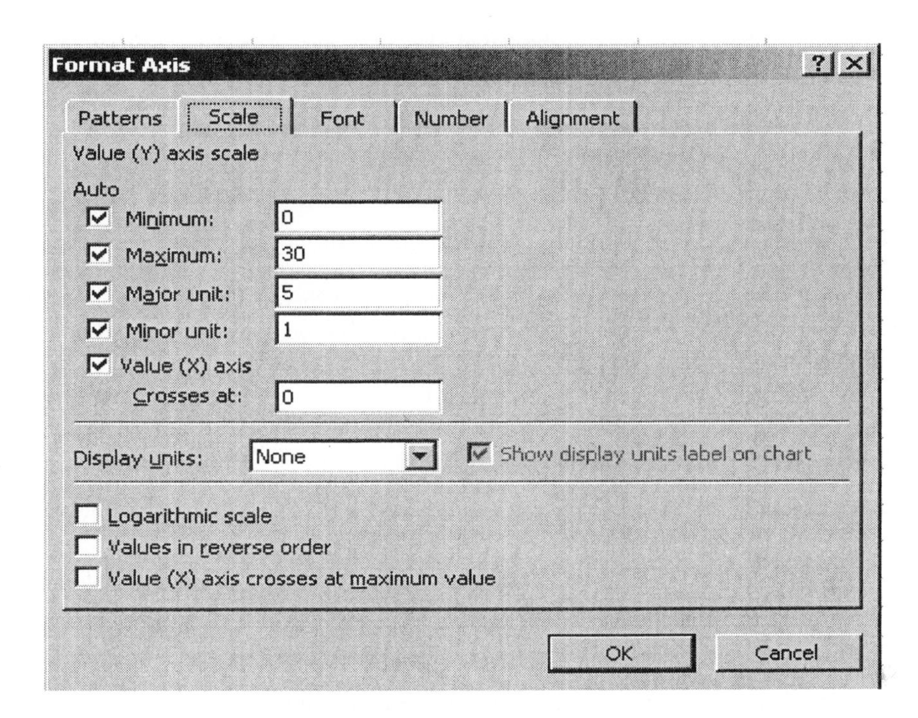

Figure 2-11. The **Format Axis** Dialog Box

Relocating the Chart

The chart is finished. If you would like to look at a larger version, you can change its location property from **As object in Sheet1** to **As new sheet**. To do this:

- Click in an empty area of the chart to have the **Chart** menu appear in the menu bar.
- On the **Chart** menu, point to **Location**.
- Click **As new sheet**.

The chart appears occupying the entirety of its own sheet, which has the default name Chart 1. You can name the chart if you wish. The chart in this location can be edited exactly as we have already done for the chart **As object in Sheet1**. To relocate the chart back to **As object in Sheet1**, use the same procedure. You will find that the size of the chart changes slightly, but a couple of tugs on the size handles will quickly bring it back to the size you previously had.

Excel Example 4

Calculation of Vapor Pressures with the Antoine Equation: Plotting a Function

Review of Vapor Pressures

The equilibrium between pure A in the liquid phase and pure A in the gas phase may be written:

$$\mathrm{A}(\ell, P) = \mathrm{A}(g, P) \tag{2-16}$$

where P is the equilibrium vapor pressure. Measurements of vapor pressures are relatively simple and are valuable because their measurement over a range of temperatures permits the calculation of several important thermodynamic quantities: the enthalpy, entropy, and free energy of vaporization. The equilibrium constant for the equilibrium vaporization is given by

$$K = \frac{a_A(g)}{a_A(\ell)} \tag{2-17}$$

The activity of the pure liquid $a_A(\ell)$ equals unity, and the activity of the gas phase $a_A(g)$ equals P_A / P_A^0. Usually the standard state P_A^0 is chosen to be the ideal gas at 1 atm or 1 bar. The expression for the equilibrium constant then is simply:

$$K = P_A \tag{2-18}$$

The free energy of vaporization is related to the equilibrium constant by

$$-RT \ln K = \Delta G^0 = \Delta H^0 - T\Delta S^0 \tag{2-19}$$

This relationship is valid if the difference in heat capacity between the gas phase and the liquid phase ΔC_p is zero or small and may be rearranged to:

$$\ln P = \frac{-\Delta H^0}{RT} + \frac{\Delta S^0}{R} \tag{2-20}$$

A plot of *lnP* versus *1/T* gives a straight line, the slope of which equals $-\Delta H^0 / R$ and the intercept of which equals $\Delta S^0 / R$. These relationships suggest that vapor-pressure data can be summarized in an equation of the form

$$\log P = \frac{m}{T} + b \tag{2-21}$$

Another equation, and one that is used here, is the empirical Antoine equation, which is of the form

$$\log P = A - \frac{B}{t + C} \tag{2-22}$$

where *A, B,* and *C* are empirical constants and *t* is the temperature in degrees Celsius. Selected constants are shown in Table 2-2.

Table 2-2

Antoine Constants for $\log P = A - \frac{B}{t + C}$
(Pressures in torr and temperatures in degrees C)

BP	Formula	Name	*A*	*B*	*C*
76.72	CCl_4	carbon tetrachloride	6.92180	1235.172	228.93
61.204	$CHCl_3$	chloroform	6.95465	1170.966	226.25
64.55	CH_3OH	methanol	8.08097	1582.27	239.726
117.897	CH_3COOH	acetic acid	7.38782	1533.313	222.30
78.298	C_2H_5OH	ethanol	8.11220	1592.864	226.18
56.707	CH_3COCHH_3	acetone	7.11714	1210.595	229.66
79.589	$CH_3COC_2H_5$	methyl ethyl ketone	7.06356	1261.339	221.96
77.063	$C_4H_8O_2$	ethyl acetate	7.10179	1244.951	217.88
65.965	C_4H_8O	tetrahydrofuran	6.99515	1202.296	226.25
80.102	C_6H_6	benzene	6.89272	1203.531	219.88
80.737	C_6H_{12}	cyclohexane	6.84941	1206.001	223.14
68.740	$C6H_{14}$	hexane	6.88555	1175.817	224.86
110.622	C_7H_8	toluene	6.95805	1346.773	219.69
97.153	C_3H_8O	1-propanol	7.74416	1437.686	198.46

Source: T. Boublik, V. Fried, and E. Hála, *The Vapor Pressures of Pure Substances*, Amsterdam: Elsevier, 1973

Excel Calculation of the Vapor Pressure of Acetone

Solving the Antoine equation for P (in torr) allows the direct calculation of vapor pressures at a variety of temperatures t (in degrees C). Earlier we used Excel's functions. This example gives us the opportunity to calculate the vapor pressure with a user-defined function.

$$P = 10^{A - \frac{B}{t+C}} \tag{2-23}$$

Entering the Title and Labels

- On the **Drawing** toolbar, click the **Text Box** icon; then click and drag over the range B2:E5.
- In the text box, type **Vapor Pressures Calculated with the Antoine Equation** as the title.
- Center the title, and change it to 12 point bold.
- Create another text box in the range C7:D10.
- On the **Insert** menu, point to **Object**, and click **Microsoft Equation**, and enter as text:

$$\log P = A - \frac{B}{t+C}$$

- Select cell B12, type eight spaces, and type **For Acetone:** as the label.
- In the range D12:D14, type the labels **A =**, **B =**, and **C =**, and right-align them.
- In the range E12:E14, type **7.11714**, **1210.595**, and **229.664**, and left-align these numbers.
- Select cell E12. On the **Insert** menu, click **Name**, and type **A** as the entry name. Name cell E13 **B** and name cell E14 **Z**. (Excel does not permit the use of **C** as a name.)
- In cell A16, type **t/(deg. C)** and in cell B16, type **P/(torr)** as the column headings.

Entering the Data

- In cell A17, type **10** as the value.
- In cell A18, type **= A17 + 5** as the formula.
- Select cell A18, and drag the fill handle to cell A33.
- Select the range A17:A33; on the **Insert** menu, click **Name**, and type **t** as the name of the entry.
- Select cell B17, and type **= 10^(A – B/(t+Z))** as the formula.
- Select cell B17, and drag the fill handle to cell B33.

Creating the Vapor-Pressure Chart

- Select the range A17:B33; then click the **Chart Wizard** icon.
- In the **Chart Wizard – Step 1 of 4 – Chart Type** dialog box (Figure 2-2), select **XY (Scatter)** with lines but no points. Click **Next**.
- In the **Chart Wizard – Step 2 of 4 – Chart Source Data** dialog box (Figure 2-3), click **Next**.
- In the **Chart Wizard – Step 3 of 4 – Chart Options** dialog box, click the **Titles** tab (Figure 2-5):
  1. In the **Chart title** box, type **Vapor Pressure of Acetone** for the title.
  2. In the **Value (X) axis** box, type **t/(deg C)** as the label.
  3. In the **Value (Y) axis** box, type **P/(torr)** as the label.
  4. Click **Next**.
- In the **Chart Wizard – Step 4 of 4 – Chart Location** dialog box (Figure 2-6), Select **As Object in Sheet1**, and click **Next**.

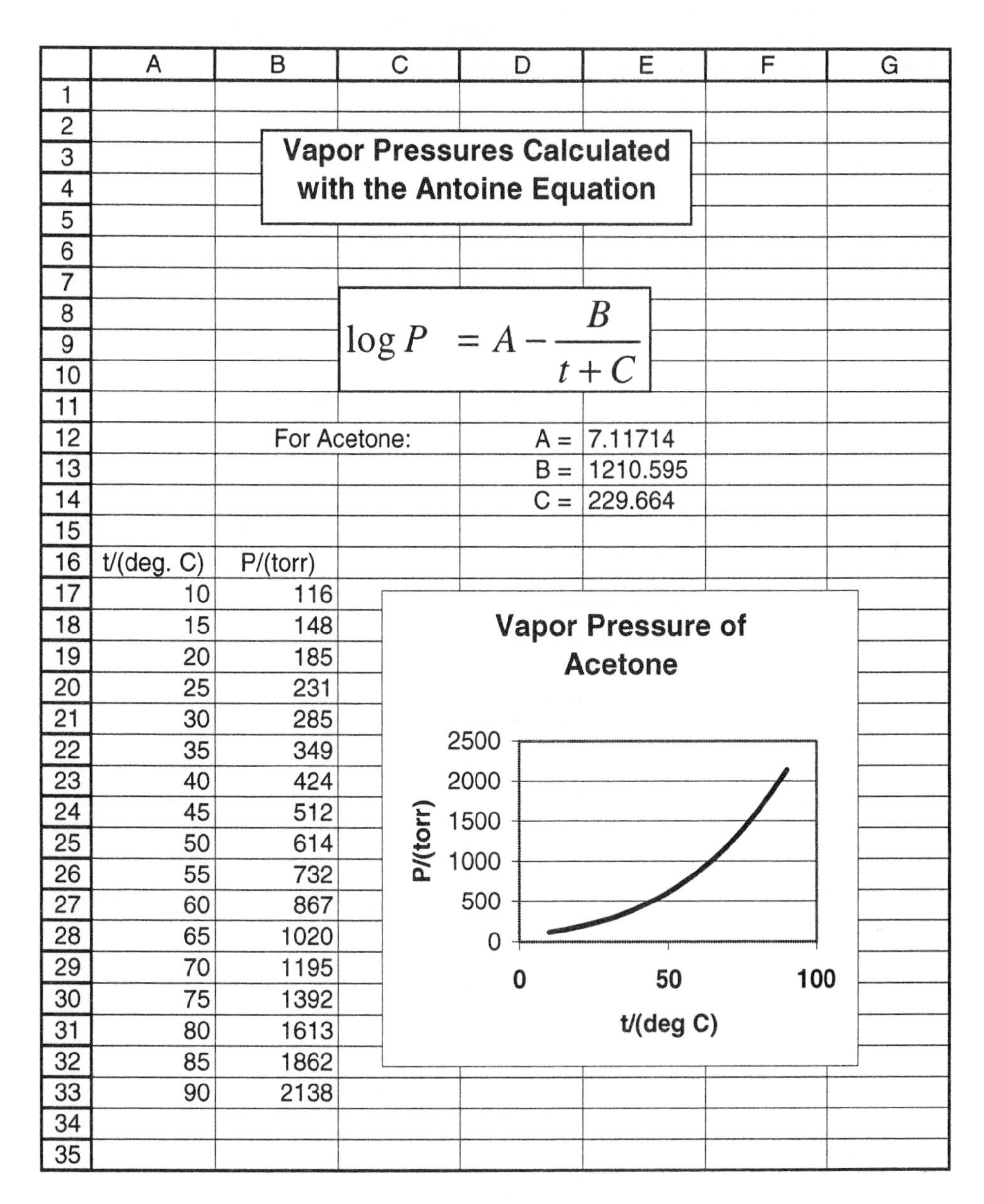

Vapor Pressures Calculated with the Antoine Equation

$$\log P = A - \frac{B}{t + C}$$

For Acetone:

A =	7.11714
B =	1210.595
C =	229.664

t/(deg. C)	P/(torr)
10	116
15	148
20	185
25	231
30	285
35	349
40	424
45	512
50	614
55	732
60	867
65	1020
70	1195
75	1392
80	1613
85	1862
90	2138

Figure 2-12. The Worksheet for the Vapor Pressure of Acetone

Editing the Vapor-Pressure Chart

- Right-click in the plot area, and click **Format Plot Area**. On the **Area** side of the **Format Plot Area** dialog box, select the white square.
- Right-click the title, and click **Format Title**. Set the font to 10 point bold.
- Click the legend box, and delete it.
- Right-click the plotted line, and in the **Format Data Series** dialog box, click the **Patterns** tab. In the **Weight** box, select a weight a little heavier than the default weight.

Excel Example 5

Multiple Point-Plots with Nonidentical *x* Values: The Vapor Pressures of Benzene and Chloroform

Scientists and engineers frequently need to make simple plots of *x,y* pairs to display visually the nature of the relationship between the independent variable *x* and the dependent variable *y*. In spreadsheet jargon, these point plots are called XY (scatter) plots. The points may be joined by line segments; a single line or no line may be drawn through them.

Creating multiple plots for several dependent variables is straightforward as long as each dependent variable is paired with the same value of the independent variable. Occasionally, it is necessary to display multiple XY (scatter) plots with data having nonidentical *x* values as shown in Table 2-3. The pressures are the dependent variable, and the temperatures are the independent variable.

Table 2-3

The Vapor Pressures of Benzene and Chloroform

Vapor Pressure of Benzene		Vapor Pressure of Chloroform	
t/deg. C	**P/mm**	**t/deg. C**	**P/mm**
3.0	30	10	100.5
21.3	80	20	159.6
35.3	150	25	199.1
52.3	300	30	246.6
72.6	600	35	301.3
78.4	720	40	366.4
80.1	760	45	439.0
		50	526.0
		55	625.2
		60	739.6
		61.2	760

Preparing the Worksheet

The position of the titles and labels in this worksheet does not appear different from worksheets described earlier. The array of *x,y* data (temperatures and pressures) may appear unusual. The particular arrangement is, however, necessary to plot data with unequal *x* coordinates (independent variable).

Entering the Data for XY (Scatter) Plots

In order for Excel to treat the two data sets for the vapor pressures of two different substances independently, it is necessary to enter the data as shown in Figure 2-13.

- Beginning in cell C1, type **Vapor Pressure vs. Temperature for Benzene and Chloroform** as the title.
- The *x* values for both plots (the temperatures in this case) are entered as two contiguous sets in the range A5:A22. The benzene temperatures are entered in the range A5:A11 and the chloroform temperatures are entered in the range A12:A22.
- The benzene vapor pressures are entered in the range B5:B11.
- The chloroform vapor pressures are displaced one column and entered in the range C12:C22.

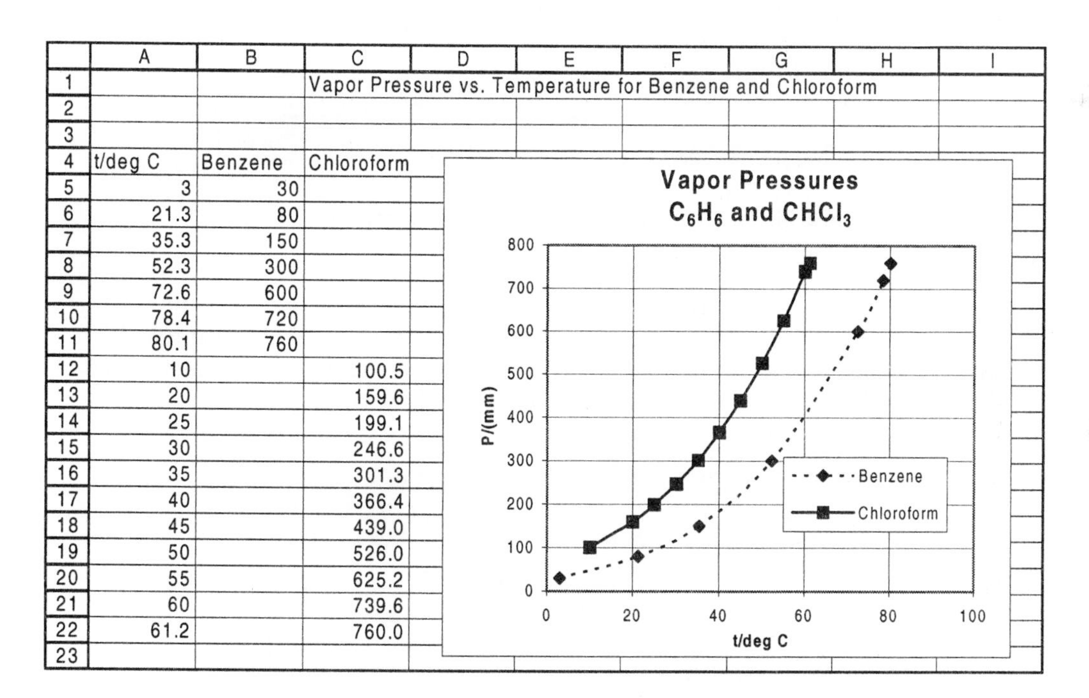

	A	B	C	D	E	F	G	H	I
1			Vapor Pressure vs. Temperature for Benzene and Chloroform						
2									
3									
4	t/deg C	Benzene	Chloroform						
5	3	30							
6	21.3	80							
7	35.3	150							
8	52.3	300							
9	72.6	600							
10	78.4	720							
11	80.1	760							
12	10		100.5						
13	20		159.6						
14	25		199.1						
15	30		246.6						
16	35		301.3						
17	40		366.4						
18	45		439.0						
19	50		526.0						
20	55		625.2						
21	60		739.6						
22	61.2		760.0						
23									

Figure 2-13. The Spreadsheet for the Vapor Pressures of Benzene and Chloroform

Creating the Chart

With the Chart Wizard (Figure 2-1), Excel automates the creation of the chart. While the immediate result is a little rough, it is a simple matter to edit the chart and make it look good. We have already done some chart editing in previous Excel Examples.

- Select cell A4 (t/deg C), and drag to cell C22 to select the entire block for the chart.

- With this block selected, click the **Chart Wizard** icon to open the Chart Wizard.

1. In the **Chart Wizard – Step 1 of 4 – Chart Type** dialog box, in the **Chart type** list, select **XY (Scatter)** (Figure 2-14). For **Chart sub-type**, select **Scatter with data points connected by smoothed lines**. Then click **Next**.

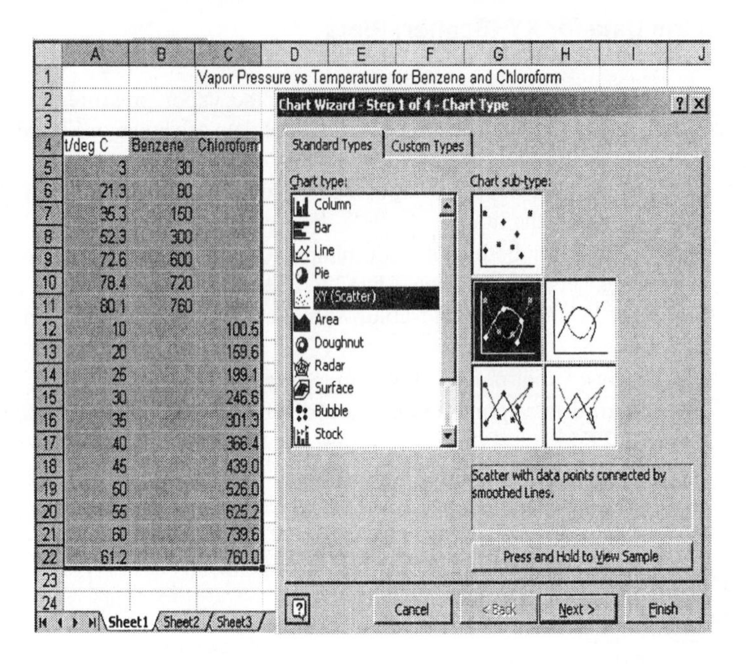

Figure 2-14. The **Chart Wizard – Step 1 of 4 – Chart Type** Dialog Box for Vapor Pressures

2. The **Chart Wizard – Step 2 of 4 – Chart Source Data** dialog box (Figure 2-15) displays the data range and a first look at the graph. Notice that Excel picks up the labels for the two dependent variables and places them in a legend box. There is no entry in this dialog box, so click **Next**.

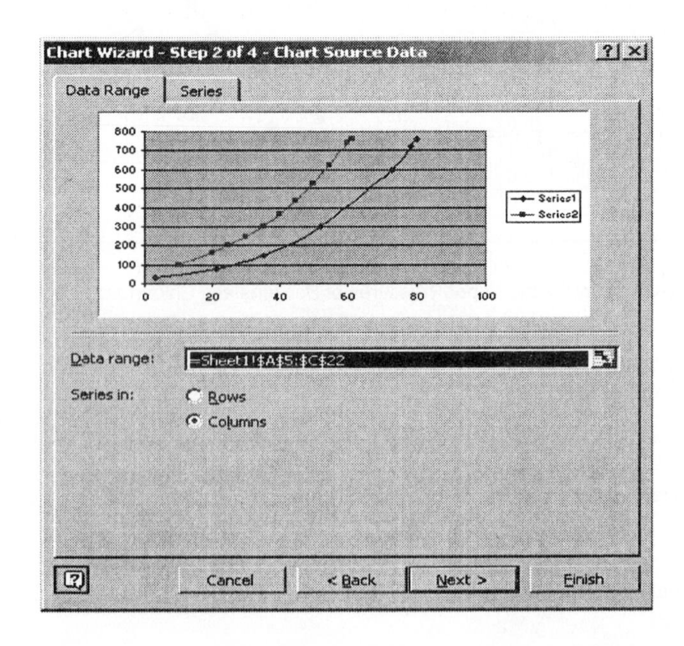

Figure 2-15. The **Chart Wizard – Step 2 of 4 – Chart Source Data** Dialog Box for Vapor Pressures

3. In the **Chart Wizard – Step 3 of 4 – Chart Options** dialog box (Figure 2-4):
    a. Click the **Gridlines** tab, and for both the **Value (X) axis** and the **Value (Y) axis** boxes, select **Major gridlines**.
    b. Click the **Titles** tab (Figure 2-5):
        - In the **Chart title** box, type **Vapor Pressure** for the title.
        - In the **Value (X) axis** box, type **t/(deg. C)** as the label.
        - In the **Value (Y) axis** box, type **P/(torr)** as the label.
        - Click **Next.**
4. In the **Chart Wizard – Step 4 of 4 – Chart Location** dialog box, select **As object in Sheet1**, and then click **Finish** (Figure 2-6).

Selecting **As object in Sheet1** results in the chart being embedded in, or floating on, the worksheet. Selecting **As new sheet** results in the chart filling a new worksheet. By selecting either chart (by clicking in it), and then, on the **Chart** menu, clicking **Location**, the location can be switched back and forth.

Editing the Chart

- Click the title twice (but do not double-click) to get a pointer in the title.
    1. Press the END key to move the pointer to the end of the title line.
    2. Press ENTER to generate a second title line. Type **C6H6 and CHCl3** as the second line.
    3. Select the first 6, and then right-click to open a shortcut menu.
    4. Click **Format [*selected chart title*]** to open the **Format Chart Title** dialog box (Figure 2-8).
        a. With the first 6 in the title selected, select the **Subscript** check box, then click **OK**.
        b. Repeat with the other 6 and the 3.
- Right-click the axis labels and numbers to edit them to your liking.
- Right-click the plotted chloroform line to open a shortcut menu.
- Click **Format Data Series** to open the **Format Data Series** dialog box (Figure 2-10).
    1. Scroll the **Style** list until it displays a dotted line. Click **OK**.
    2. Save your work and then experiment with other options in the dialog box.
- Click in a blank chart area to select the chart. Drag the size handles to resize it.
- Click in the plot area in order to size this area to your liking.
- Right-click in the plot area between the horizontal lines to open a shortcut menu.
- Click **Format Plot Area** to open the **Format Plot Area** dialog box (Figure 2-9). In the **Area** side (the right side of the dialog box), change the gray background to white by clicking a white square in the palette.

Notice that the labels in the legend arise from the column labels in cells B4 and C4. If you select the block A5:B22, the legend labels by default will be simply series 1 and series 2. You can delete the legend box by selecting it and pressing the DELETE key.

The arrangement and selection of data for this chart is not intuitive. The first guess of most spreadsheet users would be to place the chloroform data directly under the benzene data in column B, with these data contiguous. With this arrangement, the spreadsheet draws a spurious line from the last benzene data pair (80.1, 760) to the first chloroform data-pair (10, 100.5). Placing the chloroform data in its own column cures this problem (Stinson, 1992).

Chapter 3

Quantum Mechanics

Excel Example 6

Planck's Distribution of Wavelengths

A Brief History of Planck's Radiation Law

That color and temperature are connected is common knowledge as evidenced by expressions like "red hot" and "white hot." In the late 1800s physicists were making careful measurements on heat and color, that is, on temperature and wavelength. It was found that the experimental observations were reproducible, and the measurements were independent of material if the source of radiation were a black box.

A black box consists of a small, hollow and refractory object with a tiny hole drilled in it. When heated to incandescence, the color of the radiation emitted from the hole is brighter than the radiation emitted from the outside surface and is the same whether the object is made of gold, iron, carbon, aluminum oxide, or any other refractory material. By 1900 these experimental observations were carried out extremely reliably by Lummer and Pringsheim and by Rubens and Kurlbaum.

Early attempts by Raleigh-Jeans to construct a theory to account for these observations were a failure: his theory succeeded at long wavelengths but predicted that the intensity of radiation increased indefinitely as the wavelength became shorter and approached the ultraviolet. Because Raleigh-Jeans's theory was competently based on contemporary physics, his theory's failure became known as "the ultraviolet catastrophe."

In 1901 Max Planck successfully devised a theory that fit Lummer and Pringsheim's observations exactly. In applying statistical mechanics to the problem, he made the bold assumption that the oscillators (oscillating atoms in the black body) could have a discrete set of possible energy values, or energy levels, which were an integral multiple of $h\nu$, where ν is the frequency of the emitted radiation (s^{-1}) and, in his honor, h is named Planck's constant ($6.6260755 \times 10^{-34}$ J s). His assumption was bold because physicists at the time believed that energy must vary continuously. His assumption is important because it marks the birth of the quantum theory.

Planck's revolutionary 1901 paper was published in *Annalen der Physik*, volume 4, 553. The result is the Planck distribution of wavelengths:

$$\rho = \frac{8\pi hc}{\lambda^5}\left(\frac{1}{e^{hc/\lambda kT}-1}\right) \tag{3-1}$$

In this equation, ρ is the energy per unit of volume per unit of wavelength (J m^{-4}), h is Planck's constant (6.626×10^{-34} J s), c is the speed of light (2.998×10^{-8} m s^{-1}), λ is the wavelength (m), and k is the Boltzmann constant (1.381×10^{-23} J K^{-1}).

Entering the Worksheet Data

In Equation 3-1, the parameters have their usual symbols. The dependent variable is ρ, and the independent variable is λ. All the other parameters are constants, which we will enter into the worksheet (Figure 3-1) and give names along with units and labels to identify them.

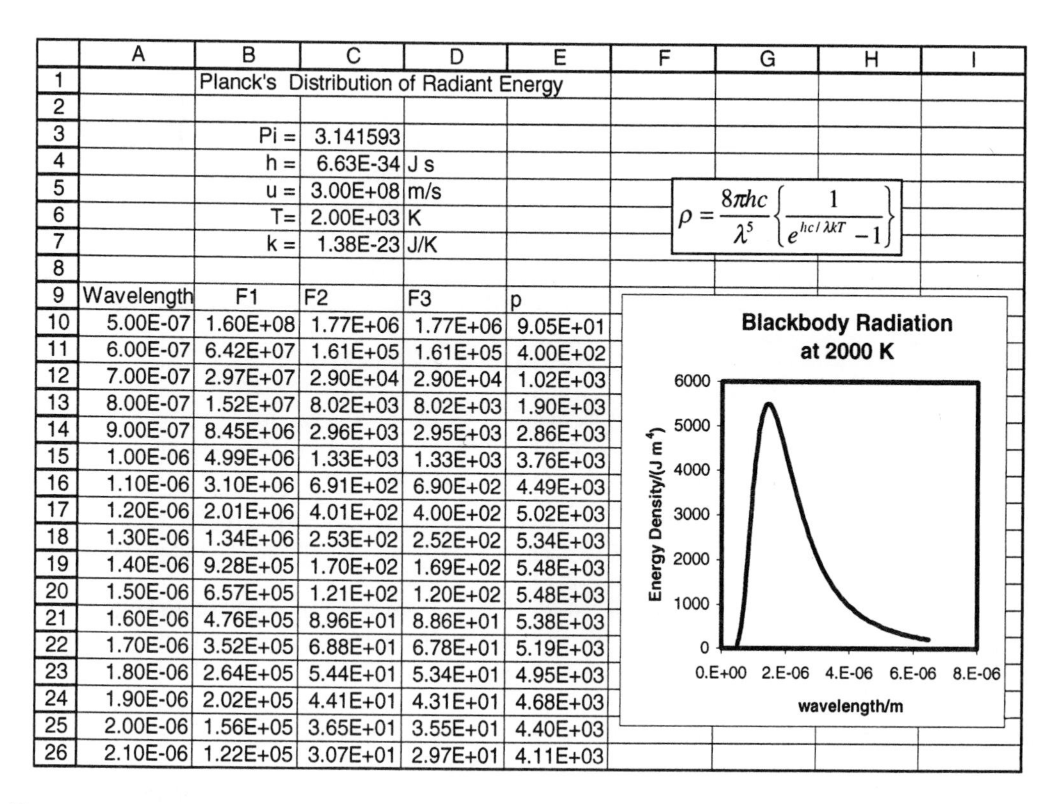

	A	B	C	D	E
1		Planck's Distribution of Radiant Energy			
2					
3		Pi =	3.141593		
4		h =	6.63E-34	J s	
5		u =	3.00E+08	m/s	
6		T=	2.00E+03	K	
7		k =	1.38E-23	J/K	
8					
9	Wavelength	F1	F2	F3	p
10	5.00E-07	1.60E+08	1.77E+06	1.77E+06	9.05E+01
11	6.00E-07	6.42E+07	1.61E+05	1.61E+05	4.00E+02
12	7.00E-07	2.97E+07	2.90E+04	2.90E+04	1.02E+03
13	8.00E-07	1.52E+07	8.02E+03	8.02E+03	1.90E+03
14	9.00E-07	8.45E+06	2.96E+03	2.95E+03	2.86E+03
15	1.00E-06	4.99E+06	1.33E+03	1.33E+03	3.76E+03
16	1.10E-06	3.10E+06	6.91E+02	6.90E+02	4.49E+03
17	1.20E-06	2.01E+06	4.01E+02	4.00E+02	5.02E+03
18	1.30E-06	1.34E+06	2.53E+02	2.52E+02	5.34E+03
19	1.40E-06	9.28E+05	1.70E+02	1.69E+02	5.48E+03
20	1.50E-06	6.57E+05	1.21E+02	1.20E+02	5.48E+03
21	1.60E-06	4.76E+05	8.96E+01	8.86E+01	5.38E+03
22	1.70E-06	3.52E+05	6.88E+01	6.78E+01	5.19E+03
23	1.80E-06	2.64E+05	5.44E+01	5.34E+01	4.95E+03
24	1.90E-06	2.02E+05	4.41E+01	4.31E+01	4.68E+03
25	2.00E-06	1.56E+05	3.65E+01	3.55E+01	4.40E+03
26	2.10E-06	1.22E+05	3.07E+01	2.97E+01	4.11E+03

Figure 3-1. The Blackbody Radiation Spreadsheet

Entering the Title, Constants, Names, and Labels

- Beginning in cell B1, type **Planck's Distribution of Radiant Energy** as the title.
- In the range B3:B7, type **Pi =**, **h =** , **u =** , **T =**, and **k =** as the labels.
- In the range D4:D7, type **J s**, **m/s**, **K**, and **J/K** as the labels.
- In cell C3, type **=Pi()** as the formula. Only seven digits are displayed, but pi is stored here as a fifteen-digit number. In Excel, **Pi** is a function without parameters.
- Define the name for the value in cell C3 to be **Pi**. To name a cell, select the cell by clicking in it, and then on the **Insert** menu, point to **Name**, and click **Define** to open the **Define Name** dialog box (Figure 3-2). Type **Pi** for the name. See also Excel Example 1.
- In cell C4, type **6.626E-34** and then, on the **Insert** menu, point to **Name**, and click **Define** to open the **Define Name** dialog box; type **h** as the name of the cell.
- In cell C5, type **2.998E8** and then, on the **Insert** menu, point to **Name**, and click **Define** to open the **Define Name** dialog box; type **u** as the name of the cell.
- In cell C6, type **2000** and then, on the **Insert** menu, point to **Name**, and click **Define** to open the **Define Name** dialog box; type **T** as the name of the cell.

- In cell C7, type **1.381E-23** and then, on the **Insert** menu, point to **Name**, and click **Define** to open the **Define Name** dialog box; type **k** as the name of the cell.
- Select cell F4. On the **Insert** menu, point to **Object**, click **Microsoft Equation Editor,** and enter as a text equation object:

$$\rho = \frac{8\pi hc}{\lambda^5}\left\{\frac{1}{e^{hc/\lambda kT} - 1}\right\}$$

By naming the constants, we make the functions that we are about to enter much more readable than if we had used cell addresses. Notice that we named cell C5 as **u** instead of **c**, the usual symbol for the speed of light. The reason? Excel does not accept **c** as a name (Figure 3-3). If Excel does not like the name you use, just change it.

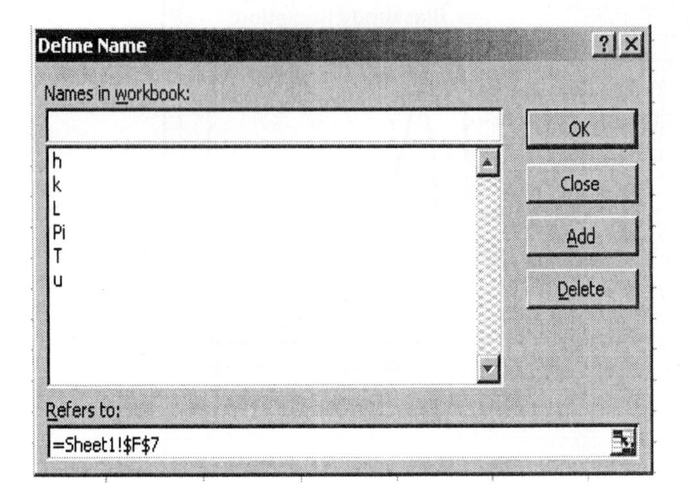

Figure 3-2. The **Define Name** Dialog Box

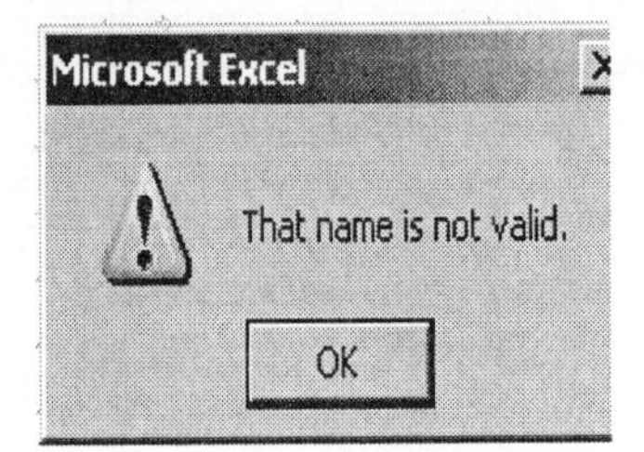

Figure 3-3. The **Invalid Name** Dialog Box.

Entering the Numerical Data and Functions. We want to plot the Planck distribution from 500 nm (5×10^{-7} m) to 6500 nm (6.5×10^{-6} nm), so we'll begin with 5×10^{-7} m and increment it by 1×10^{-7} m until we reach 6.5×10^{-6} m. The function (Equation 3-1) that we want to enter is fairly complex, so the entry is more readable if we break the function down into some factors (*F1, F2,* and *F3*) whose product equals the complete function ρ.

- In the range A9:E9, type **Wavelength**, **F1**, **F2**, **F3**, and **p** (for ρ) as the labels.
- In cell A10, type **5E-7** as the value.
- In cell A11, type **=A10+1E-7** as the formula. The value in cell A11 is now **6.00E-07**.
- Select cell A11, and drag the fill handle in its lower right corner to cell A70.
- In cell B10, type **=(8*Pi*h*u/A10^5)** and drag the fill handle to cell B70.
- In cell C10, type **=EXP(h*u/(k*T*A10)** and drag the fill handle to cell C70.
- In cell D10, type **=C10 - 1** and drag the fill handle to cell D70.
- In cell E10, type **=B10/D10** and drag the fill handle to cell E70.

Creating the Chart

Creating the plot of a function is essentially the same as creating the plot of a set of x,y point pairs. The initial step in creating an XY (scatter) chart is the selection of the x values and the y values in the columns of data. When two columns are selected, Excel by default assigns the left column to the x values and the right column to the y values. It usually simplifies the data selection if this characteristic of Excel is recognized when the programmer initially decides in which columns to locate the data.

The x,y columns do not need be adjacent. If the two columns are adjacent, the desired range can be selected by clicking and dragging from the upper left corner of the desired cells to the lower right corner. It is optional to include the column labels in the selected block. If a title is located in a row above the labels, Excel will try to use it as the title for the chart. These choices can always be edited.

If the columns are not adjacent, the left (x values) column is selected first by clicking in the topmost cell and dragging down over the desired range. The second column, on the right (y values), is selected by CTRL-clicking its topmost cell and dragging down over the desired range. In both cases, the columns may be selected by clicking the appropriate column letter at the top of the worksheet, if the columns have no text labels. Even when labels lie between the numerical data and the column letters, Excel does a quite good job of selecting the data for producing the chart. If it doesn't work satisfactorily, you can always drag over the specifically desired data columns.

Our x values lie in the A column in the range A10:A70. Our y values lie in the E column in the range E10:E70. You could click cell A10 and drag to cell A70 and then CTRL-click cell E10 and drag to cell E70. Then, by clicking the **Chart Wizard** icon, you could proceed. Instead, let's challenge Excel by clicking the A and E columns.

- Select the title and label block in the range A1:E9. Cut and paste it far to the right somewhere, for example, in the range J1:N9. We'll retrieve it later.
- Click the column A label. The entire column is selected.
- CTRL-click the column E label. The entire column is selected, along with column A.

Both column A and column E darken to indicate that they have been selected. This is clear in Figure 3-4.

- Click the **Chart Wizard** icon on the standard toolbar.
- In the **Chart Wizard – Step 1 of 4 – Chart Type** dialog box, select **XY (Scatter)** and a graph with a smooth line and no markers. Then click **Next.**
- In the **Chart Wizard – Step 2 of 4 – Chart Source Data** dialog box (Figure 3-4), which gives us our first glimpse of the chart, no action is needed. Click **Next.**
- In the **Chart Wizard – Step 3 of 4 – Chart Options** dialog box:
  1. In the **Chart title** box, type **Blackbody Radiation** for the title.
  2. In the **Value (X) axis** box, type **Wavelength/m** as the label.
  3. In the **Value (Y) axis** box, type **Energy Density/(J m4)** as the label.
  4. Click **Next.**
- In the **Chart Wizard – Step 4 of 4 – Chart Location** dialog box, select **As object in Sheet1**; click **Finish.**
- Go back to the worksheet and retrieve the label block by moving the labels and titles placed temporarily in range J1:N9. Cut and paste to A1:E9.

Editing the Chart

Edit the Title

- With the chart **As object in Sheet1,** click the title twice (but do not double-click) to get a pointer in the title.
  1. Press the END key to move the pointer to the end of the title line.
  2. Press ENTER to generate a second title line. Type **at 2000 K** for the second line of the title.
- Right-click the title, then click **Format Chart Title**.
- In the **Format Chart Title** dialog box (Figure 2-8), select 8 point bold; then click **OK**.

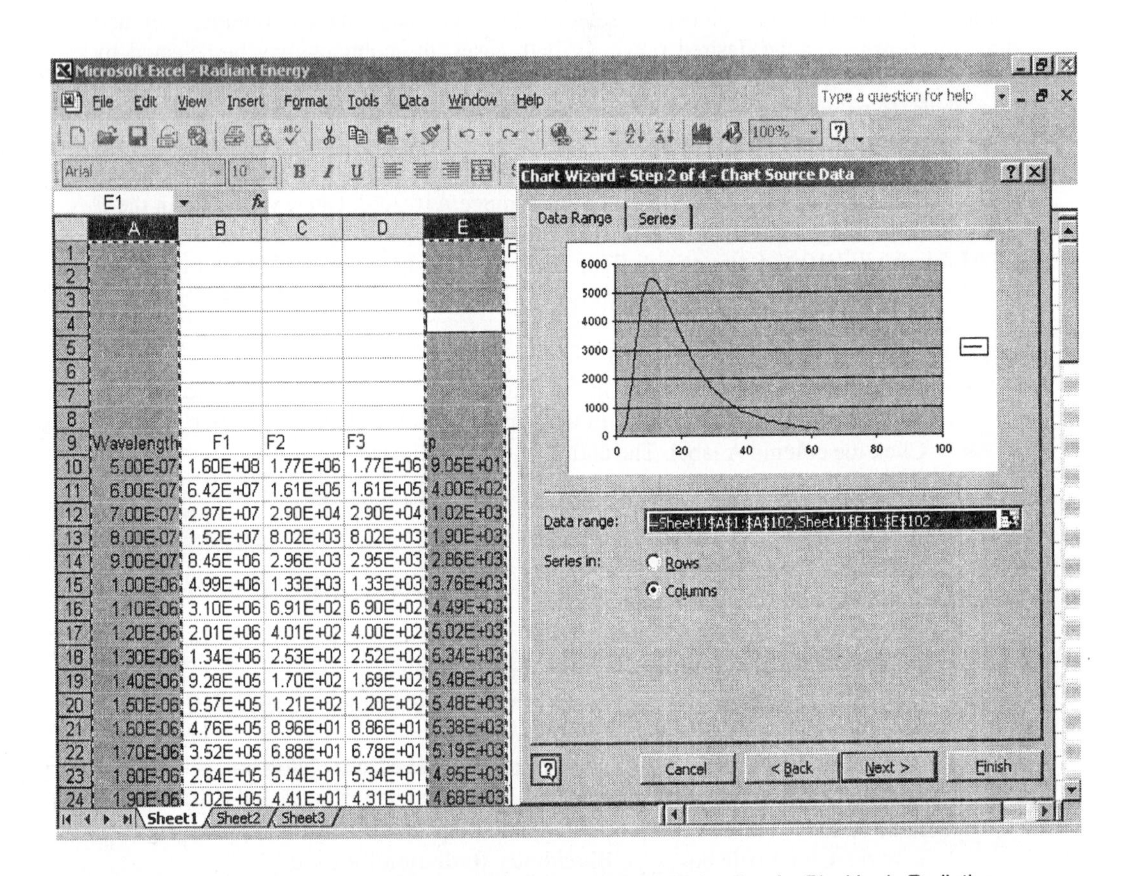

Figure 3-4. The **Chart Wizard – Step 2 of 4 – Chart Source Data** Dialog Box for Blackbody Radiation

Edit the Axes

- Click the y-axis label twice (but do not double-click) to get a pointer in the axis title. Select the **4**, and on the **Format** menu, click **Selected Axis Title.**
- In the **Format Axis Title** dialog box, select the **Superscript** check box.
- Right-Click the plotted line; then click **Format Data Series.**
- In the **Format Data Series** dialog box, change the line color to black; in the **Weight** box, select a medium heavy line.
- Right-click any x-axis number; then click **Format Axis.**

- In the **Format Axis** dialog box:
  1. Click the **Scale** tab, type **0.000008** in the **Maximum** box, and type **0** in the **Minimum** box.
  2. Click the **Number** tab, select **Number**, and type **0** in the **Decimal places** box.
  3. Repeat for the y-axis number. Leave the scale as is, but change the number of decimal places to 0.

Edit the Plot Area

- Select the chart; on the **Chart** menu, click **Chart Options**.
- Click the **Gridlines** tab, and for both **Value (X) axis** and **Value (Y) axis**, clear all four check boxes.
- Right-click anywhere in the plot area, then click **Format Plot Area**.
- In the **Format Plot Area** dialog box:
  1. In the **Area** section (the right side), click a white square on the palette to change the plot area from gray to white.
  2. In the **Border** section, click **Color**, and select black; click **Weight**, and select a medium weight line.

Excel Example 7

Radial Wave Functions: 3*s*, 3*p*, and 3*d*

Review of the Science

A hydrogen atom consists of a proton and an electron. The proton is the nucleus of the atom about which the electron travels at high speed. The region in space surrounding the nucleus in which the electron travels is called an orbital. The mathematical function that describes this three-dimensional space is also called an orbital, or a wave function. The negatively charged electron does not fall into the positively charged nucleus because of a balance between electrostatic and centripetal forces. This same model applies to hydrogen-like ions having a positively charged nucleus and a single outer electron.

The Coordinate System. The nucleus of the hydrogen atom or hydrogen-like ion is located at the origin O (Figure 3-5). The electron is located at Cartesian coordinates x, y, and z. Alternatively, the position of the electron can be expressed in polar coordinates r,θ, and ϕ. The electron is located at the end of the vector of length r.

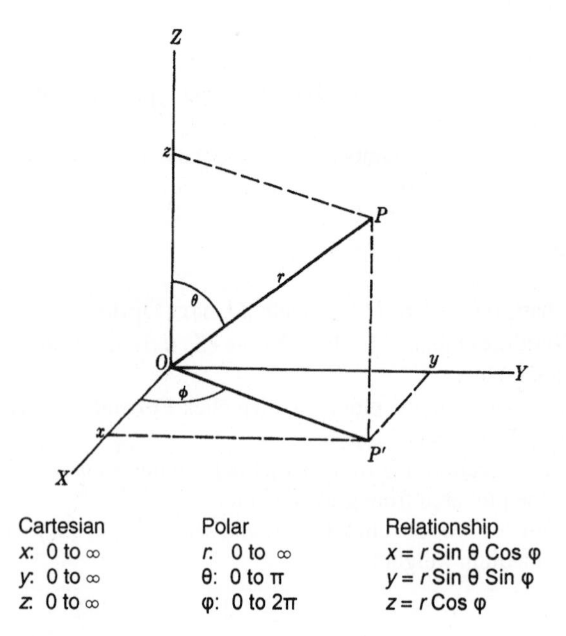

Figure 3-5. Cartesian and Polar Coordinates

The Cartesian coordinates range from zero to infinity. The polar coordinate r ranges from zero to infinity, while θ ranges from zero to π. A value of zero for θ corresponds to colinearity with the z-axis, which is vertical. The ϕ coordinate ranges from zero to 2π. A value of zero for ϕ corresponds to colinearity with the x-axis.

The Wave Equation. For a brief review of the postulates of quantum mechanics, see Excel Example 26 in Chapter 8. The quantum mechanical description of the hydrogen atom begins with the Schrödinger wave equation:

$$\hat{H}\psi = E\psi \tag{3-2}$$

where $\hat{H}$ is the Hamiltonian operator, E is the total energy of the system, and ψ is the wave function. For the hydrogen atom

$$\hat{H} = \frac{-h^2}{8\pi^2 m}\nabla^2 + U \tag{3-3}$$

U is the potential energy of an electron in the electrostatic force field of a positively charged nucleus and equals $e^2 / 4\pi\varepsilon_0 r$, m is the mass of the electron, and ∇ is the Laplacian operator.

$$\nabla = \frac{\partial}{\partial x} + \frac{\partial}{\partial y} + \frac{\partial}{\partial z} \tag{3-4}$$

This yields the wave equation for a hydrogen atom in Cartesian coordinates.

$$\frac{\partial^2\psi}{\partial x^2}+\frac{\partial^2\psi}{\partial y^2}+\frac{\partial^2\psi}{\partial y^2}+\frac{8\pi^2 m}{h^2}\left(E+\frac{e^2}{4\pi\varepsilon_0 r}\right)\psi=0 \tag{3-5}$$

The quantity E turns out to be a set of allowed energies (eigenvalues) characterized by a principal quantum number n. The associated Ψs, characterized by three quantum numbers, n, ℓ, and m, are the solutions to the wave equation and are called wave functions or eigenfunctions (after the German, *Eigenfunktionen).*

In order to solve the wave equation, it is necessary to convert Equation 3-4 and Equation 3-5 from Cartesian coordinates to polar coordinates. This conversion, described in most monographs on quantum mechanics, leads to Equation 3-6. Equation 3-6 is the Schrödinger wave equation for the hydrogen atom in polar coordinates:

$$\frac{1}{r^2}\frac{\partial}{\partial r}\left(r^2\frac{\partial\psi}{\partial r}\right)+\frac{1}{r^2\sin\theta}\frac{\partial}{\partial\theta}\left(\sin\theta\frac{\partial\psi}{\partial\theta}\right)+\frac{1}{r^2\sin^2\theta}\frac{\partial^2\psi}{\partial\phi^2}+\frac{8\pi^2 m}{h^2}\left(E+\frac{e^2}{4\pi\varepsilon_0 r}\right)\psi=0 \tag{3-6}$$

Solutions to the Wave Equation. The reason that it is worth the effort to get the Schrödinger wave equation into polar coordinates is because this equation can be factored into three much simpler equations, one a function of r only, a second equation a function of θ only, and a third equation a function of ϕ only. That last statement may be expressed as follows:

$$\psi_{n,\ell,m}(r,\theta,\varphi)=R_{n,\ell}(r)\Theta_{\ell,m}(\theta)\Phi_m(\varphi) \tag{3-7}$$

These three equations are relatively easily solved, and the product of the three solutions gives the complete solution to the Schrödinger wave equation. In examining the spatial properties of $\psi_{n,\ell,m}(r,\theta,\varphi)$, it is convenient to look at the radial part and the angular part separately. In shortened notation $\psi = R\Theta\Phi$, where R is the radial part and is a function of r only and does not depend on direction; and $\Theta\Phi$ is the angular part that is independent of r, but depends on direction, that is, on θ and ϕ. Refer to Figure 3-5. The three equations into which Equation 3-7 can be separated are

The Φ equation:

$$\frac{1}{\Phi}\frac{d^2\Phi}{d\varphi^2}=-m_l^2 \tag{3-8}$$

The Θ equation:

$$\frac{1}{\sin\theta}\frac{d}{d\theta}\left(\sin\theta\frac{d\Theta}{d\theta}\right)-\frac{-m_l^2\Theta}{\sin^2\theta}+l(l+1)\Theta=0 \tag{3-9}$$

The *R* equation:

$$\frac{1}{R}\frac{d}{dr}\left(r^2\frac{dR}{dr}\right)+\frac{8\pi^2\mu}{h^2}\left(E+\frac{Ze^2}{4\pi^2\varepsilon_0 r}\right)r^2=l(l+1) \tag{3-10}$$

The complete solution to the Schrödinger wave equation is the product of the solution to each of the preceding equations, and it depends on *r,* θ, and φ and three quantum numbers *n,* ℓ, and *m* where:

$$\psi_{n,\ell,m}(r,\theta,\varphi)=R_{n,\ell}(r)\Theta_{\ell,m}(\theta)\Phi_m(\varphi) \tag{3-11}$$

In a very shorthand form, Equation 3-11 is frequently written

$$\psi=R\Theta\Phi \tag{3-12}$$

where *R* is the radial part and is a function of *r* only and does not depend on direction. ΘΦ is the angular part that is independent of *r* but that depends on direction, that is, on θ and φ.

In examining graphically the spatial properties of $\psi_{n,\ell,m}(r,\theta,\varphi)$, it is convenient to look at the radial part and the angular parts separately. In this Excel Example, we shall look at solutions to *R*, the radial part of the wave function (Equation 3-10). In the next section (Excel Example 8), we will examine plots of the angular wave function ΘΦ.

For the details of the solution of the Schrödinger wave equation, the reader is referred to any textbook on quantum mechanics.

Some Radial Wave Functions. For the purpose of exploring Excel plots of radial wave functions, it is interesting to plot some wave functions of higher principal quantum number *n*. For principal quantum number *n* equals *3*, the first three radial wave functions are:

$n = 3$, $\ell = 0$, (3*s*)

$$R_{30}(r)=\frac{2}{81\sqrt{3}}a_0^{-3/2}[27-18\rho+2\rho^2]e^{\frac{-\rho}{3}} \tag{3-13}$$

$n = 3$, $\ell = 1$, (3*p*)

$$R_{31}(r)=\frac{4}{81\sqrt{6}}a_0^{-3/2}[6\rho-\rho^2]e^{\frac{-\rho}{3}} \tag{3-14}$$

$$n = 3,\ \lambda = 2,\ (3d) \qquad R_{32}(r) = \frac{4}{81\sqrt{30}} a_0^{-3/2} \rho^2 e^{\frac{-\rho}{3}} \tag{3-15}$$

where

$$\rho = \frac{Zr}{a_0} \tag{3-16}$$

The units of these and all other radial wave functions $R_{n\alpha}$ are $a_0^{-3/2}$. The Bohr radius is a_0, Z is the charge on the nucleus, n is the principal quantum number, and r is the distance from the nucleus. Since Z equals 1 for the H atom, a plot of R versus ρ is a plot of R versus r/a_0, a dimensionless number. For most wave function plots, r/a_0 ranges from zero to about ten or so, and is just the number of Bohr radii out from the nucleus. The worksheet with an embedded chart after editing is shown in Figure 3-6.

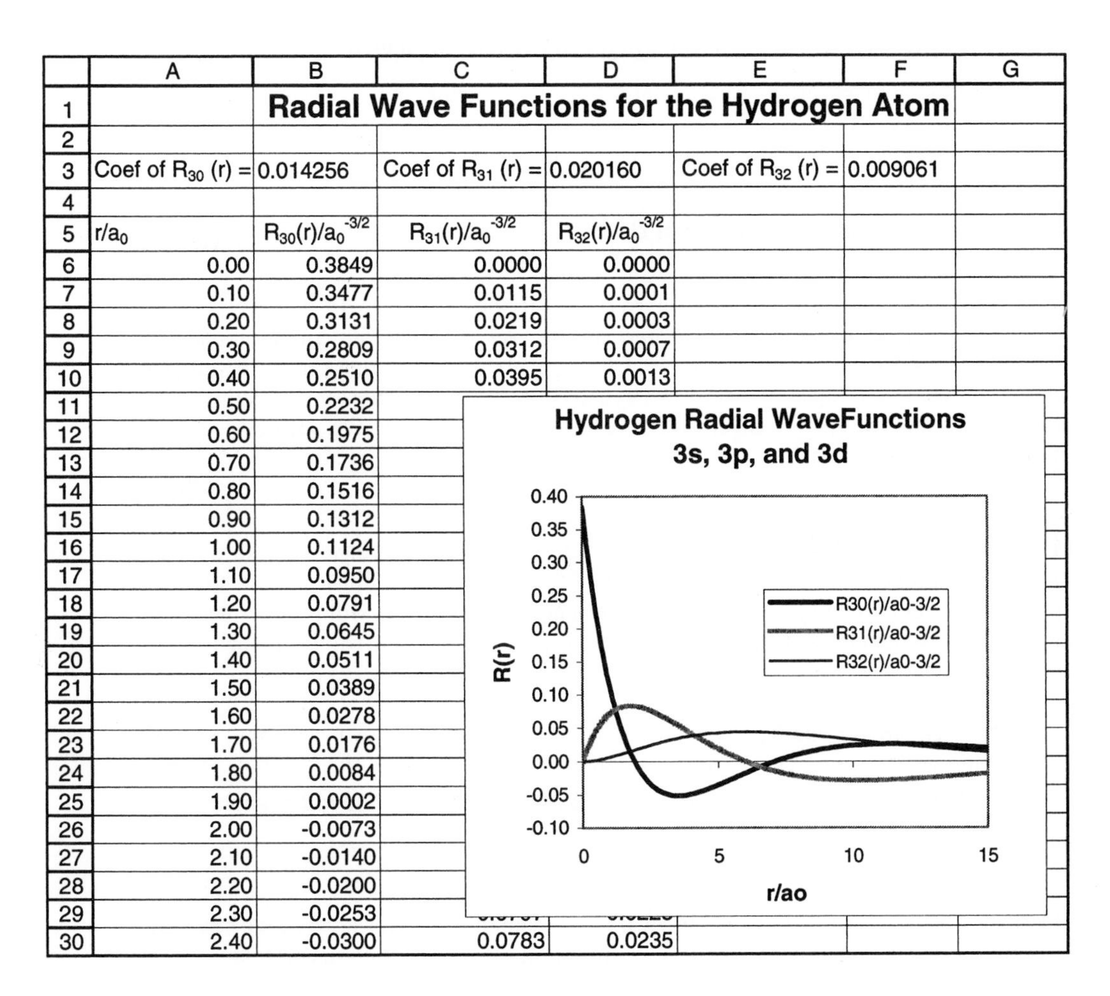

	A	B	C	D	E	F	G
1		**Radial Wave Functions for the Hydrogen Atom**					
2							
3	Coef of R_{30} (r) =	0.014256	Coef of R_{31} (r) =	0.020160	Coef of R_{32} (r) =	0.009061	
4							
5	r/a_0	$R_{30}(r)/a_0^{-3/2}$	$R_{31}(r)/a_0^{-3/2}$	$R_{32}(r)/a_0^{-3/2}$			
6	0.00	0.3849	0.0000	0.0000			
7	0.10	0.3477	0.0115	0.0001			
8	0.20	0.3131	0.0219	0.0003			
9	0.30	0.2809	0.0312	0.0007			
10	0.40	0.2510	0.0395	0.0013			
11	0.50	0.2232					
12	0.60	0.1975					
13	0.70	0.1736					
14	0.80	0.1516					
15	0.90	0.1312					
16	1.00	0.1124					
17	1.10	0.0950					
18	1.20	0.0791					
19	1.30	0.0645					
20	1.40	0.0511					
21	1.50	0.0389					
22	1.60	0.0278					
23	1.70	0.0176					
24	1.80	0.0084					
25	1.90	0.0002					
26	2.00	-0.0073					
27	2.10	-0.0140					
28	2.20	-0.0200					
29	2.30	-0.0253					
30	2.40	-0.0300	0.0783	0.0235			

Figure 3-6. The Worksheet for the 3*s*, 3*p*, and 3*d* Radial Wave Functions

Preparing the Worksheet

- Beginning in cell B1, type **Radial Wave Functions for the Hydrogen Atom** as the worksheet title (Figure 3-6).
- In cells A3, C3, and E3, type **Coef of R_{30} (r) =**, **Coef of R_{31} (r) =**, and **Coef of R_{32} (r) =** as the text labels.
  1. Select the group of characters to be superscripted or subscripted. Use CTRL-click to select more than one group of cells for formatting.
  2. On the **Format** menu, click **Cells** to open the **Format Cells** dialog box (similar to Figure 1-5); click the **Font** tab.
  3. In the **Effect** area, select the **Subscript** or **Superscript** check box, whichever is needed; then click **OK**.
- Use the preceding procedure for the four labels in row 5.
  1. In cell A5, type **r/a_0** for the label.
  2. In cell B5, type **$R_{30}(r)/a_0^{-3/2}$** for the label.
  3. In cell C5, type **$R_{31}(r)/a_0^{-3/2}$** for the label.
  4. In cell D5, type **$R_{32}(r)/a_0^{-3/2}$** for the label.
- To size these labels within the cells, click the line between the column letter headings, and drag the cell boundaries to the right until the cell fits the label. Repeat for all the boundaries that need to be widened.
- In cells B3, D3, and F3, type **0.014256**, **0.020160**, and **0.009061** and left-justify. These numbers are the coefficients of the $a_0^{3/2}$ factors in Equations 3-13, 3-14, and 3-15.

Entering the Worksheet Numerical Data

To plot the radial wave function for values of r/a_0 in the range 0 to 15 in increments of 0.1, these 150 numbers must be placed in the series A6:A156. In addition, the corresponding cells in the series B6:B156, C6:C156, and D6:D156 must be filled with the relative formulas for Equations 3-18, 3-19, and 3-20. Excel provides three methods for incrementing and filling this range, which we have used previously. We will review these three methods at this time.

Method I. The Fill-Handle Method. This method is quick and convenient for a range that is not too long, say, a page length or about 50 numbers. The Fill-Handle method is applicable for both relative formulas and incremented numbers. We will not use method I in this Excel Example. But just for comparison, here is how you would use it for the column A values:

- In cell A6, type **0.00** as the value.
- In cell A7, type **=A6+0.1** as the formula.
- Click in the lower right corner of cell A7, and drag to cell A156; the numbers from 0 to 15 incremented in steps of 0.1 appear.

Method II: The Edit Fill Series Method. This method is best for a very long series of incremented numbers but does *not* work for filling in relative formulas. Use this method in this Excel Example for the column A numbers.

- In cell A6, type **0** and press ENTER. Then select cell A6 again.
- On the **Edit** menu, point to **Fill** (Figure 3-7), and then click **Series** to open the **Series** dialog box (Figure 3-8). In the **Series** dialog box:
  1. For **Series in**, select **Columns**.

2. For **Type**, select **Linear**.
3. In the **Step value** box, type **0.1** for the step value.
4. In the **Stop value** box, type **15** for the stop value.
5. Click **OK**. The numbers from 0 to 15 incremented in steps of 0.1 appear and fill the series A6:A156.

Method III: The Edit Fill Down Method. This method works for both a very long series of relative formulas or incremented numbers. You select the range to be filled before applying the Edit Fill Down method. This method is applicable to filling down with incremented numbers or with relative formulas. Here's how you use this method to fill the column A series A6:A156:

- In cell A6, type **0.00** as the value.
- In cell A7, type **=A6+0.1** as the formula. The number 0.1 appears in cell A7, and =A6+0.1 appears in the formula bar (prior to pressing the ENTER key).

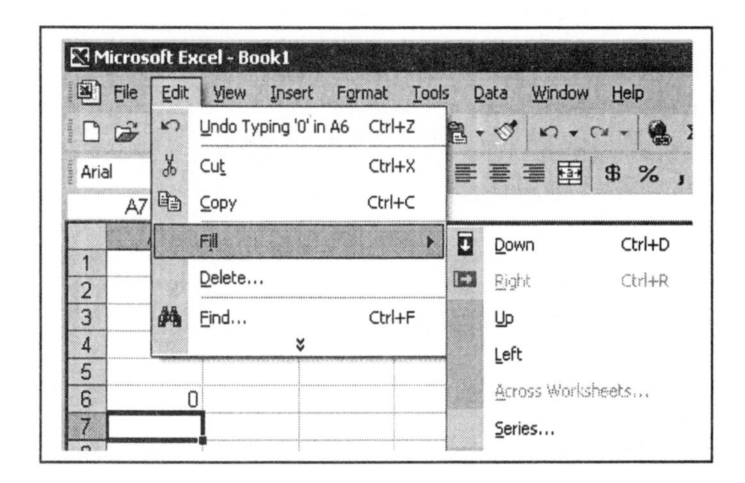

Figure 3- 7. The **Edit** Menu and the **Fill** Submenu with Filling Options

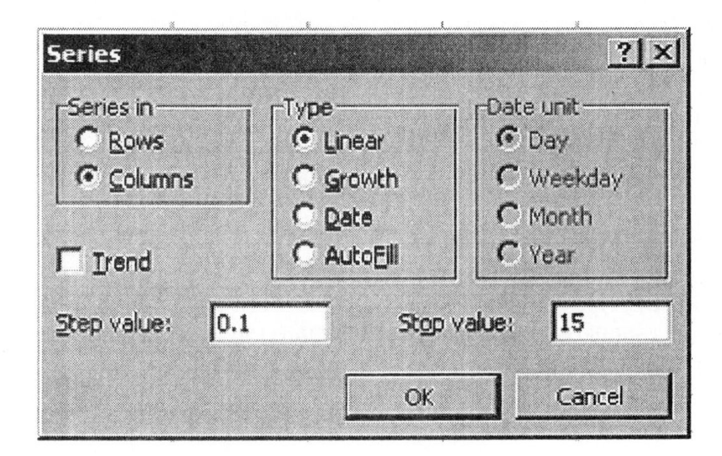

Figure 3-8. The **Edit_Fill_ Series** Dialog Box

- Click the split screen bar (Figure 1-1) and drag it to the middle of the screen. Scroll to the bottom of the split screen until cell A156 is visible.
- Click in cell A7 to select it; then hold the SHIFT key down and select cell A156. This action selects the entire range A7:A156, which blackens.
- While this range is selected, on the **Edit** menu, point to **Fill**, and click **Down**, and the range from A6 to A156 is instantly filled with values from 0 to 15 incremented in steps of 0.1.
- Double-click the split screen bar to get back to a single screen.

Although the Edit Fill Down method can be used to fill incremented values as just described, the Edit Fill Series method is generally easier for this purpose. The Edit Fill Series method cannot be used to fill in a range with relative formulas; for this purpose the Edit Fill Down method is used for long series, or the Fill-Handle method can be used for short series. Begin either method by clicking the **Edit** menu and pointing to **Fill**. For the Edit Fill Series method, there is a dialog box; for the Edit Fill Down method, there is not.

Entering the Formulas for the Wave Functions

The coefficients for each radial wave function are set up in cells B3, D3, and F3. In the formulas, they are referred to as $B$3, $D$3, and $F$3 because we want to use their absolute values, not their relative values. Remember, the function to be entered is always preceded by an = sign and is displayed in the formula box. It is convenient to use method II, the Edit Fill Down method, to fill in the cells with the relative formulas.

Entering Radial Equation 3-13: $R_{30}(r)/a_0^{-3/2}$

- In cell B6, type **=$B$3*(27-18*A6+2*A6^2)*@EXP(-A6/3)** (Equation 3-13) as the formula.
- As in method II, described previously, split the screen, select cell B6, SHIFT-click in cell B156 to select the series B6:B156, and then, on the **Edit** menu, point to **Fill**, and click **Down**.

Entering Radial Equation 3-14: $R_{31}(r)/a_0^{-3/2}$

- In cell C6, type = **$D$3*(6*A6-A6^2)*@EXP(-A6/3)** (Equation 3-14) as the formula.
- Again, split the screen, select the series C6:C156, and, on the **Edit** menu, point to **Fill**, and click **Down**.

Entering Radial Equation 3-15: $R_{32}(r)/a_0^{-3/2}$

- In cell D6, type **=$F$3*A6^2*@EXP(-A6/3)** (Equation 3-15) as the formula.
- Again, split the screen, select the series D6:D156, and, on the **Edit** menu, point to **Fill**, and click **Down**.

Creating the Chart

- Click the split screen bar, and drag it to split the screen so that cell A5 is visible on the top half and cell D156 is visible on the bottom half.

- Select cell A5 (containing the label r/a_0).
- SHIFT-click in cell D156 to select the entire block A5:D156.
- With the block A5:D156 selected, click the **Chart Wizard** icon (Figure 2-2) to open the first Chart Wizard dialog box.
- In the **Chart Wizard – Step 1 of 4 – Chart Type** dialog box, click **XY (Scatter)**, and then click the icon for smooth lines and without markers. Click **Next**.
- In the **Chart Wizard – Step 2 of 4 – Chart Source Data** dialog box, a chart should appear with three plots and a legend box with the three column labels that are in cells B5, C5, and D5. No action is necessary. Click **Next**.
- In the **Chart Wizard – Step 3 of 4 – Chart Options** dialog box:
    1. Click the **Titles** tab, and in the **Chart title** box, type **Hydrogen Radial Wave Functions** as the title.
    2. In the **Value (X) axis** box, type **r/ao**as the label.
    3. In the **Value (Y) axis** box, type **R(r)** as the label, and click **Next**.
    4. Click the **Gridlines** tab, and clear all four gridlines check boxes.
    5. Click **Next**.
- In the **Chart Wizard – Step 4 of 4 – Chart Location** dialog box, leave the chart **As object in Sheet 1**.
- Click **Finish**, and the chart appears floating in the worksheet (Figure 3-6). It needs editing.

Editing the Chart

The chart produced at this point is probably not in the best position on the worksheet. It may be too small or too large. The plot area may be too small or too large. The title may need a second line or change of font size or type. The x-axis and y-axis numbers may need changing, their labels may need editing, and the scale may not be quite right. The line colors may need changing, and the lines themselves may be too thick or too thin. Excel's editing options allow all these deficiencies to be corrected. The approach to solving these problems is nearly the same in each case. Just right-click the object to be edited and then proceed. Or sometimes right-click and drag. Some examples should help clarify.

Size and Position of the Chart on the Worksheet

- Click anywhere in a blank area of the chart, and drag the chart to where you would like it positioned on the worksheet.
- Scroll to reveal a few rows under the chart, and then click the bottom edge size handle (■) and drag to increase the size of the chart vertically.
- Click anywhere in the empty part of the plot area, and then enlarge the plot area by clicking and dragging the side and bottom size handles.
- Click anywhere in the legend box, and drag it to a suitable position in the plot area.

Tidying Up the Plot Area

- Right-click in the plot area to open a shortcut menu.
- Click **Format Chart Area** (similar to Figure 3-9), and change the gray background to white.
- Right-click the 1*s* plot line, click **Format Data Series**, and in the **Format Data Series** dialog box (Figure 2-10), select the **Color** box, and change the color to black.

- click the 2*s* plot line, click **Format Data Series**, and in the **Format Data Series** dialog box (Figure 2-10), select the **Color** box, and change the color to black; select the **Weight** box, and change the line weight to a medium heavy line.
- Right-click the 3*s* plot line, click **Format Data Series**, and in the **Format Data Series** dialog box (Figure 2-10), select the **Color** box, and change the color to black; select the **Style** box, and change the style to a light-gray cross-hatch; select the **Weight** box, and change the line weight to a medium heavy line.

Customizing the Y-Axis

- Right-click a y-axis number, then click **Format Axis** to open the **Format Axis** dialog box (Figure 2-11).
- Click the **Scale** tab, and in the **Maximum** box, type **0.4** as the value.
- Click the **Font** tab, and change to 8 point size.
- Click the **Number** tab, and type **2** in the **Decimal places** box.

Customizing the X-Axis

- Right-click an x-axis number, then click **Format Axis** to open the **Format Axis** dialog box (Figure 2-11).
- Click the **Scale** tab, and in the **Maximum** box, type **15** as the value.
- Click the **Font** tab, and change to 8 point size.
- Click the **Number** tab, select the **Number** category, and type **0** in the **Decimal places** box.
- Click the **Patterns** tab, and for **Tick mark labels**, select **Low** to place the tick marks and numbers at the bottom of the chart area.

Editing the Title

- Click twice anywhere in the chart title (but don't double-click). A vertical-line insertion point will appear.
- Move the insertion point to the right end of the title by pressing the END key. You could write more at this point, but don't.
- Instead, press the ENTER key, and the pointer drops to the next line below. Now type ***3s*, *3p*, and *3d*** for the second line. You could enter several lines of title if you wanted.
- Right-click in the chart title, click **Format Chart Title** (Figure 2-8), then click the **Font** tab, and change the size to 12 or whatever looks nice to you.

Excel Example 8

An Angular Wave Function: The $2p_x$ Orbital

Review of the Science

In examining the spatial properties of $\psi_{n,\ell,m}(r,\theta,\varphi)$, it is convenient to look at the radial wave function R separately from the angular wave function ΘΦ. In Excel Example 7, we examined some of the radial wave functions for the hydrogen atom. In this Excel Example, we shall examine an angular wave function ΘΦ that is independent of r, but depends on direction, that is, on θ and φ. We are using here the shorthand notation for the wave functions, R and ΘΦ.

Angular Wave Functions. The first few angular wave functions for the hydrogen atom are:

$$\Theta\Phi_{1s} = \left(\frac{1}{4\pi}\right)^{1/2} \tag{3-17}$$

$$\Theta\Phi_{2p_x} = \left(\frac{3}{4\pi}\right)^{1/2} \sin\theta\cos\phi \tag{3-18}$$

$$\Theta\Phi_{2p_y} = \left(\frac{3}{4\pi}\right)^{1/2} \sin\theta\sin\phi \tag{3-19}$$

$$\Theta\Phi_{2p_z} = \left(\frac{3}{4\pi}\right)^{1/2} \cos\theta \tag{3-20}$$

Equations 3-17, 3-18, 3-19, and 3-20 are wave functions. Sometimes they are called eigenfunctions from the German. They are also referred to as orbitals, for example, the $2p_z$ orbital. The square of the wave function is also referred to as an orbital. This loose nomenclature also applies to the radial wave function, for example, the $1s$ wave function or $1s$ orbital.

Spatial Characteristics. The $2p_z$ orbital is independent of φ and depends only on θ. It is usually convenient to plot an orbital in some particular plane. For example, if we choose φ to equal zero, then the plot of p_x would be in the XZ plane according to Figure 3-5. Similarly, if we choose θ to equal 90 degrees, the plot of p_x would be in the XY plane. Examination of Figure 3-5 leads to the conclusions in Table 3-1.

Table 3-1

Angular Wave Function Plots

$2p_x = \sin\theta\cos\varphi$	$2p_y = \sin\theta\sin\varphi$	$2p_z = \cos\theta$
When $\theta = 90^\circ$ the plot is in the XY plane	When $\theta = 90^\circ$ the plot is in the XY plane	When $\phi = 0^\circ$ the plot is in the YZ plane
When $\phi = 0^\circ$ the plot is in the XZ Plane	When $\phi = 90^\circ$ the plot is in the YZ plane	When $\phi = 90^\circ$ the plot is in the YZ plane
$2p_x = 0$ everywhere in the YZ plane ($\theta = 0^\circ$)	$2p_y = 0$ everywhere in the XZ plane ($\theta = 90^\circ$)	$2p_z = 0$ everywhere in The XY plane ($\theta = 90^\circ$)

Let us plot the $2p_x$ orbital (Figure 3-9). Since the angular properties of the wave function depend only on the trigonometric part, the preceding constant may be omitted so that $p_x = \sin\theta\cos\phi$. Reference to Figure 3-5 and Table 3-1 reveals that setting θ equal to 90 degrees results in the wave function lying in the XY plane, and its magnitude is given by

$$p_x = |\cos\varphi| \tag{3-21}$$

The Excel screen coordinates x (horizontal) and y (vertical) are given by

$$x = p_x \cos\varphi \tag{3-22}$$

$$y = p_x \sin\varphi \tag{3-23}$$

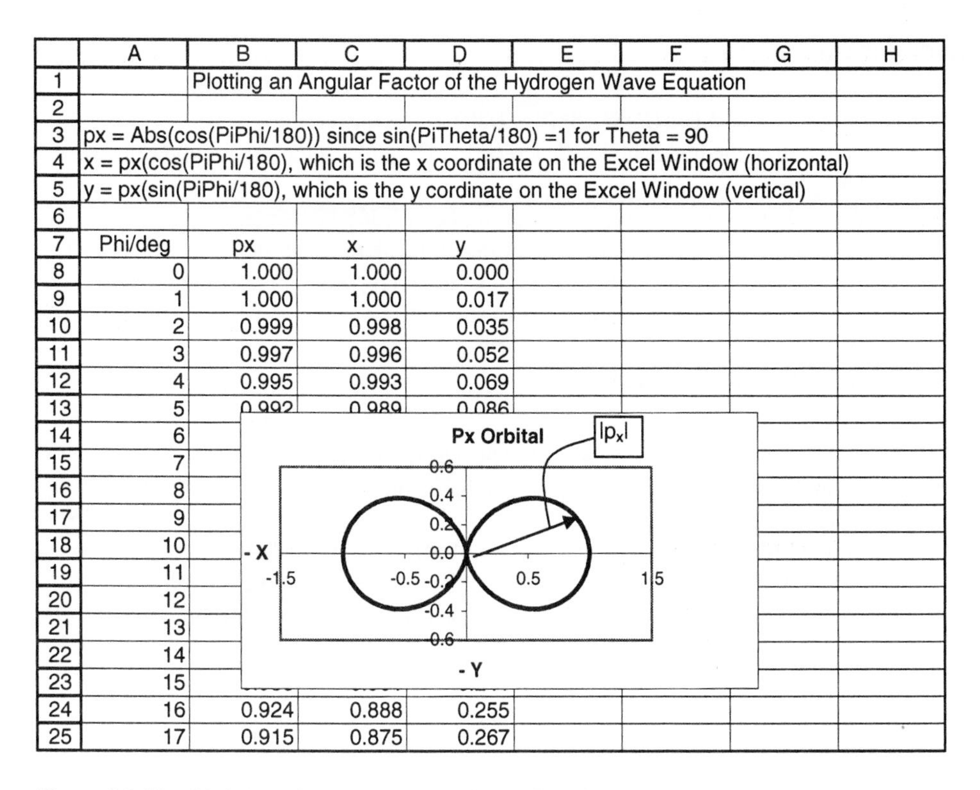

	A	B	C	D	E	F	G	H
1		Plotting an Angular Factor of the Hydrogen Wave Equation						
2								
3	px = Abs(cos(PiPhi/180)) since sin(PiTheta/180) =1 for Theta = 90							
4	x = px(cos(PiPhi/180), which is the x coordinate on the Excel Window (horizontal)							
5	y = px(sin(PiPhi/180), which is the y cordinate on the Excel Window (vertical)							
6								
7	Phi/deg	px	x	y				
8	0	1.000	1.000	0.000				
9	1	1.000	1.000	0.017				
10	2	0.999	0.998	0.035				
11	3	0.997	0.996	0.052				
12	4	0.995	0.993	0.069				
13	5	0.992	0.989	0.086				
14	6							
15	7							
16	8							
17	9							
18	10							
19	11							
20	12							
21	13							
22	14							
23	15							
24	16	0.924	0.888	0.255				
25	17	0.915	0.875	0.267				

Figure 3-9. The Worksheet for the $2p_x$ Angular Wave Function

Preparing the Worksheet

Entering the Title and Documentation for the Worksheet

- In row 1, type **Plotting an Angular Factor of the Hydrogen Wave Equation** as the title.
- In row 3, type **px = Abs(cos(Pi*Phi/180)) since sin(Pi*Theta/180) = 1 for Theta = 90** as a label.
- In row 4, type **x = px(cos(Pi*Phi/180)), which is the x coordinate on the Excel window (horizontal)** as a label.
- In row 5, type **y = px(sin(Pi*Phi/180)), which is the y coordinate on the Excel window (vertical)** as a label.
- In the range A7:D7, type **Phi/deg**, **px**, and **y** as the column labels.

Entering the Phi Values

- In cell A8, type **0** and press Enter.
- On the **Edit** menu, point to **Fill** (Figure 3-7), and then click **Series** to open the **Series** dialog box (Figure 3-8). In the **Series** dialog box:
  1. For **Series in**, select **Columns**.
  2. For **Type**, select **Linear**.

3. In the **Step value** box, type **1** for the step value.
4. In the **Stop value** box, type **360** for the stop value.

Entering the p_x Values

- In cell B8, type **=ABS(COS(A8*Pi()/180))** as the formula.
- Click and drag the horizontal split screen bar (Figure 1-1) to the middle of the screen.
- Scroll the lower half of the split screen until cell B368 is visible.
- Click in cell B8, and then SHIFT-click in cell B368 to select the range B8:B368.
- On the **Edit** menu, point to **Fill**, and click **Down** (Figure 3-7) to fill the range B8:B368 with the copies of the wave function p_x.

Entering the Excel Screen *x* Values

- In cell C8, type **=ABS(COS(B8*Pi()/180))** as the formula.
- Click and drag the horizontal split screen bar (Figure 1-1) to the middle of the screen.
- Scroll the lower half of the split screen until cell C368 is visible.
- Click in cell C8, and then SHIFT-click in cell C368 to select the range C8:C368.
- On the **Edit** menu, point to **Fill**, and click **Down** (Figure 3-7) to fill the range C8:C368 with the Excel screen *x* values.

Entering the Excel Screen *y* Values

- In cell D8, type **=ABS(SIN(B8*Pi()/180))** as the formula.
- Click and drag the horizontal split screen bar (Figure 1-1) to the middle of the screen.
- Scroll the lower half of the split screen until cell D368 is visible.
- Click in cell D8, and then SHIFT-click in cell D368 to select the range D8:D368.
- On the **Edit** menu, point to **Fill**, and click **Down** (Figure 3-7) to fill the range D8:D368 with the Excel screen *y* values.

Creating the Chart

Plotting Strategy. In Figure 3-9 the magnitude of p_x is the length of the arrow-tipped line. As φ increases, the locus of the arrow tip generates the upper right half lobe, then the upper left half lobe, and then the lower left half lobe, and finally the lower right half lobe. If $2p_x$ is plotted, each lobe is a perfect circle. If the square of $2p_x$ is plotted, the circle becomes a lobe, the characteristic shape of the *p* orbitals.

- Click and drag the horizontal split screen bar (Figure 1-1) to the middle of the screen.
- Scroll the lower half of the split screen until cell D368 is visible
- Click in cell C8, and SHIFT-click in cell D368 to select the range C8:D368.
- Click the **Chart Wizard** icon.
- In the **Chart Wizard – Step 1 of 4 – Chart Type** dialog box, select **XY (Scatter)** and the icon for smooth line, no markers; click **Next**.
- In the **Chart Wizard – Step 2 of 4 – Chart Source Data** dialog box, click **Next**.
- In the **Chart Wizard – Step 3 of 4 – Chart Options** dialog box:

1. In the **Chart title** box, type **$2p_x$ Orbital** (lowercase p, subscript x) for the title.
2. In the **Value (X) axis** box, type **< - X - >** as the label.
3. In the **Value (Y) axis** box, type **< - Y - >** as the label.
4. Click **Next.**

- In the **Chart Wizard – Step 4 of 4 – Chart Options** dialog box, Select **As object in Sheet1**; click **Finish.**

Editing the Chart

- Right-click an x-axis number, and then click **Format Axis** to open the **Format Axis** dialog box (Figure 2-10). Select the **Number** tab, and set decimals to 1.
- Repeat for the y-axis.
- Click *View_Toolbars_Drawing*
  1. In the **Drawing Toolbar** click the arrow. Then click and drag the arrow line from the origin of the plot (0,0) until it touches the plot (Figure 3-9).
  2. In the Drawing Toolbar, click the **Text Box** icon (a tiny square with the letter A). Click in its approximate location in the plot area (Figure 3-9), and size it by dragging. Type **Px** in the text box.
- Right-click anywhere along the plotted line, then click **Format Axis** to open the **Format Axis d**ialog box (Figure 2-9). In the **Color** box, select black; and in the **Weight** box, select a medium heavy line.
- Right click anywhere an the plotted line, then click on **Format Axes** to get the ***Format Axis*** dialog box (Figure 2-11). For *color* select black and for *weight,* select a medium heavy line.

Excel Example 9

The Hydrogen Molecule Ion

Review of the Science

The Model. The hydrogen molecule ion H_2^+ consists of two positively charged protons, separated by a fixed bond distance, and a single negatively charged electron at some distance away from the two protons. The bond distance, 106 pm, is known from spectroscopic measurements. It is convenient to use the one-dimensional coordinate system shown in Figure 3-10 to describe the positions of the three particles. Proton A is placed on the origin where r equals zero. Proton B is placed on the r-axis where r equals 106 pm. The electron is placed at any arbitrary position on the r-axis where its distance from proton A is r_A and its distance from proton B is r_B. Regardless of the electron's position on the r-axis, it is true that

$$r_A + r_B = 106pm \tag{3-24}$$

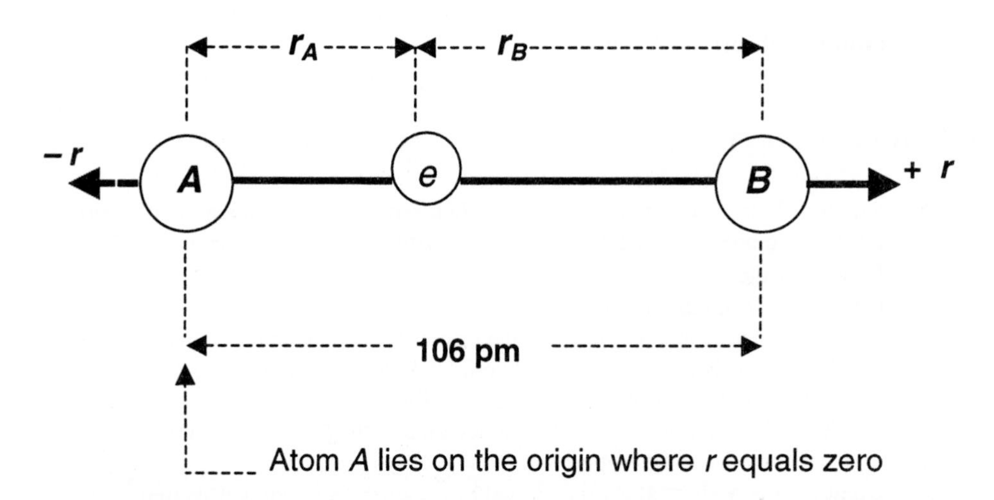

Figure 3-10. The One-Dimensional Coordinate System for the Hydrogen Molecule Ion

The electron can range anywhere from $-\infty$ to $+\infty$, but probably spends most of it time fairly close to the positive field furnished by the two protons.

Orbital Nomenclature. The complete nomenclature for hydrogen atom wave equations is described in Excel Example 7:

$$\psi_{n,\ell,m}(r,\theta,\varphi) = R_{n,\ell}(r)\Theta_{\ell,m}(\theta)\Phi_m(\varphi) \tag{3-25}$$

The radial wave function is dependent on r only and is independent of θ and φ, so Equation 3-24 becomes

$$\psi_{n,\ell}(r) = R_{n,\ell}(r) \tag{3-26}$$

The n quantum number has values from 1, 2, 3, ... and the ℓ quantum number has values from 0, 1, 2, ... , which are customarily represented by the letters *s, p, d, f,* .... The ground state of the hydrogen atom has $n = 1$ and $\ell = 0$. It is a further simplification to use $1s$ to stand in for $\psi_{1,0}(r)$.

Suppose that the electron is positioned near to, and just left of, proton A. It is virtually in the same electrical environment as an electron in a hydrogen atom, since proton B is quite distant and the electron is partially shielded from proton B by proton A. In this position it is reasonable to assume that the atomic orbital for the electron is approximately given by the wave function for an electron in the ground state of the hydrogen atom, for which $n = 1$, $\ell = 0$ and $m = 0$. The same reasoning can be applied for an electron to the right of proton B.

With these customary simplifications in mind, the equation for the ground state wave function of the hydrogen atom becomes

$$1s_A = \frac{1}{\sqrt{\pi}}\left(\frac{1}{a_0}\right)^{\frac{3}{2}} e^{\frac{-r_A}{a_o}} \tag{3-27}$$

This is the customary way the nomenclature is simplified: The ψ symbol is omitted, the value of n is given, followed by the letter (s, p, d, or f) corresponding to the value of ℓ: $1s$. The $1s$ orbital in the vicinity of proton A is the wave function given by Equation 3-27. The value of the Bohr radius a_0 is 52.9 pm. It is reasonable to assume that the $1s$ atomic orbital for the electron in the vicinity of proton B is approximately given by the same equation, but with different labels:

$$1s_B = \frac{1}{\sqrt{\pi}}\left(\frac{1}{a_0}\right)^{\frac{3}{2}} e^{\frac{-r_B}{a_o}} \tag{3-28}$$

Molecular Orbital for the H_2^+ Ion. While the Schrödinger wave equation can be solved exactly for a hydrogen atom (two particles), it cannot be solved for a hydrogen molecule ion (three particles). Consequently it is necessary to guess at the nature of a molecular orbital for H_2^+. It seems reasonable that the molecular orbital should reflect some of the characteristics of the hydrogen atom ground state atomic orbital, and indeed, it is found that a suitable wave function φ for a molecular orbital is given by a linear combination of atomic orbitals (LCAO)

$$\varphi = \frac{1}{\sqrt{2}}(1s_A + 1s_B) \tag{3-29}$$

where $\frac{1}{\sqrt{2}}$ is a normalization factor, and $1s_A$ and $1s_B$ represent equations 3-27 and 3-28. Thus the wave function for the molecular orbital (MO) is:

$$\varphi = \frac{1}{\sqrt{2}}\left[\frac{1}{\sqrt{\pi}}\left(\frac{1}{a_0}\right)^{3/2} e^{-r_A/a_0} + \frac{1}{\sqrt{\pi}}\left(\frac{1}{a_0}\right)^{3/2} e^{-r_B/a_0}\right] \tag{3-30}$$

Units. The SI units for both the MO and the AO are $m^{-3/2}$, but it is customary to plot the MO (or AO) in units of $a_0^{-3/2}$ versus either x (= $-r/a_0$) or r with a_0 given a value of 0.5292 Å or 529.2 pm. It should be noted that for hydrogen, *all* radial functions R have the dimension $a_0^{-3/2}$, electron density functions R^2 have the dimension a_0^{-3}, and all probability functions r^2R^2 have the dimension a_0^{-1}. All angular functions are dimensionless. We choose here to plot the MO φ in units of $a_0^{-3/2}$ versus r in picometers (pm) and set the interatomic distance between atom A and atom B to equal the experimental value of 106 pm, as shown in Figure 3-10.

Choosing the Origin. With this one dimensional coordinate system, we choose the origin to be on atom A. Then:

$$r_A = |r| \tag{3-31}$$

$$r_B = |r - 106| \tag{3-32}$$

where $|r|$ is the absolute value of r, the position of the electron, 106 is the experimental internuclear distance, and r is a one-dimensional coordinate running through the two nuclei with the origin at nucleus A. When r equals zero, r_A equals zero, which means that the origin has been arbitrarily placed on nucleus A. When r equals zero, r_B equals 106 pm, which means that nucleus B is 106 pm away from the origin and from nucleus A. With these substitutions, the wave function (Equation 3-30) may be written

$$\frac{\varphi}{a_0^{-3/2}} = \frac{1}{\sqrt{2\pi}}\left(e^{\frac{-|r|}{a_0}} + e^{\frac{-|r-106|}{a_0}} \right) \tag{3-33}$$

With distances expressed in picometers, Equation 3-33 becomes

$$\frac{\varphi}{a_0^{-3/2}} = 0.39894\left(e^{\frac{-|r|}{52.92}} + e^{\frac{-|r-106|}{52.92}} \right) \tag{3-34}$$

$$\frac{\varphi}{a_0^{-3/2}} = 0.3984*(\exp(-|r/52.92|)+\exp((-|r-106|)/52.92)) \tag{3-35}$$

Equation 3-35 shows that the units of a $1s$ wave function are $a_0^{3/2}$, since both sides of Equation 3-35 are dimensionless as written. Because the molecular orbital for the hydrogen molecule ion is based on the sum of two $1s$ wave functions, its units are also $a_0^{3/2}$. The coefficient of the MO is $1/\sqrt{2\pi}$, which equals 0.39894 (the normalization factor). The exponents in Equation 3-35 are dimensionless (r/a_0). By expressing a_0 in picometers, we must also express the independent variable r in picometers.

Entering the Worksheet Data

Figure 3-11 shows the appearance of the worksheet after the chart is edited. Prepare an identical worksheet. First, enter the various labels and constants. Next, enter the numerical data required for the calculations, and then calculate the data for plotting the molecular orbital. Prepare the chart and edit it. Use Figure 3-11 as a guide for entering the data in the worksheet.

Entering the Title and Labels

- Beginning in cell D1, type **The Hydrogen Molecule Ion** as the title, and make it 12 point bold.
- In cell C3, enter Equation 3-35 as an informative label:

Psi = 0.39894*(EXP(-ABS(rA/52.92))+EXP(-ABS (rA-106)/52.92))

- In the label, select the **A** in **rA**, and on the **Format** menu, click **Cells**. In the **Format Cells** dialog box, select the **Subscript** check box. Repeat for the other **A**.
- In cell A5, type **r_A/pm** (make the subscript as you did for the equation in cell C3) as the label.
- In cell B5, type **Psi/a$^{3/2}$** (make the superscript similarly) as the label.

Entering the Incremented Value of *r*

A range of *r* from –300 to +400 pm is suitable. The wave function plot shown in Figure 3-11 begins at *r* equals –300 pm and ranges to *r* equals +400 pm in increments of *r* equal to 5 pm.

- In cell A6, type **-300** and click ENTER.
- Click the -300 in cell A6. On the **Edit** menu, point to **Fill** (Figure 3-7), and then click **Series** to open the **Series** dialog box (Figure 3-8). In the **Series** dialog box:
  1. For **Series in**, select **Columns**.
  2. For **Type**, select **Linear**.
  3. In the **Step value** box, type **5** for the step value.
  4. In the **Stop value** box, type **400** for the stop value.

The last cell, A146, will contain the value 400, and the values in the intermediate cells will increment by 5 from –300 to +400. Next, enter the value of the wave function in units of $a_0^{-3/2}$.

Entering the Incremented Values of the Psi Equation

- In cell B6, type the formula: **=0.39894*(EXP(-ABS(A6/52.92))+ EXP(-ABS(A6-106)/59.92))**
- Click and drag the horizontal split screen bar (Figure 1-1) to the middle of the screen.
- Scroll the lower half of the split screen until cell B146 is visible.
- Click in cell B6, and then SHIFT-click in cell B146 to select the range B6:B146.
- On the **Edit** menu, point to **Fill**, and click **Down** (Figure 3-7) to fill the range B6:B146 with the copies of the wave function Psi.

Plotting the Molecular Orbital

- Click and drag the horizontal split screen bar (Figure 1-1) to the middle of the screen.
- Scroll the lower half of the split screen until cell B146 is visible.
- Click in cell A6, and then SHIFT-click in cell B146 to select the range A6:B146.
- Click the **Chart Wizard** icon.
- In the **Chart Wizard – Step 1 of 4 – Chart Type** dialog box, select **XY (Scatter)** and a graph with a smooth line and no markers. Then click **Next**.

- In the **Chart Wizard – Step 2 of 4 – Chart Source Data** dialog box, no action is needed. Click **Next.**
- In the **Chart Wizard – Step 3 of 4 – Chart Options** dialog box:
  1. In the **Chart title** box, type **Hydrogen Molecule Ion: H2+** for the title (subscript the 2 and superscript the + later).
  2. In the **Value (X) axis** box, type **r/pm** as the label .
  3. In the **Value (Y) axis** box, type **Psi/a3/2** as the label (superscript the 3/2 later).
  4. Click **Next.**
- In the **Chart Wizard – Step 4 of 4 – Chart Location** dialog box, select **As object in Sheet1**; click **Finish.**

Editing the Chart

- Editing the title:
  1. Click the title twice slowly (do not double-click).
  2. Select the 2 in H2+, then on the **Format** menu, click **Selected Chart Title.** Select the **Subscript** check box.
  3. Select the + in H+, then on the **Format** menu, click **Selected Chart Title.** Select the **Superscript** check box.
  4. Press the END key to move the pointer to the end of the title; press ENTER to get a new line.
  5. Type **Psi = (1/N)(1sA + 1sB)** for the second line.
  6. Make the **A** and **B** subscripts, as previously described. Press ENTER to get a new line.
  7. Type **N = 21/2** and make the **1/2** a superscript, as previously described.
- Right-click any y-axis number, then click **Format Axis.** In the **Format Axis** dialog box, select **Number**, and in the **Decimal places** box type **1** as the value.
- Right-click anywhere along the plotted line, then click **Format Data Series** to open the **Format Data Series** dialog box (Figure 2-10). Change the color to black, and make the line heavier (increase the weight).
- Click in the plot area, and make the plot area border thicker, if you wish.
- Click in the chart area, and make the chart border thicker than the plot area, if you wish.
- Add subscripts to the labels in cells A5 and B5:
  1. In cell A5, make the label **r_A/pm.**
  2. In cell B5, make the label **Psi/$a^{3/2}$.**
- On the **View** menu, point to **Toolbars**, and click **Drawing**; then click the **Text Box** icon (a tiny square with the letter A on it).
- In the upper right corner of the plot area, drag to create a text box, and type **r = 106 pm** as the label. Adjust the size of the text box if necessary, and line it up with the curve's maxima.
- On the **Drawing** toolbar, click the line with an arrow (→). Click adjacent to the text box, and drag the arrow so that it nearly touches the maximum on the right.

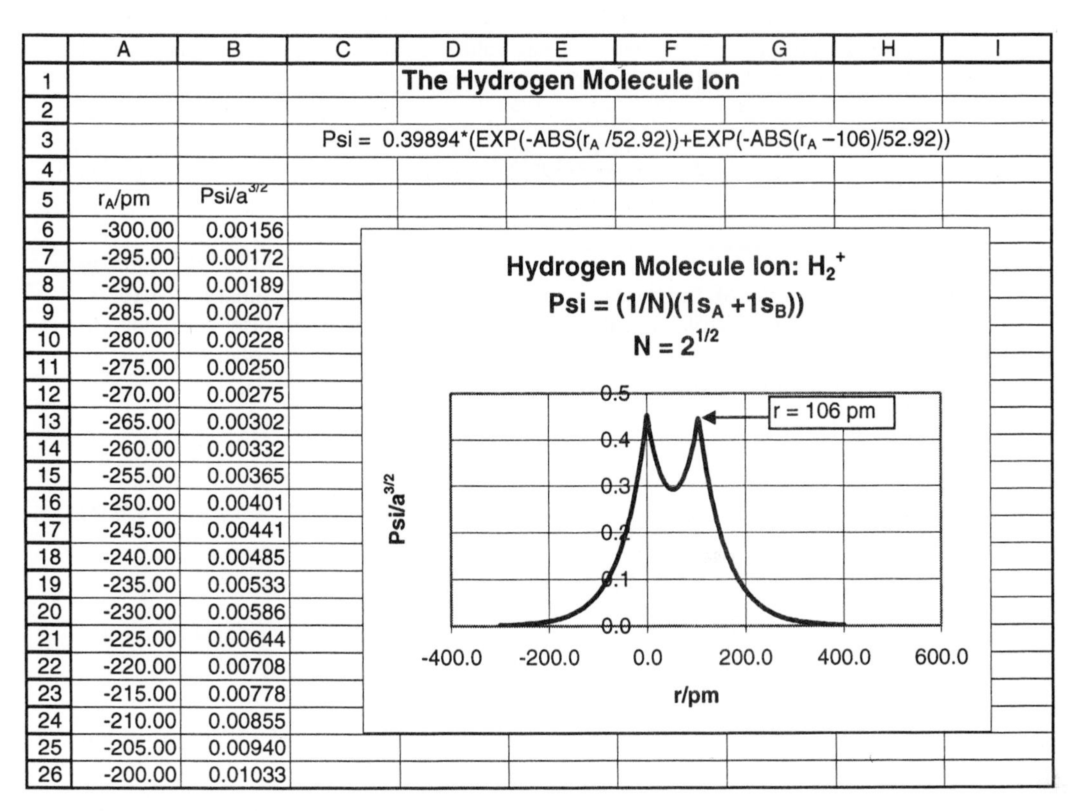

	A	B	C	D	E	F	G	H	I
1				The Hydrogen Molecule Ion					
2									
3			Psi = 0.39894*(EXP(-ABS(rA /52.92))+EXP(-ABS(rA −106)/52.92))						
4									
5	rA/pm	Psi/a$^{3/2}$							
6	-300.00	0.00156							
7	-295.00	0.00172							
8	-290.00	0.00189							
9	-285.00	0.00207							
10	-280.00	0.00228							
11	-275.00	0.00250							
12	-270.00	0.00275							
13	-265.00	0.00302							
14	-260.00	0.00332							
15	-255.00	0.00365							
16	-250.00	0.00401							
17	-245.00	0.00441							
18	-240.00	0.00485							
19	-235.00	0.00533							
20	-230.00	0.00586							
21	-225.00	0.00644							
22	-220.00	0.00708							
23	-215.00	0.00778							
24	-210.00	0.00855							
25	-205.00	0.00940							
26	-200.00	0.01033							

Figure 3-11. The Spreadsheet for the Hydrogen Molecule Ion

Excel Example 10

Rotational Energies and Rotational Spectrum

Review

A diatomic molecule consists of two atoms connected by a chemical bond, so that its shape resembles a dumbbell. Like any molecule, it has a total energy that is the sum of its translational, rotational, vibrational, and electronic energies. In this program we focus on the rotational energy of a rigidly rotating diatomic molecule. It is a simplifying approximation to consider a diatomic molecule to be rigid, which means that it is not vibrating along the bonding axis. For diatomic molecules with very strong bonds, such as N_2 or HCl, this is a good approximation; but for a molecule with a weak bond, such as Na_2 or I_2, it is not a good approximation.

A diatomic molecule cannot rotate with any arbitrary energy; its rotational energies are quantized. A solution to the Schrödinger wave equation gives a rather simple equation for the energies of a rigidly rotating diatomic molecule:

$$E_{rot} = J(J+1)\frac{h^2}{8\pi^2 I} \tag{3-36}$$

where h is Planck's constant (6.626 x 10^{-34} J s), π = 3.14159, I is the moment of inertia, and J is a rotational quantum number that has values 0, 1, 2, 3, 4, ... The moment of inertia I is a classical property of a rigid rotor, given by

$$I = \frac{m_1 m_2}{m_1 + m_2} r^2 \tag{3-37}$$

where r is the interatomic distance in meters and m_i are the masses of the two atoms in kilograms. The moment of inertia I has units of kg m^2. It is customary to express rotational energies in units of wavenumber (cm^{-1}) a cgs unit. It is also customary to express the physical properties of the diatomic molecule in terms of its rotational constant B, which also is usually reported in units of wavenumbers (Table 3-2). The rotational constant B is defined as:

$$B = \frac{h}{8\pi^2 Ic} \tag{3-38}$$

where c is the speed of light: c = 2.998 x 10^8 *m/s*. The allowed energies of a rigid rotor can be expressed in terms of B:

$$E_{rot}(cm^{-1}) = J(J+1)B \tag{3-39}$$

In this Excel Example we will create a worksheet (Figure 3-12) to calculate and chart the rotational energies from Equation 3-39.

Preparing the Worksheet

Entering the Title and Labels

- On the **Drawing** toolbar, click the **Text Box** icon; then click and drag to select the range C3:G5.
- In the text box, type **Rotational Energy Levels and Spectrum of HCl** as the title.
- Center the title, and change it to 12 point bold.
- Select cell C7, and type **The rotational constant B =** as the label.
- In cell E7, type **10.59** and center it.
- Select cell E7, and on the **Insert** menu, click **Name**; type **B** as the name of the entry.
- In cell F7, type **cm^{-1}** as the label.
- In cell A9, B9, C9, D9, and E9, type **J**, **Intensity**, **E_{rot} / (cm^{-1})**, **(E_{J+1} - E_J) / (cm^{-1})**, and **2nd Diff / (cm^{-1})** as the labels. Select the characters to be superscripted or subscripted, and on the **Format** menu, click **Cells**; select either the **Subscript** or the **Superscript** check box, as appropriate.

Table 3-2

Some Rotational Constants B (cm^{-1})

Diatomic Molecule	Symbol	B (cm^{-1})
Chlorine	$^{35}_{2}Cl$	0.2438
Carbon Monoxide	$^{12}C^{16}O$	1.9314
Hydrogen	$^{1}_{2}H$	60.809
Deuterium	$^{2}_{2}D$	30.429
Hydrogen Chloride	$^{1}H^{35}Cl$	10.5909
Iodine	$^{127}_{2}I$	0.03735

Entering Data and Formulas

- In the range A10:A16, enter **0**, **1**, **2**, **3**, **4**, **5**, and **6** as the values.
- Select the range A10:A16; on the **Insert** menu, click **Name**, and type **J** as the name for the selection.
- In the range B10:B16, type **1** in each cell.
- In cell C10, type **= J*(J+1)*B** as the formula. A 0 appears in cell C10 when you press ENTER.
- Select cell C10, and drag the fill handle to cell C16.
- In cell D11, type **=C11 - C10** as the formula.
- Select cell D11, and drag the fill handle to cell D16.
- In cell E12, type **=D12 - D11** as the formula.
- Select cell E12, and drag the fill handle to cell E16.

Creating the Rotational Energy Levels Chart

- Click and drag to select the range B10:C16; then click the **Chart Wizard** icon.
- In the **Chart Wizard – Step 1 of 4 – Chart Type** dialog box (Figure 2-3), select **XY (Scatter)** with points but no lines. Click **Next.**
- In the **Chart Wizard – Step 2 of 4 – Chart Source Data** dialog box (Figure 2-4), click **Next.**
- In the **Chart Wizard – Step 3 of 4 – Chart Options** dialog box, click the **Titles** Tab (Figure 2-6):
  1. In the **Chart title** box, type **Rotational Energy Levels** for the title.
  2. Leave the **Value (X) axis** box blank.
  3. In the **Value (Y) axis** box, type **Rotational Energy (cm-1)** as the label (superscript the -1 after cm).
  4. Click **Next.**
- In the **Chart Wizard – Step 4 of 4 – Chart Location** dialog box (Figure 2-7), select **As object in Sheet1,** and click **Finish.**

Editing the Rotational Energy Levels Chart

- Right-click the plot area, and click **Format Plot Area**. In the palette in the **Area** section of the dialog box, select the white square; click **OK**.
- Right-click a y-axis value, and click **Format Axis**. Click the **Number** tab; in the **Category** list, select **Number**, and type **0** in the **Decimal places** box.
- Click the **Font** tab, and set the font to 8 point bold.
- Right-click an x-axis value, and click **Format Axis**. Click the **Scale** tab, and type **1.0** in the **Maximum** box; click **OK**.
- Right-click an x-axis value, click **Format Axis**, and click clear.
- Click in the legend box, and delete it.
- Click the y-axis title and y-axis labels, and make them 10 point bold.
- On the **Drawing** toolbar, click the **Line** icon (\), and draw horizontal lines across the plot area from each series point.
- To change the weight of a line, select the line, and then click the **Line Weight** icon on the **Drawing** toolbar. Select the 2¼ point line.
- On the **Chart** menu, click **Chart Options**. Click the **Gridlines** tab, and clear all the gridlines check boxes.

Note: You could create an Excel bar chart for the rotational energy level chart, but Excel makes the energy axis nonlinear so that the rotational levels appear to be equally spaced.

Creating the Rotational Spectrum Chart

- Select the range B11:B16. While holding down the CTRL key, select the range D11:D16. Then click the **Chart Wizard** icon.
- In the **Chart Wizard – Step 1 of 4 – Chart Type** dialog box (Figure 2-3), select **XY (scatter)** with points but no lines. Click **Next**.
- In the **Chart Wizard – Step 2 of 4 – Chart Source Data** dialog box (Figure 2-4), click the **Series** tab.
  1. In the **X values** box, change the **B**s to **D**s.
  2. In the **Y values** box, change the **D**s to **B**s. Click **Next**.
- In the **Chart Wizard – Step 3 of 4 – Chart Options** dialog box, click the **Titles** tab (Figure 2-6):
  1. In the **Chart title** box, type **Rotational Spectrum** for the title.
  2. In the **Value (X) axis** box, type **Transition Energy (cm-1)** as the label (superscript the -1 after cm).
  3. In the **Value (Y) axis** box, type **Absorption Intensity** as the label.
  4. Click the **Gridlines** tab (Figure 2-6), and clear all the gridlines check boxes.
  5. Click **Next**.
- In the **Chart Wizard – Step 4 of 4 – Chart Location** dialog box (Figure 2-7), select **As object in Sheet1**, and click **Finish**.

Editing the Rotational Spectrum Chart

- Click in the legend box, and delete it.
- Right-click the plot area, and click **Format Plot Area**. In the palette in the **Area** section of the dialog box, select the white square; click **OK**.
- Click in the title, and set the font to 10 point bold.
- Right-click a y-axis value, and click **Format Axis**. Click the **Scale** tab, and type **1.1** in the **Maximum** box and **0.0** in the **Minimum** box; click **OK**.
- With the chart selected, on the **Chart** menu, click **Chart Type**, and select **Column**.
- Right-click any bar (column), and then click **Format Data Series**.
- Click the **Patterns** tab, and in the **Area** section, click a black color square, and then click **OK**.

- Set the axis labels to 10 point bold.

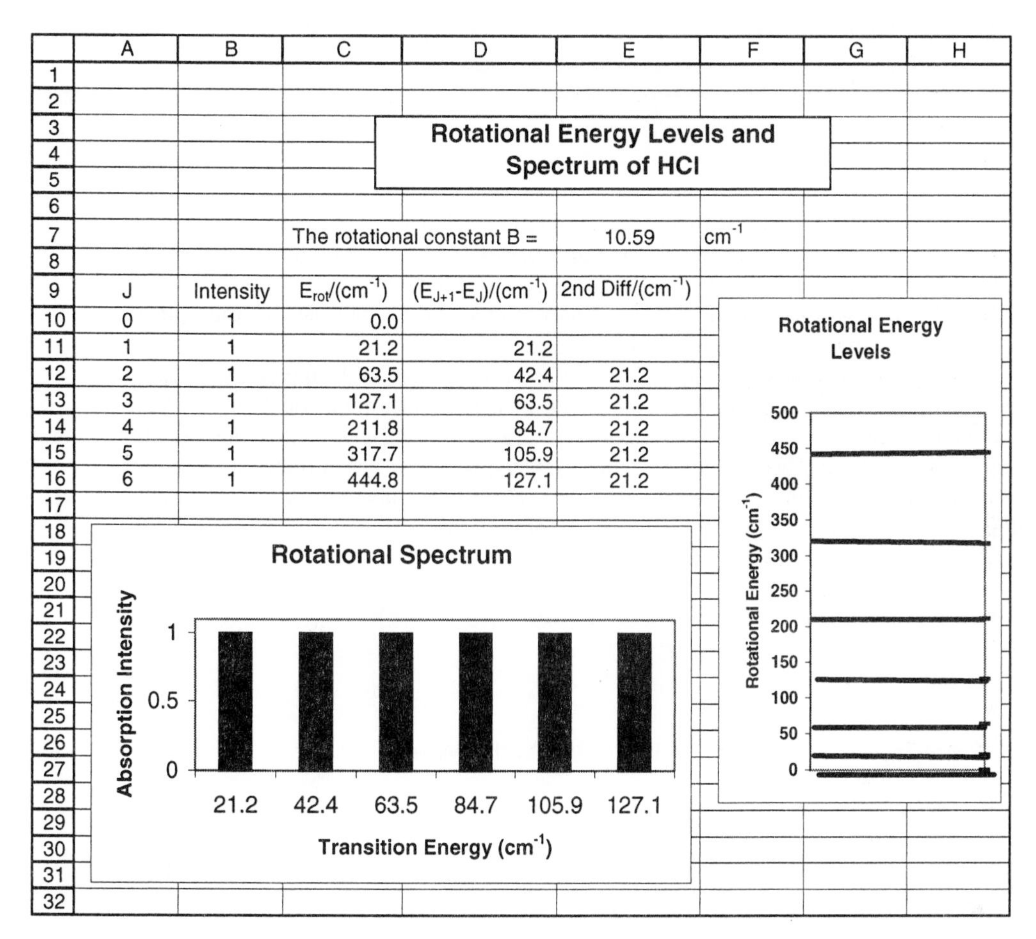

Rotational Energy Levels and Spectrum of HCl

The rotational constant B = 10.59 cm^{-1}

J	Intensity	$E_{rot}/(cm^{-1})$	$(E_{J+1}-E_J)/(cm^{-1})$	2nd Diff/(cm^{-1})
0	1	0.0		
1	1	21.2	21.2	
2	1	63.5	42.4	21.2
3	1	127.1	63.5	21.2
4	1	211.8	84.7	21.2
5	1	317.7	105.9	21.2
6	1	444.8	127.1	21.2

Figure 3-12. The Spreadsheet for Rotational Energies and the Rotational Spectrum. The actual intensities of the transitions are not equal, but depend on the population of each rotational state as shown in Chapter 4, Excel Example 15.

The rotational energy levels increase geometrically because of the J square term in Equation 3-36 Absorption transitions take place according to the selection rule that $\Delta J = +1$. These transitions are listed in the range D11:D16. The second differences are constant and equal 21.2, which equals $2B$, the spacing between the absorption lines in the rotational spectrum chart. It is assumed for this calculation that the rotating molecule is rigid.

Excel Example 11

Vibrational Wave Functions: *v* = 0, 1, and 2

Review

This section includes a brief outline of the quantum mechanics of the harmonic oscillator. The details may be found in any monograph or textbook on quantum mechanics. The graphical representation of the vibrational wave function displays the function's magnitude in one-dimensional space. What follows is an outline of the quantum mechanical origin of vibrational wave functions.

To a first approximation, the vibrations of a diatomic molecule can be treated as undergoing harmonic oscillations. The atoms in a diatomic molecule vibrate along a straight line subject to a potential energy $U = ½\, kx^2$, to the extent that the stretching motion obeys Hook's law, $F = -kx$. The displacement from an equilibrium position is given by the one-dimensional coordinate x (m). The Hook's law constant, k, has units of N/m. This gives rise to a Hamiltonian operator of the form:

$$H = -\frac{h^2}{8\pi^2\mu}\frac{\partial^2}{\partial x^2} + \frac{1}{2}kx^2 \qquad (3\text{-}40)$$

The Schrödinger wave equation, $H\psi = E\psi$, for the system is

$$\left(-\frac{h^2}{8\pi^2\mu}\frac{\partial^2}{\partial x^2} + \frac{1}{2}kx^2\right)\psi = E\psi \qquad (3\text{-}41)$$

μ is the reduced mass (in kilograms) of the diatomic molecule. A solution to the wave equation reveals that the energies are quantized so that the allowed energies E are given by

$$E = (v + 1/2)h\nu \qquad (3\text{-}42)$$

where ν is the fundamental vibrational frequency of the particular diatomic molecule, h is Planck's constant, and v is a vibrational quantum number with allowed values of 0, 1, 2, 3. If the vibrations are harmonic (obey Hook's law) then they are also described by the classical equation for the harmonic oscillator:

$$\nu = \frac{1}{2\pi}\sqrt{\frac{k}{\mu}} \qquad (3\text{-}43)$$

where v is the frequency (s^{-1}) k is the force constant (Hook's law), and μ is the reduced mass of the vibrating diatomic molecule.

$$\mu = \frac{m_1 m_2}{m_1 + m_2} \qquad (3\text{-}44)$$

Solutions to the Vibrational Wave Equation

Solutions to the vibrational wave equation are of the form

$$\psi_v(q) = N_v H_v(q) e^{\frac{-q^2}{2}} \qquad (3\text{-}45)$$

The quantity q is a generalized coordinate given by

$$q = (2a)^{1/2} x \qquad (3\text{-}46)$$

and a is a parameter of convenience defined by

$$a = 2\pi^2 v\mu / h \qquad (3\text{-}47)$$

Many monographs on quantum mechanics use α instead of a and ξ instead of q, but a and q are easier to use in a worksheet. The diatomic molecule vibrates along a line on x, displaced from zero to plus and minus values. The wave functions consist of three kinds of factors.

The Normalization Factor. The factors N_v are normalization factors and are given by

$$N_v = \left(\frac{(a/\pi)^{1/2}}{2^v v!} \right)^{1/2} \qquad (3\text{-}48)$$

The constant a is defined by equation 3-47, and v is the vibrational quantum number, which has the values 0, 1, 2, 3 ...

The Exponential Factor. The exponential factor is the same for all quantum states:

$$e^{-\frac{q^2}{2}}$$

The generalized coordinate q is defined by equations 3-46 and 3-47. For a given molecule and vibrational quantum state, it is a constant times x, the one-dimensional coordinate. The constant depends of properties of the particular molecule.

The Hermite Polynomials. The factors H_v are Hermite polynomials, well known to mathematicians. A few are listed in Table 3-3.

Table 3-3

Normalization Factors and Hermite Polynomials

Quantum Number	Normalization Factor	Hermite Polynomial
v	$N_v/(a/\pi)^{1/4}$	H_v
0	1	1
1	0.7071	$2q$
2	0.3535	$4q^2 - 2$
3	0.1443	$8q^3 - 12q$
4	0.0510	$16q^4 - 48q^2 + 12$
5	0.0161	$532q^5 - 160q^3 + 120$
6	4.66×10^{-3}	$64q^6 - 480q^4 + 720q^2 - 120$
7	1.21×10^{-3}	$128q^7 - 1344q^5 + 3360q^3 - 1680q$

Preparation of the Worksheet and Entering the Initial Data

After entering the formulas for the wave equations, we will prepare two charts. The first chart displays plots of the magnitude of the wave function versus displacement. The second chart displays plots of the square of the wave function versus displacement. Figure 3-13 shows the worksheet and charts.

Entering the Title, Labels, and Constants

As shown in Figure 3-13, $\psi_v(q)$ and $\psi_v^2(q)$ for v equal 0, 1, and 2 are graphed from q equal –4 to +4.

- In the range B1:D1, type **Vibrational Wave Functions** as the worksheet title.
- For the purpose of documentation, enter the equations for the wave functions to be plotted:
  1. Beginning in cell A3, type **Psi(v0) = EXP(-(q^2/2))** as the equation.
  2. Beginning in cell A4, type **Psi(v1) = 0.7071(EXP(-(q^2/2))*2*q** as the equation.
  3. Beginning in cell A5, type **Psi(v2) = 0.3536(EXP(-(q^2/2))*(4*q^2-2)** as the equation.
- In cells A7:H7, type the column titles shown in Figure 3-13:

q Psi(v0) Psi(v1) Psi(v2) q Psi(v0)^2 Psi(v1)^2 Psi(v2)^2

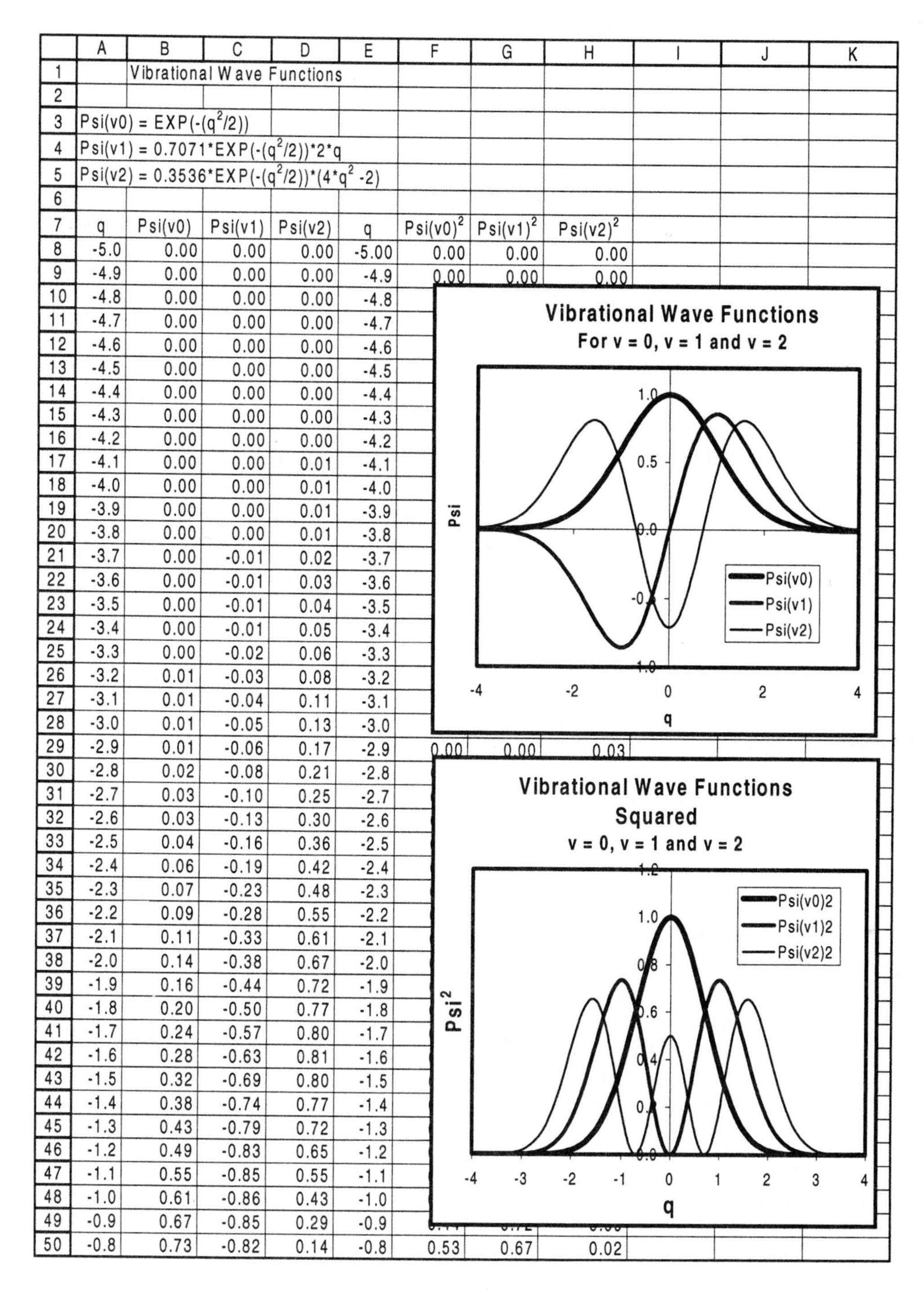

	A	B	C	D	E	F	G	H	I	J	K
1		Vibrational Wave Functions									
2											
3	Psi(v0) = EXP(-(q^2/2))										
4	Psi(v1) = 0.7071*EXP(-(q^2/2))*2*q										
5	Psi(v2) = 0.3536*EXP(-(q^2/2))*(4*q^2 -2)										
6											
7	q	Psi(v0)	Psi(v1)	Psi(v2)	q	Psi(v0)^2	Psi(v1)^2	Psi(v2)^2			
8	-5.0	0.00	0.00	0.00	-5.00	0.00	0.00	0.00			
9	-4.9	0.00	0.00	0.00	-4.9	0.00	0.00	0.00			
10	-4.8	0.00	0.00	0.00	-4.8						
11	-4.7	0.00	0.00	0.00	-4.7						
12	-4.6	0.00	0.00	0.00	-4.6						
13	-4.5	0.00	0.00	0.00	-4.5						
14	-4.4	0.00	0.00	0.00	-4.4						
15	-4.3	0.00	0.00	0.00	-4.3						
16	-4.2	0.00	0.00	0.00	-4.2						
17	-4.1	0.00	0.00	0.01	-4.1						
18	-4.0	0.00	0.00	0.01	-4.0						
19	-3.9	0.00	0.00	0.01	-3.9						
20	-3.8	0.00	0.00	0.01	-3.8						
21	-3.7	0.00	-0.01	0.02	-3.7						
22	-3.6	0.00	-0.01	0.03	-3.6						
23	-3.5	0.00	-0.01	0.04	-3.5						
24	-3.4	0.00	-0.01	0.05	-3.4						
25	-3.3	0.00	-0.02	0.06	-3.3						
26	-3.2	0.01	-0.03	0.08	-3.2						
27	-3.1	0.01	-0.04	0.11	-3.1						
28	-3.0	0.01	-0.05	0.13	-3.0						
29	-2.9	0.01	-0.06	0.17	-2.9	0.00	0.00	0.03			
30	-2.8	0.02	-0.08	0.21	-2.8						
31	-2.7	0.03	-0.10	0.25	-2.7						
32	-2.6	0.03	-0.13	0.30	-2.6						
33	-2.5	0.04	-0.16	0.36	-2.5						
34	-2.4	0.06	-0.19	0.42	-2.4						
35	-2.3	0.07	-0.23	0.48	-2.3						
36	-2.2	0.09	-0.28	0.55	-2.2						
37	-2.1	0.11	-0.33	0.61	-2.1						
38	-2.0	0.14	-0.38	0.67	-2.0						
39	-1.9	0.16	-0.44	0.72	-1.9						
40	-1.8	0.20	-0.50	0.77	-1.8						
41	-1.7	0.24	-0.57	0.80	-1.7						
42	-1.6	0.28	-0.63	0.81	-1.6						
43	-1.5	0.32	-0.69	0.80	-1.5						
44	-1.4	0.38	-0.74	0.77	-1.4						
45	-1.3	0.43	-0.79	0.72	-1.3						
46	-1.2	0.49	-0.83	0.65	-1.2						
47	-1.1	0.55	-0.85	0.55	-1.1						
48	-1.0	0.61	-0.86	0.43	-1.0						
49	-0.9	0.67	-0.85	0.29	-0.9						
50	-0.8	0.73	-0.82	0.14	-0.8	0.53	0.67	0.02			

Figure 3-13. Vibrational Wave Functions and Their Squares for *v* Equals 0, 1, and 2

Entering the Numerical Data for the First Chart

- In cell A8, type **-5** and press the ENTER key.
- Select cell A8, and on the **Edit** menu, click **Fill Series** to open the **Series** dialog box (Figure 3-8). In the **Series** dialog box:
  1. For **Series in**, select **Columns**.
  2. For **Type**, select **Linear**.

3. In the **Step value** box, type **0.1** for the step value.
4. In the **Stop value** box, type **5** for the stop value.

Next, enter the formulas for the wave functions in cells B8, C8, and D8. These are the product of $Nv/(a/\pi)^{1/4}$, Hv, and $e^{-\frac{q^2}{2}}$ for $v = 0, 1$, and 2 listed in Table 3-3.

Note: The term $(a/\pi)^{1/4}$ is not entered into the formula. This means that the units of the plot along the y-axis are $(a/\pi)^{1/4}$.

- In cell B8, type **= EXP(-(A8^2/2))** as the formula.
- In cell C8, type **= 0.7071*EXP(-(A8^2/2))*2*A8** as the formula.
- In cell D8, type **= 0.3535*EXP(-(A8^2/2))*(4*A8^2-2)** as the formula.

The values of q have already been filled in from cell A8 to Cell A109 with the use of the Edit Fill Series method. Next we shall use the Edit Fill Down method to fill in the parallel formula ranges.

- Click and drag the horizontal split screen bar (Figure 1-1) to the middle of the screen.
- Scroll the lower half of the split screen until cell B108 is visible.
- Click in cell B8, and then SHIFT-click in cell B108 to select the range B8:B108.
- On the **Edit** menu, point to **Fill**, and click **Down** (Figure 3-7) to fill the range B8:B108 with the copies of the wave function *Psi*(v0).

Repeat these step to fill the range C8:C108 with values of *Psi*(v1) and the range D8:D108 with values of *Psi*(v2). Next we shall generate the data to prepare a second chart with a plot of the squares of the wave function versus q.

Entering the Numerical Data for the Second Chart

The square of the vibrational wave function is interesting because its value at any x (or q) is the probability density or probability per unit length in one dimension at x. The probability from large negative q to large positive q (say from q equals -10 to $+10$) equals 1 for a normalized wave function. It is straightforward to plot the square of the wave functions on the same worksheet.

Entering a New Column of *q* Values

- In cell E8, type **-5** as the value.
- Select cell E8, and on the Edit menu, click **Fill Series** to open the **Series** dialog box (Figure 3-8). In the **Series** dialog box:
  1. For **Series in**, select **Columns**.
  2. For **Type**, select **Linear**.
  3. In the **Step value** box, type **0.1** for the step value.
  4. In the **Stop value** box, type **5** for the stop value.

Entering Three New Columns Containing the Squares of the Wave Function

- Select cells B8:D8, then copy and paste to cell F8.

- Click in cell F8, and change the reference from cell A8 to cell E8.
- Click in cell G8, and change the reference from cell A8 to cell E8.
- Click in cell H8, and change the reference from cell A8 to cell E8.
- Square the functions in F8:H8 by enclosing each function with parentheses and following the enclosed function with ^2 like this: (*function*)^2.
- Click and drag the horizontal split screen bar (Figure 1-1) to the middle of the screen.
- Scroll the lower half of the split screen until cell H108 is visible.
- Click in cell F8, and then SHIFT-click in cell H108 to select the range F8:H108.
- On the **Edit** menu, point to **Fill**, and click **Down** (Figure 3-7) to fill the range F8:H108 with the copies of the squares of the wave functions.

Creating the First Chart

- Click and drag the horizontal split screen bar (Figure 1-1) to the middle of the screen.
- Scroll the lower half of the split screen until cell D108 is visible.
- Click in cell A7, and then SHIFT-click in cell D108 to select the range A7:E108.
- On the toolbar, click the **Chart Wizard** icon.
- In the **Chart Wizard – Step 1 of 4 – Chart Type** dialog box, select **XY (Scatter)**, and click the icon labeled for smooth lines and without markers. Click **Next**.
- In the **Chart Wizard – Step 2 of 4 – Chart Source Data** dialog box (Figure 2-4), click **Next**.
- In the **Chart Wizard – Step 3 of 4 – Chart Options** dialog box:
    1. In the **Chart title** box, type **Vibrational Wave Functions** for the title.
    2. In the **Value (X) axis** box, type **q** as the label.
    3. In the **Value (Y) axis** box, type **Psi** as the label.
    4. Click the **Gridlines** tab, and clear the gridlines check boxes.
    5. Click the **Legend** tab, and select the **Show legend** check box.
    6. Click **Next**.
- In the **Chart Wizard – Step 4 of 4 – Chart Location** dialog box, select **As object in Sheet1**.

Creating the Second Chart

- Click and drag the horizontal split screen bar (Figure 1-1) to the middle of the screen.
- Scroll the lower half of the split screen until cell H108 is visible.
- Click in cell E7, and then SHIFT-click in cell H108 to select the range E7:H108.
- On the toolbar, click the **Chart Wizard** icon.
- In the **Chart Wizard – Step 1 of 4 – Chart Type** dialog box, select **XY (Scatter)**, and click the icon labeled for smooth lines and without markers. Click **Next**.
- In the **Chart Wizard – Step 2 of 4 – Chart Source Data** dialog box (Figure 2-4), click **Next**.
- In the **Chart Wizard – Step 3 of 4 – Chart Options** dialog box:
    1. In the **Chart title** box, type **Vibrational Wave Functions Squared** for the title.
    2. In the **Value (X) axis** box, type **q** as the label.
    3. In the **Value (Y) axis** box, type **Psi2** as the label.
    4. Click the **Gridlines** tab, and clear the gridlines check boxes.
    5. Click the **Legend** tab, and select the **Show legend** check box.
    6. Click **Next**.
- In the **Chart Wizard – Step 4 of 4 – Chart Location** dialog box, select **As object in Sheet1**.

Editing the First Chart

The Plot and Chart Area

- Right-click in the plot area to open a shortcut menu.
- Click **Format Plot Area** (similar to Figure 2-9), and change the gray background to white.
- Change the border weight to medium weight.
- Right-click in the chart area to open a shortcut menu.
- Click **Format Chart Area**, and change the border weight to medium weight.

The Title

- Click twice anywhere in the chart title (but don't double-click). A vertical-line insertion point will appear.
- Move the insertion point to the right end of the title by pressing the END key. Press the ENTER key, and the pointer drops to the next line below. Now type **For v = 0, v = 1, and v = 2** as the second line.
- Right-click in the chart title, and then click **Format Chart Title** (Figure 2-8). Click the **Font** tab, and change the size to 12 points.

The Plotted Lines

- Right-click the $v = 0$ plot line, click **Format Data Series**, and in the **Format Data Series** dialog box (Figure 2-10), select the **Color** box, and change the color to black; select the **Weight** box, and select the heaviest line weight.
- Right-click the $v = 1$ plot line, click **Format Data Series**, and in the **Format Data Series** dialog box (Figure 2-10), select the **Color** box, and change the color to black; select the **Weight** box, and select a medium heavy line.

Customizing the Y-Axis

- Right-click a y-axis number, then click **Format Axis** to open the **Format Axis** dialog box (Figure 2-11).
- Click the **Scale** tab, and type **1.2** in the **Maximum** box, and type **-1** in the **Minimum** box.
- Click the **Number** tab, select the **Number** category, and change the number of decimal places to 1.

Customizing the X-Axis

- Right click an x-axis number, then click **Format Axis** to open the **Format Axis** dialog box (Figure 2-11).
- Click the **Scale** tab, and type **4** in the **Maximum** box, and type **-4** in the **Minimum** box.
- Click the **Number** tab, select the **Number** category, and change the number of decimal places to 0.

Editing the Second Chart

- Edit the second chart exactly like the first chart.
- In the y-axis label, change **Psi2** to **Psi^2** as follows:

1. Click twice anywhere in the chart title (but don't double-click). A vertical-line insertion point will appear.
2. Select the **2** in **Psi2**.
3. On the **Format** menu, click **Selected Axis Title**, and select the **Superscript** check box.

Meaning of the Vibrational Wave Function Plots

As shown in Figure 3-13, $\psi_v(q)$ and $\psi_v^2(q)$ for v equals 0, 1, and 2 are graphed from q equals –4 to +4. The probability density of finding the atom at a displacement q from the origin is proportional to the square to the wave function. Examination of the plots in Figure 3-13 demonstrates that the probability tends to pile up at the extremes of the oscillation as the vibrational quantum number reaches high values. The system approaches classical behavior at very high quantum numbers. To compare quantum mechanical behavior of the vibrating molecule with classical behavior, visualize the shadow on the floor of a pendulum illuminated from above. The shadow of the pendulum oscillates back and forth along a straight line. The shadow spends most of its time at the extremes of the oscillation, so the highest probability of locating the shadow is at the extremes of its oscillation. The lowest probability is at its equilibrium position where the shadow is traveling fastest. This is just the opposite behavior of a vibrating molecule in its ground state.

For the ground vibrational state, v equals 0, H_V equals 1, N_V equals 1, and the shape of the wave function is just the shape of $e^{-q^2/2}$, a bell-shaped Gaussian probability distribution. For the ground state the probability reaches a maximum at the origin, not at the extremes. This behavior is least classical.

Excel Example 12

The Infrared Spectrum

Review

In Excel Example 10, we saw that the rotational energies of a diatomic molecule are given by

$$\tilde{\nu}_r = J(J+1)B \qquad (3\text{-}49)$$

where B is the rotational constant for the particular molecule (cm^{-1}) and J is the rotational quantum number, which can have integer values 0, 1, 2, 3 ...

Diatomic molecules also vibrate. The atoms in a diatomic molecule vibrate along the line (the bond) binding the two atoms together. In their lower level vibrational states, the vibrations are reasonably harmonic, and the vibrational energies of a harmonic oscillator are given by

$$\tilde{\nu}_v = (v+1/2)\tilde{\nu}_0 \qquad (3\text{-}50)$$

where $\tilde{\nu}_0$ is the fundamental vibration energy of the harmonic oscillator and v is the vibrational quantum number, which can have values of 0, 1, 2, 3 ....

In each vibrational state, the molecule is also rotating, with a rotational constant that is slightly different in each vibrational state. Figure 3-14 shows the ground vibrational state ($v'' = 0$) and the first excited vibrational state ($v' = 1$). By convention the upper state is denoted with a single prime and the lower state by a double prime. Added to each vibrational state are the rotational energies, given by $\tilde{\nu}_r' = J(J+1)B'$ for the upper state and by $\tilde{\nu}_r'' = J(J+1)B''$ for the lower state.

In the vibration-rotation absorption spectrum, molecules absorb infrared radiation and are excited from the various rotational energy levels in the ground vibrational state ($v = 0$) to various rotational levels in the $v = 1$ vibrational state. The change in the vibrational quantum number is +1 for the *absorption* of electromagnetic radiation and −1 for the *emission* of electromagnetic radiation. For the rotational transitions, the change in the rotational quantum number is ΔJ and it can equal ±1, but not zero. Figure 3-14 shows the forbidden $\Delta J = 0$ transition as a dotted line from the $v'' = 0$, $J'' = 0$ state to the $v' = 0$, $J' = 0$ state. If this energy were allowed, its energy would be $\tilde{\nu}_0$.

Energy Levels and Spectrum

The experimental absorption spectrum, shown below the energy level diagram in Figure 3-14, illustrates which transitions give rise to which absorption peaks. Notice that the transitions for which $\Delta J = +1$ are of higher energy than $\tilde{\nu}_0$, while the transitions for which $\Delta J = -1$ are of lower energy than $\tilde{\nu}_0$. Notice also that the absence of the forbidden transition ($\Delta J = 0$) gives rise to a wide space in the middle of the spectrum. The absorption peaks to the right (higher energy) are called the *R*-branch peaks, while the absorption peaks to the left (lower energy) are called the *P*-branch peaks. The energy of a transition between the lower and the upper vibrational state is given by:

$$\tilde{\nu} = \tilde{\nu}_0 = B_v' J'(J'+1) - B_v'' J''(J''+1) \tag{3-51}$$

With some rearranging, it may be shown that the two branches can be given by

$$\tilde{\nu} = \tilde{\nu}_0 + (B_v' + B_v'')m + (B_v' - B_v'')m^2 \tag{3-52}$$

where m has the values 1, 2, 3, ... for the R branch (that is, $m = J + 1$), and m has the values −1, −2, −3 ... for the P branch (that is, $m = -J$). For the HCl molecule, the transitions are represented by (Herzberg, 56–57, 113–114):

$$\tilde{\nu} = 2885.9 + 20.577m - 0.3034m^2 - 0.00222m^3 \tag{3-53}$$

The very small cubic term arises from considering the small centrifugal distortion of the rotating molecule. It is with this equation that the infrared (IR) spectrum worksheet calculates the absorption peaks in the infrared spectrum of the hydrogen chloride molecule. Figure 3-15 shows the worksheet with the embedded chart.

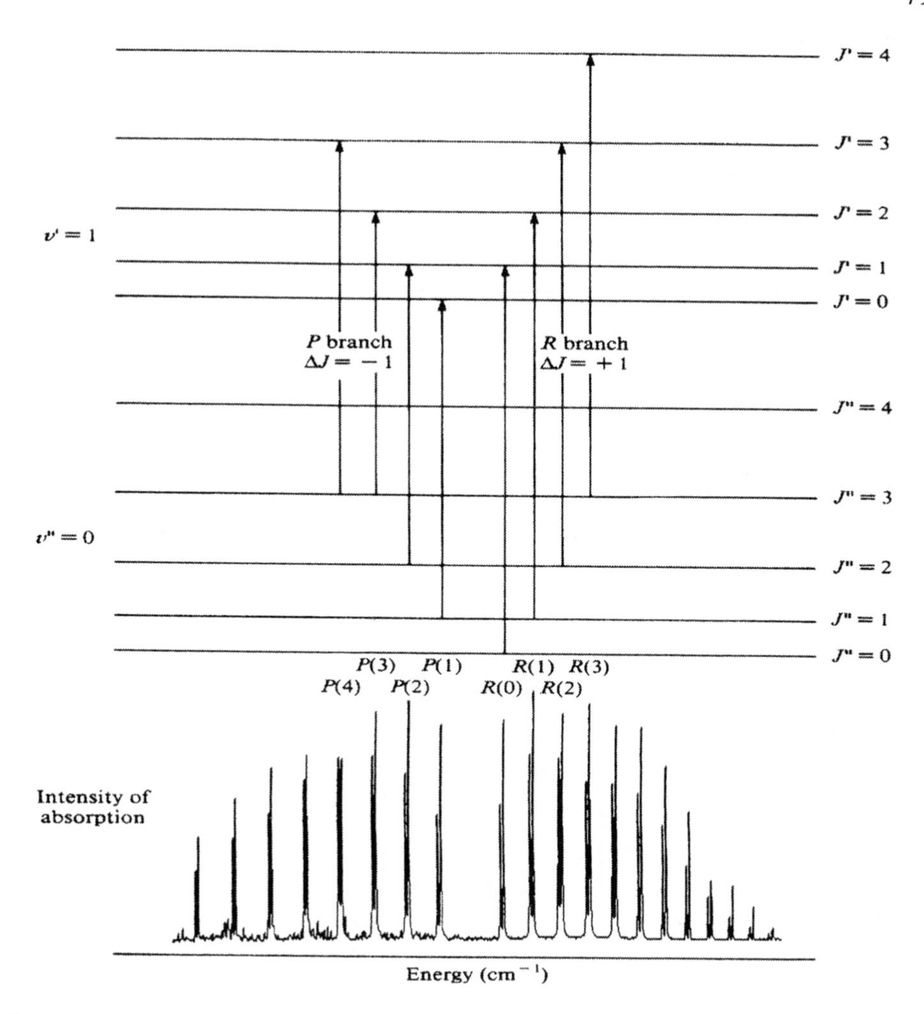

Figure 3-14. The Energy Level Diagram and Infrared Spectrum of HCl

Entering the Title and Labels

- On the **Drawing** toolbar, click the **Text Box** icon; then click and drag over the range D2:G4.
- In the text box, type **Infrared Spectrum of HCl** as the title.
- Center the title, and change it to 12 point bold; then size the box.
- Select cell D6, and type **NuBar = 2885.9 + 20.577m – 0.0303m^2 – 0.00222m^3** as the label.
- In cells A8, B8, and C8, type **m**, **Intensity**, and **NuBar** as the labels, and center them.

Entering Data and Formulas

- In cell A9, type **-6** and press ENTER.
- Select cell A9, and on the **Edit** menu, point to **Fill** (Figure 3-7), and then click **Series** to open the **Series** dialog box (Figure 3-8). In the Series dialog box:
  1. For **Series in**, select **Columns**.
  2. For **Type**, select **Linear**.
  3. In the **Step value** box, type **1** for the step value.
  4. In the **Stop value** box, type **6** for the stop value.
  5. Click **OK**. The numbers from –6 to 6, incremented in steps of 1, appear and fill the series A9:A21.

- Fill the series B9:B21 with **1**s.
- Select cell C9, and type **=288.9 + 20.577*A9 – 0.0303*A9^2 – 0.00222*A9^3** as the formula.
- Select cell C9, and drag the fill handle to cell C21.
- Select the range A15:C15, and press the DELETE key.

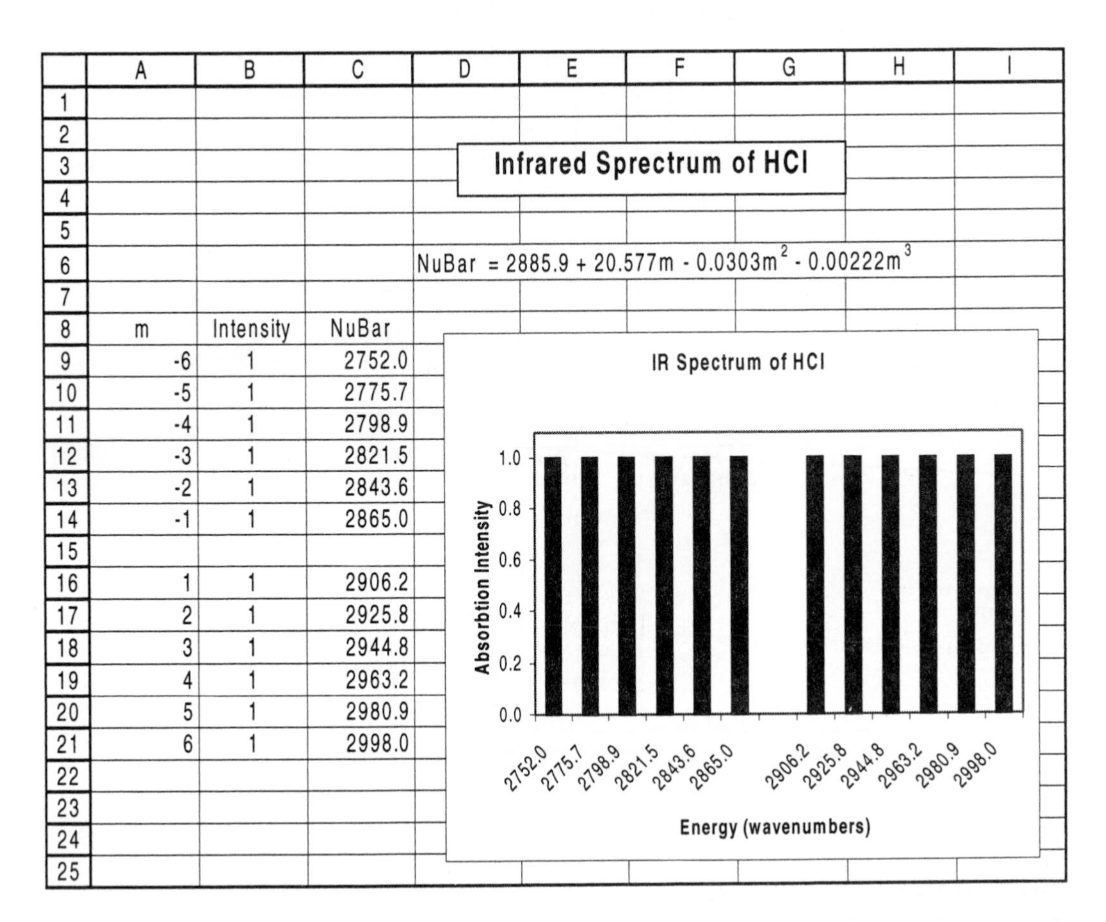

Infrared Sprectrum of HCl

NuBar = 2885.9 + 20.577m - 0.0303m^2 - 0.00222m^3

m	Intensity	NuBar
-6	1	2752.0
-5	1	2775.7
-4	1	2798.9
-3	1	2821.5
-2	1	2843.6
-1	1	2865.0
1	1	2906.2
2	1	2925.8
3	1	2944.8
4	1	2963.2
5	1	2980.9
6	1	2998.0

Figure 3-15. The Spreadsheet for the Infrared Spectrum of HCl. The actual intensities of the transitions are not equal, but depend on the population of each rotational state as shown in Chapter 4, Excel Example 15).

Creating the Infrared Spectrum of HCl Chart

- Select the range B9:B21, then click the **Chart Wizard** icon.
- In the **Chart Wizard – Step 1 of 4 – Chart Type** dialog box (Figure 2-3), select **XY (Scatter)** with points but no lines. Click **Next**.
- In the **Chart Wizard – Step 2 of 4 – Chart Source Data** dialog box (Figure 2-4), click the **Series** tab.
  1. In the **X values** box, change the **B**s to **C**s.
  2. In the **Y values** box, change the **C**s to **B**s.
- Click **Back** (to return to the **Chart Wizard – Step 1 of 4 – Chart Type** dialog box), and select **Column**, and click **Next**. In the **Chart Wizard – Step 2 of 4 – Chart Source Data** dialog box, click **Next**.
- In the **Chart Wizard – Step 3 of 4 – Chart Options** dialog box, click the **Titles** tab (Figure 2-6).
  1. In the **Chart title** box, type **IR Spectrum of HCl** for the title.
  2. In the **Value (X) axis** box, type **Energy (wavenumbers)** as the label.

3. In the **Value (Y) axis** box, type **Absorption Intensity** as the label.
4. Click the **Gridlines** tab, and clear all gridlines check boxes.
5. Click **Next.**

- In the **Chart Wizard – Step 4 of 4 – Chart Location** dialog box (Figure 2-7), select **As object in Sheet1**, and click **Finish.**

Editing the Infrared Spectrum of HCl Chart

- Right-click the plot area, click **Format Plot Area**, and in the **Area** section of the **Format Plot Area** dialog box, select the white square.
- Right-click a y-axis value, click **Format Axis**, click the **Scale** tab, and type **1.0** in the **Maximum** box.
- Click in the legend box, and delete it.
- Click the y-axis title and y-axis labels, and make them bold or larger, as you wish.

The infrared spectrum of HCl, a typical heteronuclear diatomic molecule, consists of equally spaced lines arising from transitions between rotational levels in the ground vibrational state (v = 0) to the rotational levels in the first excited vibrational state (v = 1). The selection rule is $\Delta J = \pm 1$. For m = 0, ΔJ = 0, which is forbidden. In the IR spectrum of HCl, the missing transition appears as a blank in the middle of the spectrum.

Excel Example 13

The Emission Spectrum and DesLandres Table

Vibrational Levels in Electronic States

Figure 3-16 shows a few typical electronic states of a diatomic molecule. The ground electronic state is normally called the *X* state, and successively higher electronic states are called the *A* state, *B* state, and so on. In each electronic state, the molecule also vibrates, and the vibrational energies are added to the electronic energy. Because of anharmonicity, the vibrational states become closer and closer together. Not shown in Figure 3-16 are the closely spaced rotational energy states, which are also added to the electronic energies. The rotational energies are small compared to the electronic energies.

The electronic emission spectrum originates in transitions from one electronic state to a lower electronic state, but not necessarily to the ground state. Two such transitions are shown in Figure 3-16. Of special interest is the transition labeled $\tilde{\upsilon}_{00}$, which takes place from the v' = 0 vibrational state in the upper electronic energy level to v'' = 0 in the lower electronic state. By convention, a double prime is always used to denote properties of the lower electronic state and a single prime for the upper electronic state.

In emission spectra of moderate resolution, the rotational fine structure is not fully resolved and appears as bands converging to a band head for each electronic transition. For this reason, electronic spectra of diatomic molecules are often referred to as *band spectra.*

It is apparent from Figure 3-16 that the energy of the spectral lines depends not only on the energy difference between the two electronic states but also on the vibrational energy level spacing in each electronic state, and that spacing appears to vary geometrically. Measurements of the emission spectrum of the PN molecule (Herzberg, 33, 41–43, 153–159) show that the vibrational energies of the upper *B* state are give by

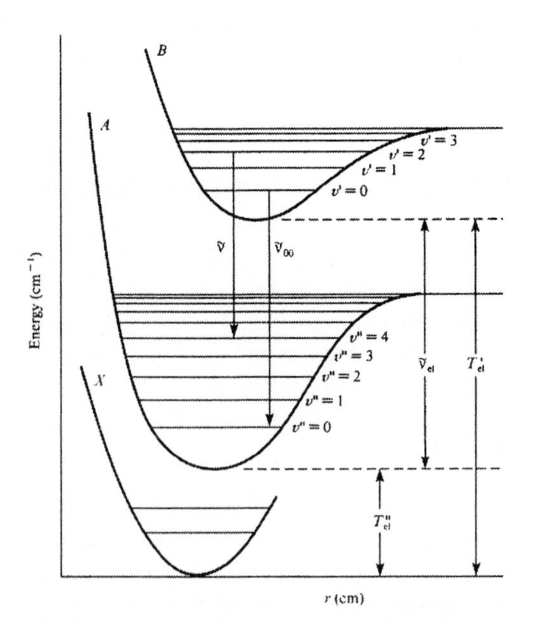

Figure 3-16. Electronic States and Transitions

$$Upper = 1094.8\, v' - 7.25\, v'^2 \qquad (3\text{-}54)$$

and lower *A* state vibrational energies are given by

$$Lower = 1329.38\, v'' - 6.98\, v''^2 \qquad (3\text{-}55)$$

The observed difference between the two electronic states is $\tilde{\upsilon}_{00}$, which equals 39699.01 cm^{-1} for the transition between $v'' = 0$ in the *B* electronic state and the $v' = 0$ vibrational level in the *A* electronic state (Figure 3-16). The $v' = 0$ to $v'' = 0$ transition corresponds to the *center* of the spectrum, The energies of the general emission spectral lines $\tilde{\upsilon}$ between the *B* and *A* states are given by

$$\tilde{\upsilon} = \tilde{\upsilon}_{00} + Upper - Lower \qquad (3\text{-}56)$$

$$\tilde{\upsilon} = 39699.01 + 1094.8\, v' - 7.25\, v'^2 - 1329.38\, v'' + 6.98\, v''^2 \qquad (3\text{-}57)$$

The DesLandres Table

The DesLandres table is a way of organizing a display of the electronic transitions between upper electronic state vibrational levels and lower electronic state vibrational levels. All that information lies in Equation 3-57, with which we will calculate a DesLandres table.

This task is simple with Equation 3-57, especially compared to the task of 19th-century spectroscopists, who constructed DesLandres tables from observed transitions in order to assign vibrational quantum numbers to each observed transition.

	A	B	C	D	E	F	G	H
1								
2								
3			**Delandres Table for the PN Molecule**					
4								
5			NuBar = NuBar00 +(1094.8v' - 7.25v'^2) - (1329.38v" - 6.98v''^2)					
6								
7				NuBar00=	39699	(wavenumbers)		
8								
9								
10				Upper State Quantum Numbers v"				
11								
12			**0**	**1**	**2**	**3**	**4**	**5**
13		0	39699.0	38376.6	37068.2	35773.7	34493.2	33226.6
14	Lower	1	40786.6	39464.2	38155.7	36861.2	35580.7	34314.2
15	State	2	41859.6	40537.2	39228.8	37934.3	36653.8	35387.2
16	Quantum	3	42918.2	41595.8	40287.3	38992.8	37712.3	36445.8
17	Numbers	4	43962.2	42639.8	41331.4	40036.9	38756.4	37489.8
18	v'	5	44991.8	43669.4	42360.9	41066.4	39785.9	38519.4
19		6	46006.8	44684.4	43376.0	42081.5	40801.0	39534.4
20								
21								
22								

Figure 3-17. Worksheet for the DesLandres Table for the PN Molecule

Preparing the Worksheet for the DesLandres Table

As shown in the worksheet in Figure 3-17, after entering suitable labels, the equation for the spectral band heads (Equation 3-57) is entered. The first column of lines is calculated by filled the cells with the simple click and drag method, with *v"* values held constant. Then, with the *v'* values held constant, the first row is filled with the click and drag method. The next six rows are filled the same way.

Entering the Labels

- Open the **Drawing** toolbar, and then click the **Text Box** icon.
- Click and drag approximately from cell B2 to cell H4, and type **DesLandres Table for the PN Molecule** as the title.
- Starting at cell C5, type Equation 3-57 as a label in the following form:
  NuBar = NuBar00 + (1094.8v’ - 7.25v’2) – (1329.38v” – 6.98v”2)
- Select the final **2**s, and on the **Format** menu, click **Cells**; select the **Superscript** check box.
- In cell D7, type **NuBar00 =** as the label; in cell E7, type **39699** as the label; and in cell F7, type **(wavenumbers)** as the label.
- As you have done previously, place a text box in the range D9:G10, and type **Upper State Quantum Numbers v”** as the label.
- In the range C12:H12, enter **0**, **1**, **2**, **3**, **4**, and **5** and make these numbers bold and centered.

- In the range A13:B18, add a text box, and type **Lower State Quantum Numbers v'** as the label. Make the text box one word wide. Center the label.
- In the range B13:B19, type **0**, **1**, **2**, **3**, **4**, **5**, and **6** and make these numbers bold.

Entering the Formulas

- Select cell C13 and type the formula:
  = $E$7 + 1094.8*B13 - 7.25*B13^2 - 1329.38*C$12 + 6.98*$C$12^2
- Click and drag from cell C13 (which should read 39699.0) to cell C19.
- Select cell C13.
  1. Change all occurrences of **B13** to **$B$13**.
  2. Change all occurrences of **$C$13** to **C13**.
- Select the range C13:H13.
- In a similar manner, fill the next six rows. The worksheet should appear as in Figure 3-16.

Discussion

As an aid in indexing emission spectral lines, that is, in determining the two vibrational quantum numbers (v' and v'') associated with each transition, spectroscopists find it useful to prepare a DesLandres table from the observed transitions. Differences in the energies of transitions in neighboring rows or columns in the DesLandres table recur. The recurring differences can be identified as members of a *progression* or a *sequence*.

Progressions. Figure 3-18 shows how the electronic transition may be grouped. For the sake of simplicity, the effect of anharmonicity is omitted. The groups in the same progression all have either a common lower level vibrational quantum number v'' or a common upper level quantum number v' (Figure 3-18a). Careful inspection of the progressions reveals that the difference in energy of selected adjacent pairs of lines is constant and corresponds to the energy difference between certain vibrational levels in the same electronic state. Grouping the lines into progressions and sequences is of great assistance in indexing the lines, that is, in assigning vibrational quantum numbers (v' and v'') to each line in the electronic spectrum.

Examination of the DesLandres table in Figure 3-17 reveals that the first column corresponds to the progression in which $v'' = 0$. The first row corresponds to the progression in which $v' = 0$.

Sequences. Figure 3-18b displays the bands heads as lines (omitting the rotational band structure) increasing in energy from left to right. The energy level diagram above the spectrum shows the origin of the spectral lines. The transitions above the spectrum are arranged in *sequences*. Because each member of a sequence has the same value of Δv, their energies are fairly close together. In the DesLandres table of Figure 3-18 transitions lying on the diagonal from upper left to lower right correspond to the sequence in which Δv equals zero.

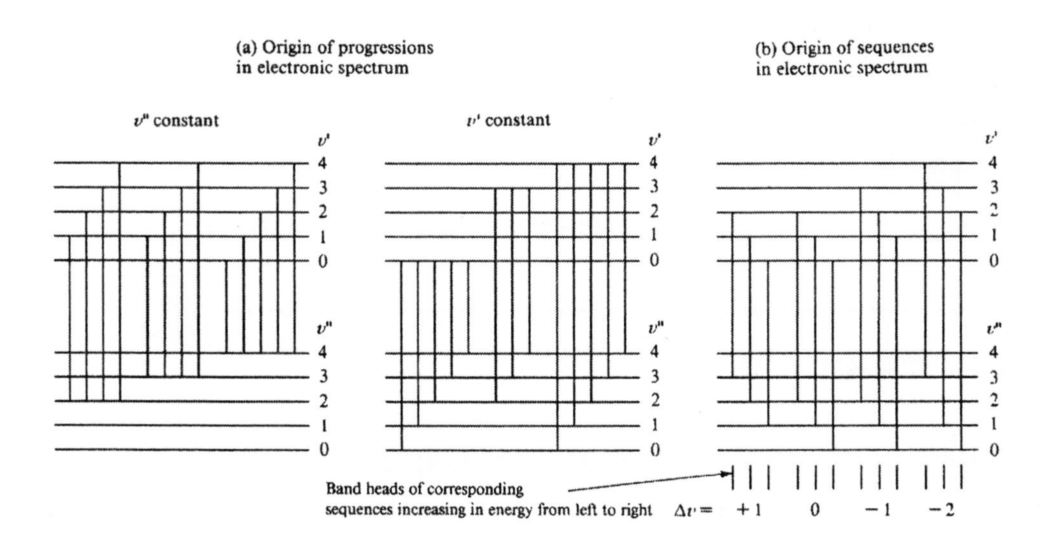

Figure 3-18. Origin of Progressions (a) and Sequences (b). The arrow points to a simulated band spectrum with only the band head locations showing.

Chapter 4

Statistical Thermodynamics

Excel Example 14

Calculation of the Vibrational Heat Capacities of I_2, Cl_2, and N_2

Creating this worksheet provides the opportunity to review some of the concepts and applications of statistical thermodynamics and further practice in creating multiple plot graphics.

Review of the Science

One of the goals of statistical thermodynamics is the calculation of macroscopic thermodynamic properties of a system from microscopic properties of individual molecules, especially their energies. Gaseous polyatomic molecules have energies arising from their translational, rotational, vibrational, and electronic degrees of freedom. These energies are quantized, and quantum mechanics provides a description of these energies.

Molecular Energies

Solution of the Schrödinger wave equation gives the various energies of an individual molecule.

Translational $$\varepsilon_n = \frac{n^2h^2}{8ma^2} \qquad n = 1, 2, 3, \ldots \tag{4-1}$$

Rotational $$\varepsilon_J = J(J+1)\frac{h^2}{8\pi^2 I} \qquad J = 0, 1, 2, \ldots \tag{4-2}$$

Vibrational $$\varepsilon_v = (v+\frac{1}{2})h\nu_{vib} \qquad v = 0, 1, 2, .. \tag{4-3}$$

No simple equation is available for the electronic energies ε_{elec}, but in favorable cases spectroscopic measurements can give information about the electronic energy of a molecule. The symbols for these equations are listed in Table 4-1.

Table 4-1

Symbols and Units in the Molecular Energy Equations

Symbol	SI Units	Description
a	m	Length
ε_n	J	Translational energy
ε_J	J	Rotational energy
ε_v	J	Vibrational energy
ε_{elec}	J	Electronic energy
g_i	—	Degeneracy
h	J s	Planck's constant
I	kg m^2	Moment of inertia
J	—	Rotational quantum number
k	J K^{-1} mol^{-1}	Boltzmann constant
m	kg	Mass
n	—	Translational quantum number
T	K	Temperature
v	—	Vibrational quantum number
ν_{vib}	s^{-1}	Classical vibration frequency

Tabulations of vibrational energies usually list $\tilde{\upsilon}$ in units of wave numbers (cm^{-1}), which often have the symbol ω. Rotational energies are often tabulated in terms of B, the rotational constant, which also has the unit of wavenumbers (Table 4-2).

$$B = \frac{h}{8\pi^2 Ic} \tag{4-4}$$

The energy ε_J in terms of the rotational constant B is given by

$$\varepsilon_J = \frac{J(J+1)hcB}{kT} \tag{4-5}$$

The total energy of a single molecule is the sum of its translational, rotational, vibrational, and electronic energies:

$$\varepsilon_{tot} = \varepsilon_{tran} + \varepsilon_{rot} + \varepsilon_{vib} + \varepsilon_{elect} \tag{4-6}$$

Similarly, on a macroscopic scale the total energy of a collection of molecules is

$$E^0 - E_0^0 = (E^0 - E_0^0)_{tran} + (E^0 - E_0^0)_{rot} + (E^0 - E_0^0)_{vib} + (E^0 - E_0^0)_{elect} \quad (4\text{-}7)$$

Table 4-2

Molecular Constants of Diatomic Molecules

Diatomic Molecule	Symbol	B/(cm^{-1})	$\tilde{\nu}$ /(cm^{-1})
Carbon monoxide	$^{12}C^{16}O$	1.9313	2170.21
Chlorine	$^{35}Cl_2$	0.2438	564.9
Deuterium	$^{2}D_2$	30.429	3118.4
Hydrogen	$^{1}H_2$	60.80	4395.2
Hydrogen chloride	$^{1}H^{35}Cl$	10.5909	2898.74
Iodine	$^{127}I_2$	0.03735	214.57
Nitrogen	$^{14}N_2$	2.010	2359.61

Energy and the Partition Functions

The calculation of thermodynamic quantities in statistical thermodynamics begins with the Boltzmann distribution law and the definition of the partition function. The following review deals with the translational, rotational and vibrational energies of homonuclear diatomic molecules. Determining separate partition functions for translational, rotational, and vibrational energy levels facilitates calculation of the translational, rotational, and vibrational contributions to the total energy and other thermodynamic properties of a collection of molecules.

The Boltzmann Distribution Function. Different numbers of molecules are distributed among all the available energy levels. The Boltzmann distribution function describes how the individual molecules populate the available energy states or levels.

$$n_i = \frac{N g_i e^{-\varepsilon_i / kT}}{\sum_i g_i e^{-\varepsilon_i / kT}} \quad (4\text{-}8)$$

The n_i are the number of molecules occupying the ε_i energy levels. N is the total number of molecules, so n_i/N is the fraction of all molecules that occupy the ith energy level. The degeneracy of an energy level is g_i.

The Total Energy from the Boltzmann Distribution. The total energy of a collection of molecules depends on the sum of the energy arising from the molecules populating each energy level.

$$E^0 - E_0^0 = \sum_i n_i \varepsilon_i \tag{4-9}$$

Replacing n_i in Equation 4-9 with the right side of Equation 4-8 gives

$$E^0 - E_0^0 = \frac{N \sum_i \varepsilon_i g_i e^{-\varepsilon_i / kT}}{\sum_i g_i e^{-\varepsilon_i / kT}} \tag{4-10}$$

The Total Energy from the Partition Function. The denominator of Equation 4-8 is the *partition function*:

$$Q = \sum_i g_i e^{-\varepsilon_i / kT} \tag{4-11}$$

The partition function, together with its derivative (Equation 4-12), simplifies the form of Equation 4-8. The derivative of the partition function is

$$\frac{\partial Q_t}{\partial T} = \frac{\sum_i \varepsilon_i g_i e^{-\varepsilon_i kT}}{kT^2} \tag{4-12}$$

Combining equations 4-10, 4-11 and 4-12 results in an equation for the energy in terms of the partition function:

$$E^0 - E_0^0 = RT^2 \left(\frac{\partial \ln Q_{tot}}{\partial T} \right) \tag{4-13}$$

Like the total energy, the total partition function Q_{tot} can be separated into translation, rotation, vibration, and electronic contributions:

$$Q_{tot} = Q_t \, Q_{rot} \, Q_{vib} \, Q_{elec} \tag{4-14}$$

This means that the each partition function can be calculated separately, and each energy contribution (translational, rotational, vibrational, and electronic) can be calculated separately.

The Calculation of the Separate Partition Functions Q_t, Q_r, and Q_v. Equation 4-14 shows that the partition functions for translation, rotation, and vibration can be calculated separately.

Translational Partition Function. In one dimension, g_i for translation equals unity. When Equation 4-1 for the translational energy is substituted into Equation 4-11, the definition of the partition function, the result is

$$Q_{tran} = \sum_{n=1}^{\infty} e^{-\frac{n^2h^2}{8ma^2kT}} = \sum_{n=1}^{\infty} e^{-bn^2} \tag{4-15}$$

where

$$b = \frac{h^2}{8ma^2kT} \tag{4-16}$$

Since the translational levels are many, close together, and fully populated, the summation in Equation 4-15 can be replaced by an integral:

$$Q_{tran} = \int_0^{\infty} e^{-bn^2}\,dn = \frac{1}{2}\sqrt{\frac{\pi}{b}} = (2\pi mkT)^{\frac{1}{2}}\frac{a}{h} \tag{4-17}$$

In three dimensions, Equation 4-17 may be written

$$Q_{tran} = (2\pi mkT)^{3/2}\frac{V}{h^3} \tag{4-18}$$

Rotational Partition Function. The degeneracy g_i for the rotational levels is $2J + 1$. When Equation 4-2 for the rotational energy is substituted into Equation 4-11, the definition of the partition function, the result is

$$Q_{rot} = \sum_{J=0}^{\infty} (2J+1)e^{-\frac{J(J+1)h^2}{8\pi^2 IkT}} \tag{4-19}$$

As J becomes very large, $2J + 1 \approx 2J$ and $J(J + 1) \approx J^2$, so the summation may be written

$$Q_{rot} = 2\sum_{J=0}^{\infty} Je^{-\frac{J^2h^2}{8\pi^2 IkT}} = 2\int_0^{\infty} Je^{-dJ^2}\,dJ = \frac{1}{d} \tag{4-20}$$

$$Q_{rot} = \frac{8\pi^2 IkT}{h^2} \qquad (4\text{-}21)$$

Vibration Partition Function. The degeneracy of the single vibrational mode of a diatomic molecule is unity. Unlike the translational and rotation energies (Equations 4-1 and 4-2), the ground state vibrational energy (Equation 4-3) does not equal zero. To take this into account, we write the partition function (Equation 4-11) for vibration as

$$Q_{vib} = \sum_i e^{-(\varepsilon_v - \varepsilon_0)/kT} \qquad (4\text{-}22)$$

From Equation 4-3, $\varepsilon_v = (v + \frac{1}{2})\, h\upsilon_{vib}$ where $v = 0, 1, 2 \ldots$ and $\varepsilon_v - \varepsilon_0 = v\varepsilon$ where $\varepsilon = h\upsilon_{vib}$. The vibrational partition function then takes the form

$$Q_{vib} = \sum_{v=0}^{\infty} e^{\frac{-v\varepsilon}{kT}} \qquad (4\text{-}23)$$

With the substitution

$$x = \frac{\varepsilon}{kT} \qquad (4\text{-}24)$$

the vibrational partition function may be written

$$Q_{vib} = \sum_{v=0}^{\infty} e^{-vx} \qquad (4\text{-}25)$$

The summation shown in Equation 4-20 converges so that

$$Q_{vib} = (1 - e^{-x})^{-1} \qquad (4\text{-}26)$$

The Translational and Rotational Contributions to the Energy

Equation 4-13, applied to the translation and rotational partition functions for diatomic molecules, gives the translational and rotational contribution to the energy.

$$(E^0 - E_0^0)_{tran} = \frac{3}{2} RT \qquad (4\text{-}27)$$

$$(E^0 - E_0^0)_{rot} = RT \qquad (4\text{-}28)$$

These contributions to the energy are a function of the temperature only and independent of the nature of the gas. Thus, at the same temperature, the translation and rotational energies of all gases are equal. This is not true of the vibrational contribution to the energy as can be seen in the next section.

The Vibrational Contribution to the Energy

To apply Equation 4-13 by taking the derivative of the vibrational partition (Equation 4-21), it is useful to make the following substitutions:

$$x = \frac{a}{T} \text{ , and } a = \frac{h\nu}{k} = \frac{hc}{k\lambda} = \frac{hc\tilde{\nu}}{k} \text{ , since } \nu = \frac{c}{\lambda} \tag{4-29}$$

The quantity a is often given the symbol θ and is called the characteristic vibrational temperature.

With this substitution, the vibrational partition function may be written as

$$Q_{vib} = (1 - e^{\frac{-a}{T}}) \tag{4-30}$$

$$\frac{d\ln(1 - e^{\frac{-a}{T}})}{dT} = \frac{ae^{\frac{-q}{T}}}{T^2(1 - e^{\frac{-a}{T}})} = \frac{xe^{-x}}{T(1 - e^{-x})} \tag{4-31}$$

Combining Equation 4-31 with Equation 4-13 gives an equation for calculating the internal energy from molecular parameters:

$$(E^0 - E_0^0)_{vib} = RT^2 \frac{xe^{-x}}{T(1 - e^{-x})} \tag{4-32}$$

Slight rearrangement puts Equation 4-32 in a more compact form:

$$(E^0 - E_0^0)_{vib} = RT \frac{x}{(e^x - 1)} \tag{4-33}$$

The Vibrational Heat Capacity of a Gas

The vibrational heat capacity is defined in terms of the vibrational energy:

$$C_{vib} = \left(\frac{\partial (E^0 - E_0^0)_{vib}}{\partial T} \right)_V \tag{4-34}$$

where $(E^0 - E_0^0)_{vib}$ is given in Equation 4-33. Thus the heat capacity is

$$C_{vib} = \frac{d}{dT}\left(\frac{Ra}{e^{a/T}-1}\right) = \frac{Ra^2 e^{a/T}}{T^2(e^{a/T}-1)^2} \tag{4-35}$$

A short rearrangement of Equation 4-35 leads to the usual form:

$$C_{vib} = R\frac{x^2 e^x}{(e^x-1)^2} \tag{4-36}$$

The coefficients of *RT* in Equation 4-33 and of *R* in Equation 4-36 are called Einstein functions. Both of these functions approach a value of unity as *x* becomes small (low $\tilde{V}$ and high *T*), and both functions approach a value of zero as *x* becomes large (high $\tilde{V}$ and low *T*). Thus the high temperature limit of a vibrational contribution to the internal energy of all gases is *RT*. Similarly, the high temperature limit of the vibrational contribution to the heat capacity is *R*.

From the definition of C_v (Equation 4-34) and Equations 4-27 and 4-28, the translational contribution to C_v is $3R/2$, and the rotational contribution to the heat capacity is *R*. Unlike the vibrational contribution to C_v, the translational and rotational contributions to the heat capacity are independent of temperature.

Preparation of the Worksheet

This Excel example illustrates the dependence of the vibrational heat capacity on both the temperature and the fundamental vibration frequency. It does this with plots of the heat capacity of three different diatomic molecules, iodine, chlorine, and nitrogen, at temperatures from 10 to 1010 K.

Entering the Title, Labels, and Constants. Figure 4-1 shows the worksheet with the input constants and labels and displays an embedded chart containing plots of the heat capacities of gaseous iodine, chlorine, and nitrogen.

- Beginning in cell B2, type **The Vibrational Heat Capacities of Gaseous Iodine, Chlorine, and Nitrogen** as the worksheet title.
- In cell C4, enter Planck's constant; in cell E4, enter the speed of light; in cell G4, enter the Boltzmann constant; and in cell I4, enter the gas constant, all in SI units as indicated in the cells below each entry.
- Identify each constant with its conventional label in cells B4, D4, F4, and H4 (**h=**, **c=**, **k=**, and **R=**). Right-justify the labels, and left-justify the constants.
- Enter the units in row 5 under the values in row 4, and center the unit labels.
- In a similar manner, enter the fundamental vibration frequencies in SI units for iodine in cell C9, for chlorine in cell E9, and for nitrogen in cell G9. Add labels and give units as shown in Figure 4-1.

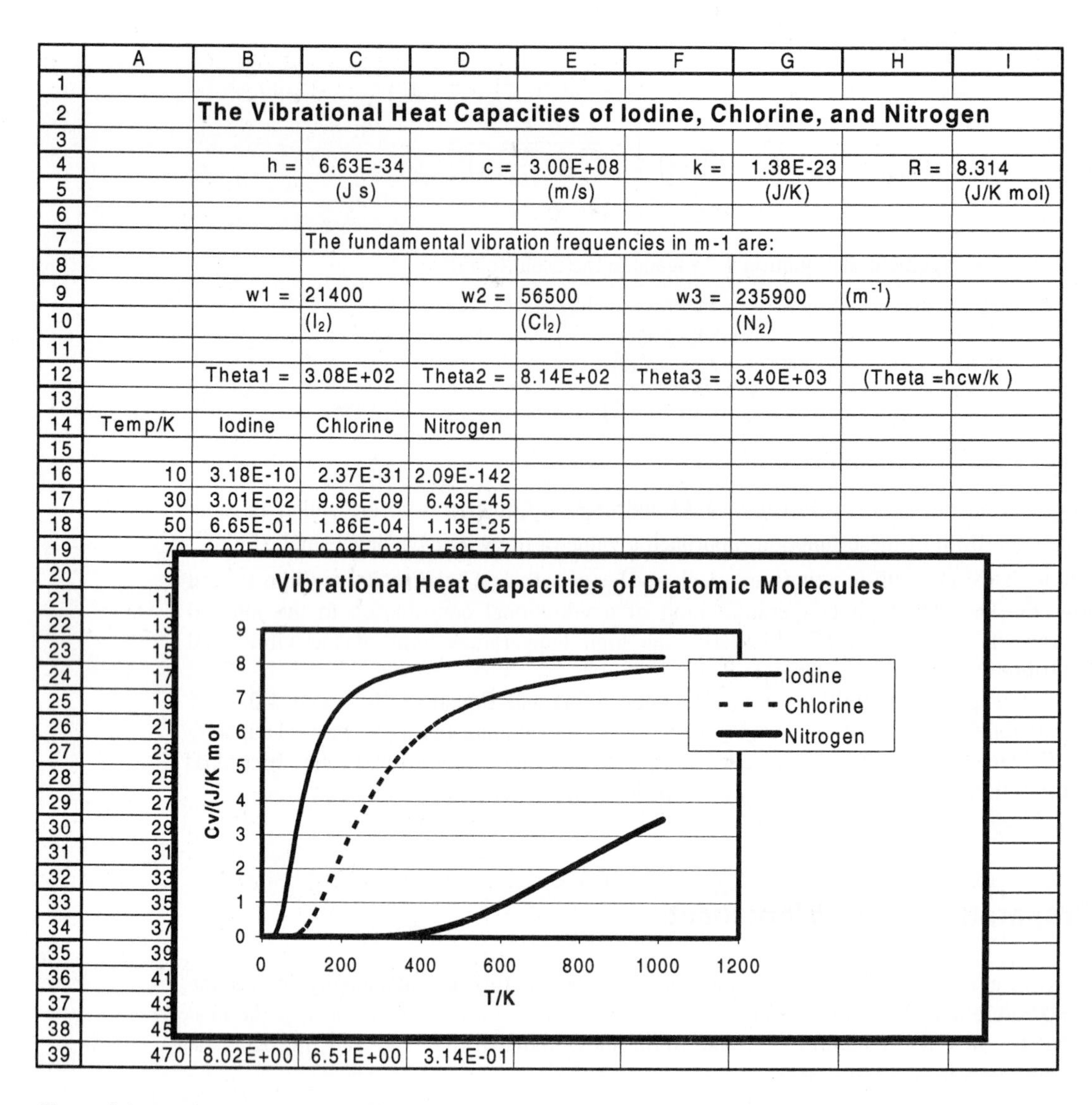

	A	B	C	D	E	F	G	H	I
1									
2		The Vibrational Heat Capacities of Iodine, Chlorine, and Nitrogen							
3									
4		h =	6.63E-34	c =	3.00E+08	k =	1.38E-23	R =	8.314
5			(J s)		(m/s)		(J/K)		(J/K mol)
6									
7			The fundamental vibration frequencies in m-1 are:						
8									
9		w1 =	21400	w2 =	56500	w3 =	235900	(m^{-1})	
10			(I_2)		(Cl_2)		(N_2)		
11									
12		Theta1 =	3.08E+02	Theta2 =	8.14E+02	Theta3 =	3.40E+03	(Theta =hcw/k)	
13									
14	Temp/K	Iodine	Chlorine	Nitrogen					
15									
16	10	3.18E-10	2.37E-31	2.09E-142					
17	30	3.01E-02	9.96E-09	6.43E-45					
18	50	6.65E-01	1.86E-04	1.13E-25					
19	70								
20	9								
21	11								
22	13								
23	15								
24	17								
25	19								
26	21								
27	23								
28	25								
29	27								
30	29								
31	31								
32	33								
33	35								
34	37								
35	39								
36	41								
37	43								
38	45								
39	470	8.02E+00	6.51E+00	3.14E-01					

Figure 4-1. Worksheet for the Heat Capacities of I_2, Cl_2, and N_2

The Vibrational Temperature. At this point, it is convenient to calculate the vibrational temperature *a* (θ) of each molecule (Equation 4-29).

- In cell C12, type **=$C$4*$E$4*$C$9/$G$4** as the formula.
- Copy and paste the contents of cell C12 to cell E12, and change the **$C$9** to **$E$9**.
- Copy and paste the contents of cell C12 to cell G12, and change the **$C$9** to **$G$9**.
- In cells B12, D12, and F12, type **Theta1 =**, **Theta2 =**, and **Theta3** = as the labels, respectively.
- Select cell C12, and on the **Insert** menu, point to **Name**, and click **Define**; type **Th1** as the name of the cell.
- Select cell E12, and on the **Insert** menu, point to **Name**, and click **Define**; type **Th2** as the name of the cell.
- Select cell G12, and on the **Insert** menu, point to **Name**, and click **Define**; type **Th3** as the name of the cell.

- Select cell I4, and on the **Insert** menu, point to **Name**, and click **Define**; type **Rg** as the name of the cell.
- In cells A14, B14, C14, and D14, type **Temp/K**, **Iodine**, **Chlorine**, and **Bromine** as the labels. These element labels will appear on the legend of the chart.

Entering the Temperatures. The heat capacities are calculated as 20-unit increments from 10 to 1010 Kelvin.

- In cell A16, type **10** as the first temperature. Press ENTER.
- Select cell A16, which contains the initial temperature value of 10.
- On the **Edit** menu, point to **Fill** (Figure 3-7), and then click **Series** to open the **Series** dialog box (Figure 3-8). In the **Series** dialog box:
  1. For **Series in**, select **Columns**.
  2. For **Type**, select **Linear**.
  3. In the **Step value** box, type **20** for the step value.
  1. 4. In the **Stop value** box, type **1010** for the stop value, and click **OK**.
- Define the name for the temperatures in the range A16:A66 by selecting the range; then, on the **Insert** menu, point to **Name**, and click **Define** to open the **Define Name** dialog box (Figure 1-6). Type **T** for the name of the list of temperatures.

Filling the Cells with the Equations. The vibrational heat capacity equation (Equation 4-35) is entered once and then copied and edited for the second and third molecule. Use the Edit Fill Down method to fill the heat capacity equations into ranges B16:B66, C16:C66, and D16:D66.

- In cell B16, type **=(Th1^2)*Rg*EXP(Th1/T)/(T^2*(EXP(Th1/T) – 1)^2)** as the formula.
- Copy and paste the contents of cell B16 to cell C16. Change all **Th1** to **Th2**.
- Copy and paste the contents of cell B16 to cell D16. Change all **Th1** to **Th3**.

Next use the Edit Fill Down method to fill in the parallel formula ranges for the vibrational heat capacities of iodine, chlorine, and nitrogen. The formulas are already entered in cells B16, C16, and D16.

- Click and drag the horizontal split screen bar (Figure 1-1) to the middle of the screen.
- Scroll the lower half of the split screen until cell B66 is visible.
- Click in cell B16, and then SHIFT-click in cell B66 to select the range B16:B66.
- On the **Edit** menu, point to **Fill**, and click **Down** (Figure 3-7) to fill the range B16:B66 with the copies of the vibrational heat capacity equation for iodine.

Repeat theses steps to fill the range C16:C66 with the heat capacity data for chlorine; repeat again for the range D16:D66 with the heat capacity data for nitrogen. On the **Windows** menu, click **Remove Split**.

Creating the Chart

Three plots, one for each element, are placed on a single chart.

- Click and drag the horizontal split screen bar (Figure 1-1) to the middle of the screen.
- Scroll the lower half of the split screen until Row 66 is visible.
- Click in cell A14, and then SHIFT-click in cell D66 to select the range A14:D66.

- On the standard toolbar, Click the **Chart Wizard** icon.
- In the **Chart Wizard – Step 1 of 4 – Chart Type** dialog box, select **XY (Scatter)**, and click the icon for smooth lines and without markers. Click **Next**.
- In the **Chart Wizard – Step 2 of 4 – Chart Source Data** dialog box (Figure 2-3), click **Next**.
- In the **Chart Wizard – Step 3 of 4 – Chart Options** dialog box:
  1. Click the **Titles** tab, and in the **Chart title** box, type **Vibrational Heat Capacities of Diatomic Molecules** as the title.
  2. In the **Value (X) axis** box, type **T/K** as the label.
  3. In the **Value (Y) axis** box, type **Cv/(J/K mol)** as the label.
  4. Click the **Legend** tab, and select the **Show legend** check box. Click **Next**.
- In the **Chart Wizard – Step 4 of 4 – Chart Location** dialog box, select **As object in Sheet1**.

The chart with three plots appears embedded in the sheet. By default, the plot area is gray, and the plots are of three different colors, but the same weight and style.

Editing the Chart

To improve the chart's appearance, several objects are edited. The object of the chart is to transmit visual information, and editing makes this transmission more effective.

The Plot and Chart Area

- Right-click in the plot area to open a shortcut menu.
- Click **Format Plot Area** (similar to Figure 2-9), and change the gray background to white.
- Change the border weight to medium weight.
- Right-click in the chart area to open a shortcut menu.
- Click **Format Chart Area**, and change the border weight to medium weight.

The Plotted Lines

- Right-click the iodine plot line, click **Format Data Series**, and in the **Format Data Series** dialog box (Figure 2-10), select **Color**, and change it to black; select **Weight**, and select the medium line weight.
- Right-click the chlorine plot line, click **Format Data Series**, and in the **Format Data Series** dialog box, select **Color**, and change it to black. Select **Weight**, and change it to a heavy line. Select **Style**, and change it to a dotted line.
- Right-click the nitrogen plot line, click **Format Data Series**, and in the **Format Data Series** dialog box, select **Color**, and change it to black. Select **Weight**, and change it to a heavy line.

Customizing the Y-Axis

- Right-click a y-axis number, then click **Format Axis** to open the **Format Axis** dialog box (Figure 2-11).
- Click the **Scale** tab, and type **0** in the **Minimum** box.

Customizing the X-Axis

- Right-click an x-axis number, then click **Format Axis** to open the **Format Axis** dialog box (Figure 2-11).
- Click the **Patterns** tab, and for **Tick mark labels**, select **Low**.

Meaning of the Chart of Heat Capacities

According to Equation 4-36, the vibrational contribution to the heat capacity is just equal to R times an Einstein function. The value of the Einstein function varies from 0 to 1. Its value approaches zero at low temperatures T and high vibrational energies $\tilde{\nu}$. Its value approaches unity at high temperatures and low vibrational energies.

At room temperature (~300 K), the vibrational contribution to the heat capacity of nitrogen is approximately zero, of chlorine about 5 J K^{-1} mol^{-1}, and of iodine about 8 J K^{-1} mol^{-1}, reflecting the different magnitudes of the Einstein function. At a constant temperature of 300 K, the variation in the magnitude of the Einstein function, and thus the contribution to the heat capacity, is due to the differences in vibrational energies $\tilde{\nu}$. This variation in turn reflects the difference in bond strengths of the three molecules.

The high temperature limit of the vibrational contribution is R or about *8.3* J K^{-1} mol^{-1}, according to Equation 4-36. At any temperature, the total heat capacity is higher by the translational contribution (3R/2) plus the rotational contribution (R) or a total of 5R/2 for a homonuclear diatomic molecule.

Excel Example 15

Population of Rotational Energy Levels

Review

Excel Example 14 includes a review of molecular energies, the Boltzmann distribution law, and partition functions. In Excel Example 15, we will use the Boltzmann distribution law to examine how molecules are distributed among available rotational energy levels.

The Rotational Energy. The rotational energy levels of a diatomic molecule are given by Equation 4-2:

$$\varepsilon_J = J(J+1)\frac{h^2}{8\pi^2 I} \tag{4-37}$$

The moment of inertia is I, h is Planck's constant, and J is the rotational quantum number. Each of the energy levels defined by J consists of $2J + 1$ quantum states. The degeneracy g_J of the Jth level is defined by

$$g_J = 2J + 1 \tag{4-38}$$

It is convenient to express the rotational energy in terms of the rotational constant B, a frequently tabulated molecular parameter. The constant B depends only on one molecular parameter, the moment of inertia I of the molecule.

$$B = \frac{h}{8\pi^2 Ic} \tag{4-39}$$

The energy ε_J in terms of the rotational constant B is given by

$$\varepsilon_J = \frac{J(J+1)hcB}{kT} \tag{4-40}$$

Collections of molecular parameters in the literature usually tabulate B in wave numbers (cm^{-1}). The SI unit of B is m^{-1}. (B in m^{-1} equals *100* x B in cm^{-1}). Table 4-2 in Excel Example 10 lists rotational constants B for several diatomic molecules.

The Rotational Partition Function. The partition function Q_{rot} for rotational energy levels is given by Equation 4-21. It is convenient to express the partition function in terms of B, the rotational constant:

$$Q_{rot} = \frac{8\pi^2 IkT}{h^2} = \frac{kT}{hcB} \tag{4-41}$$

The Boltzmann Distribution. The Boltzmann distribution law gives the fraction (n_J/N) of all molecules that populate the Jth quantum level:

$$\frac{n_J}{N} = \frac{g_J e^{-(\varepsilon_J - \varepsilon_0)/kT}}{Q_{rot}} \tag{4-42}$$

At a given temperature, Q_R is a constant, a dimensionless number, so the fraction of molecules populating an energy level is proportional to the numerator of the right side of Equation 4-42. Thus, n_i/N increases as g_J and T increase, but n_J/N decreases as $(\varepsilon_i - \varepsilon_0)$ increases. In the worksheet, we will enter this equation in terms of B, the rotational constant:

$$\frac{n_J}{N} = \frac{(2J+1)e^{\frac{(-J(J+1)Bhc}{kT}}}{Q_{rot}} \tag{4-43}$$

A bar graph serves well to demonstrate the dependence of the population of energy levels on degeneracy and energy.

It is interesting to examine the population of rotational energy states because both their energies and degeneracies play such a visible role in the population of energy levels, as shown in Figure 4-2. In Figure 4-2, SI units are used. The units of h are J s, the units of c are m/s, the units of k are J/K, and B has units of m^{-1}. The tabulated literature value of B for the HCl molecule is 10.5909 cm^{-1} (1059.09 m^{-1}).

Creating the Worksheet

The Excel worksheet for the rotational energy states of HCl(*g*) at 298 K is shown in Figure 4-2. The steps in preparing the Excel worksheet (Figure 4-2) are as follows.

- On row 2, enter a suitable title.
- On row 4, define the constants **h**, **c**, **k**, and **T**, all on one line. Add labels.
- On row 5, indicate the units of the values in row 4.

Calculating the Rotational Partition Function

- In cell B7, type **1059.09** as the value for the rotational constant *B* for HCl. Table 4-2 lists values for some other molecules.
- On the **Insert** menu, point to **Name**, and click **Define**. For the contents of cell B4, type **h** as the name; for cell D4, type **s** as the name; for cell F4, type **k** as the name; for cell H4, type **T** as the name; and for cell B7, type **B** as the name.
- In cell A7, type the label **B=**
- Calculate the partition function (Equation 4-41) by typing **=k*T/(h*s*B)** in cell I7.
- In the range, E7:H7, type the label **The partition function Qr = kT/hcB =**
- On the **Insert** menu, point to **Name**, and click **Define**, and for cell I7, type **Q** as the name of the contents.

Entering the Data for the Chart

- In cell A10, type **J** as the label, and in cell B10, type **Fraction** as the label.
- In cell A12, type **0** for the first value of *J*.
- On the **Edit** menu, point to **Fill** (Figure 3-7), and then click **Series** to fill the *J* values from 0 to 13 (step value equal 1) in the range A12:A25.
- Select the range A12:A25; on the **Insert** menu, point to **Name**, and click **Define**, and type **J** as the name of this range.
- In cell B12, type Equation 4-42, **=(2*J+1)*EXP(-J*(J+1)*B*h*s/(k*T))/Q** as the formula.
- In the range B12:B25, on the **Edit** menu, point to **Fill**, and click **Down** to fill the values for the Fraction column.

Plotting the Population Distribution as a Bar Graph

- Click and drag from cell A12 to cell B25.
- On the standard toolbar, click the **Chart Wizard** icon.
- In the **Chart Wizard – Step 1 of 4 – Chart Type** dialog box, select **XY (Scatter),** and click the icon that shows points with no line. Click **Next**.
- In the **Chart Wizard – Step 2 of 4 – Chart Source Data** dialog box (Figure 2-3), click **Next**.
- In the **Chart Wizard – Step 3 of 4 – Chart Options** dialog box:
    1. Click the **Titles** tab, and in the **Chart title** box, type **Population of Rotational Energy Levels** as the title.
    2. In the **Value (X) axis** box, type **J** as the label.
    3. In the **Value (Y) axis** box, type **ni/N** as the label. Click **Next**.
- In the **Chart Wizard – Step 4 of 4 – Chart Location** dialog box, select **As object in Sheet1**, and click **Finish**.

- Select the chart, and on the **Chart** menu, click **Chart Type**.
- In the **Chart Type** dialog box, select **Columns**. Click **OK**.

Editing the Plot Area and Chart Area

- Right-click in the plot area to open a shortcut menu.
- Click **Format Plot Area** (similar to Figure 2-9), and change the gray background to white.
- Change the border weight to medium weight.
- Right-click in the chart area to open a shortcut menu.
- Click **Format Chart Area**, and change the border weight to medium weight.
- Right-click any bar in the chart.
- Click **Format Data Series**, and in **the Format Data Series** dialog box, click a black square in the palette to make the bars black.

The Total Population

The sum of all the fractional populations of the energy levels should equal unity. One can see that the population of levels with J greater than 10 is quite small, so a sum over the first eleven states should come out close to unity.

- Select the range B12:B28.
- Click the summation icon, and the sum of the fractional population appears in cell B28.
- Add a label in Cell A28.

Discussion. Compare Figure 4-2 with Figure 3-14. Notice how the R-branch peak heights in Figure 3-14 follow the populations of rotational levels in Figure 4-2. The R-branch infrared transitions originate from rotational levels with the same J values as shown in the abscissa of the chart in Figure 4-2. The intensity of a transition depends on the population of molecules in the level from which it originates. The intensities of the P-branch transition follow the same pattern for the same reasons.

The Effect of Temperature and Rotational Constant on the Distribution

To see how increasing the temperature increases the population of higher energy rotational state, try changing the temperature T to 500 and then to 100. To see how the rotational constant B (Table 4-2) affects the population of rotational energy states, change the temperature T back to 298 K, and then try B equals 6080 m^{-1} for H_2 and 243.8 m^{-1} for CO. Then vary the temperature for H_2 and CO.

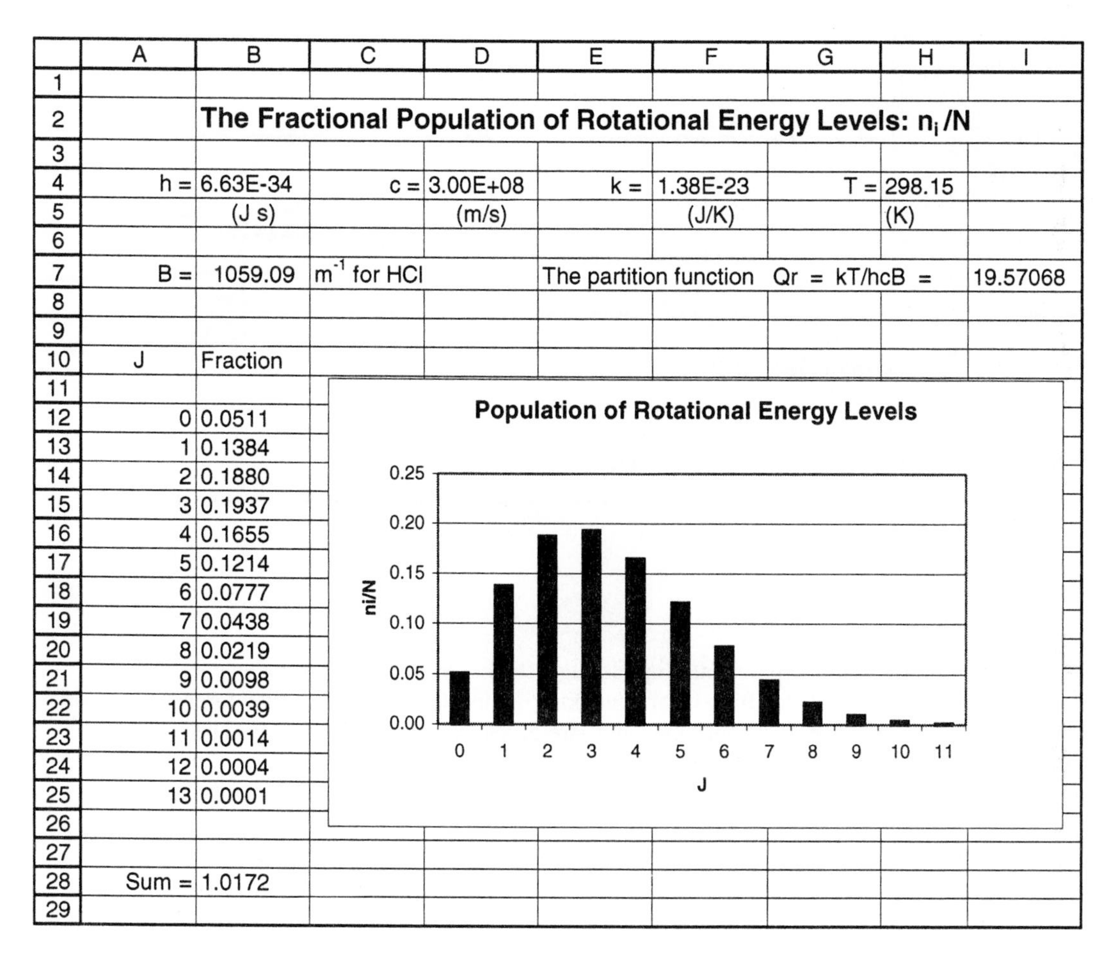

	A	B	C	D	E	F	G	H	I
1									
2		**The Fractional Population of Rotational Energy Levels: n_i /N**							
3									
4	h =	6.63E-34	c =	3.00E+08	k =	1.38E-23	T =	298.15	
5		(J s)		(m/s)		(J/K)		(K)	
6									
7	B =	1059.09	m^{-1} for HCl		The partition function	Qr = kT/hcB =			19.57068
8									
9									
10	J	Fraction							
11									
12	0	0.0511							
13	1	0.1384							
14	2	0.1880							
15	3	0.1937							
16	4	0.1655							
17	5	0.1214							
18	6	0.0777							
19	7	0.0438							
20	8	0.0219							
21	9	0.0098							
22	10	0.0039							
23	11	0.0014							
24	12	0.0004							
25	13	0.0001							
26									
27									
28	Sum =	1.0172							
29									

Figure 4-2. Worksheet for the Population of Rotational Energy Levels

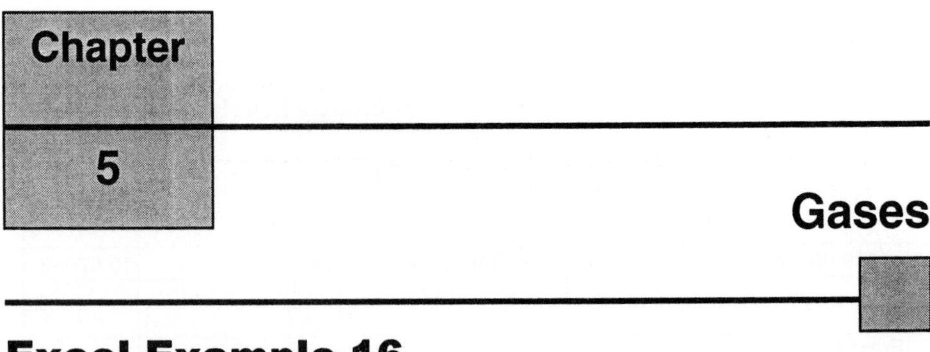

Excel Example 16

The Lennard-Jones Potential

Review

The condensation of nonpolar gases into liquids suggests that even these molecules exert attractive forces between them. It is clear that polar molecules exert attractive forces between them, but even nonpolar molecules and rare gases exert attractive forces. These forces, called London or dispersion forces, arise from the instantaneous dipole of the nonpolar molecule. The instantaneous dipole, in turn, arises from the instantaneous asymmetry of the electrons in motion about a nucleus. In addition to these attractive forces between molecules, repulsive forces arise because in close proximity the electrons of one molecule repel the electrons of another molecule, and the nuclei of one molecule repel the nuclei of another molecule. J. E. Lennard-Jones studied the tug of war between intermolecular attractive and repulsive forces and proposed an empirical function for the potential energy between molecules:

$$V = 4\varepsilon\left[\left(\frac{\sigma}{r}\right)^{12} - \left(\frac{\sigma}{r}\right)^{6}\right] \qquad (5\text{-}1)$$

In Equation 5-1, the potential energy is V. The molecular parameter ε is the measure of the interaction between molecules. It has units of *J*/molecule, but is usually tabulated as ε/k, where k is the Boltzmann constant. Thus ε/k has units of *K*. The effective diameter of the molecules is given by σ. Table 5-1 lists a few value of ε/k and σ. The Excel worksheet for the Lennard-Jones potential energy function is shown in Figure 5-1. Table 5-1 lists the data used.

Creating the Worksheet

Because the Lennard-Jones potential changes so rapidly at small separation r, it is necessary to exercise considerable care in preparing plots that convey information effectively. Experimentation with the scale is helpful.

First enter the constants and labels as shown in Figure 5-1. Next, enter the formulas for the equations. Finally, create and edit the charts.

Entering the Constants and Labels. In the worksheet the symbol *e* is used for ε/k, and the symbol *s* is used for σ.

- On row 1, type a suitable title, **The Lennard-Jones Potential** for example.
- In cell C3, type **s** as the label; in cell D3, type **e** as the label.
- In the range B4:B7, type **Hydrogen**, **Methane**, **Oxygen**, and **Chlorine** as the labels.
- In the range C4:C7, type **293**, **382**, **358**, and **440** as the *s* values.
- On the **Insert** menu, point to **Name**, and click **Define**; type the names **sh**, **sm**, **so**, and **sc** to define the preceding values.
- In the range D4:D7, type **37**, **148**, **118**, and **256** (from Table 5-1) as the *e* values.
- On the **Insert** menu, point to **Name**, and click **Define**; type the names **eh**, **em**, **eo**, and **ec** to define the preceding values.
- In the range A10:E10, type **r**, **Hydrogen**, **Methane**, **Oxygen**, and **Chlorine** as the labels.

Table 5-1

Lennard-Jones Parameters

Molecule	$\frac{\varepsilon}{k}$/(K)	σ/(pm)
Ar	120	341
Cl_2	256	440
H_2	37	293
O_2	118	358
CH_4	148	382

Entering the Equations. A suitable range for the four plots on the worksheet chart are 250 pm to 700 pm for *r*. When the chart first appears, this range does not appear promising, but a little editing will produce a satisfactory chart.

- In cell A12, type **250** as the beginning value of *r*.
- With cell A12 selected, on the **Edit** menu, point to **Fill**, and click **Series**.
  1. For **Series in**, select **Columns**.
  2. For **Type**, select **Linear**.
  3. In the **Step value** box, type **10** for the step value.
  4. In the **Stop value** box, type **700** for the stop value.
  5. Select the range A12:A70; on the **Insert** menu, point to **Name**, and click **Define**; type the name **d** to define the range.
- In cell B12, type **=4*eh*((sh/d)^12-(sh/d)^6)** as the formula.
- On the **Window** menu, click **Split**.
  1. Select cell B12.
  2. SHIFT-click in cell B57, which selects the range B12:B57.
- With the range B12:B57 selected, on the **Edit** menu, point to **Fill**, and click **Down**.
- Select cell B12, and copy and paste its contents to cells C12, D12, and E12.
  1. In cell C12, change **eh** to **em** and **sh** to **sm**.
  2. In cell D12, change **eh** to **eo** and **sh** to **so**.
  3. In cell E12 change, **eh** to **ec** and **sh** to **sc**.
- Split the window, and fill the ranges C12:C57, D12:D57, and E12:E57 in the same way that you previously filled the range B12:B57.

Creating the Chart. When you split the screen and select the entire block A10:E57, Excel includes the labels in the range B10:E10 in the legend box of the chart.

- Spit the screen, click in cell A10; then SHIFT-click in cell D57.
- With the screen split, click the **Chart Wizard** icon.
  1. In the **Chart Wizard – Step 1 of 4 – Chart Type** dialog box, select **XY (Scatter)**, and click the icon labeled for smooth lines and without markers. Click **Next**.

2. In the **Chart Wizard – Step 2 of 4 – Chart Source Data** dialog box (Figure 2-3), click **Next**.
3. In the **Chart Wizard – Step 3 of 4 – Chart Options** dialog box:
   a. Click the **Titles** tab, and in the **Chart title** box, type **The Lennard-Jones Potential Energy Function** as the title.
   b. In the **Value (X) axis** box, type **r/pm** as the label.
   c. In the **Value (Y) axis** box, type **(e/k)/K** as the label.
   d. Click the **Legend** tab, and select the **Show legend** check box.
   e. Click the **Gridlines** tab, and select both major axes. Click **Next**.
4. In the **Chart Wizard – Step 4 of 4 – Chart Location** dialog box, select **As object in Sheet1**.

The chart with four plots appears embedded in the sheet. By default, the plots are of three different colors but the same weight and style. Changing the scale quickly produces a readable chart.

Editing the Chart

The Plot and Chart Area

- Right-click in the plot area to open a shortcut menu.
- Click **Format Plot Area** (similar to Figure 2-9), and change the gray background to white.
- Change the border weight to medium weight.
- Right-click in the chart area to open a shortcut menu.
- Click **Format Chart Area**, and change the border weight to medium weight.
- Click and drag the legend box to the lower left of the plot area.

The Plotted Lines

- Right-click the Hydrogen plot line, click **Format Data Series**, and in the **Format Data Series** dialog box (Figure 2-10), select **Color**, and change the color to black; select **Weight**, and select the heavy line weight.
- Right-click the Methane plot line, click **Format Data Series**, and in the **Format Data Series** dialog box, select **Color**, and change it to black. Select **Weight**, and change it to a medium line. Select **Style**, and change it to a dotted line.
- Right-click the Oxygen plot line, click **Format Data Series**, and in the **Format Data Series** dialog box, select **Color**, and change it to black. Select **Weight**, and change it to a medium line.
- Right-click the Chlorine plot line, click **Format Data Series**, and in the **Format Data Series** dialog box, select **Color**, and change it to black. Select **Weight**, and change it to a heavy line. Select **Style**, and change it to a dashed line.

Customizing the Y-Axis

- Right-click a y-axis number, then click **Format Axis** to open the **Format Axis** dialog box (Figure 2-11).
- Click the **Scale** tab; type **–300** in the **Minimum** box, and type **150** in the **Maximum** box.

Customizing the X-Axis

- Right-click an x-axis number, then click **Format Axis** to open the **Format Axis** dialog box (Figure 2-11).
- Click the Scale tab; type **250** in the **Minimum** box, and type **700** in the **Maximum** box.

- Click the **Patterns** tab, and for **Tick mark labels**, select **Low**. This action places the x-axis numbers below the chart.

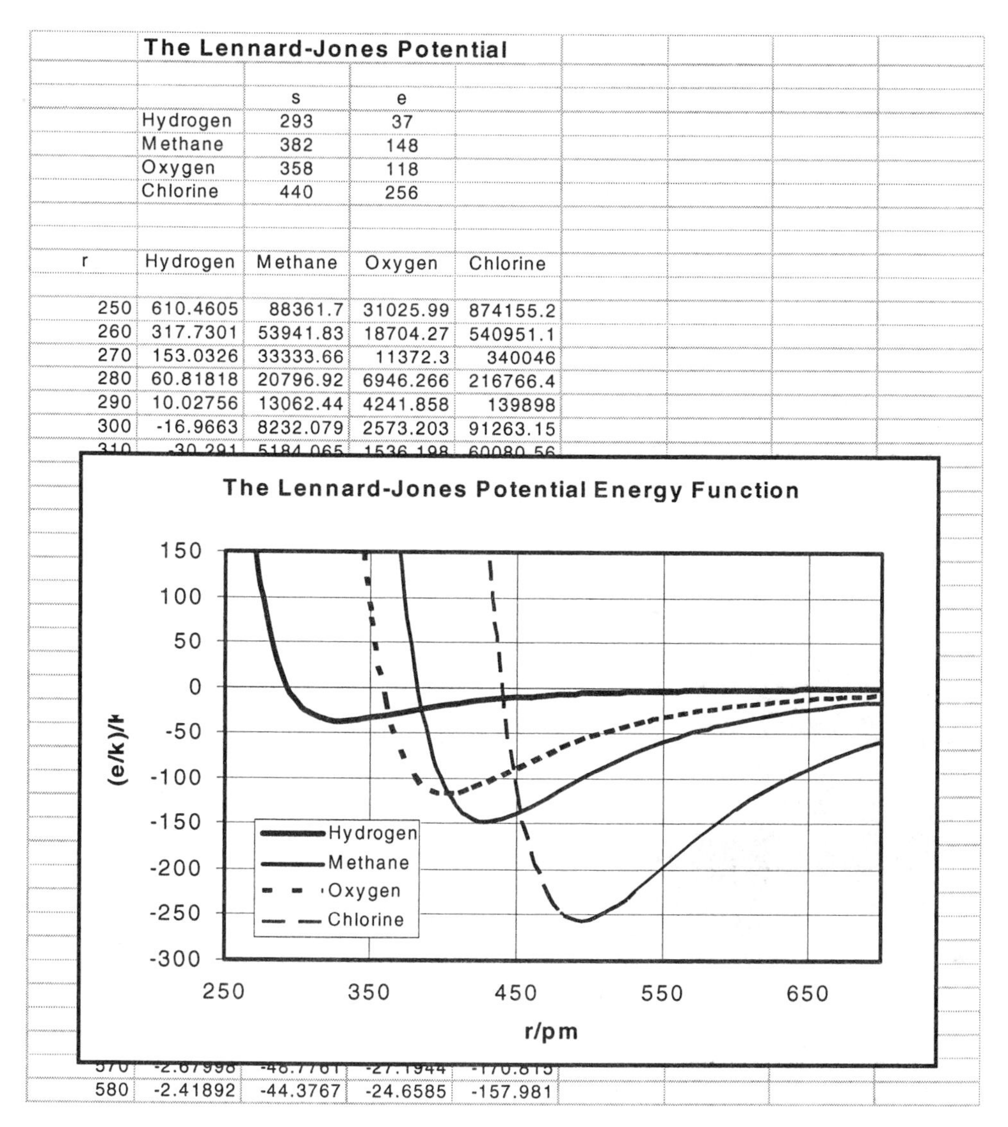

The Lennard-Jones Potential

	s	e
Hydrogen	293	37
Methane	382	148
Oxygen	358	118
Chlorine	440	256

r	Hydrogen	Methane	Oxygen	Chlorine
250	610.4605	88361.7	31025.99	874155.2
260	317.7301	53941.83	18704.27	540951.1
270	153.0326	33333.66	11372.3	340046
280	60.81818	20796.92	6946.266	216766.4
290	10.02756	13062.44	4241.858	139898
300	-16.9663	8232.079	2573.203	91263.15
310	-30.291	5184.065	1536.198	60080.56
580	-2.41892	-44.3767	-24.6585	-157.981

Figure 5-1. Worksheet for the Lennard-Jones Potential

Excel Example 17

The van der Waals Equation of State

Review

The model for the ideal gas equation of state, postulates that gas molecules have zero volume, and do not attract each other at any distance. According to van der Waals, modifications of the ideal gas equation of state should take into account the volume taken up by molecules and the interactive forces between molecules. The ideal gas equation of state is

$$P = \frac{nRT}{V} \tag{5-2}$$

According to van der Waals , the pressure should be reduced by a term that takes into account the attractive forces between molecules. Such a term should be inversely proportional to the V containing the molecules. A satisfactory empirical term has the form a/V^2. The volume available to the molecules is the container volume V reduced by the actual volume b occupied by the molecules. Table 5-2 lists some typical values of a and b, and their units. With these modifications, the ideal gas equation of state becomes the van der Waals equation of state:

Table 5-2

Van der Waals Constants

Molecule	***a*** $\frac{L^2\ atm}{mol^2}$	***b*** $\frac{L}{mol}$
Ar	1.34	0.032
He	0.034	0.024
H_2	0.24	0.027
N_2	1.4	0.039
CCl_4	2.25	0.043

$$P = \frac{nRt}{V-b} - \frac{an^2}{V^2} \tag{5-3}$$

Entering the Titles, Constants, and Their Labels (Figure 5-2)

- In row 2, enter a suitable title.
- In cells B4, B6, and B8, type **2.25**, **0.043**, and **0.08205** as the values, respectively.
- On the **Insert** menu, point to **Name**, and click **Define;** type **a**, **b**, and **Q** to define the names of these three cells (Excel does not permit *R* as a name).
- In the corresponding range C4:C8, enter labels for the units.
- In the range A4:A8, enter the labels for the constants.

Entering the Temperature for the Pressure Calculation

- In the ranges B10:B11 and D10:D11 and in cell G10, type **150**, **160**, **170**, **180**, and **190** as the five temperatures.
- Type the labels shown in the range A10:F11 of the worksheet.
- On the **Insert** menu, point to **Name**, and click **Define**, and define the names of these cells to be the same as the corresponding label (Ta, Tb, and the other names shown on the worksheet).

Entering the Volumes for the Pressure Calculations

- In cell A13, type **V** as the label; then type **P** in the next six cells of row 13.
- In the range B14:G14, type the labels for the temperature for the pressure calculations. These labels will appear in the legend box of the chart (150, 160, 170, 180, 190, and 190_{Ideal}).
- In cell A16, type **0.060** as the value.
- With cell A16 selected, on the **Edit** menu, point to **Fill**, and then click **Series** to open the **Series** dialog box.

1. For **Series in**, select **Columns**.
2. For **Type**, select **Linear**.
3. In the **Step value** box, type **0.005** for the step value.
4. In the **Stop value** box, type **0.320** for the stop value, and Click **OK**.

- Select C16:C68; on the **Insert** menu, point to **Name**, and click **Define**, and type **V** as the name for the range.

Entering the Formulas for the van der Waals Equation

- In cell B16, type = **((Q*Ta)/(V-b))-a/V^2** as the formula.
- On the **Window** menu, click **Split**.
  1. Click cell B16 to select it.
  2. SHIFT-click cell B68, which selects the range B16:B68.
- With the range B16:B68 selected, on the **Edit** menu, point to **Fill**, and click **Down**.
- Select cell B16, and copy and paste its contents to cells C16, D16, E16, and F16.
- In cell G16 type **=Q*Te/V** as the formula.
  1. In cell C16, change **Ta** to **Tb**.
  2. In cell D16, change **Ta** to **Tc**.
  3. In cell E16, change **Ta** to **Td**.
  4. In cell F16, change **Ta** to **Te**.
  5. In cell G16, change **Ta** to **Te**.
- Split the window, and fill the range C16:C68 in the same way that you previously filled the range B12:G57.
- Similarly, fill the remaining ranges.

Creating the Chart. When you split the screen and select the entire block A14:G68, Excel includes the labels in the range B14:G14 in the legend box of the chart.

- Spit the screen, click in cell A14; then SHIFT-click in cell G68.
- With the screen split, click the **Chart Wizard** icon.
  1. In the **Chart Wizard – Step 1 of 4 – Chart Type** dialog box, select **XY (Scatter)**, and click the icon for smooth lines and without markers. Click **Next**.
  2. In the **Chart Wizard – Step 2 of 4 – Chart Source Data** dialog box (Figure 2-3), click **Next**.
  3. In the **Chart Wizard – Step 3 of 4 – Chart Options** dialog box:
     a. Click the **Titles** tab, and in the **Chart title** box, type **van der Waals Equation of State for Methane** as the title.
     b. In the **Value (X) axis** box, type **V/L** as the label.
     c. In the **Value (Y) axis** box, type **P/atm** as the label.
     d. Click the **Legend** tab, and select the **Show legend** check box.
     e. Click the **Gridlines** tab, and select both major axes. Click **Next**.
  4. In the **Chart Wizard – Step 4 of 4 – Chart Location** dialog box, select **As object in Sheet1**.

The chart with four plots appears embedded in the sheet. By default, the plots are of different colors, but the same weight and style. Changing the style quickly produces a readable chart.

Editing the Chart

The Plotted Lines

- Right-click the 150 plot line, click **Format Data Series**, and in the **Format Data Series** dialog box (Figure 2-10), click **Style**, and select dashed; click **Color**, and select black; click **Weight**, and select the heavy line weight.
- In the same manner, select each of the remaining five lines and give them a unique combination of style and weight, as shown in Figure 5-2.

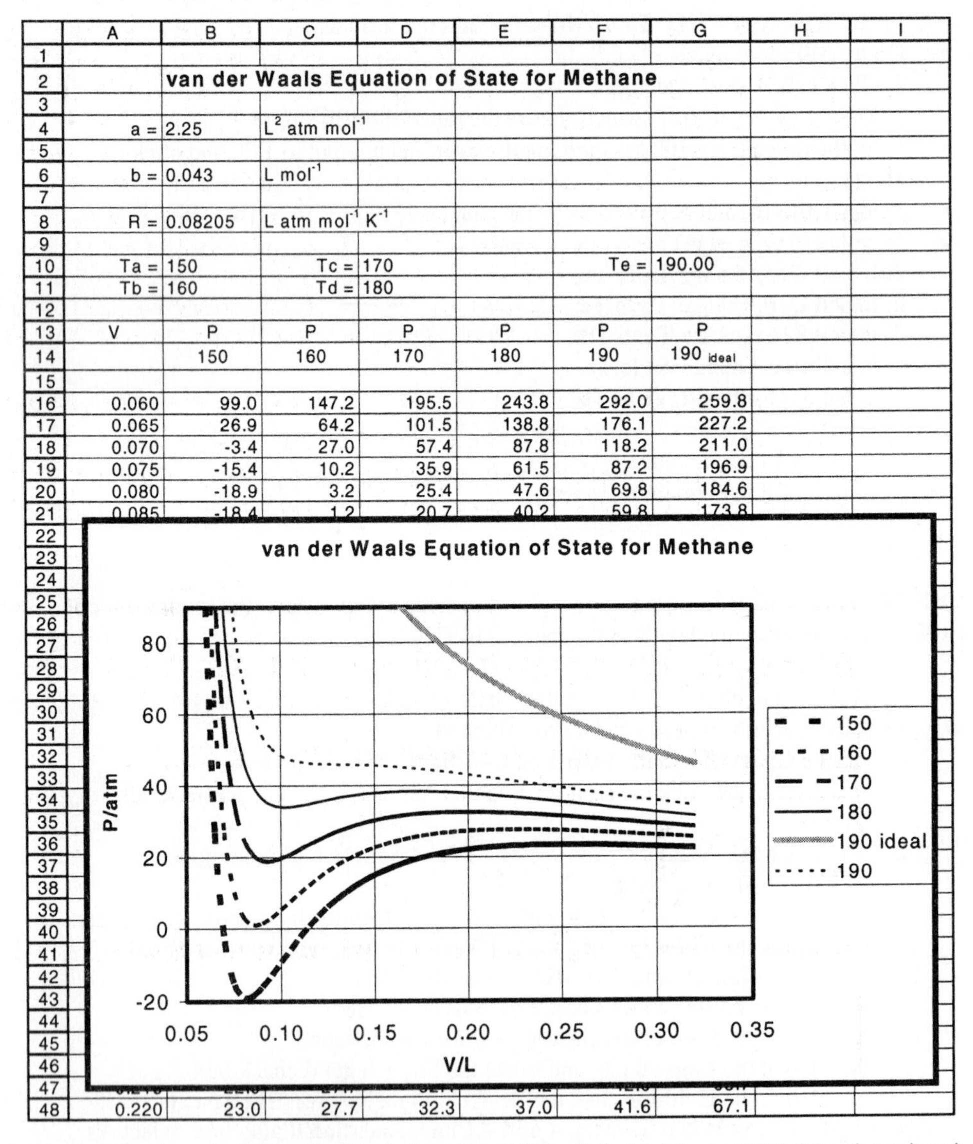

	A	B	C	D	E	F	G	H	I
1									
2		van der Waals Equation of State for Methane							
3									
4	a =	2.25	L^2 atm mol^{-1}						
5									
6	b =	0.043	L mol^{-1}						
7									
8	R =	0.08205	L atm mol^{-1} K^{-1}						
9									
10	Ta =	150	Tc =	170		Te =	190.00		
11	Tb =	160	Td =	180					
12									
13	V	P	P	P	P	P	P		
14		150	160	170	180	190	190 ideal		
15									
16	0.060	99.0	147.2	195.5	243.8	292.0	259.8		
17	0.065	26.9	64.2	101.5	138.8	176.1	227.2		
18	0.070	-3.4	27.0	57.4	87.8	118.2	211.0		
19	0.075	-15.4	10.2	35.9	61.5	87.2	196.9		
20	0.080	-18.9	3.2	25.4	47.6	69.8	184.6		
21	0.085	-18.4	1.2	20.7	40.2	59.8	173.8		
48	0.220	23.0	27.7	32.3	37.0	41.6	67.1		

Figure 5-2. Worksheet for the van der Waals Equation of State. Note the failure at 150k (negative volume).

The Plot and Chart Area

- Right-click in the plot area to open a shortcut menu.
- Click **Format Plot Area** (similar to Figure 2-9), and change the gray background to white.
- Change the border weight to medium weight.

- Right-click in the chart area to open a shortcut menu.
- Click **Format Chart Area**, and change the border weight to medium weight.

Customizing the Y-Axis

- Right-click a y-axis number, then click **Format Axis** to open the **Format Axis** dialog box (Figure 2-11).
- Click the **Scale** tab; type **–20** in the **Minimum** box, and type **90** in the **Maximum** box.

Customizing the X-Axis

- Right click an x-axis number, then click **Format Axis** to open the **Format Axis** dialog box (Figure 2-11).
- Click the **Scale** tab; type **0.05** in the **Minimum** box, and type **0.35** in the **Maximum** box.
- Click the **Patterns** tab, and for **Tick mark labels**, select **Low**. This action places the x-axis numbers below the chart.

Excel Example 18

Velocity and Speed Distribution Functions

Review

It is useful to examine both one-dimensional and three-dimensional gases in order to visualize more clearly the meaning of the distribution of velocities, the probability and the probability density.

One-Dimensional Velocity

It is postulated in the kinetic theory of gases that molecules are in rapid, random motion, and thus have random molecular velocities. It seems reasonable to assume that

- High velocities (and energies) are less probable than low velocities.

- The probability that the velocity of a molecule lies between u_x and $u_x + du_x$ is inverse-exponentially proportional to its energy; that is, the probability density follows a Boltzmann distribution.

Figure 5-3 shows the coordinate system for a one-dimensional gas. The gas can have a velocity in a positive or negative direction, represented by the horizontal line in Figure 5-3. The probability that a molecule has a certain velocity u_x is dN/N, the fraction of molecules having the velocity u_x. The probability that a molecule has a velocity between u_x and $u_x + du_x$ is the *probability density,* and it equals $(1/N)(dN/du)_x$.

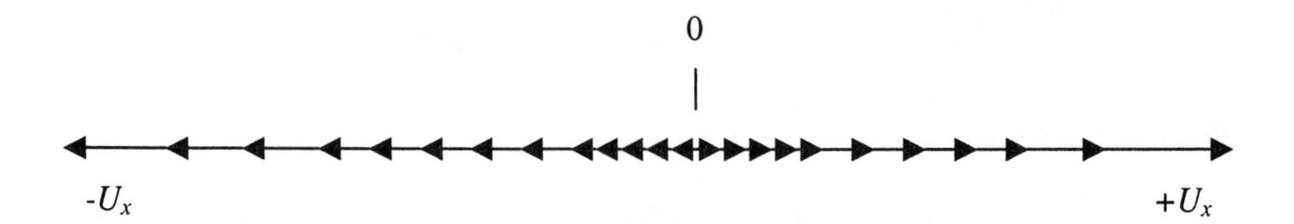

Figure 5-3. Coordinate System for a One-Dimensional Gas. The figure represents a one-dimensional velocity space. Molecules are confined to traveling back and forth on the u_x axis. Each vector represents the velocity of a single molecule. The fraction of molecules per units distance (in velocity space) ($dN/N/du_x$) falls off exponentially in both the positive and negative direction.

If the probability density decreases inverse exponentially according to the Boltzmann distribution, then

$$\frac{1}{N}\frac{dN}{du_x} = Ae^{\frac{-mu_x^2}{2kT}} \tag{5-4}$$

The constant A is a proportionality constant, which can be determined since the probability that the molecules has a velocity between $-\infty$ and $+\infty$ must equal unity:

$$\int_{-\infty}^{+\infty} Ae^{\frac{-mu_x^2}{2kT}}\, du_x = 1 \tag{5-5}$$

The proportionality constant A is found from

$$A = \frac{1}{\int_{-\infty}^{+\infty} e^{\frac{-mu_x^2}{2kT}}\, du_x} \tag{5-6}$$

This integral is of the form

$$\int_{-\infty}^{+\infty} e^{-au}\, du = \sqrt{\frac{\pi}{a}} \tag{5-7}$$

where $a = m/2kT$. It follows that

$$A = \sqrt{\frac{m}{2\pi kT}} \tag{5-8}$$

With the constant A determined, the distribution of molecular velocities (probability density) for a one-dimensional gas is

$$\frac{1}{N}\frac{dN}{du_x} = \sqrt{\frac{m}{2\pi kT}}\, e^{\frac{-mu_x^2}{2kT}} \tag{5-9}$$

Figure 5-4 shows a plot of the probability density dN/Ndu_x versus velocity u_x. Compare it with Figure 5-3. The probability density falls off exponentially in both the positive and negative u_x directions. The most probable and average velocity is zero.

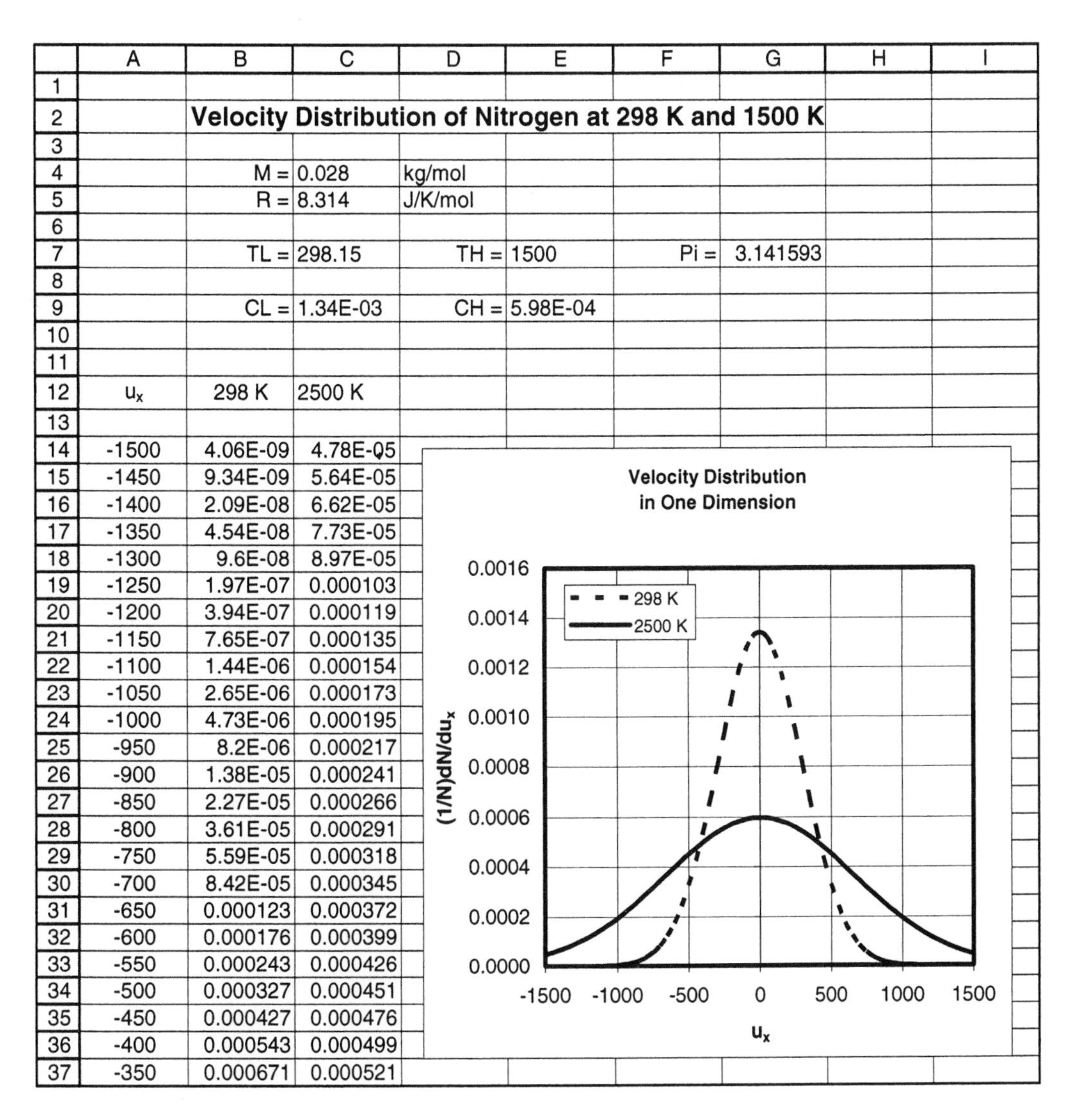

	A	B	C	D	E	F	G	H	I
1									
2		Velocity Distribution of Nitrogen at 298 K and 1500 K							
3									
4		M =	0.028	kg/mol					
5		R =	8.314	J/K/mol					
6									
7		TL =	298.15	TH =	1500	Pi =	3.141593		
8									
9		CL =	1.34E-03	CH =	5.98E-04				
10									
11									
12	u_x	298 K	2500 K						
13									
14	-1500	4.06E-09	4.78E-05						
15	-1450	9.34E-09	5.64E-05						
16	-1400	2.09E-08	6.62E-05						
17	-1350	4.54E-08	7.73E-05						
18	-1300	9.6E-08	8.97E-05						
19	-1250	1.97E-07	0.000103						
20	-1200	3.94E-07	0.000119						
21	-1150	7.65E-07	0.000135						
22	-1100	1.44E-06	0.000154						
23	-1050	2.65E-06	0.000173						
24	-1000	4.73E-06	0.000195						
25	-950	8.2E-06	0.000217						
26	-900	1.38E-05	0.000241						
27	-850	2.27E-05	0.000266						
28	-800	3.61E-05	0.000291						
29	-750	5.59E-05	0.000318						
30	-700	8.42E-05	0.000345						
31	-650	0.000123	0.000372						
32	-600	0.000176	0.000399						
33	-550	0.000243	0.000426						
34	-500	0.000327	0.000451						
35	-450	0.000427	0.000476						
36	-400	0.000543	0.000499						
37	-350	0.000671	0.000521						

Figure 5-4. The Distribution of Velocities in the *x* Direction

Entering the Constants and Labels for the Velocity Distribution

- In cell B2, type **Velocity Distribution of Nitrogen at 298 K and 1500 K** as the title.
- In cell B4, type **M =** as the label.
- In cell B5, type **R =** as the label.
- In cell B7, type **TL =** as the label.
- In cell B9, type **CL =** as the label.
- In cell D7, type **TH =** as the label.
- In cell D9, type **CH =** as the label.
- In cell F7, type **Pi =** as the label, and right align all seven of the preceding labels.
- In cell C4, type **0.028** as the value. On the **Insert** menu, point to **Name**, and click **Define**, and type **M** as its name.
- In cell C5, type **8.314** as the value. On the **Insert** menu, point to **Name**, and click **Define**, and type **Rg** as its name.
- In cell C7, type **298.15** as the value. On the **Insert** menu, point to **Name**, and click **Define**, and type **TL** as its name.
- In cell E7, type **1500** as the value. On the **Insert** menu, point to **Name**, and click **Define**, and type **TH** as its name.
- In cell G5, type **PI()** as the value. On the **Insert** menu, point to **Name**, and click **Define**, and type **Pi** as its name.
- In cells D4 and D5, type **kg/mol** and **J/K/mol** as the labels, respectively.

Next calculate the constant terms *CL* and *CH*.

- In cell C9, type **=SQRT(M/(2*Pi*Rg*TL))** as the formula.
- Copy and paste this formula to cell E9, and change the **TL** to **TH**.
- On the **Insert** menu, point to **Name**, and click **Define**, and in cell C9, type **CL** as the name, and in cell E9, type **CH** as the name.
- In cells A12, B12, and C12, type $\mathbf{u_x}$, **298 K**, and **2500 K** as the labels.

Now fill the values of the velocity *u* from –1500 to 1500 in steps of 50.

- In cell A14, type **-1500** and press **Enter**.
- With cell A14 selected, on the **Edit** menu, point to **Fill** (Figure 3-7), and then click **Series** to open the **Series** dialog box (Figure 3-8). In the **Series** dialog box:
  1. For **Series in**, select **Columns**.
  2. For **Type**, select **Linear**.
  3. In the **Step value** box, type **50** for the step value.
  4. In the **Stop value** box, type **1500** for the stop value.
  5. Click **OK**. The numbers from –1500 to 1500 incremented in steps of 50 appear and fill the series A14:A74.
- On the **Window** menu, click **Split**. Then click cell A14, and SHIFT-click cell A74.
- With the range A14:A74 selected, on the **Insert** menu, point to **Name**, and click **Define**; type **u** to define the name of the velocity range.

Now enter the distribution formula.

- In cell B14, type **=CL*EXP((-M*u^2)/(2*Rg*TL))** as the formula.
- On the **Window** menu, click **Split**. Then click cell B14, and SHIFT-click cell B64.
- With the range B14:B74 selected, on the **Edit** menu, point to **Fill**, and click **Down**, and the range B14:B74 fills with the values calculated with the distribution function formula.

- Copy and paste the formula in cell B14 to cell C14.
- In cell C14, change **CL** to **CH**, and change **TL** to TH.
- As in the preceding, use the Edit Fill Down method to fill the range C14:C74 with values calculated with the distribution function at the high temperature.

Creating the Chart. When you split the screen and select the entire block A12:C64, the low and high temperature plots can be created on one chart with labels for the temperatures.

- On the **Window** menu, click **Split**. Then click cell A12, and SHIFT-click cell C74.
- With the range A12:C74 selected, click the **Chart Wizard** icon.
- In the **Chart Wizard – Step 1 of 4 – Chart Type** dialog box (Figure 2-2):
    1. Click **XY (Scatter)**, then select the plot with lines and no points.
    2. Click **Next**.
- In the **Chart Wizard – Step 2 of 4 – Chart Source Data** dialog box (Figure 2-2):
    1. You get a first look at the chart in this step.
    2. The data range you chose is displayed, and **Series in Columns** is selected.
    3. Click **Next** if the chart appears to be what you expected.
- In the **Chart Wizard – Step 3 of 4 – Chart Options** dialog box:
    1. Click the **Gridlines** tab (Figure 2-4), and for both **Value (X) axis** and **Value (Y) axis**, select **Major gridlines**.
    2. Click the **Titles** tab (Figure 2-5).
        a. In the **Chart title** box, type **Velocity Distribution in One Dimension** as the title.
        b. In the **Value (X) axis** box, type **ux** as the label, and subscript the **x**.
        c. In the **Value (Y) axis** box, type **(1/N)dN/du$_x$** as the label.
        d. Click **Next**.
- In the **Chart Wizard – Step 4 of 4 – Chart Location** dialog box (Figure 2-6):
    1. Select **As object in Sheet1**.
    2. Selecting **As new sheet** covers a whole new worksheet with the chart and names the worksheet Chart1 (unless you eventually rename it). Keep your selection **As object in Sheet1** for now.

Editing the Chart

The Plot and Chart Area

- Right-click in the plot area to open a shortcut menu.
- Click **Format Plot Area** (similar to Figure 2-9), and change the gray background to white.
- Change the border weight to medium weight.
- Right-click in the chart area to open a shortcut menu.
- Click **Format Chart Area**, and change the border weight to medium weight.

The Plotted Lines

- Right-click the 298 K plotted line, click **Format Data Series**, and in the **Format Data Series** dialog box, select **Color**, and change it to black. Select **Weight**, and change it to a heavy line. Select **Style**, and choose a dashed line.
- Right-click the 2500 K plot line, click **Format Data Series** (Figure 2-10), and in the **Format Data Series** dialog box, select **Color**, and change it to black. Select **Weight**, and change it to a heavy line.

Customizing the Y-Axis

- Right-click a y-axis number, then click **Format Axis** to open the **Format Axis** dialog box (Figure 2-11).
- Click the Number tab, and type **4** in the **Decimal places** box.

Because of the symmetric distribution, the most probable velocity and the average velocity equal zero. Figure 5-3 shows that the probability density is equal in the $-x$ and in the $+x$ direction.

Three-Dimensional Speed Distribution

In three dimensions (Figure 5-5), a molecule has a component of velocity in three directions: x, y, and z. The probability that a molecule has a velocity between u_x and $u_x + du_x$ is given by $(dN/N)_x$. The probability that a molecule has a velocity between u_y and $u_y + du_y$ is given by $(dN/N)_y$. The probability that a molecule has a velocity between u_z and $u_z + du_z$ is given by $(dN/N)_z$. The probability dN/N that a molecule simultaneously has between u_x and $u_x + du_x$, between u_y and $u_y + du_y$, and between u_z and $u_z + du_z$ is given by the product of the probabilities in each direction:

$$\left(\frac{dN}{N}\right)=\left(\frac{dN}{N}\right)_x\left(\frac{dN}{N}\right)_y\left(\frac{dN}{N}\right)_z=$$

$$\left(\frac{m}{2\pi kT}\right)^{1/2} e^{\frac{-mu_x^2}{2kT}} du_x \left(\frac{m}{2\pi kT}\right)^{1/2} e^{\frac{-mu_y^2}{2kT}} du_y \left(\frac{m}{2\pi kT}\right)^{1/2} du_z \qquad (5\text{-}10)$$

The probabilities are multiplicative, just as the probability for tossing three coins and having all three of them come up heads equals ½ x ½ x ½ equals 1/8. The probability density for a three-dimensional gas is

$$\frac{dN/N}{du_x du_y du_z}=\left(\frac{m}{2\pi kt}\right)^{3/2} e^{\frac{-m(u_x^2+u_y^2+u_x^2)}{2kT}} \qquad (5\text{-}11)$$

The probability density (left side of Equation 5-11) is the probability per unit volume and has dimensions of m^{-3}. Figure 5-5 permits visualization of N velocity vectors. The unit volume is $du_x\, du_y\, du_z$. The arrows represent the N velocity vectors, and dN represents the number of velocity vectors lying in the differential volume $du_x\, du_y\, du_z$. Figure 5-4 displays a sphere of radius u, the center of which lies at the origin of a three-dimensional coordinate system. A molecule having velocity u has components of velocity in the u_x (horizontal direction), u_y (vertical direction), and u_x (perpendicular of the plane of the page).

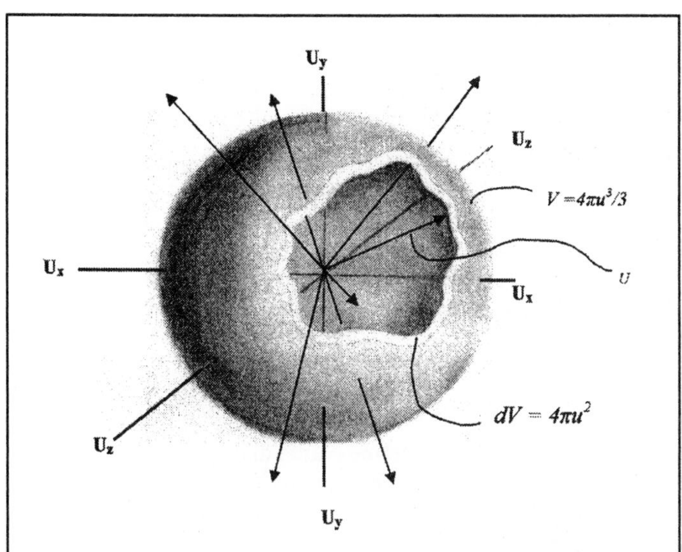

Figure 5-5. Three-dimensional Velocity Space. The coordinate axes U_x, U_y, and U_z are the velocities in the *x, y,* and *z* directions. The radius of the sphere is *u*. The *probability density*, the velocity vectors per unit volume within the shell, is $(dN/N)/du_x du_y du_z$.

Three-Dimensional Speed Distribution. Equation 5-11 gives the probability density, the number of velocity vectors of length u in the direction that terminates in the volume element $du_x\ du_y\ du_z$. Velocity vectors of the same magnitude (length) u, but terminating in other differential volume elements have the same speed but are of different velocities.

To find the speed distribution, we must consider all the vectors of length u, but which have any direction. These vectors lie within a shell of volume dV and lie a distance u from the origin. For a sphere of radius u (Figure 5-5)

$$dV = 4\pi u^2 du \tag{5-12}$$

To find the number of molecules that have the same speed, we multiply the probability density by the differential volume:

[Total vectors in the shell] = [Vectors per unit volume] x [Volume of shell]

$$dN \quad = \quad \frac{dN}{du_x du_y du_z} \quad \text{x} \quad 4\pi u^2 du$$

$$dN = N\left(\frac{m}{2\pi kT}\right)^{3/2} e^{\frac{-m(u_x^2+u_y^2+u_z^2)}{2kT}} (4\pi u^2 du) \tag{5-13}$$

$$\frac{dN/N}{du} = 4\pi\left(\frac{m}{2\pi kt)}\right)^{3/2} u^2 e^{\frac{-mu^2}{2kT}} \qquad (5\text{-}14)$$

Equation 5-14 gives the distribution of speeds for a three-dimensional gas. It is of interest to graph this function at various temperatures to see how the distribution responds to low and high temperatures.

Creating the Worksheet

The Excel worksheet for the speed distribution function is shown in Figure 5-6. After entering a suitable title, a number of constants are defined. On the **Insert** menu, point to **Name**, and click **Define** to open the **Define Name** dialog box, where you give all of these names to clarify writing the distribution equation itself. At each temperature, the term proceeding the exponential term is a constant. These terms are evaluated separately and given the names *CL* and *CH*.

Entering the Constants and Labels for the Speed Distribution

Change the title, but enter the constants and labels for the speed distribution the same as for the velocity distribution (Figure 5-4). The groups of constants labeled **CL** and **CH** have different values in Figure 5-4 and Figure 5-6.

Next calculate the constant terms CL and CH.

- In cell C9, type **=4*Pi*(M/(2*Pi*Rg*TL))^(3/2)** as the formula.
- Copy and paste this formula to cell E9, and change the **TL** to **TH**.
- On the **Insert** menu, point to **Name**, and click **Define**, for cell C9, type **CL** as the name, and for cell E9, type **CH** as the name.
- In cells A12, B12, and C12, type **u**, **298 K**, and **2500 K** as the labels.

Now fill the range A14:A64 with values of the velocity *u* from 0 to 2500 in steps of 50.

- In cell A14, type **0** and press **Enter**.
- With cell A14 selected, on the **Edit** menu, point to **Fill** (Figure 3-7), and then click **Series** to open the **Series** dialog box (Figure 3-8). In the **Series** dialog box:
    1. For **Series in**, select **Columns**.
    2. For **Type**, select **Linear**.
    3. In the **Step value** box, type **50** for the step value.
    4. In the **Stop value** box, type **2500** for the stop value.
    5. Click **OK**. The numbers from 0 to 2500 incremented in steps of 50 appear and fill the series A14:A64.
- On the **Window** menu, click **Split**. Then click cell A14, and SHIFT-click cell A64.
- With the range A14:A64 selected, on the **Insert** menu, point to **Name**, and click **Define**, and type **u** to define the name of the velocity range.

Now enter the distribution formula.

- In cell B14, type **=CL*(u^2)*EXP((-M*u^2)/(2*Rg*TL))** as the formula.
- On the **Window** menu, click **Split**. Then click cell B14, and SHIFT-click cell B64.

- With the range B14:B64 selected, on the **Edit** menu, point to **Fill**, and click **Down**, and the range B14:B64 fills with the values calculated with the distribution function formula.
- Copy and paste the formula in cell B14 to cell C14.
- Change **CL** to **CH**, and change **TL** to **TH**.
- As in the preceding, on the **Edit** menu, point to **Fill**, and click **Down** to fill the range C14:C64 with values calculated with the distribution function at the high temperature.

Creating the Chart. When you split the screen and select the entire block A12:C64, the low and high temperature plots can be created on one chart.

- On the **Window** menu, click **Split**. Then click cell A12, and SHIFT-click cell B64.
- With the range A12:C64 selected, click the **Chart Wizard** icon.
- In the **Chart Wizard – Step 1 of 4 – Chart Type** dialog box (Figure 2-2):
  1. Click **XY (Scatter)**, then select the plot with lines and no points.
  2. Click **Next**.
- In the **Chart Wizard – Step 2 of 4 – Chart Source Data** dialog box (Figure 2-2):
  1. You get a first look at the chart in this step.
  2. The data range you chose is displayed, and **Series in Column** is selected.
  3. Click **Next** if the chart appears to be what you expected.
- In the **Chart Wizard – Step 3 of 4 – Chart Options** dialog box:
  1. Click the **Gridlines** tab (Figure 2-4), and for both **Value (X) axis** and **Value (Y) axis**, select **Major gridlines**.
  2. Click the **Titles** tab (Figure 2-5).
     a. In the **Chart title** box, type **Nitrogen Speed Distribution** as the title.
     b. In the **Value (X) axis** box, type **u/(m/s)** as the label.
     c. In the **Value (Y) axis** box, type **(dN/N)du/(s/m)** as the label.
     d. Click **Next**.
- In the **Chart Wizard – Step 4 of 4 – Chart Location** dialog box (Figure 2-6):
  1. Select **As object in Sheet1**.
  2. Selecting **As new sheet** covers a whole new worksheet with the chart and names the worksheet Chart1 (unless you eventually rename it). Keep your selection **As object in Sheet1** for now.

The chart appears similar to Figure 5-6.

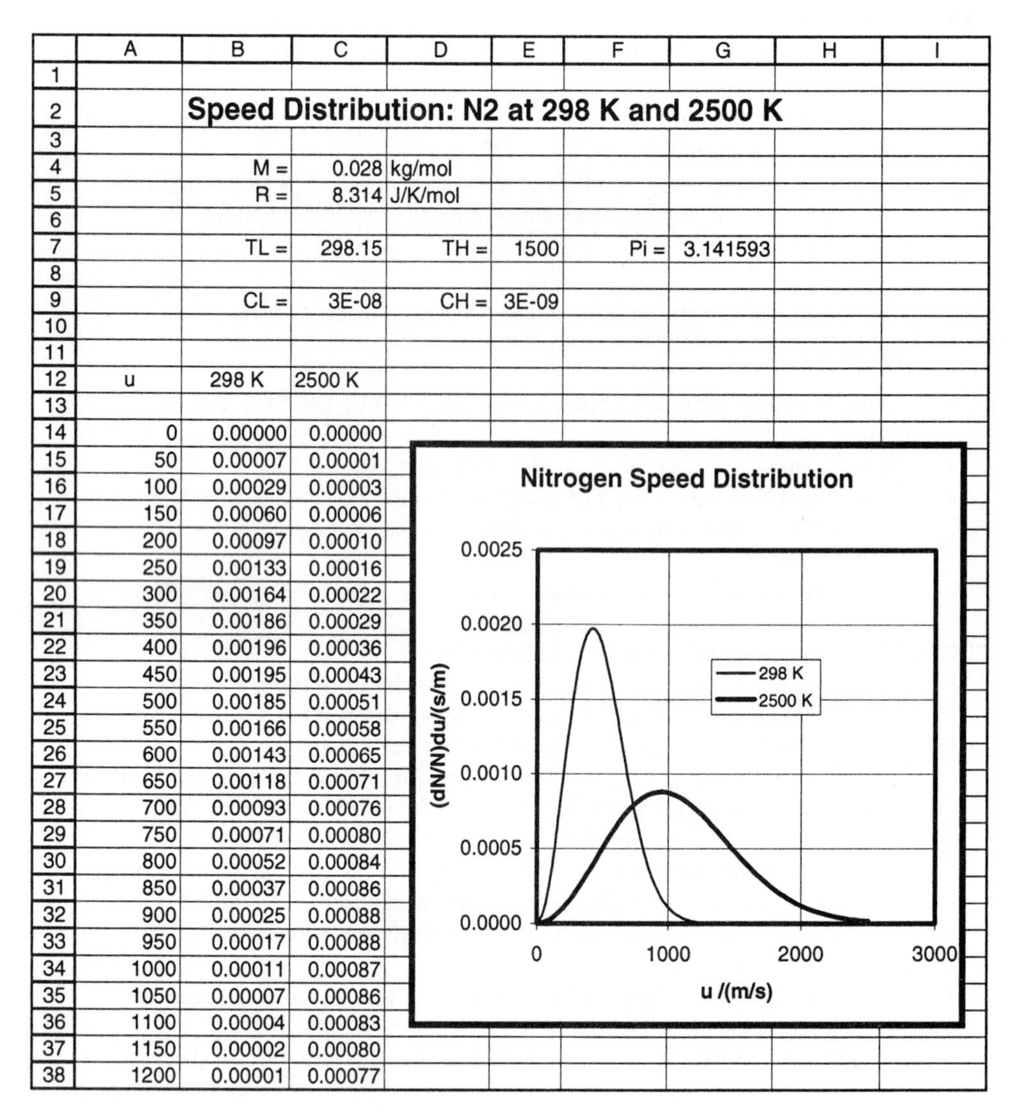

	A	B	C	D	E	F	G	H	I
1									
2		Speed Distribution: N2 at 298 K and 2500 K							
3									
4		M =	0.028	kg/mol					
5		R =	8.314	J/K/mol					
6									
7		TL =	298.15	TH =	1500	Pi =	3.141593		
8									
9		CL =	3E-08	CH =	3E-09				
10									
11									
12	u	298 K	2500 K						
13									
14	0	0.00000	0.00000						
15	50	0.00007	0.00001						
16	100	0.00029	0.00003						
17	150	0.00060	0.00006						
18	200	0.00097	0.00010						
19	250	0.00133	0.00016						
20	300	0.00164	0.00022						
21	350	0.00186	0.00029						
22	400	0.00196	0.00036						
23	450	0.00195	0.00043						
24	500	0.00185	0.00051						
25	550	0.00166	0.00058						
26	600	0.00143	0.00065						
27	650	0.00118	0.00071						
28	700	0.00093	0.00076						
29	750	0.00071	0.00080						
30	800	0.00052	0.00084						
31	850	0.00037	0.00086						
32	900	0.00025	0.00088						
33	950	0.00017	0.00088						
34	1000	0.00011	0.00087						
35	1050	0.00007	0.00086						
36	1100	0.00004	0.00083						
37	1150	0.00002	0.00080						
38	1200	0.00001	0.00077						

Figure 5-6. The Speed Distribution for Nitrogen at 298 K and 2500 K

Editing the Chart

The Plot and Chart Area

- Right-click in the plot area to open a shortcut menu.
- Click **Format Plot Area** (similar to Figure 2-9), and change the gray background to white.
- Change the border weight to medium weight.
- Right-click in the chart area to open a shortcut menu.
- Click **Format Chart Area**, and change the border weight to medium weight.

The Plotted Lines

- Right-click the 298 K plot line, click **Format Data Series**, and in the **Format Data Series** dialog box, select **Color**, and change it to black. Select **Weight**, and change it to a heavy line.
- Right-click the 2500 K plot line, click **Format Data Series** (Figure 2-10), and in the **Format Data Series** dialog box, select **Color**, and change it to black. Select **Weight**, and change it to a heavy line. Select **Style**, and choose a dashed line.

Customizing the Y-Axis

- Right-click a y-axis number, then click **Format Axis** to open the **Format Axis** dialog box (Figure 2-11).
- Click the **Number** tab, and type **4** in the **Decimal places** box.

Because of the asymmetric distribution, the most probable speed and the average speed are not equal. Since the distribution is skewed toward higher speeds, the average speed is slightly greater than the most probable speed. This difference is shown in the next section.

The Most Probable Speed. The most probable speed corresponds to the speed at the maximum in Equation 5-14, which is of the form

$$f(u) = au^2 e^{-bu^2} \tag{5-15}$$

At the maximum, the derivative *df(u)/du* equals zero

$$-bu_{MP}^2 + 1 = 0$$

so that

$$u_{MP} = \left(\frac{1}{b}\right)^{1/2}$$

$$u_{MP} = \sqrt{\frac{2RT}{M}} \tag{5-16}$$

With R, T, and M in SI units, u_{MP} has units of m/s.

Average Quantities

If the probability of x_i is P_i, then the average value of x is given by

$$\bar{x} = \sum_i P_i x_i \tag{5-17}$$

Example 1. The average throws of a die. The possible values x_i of throwing a die are 1, 2, 3, 4, 5 and 6. The probability of throwing each value equals 1/6.

$$\bar{x} = \sum_i P_i x_i = \frac{1}{6}1 + \frac{1}{6}2 + \frac{1}{6}3 + \frac{1}{6}4 + \frac{1}{6}5 + \frac{1}{6}6 = 3.5 \tag{5-18}$$

Example 2. Average atomic mass. The natural isotopic abundance of each isotope of an element equals the probability of occurrence of that isotope. Neon has three isotopes, one with a mass of 19.99244 u and an abundance of 0.9094, a second with a mass of 20.99395 u and an abundance of 0.00257, and a third with a mass of 21.99138 u and an abundance of 0.0882.The average atomic weight is $u_{ave.}$.

$$u_{\text{ave}} = 0.9094\ (19.99244\ u) + 0.00257\ (20.99395\ u) + 0.0882\ (21.99138\ u) = 20.17\ u$$

The Average Speed. For gas velocities, the velocity distribution function $f(u)$ provides the probabilities of the occurrence of a velocity u.

$$\bar{u} = \sum_{\textit{all velocities}} f(u)u = \int_0^\infty f(u)u\, du \tag{5-19}$$

$$\bar{u} = 4\pi\left(\frac{m}{2\pi kT}\right)^{3/2} \int_0^\infty e^{\frac{-mu^2}{2kT}} u^3 du \tag{5-20}$$

The integral is of the form

$$\int_0^\infty e^{-au^2} u^3 du = \frac{1}{2a^2} \tag{5-21}$$

where $a = -m/2kT$. The result is

$$\bar{u} = \sqrt{\frac{8kT}{\pi m}} = \sqrt{\frac{8RT}{\pi M}} \tag{5-22}$$

The Root-Mean-Square (RMS) Velocity. The average of the squares of the velocities is obtained in a similar fashion:

$$u^2 = 4\pi\left(\frac{m}{2\pi kT}\right)^{3/2} \int_0^\infty e^{\frac{-mu^2}{2kT}} u^4 du \tag{5-23}$$

The integral is of the form

$$\int_0^\infty e^{-au^2} u^4 du = \frac{3}{8}\sqrt{\frac{\pi}{a^5}} \tag{5-24}$$

where $a = -m/2kT$. The result is:

$$u^2 = \frac{3kT}{m} = \frac{3RT}{M} \tag{5-25}$$

$$u_{MP} : \overline{u} : \sqrt{\overline{u^2}} = 1 : 1.128 : 1.224 \tag{5-26}$$

Because the distribution is skewed to higher speeds, both the average speed and the RMS are greater than the most probable velocity.

Chapter 6

Kinetics

Excel Example 19

Kinetics: Consecutive rate processes: A → B → C

Review of the Science

Rate processes of the type A → B → C frequently occur in chemistry, physics, biology, and engineering. Radioactive decay, population growth, and fluid flow are all described by sequential rate processes. In chemistry, a plot of the changing concentrations of A, B, and C on a single chart displays the relative changes taking place. The relative amounts of reactants and products present at any time for a consecutive reaction depend on the relative magnitude of the rate constants k_1 and k_2:

$$A \xrightarrow{k_1} B \xrightarrow{k_2} C \tag{6-1}$$

If [A*o*] is the concentration of A at time zero, then it may be shown that the concentrations [A], [B], and [C] present at any time t are given by

$$[A] = [A_0]e^{-k_1 t} \tag{6-2}$$

$$[B] = [A_0]\frac{k_1}{k_2 - k_1}(e^{-k_1 t} - e^{-k_2 t}) \tag{6-3}$$

$$[C] = \frac{[A_0]}{k_2 - k_1}\left[k_2(1 - e^{-k_1 t}) - k_1(1 - e^{-k_2 t})\right] \tag{6-4}$$

Plots of [A], [B], and [C] versus t on the same chart (Figure 6-1) provide a display of the simultaneous progress of the *reaction* sequence $A \xrightarrow{k_1} B \xrightarrow{k_2} C$. In Chapter 11, we will write macros (Visual Basic procedures) that provide an even more dramatic presentation of this reaction sequence as the ratio of k_1 to k_2 is dynamically varied.

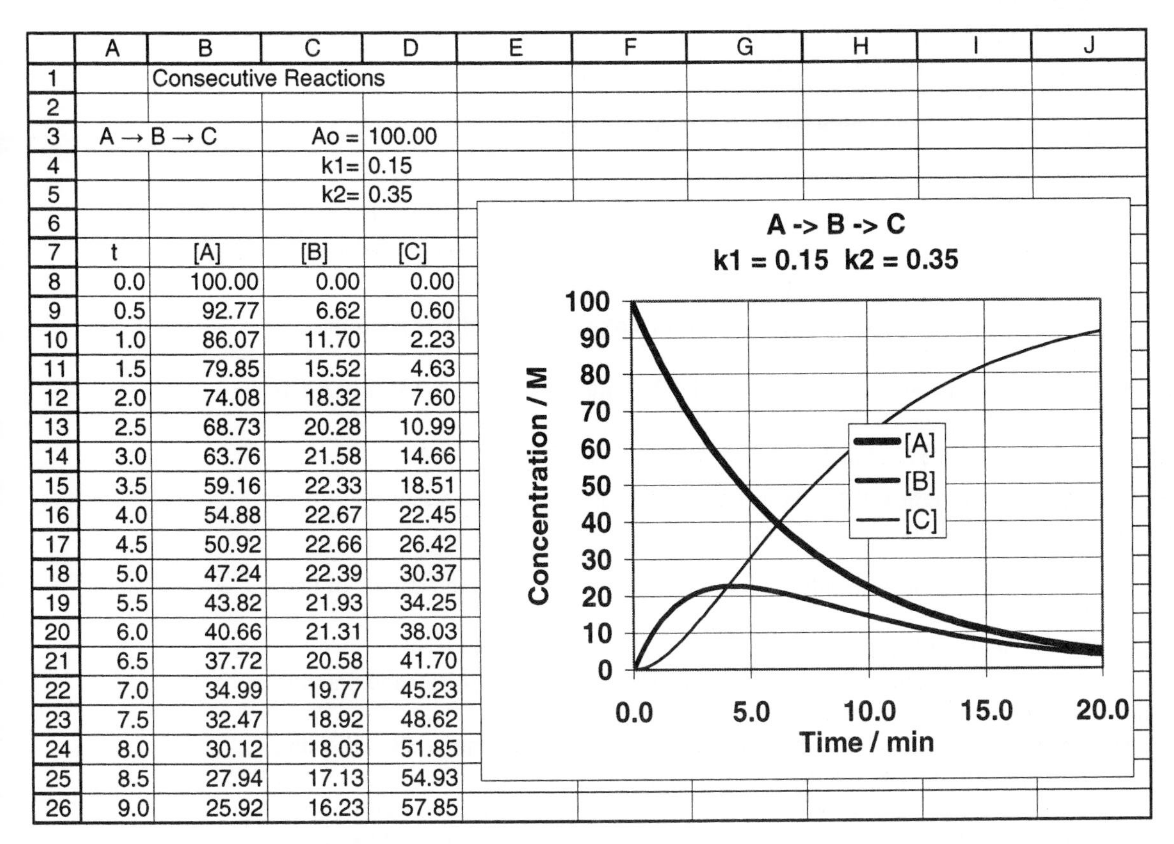

	A	B	C	D
1		Consecutive Reactions		
2				
3	A → B → C		Ao =	100.00
4			k1=	0.15
5			k2=	0.35
6				
7	t	[A]	[B]	[C]
8	0.0	100.00	0.00	0.00
9	0.5	92.77	6.62	0.60
10	1.0	86.07	11.70	2.23
11	1.5	79.85	15.52	4.63
12	2.0	74.08	18.32	7.60
13	2.5	68.73	20.28	10.99
14	3.0	63.76	21.58	14.66
15	3.5	59.16	22.33	18.51
16	4.0	54.88	22.67	22.45
17	4.5	50.92	22.66	26.42
18	5.0	47.24	22.39	30.37
19	5.5	43.82	21.93	34.25
20	6.0	40.66	21.31	38.03
21	6.5	37.72	20.58	41.70
22	7.0	34.99	19.77	45.23
23	7.5	32.47	18.92	48.62
24	8.0	30.12	18.03	51.85
25	8.5	27.94	17.13	54.93
26	9.0	25.92	16.23	57.85

Figure 6-1. The Worksheet for Consecutive Reactions

Entering the Title, Labels, and Constants

Enter the data as shown in Figure 6-1.

- Beginning in cell B1, type **Consecutive Reactions** as the title.
- In cell A1, type **A - > B - > C** as the label.
- In cells C3:C5, type **Ao =**, **k1 =**, and **k2 =** and right-justify these entries.
- In cells A7:D7, type **t**, **[A]**, **[B]**, and **[C]** and center these entries.
- In cells D3:D5, type **100**, **0.15**, and **0.35** and left-justify these entries.
- Select cell D3; on the **Insert** menu, point to **Name**, and click **Define**; type **Ao** for the name.
- In the same way, define the name of cell D4 as **ka** and the name of cell D5 as **kb**.
  Note: Excel does not allow **k1** and **k2** as defined names.

Entering the Numerical Data

- In cell A8, type **0** and press the ENTER key.
- Select cell A8; on the **Edit** menu, point to **Fill** (Figure 3-7), and then click **Series** to open the **Series** dialog box (Figure 3-8). In the **Series** dialog box:
  1. For **Series in**, select **Columns**.
  2. For **Type**, select **Linear**.
  3. In the **Step value** box, type **0.5** for the step value.
  4. In the **Stop value** box, type **20** for the stop value, and click **OK**.

Enter the formulas for the rate equations in cells B8 (Equation 6-2), C8 (Equation 6-3) and D8 (Equation 6-4):

- In cell B8, type = **Ao*EXP(-ka*A8)** as the formula.
- In cell C8, type = **(Ao*ka/(kb-ka)* (EXP(-ka*A8)-EXP(-kb*A8)))** as the formula.
- In cell D8, type =**(Ao/(kb-ka))*(kb*(1-EXP(-ka*A8))-ka*(1-EXP(-kb*A8)))** as the formula.

Next we shall use the Edit Fill Down method to fill in the parallel formula ranges. All three equations can be filled in one operation.

- Click and drag the horizontal split screen bar (Figure 1-1) to the middle of the screen.
- Scroll the lower half of the split screen until cell D50 is visible.
- Click in cell B8 and drag to cell D8; then SHIFT-click in cell B50 and drag to cell D50 to select the range B8:D50.
- On the **Edit** menu, point to **Fill**, and click **Down** to fill the range B8:D50 with the copies of the rate equations.
- Right-click anywhere in the chart; on the **Chart** menu, click **Chart Options**. Click the **Legend** tab, and select the **Show legend** check box. Click **OK**, and drag the legend box into the plot area.
- Select column A. On the **Format** menu, click **Cells**. Click the **Number** tab, select **Number**, and type **1** in the **Decimal places** box.
- Select columns B, C, and D. On the **Format** menu, click **Cells**. Click the **Number** tab, select **Number**, and type **2** in the **Decimal places** box.

Creating the Chart

- Select the data block to be plotted by clicking in cell A7 and dragging to cell D50.
- With the block selected (darkened), click the **Chart Wizard** icon.
- In the **Chart Wizard – Step 1 of 4 – Chart Type** dialog box (Figure 2-2), click **XY (Scatter)**.
- Click the icon showing curved lines without markers; click **Next**.
- In the **Chart Wizard – Step 2 of 4 – Chart Source Data** dialog box, no action is necessary. Click **Next**.
- In the **Chart Wizard – Step 3 of 4 – Chart Options** dialog box:
  1. In the **Chart title** box, type **A -> B -> C** as the title.
  2. In the **Value (X) axis** box, type **Time / min** as the label.
  3. In the **Value (Y) axis** box, type **Concentration / M** as the label. Click **Next**.
- In the **Chart Wizard – Step 4 of 4 – Chart Location** dialog box, select **As object in Sheet1**, and click **Finish**.

Editing the Chart

The chart produced by the Chart Wizard invariably needs some editing. Both the display of text in labels and titles and the axial numbers are improved with editing.

Editing the Title

- Somewhere in the title, click twice slowly. (Do not double-click.)
  1. Press the END key to move the insertion point that appears to the right end of the line, and press ENTER to move the insertion point one line down.
  2. Type **k1 = 0.15, k2 = 0.35** as the second line of the title.

Editing the Chart and Plot Areas

- Click in an empty white area of the chart to adjust the size of the chart with the size handles. Click in the plot area, and also adjust the size.
- Right-click in the plot area; click **Format Plot Area**, and in the **Format Plot Area** dialog box:
  1. In the **Border** section of the dialog box, click **Color**, and change it to white; click **Weight**, and change it to medium heavy.
  2. In the **Area** section of the dialog box, select the white square in the color palette, and click **OK**.
  3. Right-click in the plot area, click **Chart Options**, and in the **Chart Options** dialog box, click the **Gridlines** tab; for both **Value (X) axis** and **Value (Y) axis**, select **Major gridlines**.
- Right-click each of the plotted lines, and click **Format Data Series** to open the **Format Data Series** dialog box. For all of the lines, select black for the color.
  1. For the [A] line, click **Weight**, and select the heaviest weight.
  2. For the [B] line, click **Weight**, and select the medium weight.
  3. For the [C] line, click **Weight**, and select the lightest weight.
- Right-click the legend box, click **Format Legend**, and in the **Format Legend** dialog box, click the **Font** tab and select 10 point bold.

Editing the Axes

- Right-click any number on the y (Concentration / M) axis; click **Format Axis** to open the **Format Axis** dialog box.
  1. Click the **Number** tab, select **Number**, and type **0** in the **Decimal places** box.
  2. Click the **Font** tab and set the font to 10 point bold.
- Right-Click any number on the x (Time / min) axis; click **Format Axis** to open the **Format Axis** dialog box.
  1. Click the **Number** tab, select **Number**, and type **0** in the **Decimal places** box.
  2. Click the **Font** tab and set the font to 10 point bold.

Results

The plot in Figure 6-1 shows how the concentrations [A], [B], and [C] change with time as the reaction sequence $A \xrightarrow{k_1} B \xrightarrow{k_2} C$ progresses. At any given time, the slope of the curve gives the rate of change of concentration. It can be seen that the maximum rate of formation of C ($d[C]/dt$) occurs at the time at which $[B]$ is at a maximum. After a sufficiently long period of time, A and B have disappeared, and only C remains.

By changing the values of k_1 and k_2 in rows 4 and 5 of the worksheet, you can investigate the effect of different k_1 / k_2 ratios on the appearance of the plotted concentrations. Visual Basic for Applications provides sophisticated and convenient ways to investigate the effect of different k_1 / k_2 ratios. Compare Figure 6-1 with Figure 18-18.

Excel Example 20

Linear Regression and the Trendline: Enzyme Kinetics

Review of the Science

Enzymes are proteins that act as catalysts for biological reactions. A reactant participating in an enzyme-catalyzed reaction is called a *substrate*. The protein part of an enzyme provides functional groups in a three-dimensional pattern that allows the enzymes to bind specifically to a particular substrate permitting a biological reaction to occur.

In laboratory studies, the concentration of enzyme is much less than the concentration of substrate. Because the products themselves may inhibit the reaction, it is convenient to study the initial steady-state reaction velocity. The reaction velocity (v) depends on the substrate concentration $[S]$ as shown in Figure 6-2. At low substrate concentrations, the reaction is first order with respect to the substrate concentration, but with increasing substrate concentration, the reaction velocity levels out, reaching a maximum velocity V_{max}, and becomes zero order with respect to the substrate concentration.

The Michaelis and Menten Mechanism

Although most biochemical reactions involve two or more substrates, it is instructive to consider first a single substrate reaction of overall stoichiometry:

$$S \rightarrow P \tag{6-5}$$

A kinetic mechanism for this enzyme-catalyzed reaction was first given by Michaelis and Menten (1913):

$$\mathrm{E+S} \underset{k_{-1}}{\overset{k_1}{\rightleftarrows}} \mathrm{ES} \tag{6-6}$$

$$\mathrm{ES} \xrightarrow{k_2} \mathrm{E+P} \tag{6-7}$$

E is the free enzyme (catalyst).
S is the substrate: the substance in a living cell undergoing chemical change.
E*S* is an intermediate addition complex: the bound enzyme.
P is the product of the reaction.

Notice that the sum of the preceding two reactions (6-5 and 6-7) is equal to Equation 6-5 which is just the chemical equation for one substance, the substrate, *S*, changing to some product, *P*. Since the enzyme *E* is a catalyst, not consumed in the overall reaction, it does not appear in the overall balanced chemical equation. The simple mechanism described is sometimes displayed with a Cleland diagram (Cleland 1963):

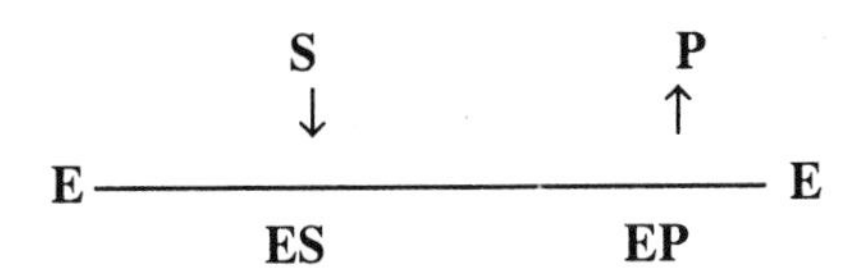

Reactants and products are shown above the line; intermediates, usually bound enzymes, are shown below the line. The initial enzyme is at the left end, the regenerated enzyme at the right end. Reactant participation is shown with a down arrow, product formation with an up arrow. When the mechanism is more complicated, the Cleland diagram often clarifies the nature of the mechanism.

Ordinarily, enzymes (catalysts) are present in relatively low concentration. Thus, [E]<<[S], so it is also true that [E*S*]<<[S]. It is also believed that under steady-state conditions, the concentration of the bound enzyme [*ES*] is not only small, but constant. Consequently, application of the steady-state approximation to Equations 6-6 and 6-7 gives

$$\frac{d[\mathrm{ES}]}{dt} = k_1[\mathrm{E}][\mathrm{S}] - k_{-1}[\mathrm{ES}] - k_2[\mathrm{ES}] = 0 \tag{6-8}$$

Since the total enzyme present, E_0, must equal the sum of the free enzyme *E* and the bound enzyme *ES*, we can write

$$[\mathrm{E_0}] = [\mathrm{E}] + [\mathrm{ES}] \tag{6-9}$$

Solving this for $[E]$, substituting in Equation 6-8, simplifying and rearranging results in an explicit expression for the steady-state concentration of the bound enzyme $[ES]$:

$$[\mathrm{ES}] = \frac{k_1[\mathrm{S}][\mathrm{E}_0]}{\mathrm{k}_{-1} + k_2 + k_1[S]} \tag{6-10}$$

This mechanism applies only to the initial stages of the reaction, where the reaction velocity is v_0. The rate of the reaction is just the rate of formation of the product P of the reaction.

$$v_0 = \frac{dP}{dt} \tag{6-11}$$

According to Equation 6-12, the rate of production of the product **P** depends on k_2 and the steady-state concentration of the bound enzyme $[ES]$:

$$\frac{dP}{dt} = v_0 = k_2[\mathrm{ES}] \tag{6-12}$$

Consequently, as a result of substituting Equation 6-12 in Equation 6-10:

$$v_0 = \frac{k_1 k_2[\mathrm{S}][\mathrm{E}_0]}{k_{-1} + k_2 + k_1[\mathrm{S}]} = \frac{k_2[\mathrm{S}][\mathrm{E}_0]}{\frac{(k_{-1} + k_2)}{k_1} + [\mathrm{S}]} \tag{6-13}$$

$$v_0 = \frac{V_{\max}[\mathrm{S}]}{\mathrm{K}_m + [\mathrm{S}]} \tag{6-14}$$

where

$$V_{\max} = k_2[\mathrm{E}]_0 \tag{6-15}$$

$$K_m = \frac{k_{-1} + k_2}{k_1} \tag{6-16}$$

where K_m is the Michaelis constant. Equation 6-14 is known as the Michaelis-Menten equation. Examination of Equation 6-14 reveals that as [S] becomes large, v_0 becomes equal to V_{max}, which corresponds to the constant velocity approached at high substrate concentrations revealed in Figure 6-2 (upper plot).

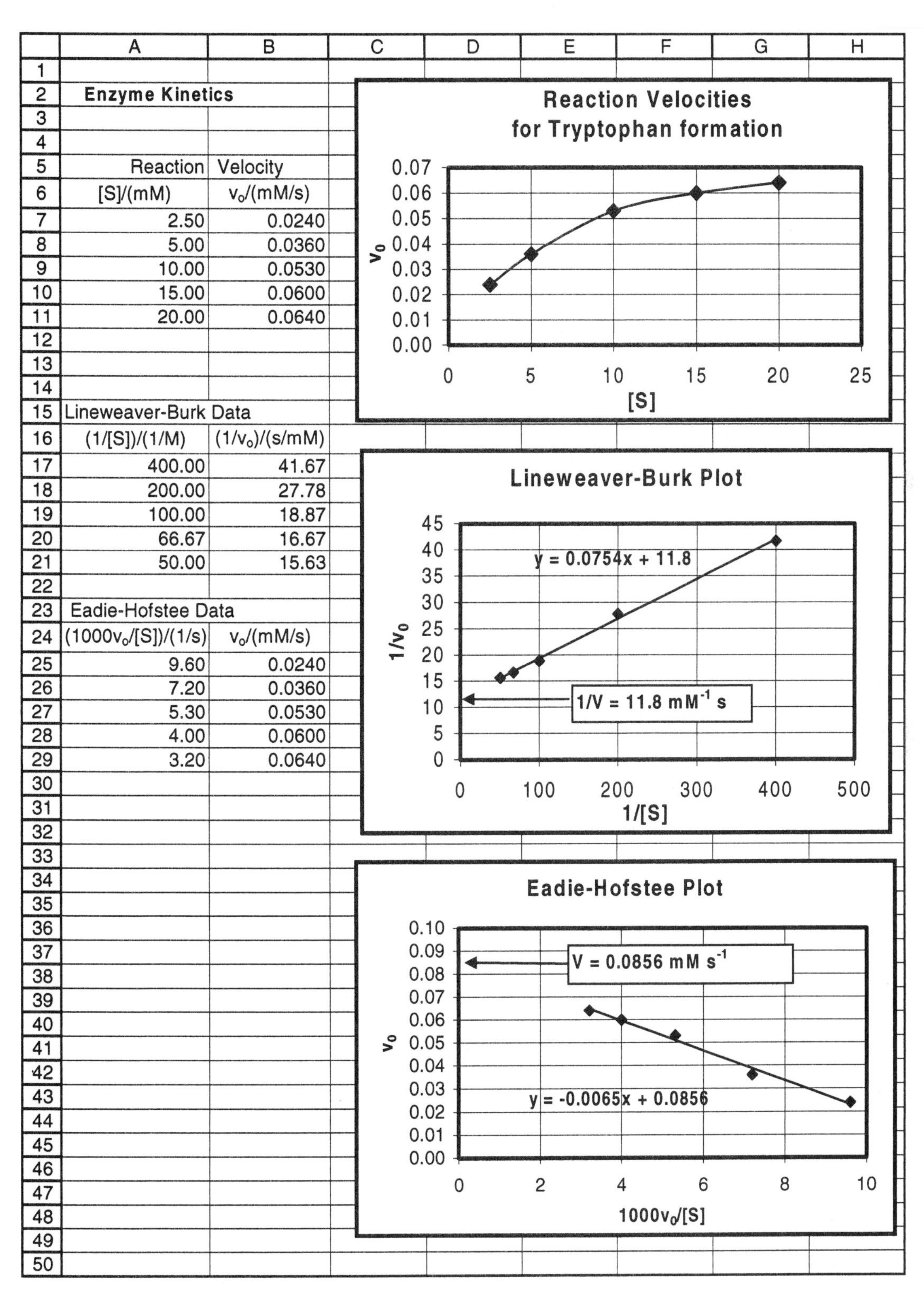

Figure 6-2. The Worksheet for Enzyme Kinetics: The Trendline and Linear Regression

Determination of V_{max} and K_m

The Lineweaver-Burk Plot. Taking reciprocals of both sides of the Michaelis- Menten equation (Equation 6-14) results in

$$\frac{1}{v_0} = \frac{K_m + [S]}{V_{max}[S]} \tag{6-17}$$

which rearranges to

$$\frac{1}{v_0} = \frac{K_m}{V_{max}[S]} + \frac{1}{V_{max}} \tag{6-18}$$

an equation known as the Lineweaver-Burk equation. For systems obeying the Michaelis-Menten equation, a plot of $1/v_0$ against $1/[S]$ (a Lineweaver-Burk plot) results in a straight line of slope equal to K_m/V_{max} and intercept equal to $1/V_{max}$ (Figure 6-2, middle plot). The intercept on the $1/[S]$ axis (where $1/v_0 = 0$) equals $-1/K_m$. The constant k_2, which equals $V_{max}/[E]_o$ (equation 6-15), is frequently given the symbol k_{cat} and is called the turnover number. The turnover number is the number of product molecules produced per enzyme molecule per second and thus has units of s^{-1}.

The Eadie-Hofstee Plot. Although the most frequently used plot is the Lineweaver-Burk plot, it has the disadvantage that the plotted points tend to bunch up at small values of $1/[S]$. Because the data points plotted with the Eadie-Hofstee method tend to remain more evenly spaced, it is arguably superior to the Lineweaver-Burk method (Alberty 1987). The Eadie-Hofstee plot is a linear transformation of the Michaelis-Menten equation, written as

$$v_0 = -K_m \cdot \frac{v_0}{[S]} + V \tag{6-19}$$

The Eadie-Hofstee plot consists of a plot of v_0 versus $v_0/[S]$. The slope of the Eadie-Hofstee plot equals $-K_m$, and the intercept equals V as shown in Figure 6-2, lower plot. Table 6-1 lists some typical data for an enzyme-catalyzed reaction:

carbobenzoxyglycyl-L-tryptophan + H_2O → carbobenzoxyglycine + L-tryptophan

This reaction is catalyzed by the enzyme pancreatic carboxypeptidas. The concentration of pancreatic carboxypeptidas is given by [E], and the concentration of carbo-benzoxyglycyl-L-tryptophan is given by [S].

A plot of v_0 as the dependent variable on the y-axis against [S] as the independent variable on the x-axis generally shows that v_0 rises rapidly at first. Then v_0 levels off as [S] is further increased, as shown in the upper plot in Figure 6-2.

Entering the Worksheet Data

Excel expects the independent variable x to be in a worksheet column to the left of the column containing the dependent variable y. With this in mind, enter the data from Table 6-1 for a plot of reaction velocities (v_0) versus substrate concentrations.

- In cell A2, type **Enzyme Kinetics** as the worksheet title.
- In cell A5, type **Reaction Velocity** as the title.
- In the range A6:A11, type **[S]/(mM)** as the label, and enter the corresponding data from Table 6-1.

- In the range B6:B11, type **v_0 / (mM/s)** as the label, and enter the corresponding data from Table 6-1.

Table 6-1

Dependence of Initial velocity v_0 on Substrate Concentration [S]

v_0 / (mM/s)	[S] / (mM)
0.024	2.50
0.036	5.00
0.053	10.00
0.060	15.00
0.064	20.00

Lumry, R., E. L. Smith, and R. R. Glantz, *J. Amer. Chem. Soc. 73,* 4330 (1951)

Enter the data for the Lineweaver-Burk Plot, which are calculated from the reaction velocity data just entered.

- In cell A15, type **Lineweaver-Burk Data** as the title.
- In cell A16, type **(1/[S])/(1/M)** as the label.
- In cell A17, type **=1000/A7** and press the ENTER key.
- Select cell A17, and drag the fill handle to cell A21.
- In cell B16, type **(1/v_0)/(s/mM)** as the label.
- In cell B17, type **=1/B7** and press the ENTER key.
- Select cell B17, and drag the fill handle to cell B21.

Next enter the data for the Eadie-Hofstee plot, which are calculated from the reaction velocity data previously entered.

- In cell A23, type **Eadie-Hofstee Data** as the title.
- In cell A24, type **(1000v_0/[S])/(1/s)** as the label.
- In cell A25, type **=1000*B7/A7** and press the ENTER key.
- Select cell A25, and drag the size-handle to cell A29.
- In cell B24, type **v_0/(mM/s)** as the label.
- Click and drag from cell B7 to cell B11.
- Copy and paste to cell B25.
- In cells B6, B16, A24, and B24, subscript the zeros in the **v0** terms:
  1. Click in the cell, select the **0**, and on the **Format** menu, click **Cells**.
  2. In the **Format Cells** dialog box, select the **Subscript** check box.

Creating the Charts and Adding the Trendline

- For the first chart, select the range A7:B11; then click the **Chart Wizard** icon.
  1. In the **Chart Wizard – Step 1 of 4 – Chart Type** dialog box, select **XY (Scatter)**, and click the icon for scatter with data points connected by smoothed lines without markers. Click **Next.**
  2. In the **Chart Wizard – Step 2 of 4 – Chart Source Data** dialog box, no action is needed. Click **Next.**
  3. In the **Chart Wizard – Step 3 of 4 – Chart Options** dialog box:
     a. In the **Chart title** box, type **Reaction Velocities for Tryptophan Formation** as the title.
     b. In the **Value (X) axis** box, type **[S]** as the label.
     c. In the **Value (Y) axis** box, type **v0** as the label. Click **Next.**
  4. In the **Chart Wizard – Step 4 of 4 – Chart Location** dialog box, select **As object in Sheet1**, and click **Finish.**
- For the second chart, select the range A17:B21; then click the **Chart Wizard** icon.

1. In the **Chart Wizard – Step 1 of 4 – Chart Type** dialog box, select **XY (Scatter)**, and click the icon for compares pairs of values (points, but no lines. Click **Next**.
2. In the **Chart Wizard – Step 2 of 4 – Chart Source Data** dialog box, no action is needed. Click **Next**.
3. In the **Chart Wizard – Step 3 of 4 – Chart Options** dialog box:
    a. In the **Chart title** box, type **Lineweaver-Burk Plot** as the title.
    b. In the **Value (X) axis** box, type **1/[S]** as the label.
    c. In the **Value (Y) axis** box, type **1/v0** as the label. Click **Next**.
4. In the **Chart Wizard – Step 4 of 4 – Chart Location** dialog box, select **As object in Sheet1**, and click **Finish**.

- For the third chart, select the range A25:B29; then click the **Chart Wizard** icon.
  1. In the **Chart Wizard – Step 1 of 4 – Chart Type** dialog box, select the icon for compares pairs of values.
  2. In the **Chart Wizard – Step 2 of 4 – Chart Source Data** dialog box, no action is needed. Click **Next**.
  3. In the **Chart Wizard – Step 3 of 4 – Chart Options** dialog box:
      a. In the **Chart title** box, type **Eadie-Hofstee Plot** as the title.
      b. In the **Value (X) axis** box, type **1000v0/[S]** as the label.
      c. In the **Value (Y) axis** box, type **v0** as the label. Click **Next**.
  4. In the **Chart Wizard – Step 4 of 4 – Chart Location** dialog box, select **As object in Sheet1**, and click **Finish**.
- In the Lineweaver-Burk plot and the Eadie-Hofstee plot, add the trendline, which we normally call the linear regression line.
- Right-click any data point (marker) in the plot area, and then in the shortcut menu, click **Add Trendline** to open the **Add Trendline** dialog box.
  1. In the **Add Trendline** dialog box, click the **Type** tab, then select **Linear**.
  2. Click the **Options** tab, and select the **Display Equation on Chart** check box.
- In the plot area of the Lineweaver-Burk and Eadie-Hofstee plots, place a text box and an arrow-tipped line as shown in Figure 6-2.
- On the **View** menu, point to **Toolbars**, and click **Drawing**.
  1. In the **Drawing** toolbar, click the **Text Box** icon (a square with the letter A).
  2. Place the pointer in the plot area, and click and drag from the upper left corner of the text box to its lower right corner.
  3. Place the pointer in the text box, and enter the text:
      a. For the Lineweaver-Burk plot, type **1/V = 11.8 mM-1 s** as the text.
      b. For the Eadie-Hofstee plot, type **V = 0.0856 mM s-1** as the text.
  5. In the **Drawing** tool bar, click the arrow-tipped line. Place the pointer on the left edge of the text box, and drag to the y-axis as shown in Figure 6-2.

Chart Versus Graph and Regression Line Versus Trendline. Since business people outnumber scientists in the use of Excel and other spreadsheets, the nomenclature of spreadsheets is a bit different from that used in scientific software. A *chart* is what scientists and engineers would call a graph or a plot. A *trendline* is what scientists and engineers would call a *regression line*. Business people talk about trends in sales and profits and probably feel more comfortable with a term like *trendline* than with *linear regression*.

Editing the Charts

- In all three charts, subscript the zeros in the **v0** terms:
  1. Click twice slowly on the axis label with a **v0** in it.
  2. Select the **0**, and, on the **Format** menu, click **Selected Axis Title**.
  3. In the **Format Axis Title** dialog box, select the **Subscript** check box.
- Make all the chart borders and plot area borders medium heavy:
  1. Right-click the chart or plot area, then click **Format Chart Area** or **Format Plot Area**.
  2. For the plot area, in the **Format Plot Area** dialog box:
     a. In the **Area** section, click the white square.
     b. In the **Border** section, click **Color**, and select black; click **Weight**, and select the medium weight.
  3. In the **Format Chart Area** dialog box, click **Color**, and select black; click **Weight**, and select the medium weight.
- Make all axis labels, axis numbers, titles, equations, and text box contents 10 point bold:
  1. Right-click the object whose font is to be changed, and then click **Format [*object name*]** to open the **Format [*object name*]** dialog box.
  2. Click the **Font** tab, and select 10 point bold.
- Change the number of decimal places in the axis numbers to conform to those shown in Figure 6-2:
  1. Right-click any number, and then click **Format Axis** to open the **Format Axis** dialog box.
  2. Select the **Number** tab, select the **Number** category, and type **0** in the **Decimal places** box for all the x-axes and for the y-axis of the Lineweaver-Burk chart. Set the y-axis decimal to **2** for the reaction velocity chart and the Eadie-Hofstee chart.

Using the Linear Regression Data. From the Lineweaver-Burk plot, the intercept is $1/V$, and the slope is K_M:

$$1/V = 11.8 \text{ s mM}^{-1} \text{ and } V = 0.0847 \text{ mM s}^{-1}$$

$$K_m/V = 75.4 \text{ mM}$$

$$K_\mathrm{m} = \frac{K_m}{V}V = (75.4)(0.0847) = 6.38 \text{ mM}$$

From the Eadie-Hofstee plot, the intercept is V, and the slope is K_m:

$$V = 0.0856 \text{ mM s}^{-1}$$

$$K_m = 6.50 \text{ mM}$$

If the data were free of experimental error, the values for the Lineweaver-Burk and the Eadie-Hofstee plots would give identical values. In the next example, we will demonstrate how Excel can treat the experimental error statistically.

Chapter 7

Statistics

Excel Example 21

Linear Regression with Data Analysis: The Hydrogen Sulfide Equilibrium

Calculating and plotting a linear regression line is so common that it is illustrated for Excel in this chapter and for Lotus 1-2-3 and Quattro Pro in Chapter 9. For each of these spreadsheets the goal is to generate an XY (scatter) plot of the experimental points and then to plot the linear regression line calculated from these points. The result should be a straight line, the regression line, with the *x,y* points scattered randomly about the line. With Excel and Lotus 1-2-3 this task is essentially automatic. With Quattro Pro, the task is slightly more manual. For the purpose of illustration, some equilibrium constant data serve as a realistic example of linear regression with experimental data.

Review of the Science

Measurements of equilibrium constants over a range of temperatures provide one important method for determining thermodynamic data for a chemical change of state. The equilibrium constant K is related to the change in free energy ΔG^0, which in turn is related to the change in enthalpy ΔH^0, and to the change in entropy ΔS^0 by the following equation:

$$-RT\ln K = \Delta G^0 = \Delta H^0 - T\Delta S^0 \tag{7-1}$$

The free energy ΔG^0 is temperature dependent and can be calculated at any temperature for which an equilibrium constant K has been measured. Rearrangement of Equation (7-1) gives

$$R\ln K = \frac{-\Delta H^0}{T} + \Delta S^0 \tag{7-2}$$

A plot of $R \ln K$ versus $1/T$ gives a straight line, for which the slope equals $-\Delta H^0$ and the intercept equals ΔS^0. Experimental equilibrium constants for the equilibrium between hydrogen, sulfur, and hydrogen sulfide (Table 7-1) are suitable for our use. The equilibrium in question is for

$$H_2\ (g) + \tfrac{1}{2}\ S_2\ (g) \rightleftharpoons H_2S\ (g) \quad (7\text{-}3)$$

$$K = \frac{[H_2S]}{[H_2][S_2]^{1/2}} \quad (7\text{-}4)$$

Table 7-1.

Temperatures and Equilibrium Constants for the Hydrogen, Sulfur, and Hydrogen Sulfide Equilibrium

t°C	K
750	106.2
830	45.0
945	23.5
1065	8.9
1089	8.2
1132	6.2
1200	5.0
1264	3.2
1394	1.6

Data from: Pitzer, K. S., and L. Brewer, *Thermodynamics,* 3rd ed., McGraw-Hill, 1995.

Entering the Data

Entering the Labels and Constants

- Beginning in cell B1 (see Figure 7-1), type **K = [H2S]/[H2][S2]1/2** for the title. Subscript the **2** in **[H2]** and the **2** in **[S2]**, and superscript the **1/2**:
  1. Select the group of characters to be superscripted or subscripted by clicking and dragging. CRTL-click and drag to select more than one group of cells for formatting.
  2. On the **Format** menu, click **Cells** to open the **Format Cells** dialog box (similar to Figure 1-5). Click the **Font** tab.

3. In the **Effects** section, select the **Subscript** or **Superscript** check box, as appropriate; then click **OK**.

- In cell C3, type **T = t deg C + 273** as the label.
- In cell C4, type **8.314** as the gas constant, and type the labels, as shown in Figure 7-1, in cells B4 and D4.
- Type the data column labels shown in Figure7-1 in the range A6:E6.

Entering the Numerical Data

- From Table 7-1, in the range A7:A15, type the temperatures (t/deg C) in descending order.
- Leave column B blank for now.
- From Table 7-1, in the range C7:C15, enter the equilibrium constants *K*, in ascending order.

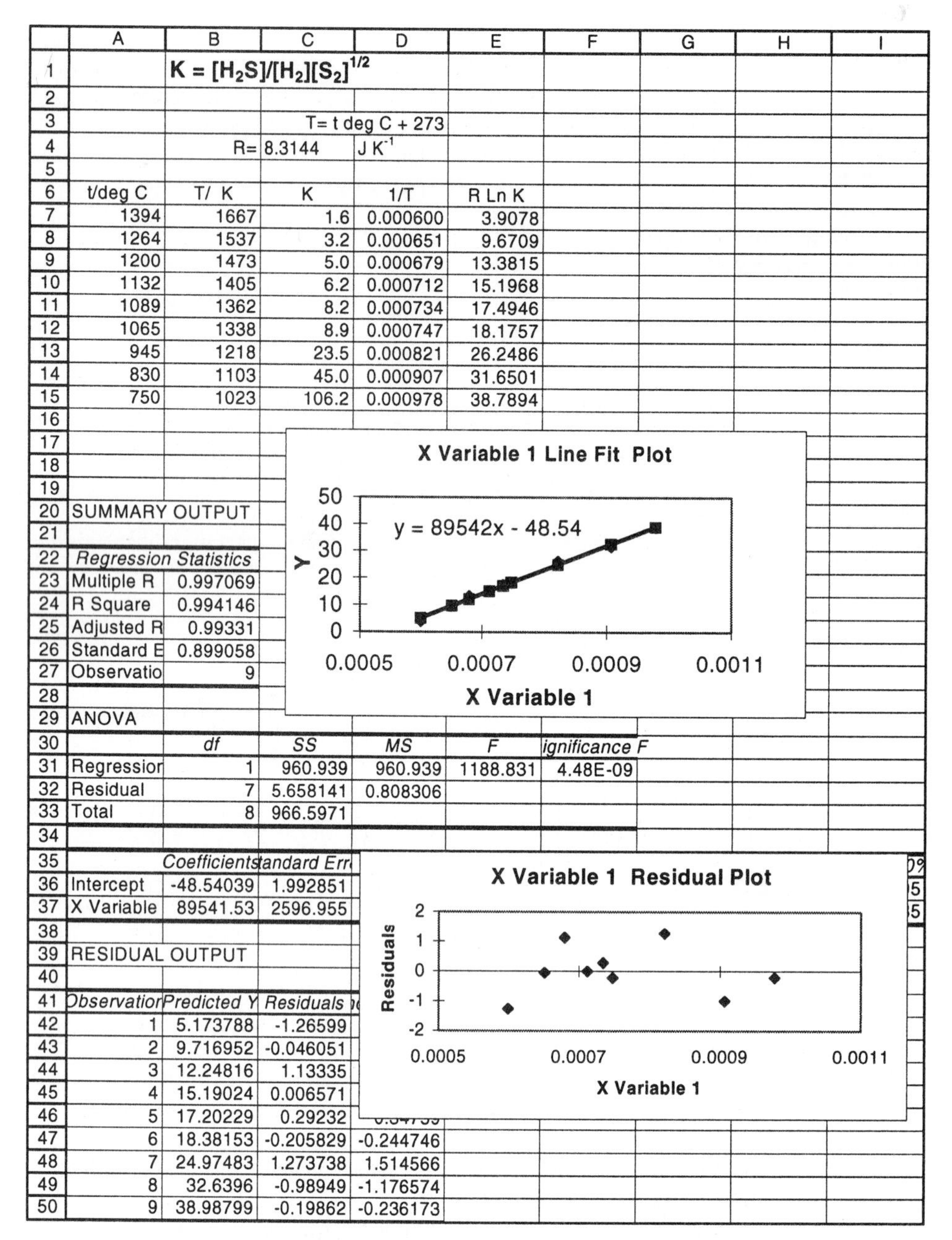

	A	B	C	D	E
1		$K = [H_2S]/[H_2][S_2]^{1/2}$			
2					
3			T= t deg C + 273		
4		R=	8.3144	J K^{-1}	
5					
6	t/deg C	T/ K	K	1/T	R Ln K
7	1394	1667	1.6	0.000600	3.9078
8	1264	1537	3.2	0.000651	9.6709
9	1200	1473	5.0	0.000679	13.3815
10	1132	1405	6.2	0.000712	15.1968
11	1089	1362	8.2	0.000734	17.4946
12	1065	1338	8.9	0.000747	18.1757
13	945	1218	23.5	0.000821	26.2486
14	830	1103	45.0	0.000907	31.6501
15	750	1023	106.2	0.000978	38.7894
16					
17					
18					
19					
20	SUMMARY OUTPUT				
21					
22	*Regression Statistics*				
23	Multiple R	0.997069			
24	R Square	0.994146			
25	Adjusted R	0.99331			
26	Standard E	0.899058			
27	Observatio	9			
28					
29	ANOVA				
30		*df*	*SS*	*MS*	*F*	*ignificance F*
31	Regressior	1	960.939	960.939	1188.831	4.48E-09
32	Residual	7	5.658141	0.808306		
33	Total	8	966.5971			
34						
35		*Coefficients*	*tandard Err*			
36	Intercept	-48.54039	1.992851			
37	X Variable	89541.53	2596.955			
38						
39	RESIDUAL OUTPUT					
40						
41	*Observatior*	*Predicted Y*	*Residuals*			
42	1	5.173788	-1.26599			
43	2	9.716952	-0.046051			
44	3	12.24816	1.13335			
45	4	15.19024	0.006571			
46	5	17.20229	0.29232			
47	6	18.38153	-0.205829	-0.244746		
48	7	24.97483	1.273738	1.514566		
49	8	32.6396	-0.98949	-1.176574		
50	9	38.98799	-0.19862	-0.236173		

Figure 7-1. The Worksheet for the H_2S Equilibrium: Linear Regression with *Tools_Data Analysis*

Preliminary Calculations

- In cell B7, type **=A7+273** as the formula; press the ENTER key.
- Select cell B7, and drag the fill handle to cell B15.
- In cell D7, type **=1/B7** as the formula; then press the ENTER key.
- Select cell D7, and drag the fill handle to cell D15.
- In cell E7, type **=$C$4*LN(C7)** as the formula.
- Select cell E7, and drag the fill handle to cell E15.

The worksheet should look like the range A1:E15 in Figure 7-1. The chart has not yet been created.

Creating the Chart

With all previous charts, we used the Chart Wizard's four steps to create the chart. For this example we'll create the chart with a new method that uses the **Regression** dialog box. This method not only creates the chart with data points (X_i, Y_i) and (X_i, Y_i(calc)) but also provides a complete statistical analysis of the data.

- On the **Tools** menu, click **Data Analysis** to open the **Data Analysis** dialog box (Figure 1-13).
- In the **Analysis Tools** list, scroll to **Regression**, and click **OK** to open the **Regression** dialog box (Figure 7-2).

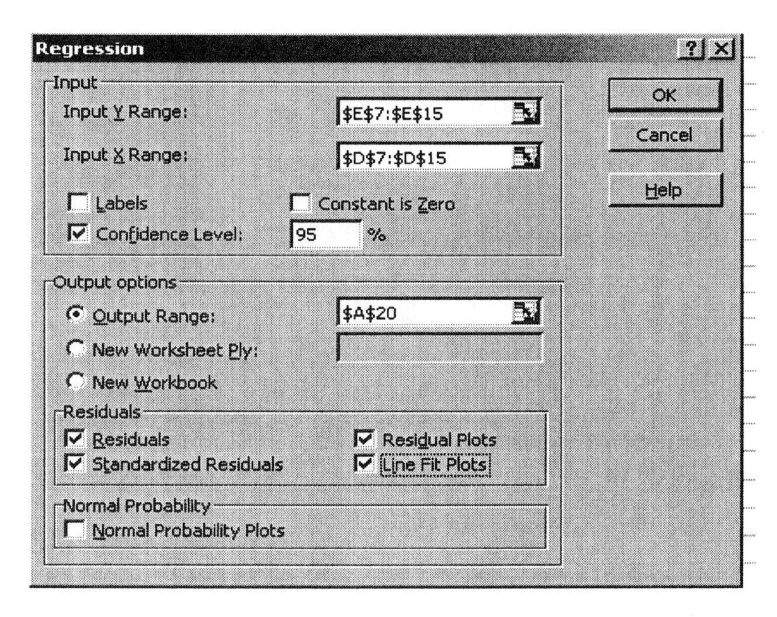

Figure 7-2. The **Regression** Dialog Box

- In the **Input** section of the **Regression** dialog box:
  1. With the insertion point in the **Input Y Range** box, click and drag to select the range E7:E15 on the worksheet (Excel adds the dollar signs).
  2. With the insertion point in the **Input X Range** box, click and drag to select the range D7:D15 on the worksheet (Excel adds the dollar signs).
  3. Select the **Confidence Level** check box, and type **95** in the **%** box.
- In the **Output options** section of the **Regression** dialog box:
  1. Select **Output Range**, and and type **A20** (Excel adds the dollar signs). The upper left corner of the statistical output will begin at this cell.
  2. In the **Residuals** section, select all four check boxes.
- Click **OK**.
  1. The label **SUMMARY OUTPUT** appears in cell A20 along with four tables of statistical data below it.
  2. A plot appears displaying the data (X_i, Y_i) and (X_i, Y_i(calc)) and the title **X Variable 1 Line Fit Plot**. The chart is almost like the chart in Figure 7-1, except that no regression line has been added yet. This chart, its plot area, title, and labels can be edited by the methods previously described. We will leave the charts with their default labels.
  3. A second plot appears displaying the residuals and the title **X Variable 1 Residual Plot**. It also can be edited and retitled.

Adding the Trendline

The chart, generated through the **Regression** dialog box, has a title, axis titles, and numbers (X_i,Y_i , and X_i,Y_i (calc)), but no regression line (trendline). We want to eliminate the (X_i,Y_i (calc)) markers and add a trendline that represents a linear regression line drawn through the (X_i, Y_i),markers.

- Click anywhere in the chart to have the **Chart** menu appear in the menu bar.
  1. On the **Chart** menu, click **Source Data** to open the **Source Data** dialog box; click the **Series** tab (Figure 7-3).
  2. In the box in the lower left corner, select **Predicted Y**, and click **Remove**.
- Right-click any data point in the plot area.
  1. In the shortcut menu, click **Add Trendline** to open the **Add Trendline** dialog box.
  2. Click the **Type** tab, and select **Linear** (Figure 7-4).
  3. Click the **Options** tab, and select the **Display equation on chart** check box (Figure 7-5); click **OK.**

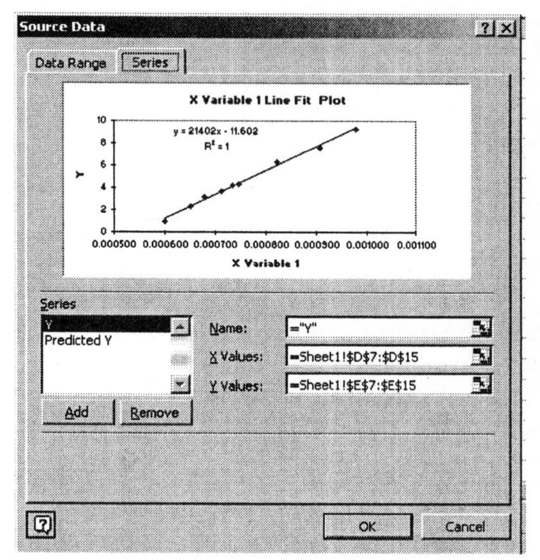

Figure 7-3. The **Source Data Series** Dialog Box

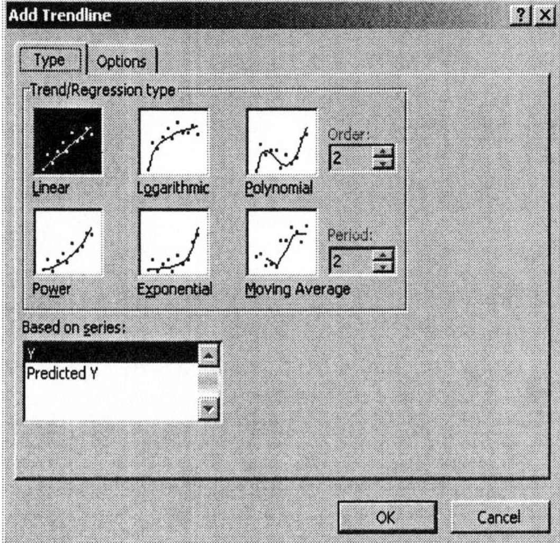

Figure 7-4. The **Add Trendline Type** Dialog Box

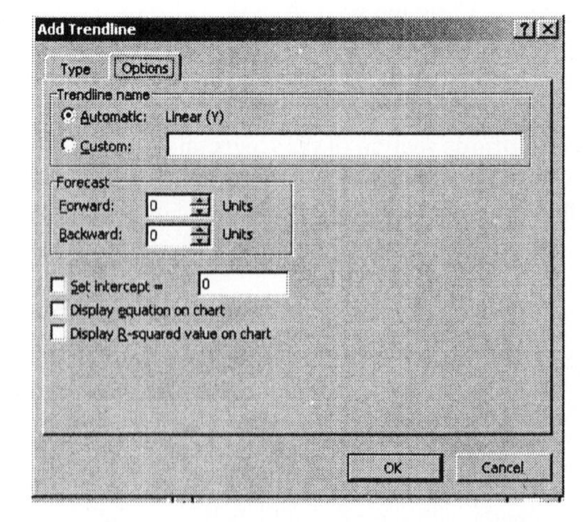

Figure 7-5. The **Add Trendline Options** Dialog Box

Editing the Chart

At this point, the chart titles appear as shown in Figure 7-1. The trendlline has been added, and the predicted *y* values have been removed.

- Click in the chart, and use the size handles to squeeze the line fit plot between data tables as shown in Figure 7-1. Do the same with the residual plot chart.
- Right-click the legend box, then click clear.

Next, edit the chart title and axis titles. In Figure 7-1, the default titles and labels were left in place.

- Click the default chart title, **X Variable 1 Line Fit Plot**, twice slowly (don't double-click). The insertion point remains in the title.
- Edit the title as you would with a word processor.
- Repeat for the y-axis title, and change **Y** to **R Ln K**.
- Repeat for the x-axis title, and change **X Variable 1** to **1/T**.

We could do some more editing, but we'll come back to chart editing later. However, if you wish to experiment with the chart at this time, it's useful to become acquainted with several right-click operations even though we'll look at them in more detail later.

- Right-click the chart title or axis titles to change font and colors.
- Right-click in the chart or plot area background to change their colors or borders.
- Right-click on the x-axis or y-axis numbers to change:

  1. Patterns (borders and tic marks, line and marker color)
  2. Scale (for example, minimum and maximum)
  3. Font (type, size, italic, bold, color)
  4. Number (general, currency, scientific, decimal points)
  5. Alignment (parallel, perpendicular, or slanted)

Using the Linear Regression Data

The table entries in lines 36 and 37 give the intercept and the slope of the plot of $R \ln K$ versus $1/T$. From Equation 7-2, the slope of this plot equals $-\Delta H^0$ and the intercept equals ΔS^0, so for the equilibrium

$$H_2\,(g) + \tfrac{1}{2}\,S_2\,(g) = H_2S\,(g) \qquad (7\text{-}5)$$

$$\Delta H^0 = -89{,}542\ J \text{ and } \Delta S^0 = -48.54\ J/K$$

ΔH^0 is negative, so the formation of H_2S (g) from the elements is exothermic, which is to be expected when more bonds are formed than broken in the reaction process. A decrease in the number of moles (1½ to 1) for the reaction corresponds to a decrease in the entropy of reaction, consistent with the negative sign for ΔS^0. These values represent the values calculated by the program, not taking into account realistic significant figures. Equation 7-2 may be rewritten to show the explicit dependence of the equilibrium constant K on temperature T, with more realistic significant figures.

$$R \ln K = \frac{89{,}500}{T} - 48.5 \tag{7-6}$$

Related Excel Regression Functions: SLOPE, INTERCEPT, and LINEST

Excel's **SLOPE** function returns the slope of the linear regression line through a set of *x,y* data pairs. The **INTERCEPT** function returns the intercept of the linear regression line through a set of *x,y* data pairs. The **LINEST** function returns an array of values, including the slope and intercept of the linear regression line through a set of *x,y* data pairs, along with regression statistics. You select all these functions by pointing to the **Insert** menu and clicking **Function**, selecting **Statistical** in the category list, and selecting the specific function, for example, **SLOPE**, **INTERCEPT**, or **LINEST**, in the function list.

To compare these regression functions with the **Regression** dialog box method, copy the data in Sheet1 (Figure 7-1) to Sheet2. In the original worksheet (Figure 7-1), select the block A1:E15, and click the **Copy** icon. Then click the **Sheet2** tab to open it. Select cell A1, and click the **Paste** icon to copy the data from Sheet1 to Sheet2.

The SLOPE Function

- In Sheet2, select cell D17, and type **Slope** as the label; select cell E17, and type **Intercept** as the label.
- Select cell D18, and, on the **Insert** menu, click **Function** to open the **Insert Function** dialog box (Figure 1-2).
- In the **Insert Function** dialog box, select **Statistical** as the category, and select **SLOPE** as the function to open the **Function Arguments** dialog box (Figure 7-6).
- With the insertion point in the **Known_y's** box, select the range E7:E15.
- With the insertion point in the **Known_x's** box, select the range D7:D15, and click **OK**. The slope is returned to cell D18 (Figure 7-7).

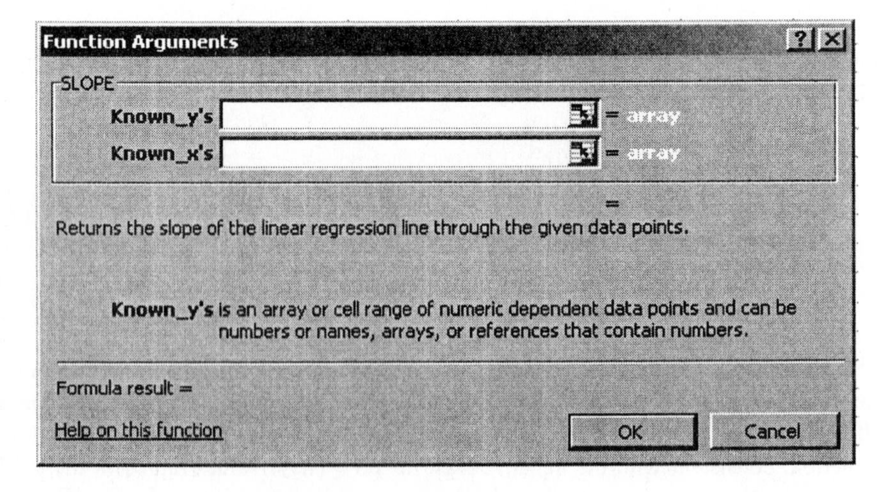

Figure 7-6. The **Function Arguments** Dialog Box

The INTERCEPT Function

- Select cell E18, and on the **Insert** menu, click **Function** to open the **Insert Function** dialog box (Figure 1-2).
- Repeat the preceding steps, but select the **INTERCEPT** function.
- The slope is returned to cell E18 (Figure 7-7).

The LINEST Function. Because the **LINEST** function returns an array of values, it is necessary to select a block of empty cells on the worksheet that is congruent with the returned array. The **LINEST** function returns a 5 x 2 array of values, so a worksheet block consisting of five rows and two columns matches the returned array.

- In Sheet2, select the range D20:E24.
- Repeat the preceding steps, but select the **LINEST** function.
- An array of values is returned to the range D20:E24, but with no labels (Figure 7-7).
- Enter the following labels in the range F20:G24:
  1. In cell F20, type **Intercept** as the label.
  2. In cell F21, type **Std. Err. of Intercept** as the label.
  3. In cell F22, type **Std. Err. of Y est.** as the label.
  4. In cell F23, type **Degrees of Freedom** as the label.
  5. In cell F24, type **Sum Sq** as the label.
- Enter the following labels in the range B20:C24:
  1. In cell C20, type **Slope** as the label.
  2. In cell B21, type **Std. Err. of Slope** as the label.
  3. In cell C22, type **r2** as the label.
  4. In cell C23, type **F** as the label.
  5. In cell C24, type **Sum Sq** as the label.

Figure 7-7 displays Sheet2. Excel's **TREND**, **GROWTH**, and **LOGEST** functions calculate regression formulas similar to those of **LINEST** and are accessed in the same way as the **LINEST** function.

Other Spreadsheets

Lotus 1-2-3 and Quattro Pro have similar commands for regression analysis, and it is useful at this time to anticipate and compare these commands.

- With Excel, on the **Chart** menu, click **Add Trendline**, and on the **Type** tab, select **Linear**; or on the **Tools** menu, click **Data Analysis**, and in the **Data Analysis** dialog box, click **Regression**.

- With Lotus 1-2-3, on the **Chart** menu, click **Properties for Series**, and click the **Series Trend** tab to create a regression line, and on the **Range** menu, point to **Analyze**, and click **Regression** to analyze the regression line.

- With Quattro Pro, on the **Tools** menu, point to **Numeric Tools**, and click **Regression**, and ***Analysis_Linear Fit*** (accessible from the **Line Series** Dialog Box).

The commands for Lotus 1-2-3 and Quattro Pro are illustrated in Chapter 9. It is relatively straightforward to shift between spreadsheets because the commands and their application are so similar.

	A	B	C	D	E	F	G	H
1		$K = [H_2S]/[H_2][S_2]^{1/2}$						
2								
3			T= t deg C + 273					
4		R=	8.3144	J/K mol				
5								
6	t/deg C	T/ deg K	K	1/T	R Ln K			
7	1394	1667	1.6	0.000600	3.907798			
8	1264	1537	3.2	0.000651	9.670901			
9	1200	1473	5.0	0.000679	13.38151			
10	1132	1405	6.2	0.000712	15.19681			
11	1089	1362	8.2	0.000734	17.49461			
12	1065	1338	8.9	0.000747	18.1757			
13	945	1218	23.5	0.000821	26.24856			
14	830	1103	45.0	0.000907	31.65011			
15	750	1023	106.2	0.000978	38.78937			
16								
17				Slope	Intercept			
18				89541.53	-48.54039			
19								
20			Slope	89541.53	-48.54039	Intercept		
21		Std. Err. Of Slope		2596.955	1.992851	Std. Err. of Intercept		
22			r2	0.994146	0.899058	Std. Err. of Y est.		
23			F	1188.831	7	Degrees of Freedom		
24			Sum Sq	960.939	5.658141	Sum Sq		
25								
26								

Figure 7-7. Sheet2 Showing the Result for the **SLOPE()**, **INTERCEPT()**, and **LINEST()** Functions

Excel Example 22

Multiple Linear Regression Lines: emf versus T

Excel allows the user to create a chart with more than one regression line. For example the linear regression data for two independent data sets can be placed on the same chart, providing of course, that the scale ranges are reasonably close together. In this Excel example, we shall tabulate the emf (cell potential) versus temperature data for two different electrochemical cells, plot the data on a chart, and add the linear regression lines for both data sets with the **Add Trendline** command. We will also use the **Regression** dialog box to calculate the regression line parameters and statistics.

In Example 21 we used the **Regression** dialog box both to create the chart and to calculate the regression statistics. In the present example we will use the Chart Wizard to create the chart and the **Regression** dialog box to calculate the regression statistics. These two tools can be used in any combination, together or alone.

Review of the Science: The Thermodynamics of Electrochemical Cells

Measurement of the emf (electromotive force) E of an electrochemical cell permits the evaluation of the free energy ΔG, the entropy ΔS, and enthalpy ΔH for the overall cell reaction. At each temperature, the free energy change for the cell reaction is given by

$$\Delta G = -nFE \tag{7-7}$$

where n is the number of moles and F is the Faraday constant (96485.3 C mol^{-1}). Since the free energy depends upon temperature according to

$$\left(\frac{\partial G}{\partial T}\right)_P = -\Delta S \tag{7-8}$$

it follows from Equation 7-7 and Equation 7-8 that the entropy change is given by

$$\Delta S = nF\left(\frac{\partial E}{\partial T}\right)_P \tag{7-9}$$

where $(\partial E/\partial T)_P$ is the slope of a plot of E versus T at constant pressure. Then, since $\Delta G = \Delta H - T\,\Delta S$, ΔH can be calculated from

$$\Delta H = -nFE + nFT\left(\frac{\partial E}{\partial T}\right)_P \tag{7-10}$$

The data of R. H. Gerke (*J. Am. Chem. Soc. 44*:1684–1704, 1922) illustrate the linear dependence of emf on temperature. Some of his data are summarized in Table 7-2. In this Excel Example, we will duplicate the worksheet shown in Figure 7-9 by using the Chart Wizard and the trendline feature of Excel.

Preparing the Worksheet

Entering a Title in a Text Box

- On the **View** menu, point to **Toolbars**, and click **Drawing**; click the **Text Box** icon.
  1. Beginning in the middle of cell A1, click and drag to about cell G3.
  2. Type **The Dependence of emf on Temperature** as the title.
  3. Use the icons on the formatting toolbar to center the text or change fonts.

  Note: To delete a text box, select it, click its fuzzy edge, and press the DELETE key.
- In the Range A5:C5, type **T/K**, **Pb/AgCl**, and **Pb/HgCl** as the three column headings.

Table 7-2

emf and Temperature Data for Two Electrochemical Cell Reactions

$Pb(s) + 2HgCl(s) \rightarrow PbCl_2(s) + 2Hg(\ell)$		$Pb(s) + 2AgCl(s \rightarrow PbCl_2(s) + 2Ag(s)$	
T/K	**Volts**	**T/K**	**Volts**
278.15	0.52788	278.15	0.48837
288.15	0.52856	287.71	0.48637
298.15	0.52987	298.01	0.48434
308.11	0.53120	308.11	0.48231
318.35	0.53257	318.20	0.48025

Entering the Data

- In the range A6:A15, enter the temperatures from Table 7-2, first the Pb/AgCl temperature data followed by the Pb/HgCl temperature data.
- In the range B6:B10, enter the AgCl emf data.
- In the range C11:C15, enter the HgCl emf data.

Notice that these data are entered in same pattern shown in Excel Example 5 for the multiple XY (scatter) plots for the vapor pressure of benzene and chloroform. If the temperatures were identical for both emf data sets, then it would not be necessary to use the staggered two-column pattern of data entry.

Creating the Chart

- Select the block A5:C15. This selection will allow Excel to plot two graphs on one chart and distinguish between them in the legend box.
- With this block selected, click the **Chart Wizard** icon.
  1. In the **Chart Wizard – Step 1 of 4 – Chart Type** dialog box, for the chart type, click **XY (Scatter)**. For the chart subtype, select the icon for scatter (points, no line), then click **Next**.
  2. In the **Chart Wizard – Step 2 of 4 – Chart Source Data** dialog box, no action is necessary. Click **Next**.
  3. In the **Chart Wizard – Step 3 of 4 – Chart Options** dialog box:
     a. In the **Chart title** box, type **emf vs Temperature** as the title.
     b. In the **Value (X) axis** box, type **T/K** as the label.
     c. In the **Value (Y) axis** box, type **emf/Volt** as the label.
     d. Click **Next**.
- In the **Chart Wizard – Step 4 of 4 – Chart Location** dialog box, select **As object in Sheet1**, and click **Finish**.

The chart created at this point (Figure 7-8), appears with default formats and styles. No regression lines are drawn through the markers. After editing the chart, trendlines will be drawn through both sets of data pairs. Equations for the trendlines will be embedded in the chart next to each trendlline.

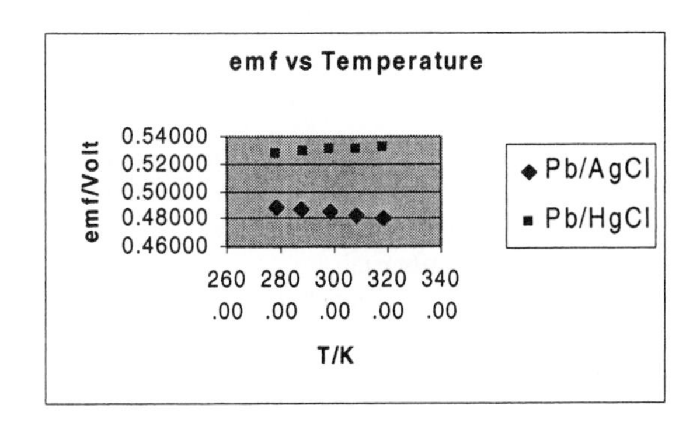

Figure 7-8. The Unedited emf vs Temperature Chart Created with the Chart Wizard. Figure 7-9 shows the chart after editing.

Editing the Chart

In Step 1 of 4 of the Chart Wizard, we chose the icon for the scatter (points, no lines) chart subtype. If we had chosen scatter (with data points connected by smooth lines), Excel would have drawn line segments between the lines. Instead, we want the linear regression line drawn through the points. We get this line with the **Add Trendline** command. We'll do this after some initial chart editing.

- Right-click in the chart or plot area background to change their colors or borders.
- Right-click the chart title or axis titles to change their font and colors.
- Click a title or label slowly twice (don't double-click) to edit it.
- Right-click the x-axis or y-axis numbers to change:
  1. Patterns (borders and tic marks)
  2. Scale (for example, minimum and maximum)
  3. Font (type, size, italic, bold, color)
  4. Number (general, currency, scientific, or other format)
  5. Alignment (parallel, perpendicular, or slanted)

Plotting the Linear Regression with Trendlines

As already noted, the chart up to this point has no linear regression lines drawn through the points. In the **Chart Wizard – Step 1 of 4 – Chart Type** dialog box, we selected the XY (scatter) type and the scatter (points, no line) subtype. The two plots should consist only of unconnected points and no line (Figure 7-8). We now add the linear regression lines through the two plots.

- Right-click any data point (marker) in the plot area, and then in the shortcut menu, click **Add Trendline** to open the **Add Trendline** dialog box.
  1. In the **Add Trendline** dialog box, select the **Type** tab, then select **Linear**.
  2. Click the **Options** tab, and select the **Display equation on chart** check box.
- Click any data marker in the second set, and repeat the preceding operations to insert a second trendline.

Notice that on Figure 7-9, the equation coefficients of the embedded regression equations are shown in their default format in the lower (second) equation and in scientific notation with four decimal places in the upper (first) equation. Often the default format displays too few significant figures: the slope of the lower equation is given as –0.0002. The format of these numbers can be changed to give the desired number of significant figures. To do so:

- Right-click the first equation, click **Format Data Labels** to open the **Format Data Labels** dialog box (similar to the **Format Cells** dialog box, Figure 1-5, but with fewer tabs from which to select).
- Click the **Number** tab.
- For the category, select **Scientific**.
- Type **4** in the **Decimal places** box, and click **OK**. The number formats of both equations are now in scientific notation with four decimal places.
- You can click and drag these equations to place them in the most suitable position on the chart.

The chart, embedded in the worksheet, appears as shown in Figure 7-9 after sizing and formatting to taste.

Calculating the Thermodynamic Functions

For the cell reaction Pb(s) + 2HgCl(s) → $PbCl_2$(s) + 2Hg(ℓ), the Excel worksheet (Figure 7-9) shows that the slope of E versus T is 1.1980 x 10^{-4} V/K. Faraday's constant F equals 96,484 J V^{-1} mol^{-1}, and at 298.15 K, E equals 0.52987 V. With equation 7-9, we can calculate the entropy change for the cell reaction:

$$\Delta S = nF\left(\frac{\partial E}{\partial T}\right)_P = (2\text{ mol})\,(96{,}485\text{ J V}^{-1}\text{ mol}^{-1})\,(1.1980 \times 10^{-4}\text{ V K}^{-1}) = 23.12\text{ J K}^{-1}$$

Combining Equation 7-9 and Equation 7-10, we can calculate the enthalpy change for the cell reaction:

$$\Delta H = -nFE + nFT\left(\frac{\partial E}{\partial T}\right)_P = -nFE + T\Delta S$$

$\Delta H = (-2 \text{ mol})\,(96{,}484 \text{ J V}^{-1} \text{ mol}^{-1})(0.52987 \text{ V}) + (298.15 \text{ K})(\,23.12 \text{ J K}^{-1}) = -95{,}563 \text{ J}$

It is readily apparent from the plots that one of the two cell reactions has a positive slope (positive entropy change) and the other has a negative slope (negative entropy change). Qualitative statistical thermodynamics offers an explanation. For the cell reaction

$$Pb(s) + 2HgCl(s) \rightarrow PbCl_2(s) + 2Hg(\ell)$$

the change of state involves three moles of solid going to one mole of solid and two moles of liquid mercury. The liquid state is a more random state than a solid, and so it has larger entropy, leading to an increase in the entropy for the cell reaction change of state.

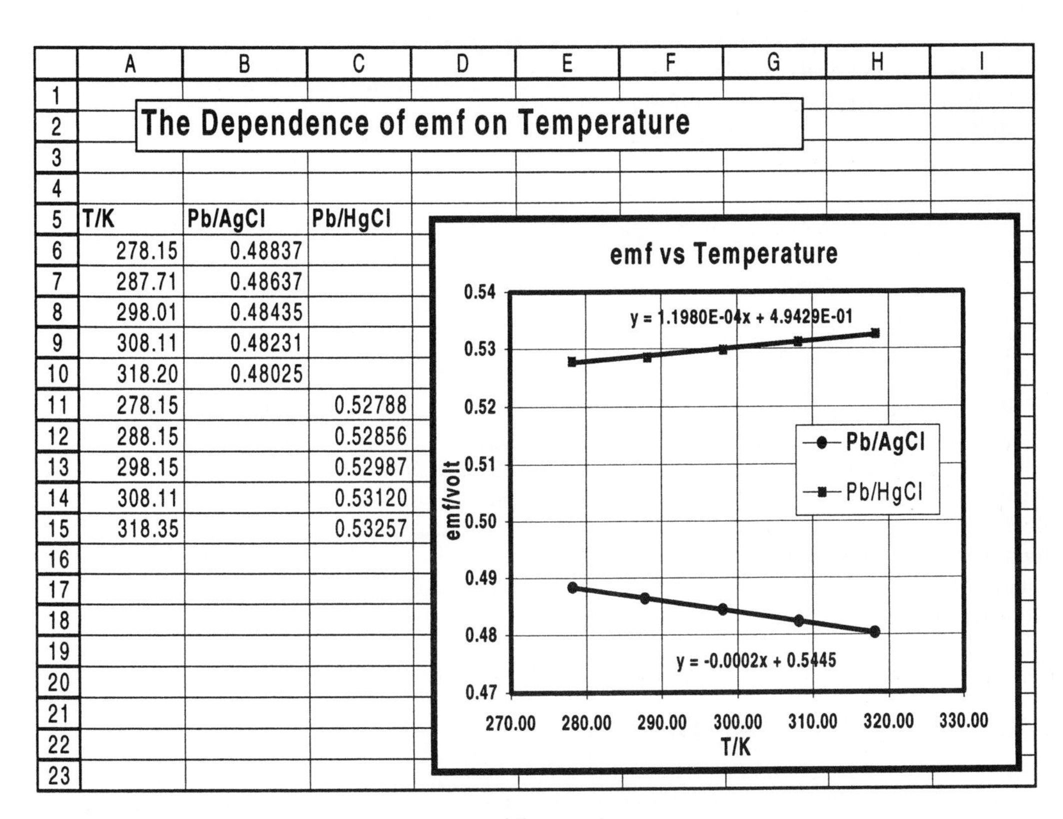

The Dependence of emf on Temperature

T/K	Pb/AgCl	Pb/HgCl
278.15	0.48837	
287.71	0.48637	
298.01	0.48435	
308.11	0.48231	
318.20	0.48025	
278.15		0.52788
288.15		0.52856
298.15		0.52987
308.11		0.53120
318.35		0.53257

Figure 7-9. Worksheet for the emf as a Function of Temperature

Excel Example 23

Nonlinear Regression: A Calibration Curve

Scientists and engineers frequently use calibration curves for a variety of purposes. As an example, chemists sometimes use the measurement of some physical property of a mixture to analyze the mixture. Chloroform and acetone have quite different refractive indices, so it is possible to analyze a mixture of these two components by simply measuring the refractive index of a mixture. Table 7-3 displays the refractive indices of mixtures of chloroform and acetone.

In this Excel Example, we will use Excel's **Regression** dialog box to carry out a linear regression with the data in Table 7-3. This is shown in Figure 7-10, which shows evidence of nonlinearity.

Second, we will carry out a second-order regression and again plot the residuals to see if the data in Table 7-3 fit the second order line better.

You cannot use Excel's **Regression** dialog box to carry out a nonlinear regression, analyze the data, or create a residuals plot (except for the linear regression). For these reasons, we will plot the data in Table 7-3 with the Excel Chart Wizard and then carry out a second-order regression with the trendline. The trendline also furnishes an equation for the second-order plot. The trendline does not calculate the residuals, so we will carry out our own calculation of the residuals and plot them with the Chart Wizard.

Table 7-3

The Refractive Indices of Mixtures of Chloroform and Acetone.

Refractive Index	X_{chlor}
1.4470	0.0000
1.4272	0.2509
1.4061	0.4992
1.3835	0.7503
1.3605	1.0000

Entering the Worksheet Title and Data

- Open a new workbook with Sheet1 the active sheet.
- On the **View** menu, point to **Toolbars**, and click **Drawing**; then click the **Text Box** icon (a tiny square with the letter A on it).
- Drag from cell A2 to about Cell G4.
  1. Type **A Linear Calibration Curve** as the first line of the title. Press ENTER.
  2. Type **Mole Fraction Chloroform vs Refractive Index** as the title's second line.
  3. Change the first title line to 12 point, the second to 10 point, and make both lines bold.
- Enter the data column labels and data:
  1. In cell A6, type **Refr. Index** as the label.
  2. In cell B6, **Xclor** as the label.
  3. In the range A7:B11, enter the numerical data shown in Table 7-3 and in Figures 7-11 and 7-12.

Calculating a Linear Regression

- On the **Tools** menu in the menu bar, click **Data Analysis** to open the **Data Analysis** dialog box (Figure 1-13).
- Scroll the list and select **Regression** to open the **Regression** dialog box (Figure 7-2).
  1. In the **Input Y range** box, type **B7:B11** as the range.
  2. In the **Input X range** box, type **A7:A11** as the range.
  3. In the **Output options** section, select **Output range**, and type **A15** to place the output under the input data.
  4. In the **Residuals** section, select the **Residuals, Residual plots**, and **Line fit plots** check boxes.

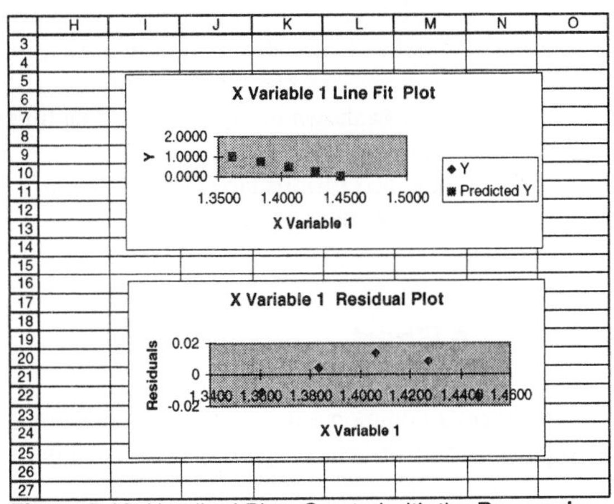

Figure 7-10. Unedited Plots Created with the **Regression** Dialog Box

Beginning at cell A15, the output appears. The output includes two graphs (Figure 7-10) and several tables of statistical data (deleted from Figure 7-10). The plots are automatically prepared, but with no line drawn through the markers. The upper plot, labeled X Variable 1 Line Fit Plot, appears on first examination to be a linear fit. However, the lower plot, labeled X Variable 1 Residual Plot, displays a systematic deviation of the residuals. If the regression were linear, the deviation markers would be randomly and equally scattered above and below the zero line. The residual plot strongly suggests that the relationship between mole fraction and refractive index is not linear for chloroform-acetone mixtures. Our initial assumption of linearity appears to be wrong.

Even when enlarged and edited (Figure 7-11), the upper graph appears to be linear. In Figure 7-11 the upper plot has been subtitled "A Linear Calibration Curve," the linear regression line added, and an equation for the regression displayed.

Calculating a Nonlinear Regression

Now return to the original worksheet with the data in the range A7:B11.

- Select the block A7:B11.
- With this block selected, click the Chart Wizard icon.

1. In the **Chart Wizard – Step 1 of 4 – Chart Type** dialog box, click **XY (Scatter)**. For the chart subtype, select the icon for scatter (points, no line), then click **Next.**
2. In the **Chart Wizard – Step 2 of 4 – Chart Source Data** dialog box, no action is necessary. Click **Next.**
3. In the **Chart Wizard – Step 3 of 4 – Chart Options** dialog box:
   a. In the **Chart title** box, type **2nd Order Calibration Curve** as the title.
   b. In the **Value (X) axis** box, type **Refractive Index** as the label.

c. In the **Value (Y) axis** box, type **(Xc) Mole Fraction Chloroform** as the label, and click **Next.**

4. In the **Chart Wizard – Step 4 of 4 – Chart Location** dialog box, select **As object in Sheet1**, and click **Finish**.

- Resize and reposition the chart so that it appears approximately like the chart in Figure 7-12. The chart still lacks a line drawn through the markers.

Adding a Polynomial (Nonlinear) Trendline

Since a plot of the residuals shows that the plot is nonlinear, we will now add a polynomial regression line through the markers.

- Right-click any marker to open a shortcut menu, and click **Format Data Series.**
- Click **Add Trendline** to open the **Add Trendline** dialog box.
  1. Click the **Type** tab (Figure 7-4), click **Polynomial**, Type **2** in the **Order** box, then click **OK**, and the polynomial regression line appears through the points as shown in Figure 7-12. The line appears nearly linear but in fact is a polynomial of order 2.
  2. Click the **Options** tab (Figure 7-5), and select the **Display equation on chart** check box.

Editing the Chart

Many of Excel's common editing features have already been covered in earlier examples. To create the chart shown in Figure 7-12, several parts were edited. In way of review:

- To edit the text of the title and axis labels, click twice slowly anywhere in the title, and make any necessary changes. To change the font in the titles, right-click the title, and use the options in the **Format Chart Title** dialog box (Figure 2-8).
- To edit the axis numbers, right-click an axis number, and use the options in the **Format Axis** dialog box to change the scale, font, or numbers (Figure 2-11).
- To edit the plot area, right-click in any blank part of the plot area.
  1. Click **Format Plot Area** to change:
     a. Border lines, style, color, or weight
     b. Plot area color (the default is gray; white usually looks better)
  2. Click **Chart Options** to add or eliminate grid lines, then select the **Gridlines** tab.

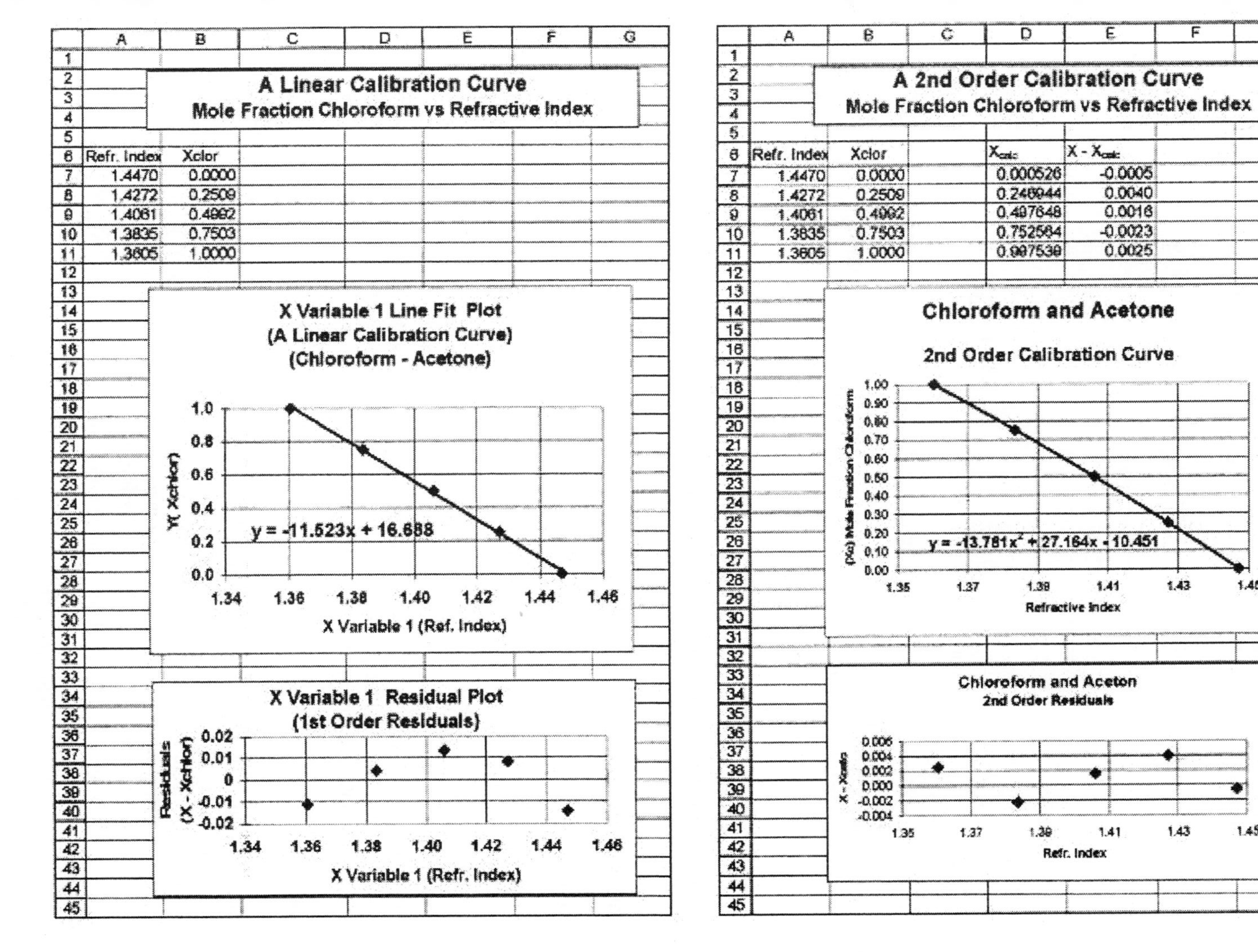

Figure 7-11. Worksheet for a Linear Calibration. Compare the linearity and residuals with Figure 7-12

Figure 7-12. Worksheet for a Second-Order Calibration Curve. The first-order residuals show random variation and the residuals about one-fifth as large as those in Figure 7-11.

Editing the Trendline

- To edit the trendline, right-click it to open the **Format Trendline** dialog box (Figure 7-13).
  1. Click the **Patterns** tab, and change the style or weight of the trendline.
  2. Click the **Type** tab, and change from linear to nonlinear, polynomial, logarithmic, and so on.
  3. Click the **Options** tab. Display the equation or *R*-squared value if you forgot earlier.

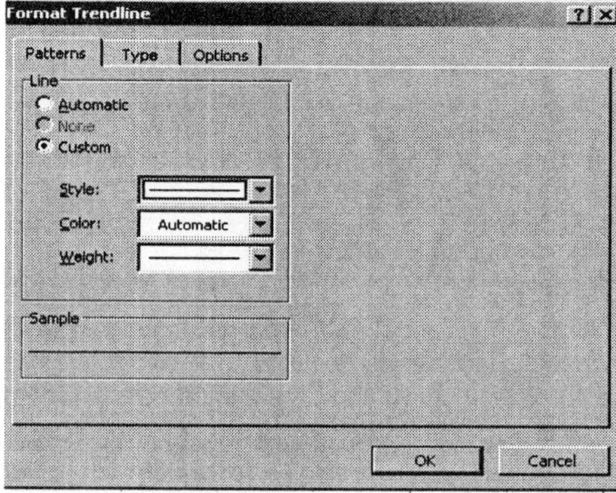

Figure 7-13. Trendline Dialog Box

- To edit the markers, right-click any marker to open the **Format Data Series** dialog box (Figure 2-10). When all the markers fall very nearly on the trendline, it may be difficult to select a marker and not the trendline. One solution is to edit the marker before adding the trendlline. Another is to delete the trendline, edit the markers, and then add the trendline. In any case, right-clicking a marker opens the **Format Data Series** dialog box (Figure 2-10) with several tabs:
  1. The **Patterns** tab shows some of the options for the patterns for the trendline and markers. The term *patterns* includes line style, line color, and line weight, and for markers it includes size (adjustable), shape, and color.
  2. Other tabs on this dialog box, less often used, are:
     a. **Data Labels**
     b. **Series Order**
     c. **Options**
     d. **Axis**
     e. **X Error Bars**
     f. **Y Error Bars**

Excel omits few options. If you have a graphics need, chances are that the designers of Excel have already thought of it. These Excel Examples have touched upon most of the common features. For graphics editing, it is useful to create some of the charts described so far, and then click and right-click and experiment and explore the nature of Excel's possibilities. When in doubt, save your work before clicking.

Discussion

Figure 7-11 displays a linear regression of the data in Table 7-3. The data pairs in the upper chart appear to fall closely on a linear regression line. However, the lower chart in Figure 7-11 shows a systematic deviation of the residuals ($X - X_{chlor}$) over the range of x variable (refractive index). The residuals show a pronounced maximum in the middle range. Both of these charts were produced by the **Regression** dialog box method. The tables output by the **Regression** dialog box method (shown in Figure 7-1) were omitted for the sake of clarity.

Figure 7-12 displays a second-order regression of the same data. Again the data pairs in the upper chart appear to fall closely on the second-order regression line. However the residuals are much smaller and more random than the residuals with the linear regression. The range of the residuals with the linear regression is about five times as large as with the second-order regression. Both charts in Figure 7-12 were made with the Chart Wizard and the **Add Trendline** command.

Excel Example 24

Statistics: Errors of Measurement

Review

An error in the measurement of a sample is the difference between the sample value and the true value so that

$$\varepsilon_i = x_i - \text{ true value} \qquad (7\text{-}11)$$

where ε_i is the error in the ith sample measurement and x_i is the ith observed sample value. What, however, is the true value of a physical measurement? The true value is never known; consequently the error of a measurement cannot be known. If all the errors in a set of sample measurements are random, then the sample mean is taken as the best estimate of the true value. If the errors are random, and an infinite number of samples are measured, then the true value does indeed equal the mean of the infinite set of samples in the absence of systematic errors. We shall distinguish between the real finite sample mean (m) and the mean of an infinitely large sample set:

$$m = \frac{\sum x_i}{n} = \text{mean of a finite sample set} \qquad (7\text{-}12)$$

$$\mu = \text{the mean of an infinitely large sample} \qquad (7\text{-}13)$$

The Excel **AVERAGE()** function returns the average, that is, m, of its argument, which can be a list of numbers, a range, or the name of a range.

The Normal Error Distribution

As the sample set becomes increasingly large, m becomes a better and better estimate of μ, the true value:

$$\lim_{n \to \infty} m = \mu \qquad (7\text{-}14)$$

With these concepts in mind, we can define the error as

$$\varepsilon_i = x_i - \mu \qquad (7\text{-}15)$$

If observed experimental results are obtained by independent measurements, the errors tend to be distributed with equal probability above and below the mean *m*, with a maximum probability at the mean. The Gaussian or normal error distribution is given by

$$P(\varepsilon) = \frac{1}{\sigma\sqrt{2\pi}} e^{-(x-\mu)^2/2\sigma^2} \tag{7-16}$$

where $P(\varepsilon)$ is the error probability function, σ is the standard deviation, ε is the error, π = 3.14159, and e = 2.71828. Excel's **NORMDIST (x, mean, standard_dev, cumulative)** function returns the normal distribution for specified x, μ, and ε. Equation 7-16 is normalized so that the probability that an error ε lies between $-\infty$ and $+\infty$ is one. Thus

$$P = \int_{-\infty}^{+\infty} P(\varepsilon)\, d\varepsilon = 1$$

where P is the probability that the error ε lies between $-\infty$ and $+\infty$. For any x, the error is the difference between the quantity measured, x, and the mean of an infinite number of measurements, μ (Equation 7-13). The standard deviation σ is defined by

$$\sigma = \sqrt{\frac{\sum (x_i - \mu)^2}{n}} = \sqrt{\frac{\sum \varepsilon_i^2}{n}} \tag{7-17}$$

where x_i is the value of the ith sample measurement, μ is the mean of the infinite sample set, and n is the number of observations.

The Standard Deviation

The sample standard deviation s is defined by

$$s = \sqrt{\frac{\sum d_i^2}{n-1}} = \sqrt{\frac{\sum (x_i - m)^2}{n-1}} \tag{7-18}$$

Excel's **STDEVP()** function returns the s of its argument, which can be a list of numbers, a range, or the name of a range. The variance is s^2, and the sample standard deviation, s, is the square root of the variance. The deviation from average d_i is given by

$$d_i = x_i - m \tag{7-19}$$

Significance of the Standard Deviation. The area under the normal error distribution curve (Equation 7-16) between a range of errors gives the probability that the error falls within this range (Figure 7-14). The area under the curve between $-\infty$ and $+\infty$

equals 1.00; that is, it is certain that the error must lie between $-\infty$ and $+\infty$. The probability P that an error lies between $+\sigma$ and $-\sigma$ is given by

$$P = \int_{-\sigma}^{+\sigma} P(\varepsilon)d\varepsilon = 0.683 \tag{7-20}$$

This means that there is a 68.3 percent chance that the error lies between $+\sigma$ and $-\sigma$. Similarly, the probability that an error lies between -2σ and $+2\sigma$ is 95.45 percent. In other words, the probability that a measurement will deviate from the average by more than two standard deviations is only 4.55 percent.

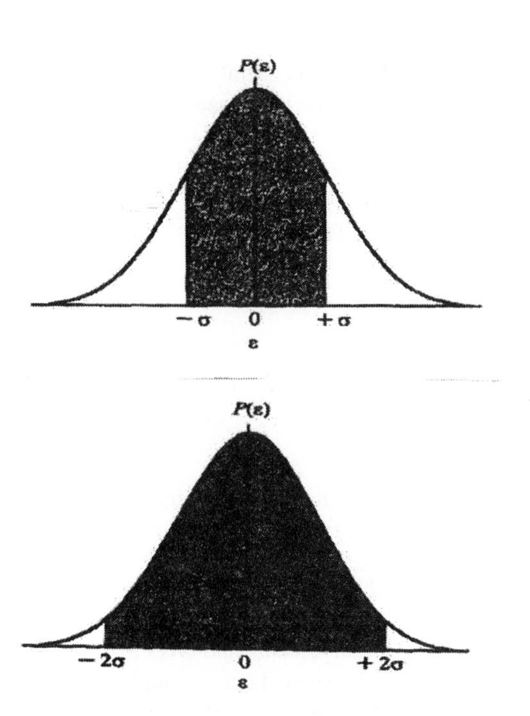

Figure 7-14. Upper, the Probability of an Error between $-\sigma$ and $+\sigma$, and Lower, between -2σ and $+2\sigma$

Confidence Levels and Confidence Factors. Let us rephrase the meaning of a standard deviation. From the preceding discussion it follows that we can be 68.3 percent confident that the error lies between true value plus one standard deviation and the true value minus one standard deviation. The confidence level in this example is 68.3 percent, and the confidence factor is one, that is, one standard deviation. If we choose a larger confidence factor, for example, two (standard deviations), we can be more confident that the true value lies within our reported confidence limit. The confidence levels for integral confidence factors are listed in Table 7-4. It is perhaps more usual to describe errors in terms of integral confidence levels, especially the 95 or 99 percent confidence level. These values are also shown in Table 7-4.

Table 7-4

Confidence Factors

Integral Confidence Factors

Confidence Factor	1σ	2σ	3σ	4σ
Confidence Level (%)	68.268	95.450	99.730	99.994

Integral Confidence Levels

Confidence Factor	0.674σ	1.28σ	1.96σ	2.58σ
Confidence Level (%)	50	80	95	99

These assignments of errors are valid when n, the number of observations, is infinite, the only errors are random, and the error distribution is Gaussian (normal). In principle, then, none of the confidence levels and confidence factors can be calculated, since σ is not known except for an infinitely large sample set. In practice, the situation is not so dismal because it is possible to calculate approximate values for the standard deviation σ of the infinite sample set.

The Standard Deviation of the Mean

In the previous paragraph we have interpreted s in terms of probabilities associated with a single measurement. Now suppose our sample size is finite, but large, say, between 20 or 200. It is an important concept of statistics that every finite set of measurements may be considered a sample from an infinite set of similar data. Up to this point we have discussed ways to estimate σ and μ, parameters of the infinite set. Now let us see how to handle large but finite sample sets.

Suppose we measure several finite sample sets of, say, 20 samples each. It turns out that the distribution of the *m*s of these sets also follows a normal distribution with its sample standard deviation, s_m, called the standard deviation of the mean. It can be shown (Baird 1962, p. 64; Young 1962, p. 95) that

$$s_m = \frac{s}{\sqrt{n}} \tag{7-21}$$

where n is the number of samples, s is the standard deviation as calculated with Equation 7-18, and s_m is the standard deviation of the mean as calculated with Equation 7-21.

It makes sense that the standard deviation of the mean s_m is less than the sample standard deviation s. This is consistent with our intuitive idea that the more measurements we make the more confidence we can have in the results. Notice that the number of samples enters as the square root. If you make four measurements and wish to decrease s_m by one half, you would be required to make 16 measurements. Thus, a couple of extra runs will have little effect on the experimental precision (as measured by s_m). Your time would be spent more advantageously improving the design of the experiment.

Sample Calculations with Data from Excel Example 1. In Excel Example 1 we calculated the standard deviation of some measurements of the lattice dimension of a cubic crystal lattice. The results are shown in Figure 1-4. Here, several of Excel's statistical function were used: **AVERAGE()**, **STDEV()**, **MAXIMUM()**, and **MINIMUM()**. From the standard deviation s already calculated, we can calculate the standard deviation of the mean s_m with Equation 7-21:

$$s_m = \frac{s}{\sqrt{n}} = \frac{0.0068}{\sqrt{6}} = 0.0028 \tag{7-22}$$

We would then report our measurements: $a_0 = 3.604(28)$ angstroms.

Confidence Limits. Instead of reporting s_m as a measure of precision, the value of s_m times a factor required to give a desired confidence level can be reported. The value of s_m times a factor is called the confidence limit, and is given the symbol λ.

$$\lambda_{95} = F_C \times s_m \tag{7-23}$$

Thus in the previous example, we reported $a_0 \pm s_m$. If we wish to report $a_0 \pm \lambda_{95}$, we calculate λ_{95} using the confidence factors F_C listed in Table 7-5. For the crystallographic data, let us calculate λ_{95}:

$$\lambda_{95} = F_c \times s_m = 1.96 \times 0.0028 = 0.0055$$

and we report $a_0 = 3.604 \pm 0.0055$, where 0.0055 equals λ_{95}, the 95 percent confidence limit. This might also be written as $a_0 = 3.604(55)$, where (55) refers to the uncertainty in the last digits. One could argue that two significant figures in the confidence limit is meaningless, so that what really should be reported is 3.604(6). However, it is customary to retain one extra significant figure.

Table 7-5

Confidence Factors

F_C	%
1.00	68.268
1.28	80.000
1.96	95.000
2.00	95.450
2.58	99.000
3.00	99.730
3.29	99.900
4.00	99.994

In any case, it is important to explain F_c or whatever measure of precision is being reported. If it is the sample standard deviation, s_m, then that should be stated. If it is the 95 percent confidence limit, that fact must be explicitly stated by the writer or experimenter.

Excel's TINV() Statistical Function: Student's *t*-Distribution

For a small sample, say fewer than 20 or so, Student's *t*-distribution provides a better description of the errors in a series of measurements than the previously described confidence factors.

Scientists and engineers constantly face the problem of evaluating their work and communicating some kind of measure of the precision of the average value of the measurements that they make. Although the average deviation $\overline{d}$, the standard deviation s, or the standard deviation of the mean s_m is frequently used, it is often desirable to use confidence limits λ_P calculated from Student's t factor

$$\lambda_P = ts_m = t\frac{s}{\sqrt{n}} \tag{7-24}$$

where n is the number of measurements; t is a factor that lies between about 2 and 3 for much experimental work. Equation 7-24 is similar to Equation 7-23, with t replacing the confidence factor F_C. For large sample sets, F_C is applicable, but Student's t is better for a small sample set. Excel's **TINV**() statistical function makes it easy to prepare a table of critical values of t (Figure 7-15).

Each row in the table, shown in Figure 7-15, displays values of t for various confidence limits and a given number n of experiment trials. The number of degrees of freedom equal $n - 1$.

	A	B	C	D	E	F	G	H
1				**Table of Critical Values of t**				
2								
3				**Confidence Limits (%)**				
4								
5	**n - 1**	**60.0**	**70.0**	**80.0**	**90.0**	**95.0**	**99.0**	**99.9**
6	**0**							
7	**1**	1.376	1.963	3.078	6.314	12.706	63.656	636.578
8	**2**	1.061	1.386	1.886	2.920	4.303	9.925	31.600
9	**3**	0.978	1.250	1.638	2.353	3.182	5.841	12.924
10	**4**	0.941	1.190	1.533	2.132	2.776	4.604	8.610
11	**5**	0.920	1.156	1.476	2.015	2.571	4.032	6.869
12	**6**	0.906	1.134	1.440	1.943	2.447	3.707	5.959
13	**7**	0.896	1.119	1.415	1.895	2.365	3.499	5.408
14	**8**	0.889	1.108	1.397	1.860	2.306	3.355	5.041
15	**9**	0.883	1.100	1.383	1.833	2.262	3.250	4.781
16	**10**	0.879	1.093	1.372	1.812	2.228	3.169	4.587
17	**11**	0.876	1.088	1.363	1.796	2.201	3.106	4.437
18	**12**	0.873	1.083	1.356	1.782	2.179	3.055	4.318
19	**13**	0.870	1.079	1.350	1.771	2.160	3.012	4.221
20	**14**	0.868	1.076	1.345	1.761	2.145	2.977	4.140
21	**15**	0.866	1.074	1.341	1.753	2.131	2.947	4.073
22	**16**	0.865	1.071	1.337	1.746	2.120	2.921	4.015
23	**17**	0.863	1.069	1.333	1.740	2.110	2.898	3.965
24	**18**	0.862	1.067	1.330	1.734	2.101	2.878	3.922
25	**19**	0.861	1.066	1.328	1.729	2.093	2.861	3.883
26	**20**	0.860	1.064	1.325	1.725	2.086	2.845	3.850
27	**21**	0.859	1.063	1.323	1.721	2.080	2.831	3.819
28	**22**	0.858	1.061	1.321	1.717	2.074	2.819	3.792
29	**23**	0.858	1.060	1.319	1.714	2.069	2.807	3.768
30	**24**	0.857	1.059	1.318	1.711	2.064	2.797	3.745
31	**25**	0.856	1.058	1.316	1.708	2.060	2.787	3.725
32	**26**	0.856	1.058	1.315	1.706	2.056	2.779	3.707
33	**27**	0.855	1.057	1.314	1.703	2.052	2.771	3.689
34	**28**	0.855	1.056	1.313	1.701	2.048	2.763	3.674
35	**29**	0.854	1.055	1.311	1.699	2.045	2.756	3.660
36	**30**	0.854	1.055	1.310	1.697	2.042	2.750	3.646
37	**40**	0.851	1.050	1.303	1.684	2.021	2.704	3.551
38	**50**	0.849	1.047	1.299	1.676	2.009	2.678	3.496
39	**60**	0.848	1.045	1.296	1.671	2.000	2.660	3.460
40	**70**	0.847	1.044	1.294	1.667	1.994	2.648	3.435
41	**120**	0.845	1.041	1.289	1.658	1.980	2.617	3.373
42	**10000**	0.842	1.036	1.282	1.645	1.960	2.576	3.291

Figure 7-15. The Worksheet for Student's *t*-Distribution

Creating a Table of Critical Values of *t*

With Excel's **TINV()** statistical function, we can prepare an Excel worksheet (Figure 7-15) with a table of critical values of *t*.

Entering the Labels

- Open a new workbook with Sheet1 the active sheet.
- Enter the title and subtitle shown in Figure 7-15 in rows 1 and 3.
- In cell A5, type **n-1** as the label.
- In the range B5:BH5, enter the confidence limits, *P*, in percent.

Filling the Table with Degrees of Freedom. The Fill-Handle method is appropriate for filling the required range with these values.

- In cell A6, type **0** as the value.
- In cell A7, type **=A6 +1** as the formula, which causes a 1 to appear in cell A7.
- Click the fill-handle in cell A7 and drag until the value **30** is reached.
- In cell A37, type **=A36 + 10** as the formula, and drag until the value **70** is reached.
- In the next cell (A41), type **120** as the value, and in the next cell (A42), type **10000** as the value.

Filling in the Student's *t* Values. Excel's **TINV(p,f)** function returns the critical value of *t* for the desired probability, *p*, and degrees of freedom, *f*, which equals $n - 1$ where *n* is the number of measurements. Since Excel's **TINV(p,f)** function returns the inverse of the Student's *t*-distribution for the specified degrees of freedom, its probability argument is given by $p = 1 - P/100$, where *P* is the probability (confidence limit in percent) listed in Figure 7-15.

- In cell B7, type = **TINV(0.4, A7)** as the formula.
  Note: $p = 1 - P/100 = 1 - 60/100 = 0.4$.
- In cell C7, type = **TINV(0.3,A7)** as the formula.
- Continue entering the function in the range D7:H7. In H7, type **=TINV(0.001,A7)** as the function.
- Select cell B7, and drag the fill handle to cell B42.
- Similarly, select cells C7, D7, E7, F7, G7, and H7, and drag the fill handles to row 42.
- Select the block E7:H26 for shading, and on the **Format** menu, click **Cells** to open the **Format Cells** dialog box (Figure 1-5)
- Click the **Patterns** tab, then click the down arrow in the **Patterns** box, and select **12.5 % gray** (the second choice from the top right).

Using the Student's *t* Value. In Excel Exercise 1, *n* equals 6, so *f* equals 5. Now for each probability, let us tabulate *t* from $n - 1$ equals 1 down to 10000 at the selected increments. The unit cell dimension a_0 calculated from experimentally measured diffraction angles of X-rays is 3.604 angstroms, the standard deviation *s* is 0.0068, and the sample standard deviation s_m equals 00.55. The number of data *n* equals 6, so the number of degrees of freedom $(n - 1)$ equals 5. Let us calculate confidence limits for 95 percent confidence (λ_{95}). That means the value we report for a_0 lies within the confidence limits with 95 percent certainty. From Figure 7-15, the critical value of *t* for 95 percent confidence and 5 degrees of freedom equals 2.571

$$\lambda_P = t_P s_m = t_P \frac{s}{\sqrt{n}} = 2.571 \frac{0.0068}{\sqrt{6}} = 0.014 \qquad (7\text{-}25)$$

The experimenter could report with 95 percent certainty that $a_0 = 3.604\ \text{Å} \pm 0.014$.

The gray area in Figure 7-15 indicates the most commonly used t values. For most repetitive experiments, the maximum n is probably fewer than 10 or 20. The desired confidence interval probably ranges from 90 to 99.99 with 95 and 99 the most commonly reported. It is the obligation of the experimenter not only to report the confidence interval but also to report the percent confidence interval.

Equation 7-25 reveals that the reported confidence interval decreases as the number of repetitive experiments increases. That is not surprising, as you would expect that the more often you repeat an experiment, the more reliable the average result would be. Notice, however, that the decrease is slow because of the square root dependence. You need to quadruple the number of repetitions to halve the confidence interval. In the diffraction experiment, that means that 24 diffraction angles would need to be measured to decrease λ_{95} from 0.014 to 0.0007.

It is also clear from Figure 7-15 that the critical values of t increase as the confidence limit is increased. The increase is quite rapid for small sample sizes and large confidence limits. This means that for a given data set, the confidence interval is larger for higher confidence intervals.

The t-distribution was first published by William Sealy Gosset writing under the pseudonym Student. The distribution was and is referred to as Student's t-distribution. Gosset, an employee of the Guinness brewery in Dublin in the early 1900s, was interested in the quality control of beer.

Chapter 8

Three-Dimensional Plots

Excel Example 25

Connected Points in Space

Two-dimensional point plots were described in Excel Examples 3 and 4. Example 25 furnishes an example of a three-dimensional (3-D) point plot. The points for a 3-D point plot could be experimental points z_i (a dependent variable) and x_i, y_i (independent variables). The points in Excel Example 25 are, however, fictitious.

Entering the Worksheet Title and Data

In this section we shall create on the worksheet an *x,y* coordinate system, with *x* values ranging from *x* equals 0 to *x* equals 14, and with *y* values ranging from *y* equals 0 to *y* equals 14. In the cells at the intersection of the *x* and *y* values, we shall enter some *z* values that represent the distance of *z* above the XY plane at that *x,y* value.

- Open a new workbook with Sheet1 the active sheet.
- In cell I2, type **Y Values** as the label and make it bold.
- In cell A10, type **X** as the label and make it bold.
- In cell C3, type **0.00** as the value.
- In cell B4, type **0.00** as the value.

Next, use the Edit Fill Series method to fill in the *x* coordinates in the range B4:B18 and the *y* coordinates in the range C3:Q3.

- Select cell B4; then, on the **Edit** menu, point to **Fill**, and click **Series**.
- In the **Series** dialog box, for **Series in**, select **Columns**; for **Type**, select **Linear**; in the **Step value** box, type **1** for the step value; and in the **Stop value** box, type **14** for the stop value. Then Click **OK**.
- Select cell C3; then, on the **Edit** menu, point to **Fill**, and click **Series**.
- In the **Series** dialog box, for **Series in**, select **Rows**; for **Type**, select **Linear**; in the **Step value** box, type **1** for the step value; and in the **Stop value** box, type **14** for the stop value. Then Click **OK**.

The range B4:B18 fills with the *x* values from 0 to 14 in increments of 1, and the range C3:Q3 fills with the *y* values from 0 to 14 in increments of 1.

Enter the values of *z*, shown in Figure 8-1 in the range C4:Q18. To ease the task, first fill in the range C4:J18. Then copy and paste range I4:I18 to the range K4:K18, H4:H18 to L4:L18, and copy the remaining columns similarly.

	A	B	C	D	E	F	G	H	I	J	K	L	M	N	O	P	Q
1																	
2									Y Values								
3			0	1	2	3	4	5	6	7	8	9	10	11	12	13	14
4		0	0	0	0	0	0	0	0	0	0	0	0	0	0	0	0
5		1	0	10	10	10	10	10	10	20	10	10	10	10	10	10	0
6		2	0	10	20	20	20	20	20	40	20	20	20	20	20	10	0
7		3	0	10	20	30	30	30	40	60	40	30	30	30	20	10	0
8		4	0	10	20	30	40	40	60	80	60	40	40	30	20	10	0
9		5	0	10	20	30	40	50	60	80	60	50	40	30	20	10	0
10	X	6	0	10	20	30	40	60	80	80	80	60	40	30	20	10	0
11		7	0	10	20	30	40	60	80	100	80	60	40	30	20	10	0
12		8	0	10	20	30	40	60	80	80	80	60	40	30	20	10	0
13		9	0	10	20	30	40	50	60	80	60	50	40	30	20	10	0
14		10	0	10	20	30	40	40	60	80	60	40	40	30	20	10	0
15		11	0	10	20	30	30	30	40	60	40	30	30	30	20	10	0
16		12	0	10	20	20	20	20	20	40	20	20	20	20	20	10	0
17		13	0	10	10	10	10	10	10	20	10	10	10	10	10	10	0
18		14	0	0	0	0	0	0	0	0	0	0	0	0	0	0	0

A Tower over the XY Plane

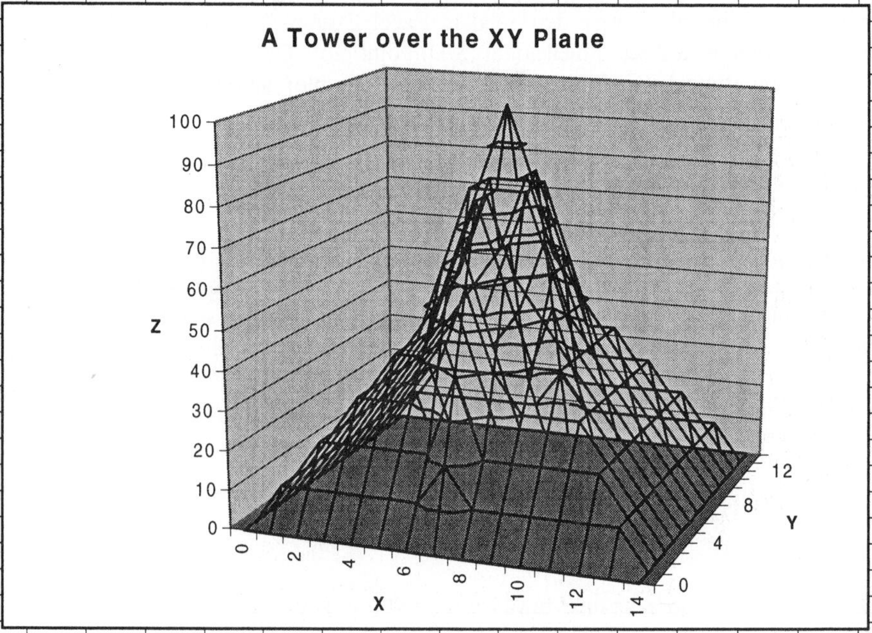

Figure 8-1. Worksheet for a 3-D Point Chart

Creating the 3-D Chart

The data have now been entered so that a three-dimensional chart can be created. The Chart Wizard is employed as usual.

- Select the range C4:Q18.
- Click the **Chart Wizard** icon.

1. In the **Chart Wizard – Step 1 of 4 – Chart Type** dialog box, select **Surface** and the icon for wire frame 3-D surface chart without color; click **Next**.

2. In the **Chart Wizard – Step 2 of 4 – Chart Source Data** dialog box, click **Next.**
3. In the **Chart Wizard – Step 3 of 4 – Chart Options** dialog box:
   a. In the **Chart title** box, type **A Tower over the XY Plane** as the title.
   b. In the **(X) axis** box, type **X** as the label; in the **(Y) axis** box, type **Y** as the label; in the **(Z) axis** box, type **Z** as the label; click **Next.**
4. In the **Chart Wizard – Step 4 of 4 – Chart Options** dialog box, select **As object in Sheet1**; click **Finish.**

- Select the legend box on the right side of the chart, and press the DELETE key.
- Right-click the title, and change the font to 12 point.

The chart appears as shown in Figure 8-1. It needs little or no editing. To change the orientation of the chart, click and drag a corner of the plot area. To change the view of the 3-D chart, click in the chart to select it, and then on the **Chart** menu, click **3-D View** to open the **3-D View** dialog box (Figure 8-2). The **3-D View** dialog box provides a number of options.

Plotting a 3-D wave function is similar to building the tower. Instead of filling the z values in the worksheet cells by hand, the z values are just the value of the wave function at each cell whose coordinates are x and y.

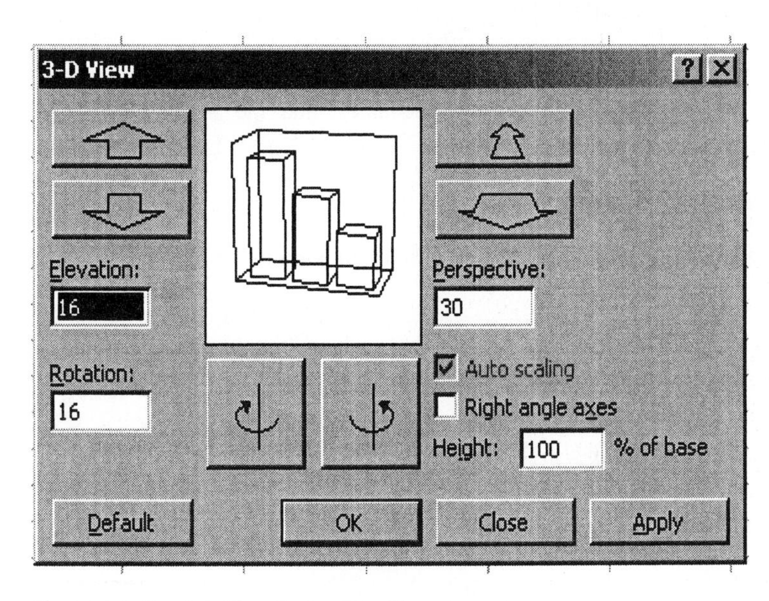

Figure 8-2. The **3-D View** Dialog Box. The programmer can change the orientation of the 3-D plot.

Excel Example 26

Three-Dimensional Charts: Wave Functions

In Excel Example 25, it is demonstrated that Excel can produce quite acceptable three-dimensional charts. In this example, we shall use Excel to produce a three-dimensional chart of a wave function. Any function of the form $z = f(x,y)$ would suffice, but it is of special interest to create a 3-D chart of the wave function, for example, for a particle in a two-dimensional box. Plotting a 3-D wave function is similar to building the tower. Instead of filling the z values in the worksheet cells by hand, the z values are just the value of the wave function at points in the XY plane, whose coordinates are x and y.

Review of Quantum Mechanics

The simplicity of the particle-in-a-box problem offers a low-key opportunity to review operator equations, the postulates of quantum mechanics, and the Schrödinger wave equation and its application to a simple system.

Operator Equations. The general form of an operator equation is

$$\hat{a}\psi = a\psi \tag{8-1}$$

This equation can be read as follows, from left to right: If ψ is a solution to this operator equation, then operation with the operator $\hat{a}$ on ψ (an *eigenfunction*) gives back the same function ψ, multiplied by a constant a (an *eigenvalue*).

Example 1. Given the operator $\hat{a} = \dfrac{\partial}{\partial x}$, is $\psi_1 = kx^2$ a solution to the operator equation, Equation 8-1?

$$\frac{\partial}{\partial x}(kx^2) = 2kx \tag{8-2}$$

Since $x \neq x^2$, ψ_1 is not a solution (eigenfunction) of Equation 8-1. Operation with the operator does not give back the function multiplied by a constant.

Example 2. Given the same operator $\hat{a} = \dfrac{\partial}{\partial x}$, is $\psi_2 = e^{kx}$ a solution to the operator equation, Equation 8-1?

$$\frac{\partial}{\partial x}(e^{kx}) = ke^{kx} \tag{8-3}$$

Since ke^{kx} equals ψ_2 multiplied by a constant, ψ_2 is a solution (eigenfunction) of Equation 8-1. Operation with the operator does give back the function multiplied by a constant.

Example 3. Given the same operator $\hat{a} = \frac{\partial}{\partial x}$, is $\psi_3 = e^{kx^2}$ a solution to the operator equation, Equation 8-1?

$$\frac{\partial}{\partial x}(e^{kx^2}) = 2kxe^{kx^2} \tag{8-4}$$

Since $xe^{kx} \neq e^{kx}$, ψ_3 is not a solution (eigenfunction) of Equation 8-1. Operation with the operator does not give back the function multiplied by a constant.

The Postulates of Quantum Mechanics. The postulates of quantum mechanics provide the guidelines, rules, and meaning of the Schrödinger wave equation, which is also an operator equation, similar to Equation 8-1. The postulates show how to write the Schrödinger wave equation for a system of interest and give guidelines for finding a solution (eigenfunction) to the Schrödinger wave equation. The physical systems to which the Schrödinger wave equation applies usually consist of a particle or a few particles, such as electrons and protons.

Postulate 1 tells how the state of a system is described. The physical state of a system is described by a function of the spatial coordinates of the particles that make up the system.

a. The function is usually denoted by the symbol ψ, or less commonly by $\psi(x, y, z, t)$. The function ψ is called a wave function or an eigenfunction, and its complex conjugate is ψ *.

b. The probability of finding the particle between x and $x + dx$ is proportional to $\psi^* \psi$, the probability density. Since the probability of finding the particle anywhere on the x-axis is unity

$$\int_{-\infty}^{+\infty} \psi^* \psi dx = 1 \tag{8-5}$$

if the wave function is normalized.

c. If the system is independent of time, the state of the system is called a stationary state.

d. The average value or expectation value $<a>$ of an observable a is given by

$$< a > = \int_{-\infty}^{+\infty} \psi^* \hat{a} \psi d\tau \qquad (8\text{-}6)$$

Postulate 2 introduces the Schrödinger wave equation. The wave functions ψ are obtained by solving the Schrödinger wave equation, an operator equation of the form of Equation 8-1. It is written

$$\hat{H}\psi = E\psi \qquad (8\text{-}7)$$

The symbol $\hat{H}$ represents the Hamiltonian operator, ψ is a wave function, and E is a constant (an observable, in this case, the energy).

Postulate 3 tells how to construct the Hamiltonian operator. Quantum mechanical operators for *observables* (such as energy, position, momentum) are obtained from the classical expression for the observable with the use of certain established substitutions. An abbreviated list of these substitutions is given in Table 8-1. This table is itself an integral part of Postulate 3.

The Particle in a Two-dimensional Box: ψ_{22} and ψ_{22}^2

The wave function for the particle in a two-dimensional (2-D) box problem provides a suitable example of a three-dimensional surface plot.

The System. First, let us review the model for the particle in a box in *one* dimension. In a one-dimensional system, the particle is constrained to a line with coordinates that range from $x = 0$ to $x = a$. This means that the potential energy $V(x) = 0$ for $0 \leq x \leq a$ and $V(x) = \infty$ for any other value of x.

$$E = T + V(x) \qquad (8\text{-}8)$$

$$E = \frac{p_x^2}{2m} + V(x) \qquad (8\text{-}9)$$

Constructing the Hamiltonian Operator. The classical total energy E of a particle in one dimension is the sum of the kinetic and potential energies. According to Postulate 3, Table 8-1, the Hamiltonian operator for the Schrödinger wave equation, $H\psi = E\psi$ (Equation 8-7), is of the form

$$\hat{H} = \frac{-\hbar^2}{2m}\frac{\partial^2}{\partial x^2} \qquad (8\text{-}10)$$

Table 8-1

Classical Observables and Their Corresponding Quantum Mechanical Operators

Observables		Corresponding Operators	
Name	**Symbol**	**Symbol**	**Operation**
Position	x	$\hat{x}$	Multiply by x
Position squared	x^2	$\hat{x}^2$	Multiply by x^2
Momentum	p_x	$\hat{p}_x$	Operate with $\frac{\hbar}{i}\frac{\partial}{\partial x}$
Momentum squared	p_x^2	$\hat{p}_x^2$	Operate with $\hbar^2 \frac{\partial^2}{\partial x^2}$
Potential energy	$V(x)$	$\hat{V}(x)$	Multiply by $V(x)$
Kinetic energy	$\frac{p_x^2}{2m}$	$\hat{T}_x$	Operate with $\frac{-\hbar^2}{2m}\frac{\partial^2}{\partial x^2}$
Total energy	$\frac{p_x^2}{2m} + V(x)$	$\hat{H}$	Operate with $\frac{-\hbar^2}{2m}\frac{\partial^2}{\partial x^2} + V(x)$

The symbol $\hbar$ is often used to represent $h/2\pi$.

Notice that the potential energy $V(x)$ of the particle in a box equals zero. With this operator, the Schrödinger wave equation (Equation 8-7) becomes

$$\frac{-\hbar^2}{2m}\frac{\partial^2 \psi}{\partial x^2} = E\psi \tag{8-11}$$

or

$$\frac{\partial^2 \psi}{\partial x^2} = -\frac{2mE}{\hbar^2}\psi \tag{8-12}$$

Notice the similarity of this operator equation and Equation 8-1. To find a solution to this operator equation is to find a function ψ. For an operator equation like Equation 8-1, ψ is a function, which when operated on twice gives back itself times a constant.

Solving the Schrödinger Wave Equation. We seek a solution ψ to this equation, which

1. reproduces itself (times a constant) when differentiated twice,
2. and makes sense according to the previously described model.

Intuitively (meaning we try a few common functions we have had experience with), we see that $\psi = \sin kx$, $\psi = \cos kx$, and $\psi = e^{kx}$ fit criterion number one. Operate on these functions with $\partial^2 / \partial x^2$ to verify this assertion.

But only sin kx makes sense. When $x = 0$, cos kx and e^{kx} both equal one. When $x = 0$, sin $kx = 0$, which it should at the end of the "box." Why must ψ equal zero when x equals zero? Because at x less than zero, ψ must be zero because the probability of the particle being outside the box is zero: our model puts it in the box! The same argument must be made at the other end of the box, where x equals a. At x greater than a, ψ must be zero because the particle must be inside the box. In order that sin $kx = 0$ when $x = a$, it must be true that

$$ka = \pi, 2\pi, \ldots n\pi \tag{8-13}$$

$$k = \frac{n\pi}{a} \tag{8-14}$$

$$\psi = A \sin \frac{n\pi x}{a} \tag{8-15}$$

Here, A is the proportionality constant to be evaluated shortly. Double differentiation of this wave function shows that it is an eigenfunction of the operator equation if

$$k^2 = \frac{2mE}{\hbar^2} \tag{8-16}$$

Since $ka = n\pi$ (Equation 8-14) and $\hbar = h / 2\pi$, then E from Equation 8-16 is

$$E = \frac{n^2 h^2}{8ma^2} \tag{8-17}$$

where $n = 1, 2, 3, \ldots$. The value of $n = 0$ is not allowed, because then ψ would equal zero everywhere in the box, the probability would equal zero everywhere in the box, and the particle would be nowhere in the box. But that is not in accord with our model: the particle is in the box.

Normalizing the Wave Function. The constant A is the normalization factor, required so that

$$\int_0^a \psi^2 dx = 1 \qquad (8\text{-}18)$$

This equation gives the probability of finding the particle between $x = 0$ and $x = a$. This probability must equal 1 (corresponding to certainty), since the model requires the particle to be between $x = 0$ and $x = a$. By substituting for ψ and getting the equation in the form of sin u du, we have

$$\frac{aA^2}{n\pi}\int_0^a \sin^2\left(\frac{n\pi x}{a}\right) d\left(\frac{n\pi x}{a}\right) = 1 \qquad (8\text{-}19)$$

The integral of sin u du is in any table of integrals. After integrating and simplifying

$$A = \left(\frac{2}{a}\right)^{1/2} \qquad (8\text{-}20)$$

and so the normalized wave function is

$$\psi = \left(\frac{2}{a}\right)^{1/2} \sin\frac{n\pi x}{a} \qquad (8\text{-}21)$$

This problem may be extended to two or three dimensions. For the three-dimensional case in a cubic box, the energies and wave functions are

$$E = (n_x^2 + n_y^2 + n_z^2)\frac{h}{8ma^2} \qquad (8\text{-}22)$$

$$\psi(x, y, z) = \left(\frac{2}{a}\right)^{3/2} \sin\frac{n_x\pi x}{a}\sin\frac{n_y\pi y}{a}\sin\frac{n_z\pi z}{a} \qquad (8\text{-}23)$$

A surface plot of a three-dimensional wave function requires a four-dimensional plot, so we will settle for two-dimensional plots of ψ_{22} and ψ_{22}^2, for which both n_x and $n_y = 2$. Figure 8-4 and Figure 8-5 show these two plots.

Entering the Worksheet Title and Data

In the previous Excel Example we entered a numerical value at each x,y point in the grid of x,y values. In this section we shall enter the formulas for the ψ_{22}^2 wave function at each *x,y* point and use the values calculated from the formulas to create the chart of a wave function.

- Open a new workbook with Sheet1 the active sheet.
- Leave row 1 and row 2 blank for the later entry of a title.
- In cell B3, type **0.00** as the value.
- In cell A4, type **0.00** as the value.
- In cell B4, type **=(SIN(2*PI()*A4)*SIN(2*PI()*$B$3))^2** as the formula.

The wave function is calculated from *x* equals 0.00 to 1.00 in increments of 0.02. Similarly in the *y* direction, the wave function is calculated from *y* equal 0.00 to 1.00 in increments of 0.02. The Excel Edit Fill Series method is convenient for filling row 3 and column A with these values.

- Select cell A4; then, on the **Edit** menu, point to **Fill**, and click **Series**.
- In the **Series** dialog box, for **Series in**, select **Columns**; for **Type**, select **Linear**; in the **Step value** box, type **0.02** for the step value; and in the **Stop value** box type **1.00** as the stop value. Then Click **OK**.
- Select Cell B3; then, on the **Edit** menu, point to **Fill**, and click **Series**.
- In the **Series** dialog box, for **Series in**, select **Rows**; for **Type**, select **Linear**; in the **Step value** box, type **0.02** for the step value; and in the **Stop value** box, type **1.00** as the stop value. Then Click **OK**.

The range A4:A54 fills with the x values from 0 to 1 in 0.02 increments, and the range B3:AZ fills with the y values from 0 to 1 in 0.02 increments. To fix the decimal points:

- Click in the blank square at the intersection of the column labels (A, B, C ...) and the row labels (1, 2, 3 ...).
- On the **Format** menu, click **Cells**; click the **Number** tab, and in the **Decimal places** box, type **3** for the number of decimal places; click **OK**.

Next, fill in the range B4:B54 with the formulas for the wave function at *y* equals 0.02 and *x* ranging from 0 to 1 in 0.02 increments.

- Select any cell in the middle of the screen.
- On the **Window** menu, click **Split**.
- Click in cell B4, and then SHIFT-click in cell B54 to select the range B4:B54.
- On the **Edit** menu, point to **Fill**, and click **Down**.

Column B displays the values of the wave function formulas. Next we want to use the Edit Fill Right method to fill the cell across the rows. To do this we need to make some changes in the formulas in column B. Click in cell B4 and notice that the displayed formula now has an A4 and a $B$3. Before filling to the right, we must change these to $A$4 and B3, respectively, in cell B4 and in all the cells below. This is conveniently accomplished with Excel's find and replace capability.

- On the **Edit** menu, click **Find**, and type **A** as the value to find.
- Click the **Replace** tab, and type **A** as the value to replace with; then click **Replace All**.
- On the **Edit** menu, click **Find**, and type **B** as the value to find.
- Click the **Replace** tab, and type **B** as the value to replace with; then click **Replace All**.

The next step is to select the range B4:AZ54 and to complete filling in the remaining cells in the range B4:AZ54 with their formulas to calculate the *z* (psi squared) values at each *x,y* point.

- Select any cell in the center of the worksheet, and on the **Window** menu, click **Split**.
- Scroll the bottom right window to cell AZ54.
- Scroll the top left window to cell B4.
- Click in cell B4; then SHIFT-click in cell AZ54 to select the range B4:AZ54.
- On the **Edit** menu, point to **Fill**, and click **Right**. The cells are now filled with the wave function formulas, and their values are displayed.
- Click anywhere in the sheet to deselect the cells.
- On the **Window** menu, click **Remove Split.**

The upper left corner of the worksheet now appears as shown in Figure 8-3. In this figure, the cell E7 has been selected, revealing that the value of the wave function (0.018) at x equals 0.060 and at y equals 0.060. The wave function that calculates the value of 0.018 at (0.060, 0.060) is shown in the formula toolbar: =(SIN(2*PI()*$A$7)*SIN(2*PI()*E3))^2.

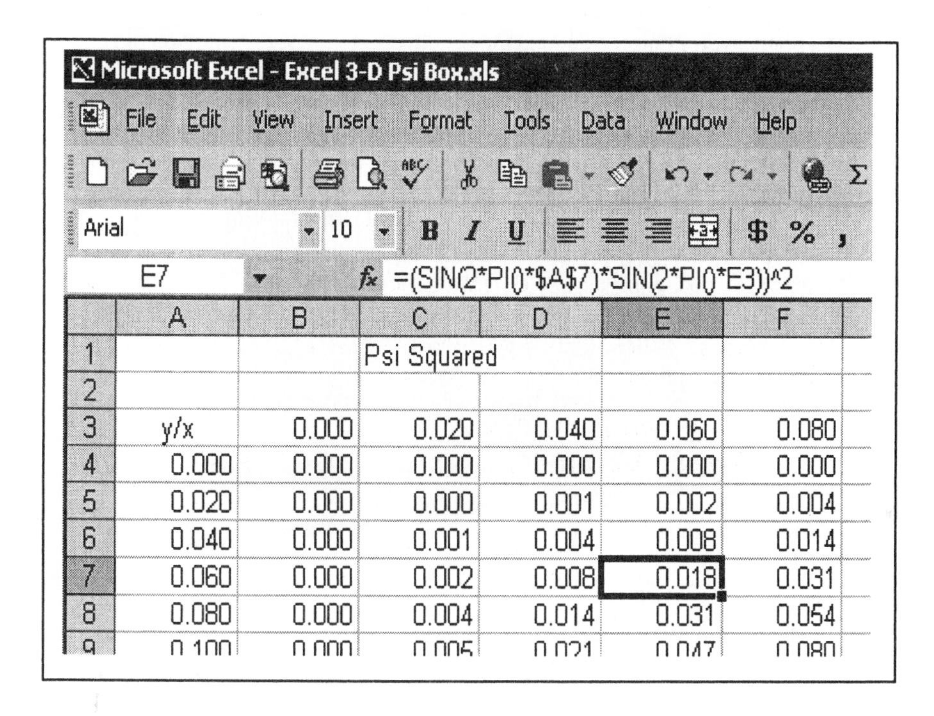

Figure 8-3. Part of the Worksheet with the Formula Toolbar Displaying the Formula for ψ_{22}

Creating the 3-D Chart

The data have now been entered so that a three-dimensional chart can be created. The Chart Wizard is employed as usual.

- Place the cursor anywhere in the middle of the sheet, and on the **Window** menu, click **Split**.
- Click in cell A3, and then SHIFT-click in cell AZ54 to select all the cells in the range A3:AZ54.
- Click the **Chart Wizard** icon.

1. In the **Chart Wizard – Step 1 of 4 – Chart Type** dialog box, select **Surface** and the icon for surface chart with color; click **Next**.

2. In the **Chart Wizard – Step 2 of 4 – Chart Source Data** dialog box, for **Series in**, select **Rows**; click **Next**.
3. In the **Chart Wizard – Step 3 of 4 – Chart Options** dialog box:
   a. In the **Chart title** box, **Psi Squared** as the title.
   b. In the **(X) axis** box, type **X** as the label; in the **(Y) axis** box, type **Y** as the label; in the **(Z) axis** box, type **Psi ^2** as the label; click **Next**.
4. In the **Chart Wizard – Step 4 of 4 – Chart Options** dialog box, select **As object in Sheet1**; click **Finish**.

- Select the legend box on the right side of the chart, and press the DELETE key.

The chart appears as shown in Figure 8-4. It needs little or no editing. This is a 3-D chart of Psi_{22}^2.

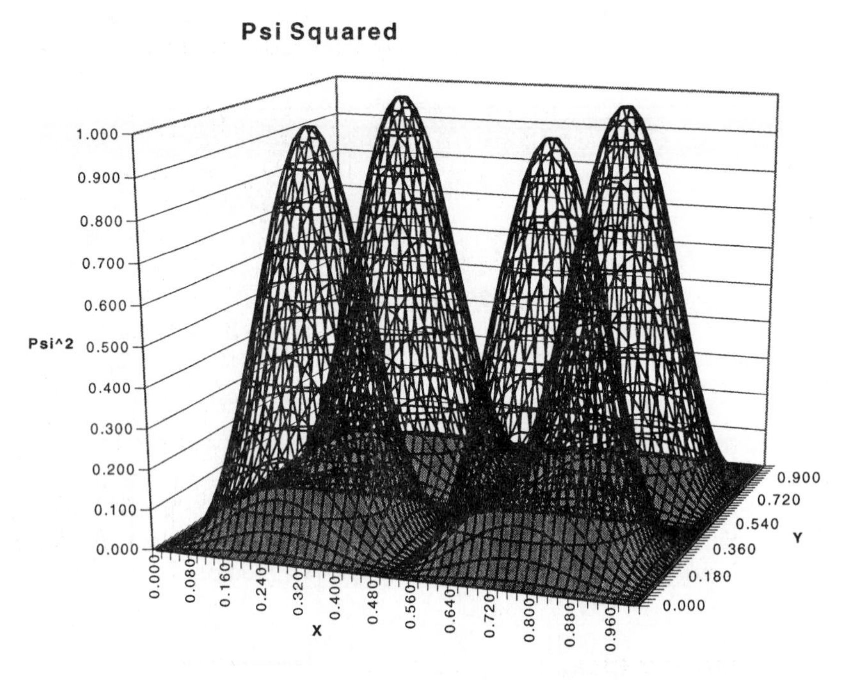

Figure 8-4. The ψ_{22}^2 Wave Function for a Particle in a Box. All four lobes are plus.

To plot psi instead of psi squared (Figure 8-5):

- On the **Edit** menu, click **Find**, and type **^2** as the value to find.
- Replace with nothing.
- Edit the title: change **Psi Squared** to **Psi**.

To replot psi squared (Figure 8-4):

- On the **Edit** menu, click **Find**, and type)) as the value to find.
- Type))**^2** as the value to replace with, and click **Replace All**.
- Edit the title: change **Psi Squared** to **Psi**.
- Edit the **(X) axis** box**:** change **Psi^2** to **Psi**

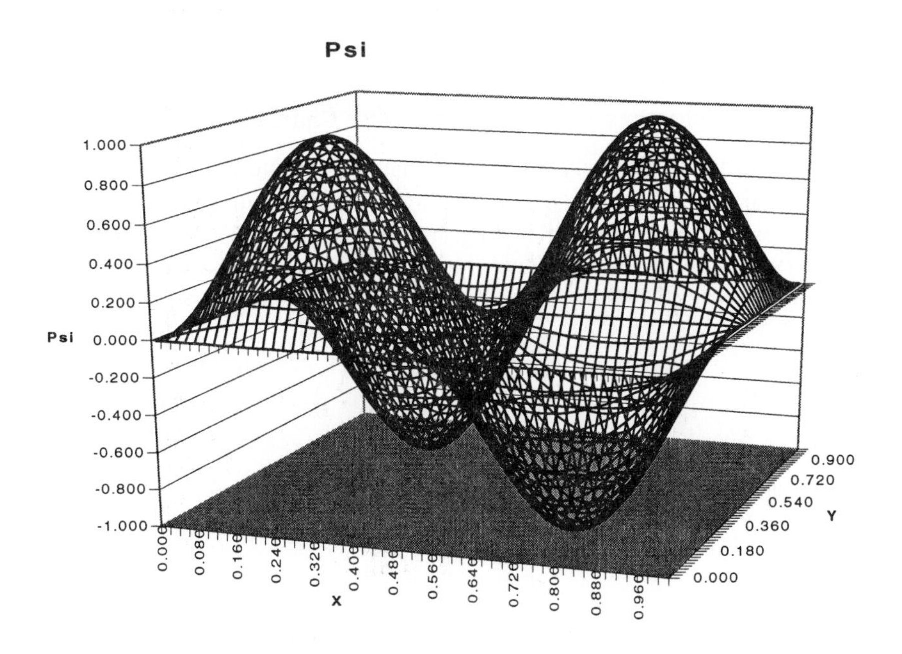

Figure 8-5. The ψ_{22} Wave Function for a Particle in a Box. Two lobes are plus, and two lobes are minus.

To create the contour chart shown in Figure 8-6, click in the chart (Figure 8-4). Then on the **Chart** menu, click **Type**. Select **Contour** as the chart subtype.

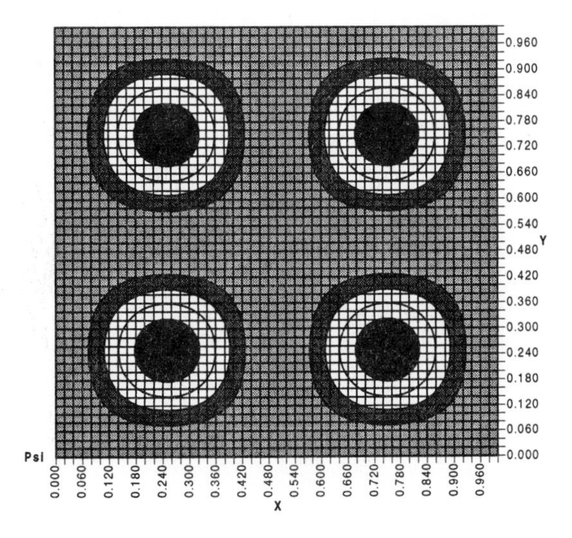

Figure 8-6. A Contour Plot of ψ_{22}^2 for a Particle in a 2-D Box

Discussion

The plot of ψ_{22}^2 in Figure 8-4 is positive everywhere but approaches values of zero at the edges of the two-dimensional box. The plot of ψ_{22} in Figure 8-5 consists of two positive peaks and two negative peaks.

Select the ψ_{22}^2 chart by clicking in it. Then on the **Chart** menu, click **Type**, and change the ψ_{22}^2 chart to a contour chart, a view from above (Figure 8-5). Plotting ψ^2 for a higher quantum number, say 5, would show 25 peaks symmetrically placed on the XY plane, spreading out the probability of finding the particle more evenly. At very high quantum numbers, the probability of finding the particle is about the same everywhere, corresponding to classical behavior.

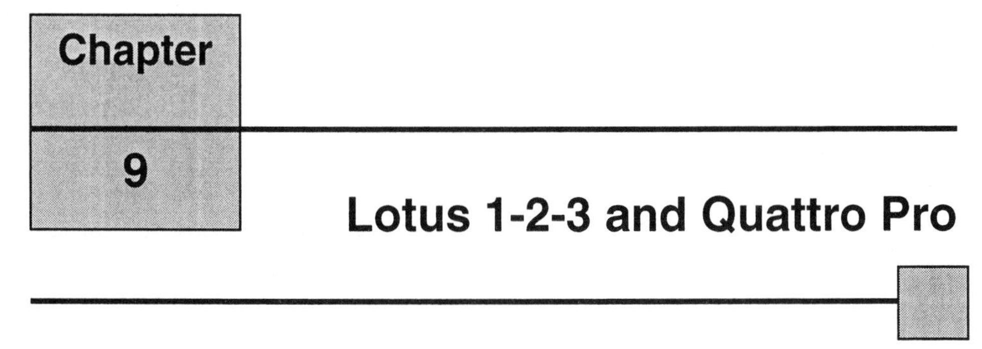

Chapter 9

Lotus 1-2-3 and Quattro Pro

Lotus 1-2-3 Introduction

Currently, IBM and a few other personal computer manufacturers include Lotus Suite with new computer sales. Like Microsoft Office, Lotus Suite includes a number of applications headed up by a word processor application and a spreadsheet application. Lotus Suite includes Lotus 1-2-3, which has been around for a number of years, and like Excel, it is a powerful spreadsheet program. Some users claim that Lotus 1-2-3 is arguably a little easier to use. Nevertheless, when one is familiar with one spreadsheet, it is not difficult to use another. If you have already chosen another spreadsheet as your favorite, you can still use Lotus 1-2-3. You will see that many commands are identical and others similar. The spreadsheet layout, shown in Figure 9-1, is similar to that of Excel, already displayed in Figure 1-1.

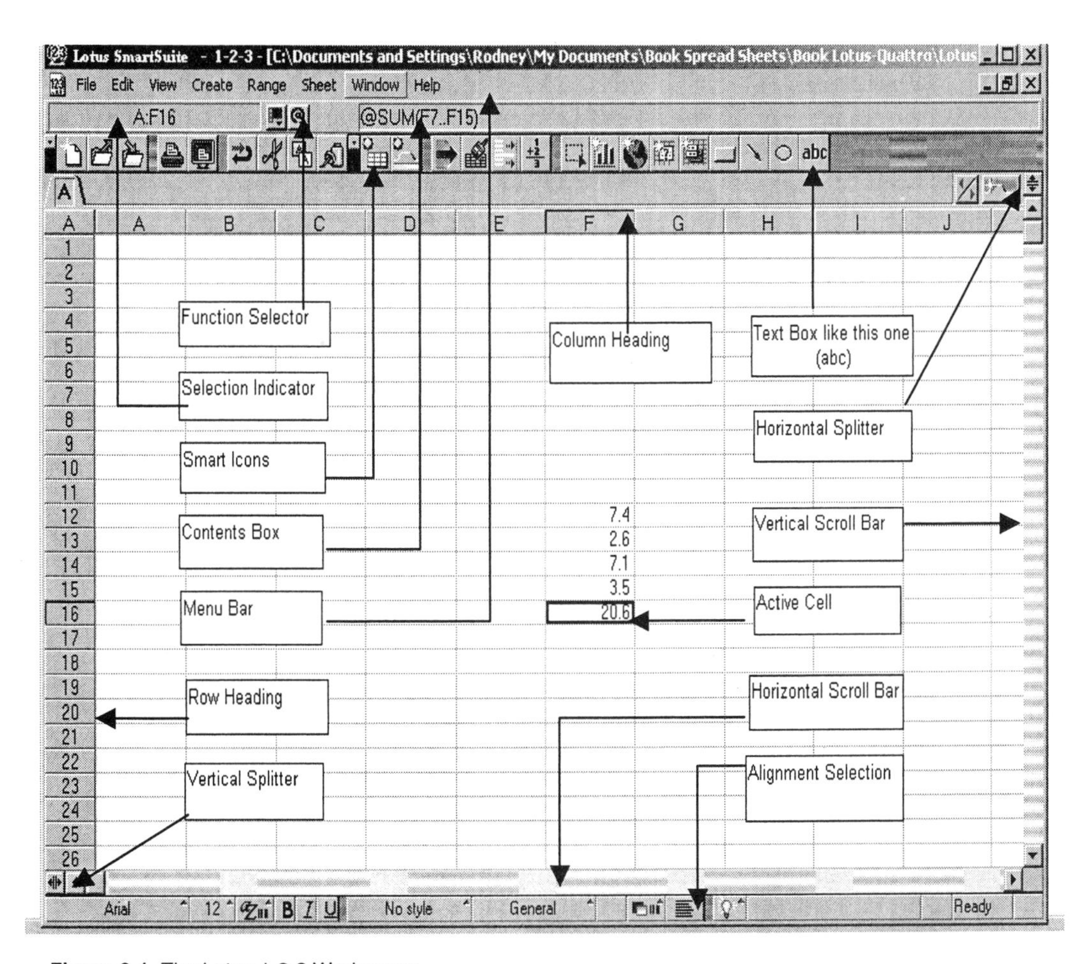

Figure 9-1. The Lotus 1-2-3 Workspace

The Lotus 1-2-3 Workspace Layout

Figure 9-1 displays the Lotus 1-2-3 screen layout, which is not very different from that for other spreadsheets and Windows applications. The menu bar at the top is standard for Windows, with **File**, **Edit**, **View**, **Insert**, and other menus. Just below the menu bar lies the contents toolbar, similar to the formula toolbar in Excel. The **Function Selector** icon (for locating and inserting functions) is on the contents toolbar. Below the contents toolbar lies the smart icons toolbar, similar to the Excel standard toolbar, with icons for **New Document**, **Open**, **Save**, **Print**, and so on. The **Create Chart** icon in Lotus looks like the analogous icon in Excel: a tiny bar graph.

The Menu Bar. The menu bar is essentially the same as in Excel or any other Windows application. It begins, as usual, with **File**, **Edit**, and **View** menus, but then includes some menu items unique to Lotus 1-2-3: **Create**, **Range**, and **Sheet**. Note that the range separator in Lotus 1-2-3 is a pair of periods (..); in Excel, it is a colon (:). The **Create_Drawing** command functions like Excel's **Drawing** toolbar.

The Contents Toolbar. The Lotus 1-2-3 contents toolbar, directly under the menu bar, is similar to the Excel formula toolbar. The first field on the left is the **Selection Indicator**, which gives the address of the selected cell. The long field on the right is the **Contents Box**, which displays the value of a selected cell or the formula contained in a selected cell. To its immediate left is the **Function Selector** icon (@), which serves like the **Insert Function** icon in Excel.

The Smart Icons. This row of icons is essentially the same as the standard toolbar in Excel. The first four icons have identical functions: **New Sheet**, **Open**, **Save**, and **Print**. The icons for **Redo**, **Cut**, **Copy**, and **Paste** follow the first four icons. The **Create a Chart** icon (looks just like the Excel Chart Wizard icon) consists of three tiny vertical bars, red, green, and blue. The button for creating a text block (the same as a text box in Excel) is located at the far right. Its icon is labeled **abc**.

The Status Bar. The status bar has many of the functions of Excel's formatting toolbar. To view the status bar, on the **View** menu, click **Show Status Bar**. The status bar is located at the bottom of the workspace.

Functions. The icon for the function selector is @, located on the contents toolbar. Clicking the @ icon causes a drop-down menu to appear, at the top of which is the selection **List All**. Selecting **List All** opens the **Function** dialog box, which lists categories of functions from which you can select. This is nearly exactly like Excel.

This Chapter

To compare Lotus 1-2-3 and Quattro Pro with Excel, this chapter includes five examples based on Excel Examples from previous chapters. Together, these programs require some of the features that scientists and engineers are likely to use: numerical calculation, charts, and statistics.

1. Lotus 1-2-3 Example 1. Linear Regression and Enzyme Kinetics
2. Lotus 1-2-3 Example 2. Plotting an Angular Wave Function
3. Lotus 1-2-3 Example 3. The Hydrogen Molecule Ion
4. Quattro Pro Example 1. Linear Regression: The H_2S Equilibrium
5. Quattro Pro Example 2. Vibrational Wave Functions

Lotus 1-2-3 Example 1

Linear Regression and Enzyme Kinetics

The science background underlying enzyme kinetics is described in Excel Example 20, Linear Regression and the Trendline: Enzyme Kinetics. In Excel Example 20, the Michaelis-Menten equation was derived and described. This equation describes how the reaction velocity v_0 depends of substrate concentration $[S]$.

$$v_0 = \frac{V_{\max}[S]}{K_m + [S]} \tag{9-1}$$

K_m is the Michaelis constant and V_{max} is the reaction velocity at high substrate concentration. Examination of Equation 9-1 reveals that as $[S]$ becomes large, v becomes equals to V_{max}, which corresponds to the constant velocity approached at high substrate concentrations revealed in Figure 9-2 (upper plot).

The Lineweaver-Burk Plot

Taking reciprocals of both sides of the Michaelis-Menten equation (described in Excel Example 20) results in

$$\frac{1}{v_0} = \frac{K_m + [S]}{V_{\max}[S]} \tag{9-2}$$

$$\frac{1}{v_0} = \frac{K_m}{V_{\max}[S]} + \frac{1}{V_{\max}} \tag{9-3}$$

Equation 9-3 is known as the Lineweaver-Burk equation (Excel Example 20). For systems obeying the Michaelis-Menten equation, a plot of $1/v_0$ against $1/[S]$ (a Lineweaver-Burk plot) results in a straight line of slope equal to K_m/V_{max} and intercept equal to $1/V_{max}$ (Figure 9-2, middle plot). The intercept on the $1/[S]$ axis (where $1/v_0 = 0$) equals $-1/K_m$.

The constant k_2, which equals $V_{max}/[E]_o$ (Equation 6-15) is frequently given the symbol k_{cat} and is called the turnover number. The turnover number is the number of product molecules produced per enzyme molecules per second and thus has units of sec^{-1}.

The Eadie-Hofstee Plot.

Although the most frequently used plot is the Lineweaver-Burk plot, it has the disadvantage that the plotted points tend to bunch up at small values of $1/[S]$. Because the data points plotted with the Eadie-Hofstee method tend to remain more evenly spaced, it is arguably superior to the Lineweaver-Burk method (Alberty1987, Dunford 1984).

The Eadie-Hofstee plot is a linear transformation of the Michaelis-Menten equation, written as

$$v_0 = -K_m \frac{v_0}{[S]} + V \tag{9-4}$$

The Eadie-Hofstee plot consists of a plot of v_0 versus $v_0/[S]$. The slope of the Eadie-Hofstee plot equals $-K_m$, and the intercept equals V_m as shown in Figure 9-2, lower plot.

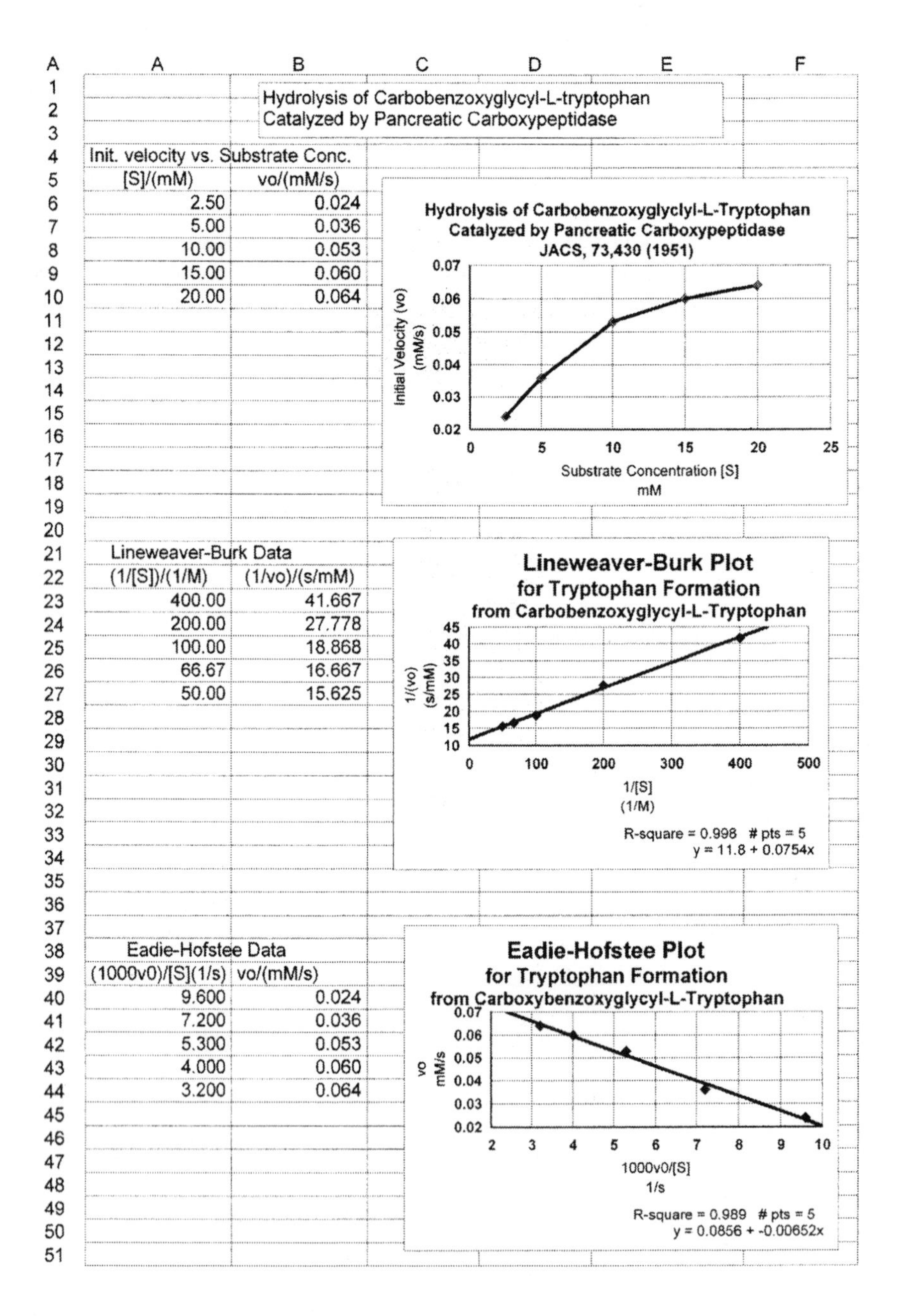

Figure 9-2. The Sheet for Enzyme Kinetics

Enzyme Kinetics and the Lotus 1-2-3 Spreadsheet

With this example, we will use Lotus 1-2-3 and the same data from Chapter 6, Excel Example 20, displayed in Table 6-1. Using the same data provides a suitable comparison between Excel and Lotus 1-2-3. The worksheets and charts from these two spreadsheet applications differ in details, in the appearance of the dialog boxes, the various options, and the editing features. But you will see that if you can use one of these applications, you can use the other with little difficulty should the occasion arise.

Entering the Experimental Data

- Create a text block by clicking the **Text Block** icon, a square button labeled **abc** located on the smart icons toolbar. (The Lotus text block is called a text box in Excel.) Place the cursor where you want the upper left corner of the text block to be located, and then drag to where you want the lower right corner to be located.

    1. Type the long label shown at the top of Figure 9-2 in the text block.
    2. Click in an empty area of the text block to select it, and change its size with the pull handles (■).
    3. Double-click the text inside the text block to edit it.

- In the cell range A4..C4, type **Init. Velocity vs. Substrate Conc.** as the label.
- In cells A5 and B5, type **[S]/(mM)** and **vo/(mM/s)** as the labels.
- From Table 6-1, in the range A6..A10, type the independent variable data [S].
- From Table 6-1, in the range B6..B10, type the dependent variable data [v_0].

Creating the Chart

- Click and drag from cell A5 (the column label) to cell B10 (the last dependent variable value).
- With this block highlighted, click the **Create Chart** icon (it looks like a tiny bar graph) on the smart icons toolbar, and position the pointer (which has become a tiny bar chart) in the middle of cell C5. Then drag it down to about cell F19. The cursor generates a dotted line around the rectangular area where the chart will appear. When you release the mouse button, Lotus 1-2-3 creates a bar chart from your data (Figure 9-3).

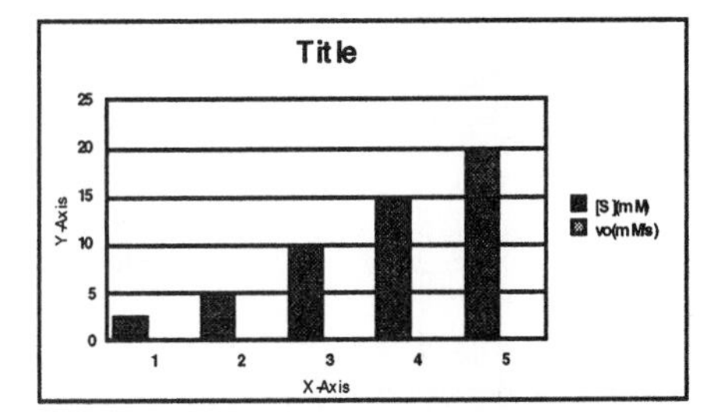

Figure 9-3. The Default Bar Chart Created by Lotus

Editing the Chart

Like most spreadsheet "Chart Wizards," Lotus 1-2-3 initially generates a bar chart. Lotus, Excel, and Quattro Pro all generate embedded charts in the spreadsheet with this procedure and, by default, they all produce a bar chart. However, it is a simple matter to change this bar chart to an XY (scatter) chart. The chart you have just produced has by default been selected, that is, it has black squares at its corners and sides. Dragging these "pull handles" permits changing the length or width of the chart. Notice that when a chart has been selected, the **Chart** item appears on the menu bar. Click anywhere on the sheet outside the chart area, and the black squares disappear, and so does the **Chart** item disappear from the menu bar. Move the cursor outside the chart, click, and the squares disappear. Repeat clicking inside and outside the chart a few times, while looking at the menu bar. You will notice that when the chart has been selected—and only then—the **Chart** item appears on the menu bar.

- With the chart selected, click **Chart** on the menu bar, and a drop-down menu appears (Figure 9-4).
- On the **Chart** menu, click **Chart Type** to open the **Properties for Chart** dialog box. Click **XY (Scatter)** once to select that type, and the bar chart instantly changes to an XY (scatter) chart, and the screen appears as shown in Figure 9-5.
- Right-click one of the markers to open a shortcut menu.
- On the shortcut menu, click **Series Properties** to open the **Properties for Series** dialog box (Figure 9-6).
    1. Click the tab with an icon (not the **Options** tab or **Series trend** tab) to open the dialog box shown in Figure 9-6.
    2. Select the **Connect points** check box.
    3. You can experiment with other tabs and choices on this dialog box if you wish.

Figure 9-4. The **Chart** Menu

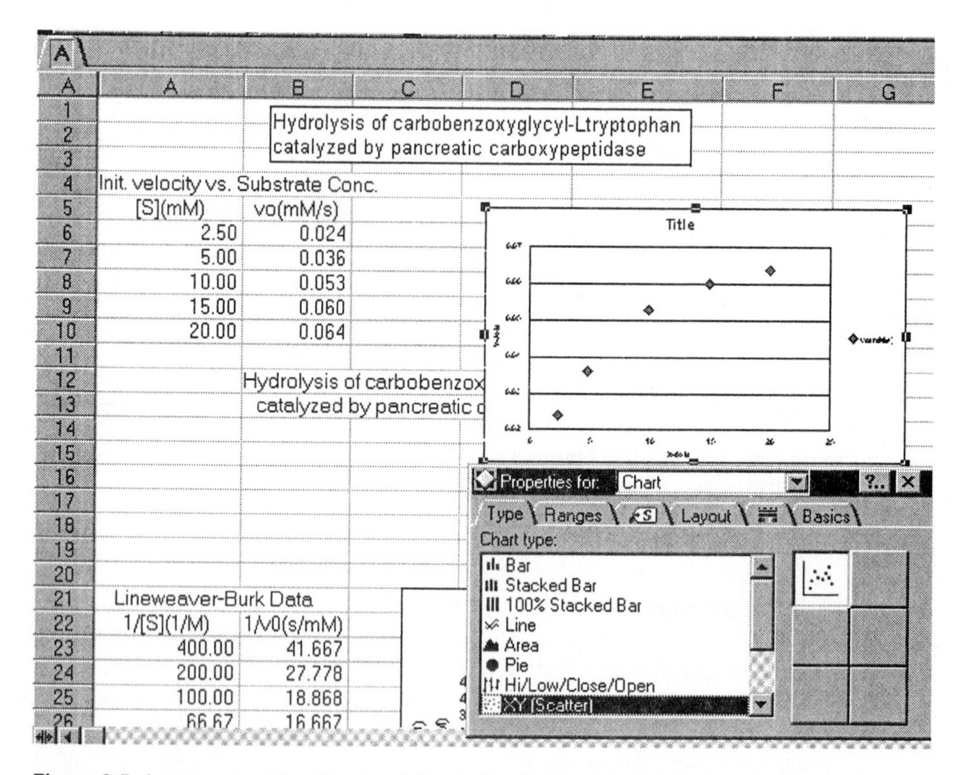

Figure 9-5. Appearance of the Sheet and Chart after Clicking **XY (Scatter)** on the **Properties for Chart** Dialog Box

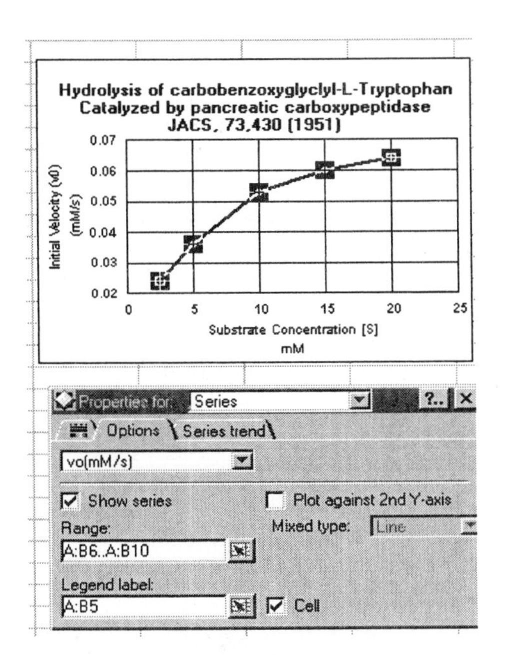

Figure 9-6. The **Properties for Series** Dialog Box for Line and Marker Appearance

Instead of using the **Chart** menu, you can right-click any object on the chart (such as a marker). This action results in a shortcut menu that gives you access to the object's **Properties** dialog box.

Try clicking successively on the chart's title, x-axis title, y-axis title, x-axis numbers, y-axis numbers, and any plot marker. The corresponding dialog box that appears for each object usually has several tabs for groups of related properties. Try also clicking these tabs to explore the way Lotus 1-2-3 allows the user to change properties of objects on the chart.

Notice that the objects selected (the series markers) are darkened.

Labeling and Editing the Lotus Chart

- Click in the chart to select it, and on the **Chart** menu, click **Title** (Figure 9-7). Fill in a title for the chart in line 1, and optional subtitles in lines 2 and 3.
- On the **Chart** menu, point to **Axes & Grids**, and click **X-axis**. The **Properties for X-axis** dialog box (Figure 9-8) with several selectable tabs opens; click the **Titles** tab, and in the x-axis **Line 1** text box, type **Substrate Concentration [S]** as the title, and in the **Line 2** text box, type **mM** as the subtitle.
- Click each of the **Properties for X-axis** dialog box tabs to explore them, but it is not necessary to make any changes at this time. These tabs allow changing the font, number format, grids, and scale.
- On the **Chart** menu, point to **Axes & Grids**, and click **Y-axis**. The **Properties for Y-axis** dialog box, similar to the **Properties for Y-axis** dialog box, opens. Click the **Titles** tab. The default title is y-axis; type **Initial Velocity (v0)** as the title. Type **(mM/s)** as the subtitle.

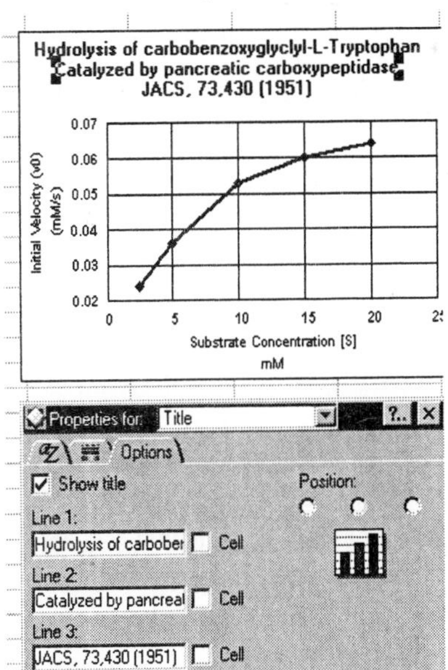

Figure 9-7. The **Properties for Title** Dialog Box

Lineweaver-Burk Plot

We will calculate data needed to create a Lineweaver-Burk plot, plot the data, calculate a linear regression for this plot, and draw the regression line through the data.

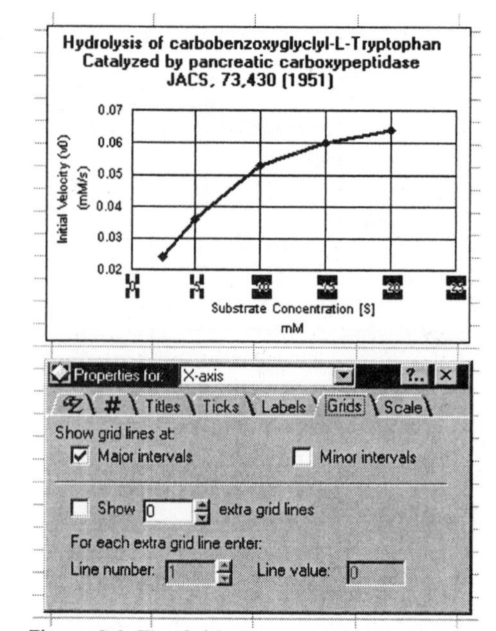

Figure 9-8. The **Grids** Tab in the **Properties for X-axis** Dialog Box

Entering the Data

- Beginning about cell A21, enter the data series title. Enter the labels for the *x* and *y* series in cells A22 and B22, as shown in Figure 9-2. Placing these numbers well below the data series for the first plot leaves room for the charts as they appear on Figure 9-2.
- In cell A23, type **=1000/A6** as the formula. In all of our spreadsheets (Lotus, Excel, and Quattro Pro), this is the format for entering a formula into a cell. After pressing the ENTER key, the contents of cell A23 reads 400. The 1000 changes mM (millimoles/liter) to M (moles/liter). Formulas can be much more complex as we will see later.
- Click in the lower right corner of cell A23. When you see three tiny arrowheads (a fill handle) pointing to the right and down, drag the cursor to cell A27. The numbers showing in Figure 9-2 (left of middle plot) will appear in the series A23..A27. When the series to be filled is long …fifty or hundred cells … and the formula is complex, this becomes a powerful tool. It is available in Excel nearly exactly the same, and in Quattro Pro in a slightly different form.
- Now repeat this procedure for the *y* series. In cell B23, type **=1/B6** as the formula. The value 41.677 appears in cell B23.
- Click in the lower right corner of cell B23 and drag to cell B27. The numbers in Figure 9-2 appear in the series B23..B27.

Creating the Chart

- Select the data for the chart block by clicking and dragging over the column heading and data, A22..B27. This selection transfers the labels and the data to the Chart Wizard.
- With the block A22..B27 selected, click the **Create Chart** icon, and then position the pointer—which has become a tiny bar chart—in the middle of cell C21. Drag it to about cell G35. The cursor generates a dotted line around the rectangular area where the chart will be generated when you release the mouse button.

Editing the Chart

- Once again Lotus produces a bar chart by default. As before, change it to an XY (scatter) chart; on the **Chart** menu, point to **Chart Type** and click **XY (Scatter)** (Figure 9-5).
- Give titles for the chart, the x-axis, and the y-axis with the **Properties for Title** dialog box (Figure 9-7).

Creating the Linear Regression Line with Lotus 1-2-3

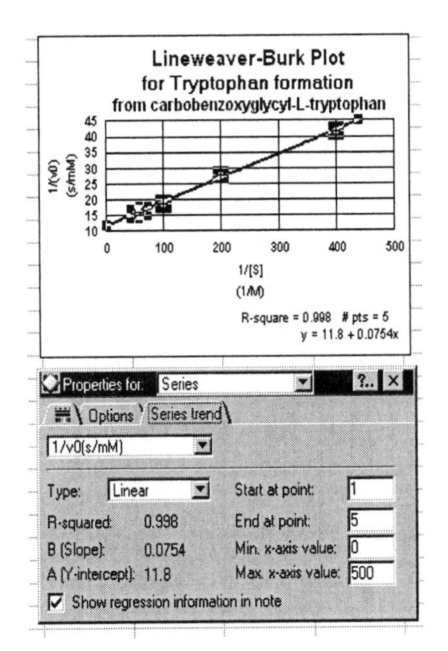

Figure 9-9. The **Series Trend Tab** of the **Properties for Series** Dialog Box

- Right-click any data point on the chart, and a shortcut menu appears. On the shortcut menu, click **Series Properties**, and the **Properties for Series** dialog box appears with three tabs (Figure 9-9).
- In the **Series trend** tab, for **Type**, select **Linear** in the drop-down list. As soon as you do this, a straight line is drawn through the chart data (Figure 9-9).
- The boxes at the right (**Start at point**, **End at point**, **Min. x-axis value**, **Max. x-axis value**) can be changed, although they are usually OK.
- Select the **Show regression information in note** check box. Selecting this check box places the equation of the linear regression line in the plot area.
- Click the tab with tiny colored lines and bars for options that deal with the line and markers (Figure 9-6).

    1. Select **Line color**, and change it to black.
    2. Select **Line width**, and make it a little heavier than the default width.
    3. Scroll down the **Marker symbol** drop-down list, and select a marker that you like.
    4. Select **Marker color**, and scroll to black.

- Click in the chart to select it, and then, on the **Chart** menu, click **Title** (Figure 9-4). Next fill in a title for the chart in line 1, and optional subtitles in lines 2 and 3, as shown in Figure 9-2, middle chart. Exactly as with the chart of raw data, click the line to edit the line and marker colors and other features.
- Close the dialog box to see the finished chart.

Eadie-Hofstee Plot

We will calculate data needed to create an Eadie-Hofstee plot, plot the data, calculate a linear regression for this plot, and draw the regression line through the data. The procedure used is essentially the same as for the Lineweaver-Burk plot.

Entering the Data. The lower plot on Figure 9-2 is the Eadie-Hofstee plot, and you can create it just as you created the Lineweaver-Burk plot.

- Enter the labels in the block A38..B39.
- In cell A40, type **=1000*B6/A6** as the formula.
- Drag the fill handle to fill the range A40..A44.
- In cell B40, type **=B6** as the formula.
- Drag the fill handle to fill the range B40..B44.
- Select the block A39..B44, then click the **Create Chart** icon, and drag to create the chart.
- Edit the chart as described for the chart of the raw data and for the Lineweaver-Burk plots. The chart should look like Figure 9-2, lower chart.

Just as with Excel, you have a couple of options for editing the chart. If you first select the chart, the **Chart** menu appears on the menu bar. Clicking **Chart** provides a drop-down menu (Figure 9-4) from which you can select the feature of the chart you wish to edit (**Chart Type**, **Chart Style**, **Chart Properties**, **Title**, **Legend**, **Axes & Grids**, **Series**, and **Series Labels**). Alternatively, as with Excel, you can right-click the object you wish to edit. Right-clicking opens a shortcut menu. Clicking a command on it opens the dialog box associated with the object that you right-clicked. The dialog boxes generally are identical to those that you arrive at by selecting the chart and then clicking a command on the **Chart** menu, so which method you use is a matter of personal preference.

Lotus 1-2-3 Example 2

Plotting an Angular Wave Function

Review

The coordinate system and general features of the wave functions for the hydrogen atom are described in Chapter 3, Excel Example 7, Radial Wave Functions: 1*s*, 2*s*, and 3*s*. The science background underlying angular wave functions are described in Excel Example 8, An Angular Wave Function: The $2p_x$ Orbital. To compare Lotus 1-2-3 with Excel, we shall again plot the p_x orbital (Equation 3-17). Figure 9-10 displays the finished spreadsheet for the p_x orbital.

Entering the Data

The text for titles and labels is entered in about the same way for Excel, Lotus 1-2-3, and Quattro Pro. The spreadsheet functions for the absolute value is written

@ABS(argument)

You can find and insert any Lotus 1-2-3 function by selecting the cell where you want the function and then clicking the ampersand (@) located on the contents toolbar below the menu bar (Figure 9-1).

Entering the Text and Title Data

- In rows 2, 4, 5, and 6, type the following five lines of documentation:
  Row 2: **The px Angular Wave Function for the Hydrogen Atom**
  Row 4: **px = Abs(cos(PiPhi/180)) since sin(PiTheta/180) = 1 for Theta = 90. (= arrow on chart)**
  Row 5: **x = px(cos(PiPhi/180)), which is the x (horizontal) coordinate on the Lotus Window**
  Row 6: **y = px(sin(PiPhi/180)), which is the y (vertical) coordinate on the Lotus Window**
- In the range A8..D8, type **Phi/deg**, **px**, **x**, and **y** as the column labels.
- Select the range A8..D8, and click the **Center Alignment** button at the bottom of the screen.

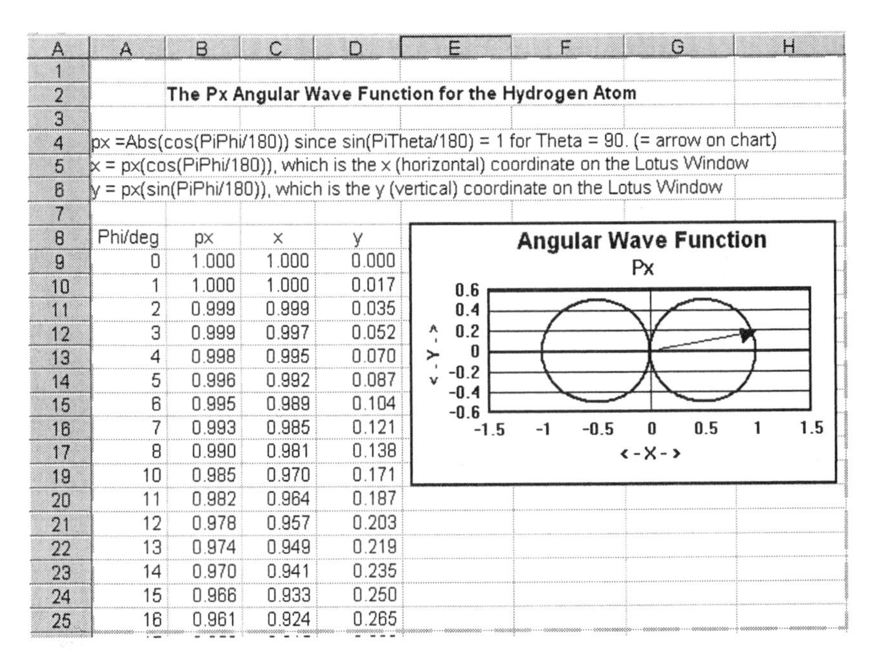

The Px Angular Wave Function for the Hydrogen Atom

px =Abs(cos(PiPhi/180)) since sin(PiTheta/180) = 1 for Theta = 90. (= arrow on chart)
x = px(cos(PiPhi/180)), which is the x (horizontal) coordinate on the Lotus Window
y = px(sin(PiPhi/180)), which is the y (vertical) coordinate on the Lotus Window

Row	Phi/deg	px	x	y
9	0	1.000	1.000	0.000
10	1	1.000	1.000	0.017
11	2	0.999	0.999	0.035
12	3	0.999	0.997	0.052
13	4	0.998	0.995	0.070
14	5	0.996	0.992	0.087
15	6	0.995	0.989	0.104
16	7	0.993	0.985	0.121
17	8	0.990	0.981	0.138
19	10	0.985	0.970	0.171
20	11	0.982	0.964	0.187
21	12	0.978	0.957	0.203
22	13	0.974	0.949	0.219
23	14	0.970	0.941	0.235
24	15	0.966	0.933	0.250
25	16	0.961	0.924	0.265

Figure 9-10. Sheet for the p_x Angular Wave Function for the Hydrogen Atom

Entering φ (The Angle between the Vector and the X-Axis in Figure 9-10)

- Select cell A9.
- On the **View** menu, click **Split**, and select **horizontal**.
- Scroll the lower screen so that cell A:A400 is visible.
- Select cell A9, and SHIFT-click in cell A400 to select the range A9..A400.
- On the **Range** menu (Figure 9-11), click **Fill** to open the **Fill** dialog box (Figure 9-12).
- For **Fill using**, select **Numbers** (Figure 9-12).
- Type **0** in the **Start at** block, **1** in the **Increment by** block, and **360** in the **Stop at** block. The final cell in the range can be cell A369 or any cell beyond cell A369. Filling will stop at the number value of 360. These values are the incremental values of φ.
- Click **OK**, and the range A9..A369 fills with the number series 0 to 360.

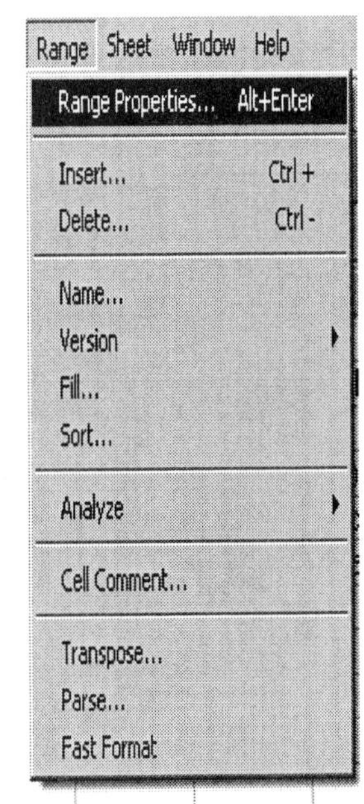

Figure 9-11. The **Range** Menu

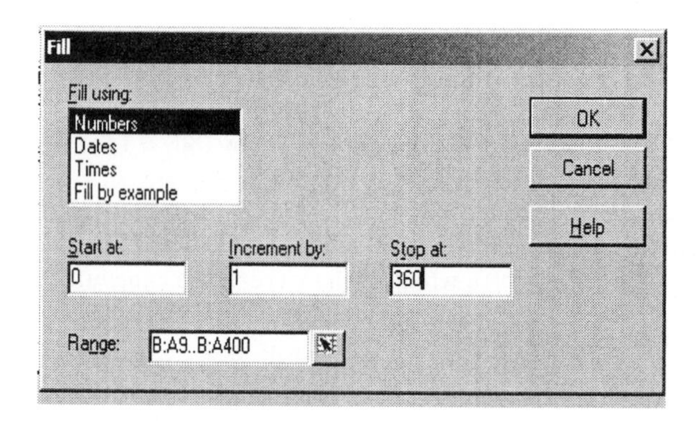

Figure 9-12. The **Fill** Dialog Box

Calculating the p_x, x, and y Data for the plot

Calculating p_x

- In cell B9, type Equation 3-20 as a Lotus function: **@ABS(@COS(@PI*A9/180))**
- Click and drag the split screen bar to the middle of the screen, or, on the **View** menu, click **Split**.
- Scroll the lower half of the slit screen until the 360 is seen in cell A369.
- Select cell B9, and then SHIFT-click in cell B369 to select the range B9..B369.
- On the **Edit** menu, click **Copy Down** (Figure 9-13) to fill the range B9..B369 with the wave equation (Equation 3-20) at each ϕ. Figure 9-13 shows the **Edit** menu just before clicking the **Copy Down** command.
- In a similar manner, type Equation 3-21 in cell C9 and Equation 3-22 in cell D9 for the x and y values, and fill in the ranges using the **Copy Down** command.

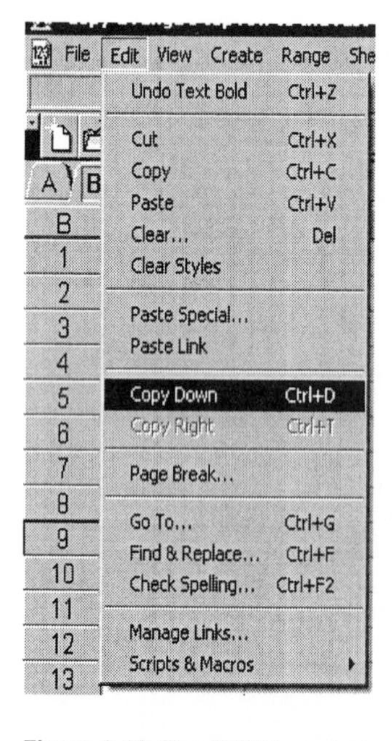

Figure 9-13. The **Edit** Menu Just before Clicking **Copy Down**

Calculating the *X* Coordinate

- Select cell C9, and type Equation 3-21: **=B9*@COS(@PI*A9/180)**
- Split the page, and as previously select the range C9..C369.
- As previously, on the **Edit** menu, click **Copy Down** to fill the range C9..C369 with the x values.

Calculating the *Y* Coordinate

- Select cell D9, and enter Equation 3-22: **=B9*@SIN(@PI*A9/180)**
- Split the page, and select the range D9..D369.
- On the **Edit** menu, click **Copy Down** (Figure 9-13) to fill the range D9..D369 with the y values.

Creating the Chart

- Split the screen as described earlier. Scroll the bottom half until cell D369 is visible.
- Click and drag across cells C9 and D9.
- SHIFT-click and drag across cells C369 and D369.
- Click the **Create Chart** icon on the smart icons toolbar.
- Place the pointer where you would like the upper left corner of the chart to be placed, and drag to where you would like the lower right corner of the chart to be. These actions create the Lotus default chart type, a bar chart, so we need to change the type and do some editing until the chart appearance is satisfactory.
- Click anywhere in a blank area of the chart to select it. The **Chart** menu appears on the menu bar.
- On the **Chart** menu, click **Chart Type**, and then select **XY (Scatter)** (Figure 9-5).

Editing the Chart

The chart needs some editing to improve its appearance. Right-clicking the feature to be changed gives access to most of the internal features of the chart. Alternatively, when the chart is selected, the **Chart** menu appears on the menu bar, and clicking it also provides access to editing the chart. Clicking an area and pulling the size handles to achieve the desired effect changes the position and size of the chart area or plot area slightly. The internal features should be changed first, since they perturb the size of the plot and chart areas slightly.

- Right-click the fuzzy *p* orbital plot to open the **Properties for Series** dialog box for line and marker appearance.
  1. Select the **Connect points** check box.
  2. Click the **Line width** block, and select a line width to your personal taste.
  3. Make the line black.
  4. Leave the line style as continuous (not dashed).
  5. Clear the **Show marker** check box (the marker properties and color can be ignored).
- Right-click the chart title, which by default is "title."
  1. In the Line 1 block, type **Angular Wave Function** as the first line.
  2. In the Line 2 block, type **px** as the second line.
- Right-click the x-axis, which has the default title "x-axis." Explore the tabs.
  1. Change the minimum to **-1.5** and change the maximum to **+1.5**
  2. Change the font to bold.
- Right-click the y-axis, which has the default title "y-axis." Change the font to bold.
- Click the legend box, and delete it.
- Select the chart, and use the size handles to size it so that the left and right *p* orbital lobes are circles. The chart should now look like the chart in Figure 9-10.
- On the **Create** menu, point to **Drawing**, and click **Arrow**.
  1. Place the cursor on the origin of the *p* orbital plot.
  2. Drag the cursor over to the right, as shown in Figure 9-10.
  3. If you are dissatisfied, select the arrow, delete it, and try again.

Lotus 1-2-3 Example 3

The Hydrogen Molecule Ion

Review

The coordinate system and theory for the hydrogen molecule ion are developed in Excel Example 7. In this Lotus 1-2-3 example we will compare the results of plotting the wave function with Lotus and Excel. As shown in Chapter 3, Excel Example 7, the function to be plotted is

$$\frac{\varphi}{a_0^{-3/2}} = 0.3984*(\exp(-|r/52.92|)+\exp((-|r-106|)/52.92)) \qquad (9\text{-}5)$$

Entering Data into the Lotus 1-2-3 Spreadsheet

The titles and labels are entered as usual and are shown in Figure 9-14, but instead of using the pull handles to fill the columns with data, we will use the Range Fill method. This method is comparable to the Edit Fill Down method used in Excel Example 7.

Entering the Labels

- Create a text block with the **Text Block** icon on the smart icons toolbar. Click the icon, then place the pointer in the block's upper left corner and drag it to the lower left corner. Type the text **The Hydrogen Molecule Ion** as shown in Figure 9-14.
- Right-click the text block, and on the resulting shortcut menu, click **Drawing properties** to open the **Properties for Draw Object** dialog box (Figure 9-15).
  1. Click the **Alignment** tab, and align the text vertically and horizontally.
  2. Click the **Color, Pattern, and Line Style** tab, and make the border line black and a bit heavy.
  3. Click the **aZ** tab, and make the text 12 point bold.
- In the same way, create another text block, and for the purpose of documentation, type the equation shown in the second text block at the top of Figure 9-14 (this is Equation 9-5):
  Psi/ao-3/2 = 0.39894*(@EXP(-@ABS(A9/52.92))+@EXP(-@ABS((A9-106)/59.92)))
- In cell A8, type **r/pm** as the column label.
- In cell B8, type **Psi/ao-3/2** as the column label.

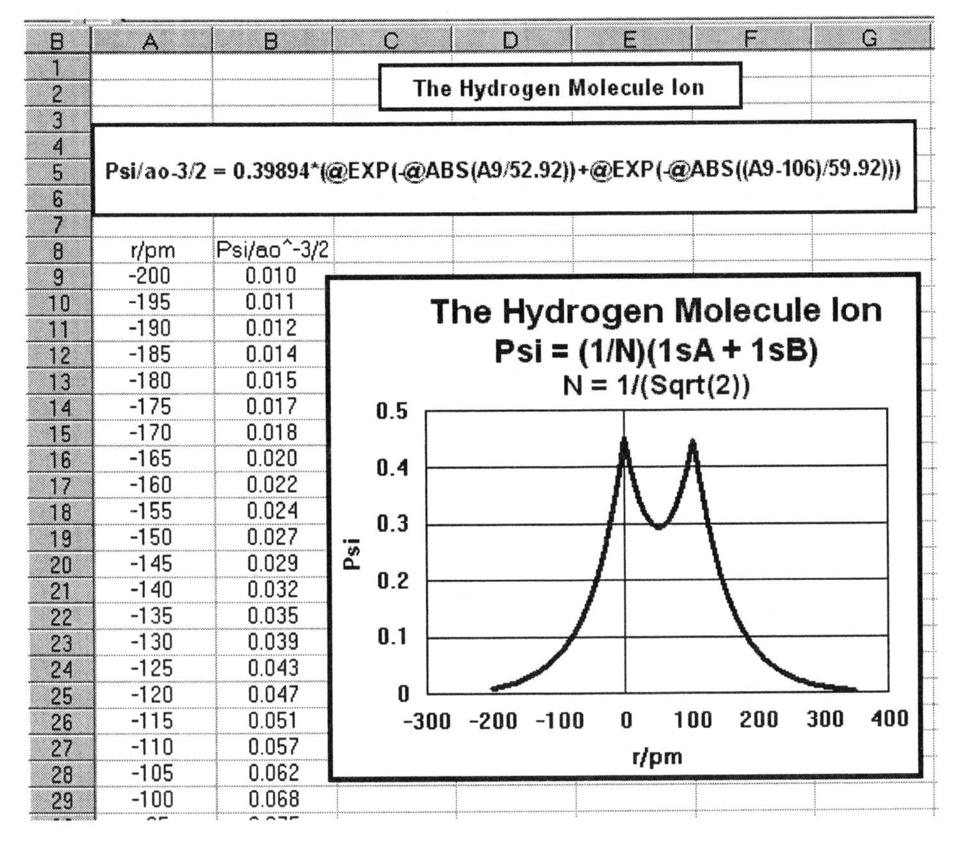

r/pm	Psi/ao^-3/2
-200	0.010
-195	0.011
-190	0.012
-185	0.014
-180	0.015
-175	0.017
-170	0.018
-165	0.020
-160	0.022
-155	0.024
-150	0.027
-145	0.029
-140	0.032
-135	0.035
-130	0.039
-125	0.043
-120	0.047
-115	0.051
-110	0.057
-105	0.062
-100	0.068

Figure 9-14. Sheet for the Hydrogen Molecule Ion

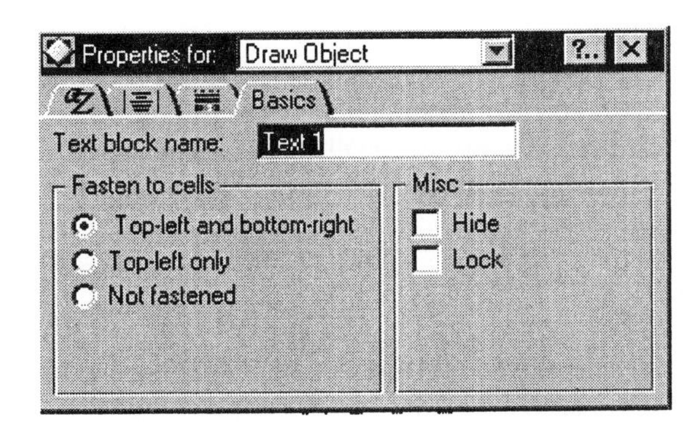

Figure 9-15. The **Properties for Draw Object** Dialog Box

Entering the Incremented Value of *r*. The wave function plot shown in Figure 9-14 begins at *r* equals –200 and ranges to *r* equals +350 in increments of *r* equal to 5.

Method I. Fill Arrows

- In cell A9, type **-200** as the value.
- In cell A10, type **=A9+5** as the formula.
- Click in the lower right corner of cell A10, and when the arrows appear, drag the pointer to cell A119. Cell A119 will contain the value 350.

Method II. Range Fill

- Select cell A9 by clicking in it.
- On the **Range** menu, click **Fill** to open the Fill dialog box (Figure 9-12).
- In the **Start at** block, type **-200** as the value.
- In the **Increment by** block, type **5** as the value.
- In the **Stop at** block, type **+350** as the value.
- In the **Range** block, type **A9..A119** as the range.

Entering the Formulas. The last cell, A119, will contain the value 350, and the values in the intermediate cells will increment by 5 from –250 to +350. If cell A119 is selected, the expression to be seen in the contents toolbar is 350 if method II is used, but is A6+5 if method I is used. Enter the value of the molecular orbital φ in units of $a_0^{-3/2}$.

- In cell B9, type the formula:
  =0.39894*(@EXP(-@ABS(A9/52.92))+@EXP(-@ABS((A9-106)/59.92)))
- Click and drag the split screen bar to the middle of the screen, or, on the **View** menu, click **Split**.
- Scroll the lower half of the split screen until cell B119 is visible.
- Select cell B9, and then SHIFT-click in cell B119 to select the range B9..B119.
- On the **Edit** menu, click **Copy Down** (Figure 9-13) to fill the range B9..B119 with the values of φ.

Plotting the Molecular Orbital

- Click in cell C9, and drag to B119 to select the series to be plotted.
- Click the **Create Chart** icon once.
- Place the pointer in cell C8 and drag to cell G27.

This produces a peculiar looking chart, which is actually a bar chart, the default chart first created by most spreadsheets. It is floating in the spreadsheet between cells C8 and G27 and is already selected as seen by the small black squares at its corners and sides. Because it has been selected, the **Chart** menu appears in the menu bar, where it does not normally appear unless a chart has been created and selected.

Editing the Chart

- With the chart selected, on the **Chart** menu, click **Chart Type**, and select the **XY (Scatter)** type (Figure 9-5).
- On the **Chart** menu, click **Title** to add a title and up to two subtitles (Figure 9-7).
- On the **Chart** menu, point to **Axes & Grids**, and click **X-axis** or **Y-axis** to label the axes (Figure 9-8).

All three of these dialog boxes have tabs that can be clicked for further editing the font, number format, grids, borders, and scale.

Quattro Pro Introduction

Borland International was the original developer of Quattro Pro (QP). Corel, best known for Corel Draw, created Corel Office by absorbing WordPerfect as its word processor and Quattro Pro as its spreadsheet. The Quattro Pro spreadsheet layout (Figure 9-16) is similar to the layouts of Excel and Lotus 1-2-3. The first Quattro Pro Example illustrates how this spreadsheet deals with linear regression.

The Menu Bar

The topmost bar, the menu bar, is close to that of Excel. If you right-click the Quattro Pro menu bar, a small menu opens that allows you to select the menu bar for Excel or Lotus 1-2-3 in place of the QP menu bar. You can then work almost as if you were working in either of these two applications.

QP:	File	Edit	View	Insert	Format	Tools	Window	Help
Excel:	File	Edit	View	Insert	Format	Tools	Data	Window Help
Lotus:	File	Edit	View	Create	Range	Sheet	Window	Help

The Standard Toolbar

The toolbar under the menu bar is simply called the toolbar. The first four icons are identical in appearance with those in Excel and identical in function with those in Lotus. To place other toolbars on top of the spreadsheet in QP, right-click the toolbar to open a menu of toolbar selections. In Excel, you perform this operation by, on the **View** menu, clicking **Toolbars**. Lotus is similar to QP in opening new toolbars, although in Lotus it is referred to as the smart icon toolbar, and the icons are called smart icons.

The Property Bar

In Excel the QP property bar is called the formatting toolbar. It displays the current font name and size and allows changing them. It also has the icons for bold, italic, and underline.

The Formula Toolbar

The toolbar under the property bar has no name in Quattro Pro, but is about the same as the formula toolbar in Excel. The first field (far left) displays the name of the currently selected cell. To its right is the icon for inserting a function, the ampersand (@). The remainder of this toolbar is a field containing the formula or value of the selected cell.

Functions

As just mentioned, the @ icon on the formula toolbar allows access to Quattro Pro's functions, which are about as numerous as those in Excel and Lotus. Clicking the @ icon opens the **Function** dialog box, which categorizes and lists the 500 or so accessible QP functions.

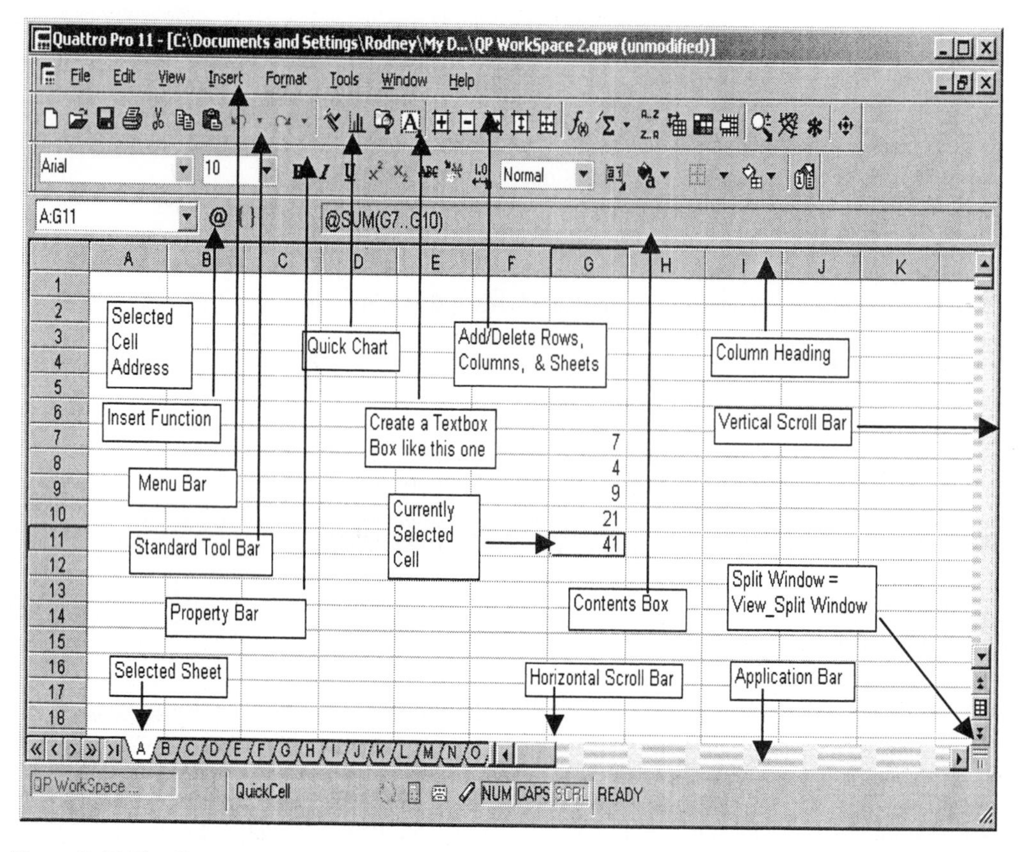

Figure 9-16. The Quattro Pro Workspace Layout

Quattro Pro Example 1

Linear Regression: The H_2S Equilibrium

The calculation and plot of a linear regression line is so common that it is illustrated for Excel, Lotus 1-2-3, and Quattro Pro. For each of these spreadsheets the goal is to generate an XY (scatter) plot of the experimental points and then to plot the linear regression line calculated from these points. The result is a straight line, the regression line, with the *x,y* points scattered randomly about the line. With Excel and Lotus 1-2-3, this task is essentially automatic. With Quattro Pro, the task is slightly more manual.

Entering Title and Labels

Enter the initial data shown in the upper left corner of the spreadsheet (Figure 9-17) so that we can reproduce the worksheet shown in Figure 9-34. Open a new worksheet, and:

- To create a text box, click the **Text Box** icon (an A), and click and drag the pointer from cell A1 to cell B3.
- Type **The Hydrogen Sulfide Equilibrium** as the title.
- Click and drag over the label to select it, and then on the **Format** menu, point to Selection Property_Text Font_Size_12_Bold.
- In cell C4, type **1.9873** as the gas constant, and in cell B4, type its label. (The original work used cgs rather than SI units for the gas constant. See Chapter 3, Excel Example 7, for SI units.)
- Enter the data column labels in the range A6..E6 as in Figure 9-17.
- Enter the temperatures (t/deg) in the range A7..A15.
- Leave column B blank for now.
- Enter the equilibrium constants *K* in the range C7..C15.

	A	B	C	D	E
1					
2		The Hydrogen Sulfide Equibrium			
3					
4		R =	1.9873		
5					
6	t/deg	T/K	K	1/T	R ln K
7	750	1023	106.20	0.0009775	9.271
8	830	1103	45.00	0.0009066	7.565
9	945	1218	23.50	0.0008210	6.274
10	1065	1338	8.90	0.0007474	4.344
11	1089	1362	8.20	0.0007342	4.182
12	1132	1405	6.20	0.0007117	3.626
13	1200	1473	5.00	0.0006789	3.198
14	1264	1537	3.20	0.0006506	2.312
15	1394	1667	1.60	0.0005999	0.934

Figure 9-17. The H_2S Spreadsheet before Creating the Chart

Entering the Formulas and Numerical Data

This completes the initial data entry to prepare Quattro Pro to do the calculations. We now need to enter some formulas to convert temperatures from Celsius to Kelvin, and to calculate *1/T* and *R ln K*. We want to plot *ln K* as the dependent variable along the y-axis and *1/T* as the independent variable along the x-axis. If these data are in adjacent selected columns, Quattro Pro uses the left column as the independent variable and the right column as the dependent variable. This characteristic of Quattro Pro (and Excel and Lotus 1-2-3) dictates the placement of the *1/T* data in column D and the *RlnK* data in column E. Although the columns selected for the x-axis and y-axis can be reversed later, it is usually simpler to anticipate the spreadsheet's choice.

- In cell B7, type **=A7 + 273** as the formula.
- Select cell B7, and drag its fill handle to cell B15.
- In cell D7, type **= 1/ B7** as the formula.
- Select cell D7, and drag its fill handle to cell D15.
- In cell E7, type **= $C$4*@LN(C7)** as the formula.
- Select cell E7, and drag its fill handle to cell D15.

The upper left corner of your spreadsheet should look like Figure 9-17. Quattro Pro requires an @ to precede a function call and capital letters for the function name. To find the required function, on the **Insert** menu, click **Function**, and the **Functions** dialog box appears, with the function category displayed on the left and the functions for each category on the right. This completes the data entry and preparation for the chart.

Creating the Chart for the H_2S Equilibrium Data

- Select the range D7..E15.
- On the **Insert** menu, click **Chart** to activate the **Expert – Step 1 of 4** dialog box.
- Click **Next**, and the **Expert – Step 2 of 4** dialog box appears.

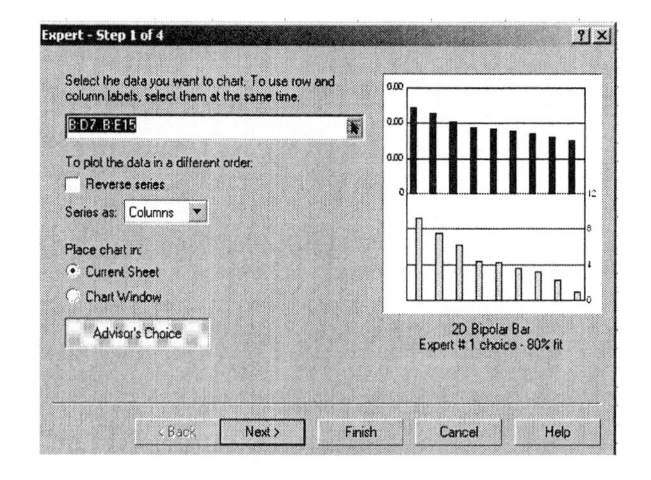

Figure 9-18. The **Expert – Step 1 of 4** Dialog Box. In this dialog box, you select the data you want to chart. Notice that a bar chart is displayed; ignore this for now.

1. Verify that the correct data range is selected.
2. Click **Next**.

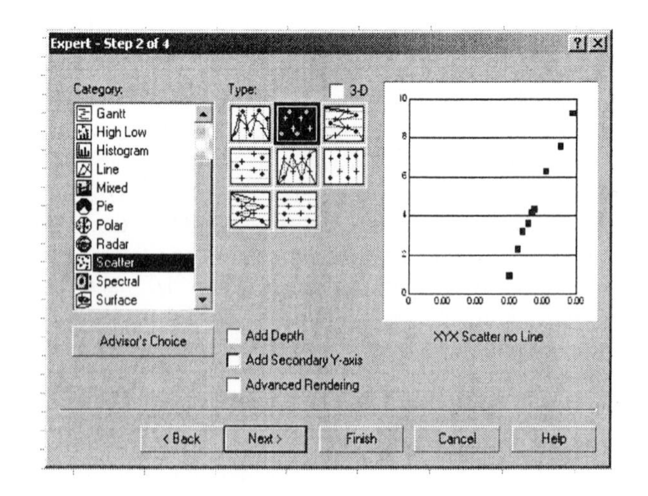

Figure 9-19. The **Expert – Step 2 of 4** Dialog Box. In this dialog box, you select the category and type.

1. In the **Category** list, scroll down and select **Scatter**.
2. If a message box appears with the message **Insufficient Data for This Chart Type**, ignore it and click **OK** in the message box.
3. Clear the **Add Secondary Y-axis** check box.
4. In the **Type** section, select **XYX Scatter no Line** (middle icon, first row).
5. Click **Next**.

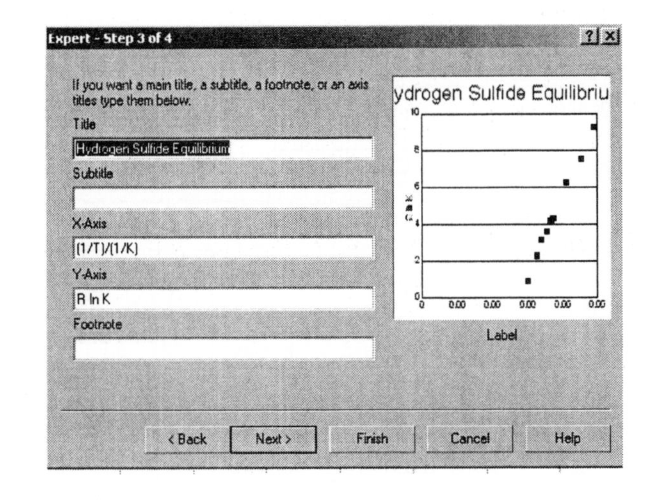

Figure 9-20. The **Expert – Step 3 of 4** Dialog Box. In this dialog box, you give titles to the chart and the axes.

1. In the **Title** box, type **Hydrogen Sulfide Equilibrium** as the title.
2. In the **X-Axis** box, type **1/(T/K)** as the label.
3. In the **Y-Axis** box, type **R ln K** as the label.
4. Click **Next**.

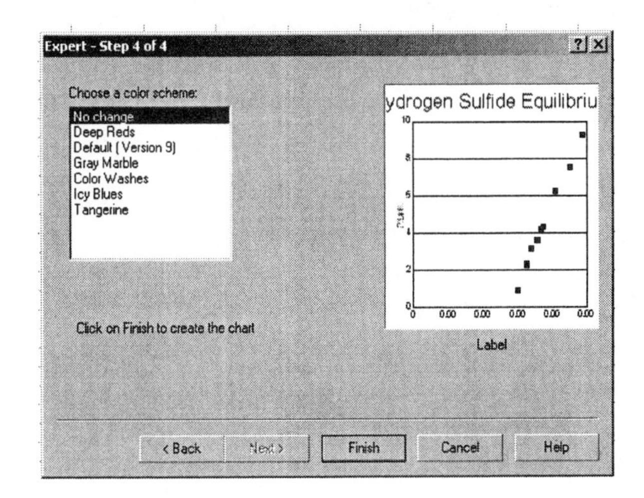

Figure 9-21. The **Expert – Step 4 of 4** Dialog Box. In this dialog box, you choose a color scheme.

- Select **No change**, and click **Finish**. The dialog box disappears, and the pointer changes to a tiny bar chart.

Place the pointer in cell F4 and drag it to cell J22 to create and size the chart. The spreadsheet with a chart floating in the range F4..J22 now appears similar to Figure 9-22. After you complete the **Expert – Step 4 of 4** dialog box, the chart shows a plot of the experimental points, with line segments drawn between them. We must delete the line segments and insert a linear regression line. In addition, it is clear that the font size of the title is too large and must be edited. Before inserting the linear regression line, let us learn how to handle editing in Quattro Pro.

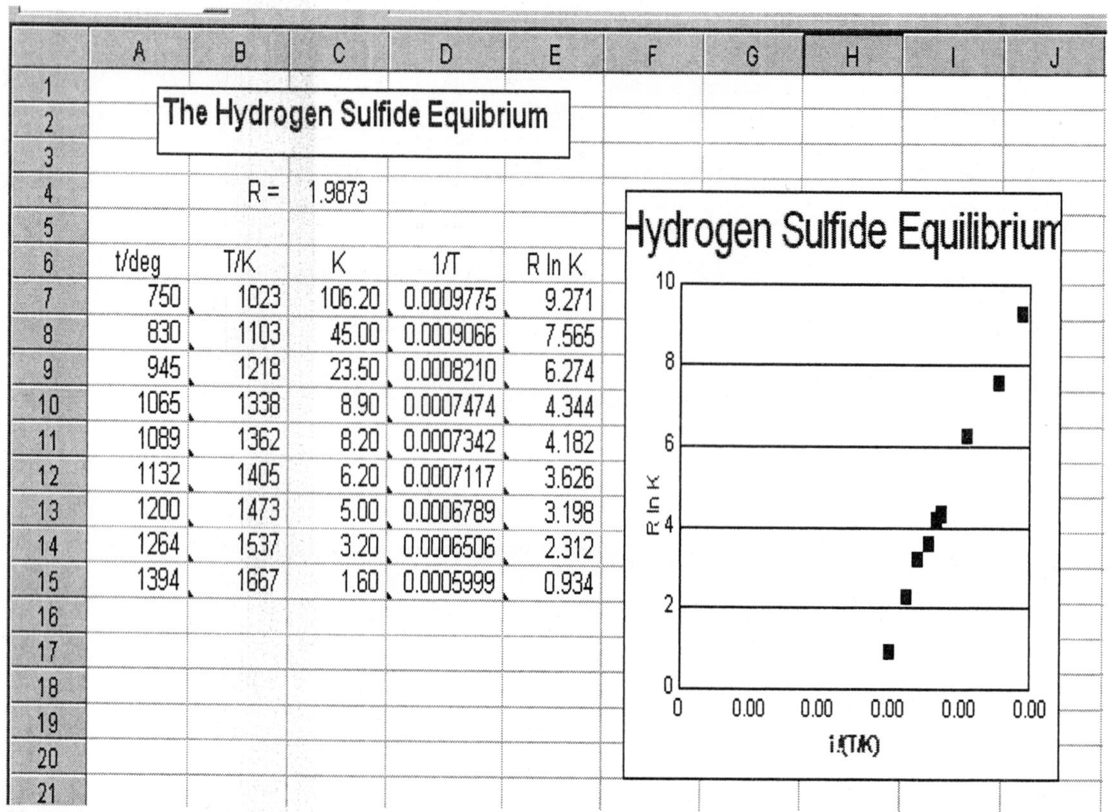

The Hydrogen Sulfide Equibrium

R = 1.9873

t/deg	T/K	K	1/T	R ln K
750	1023	106.20	0.0009775	9.271
830	1103	45.00	0.0009066	7.565
945	1218	23.50	0.0008210	6.274
1065	1338	8.90	0.0007474	4.344
1089	1362	8.20	0.0007342	4.182
1132	1405	6.20	0.0007117	3.626
1200	1473	5.00	0.0006789	3.198
1264	1537	3.20	0.0006506	2.312
1394	1667	1.60	0.0005999	0.934

Figure 9-22. The Sheet and Unedited Chart after Step 4 of 4.

Editing the Chart

The spreadsheet now looks nearly like Figure 9-22. The chart is floating on the spreadsheet. Before editing the floating chart, it's useful to explore the chart's behavior. The initially created chart appears similar to the one shown in Figure 9-23a. If you place the pointer on the chart near an edge, the pointer changes from the normal pointer ⇨ to a cross of double-tipped arrows:

You can click and drag with this pointer to move the chart around on the spreadsheet. If you click when the pointer has this crossed arrow shape, the chart appears as shown in Figure 9-23b, with sizing handles (■). When the sizing handles are visible, clicking the DELETE key deletes the chart. When the crossed arrow pointer is visible on the chart, you may drag the chart to a different position on the spreadsheet.

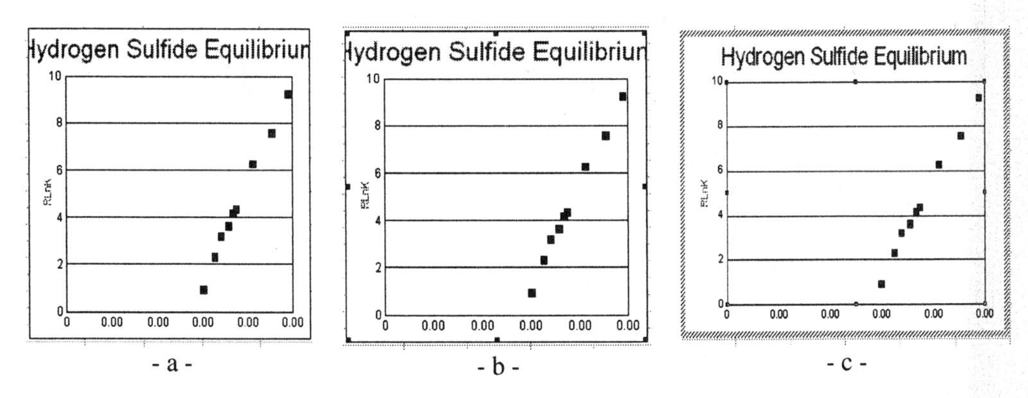

Figure 9-23. a. The Chart. b. The Chart Selected for Sizing or Moving (Has 8 ▪'s). c. The Chart Selected for Editing (Has Fuzzy Edges).

If you move the normal pointer ⇨ anywhere into the interior of the chart, it retains its shape. Clicking with the normal pointer inside the chart changes the chart's appearance so that the chart appears as shown in Figure 9-23c, a chart with fuzzy edges. At the same time, the **Chart** menu is inserted between the **Format** and **Tools** menus on the menu bar. When the chart has its fuzzy edges, it is in edit mode. Editing can be accomplished in two ways.

Editing the Chart with the Chart Menu. Click anywhere in the chart to get the **Chart** menu on the menu bar. The chart will have fuzzy edges. Then point to **Chart** to open the **Chart** drop-down menu (Figure 9-24). Then, for example, click **Titles** to open the **Titles** dialog box (Figure 9-25). This dialog box allows you to change the chart title and the labels for the x-axis and y-axis. Several other options are available in the **Chart** menu.

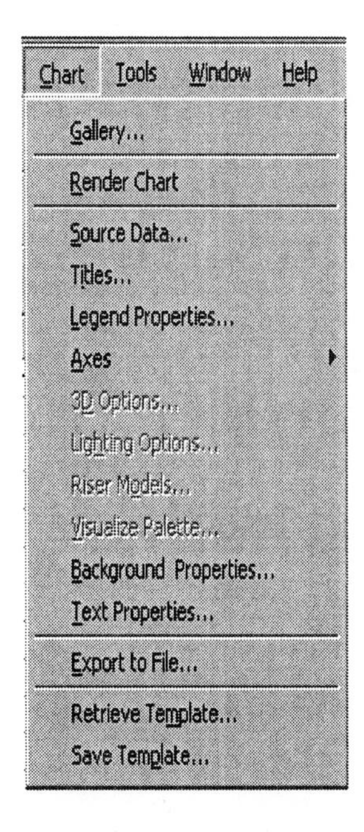

Figure 9-24. The **Chart** Menu Appears after Clicking **Chart** in the Menu Bar.

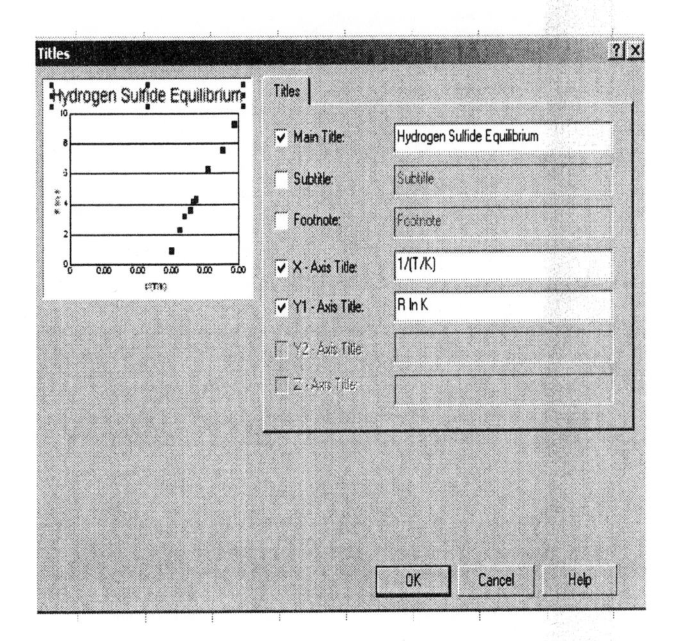

Figure 9-25. The **Titles** Dialog Box Appears after Clicking **Titles** in the **Chart Drop-Down Menu**

Editing by Right-Clicking a Chart Object. Another path to editing a chart object is to click the chart anywhere to get the chart in edit mode (fuzzy edges), and then right-click a chart object, for example the title, an axis label, or an axis number. This opens a shortcut menu. For example, right-clicking the Y-Axis when the chart is in edit mode opens a shortcut menu (Figure 9-26). In the shortcut menu, click **Y-Axis Properties** to open the **Axis Properties** dialog box (Figure 9-27).

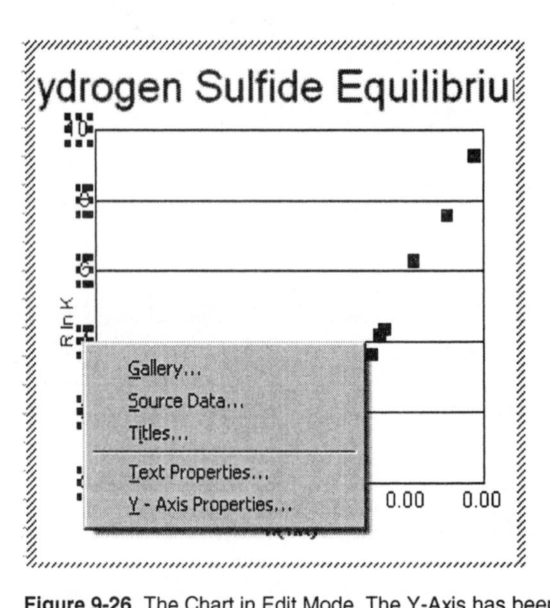

Figure 9-26. The Chart in Edit Mode. The Y-Axis has been right-clicked while in edit mode (Chart has fuzzy edges).

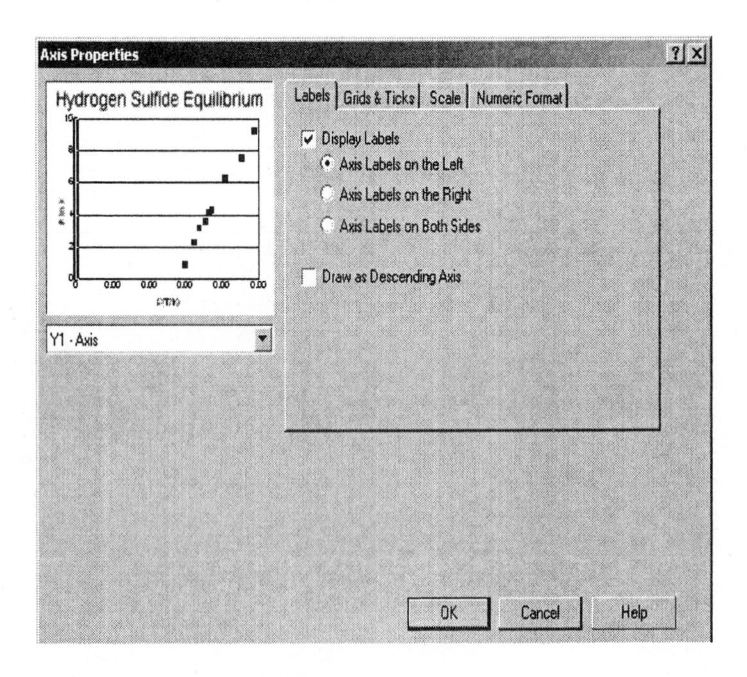

Figure 9-27. The **Axes Properties** Dialog Box. You open this dialog box by clicking **Y-Axes Properties** in the shortcut menu shown in Figure 9-26.

Summary of Chart Clicking.

To edit a chart object, click on it. Then you have two principal choices. Either click **Chart** on the **Main Menu** or **right** click the object you have just clicked.. Usually Right-clicking an chart object is the most straight forward method of editing the properties of an object. Table 9-1 summarizes some of these option.

Table 9-1

Options for Editing the Chart

Chart Object to be Edited	Click Object – then Click Chart – and then Click item in Figure 9-23	Click Object – Right Click Object – and then click item at bottom of small menu
Chart Title	Titles or Text Properties	Titles or Chart Titles db
Y-Axis Title	Axes or Text Properties	Titles or Axes Title Properties
Y-Axis	Axes or Text Properties	Text Properties or Y-Axes Properties
X-Axis Title	Axes or Text Properties	Titles or Axes Title Properties
X-Axis	Axes or Text Properties	Text Properties or X-Axes Properties
Line or Marker	Series Data	Series Properties or Data Point Properties
Plot Area	Background Properties	Titles or Frame Properties
Background	Background Properties	Titles or Background Properties

After you have created your chart, try out each of the chart edit clicking options summarized in Table 9-1. Most of the options for each object lead to the same dialog box. The important exception is the **Line or Marker**. To create a regression line, it is necessary to click a marker and then right click it to open the **Series Propertes** dialog box. This dialog box has a tab for the **Trendline** (linear regression line).

Clicking a chart object also cause the contents **Property bar** to change, allowing still another path to editing. You can also explore this option.

Creating the Linear Regression Line

Get the chart in edit mode by clicking in the chart. It should have fuzzy edges.

- Click a series marker.
- Right-click a series marker to open a shortcut menu.
- On the shortcut menu, click **Series Properties** to open the **Series Properties** dialog box (Figure 9-28).

The **Series Properties** dialog box (Figure 9-28) has several tabs. The **Trendline** tab allows creating the trendline or regression line, which may be linear or nonlinear. The **Line** tab (partially visible at the far right) enables editing of the regression line, its type (solid, dotted, or dashed), and weight. The **Type Options** tab allows the user to select different types of markers (round, square, triangular) of different sizes.

- Click the **Trendline** tab; as the method, select **Linear Fit**, and select the **Display Equation on Chart** check box.

Quattro Pro instantly draws a linear regression line through the points. The points remain located at their experimental *x,y* positions on the chart. Quattro Pro also places an equation for the line in the upper left corner of the chart. Creation of the chart is now complete and we can turn our attention to editing the chart.

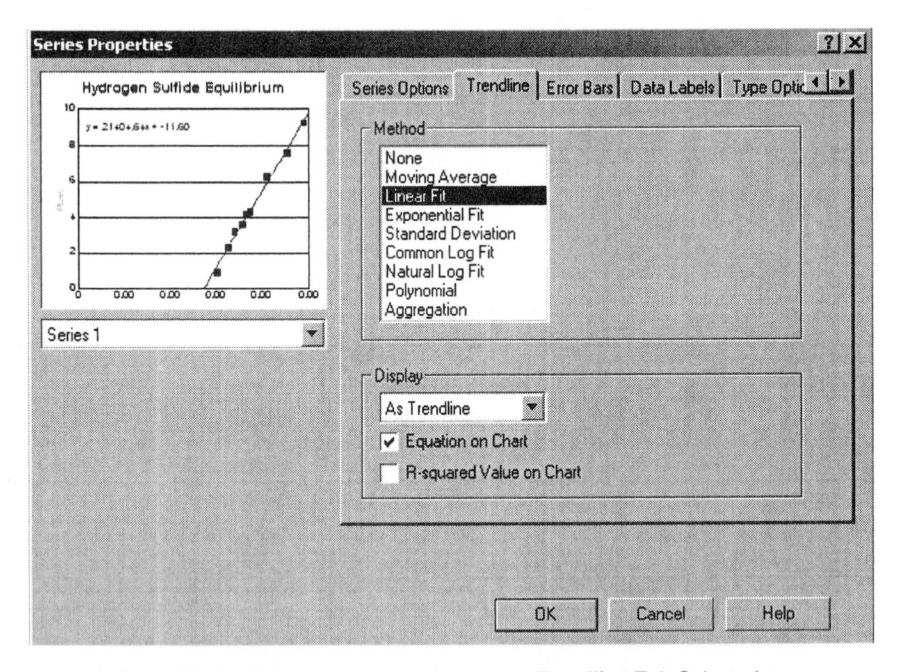

Figure 9-28. The **Series Properties** Dialog Box with the **Trendline** Tab Selected

Editing the Hydrogen Sulfide Equilibrium Chart

Editing the Chart Title. The chart title appears to be too large. We can edit the title to correct this.

- Click the title to select it. Then right-click the title, which opens a shortcut menu.

- Click **Chart Title Properties** on the shortcut menu to open the **Text Properties** dialog box (Figure 9-29).
- Change the font to Arial 10 point bold.

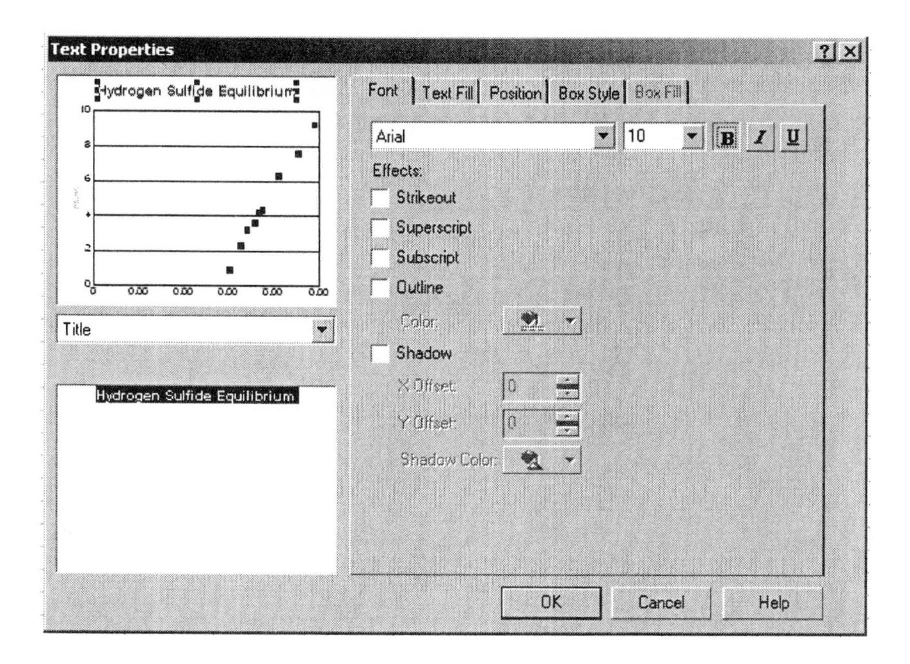

Figure 9-29. The **Text Properties** Dialog Box with the **Font** Tab Selected

Editing the X-Axis

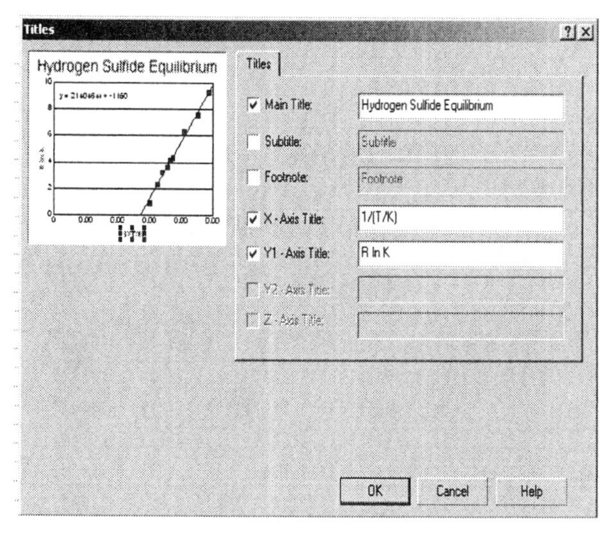

Figure 9-30. The **Titles** Dialog Box

Figure 9-29 shows that an x-axis label is missing, and the scale needs changing to fit the regression line better in the plot area. The colors of the chart object should be changed to black.

- Click any x-axis number, and then right-click any x-axis number to open a shortcut menu.
- On the shortcut menu, click **Titles** to open the **Titles** dialog box (Figure 9-30).
  1. Select the **X-Axis Title** check box.
  2. Type **1/(T/K)** for the x-axis title, and then click **OK**.
- Again open the same shortcut menu.
- Click **Axis Properties** to open the **Axis Properties** dialog box with four tabs (Figure 9-31).
- Click the **Numeric Format** tab, select **Number**, and set the number of decimal places to **4**.
- Click the **Scale** tab.
  1. In the **Max Value** box, type **0.0011** as the maximum value.

2. In the **Min Value** box, type **0.0005** as the minimum value.
3. In the **Increment** box, type **0.0001** as the increment value.
4. Click **OK**.

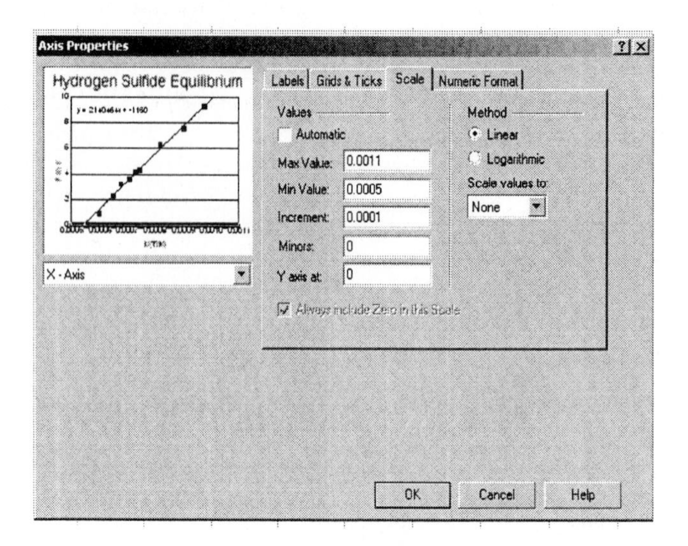

Figure 9-31. The **Axis Properties** Dialog Box

The default color of the line, markers, and equations is red, not black. To change the color to black:

- Click a marker, then right-click a marker to open a shortcut menu.
- Click **Series Properties** to open the **Series Properties** dialog Box (Figure 9-32).
- Click the **Line** tab (use the left or right arrows if necessary).
- Click **Color**, and select black from the palette.

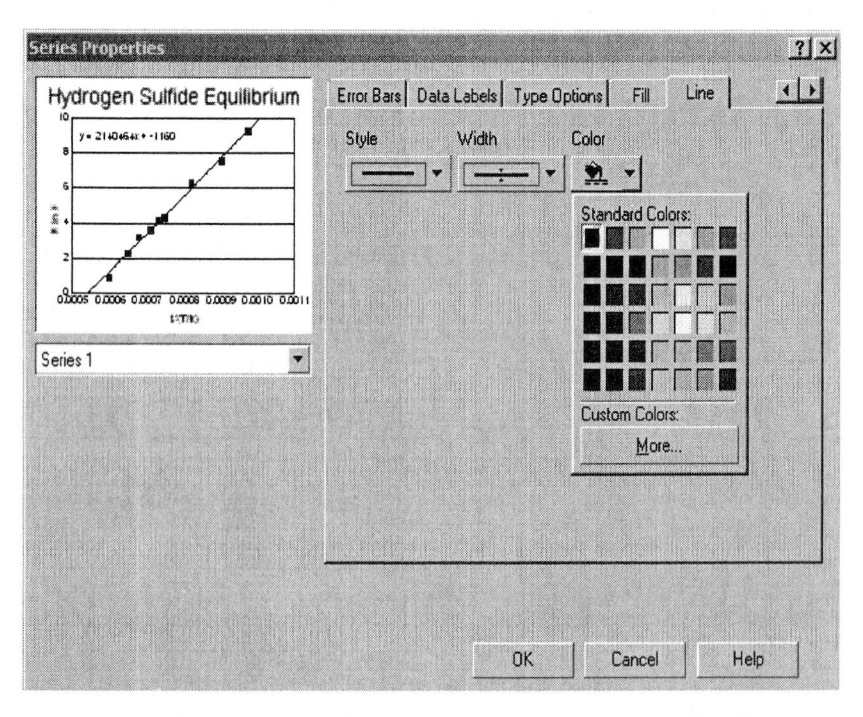

Figure 9-32. The **Series Properties** Dialog Box with the **Line** Tab Selected and **Color** Button Clicked.

Data Analysis

The equation in the upper left corner of the chart gives the slope and intercept of the linear regression line. The regression output produced by the **Linear Regression** dialog box provides more detailed statistics of the linear regression.

- Select the range D7..E15 by dragging over it.
- On the **Tools** menu (Figure 33a), point to **Numeric Tools**, and click **Regression** (Figure 9-33b).
- In the **Linear Regression** dialog box (Figure 9-33c):
    1. In the **Independent** box, type **A:D7..D15** if it is not already there.
    2. In the **Dependent** box, type **A:E7..E15** if it is not already there.
    3. In the **Output** box, type **A:A36** as the value.
    4. In the **Y Intercept** section, select **Compute**.

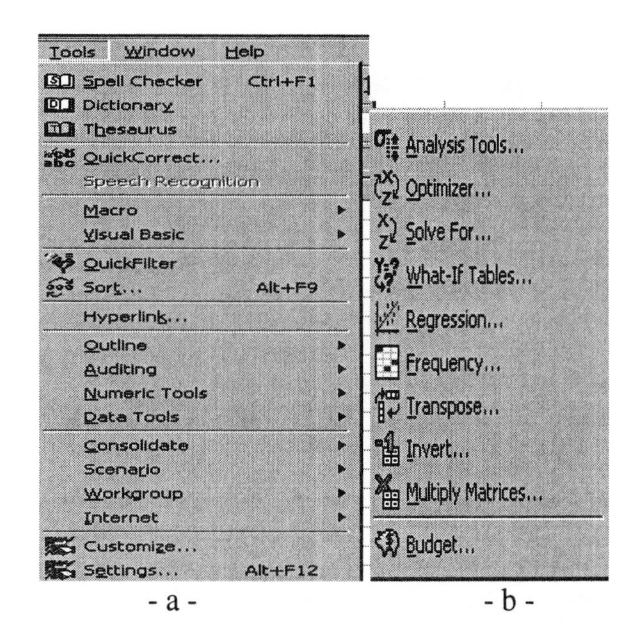

- a - - b -

The cell address entered in the **Output** box determines the upper left corner of the output block printed by **Linear Regression**

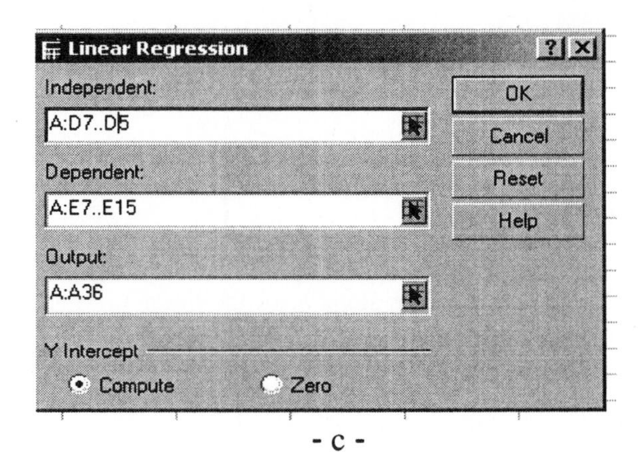

- c -

Clicking OK on the **Linear Regression** dialog box results in the regression output table shown in the bottom of the spreadsheet (Figure 9-34).

Figure 9-33. a. The **Chart** Menu
b. The **Numeric Tools** Submenu
c. The **Linear Regression** Dialog Box

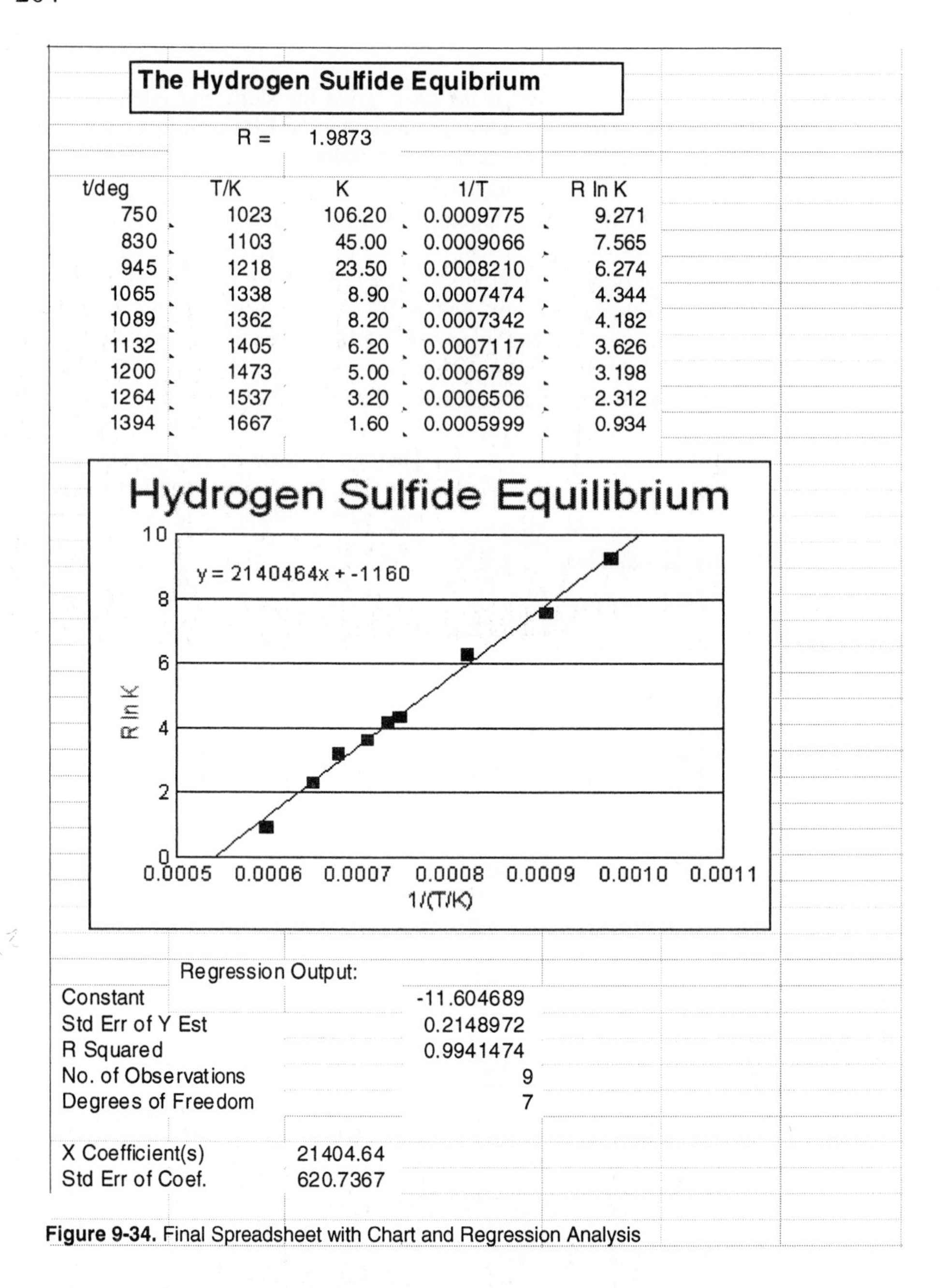

The Hydrogen Sulfide Equibrium

R = 1.9873

t/deg	T/K	K	1/T	R ln K
750	1023	106.20	0.0009775	9.271
830	1103	45.00	0.0009066	7.565
945	1218	23.50	0.0008210	6.274
1065	1338	8.90	0.0007474	4.344
1089	1362	8.20	0.0007342	4.182
1132	1405	6.20	0.0007117	3.626
1200	1473	5.00	0.0006789	3.198
1264	1537	3.20	0.0006506	2.312
1394	1667	1.60	0.0005999	0.934

Regression Output:

Constant		-11.604689
Std Err of Y Est		0.2148972
R Squared		0.9941474
No. of Observations		9
Degrees of Freedom		7
X Coefficient(s)	21404.64	
Std Err of Coef.	620.7367	

Figure 9-34. Final Spreadsheet with Chart and Regression Analysis

From the regression output table (Figure 9-34, bottom left), the slope m of the linear regression line equals 21,404, and the intercept b equals –11.6045, for the line $y = mx + b$. From a comparison with Equation 7-2, Excel Example 21, Chapter 7, it follows that ΔH^0 equals –21.4(6) kcal/mole and ΔS^0 equals –11.6(2) cal/K/mole. The equation for the equilibrium constant can be written

$$R \ln K = \frac{21{,}400}{T} - 11.60$$

Compare this cgs equation with the SI equation in Chapter 7, Excel Example 21.

Comparison of Linear Regression in Excel, Lotus, and Quattro Pro

To compare the use of Excel, Lotus and Quattro Pro in calculating a linear regression, the same data are used from Table 7-1, in Excel Example 17, Chapter 7. You will recall that Excel and Lotus 1-2-3 also have some commands for quick linear regressions:

- With Excel, on the **Chart** menu, click **Add Trendline,** and on the **Type** tab, select **Linear**; or on the **Tools** menu, click **Data Analysis**, and in the **Data Analysis** dialog box, click **Regression.**
- With Lotus 1-2-3, on the **Chart** menu, click **Properties for Series**, and click the **Series Trend** tab to create a regression line, and on the **Range** menu, point to **Analyze**, and click **Regression** to analyze the regression line.
- With Quattro Pro, on the **Tools** menu, point to **Numeric Tools**, and click **Regression**, or use **Trendline** from the **Series Properties** dialog box.

Quattro Pro Example 2

Vibrational Wave Functions

Review

With this example, we can compare how Quattro Pro handles multiple point graphs on a single chart with a similar problem already encounter with Excel Example 11, Vibrational Wave Functions: v = 0, 1, and 2, in Chapter 3. The vibrational wave function for each vibrational energy level, characterized by the vibrational quantum number v, consists of three factors: the exponential factor, a normalization factor, and a Hermite polynomial. The exponential factor is the same for all quantum states: $e^{-\frac{q^2}{2}}$. The factors H_v are Hermite polynomials, well known to mathematicians. A few normalization factors and Hermite polynomials are listed in Table 3-3.

The meaning of the vibrational wave function plots is discussed in Chapter 3, Excel Example 11. To compare Quattro Pro and Excel, the same vibrational wave functions are plotted in this Quattro Pro example as are plotted in Excel Example 11, Chapter 3.

Entering the Spreadsheet Data

As shown in Figure 9-35, $\psi_v(q)$ and $\psi_v^2(q)$ for v equals 0, 1, and 2 are graphed from q equals –5 to +5.

Entering the Text

- In the range B1..D1, type **Vibrational Wave Functions** as the spreadsheet title.

- For the purpose of documentation, enter the equations for the wave function to be plotted. Use **Insert_Symbol_Greek** on the **Menu Bar** for the Greek ψ. Use the $\mathbf{x^2}$ and $\mathbf{x_2}$ icons on the **Property Bar** for the superscripts and subscripts.
  1. Beginning in cell A3, type ψ_0 = **EXP(-(q^2/2))** as the equation.
  2. Beginning in cell A4, type ψ_1 = **0.7071*EXP(-(q^2/2))*2*q** as the equation.
  3. Beginning in cell A5, type ψ_2 = **0.3536*EXP(-(q^2/2))*(4*q^2-2)** as the equation.
- In cells A7..H7, type the column titles shown in Figure 9-35:

$q \quad \psi_0 \quad \psi_1 \quad \psi_2 \quad \psi_0^2 \quad \psi_1^2 \quad \psi_2^2$

Entering Values of *q* from –5 to +5 with the Edit Fill Series Method

- Select cell A9.
- On the **Edit** menu, click **Fill Series** to open the **Fill Series** dialog box (Figure 9-36).
  1. In the **Cells** box, type **A9..A109** as the value.
  2. In the **Start** box, type **-5** as the value.
  3. In the **Step** box, type **0.1** as the value.
  4. In the **Stop** box, type **5** as the value.
  5. In the **Order** section, select **Column.**
  6. In the **Static** section, select **Linear**; click **OK**, and the series fills with values of q from –5 to +5, incremented in steps of 0.1.
  7. Click column **A** to select it; On the **Format** menu, point to **Selection**, click **Numeric Format**, select **Number**, and for decimal places, select **4**.

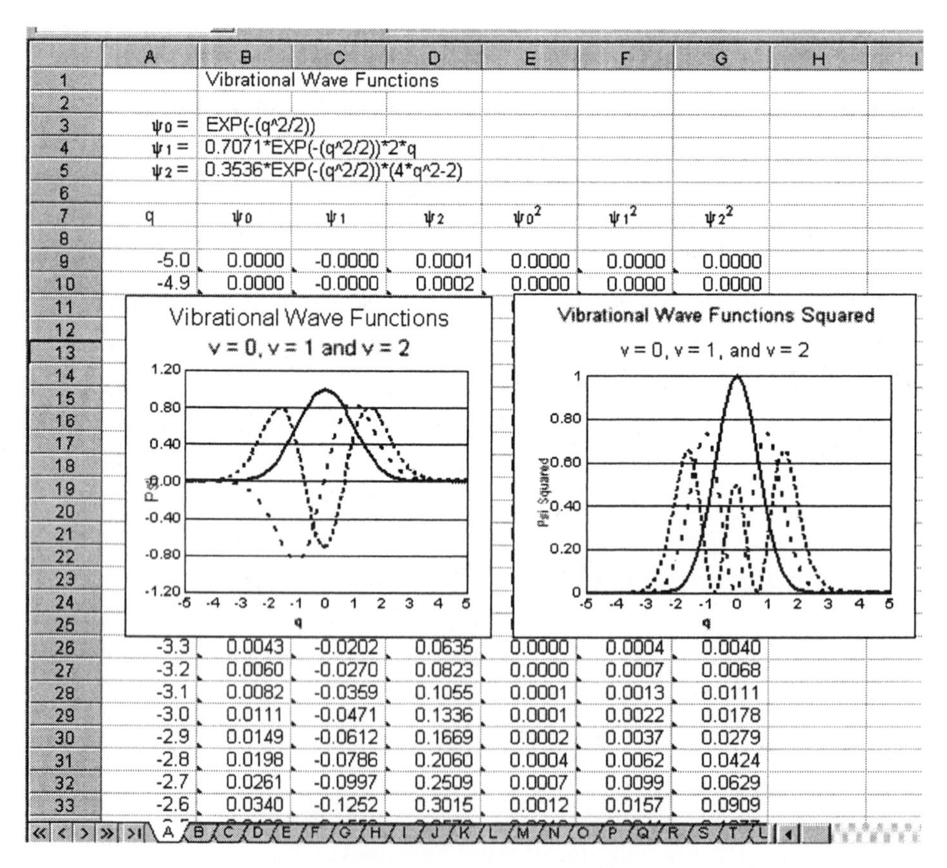
Vibrational Wave Functions

ψ0 = EXP(-(q^2/2))
ψ1 = 0.7071*EXP(-(q^2/2))*2*q
ψ2 = 0.3536*EXP(-(q^2/2))*(4*q^2-2)

q	ψ0	ψ1	ψ2	ψ0²	ψ1²	ψ2²
-5.0	0.0000	-0.0000	0.0001	0.0000	0.0000	0.0000
-4.9	0.0000	-0.0000	0.0002	0.0000	0.0000	0.0000
-3.3	0.0043	-0.0202	0.0635	0.0000	0.0004	0.0040
-3.2	0.0060	-0.0270	0.0823	0.0000	0.0007	0.0068
-3.1	0.0082	-0.0359	0.1055	0.0001	0.0013	0.0111
-3.0	0.0111	-0.0471	0.1336	0.0001	0.0022	0.0178
-2.9	0.0149	-0.0612	0.1669	0.0002	0.0037	0.0279
-2.8	0.0198	-0.0786	0.2060	0.0004	0.0062	0.0424
-2.7	0.0261	-0.0997	0.2509	0.0007	0.0099	0.0629
-2.6	0.0340	-0.1252	0.3015	0.0012	0.0157	0.0909

Figure 9-35. Spreadsheet for Vibrational Wave Functions and Their Squares for v Equals 0, 1, and 2

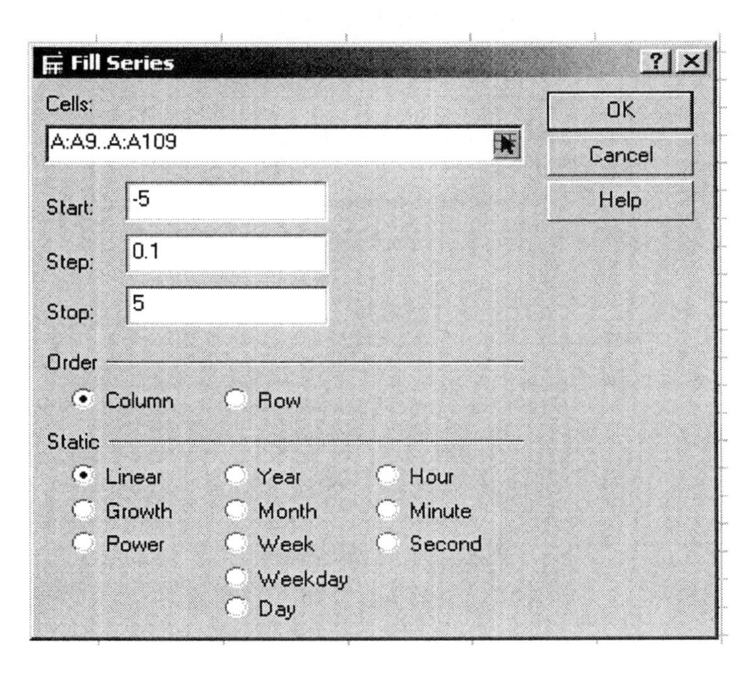

Figure 9-36. The **Fill Series** Dialog Box

Entering the Formulas with the Edit Copy Cell Method

The formulas for the wave functions are entered in cells B9, C9, and D9. These functions are the product of $Nv/(a/\pi)^{1/4}$, Hv, and $e^{-\frac{q^2}{2}}$ for $v = 0$, 1, and 2 listed in Table 3-3. **Note**: The term $(a/\pi)^{1/4}$ is not entered into the formula. This means that the unit of the plot along the y-axis is $(a/\pi)^{1/4}$. The **Copy Cell** command on the **Edit** menu is used to fill the cells with relatively referenced formulas.

Entering the Wave Functions ψ_0 ψ_1 ψ_2

- In cell B9, type **=@EXP(-(A9^2/2))** as the formula.
- On the **Edit** menu, click **Copy Cells** to open the **Copy Cells** dialog box (Figure 9-37).
  1. In the **From** box, type **A:B9..B9** as the value.
  2. In the **To** box, type **A:B9..B109** as the value.
- In cell C9, type **= 0.7071*@EXP(-(A9^2/2))*2*A9** as the formula.
- On the **Edit** menu, click **Copy Cells** to open the **Copy Cells** dialog box (Figure 9-37).
  1. In the **From** box, type **A:C9..C9** as the value.
  2. In the **To** box, type **A:C9..C109** as the value.
- In cell C9, type **= 0.3536*@EXP(-(A9^2/2))*(4*A9^2-2)** as the formula.
- On the **Edit** menu, click **Copy Cell** to open the **Copy Cells** dialog box (Figure 9-37).
  1. In the **From** box, type **A:D9..D9** as the value.
  2. In the **To** box, type **A:D9..D109** as the value.

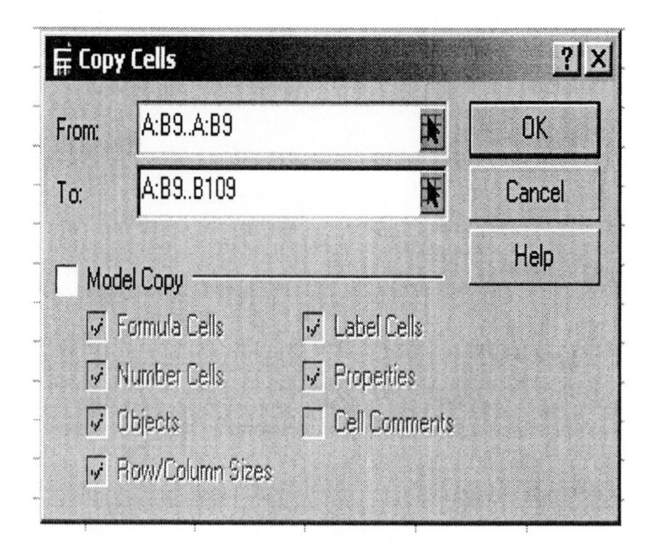

Figure 9-37. The **Copy Cells** Dialog Box

Entering the Wave Functions ψ_0^2 ψ_1^2 ψ_2^2

- Select the range B9..D9, and copy the contents to the range E9..G9.
- Select cells E9, F9, and G9, and change each occurrence of **D9** to **A9.**
- Select each of these cells, enclose in parentheses, and to square them, add **^2.**
- Fill the ranges E9..E109, F9..F109, and G9..G109 with the Edit Fill Cells method.
- To format the number, click and drag over the B, C, and D column headings.
- Select columns B, C, D, E, F, and G by dragging across them.
- On the **Format** menu, point to **Selection**, click **Numeric Format**, select **Number**, and for decimal places, select **4**.

Creating the Chart of Unsquared Wave Functions

- Select the range A9..D109.
- Move the spreadsheet so that the block I1..N40 is visible.
- On the **Insert** menu, click **Chart** to activate the Expert.
- In the **Expert – Step 1 of 4** dialog box, check that the data selected in the window is A9..D109. Click **Next.**
- In the **Expert – Step 2 of 4** dialog box:
  1. In the **Category** list, scroll down and select **Scatter**.
  2. In the **Type** section, select **XY Scatter**.
  3. Click **Next**.
- In the **Expert – Step 3 of 4** dialog box:
  1. In the **Title** box, type **Vibrational Wave Functions** as the title.
  2. In the **Subtitle** box, type **v = 0, v = 1, and v = 2** as the subtitle.
  3. In the **X-Axis** box, type **q** as the title.
  4. In the **Y-Axis** box, type **Psi** as the title.
  5. Click **Next**.
- In the **Expert – Step 4 of 4** dialog box, select **No change**, and click **Finish**.

At this point this dialog box disappears, and the pointer changes to a tiny bar graph. Place the pointer in an empty area of the spreadsheet; click and drag from upper left to lower right. The chart should display three plots

For the Y-Series one plot should have A:B9..B109, the second plot should have as the Y-Series A:C9..C109 and the third plot should have as its Y-Series A:D9..D109. To force Quattro Pro to use these series, use the following procedure.

- Click on the chart to get it into edit mode
- Click on Source Data to get the Source data Dialog box (Figure 9-38)

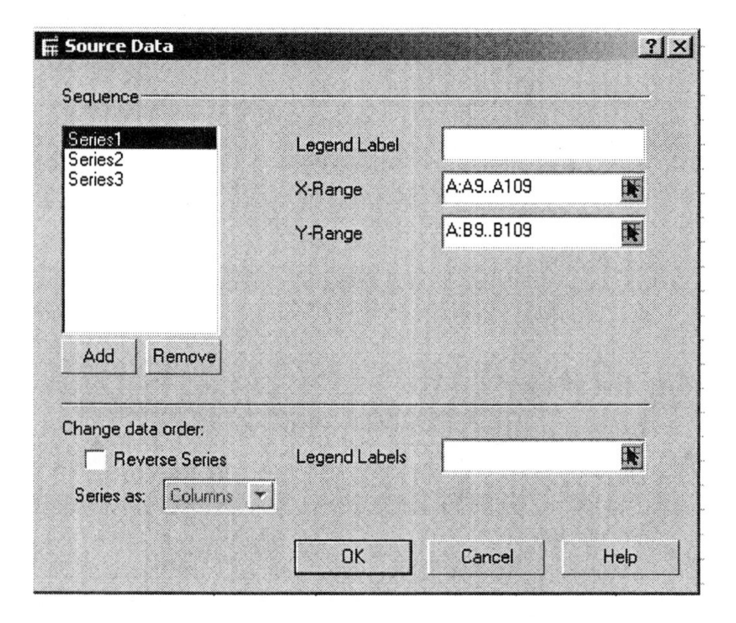

Figure 9-38. The Source Data Dialog Box.

- Select **Series1** and edit it so that
  1. the **X-Series** is **A:A9..A109**
  2. the **1st Y-Series** is **A:B9..B109**
- Select **Series2** and edit it so that
  1. the **X-Series** is **A:A9..A109**
  2. the **2nd Y-Series** is **A:C9..C109**
- Select **Series3** and edit it so that
  1. the **X-Series** is **A:A9..A109**
  2. the **3rd Y-Series** is **A:D9..D109**
- If the **Source Data Series** dialog box displays only one or two series, click the **Add** button under **Sequence** menu of series, and fill in the X-Range and Y-Range(s) as described.
- Click **OK**. The selected chart now displays three wavefunctions

Editing the Chart

Editing the Chart Titles. The chart title appears to be too large. We can edit the title to correct this.

- Click the title to select it. Then right-click the title, which opens a shortcut menu.
- Click **Chart Title Properties** on the shortcut menu to open the **Text Properties** dialog box (Figure 9-29).
- Change the font to Arial 10 point bold.
- Repeat, and change the subtitle font to Arial 8 point bold.

Editing the Plotted Lines. We want a plot consisting of a smooth line and no marker. The lines should appear different for each vibrational quantum number.

- Click the chart, then click a line marker on the v equals zero line, then right-click the line marker of the middle ($v = 0$) plot.
- On the shortcut menu that appears, click **Series Properties**.
  1. Click the **Line** tab at the far right (Figure 9-39); click **Color**, and change the color to black. Leave this line's style solid.
  2. Click the **Type Options** tab, and clear the **Marker** check box.
- Repeat for the other two lines. Make one of them dashed and the other dotted using the **Style** list in Figure 9-39).

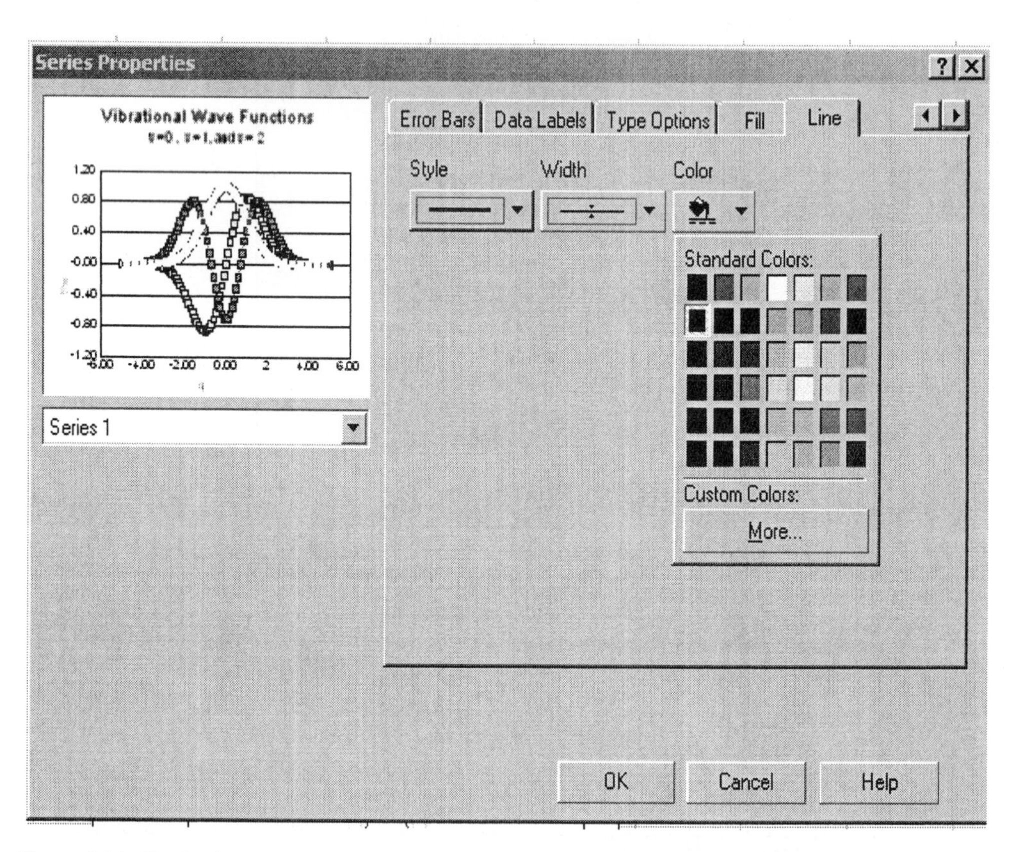

Figure 9-39. The **Series Properties** Dialog Box with the **Line** Tab and **Color** Selected

Creating the Chart of Squared Wave Functions

- Hold down the CTRL key, and click and drag from cell A9 to A109.
- Continue holding the CTRL key down, and click and drag over the range E9..G109. The numerical data in columns A, E, F, and G should now be selected.
- On the **Insert** menu, click **Chart** to activate the Expert and follow the same procedure described for the chart of the unsquared wave functions. Use the **Source Data** dialog box if necessary, as described above.

Part II. Visual Basic

Chapter 10. Visual Basic—Introduction
Chapter 11. Visual Basic Controls: Properties, Methods, and Events
Chapter 12. Visual Basic Operators, Control Structures, and Functions
Chapter 13. Visual Basic Loops without Arrays
Chapter 14. Visual Basic Loops with Arrays
Chapter 15. Visual Basic Files
Chapter 16. Visual Basic Graphics

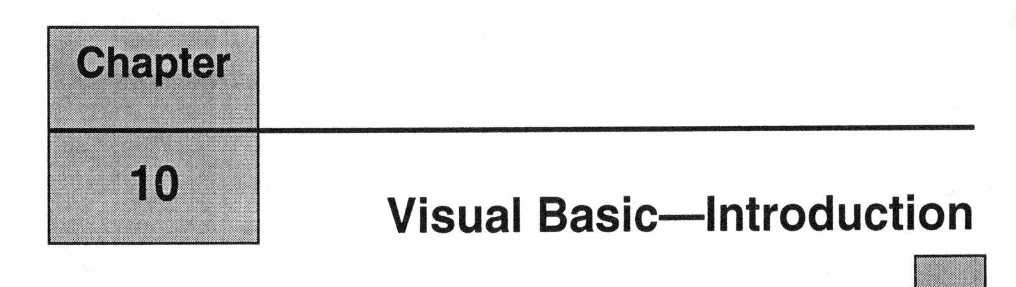

Chapter 10 Visual Basic—Introduction

Visual Basic (VB) evolved from BASIC (Beginner's All-Purpose Symbolic Instruction Code), which was developed by John Kemeny and Thomas Kuntz at Dartmouth College in 1960. Several versions followed the original, including GWBASIC (1970s) and QuickBASIC (1980s). Microsoft introduced Visual Basic 1.0 in 1990 to run on its recently developed (1985) Windows operating system.

The current version of Visual Basic is version 6.0, and it comes in three editions. The powerful, but least expensive edition is the Learning Edition, which was used for the examples in this book. The Professional Edition, for professional program developers, comes with more controls, tools, and auxiliary applications. The industrial-strength Enterprise Edition includes all of the features of the other versions and more.

As a programming language, Visual Basic is as easy to learn as Pascal, and it has power and acceptance comparable to C++. Visual Basic can make developing applications in a Windows environment so easy that it's actually fun. With Visual Basic, you can create a graphical user interface that is as slick, polished, and professional as in those applications you buy off the shelf. The Basic in Visual Basic has come a long way from the BASIC of old, but it is just as easy to learn and is much more powerful and robust. Moreover, another slight abridged version of Visual Basic is supplied with many applications marketed by Microsoft and other venders. This version, called Visual Basic for Applications (VBA), is included, for example, with Microsoft Word, Excel, Access, and PowerPoint.

This chapter and the following six chapters offer a presentation of Visual Basic that tends to concentrate on numerical calculations. These chapters are intended to stand alone as a primer on Visual Basic, but they also provide a background for Chapters 17 and Chapter 18, which introduce Visual Basic for Applications. Thus, these chapters on Visual Basic (Chapters 10 through 16) form a bridge between Excel (Chapters 1 through 9) and VBA (Chapters 17 and 18). Even without investing in Microsoft Visual Basic, you can use these chapters as a source book for using VBA, which is almost surely available with one of your applications, such as Word or Excel.

In this chapter you will learn to write a Visual Basic program that serves as an introduction: The Color Properties program.

The Nature of a Windows Program

Programs consist of a collection of objects and code. The user sees the objects on the screen window, but does not see the code. The programmer creates the code, which is invisible to the user.

What the User Sees. Look at any Windows program. It consists of a form that usually covers most of the screen. The form itself is an object. The form shown in Figure 10-1 is typical of a form in which are embedded a number of other objects. This particular form is the interface that the user encounters upon opening Hewlett-Packard's software for one of its scanners. The form in Figure 10-1 has eight command button objects. The first has the caption **Start a new scan**. To its left is a label object, with the caption **1.** Below this object is a list box object. The objects embedded in a form are called controls, even though they may or may not control anything. This label object is a control even though it controls nothing. The command button is a control and, indeed, it does function to control. A click on a control button causes some code associated with the command button to execute.

Look at the remaining controls. You recognize the list box control's down arrow that, when clicked, displays a list of options from which you can choose. To its left is another label control, this one with the caption **2.** Below it is another label control with the caption **3.** To its right is a frame control enclosing three command buttons, whose captions are iconic. Dropping down further on the right side, you find another command button control with the caption **Print the scan now**, and to its left is another label control with the caption **4.** An image box control occupies most of the right side of the form. In this program it displays the scanned image. The form holds three more command button controls below the image box, two with iconic captions and the last one with the caption

Help. This typical form contains embedded in it a total of fifteen objects (controls), and together these objects present to the program user a graphical user interface. As a Windows program user you are probably vaguely familiar with the appearance of many of the controls shown in Figure 10-1.

What the User Does. When you start a Windows program, you are in control, and the program waits for some kind of event to occur, to which it responds appropriately. As an example of an event, you might move the mouse to a new position. You might click the mouse or double-click the mouse or right-click the mouse or press a keyboard key. You might enter some data into a waiting text box. You might scroll down a list or click a command button. All these actions on the part of the user are events, and inevitably, events trigger action: something happens.

Properties of Objects. One feature of all objects in a Windows program is their set of properties. Each object has a unique set of properties. As an example, consider a command button (Figure 10-1). It has some obvious properties that you can see by looking at one: its size, its location on the form, and its caption. Actually, Visual Basic recognizes thirty other properties of a command button! It can be made visible or invisible. Its caption font can be any font that Windows supports. It can be disabled or enabled. Its default name is **Command1** (if it were the first command button to be put on the form), but the programmer can give it any name.

Some controls have properties similar to the properties of other controls. Some properties are unique to a single control. The properties of objects such as controls can be changed. Visual Basic gives every object a default set of properties when the object is first selected for use in a form. However, properties can be changed at design time or at run time or both. Writing the code for a Windows program consists in large part of changing the properties of the objects that make up the program.

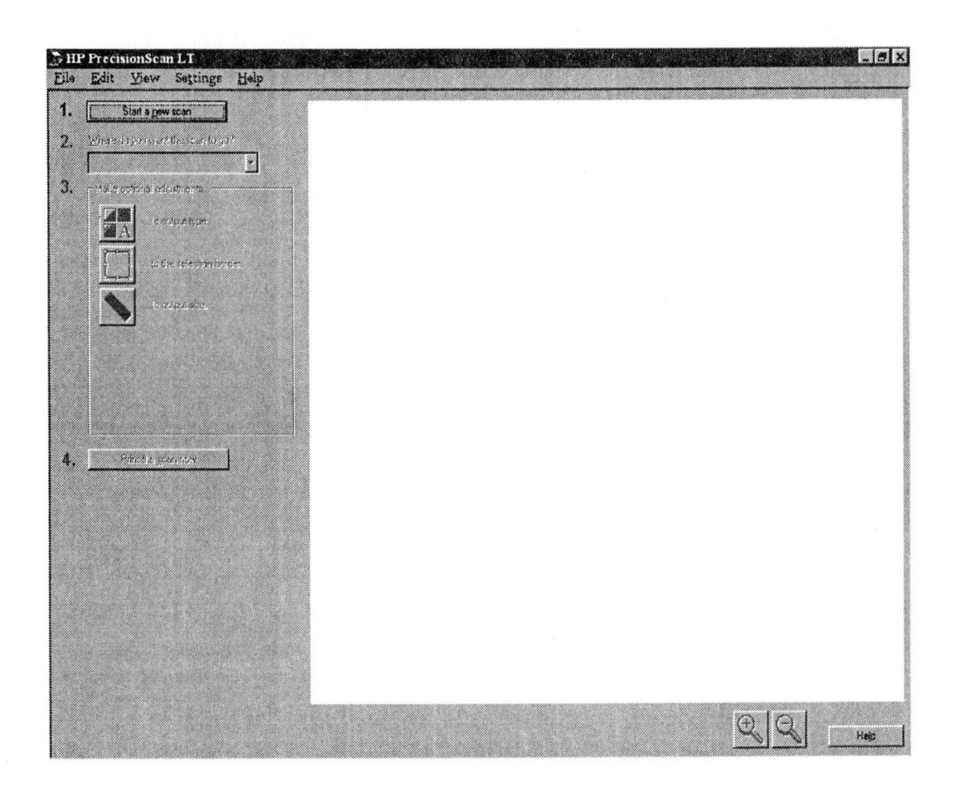

Figure 10-1. A Typical Visual Basic Form with Controls Embedded in it

What the Programmer Does. All of the aforementioned objects are visible to the user of the Windows program. The code that carries out the appropriate response is part of the program, but the code is invisible to the user. Writing a Windows program involves selecting the controls needed for the task at hand, embedding them in a form (user interface), and then writing the code that carries out the appropriate action and connects the chosen controls and the form. To summarize, the programmer is involved in:

1. **Creating the user interface** (the form and its embedded controls)
2. **Setting the properties of the controls** (caption, name, size, color, and other properties)
3. **Writing the code** (attaching the Visual Basic code to the controls)

Actually, the very first step occurs away from the computer. It involves thinking very carefully about what you want the program to do, how the form should appear to the user, and the algorithm for the code. The algorithm is an abbreviated list of steps that the code should carry out in order to do what you want the program to do.

Visual Basic provides the programmer with the form that the user will eventually see. Visual Basic provides the programmer with the controls to embed in the form. VB also provides a code editor, which is just a specialized word processor for writing the code associated with each control.

The Visual Basic Programming Environment

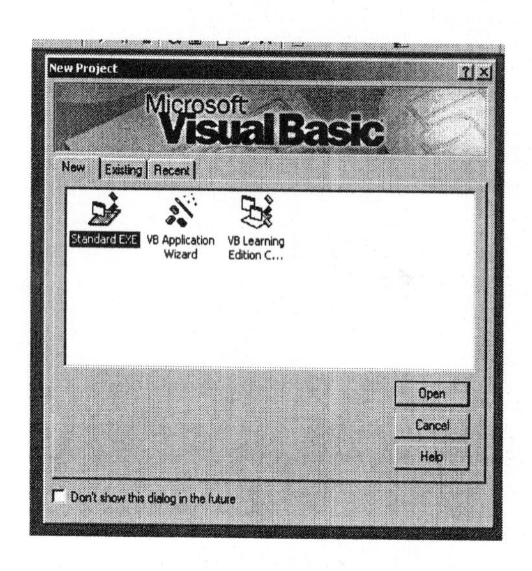

Figure 10-2. The Visual Basic New Project Dialog Box

After installing Microsoft Visual Basic, click the **Visual Basic** icon on your Windows desktop. Two windows appear (Figure 10-2). Grayed out in the background is the Visual Basic programming environment window with the title bar **Microsoft Visual Basic**. In the foreground, with the title bar **New Project**, is a window with a few icons, one of which is labeled **Standard EXE**. Either double-click it or click **Open** at the bottom of the window. The **New Project** window disappears, leaving the Visual Basic programming environment (Figure 10-3) with a title bar showing **Project 1 – Microsoft Visual Basic [design]**. From now on, we shall call this the design window.

The Design Window. Here is where the programmer works. It is in the design window that the programmer creates the interface, embeds controls in the form, and writes the code. Under the title bar you will recognize the usual Windows application menu bar, containing the **File**, **Edit**, and other menus. The design window contains four windows, and on the far left a toolbar, which is actually called the toolbox. These five objects are in constant use when you write a Visual Basic program.

Starting in the upper right of the design window (Figure 10-3) and proceeding clockwise, you will find the project window, sometimes called the Project Explorer, since it is laid out similar to the Windows Explorer. Below the project window, the properties window is located. To its left is the form window (with the complete name **Projectl – Forml [Form]**. Above the form window is the code window (with the complete name **Projectl – Forml [Code]**). In this blank region is the area where the program code is written. And at the far left is the toolbox, where the controls are available for the programmer's use. The programmer selects controls to embed in the form.

If some of these windows don't show up when you open a new project (program), don't worry. They can be opened and closed in a variety of ways that will be described shortly.

The Project Window. Take another look at the project window and see if it appears as it does in Figure 10-3. If the line **Projectl (Projectl)** has a plus sign before it, then the next two lines will be missing. Click the plus sign and the line below it appears. Try clicking the plus and minus signs to see how the list expands and contracts. The third line in the project window is **Forml (Forml)**. If the form window is missing, double-clicking this line will make the form window reappear

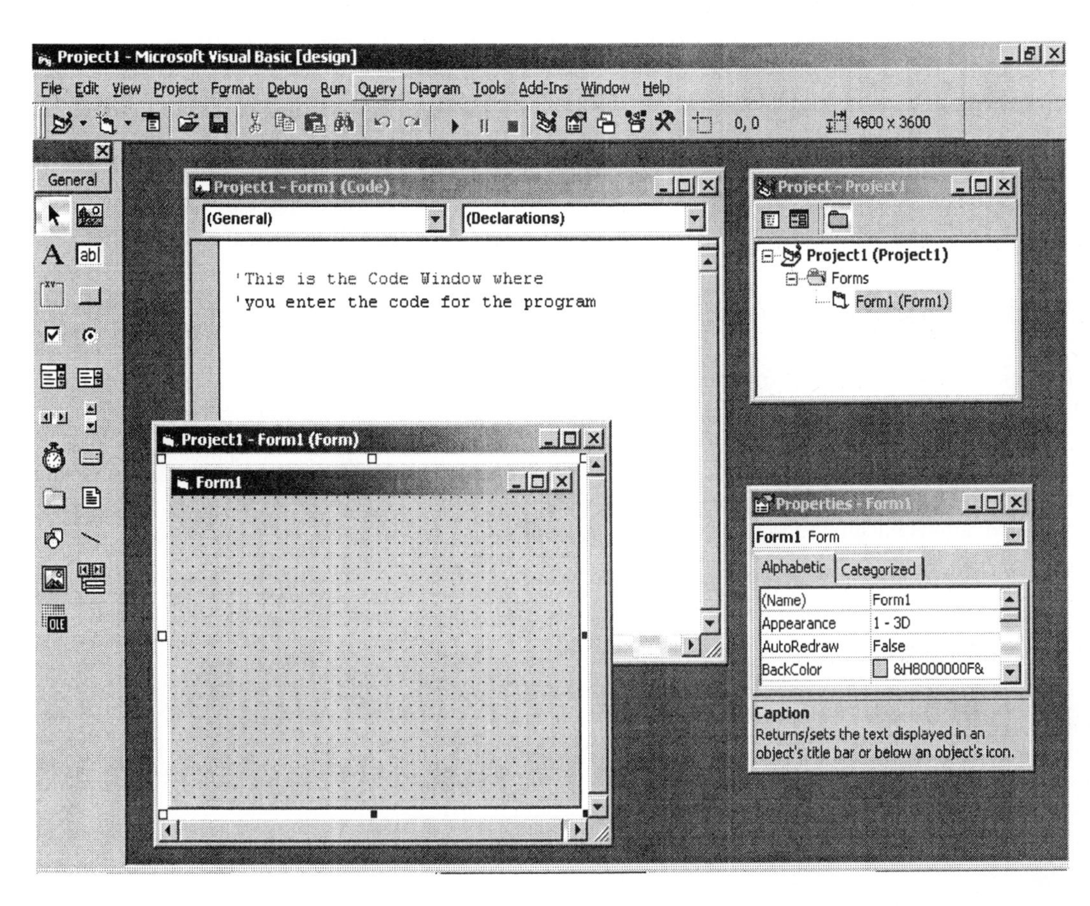

Figure 10-3. The Visual Basic Design Window, Displaying the Code Window, the Project Window, the Properties Window, and the Form

The Form Window. The form window (middle left, Figure 10-3) has two title bars (which are dark blue when the form window is selected). The upper title bar, **Project1 – Form1 (Form)**, is the default **(Name)** property of the form. When you write a program, you will usually change this name to something that is appropriate to your program and that conforms to certain naming conventions that are discussed later. For example a typical name for a form could be **frmMyFirstProgram** (one word).

The lower title bar, **Form1**, is the default caption of the form. (**Form1** is both the default caption and default name for **Form1**). When you write a program, you might change the default caption to something more appropriate; for example, **My First Program for Calculating Interest** (any number of words). For the form, the default name and caption happen to be the same, but for many controls they are different.

In Figure 10-3, you can tell that the form is the selected window, because the title bar is darkened and because the form is surrounded on its side and corners by sizing handles (tiny square boxes). Dragging these changes the size of the form. Notice that the properties window (lower right) is displaying the properties of the selected form. You can see that the default **(Name)** and **Caption** properties are **Form1**. The properties are listed alphabetically, except for the **(Name)** property, which is always at the top of the list.

The Toolbox. The tools in the toolbox (Figure 10-3, far left, and Figure 10-4) are the controls that we can select and embed in **Form1**. If you slide the pointer over each icon in the toolbox, a little yellow label appears giving the name of the control. Some of the controls really don't control anything ... like the label control. It just places a label on the form. Nevertheless, all the tools in the toolbox are called controls, even though their formal name is *programmable user interface element.* The form in which the controls are embedded has the formal name *user interface form*, but we shall refer to it simple as the form or Form1.

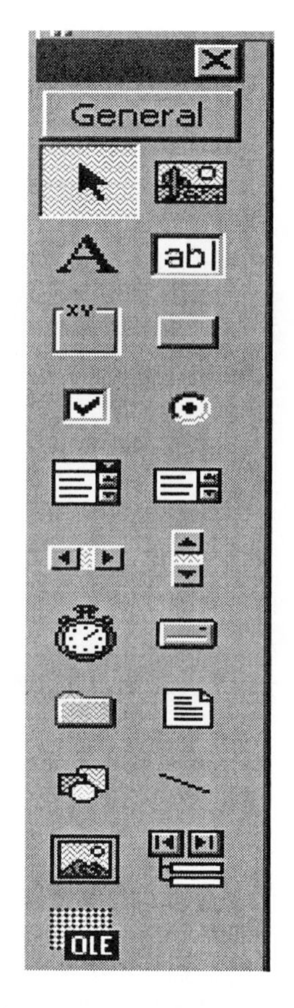

Figure 10-4. The Toolbox

The Properties Window. In Figure 10-3, this window is located in the lower right. Every control has its own set of properties. When a control is first placed on a form, the control has a set of default properties. Clicking a control at design time causes the properties window to list the default properties of the selected control. Nearly always the programmer will want to change a few properties of the selected control. The properties are listed alphabetically in the properties window, where a property may be selected and changed.

The Code Window. In Figure 10-3, the code window lies in the upper left, behind the form. If you click the code window as it appears in Figure 10-3, it would be the selected window and would come to the fore, and would lie over the form.

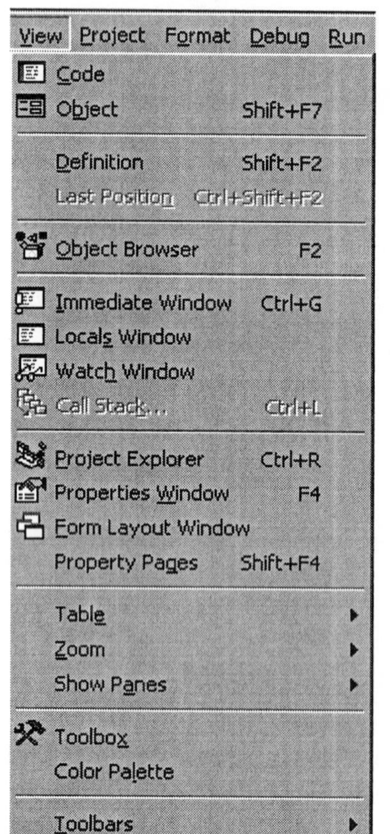

Figure 10-5. The View Menu

The code window acts as a word processor for entering code in the form of alphanumeric text. The **Edit** menu on the menu bar permits you to copy, cut, and paste just as with any word processor. The **File** menu on the menu bar permits you to print a hard copy of the code. The two lines written in the code window of Figure 10-3 are actually just comments. Any text preceded by an apostrophe is ignored by the compiler and serves as the programmer's comments.

Viewing Missing Windows. All of these windows may be positioned elsewhere on the design window by clicking and dragging their title bars. They may be resized by clicking and dragging an edge. The form is resized by selecting it and clicking and dragging its size handles (the tiny black squares that appear on its corners and edges when it is selected). And finally, one or all of these windows may be missing from the design window when you open it.

To view any missing windows or the toolbox, clicking **View** on the menu bar causes the drop-down menu shown in Figure 10-5 to appear. The menu shows an icon for each of the objects that we want to view on the design window. Beginning at the top, clicking:

- The **Code** icon makes the code window visible
- The **Object** icon makes the form window visible
- The **Project Explorer** icon makes the project window visible
- The **Properties Window** icon makes the properties window visible
- The **Toolbox** icon makes the toolbox visible.

Notice that some of the icons on the **View** menu also appear on the VB toolbar towards its right side: **Project Explorer**, **Properties Window**, and **Toolbox**. These icons provide an alternative way to make the windows and toolbox visible. After you are satisfied that all four windows and the toolbox are present and accounted for on the design window, practice resizing and repositioning them.

Docking Windows. One last characteristic of the windows in the design window should be mentioned. Each window appears to "float" around in the design window. The project window, the properties window, and the toolbox may be tied to the edge of the design window much like a boat is docked at a pier. These windows can be set to be dockable or not dockable. When an object is docked, it attaches (docks) itself to an edge of the design window. In Figure 10-3, the toolbox is docked against the left edge of the design window. You can see that the project and properties windows are not docked and can be dragged to any position in the design window. The project/form window cannot be docked.

To change the dockable state of an object, right-click the object. This opens a shortcut menu that lists three or four commands, one of which is **Dockable**. Click it to select it, and the window becomes dockable. Select **Dockable** for the toolbox, but clear **Dockable** for the project window and the properties window.

When you are confident that you can retrieve misplaced windows, we are ready to see how to write a simple program, and in doing so, you will gain insight into the roles played by the four windows and toolbox displayed on the design window.

The Command Button Control

The toolbox in Figure 10-3 displays the **Command Button** icon as the third icon down in the right column of icons. A command button allows the user to begin or end some kind of action by clicking it. It nearly always has some text written on it to prompt the user. This text is the **Caption** property of the command button. The command button has code associated with it, written by the programmer, but invisible to the user. When the user clicks the command button, the command button responds to the click event and executes the associated code. The command button is a frequently used control.

The Click() Event. One use of the command button that you have probably already experienced is a button to exit the program when clicked. The short click procedure code for a command button named **Command1** to exit the program looks like this:

```
Private Sub Command1_Click()
   End
End Sub
```

The click event is an event to which the command button responds. When the command button with the name **Command1** is clicked, the **Private Sub Command1_Click()** procedure executes. The only action that occurs in the preceding sample code is the execution of the **End** statement. Execution of the **End** statement terminates the program.

The Structure of a Procedure. This very short and simple procedure illustrates the general structure of any Visual Basic procedure.

1. **The first line.** The first line of the program is a call to the procedure named **Command1_Click()** to carry out whatever action is contained in the body of the procedure. The **Click()** event is the event that triggers the action. When you click on this command button named **Command1**, a **Click()** event occurs, and the procedure responds to it by carrying out the action in the procedure. The word **Private** identifies the scope of the procedure.

2. **The body of the procedure.** The **End** statement is the body of this procedure. Most procedures have more than one statement in the body of the procedure. The statements in the body of the procedure carry out some desired action.

3. All procedures end with the statement **End Sub.** The program needs to know when the procedure ends so that it can go on to the next procedure, if there is one.

In a Visual Basic program, a procedure carries out some action desired by the programmer. The code of procedure is attached to a control. The code, similar to BASIC, is invisible to the user, but the control is visible to the user. Hence, the name of the language: Visual Basic. The procedure executes only when it responds to an event of the control, for example, the click event of the command button.

Besides properties and events, controls also have methods. Later on, we will discuss methods in detail.

The Shape Control

With the shape control, Visual Basic makes it easy to create six predefined geometric shapes (Figure 10-6) and to place them on the form. The default shape is the rectangle. The geometric shapes are the **Shape** property of the shape control. The shape control has many properties having to do with its appearance. It can be filled with color (**FillColor** property). It can be filled in a number of styles such as solid or crosshatched (**FillStyle** property), and it can have different styles of borders (**BorderStyle** property). The shape control glories in its appearance, but it can't do anything. Unlike a command button control, the shape control responds to no events, not even the click event.

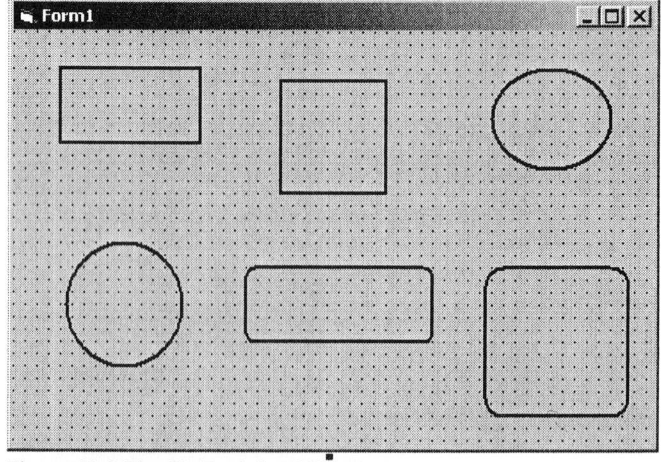

Figure 10-6. The Six Predefined Shapes

Object Browser

The Object Brower is an extremely useful feature of Visual Basic that enables you to find all the properties, methods, and events of any object. It is not directly used in writing code. It is more of a dictionary available for the convenience of the programmer. When you are in Design mode, you can, on the **View** menu, click **Object Browser**. An icon for the Object Browser is also on the toolbar. The **Object Browser** window, shown in Figure 10-7, consists mainly of two list boxes side by side. The one the right is labeled **Classes**, and in it you can find any object (plus other items).

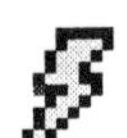

If you click an object on the left, the list box on the right lists all of that object's properties, methods, and events. A small icon to its left identifies each of these properties, methods, or events. A hand holding a piece of paper is the icon for a property; a little green book is the icon for a method; and a yellow lightening bolt is the icon for an event.

Try out the Object Browser by, on the **View** menu, clicking **Object Browser**. Then click **CommandButton** and notice its description at the bottom of the dialog box (Figure 10-7): **Looks like a push button and is used to begin, interrupt, or end a process.** Now click the **Click** property of the **CommandButton** class and notice that its description now appears at the bottom: **Occurs when the user presses and then releases a mouse button over an object.** Try this for the **Shape** class and some of its properties.

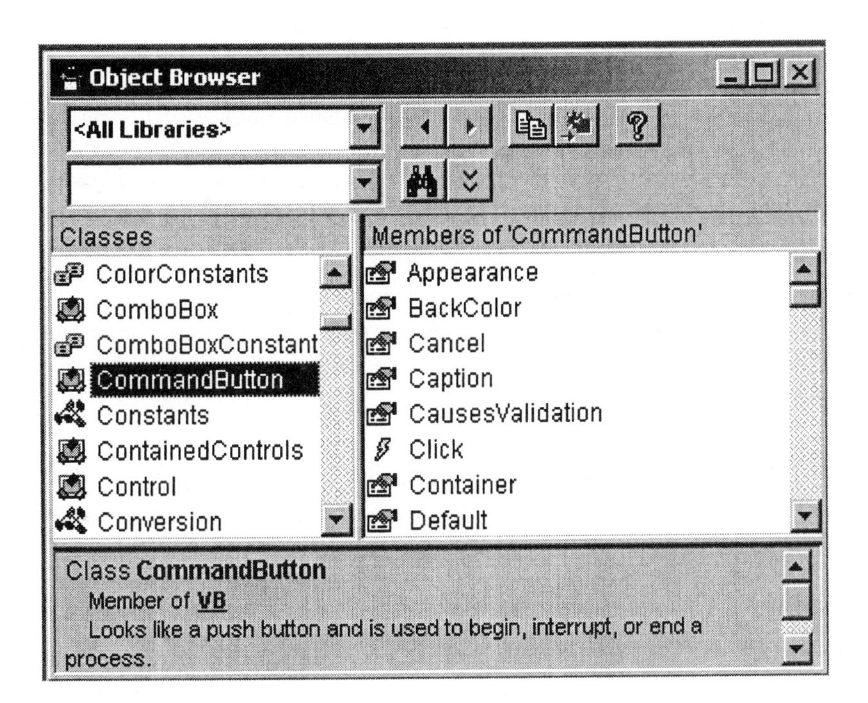

Figure 10-7. The Object Browser Dialog Box

The Color Properties Program

The Color Properties program is a short program that illustrates the Visual Basic programming environment. In addition, it demonstrates how the programmer sets the properties of objects at design time and how the program can change properties during run time. Recall that the three steps in writing a Visual Basic program are:

1. **Creating the user interface**—Embedding selected controls into the form (Figure 10-3); selecting the desired control from the toolbox (Figure 10-3)

2. **Setting properties of the controls**—Using the properties window. (Figure 10-3)

3. **Writing the code**—Using the code window and code editor (Figure 10-3)

Before Starting the Project. At some point, either while creating the Visual Basic project or upon finishing it, you will want to save your work. You might assume that saving the project creates a single file that you can name and come back to later. Actually, creating a Visual Basic project generates a set of three or four files. These files contain the code you have written and all the properties you have assigned to the controls. Saving ten projects in a single folder could result in mixing together 30 or 40 files, possibly leading to considerable confusion. For this reason it is advisable to save each project and its files in its own folder, which is given a name similar or identical to the name of the project.

Furthermore, saving all of these project folders in a single folder in your Windows My Documents folder makes it easy to locate your projects, use them, and back them up. Figure 10-8 shows how Windows Explorer displays such a file and folder organization. In the My Documents folder, the folder named Book Visual Basic Program Folders contains all the VB programs described in this book. The program folders are listed beneath the VB program folder. When Prog ColorProp is selected (Figure 10-8), the files it contains are shown in the window to the right, with their icons, their names, and their file types.

The topmost icon represents the Visual Basic project (vbp) file with the name ColorProp.vbp. The next icon represents Visual Basic workspace (vbw) with the name ColorProp.vbw. The third icon down from the top represents the form file and has the name Form1.frm. This is its default name. It could have been named, for example, Color.frm or ColorProp.frm. Every Visual Basic project has at least these three files with the extensions vbp, vbw, and frm.

The last icon (bottom right of Figure 10-8) represents the compiled file for this Visual Basic project. It is given the name Program ColorProp. Visual Basic automatically adds the extension .exe for an executable file. Visual Basic does not create the .exe file automatically. The programmer does this as described at the end of this chapter. The .exe file can be saved in the folder with the other three VB project files or in another folder that the programmer created to contain .exe files.

Normally, Windows and Visual Basic do not show file extensions. To show (or hide) file extensions, go to the **Window Control Panel**. Click **Appearance and themes_Folder Options_View**. Then clear the check box labeled **Hide Extensions for known file types**.

Starting the Project

- Start Visual Basic by clicking the **Visual Basic** icon on the desktop
- Select the **Standard EXE** icon in the **New Project** window (Figure 10-2), and click the **OK** button.
- A new project window opens with a fresh form (Figure 10-3).
- Select the form by clicking it.

- In the properties window, change the **(Name)** property to **frmColorProp**.
- Click **Save**; the file name will be frmColorProp, so click **OK**.
- The file name changes to Project1; change it to The Color Properties Program, and click the **OK** button.

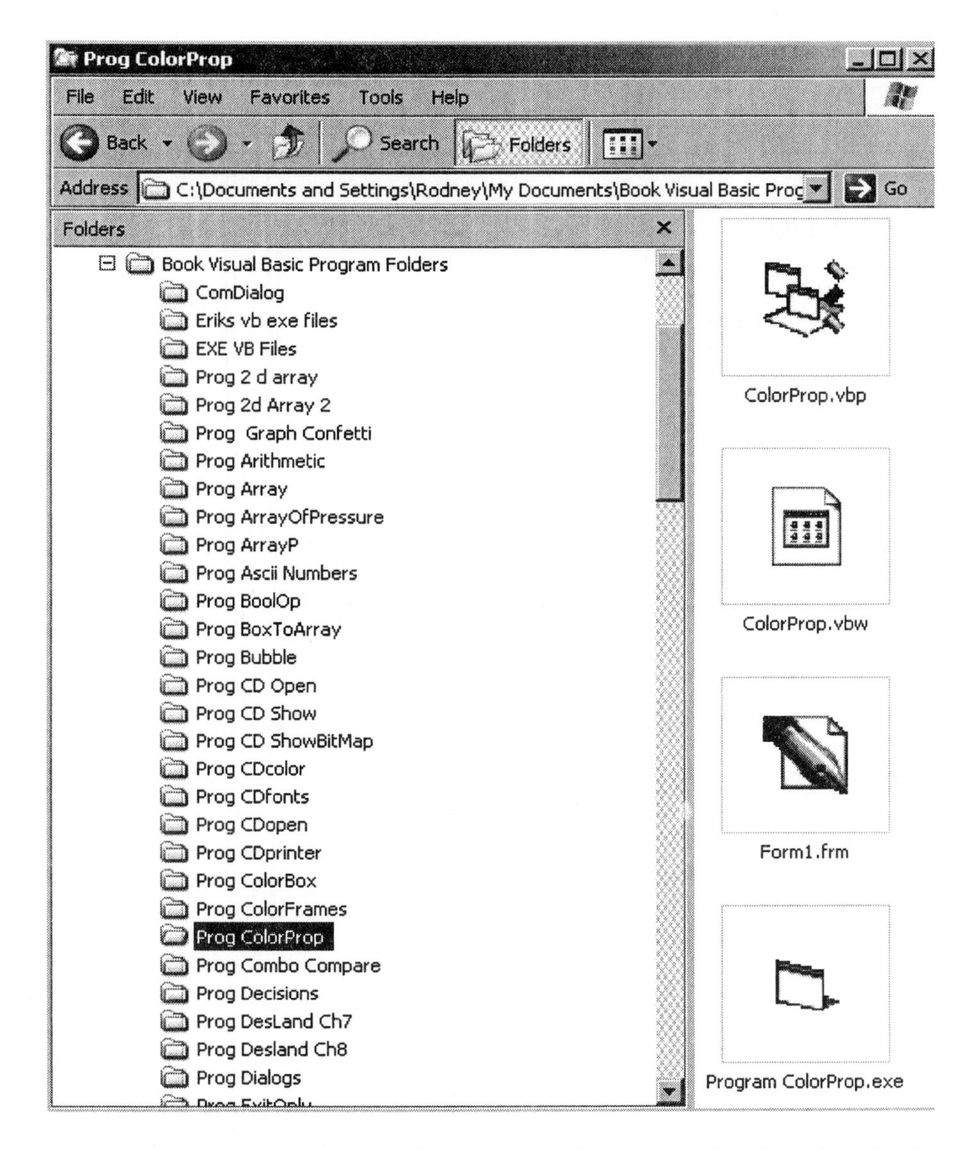

Figure 10-8. VB Program Folders Contained in a Folder. Selecting Prog ColorProp shows the file types associated with the Color Properties program.

Creating the User Interface

The toolbox (Figure 10-4) on the left side of the VB design window (Figure 10-3) displays the icons used to bring the controls to the form. For the Color Properties program, place five command button controls and three shape controls on the form positioned approximately as shown in Figure 10-9.

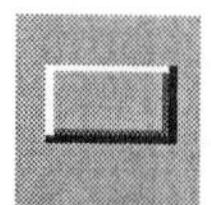

The **Command Button** icon on the toolbox looks like this.

The **Shape** icon on the toolbox looks like this.

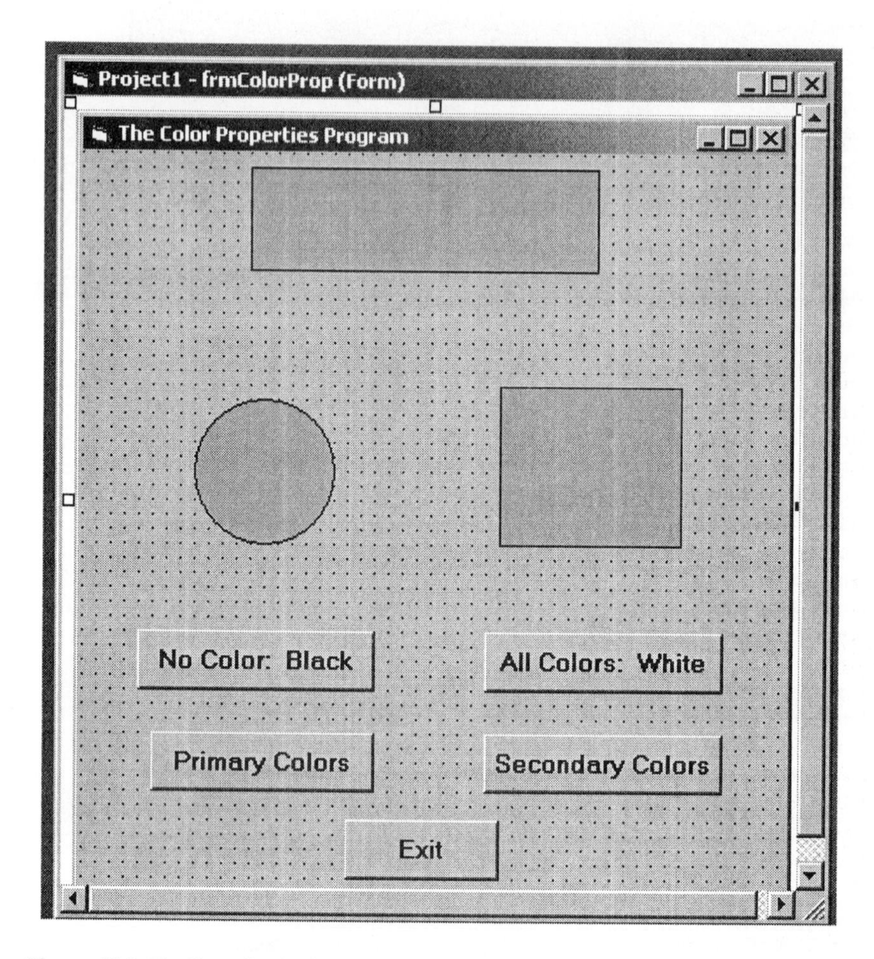

Figure 10-9. The Form for the Color Properties Program. It contains three shape controls and five command button controls.

Place Five Command Button Controls on the Form. Locate the **Command Button** icon in the toolbox (Figure 10-4). Place the five command buttons on the form, positioned as shown in Figure 10-9. To place a control on Form1:

- Click the control's icon in the toolbox.
- Move the pointer to the form, where the pointer becomes a crosshair.
- Position the crosshair where you want to place the upper left corner of the control.
- Drag the cursor until the control has the desired size.

At design time, you can select the control and resize it or reposition it. The first command button has the default **(Name)** property **Command1**. It also has the default **Caption** property **Command1**. Select the first command button and check the name and caption in the properties window (Figure 10-2). Similarly, the second command button has the default name and caption **Command2**, and so on for the remaining command buttons. Select each by clicking it, and check the (**Name**) and **Caption** properties in the properties window.

Place Three Shape Controls on the Form. Locate the **Shape** icon on the toolbox (Figure 10-4). Place the shape controls on the form using the same procedure as described previously, and place them above the command button controls, as shown in Figure 10-9. Unlike the command button control, the shape control does not have the **Caption** property. The shape control does have a **Shape** property, and the default **Shape** property of the shape control is **Rectangle**. Perhaps Microsoft should have named the shape control the geometric figure control with the default **Shape** property **Rectangle** and other **Shape** properties such as **Circle**, **Square**, and so on. As it is, we have the shape control, with the default **Shape** property **Rectangle**, and other **Shape** properties shown in Figure 10-6. After placing all eight controls on the form, practice moving and resizing them until you feel comfortable with these operations.

Setting the Properties of the Controls

Figure 10-10. The Properties Window for the Form

When a control is first placed on a form, Visual Basic gives it a set of default properties. Of the many properties that characterize a control, it is usually necessary or desirable to reset only a few at design time. The properties window greatly simplifies the task of setting control properties. The Color Properties program is short and simple, but as programs become large and complex, it is good programming practice to set up a properties table, shown in Table 10-1.

The properties table lists, in the leftmost column, the default names of the objects in the program. Notice that the first object listed is Form1, which is also an object, and it, too, has many properties, among them the (**Name**) property and the **Caption** property. Its default name is **Form1**, which we will reset to **frmColorProp**. Its default caption is also **Form1**, which we will reset to **The Color Properties Program**.

The first entry in the properties list is always (**Name**), in parentheses because it is not in alphabetical order and is always first on the properties list. The next few properties (for the form) are **Appearance**, **AutoRedraw**, **BackColor**, **BorderStyle**, and **Caption**, followed by many more. To verify these properties, click the form to select it, and inspect the properties table.

With the **Project1** window open and the form and its controls in place, open the properties window (Figure 10-3, lower right, and Figure 10-10). The properties window lists the selected object's properties and their settings, actually their default settings at this point. Select the **Alphabetic** tab on the properties window if it is not already selected. Click each object including Form1, and watch how the content of the properties window changes. After each object is selected, scroll the properties window list to the top so that the (**Name**) property is the first item on the list.

Naming Conventions for Standard Visual Basic Controls. Notice that the names of the objects bear a three-letter prefix associated with the default name of the object. We set the name of **Shapel** to **shpCircle** and it is indeed a circle, but it is also a shape control. When a form contains several controls of the same kind, then it is often confusing to the programmer if a more complete description is not used. The default names **Shapel**, **Shape2**, and **Shape3** are not as clear as **shpCircle**, **shpSquare**, and **shpRectangle**. Over time, a convention among programmers has evolved for the naming of objects, described in Table 10-2.

Table 10-1

The Properties Table for the Color Properties Program

Object	Property	Setting
Forml	**Name**	**frmColorProp**
	Caption	The Color Properties Program
Shapel	**Name**	**shpCircle**
	FillColor	Gray (row 3, col 1)
	FillStyle	0-Solid
	Shape	3-Circle
	BorderStyle	1-Solid
Shape2	**Name**	**shpSquare**
	FillColor	Gray (row 3, col 1)
	FillStyle	0-Solid
	Shape	1-Square
	BorderStyle	1-Solid
Shape3	**Name**	**shpRectangle**
	FillColor	Gray (row 3, col 1)
	FillStyle	1-Solid
	Shape	0-Rectangle
	BorderStyle	0-Transparent
Commandl	**Name**	**cmdPriColors**
	Caption	Primary Colors
Command2	**Name**	**cmdSecColors**
	Caption	Secondary Colors
Command3	**Name**	**cmdBlack**
	Caption	No Color: Black
Command4	**Name**	**cmdWhite**
	Caption	All Colors: White
Command5	**Name**	**cmdExit**
	Caption	Exit

Table 10-2

Naming Conventions for Standard Visual Basic Controls

Control	Prefix	Example
Form	frm	frmMyFirst Program
Combo box	cbo	cboMembers
Check box	chk	chkNorth
Command button	cmd	cmdExit
Data	dat	datPublishers
Directory list box	dir	dirMain
Drive list box	drv	drvUser
File list box	fil	filMain
Frame	fra	fraSection
Horizontal scroll bar	hsb	hsbSpeed
Image	img	imgMadonna
Label	lbl	lblEnergy
Line	lin	linAlpha
List box	lst	lstWavelengths
OLE	ole	oleFist
Option button	opt	optLessThan10
Picture box	pic	picNMR
Pointer	ptr	ptrSection
Shape	shp	shpCircle
Text box	txt	txtYourName
Timer	tmr	tmrInterval
Vertical scroll bar	vsb	vsbAcidity

When the form is selected, the default (**Name**) property for the form is **Form1**. Change it to **frmColorProp** and notice that the title bar for the container window now reads **Project 1 – frmColorProp**. Down about six items on the list appears the default **Caption** property, which is also **Form1**. Change the **Caption** property for the form to **The Color Properties Program**. The properties window for the form should now look like Figure 10-10. Notice the changes in the title bars for the container window and the **Project1** window.

Setting the Properties of the Form. The (**Name**) and **Caption** properties are by far the most frequently used properties, at least in the initial stages of writing a program. Notice that VB gives each control a default name, and if the form contains more than one control of the same type, it adds a number to the name. The numbering begins at zero (0, 1, 2, 3...). The default captions are often the same as the default name: (**Name**):

Command1; and **Caption**: **Command1**, and so on. The captions are visible to the program user and should be helpful, informative, and unambiguous to the program user. The **(Name)** property is invisible to the program user, but becomes part of the program's code. Names in code should be clear to the programmer and to other programmers who need to change and maintain the program. The **(Name)** property in the code is the property of a particular control. A name alone might be sensible and descriptive, but if it doesn't tell the programmer what control is it associated with, the programmer has a problem. Consequently, VB programmers use a consistent set of lowercase prefixes with the **(Name)** properties of their controls as shown in Table 10-2.

Setting the Shape Control Properties. Table 10-1 lists the four properties of each of the three shape controls that should be changed. Properties are changed in the properties window (Figure 10-11).

- Select **Shape1**.
- Change the **(Name)** property from **Shape1** to **shpCircle.**
- Change the **FillColor** property to gray.
    1. Select the **FillColor** property.
    2. Click the down arrow.
    3. Select **Palette.**
    4. Click the gray color.
- Change the **FillStyle** to **0-Solid.**
- Change the **Shape** to **3-Circle.**

In the same way, set the properties of **Shape2** and **Shape3** according to Table 10-1. Other **Shape**, **BorderStyle**, and **FillStyle** properties are listed in Table 10-3, Table 10-4, and Table 10-5.

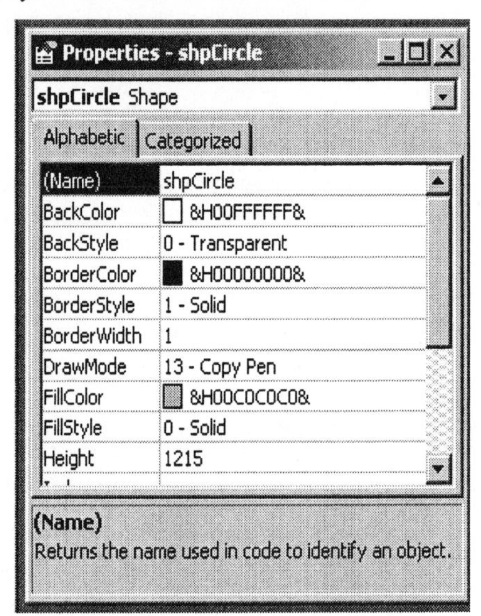

Figure 10-11. The Properties Window for a Shape Control

Table 10-3

Predefined Shapes

Shape	Value	Constant
Rectangle	0	vbShapeRectangle
Square	1	vbShapeSquare
Oval	2	vbShapeOval
Circle	3	vbShapeCircle
Rounded rectangle	4	vbShapeRoundedRectangle
Rounded square	5	vbShapeRoundedSquare

Setting the Command Button Properties. Use Table 10-1 to set the properties of each of the three command buttons on the form. When finished, resize and reposition the controls so that the form and its controls look approximately like Figure 10-9.

The program consists of several parts or, in VB terminology, several objects. These objects are the form itself, the three shapes, and the three command buttons. Objects have default names. The name is just one property of an object; objects have many properties. For example, the default name of a command button is **Command1** and if there are more of them, they are named **Command2**, **Command3**, and so on. Similarly, shapes have names **Shape1**, **Shape2**, **Shape3**, and so on. When a program has several controls of the same type, it can be confusing unless the controls are given more meaningful names.

Table 10-4

FillStyles, their Values and Constants

FillStyles	Value	Constant
Solid	0	vbFSSolid
Transparent (the default)	1	vbFSTransparent
Horizontal line	2	vbHorizontalLine
Vertical line	3	vbVerticalLine
Upward diagonal	4	vbUpwardDiagonal
Downward diagonal	5	vbDownwardDiagonal
Cross	6	vbCross
Diagonal cross	7	vbDiagonalCross

For example, it is not possible from the names **Shape1**, **Shape2**, and **Shape3** to tell which of these has the circle shape property, which has the square shape property, or which has the rectangle shape property. Similarly, is it **Command1** or is it **Command2** that changes the color to secondary colors? For this reason, it is customary in the development stage of writing a program to draw up a properties table for the form and its controls, as shown in Table 10-1 for the Color Properties program.

The name prefixes are not part of Visual Basic's syntax. In principle, you can use any name for a control. However, it is highly recommended that the conventions in Table 10-2 be followed. As you write programs that are more complex than the Color Properties program, it will become more apparent that these conventions are helpful and save time in maintaining and debugging programs.

Table 10-5

BorderStyle Property

BorderStyle	Value	Constant
Transparent	0	vbTransparent
Solid	1	vbBSSolid
Dash	2	vbBSDash
Dot	3	vbBSDot
Dash-dot	4	vbBSDashDot
Dash-dot-dot	5	vbBSDashDotDot
Inside solid	6	vbBSInsideSolid

Entering the Code for the Color Properties Program

On the **View** menu, click **Code** to open the code window. The code window is also called the Visual Basic Code Editor window. It is really just a word processor that has some special features that facilitate writing code.
It is generally good programming practice to document every program internally with comments both at the very beginning of the program and occasionally in the body of the program. A comment begins with an apostrophe (') and the program ignores everything after it on the same line. You should include the name of the program in a comment on the first line of the program, followed by a brief statement of the purpose of the program.

In a Visual Basic program, code is written in a block associated with a particular control. That block of code is called a procedure, and its name contains the name of its control. A control does not have to have an associated procedure. In the Color Properties program, the shape controls have no code written for them, but each of the five command button controls does have code.

Entering Preliminary Comments. Now enter the following three comments in the general declaration section of the program (that part of the program preceding the procedures):

```
'The Color Properties program introduces the shape control and the command
'button control. It shows how the properties of a control can be
'set at design time or at run time
```

Entering the Procedure Outline. Visual Basic provides some assistance in writing the code for the command button. Click the form or, on the **View** menu, click **Object** to get the form window in front of the code window. Then double-click the **Primary Colors** icon. As a result, the code window appears as shown in Figure 10-12. The comments just entered in the general declaration area at the top of the form are visible, and Visual Basic has furnished the first and last lines of the code for the command button control with the name **cmdPriColors**. Recall that when the form was created, the controls were embedded in it, the **(Name)** property (default: **Command1**) was set to **cmdPriColors** and the **Caption** property (default: blank) was set to **Primary Colors**. Notice the difference. The caption appears on the control and is visible to the user. The name is used in the code and is invisible to the user.

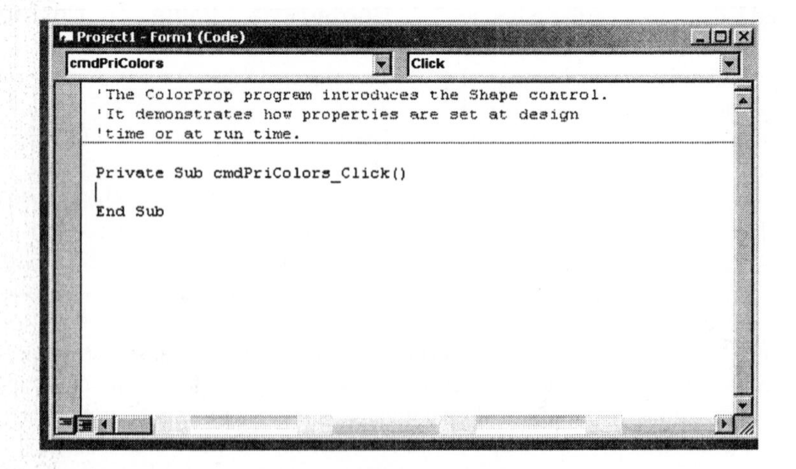

Figure 10-12. The Code Window with Comments. The window is ready for code to be entered between the first and last lines of the procedure.

Fill in the Body of the Code. Fill in the code as shown below, but stop for a moment after you type the period after **shpCircle.**

```
Private Sub cmdPriColors_Click()
    shpCircle.FillColor = RGB(255, 0, 0)        ' Assigns red to the FillColor property
    shpSquare.FillColor = RGB(255, 255, 0)      ' Assigns yellow to the FillColor property
    shpRectangle.FillColor = RGB(0, 0, 255)     ' Assigns blue to the FillColor property
End Sub
```

Automatic Code Completion. At the instant that the period after **shpCircle** is typed, the VB Code Editor inserts an automatic code completion list (Figure 10-13). The alphabetical list is for the shape control for which we have given the name **shpCircle**. As soon as a period is typed after this name, VB recognizes the name, and upon entering the period, VB knows that the user is about to type a property of that control. To lend a helping hand, VB places the automatic code completion list just below the line being typed. The list is a list of all the properties of a shape control, beginning with **BackColor**. If we forgot the exact name of the property or its spelling, we can scroll down the list and would probably recognize the property we are seeking. In this case we know that the property is **FillColor**. At the instant that the **f** or **F** is entered after the period, the list automatically scrolls to the properties beginning with **f** (Figure 10-14). If you know that **FillColor** is the property you want, just touch the SPACEBAR once, and Visual Basic's automatic code completion feature completes entering the property name for you. You save time, and you cannot make a typing or spelling error.

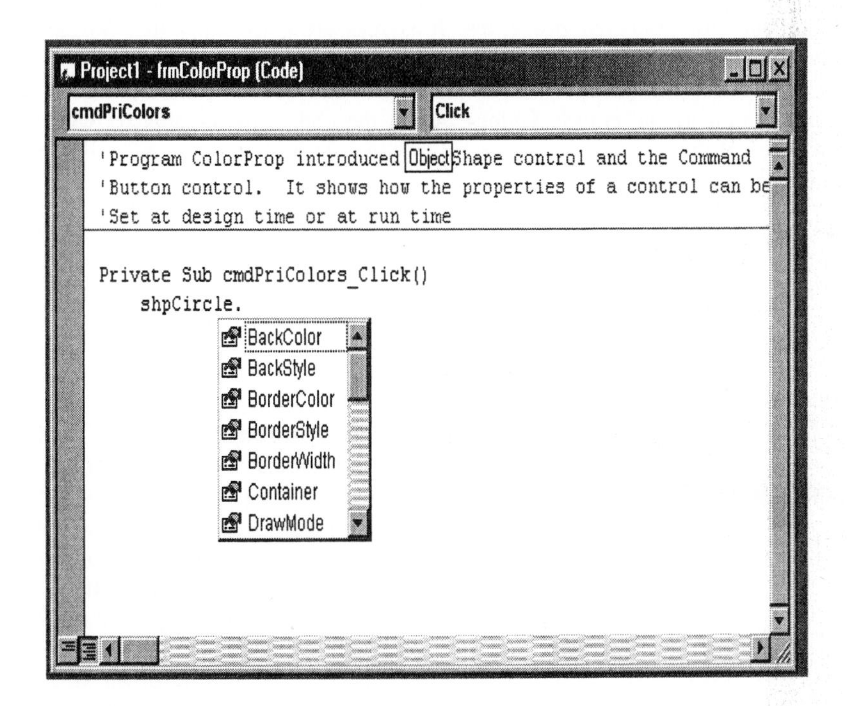

Figure 10-13. Entering a Period (.) after **shpCircle** Results in Automatic Code Completion Choices

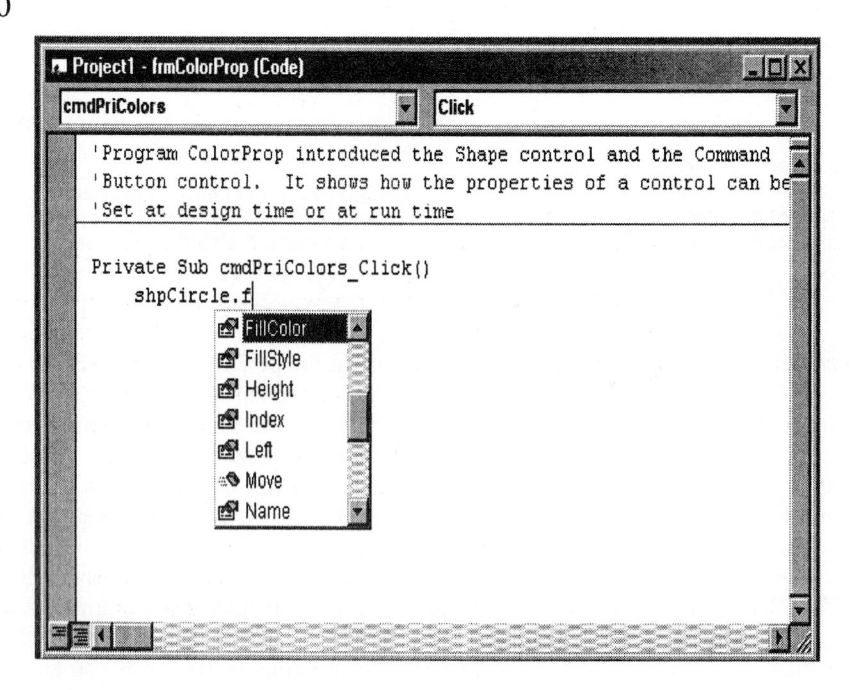

Figure 10-14. Entering an **f** after **shpCircle.** Results in Automatic Code Completion Choices in an Alphabetical List That Begins with f (**FillColor**)

The control bar lists twenty-one standard controls, each with more than twenty properties for more than four hundred properties. It is reassuring to know that Visual Basic's automatic code completion is ready to assist at the drop of a period.

Minimize the code window so that you can see the form. Double-click the command button named **cmdSecColors**. Again the code window appears with the first and last lines of the code for the command button control with the name **cmdSecColors**. Fill in the code as shown below.

```
Private Sub cmdSecColors_Click()
    shpCircle.FillColor = RGB(255, 150, 0)          'Orange
    shpSquare.FillColor = RGB(0, 255, 0)            'Green
    shpRectangle.FillColor = RGB(200, 0, 255)       'Violet
End Sub
```

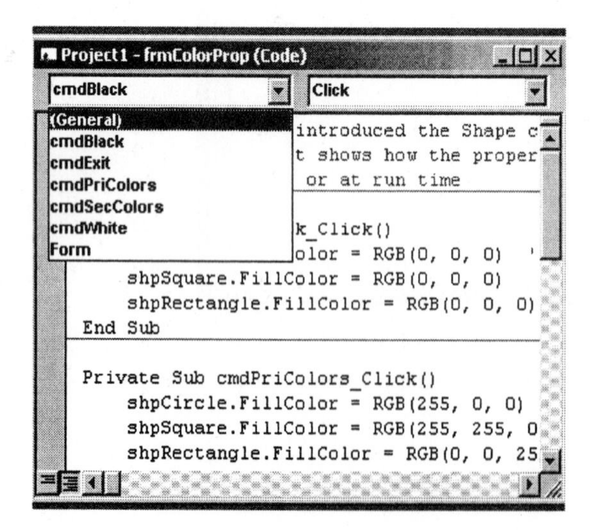

Figure 10-15. The Procedure List of the Code Window

The Procedure List. Visual Basic offers another method for getting the first and last lines of a control's procedure. Look at the top of the code window in Figure 10-12. It shows two drop-down list boxes. The one on the left has a list of the controls that you placed on the form and for which you could write code for a procedure. Click the down arrow and a procedure list drops down (Figure 10-15). On it you can see the names of the controls that you have placed on the form. In addition, **(General)** and **Form** appear on the list. **(General)** refers to the general declarations section of the program where we have only written some comments. The Color Properties program does not include any code associated with the form, but many program do. Clicking one of the control items in this list of controls causes the first and last line of a procedure for that control to be placed in the code window.

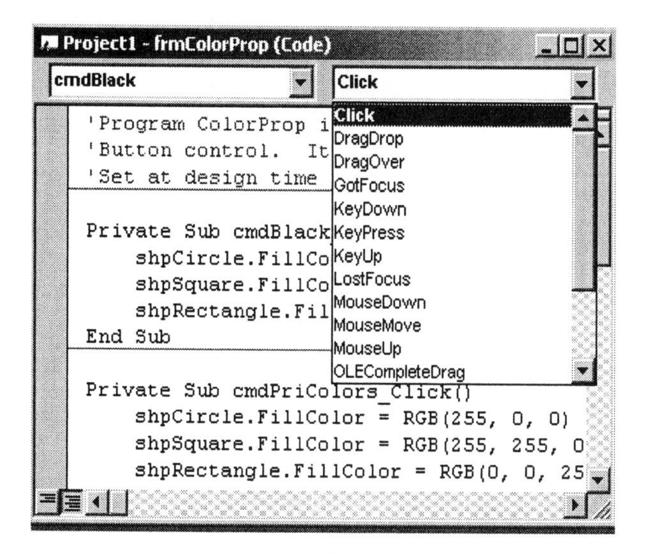

Figure 10-16. The Events List of the Code Window

The Events List. Now examine the upper right side of Figure 10-16 and notice that there is another drop-down list box. In this one, **Click** is the first entry in the list. Click the down arrow and an events list appears for the control selected in the procedure list to its left. This is the events list for the controls that you placed on the form. Many controls have events to which they respond, and the most common event is the click event. This is the event that a command button control responds to when you click the running program. The control responds by executing the code that you write in its associated procedure.

The next procedure is

```
Private Sub cmdWhite_Click()
   shpCircle.FillColor = RGB(255, 255, 255)          ' White
   shpSquare.FillColor = RGB(255, 255, 255)          ' White
   shpRectangle.FillColor = RGB(255, 255, 255)       ' White
End Sub
```

To get the first and last lines of this procedure, double-click the **cmdWhite** command button or select **cmdWhite** from the procedure list (Figure 10-15). Use these same methods for the next three procedures.

The procedure for the fourth command button is

```
Private Sub cmdBlack_Click()
   shpCircle.FillColor = RGB(0, 0, 0)                ' Black
   shpSquare.FillColor = RGB(0, 0, 0)                ' Black
   shpRectangle.FillColor = RGB(0, 0, 0)             'Black
End Sub
```

The last code for the **Exit** command button permits the user to leave the program gracefully when finished. The one-word statement **End** causes the program to close.

```
Private Sub cmdExit_Click()
   End
End Sub
```

Running the Color Properties Program

Notice the three symbols located on the VB toolbar: ▶ II ■

▶ Start	This runs the procedure
II Break	This interrupts execution
■ End	This resets your code to the beginning

The first of these is called the **Start** button, and clicking it causes your program to run. When it seems to be finished, you need to formally stop it by clicking the square button called the **End** button. The middle button, which we don't use, is called the **Break** button.

Click the **Start** icon (▶) to run the program, and observe the action (or click F5, or, on the **Run** menu, click **Start**). The toolbox and the properties window disappear (if they were visible when the program was run), and the form, with its controls that you created, appears on the Visual Basic screen. The circle, square, and rectangle are gray, because you set the **FillColor** property of these three shapes to gray at design time. The geometric figures are shown by their outline because at design time you set the **BorderStyle** property to equal **1-Solid**.

You have left design time and are now in run time. Your program is alive, running, and waiting for you to do something: it is waiting for an event. You can click one of five buttons with the captions: **No Color**, **All Colors**, **Primary Colors**, **Secondary Colors**, or **Exit**.

Click the button labeled **Secondary Colors**. The colors in the circle, square, and rectangle change to orange, green, and violet, respectively. Clicking the **Secondary Colors** command button precipitates the click event for that command button, causing execution of the code attached to it. You have changed the properties of the three shape controls during run time. When you embedded the controls in the form, you set the properties at design time (different from their default **FillColor** properties).

Now click the button labeled **Primary Colors**. The colors in the circle, square, and rectangle change to red, yellow, and blue, respectively. Once again, you have changed the properties of the three shape controls during run time. Click back and forth between the four buttons dealing with color. When you click the **Secondary Colors** button, that button responds to the click event by executing the three lines in the **Private Sub cmdSecColors_Click()** procedure.

Click the maximize button on the program's form. Your program knows how to maximize and minimize its window, even though you didn't write any code for it. Visual Basic does a lot of heavy lifting in the background so that the code that you write is quite simple. Click the minimize button, and the program returns to its original size.

It is apparent that VB furnishes a great deal of code for the various controls and windows that you don't have to bother with. That's why the code for VB programs is very small. The code for the Color Properties program at this moment is about 21.8 kilobytes. Click the **End** icon ■ to end execution (or click ALT F4 or, on the **Run** menu, select **End**).

Creating an Executable File. With the ColorProp file open, on the **File** menu, click **Make ColorProp.exe**. Visual Basic automatically supplies the name of the executable file, ColorProp.exe. This action opens the **Make Project** dialog box shown in Figure 10-17. The dialog box suggests the folder where the executable file will be saved, but you can change it if you wish. Click **OK**, and the executable file is saved as ColorProp. Go to the folder where it is saved, find it, and double-click it. It will execute. If you right-click it, and then click **Properties**, it will show that ColorProp is an application with the full name ColorProp.exe.

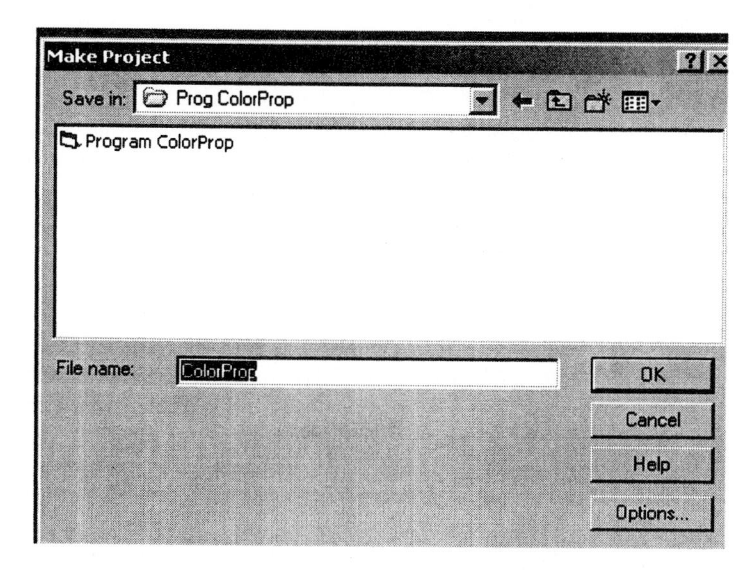

Figure 10-17. The Make Project Dialog Box

Understanding the Color Properties Program

In addition to the form, the Color Properties program has two kinds of controls: the shape control and the command button control. The shape control does not control anything, but it is still called a control. The command buttons do control in that they respond to the click event when the user clicks the mouse on a command button. The response of the clicked button is to execute the code attached to it. The shape controls just sit there on the form and look pretty and change their color property as directed by the code of the associated command button.

The Shape Control. In this program, the shapes of the shape controls are set at program design time. The shape can also be changed during run time with a statement like: **shpShape1.Shape = 3**. This statement would give **Shape1** the shape of a circle. Alternatively, this statement could be written **shpShape.Shape = vbShapeCircle**. Visual Basic has a variety of constants as a convenience for a number of uses, such as setting the shapes of shape controls and the color and other properties of certain objects. The color constants are listed in Table 10-6.

Several color properties of the shape control are changed at design time according to Table 10-1. **FillColor** sets the color that fills the shape control, and **FillStyle** sets the attributes of the **Color** property, in this case **Solid**. The shape control may have a colored border set by the **BorderColor** property, but in this program **BorderColor** is set to black. The **BorderStyle** property is set to **1-Solid**. Regardless of what the **BorderColor** setting is, setting the **BorderStyle** to **0-Transparent** makes the border and the shape control invisible. Try experimenting with the **BorderColor** and **BorderStyle** properties. Besides **0-Transparent**, the **BorderStyle** can be set to solid, dot, dash, dash-dot, and more.

The Command Button Control. Note that the word **Click** is attached to the name of each command button control in the first line of the procedure. This is the event that executes the procedure. Controls are characterized by their properties, their methods, and their events. Methods are described later in this chapter. The click event is one of the most common events and appears at the top of the event list in the code window. It is the click event to which we wish our command buttons to respond. The event list on the top right (Figure 10-16) gives a list of the events to which the control (**cmdBlack**) that is at the moment selected in the procedure list can react.

Color and the RGB Function. Clicking the command buttons in this program changes the color properties of the three shape controls. The value of the color is assigned with an assignment operator to the **FillColor** property of the shape control. The value is established with the **RGB** function. We'll deal in a more general way with functions later on. But for now, think of a function as a routine that accepts one or more values and returns a single value. Consider the **SIN (x)** function. It accepts a value of x and returns a value for the sine of x. We don't worry about how it does this. We know that if we put a value for x in one end, a value for the sine of x comes out the other end.

The **RGB** function accepts three integers (each between 0 and 255) and returns a single value, namely a color. The syntax of the **RGB** function is **RGB(red, green, blue)**. The color returned by the **RGB** function is a mixture of weighted red, green, and blue components. **RGB(255, 0, 0)** returns red, **RGB(0, 255, 0)** returns green, and **RGB(0, 0, 255)** returns blue. White light is an equal mixture of all three colors, and **RGB(255, 255, 255)** returns the color white. **RGB(0, 0, 0)** returns black, which is the absence of color. The total number of different colors is 256^3 or 16,777,216 different colors (Table 10-6).

With color constants (Table 10-6), Visual Basic provides another way of assigning a color. For example the statement `shpCircle.FillColor = vbBlack` is equivalent to the statement `shpCircle.FillColor = RGB(0, 0, 0)`. The number of color constant colors is limited, but the code in which they are used is very readable. Try some color constants in the program to see the equivalence.

Table 10-6

Selected RGB Colors and Their Color Constant Equivalents

Color	RGB Function Parameters			Color Constants
	Red Value	Green Value	Blue Value	
Black	0	0	0	vbBlack
Blue	0	0	255	vbBlue
Green	0	255	0	vbGreen
Cyan	0	255	255	vbCyan
Red	255	0	0	vbRed
Magenta	255	0	255	vbMagenta
Yellow	255	255	0	vbYellow
White	255	255	255	vbWhite

Summary and Review of the Color Properties Program

Figure 10-18 shows the Color Properties program in design mode. At the moment, the command button with the name **cmdPriColors** and caption **Primary Colors** has been selected. The black size handles around it show that is has been selected. The properties tables show a list of the properties of the selected control. At the moment, the **Caption** property has been selected, and to the right of the **Caption** property the caption has been entered: **Primary Colors**. At the top of the properties window the **(Name)** property can be seen, and the name entered is **cmdPriColors**.

The code associated with **cmdPriColors** is seen in the code window. An arrow-tipped line connects the **cmdPriColors** button to its procedure: **Private Sub cmdPriColors_Click**. An arrow connecting the rectangular shape on the form points to the line in this procedure that assigns a color value to the **shpRectangle.Fillcolor**.

Below this procedure is the procedure for the **cmdSecColors** command button. Arrow-tipped lines connect the **shpCircle** and **shpSquare** shape controls to the lines of code that assign a value to their **FillColor** property. The last procedure is the code for the **cmdExit** button with the caption **Exit**.

Finally, an arrow-tipped line beginning in the middle of the form points to its icon in the project window. The icon's label is **frmPriColors**. If you open the program and the design window is empty, double-clicking the **frmPriColors** icon in the project window opens the form to the design window. The toolbox is missing from Figure 10-17 to give room for the windows shown, but it could be opened by clicking its icon once. Figure 10-18 summarizes the steps to writing any Visual Basic program, which are:

1. Creating the user interface, embedding selected controls into the form (middle window—**Form1**).
2. Setting the properties of the controls (the properties window—lower right)
3. Writing the code (code window—left window)

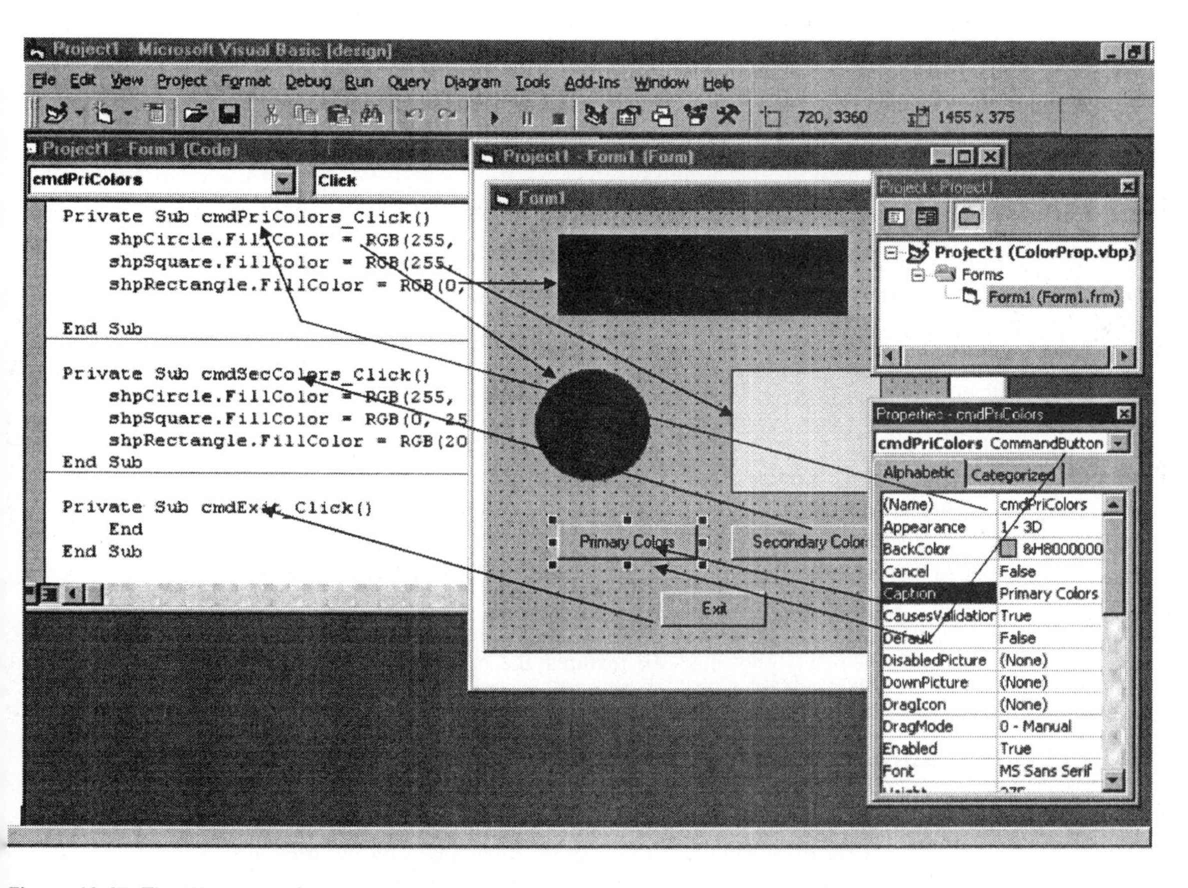

Figure 10-17. The Workspace for the Completed Color Properties Program. The **cmdPriColors** command button has been selected. Its caption is **Primary Colors**.

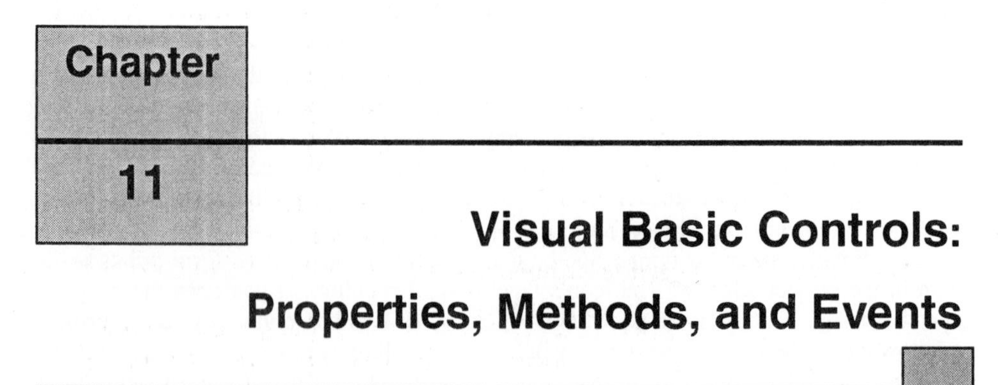

Chapter 11 Visual Basic Controls: Properties, Methods, and Events

The least that the user expects of a computer program is that the program accept input from the user, carry out some action, and then display the output in a useable form. This chapter begins the elements of these features with a program that introduces two new controls: the text box control and the label control. The text box can accept input and display output. Output can be displayed not only in a text box, but also can be printed to the form of the program or to the hard copy printer. The label control is just a label.

The appearance (format) of text and numbers is important, so we want to be able early on to control text properties such as font, size, and italics, among others, and number properties such as the number of decimal places, number formats, and scientific notation. In addition, we need to learn something about Visual Basic's data types, variables, and nomenclature. With this basic introduction, we will be prepared to move on to the nature of the Visual Basic language, its syntax, and semantics.

In Chapter 10, The Color Properties program serves as an introduction to Visual Basic and to writing a simple program in Visual Basic. It takes a little study to get familiar with the various windows and their functions, but this problem disappears with repetition. In general, a program consists of a form in which controls are embedded. Underlying each control is code in the Visual Basic language. In addition, each control has its own set of properties, methods, and events that greatly increase their power and utility.

The controls and other objects that make up a program are characterized by their properties and methods and their responses to events. The Color Properties program also demonstrated the click event, to which the command button responds. Many objects have methods. A method is an action that an object, like a form or control, can do.

In the next several chapters, new controls are introduced whose representative properties, methods, and events are illustrated in small Visual Basic programs. At the same time, the language is described, extended, and illustrated. In this chapter two VB programs serve to illustrate the newly presented VB features: the Justify program and the Print Demo program.

The Text Box Control

The text box control is another one of Visual Basic's most commonly used controls. It is important because it is useful for displaying text that can be either input by the user or output by the program. The **Text Box** icon in the toolbox looks like this:

It can also display text that is assigned to it at design time. The default name of the text box is **Text1**; the default **Text** property is also **Text1**. The **Value** property of the text box is the text that the text box displays. Besides accepting text from the user, the text box can accept the text representation of numbers, integers, decimals, and even scientific notation, for example, 3.6864E-5.

If you don't want the user to be able to input text or edit text displayed by the text box, set the **Locked** property of the text box to **True**.

Set the **Multiline** property to **True** if you want to display multiple lines of text in a text box. Scroll bars will appear with the text box whenever the **Multiline** property is set to **True**, unless the **Scrollbars** property is set to **None (0)**. If the **Multiline** property is set to **False**, then only part of a long line is visible in the text box. The text box can be small enough to accept just a short line of text or numerical data, or sized at design time to nearly cover the form. When the **Multiline** property is set to **True**, you can move through the text with the usual keys: HOME, END, CTRL-HOME, CRTL-END, and the arrow keys.

The Label Control

The **Label** icon appears as follows in the toolbox:

The label control is useful for giving information to the user, sometimes by itself and sometimes with another control to explain the control more fully to the user. It is a simple control that uses little overhead. Its principal property is its **Caption** property, but its **Visible** property allows the programmer to make the label visible or invisible as is appropriate. The **ForeColor** property (the color of the caption) and the **BackColor** property can also be specified at design time or changed at run time.

The Justify Program

The Justify program introduces two new controls: The text box control and the label control. But the main purpose of the Justify program is to illustrate how one control can change the properties of another control. Figure 11-1 shows the form for the Justify program with its embedded controls. In this program, a text box control in the form permits you to enter some text, like your name or a phrase. By default, your name is left-justified in the text box.

The program's command buttons permit you to change several text properties of your name in the text box. A label control near the text box instructs the user in what action the user should take:

- Start Visual Basic by clicking the **Visual Basic** icon on the desktop.
- Select the **Standard EXE** icon in the **New Project** window, and click the **OK** button.
- A **New Project** window opens with a fresh form.
- Select the form, and in the properties window change the **(Name)** property to **frmJustify**.
- Click **Save**; the file name will be frmJustify, so click **OK**.
- The file name changes to **Project1**; change it to **The Justify Program**.
- Click the **OK** button.
- Place controls on **frmJustify** according to Figure 11-1.
- Set the properties of the controls according to Table 11-1.

Table 11-1

The Properties Table for the Justify Program

Object	Property	Setting
Form	Name	frmJustify
	Caption	The Justify Program
Label	Name	lblInfo
	Caption	Please enter your name
	BorderStyle	1-Fixed Single
Text box	Name	txtPhrase
	Text	make it blank
	Multiline	True
Command button 1	Name	cmdRight
	Caption	Right Justify
Command button 2	Name	cmdCenter
	Caption	Center Justify
Command button 3	Name	cmdLeft
	Caption	Left Justify
Command button 4	Name	cmdBold
	Caption	Bold
Command button 5	Name	cmdItalic
	Caption	Italic
Command button 6	Name	cmdBigFont
	Caption	14 Point Font
Command button 7	Name	cmdOriginal
	Caption s	Return to Original Font
Command button 8	Name	cmdClear
	Caption	Clear for New Name
Command button 9	Name	cmdExit
	Caption	E&xit

Entering the Code of the Justify Program

To enter the procedure code, for each control either:

- Double-click the control
- Or, on the **View** menu, click **Code**, then select the desired procedure by scrolling down the procedure list of the code window shown in Figure 10-14.

In the general declarations section of the code window, VB may already have written in the statement **Option Explicit,** the meaning of which will be explained in detail later. Enter the following code in the general declaration section of the code window:

```
' The Justify program demonstrates the text box control
' and illustrates how one control can change
' the property of another control
Option Explicit
```

Enter the following code for the **Private Sub cmdBold_Click()** procedure:

```
Private Sub cmdBold_Click()
   'Set font to bold
   txtPhrase.Font.Bold = True
End Sub
```

Enter the following code for the **Private Sub cmdCenter_Click()** procedure:

```
Private Sub cmdCenter_Click()
   'Set alignment to center justify
   txtPhrase.Alignment = 2
End Sub
```

Enter the following code for the **Private Sub cmdClear_Click()** procedure:

```
Private Sub cmdClear_Click()
   txtPhrase.Text = ""                 ' The "" is an empty string"
   txtPhrase.Font.Bold = False         ' Font is no longer bold
   txtPhrase.Alignment = 0             ' Reset to Left Justify
   txtPhrase.Font.Size = 10            ' Font is reset to 10 point
   txtPhrase.SetFocus                  ' Reset focus to text box
End Sub
```

Enter the following code for the **Private Sub cmdExit_Click()** procedure:

```
Private Sub cmdExit_Click()
   End                                 ' Closes the program
End Sub
```

Enter the following code for the **Private Sub cmdItalic_Click()** procedure:

```
Private Sub cmdItalic_Click()
   ' Set font to italic
   txtPhrase.Font.Italic = True
End Sub
```

Enter the following code for the **Private Sub cmdLeft_Click()** procedure:

```
Private Sub cmdLeft_Click()
   ' Set the alignment to left justify
   txtPhrase.Alignment = 0
End Sub
```

Enter the following code for the **Private Sub cmdRight_Click()** procedure:

```
Private Sub cmdRight_Click()
   ' Change alignment to right justify
   txtPhrase.Alignment = 1
End Sub
```

Enter the following code for the **Private Sub cmdOriginal_Click()** procedure:

```
Private Sub cmdOriginal_Click()
   txtPhrase.Font.Bold = False         ' Font is no longer bold
   txtPhrase.Font.Italic = False       ' Font is no longer italic
   txtPhrase.Font.Size = 10            ' Font is reset to 10 point
End Sub
```

Enter the following code for the **Private Sub cmdBigFont_Click()** procedure:

```
Private Sub cmdBigFont_Click()
   'Set font to 14 point
   txtPhrase.Font.Size = 14
End Sub
```

Proofread your entries, correct any obvious errors, and save by clicking the **Save** button.

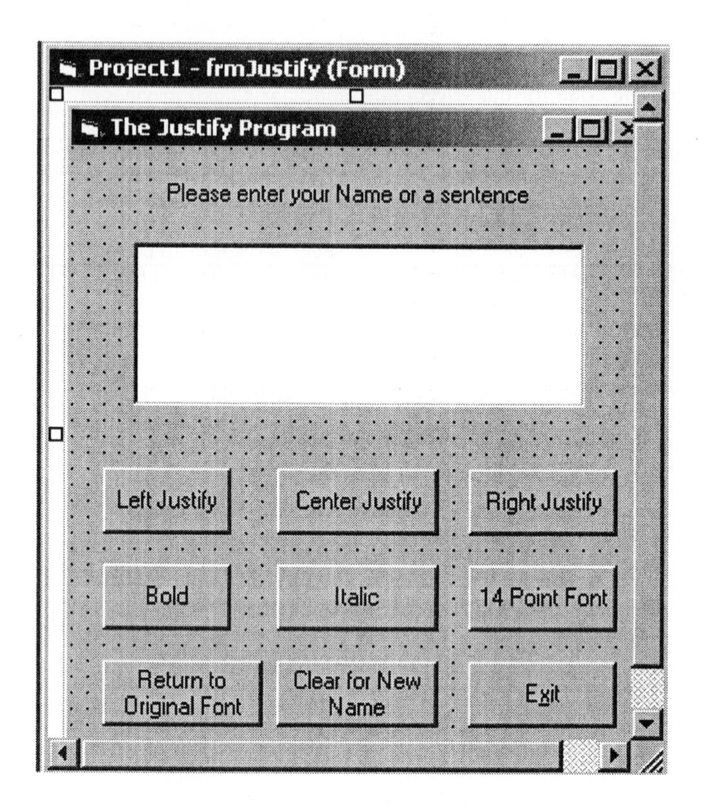

Figure 11-1. The Form for the Justify Program with Its Embedded Controls

Executing the Justify Program

- Execute the Justify program by clicking the **Start** button.
- Enter your name or a short phase in the text box.
- Click all the buttons in random order.
- Enter a sentence long enough to try out the **Multiline** property of the text box.
- Click all the buttons in random order.
- Exit the Justify program by clicking the **Exit** button.

Understanding the Code of the Justify Program. As a minimum, the general declaration section of the program should contain the title of the program. A printed listing of the program does not include the title unless you include it as a comment. Your Visual Basic may include the **Option Explicit** statement automatically. If it does not, enter it in the general declarations section as shown. The meaning of this statement is discussed in a later section on variables and the declaration of variables.

The Text Box. The Justify program introduces one of Visual Basic's most important and most commonly used controls: the text box. The Justify program uses a text box to allow the user to input text data. The **(Name)** property of the text box is set to **txtPhrase** at design time. Another property of the text box is the **Text** property. The text that is in the text box is the value of the **Text** property. When a string of text is entered from the keyboard into the text box, the value of the string is assigned to the **Text** property of **txtPhrase**. Another way of assigning text to the **Text** property of **txtPhrase** would be to use an assignment statement; for example, **property txtPhrase.Text = "George Washington"**. After this statement is executed at run time, the **Text** property of **txtPhrase** is **George Washington**.

In this program the **Text** property is changed by the user, who inputs text into the text box. The default setting of the **Text** property of the text box is **Text1**. At design time this property value is blanked out, that is, set to no text at all.

The text box has several properties that affect the way the text is displayed: **Bold**, **Italic**, **Size**, and **Alignment** (left, center, or right justification). The default setting for **Bold** is **False**, **Italic** is **False**, and **Size** is **12** (point); the **Alignment** setting is **0**, or left justification. the Justify program demonstrates how the command button controls can change the text box's properties at run time.

If the **Multiline** property is set to **True**, then the text wraps around like the display in a word processor. According to Table 11-1, the **(Name)** property of the text box is **txtPhrase**, set at design time. When the program is executed, the form and its five controls appear. The text box named **txtPhrase** has the focus, indicated by a blinking pointer at the left end of **txtPhrase**. It is waiting for text input. (The **Focus** property is discussed in more detail later in this chapter.)

One the many properties of **txtPhrase** is the **Text** property. Another property of **txtPhrase** is the **Alignment** property. When a program with a text box is executed, the default value of the **Alignment** property is **0**, which corresponds to left justification. By default then, **txtPhrase.Alignment = 0** upon execution. In this program, no procedures were written for a control named **txtPhrase**.

The Command Buttons. Procedures are written for all nine command buttons. Nearly every program has a **cmdExit** button. The code is

```
Private Sub cmdExit_Click ( )
        End      ' the program
End Sub
```

Clicking the **Exit** button terminates the program. Table 11-1 lists the changes made to this command button at design time. The **(Name)** property was changed from **Command9** to **cmdExit**, and the **Caption** property was changed from **command** to **E&xit**, which upon execution of the program, appears as **Exit** on the form. With the ampersand (**&**) in the caption before the letter **x**, pressing ALT-x has exactly the same effect as clicking the **Exit** button. The code for the next command button is

```
Private Sub cmdRight_Click()
        ' Change the alignment to right justify
        txtPhrase.Alignment = 1
End Sub
```

As usual, Visual Basic furnishes the first and last lines of code. We added a comment to explain the action, preceding it with an apostrophe. The executable line of code assigned the value **1** to the **Alignment** property of **txtPhrase**. This corresponds to right justification. The default value equals **0** and corresponds to left justification. The alignment property is set to **2** with the **Center Justify** command button and set to **0** with the **Left Justify** button. Alternatively, Visual Basic alignment constants could be used instead of **0**, **1**, and **2** as shown in Table 11-2. Using the numerical values is simple and compact, but using the alignment constants provides for more readable code. Try substituting some alignment constants for their numerical equivalents to see how they work.

Table 11-2

Alignment Constants

Constant	Value	Description
vbLeftJustify	0	Left justify
vbRightJustify	1	Right justify
vbCenter	2	Center

The important thing to realize here is that an assignment statement in one procedure can change the property of any object in the program. The controls are objects. So are the form and windows. Modern programming focuses on objects, their properties, the functions (or methods) they can do, and the events (like the click event) that trigger a control to function. This approach is called object-oriented programming, or OOP.

SetFocus Statement. The **Focus** property is a control property that indicates the currently active control. Run the program. Keep pressing the TAB key and notice how the focus goes from one control to another. When the text box has focus, it has a blinking insertion point in its upper left corner. When a command button has focus, it has a lightly dashed line running around its perimeter. The controls have focus in the order in which they were placed on the form. Notice that the label control does not have the focus control. The **SetFocus** method of a control places the focus on the control: **txtPhrase.SetFocus.** This statement sets the focus on the text box named **txtPhrase**. **SetFocus** is a method of the text box control (and many other controls).

Tab Index. Focus can also be set at design time or changed at run time with the **Tab Index** property. Close your program if it is running. Click each of the controls, and look at the **Tab Index** property in the properties window for each control. The **Tab Index** property has a value of **0** for the first control you placed on the form and then **1**, **2**, and so on for the other controls in the order in which they were placed on the form. In this program the **Tab Index** property of the text box should equal **0** so that the user sees a blinking insertion point in the text box when the program is initially run; this is an invitation to the user to enter some text. Now let's look at the code for the **cmdClear** button.

```
Private Sub cmdClear_Click()
   txtPhrase.Text = ""                    ' The "" is an empty string"
   txtPhrase.Font.Bold = False         ' Font is no longer bold
   txtPhrase.Alignment = 0              ' Reset to left justify
   txtPhrase.Font.Size = 10             ' Font is reset to 10 point
   txtPhrase.SetFocus                     ' Reset focus to text box
End Sub
```

The first line clears the text out of the text box with the empty string. Strings are placed between two quotation marks; if nothing is between two quotation marks, then the string is null and empty. The next four statements simply set the text box properties back to their default values. The last statement sets the focus back to the text box so the blinking insertion point there tells the use that the text box is ready to receive text input.

The **Private Sub cmdOriginal_Click()** procedure is similar to the **Clear** procedure, except that the **Alignment**, **Focus**, and **Text** properties are not reset.

Variables and Data

A variable has a name and a type and is used for storing data of the same type. In a program, a variable is a named storage location in the computer. That variable contains data that the program or programmer places there, and the data in that storage location may be changed by the program many times during its execution.

The first two Visual Basic programs in this section had no variables and no data. Usually a program has variables with which the programmer can store data. The data is of personal importance to the programmer or program user. The data might be numbers such as dollars of income or names of club members. The program might add up the numbers or alphabetize the names. Most programs carry out actions on data, even though our first two programs did not do this. They serve as an introduction to the Visual Basic workspace and programming with controls.

Data Types. VB recognizes several types of data. Some of the most common types are integer, string, Boolean, and two types of numbers: single and double. A more complete list of data types, their ranges, and declaration examples is given in Table 11-3. A few examples of the values of selected data types are

- Integers: 7, –4, 137, 0, –368, and 3174398
- Strings: "Cost", "D", "137", "0.00509", "Sacramento, CA 95825"
- Single: 438.6214 (unformatted), 438.6 (formatted), 4.386E2 (formatted)
- Double: 438.621384271756 (unformatted), 438.6 (formatted), 4.386E2(formatted)
- Boolean: True or False (Boolean variables have no other values)

Notice that 137 is an integer number, but "137" is a string. Visual Basic allows multiplication of 137 by 3, but Visual Basic gives an error message for the multiplication of "137" by 3. A string is just a list of characters. Words are strings. A sentence is a string that includes spaces (and other symbols, and obeys the syntax of the English language). In Visual Basic (and other languages) "1.27" is a string, but 1.27 is a number. You can square 1.27, but you can no more square "1.27" than you can square "elephant".

Table 11-3 lists Visual Basic data types. The value stored by a variable of type **Single** is stored as a seven-digit number. Formatting (treated later) makes it possible to display more meaningful precision in numeric values. A variable of type **Double** is stored as a fifteen-digit number, but it can be formatted to display fewer digits. Even when a

Single or **Double** is formatted, the variable still stores the full seven- or fifteen-digit number, so that round-off errors in calculations are independent of the variable's format.

Choosing Data Types. How do you know what data type to use? Like the name of a variable, the choice of data type is flexible (but not as flexible). If you or your program is counting, then the **Integer** type is a logical choice. If you are doing mathematical number crunching (in physics or engineering), then **Single** or **Double** might be appropriate. In financial calculations, the **Currency** data type might be the best choice.

In programming languages, a **String** is a string of characters, any character on the keyboard and several more. Some strings we might recognize as forming a word in the English language: Washington, baseball, quantum. Some strings we don't recognize: sy76dkosne, #76siddle, a;sld kfjgh, but they are still strings. If we have a program that uses **Person** as a variable name that stores the names of people, then an appropriate choice of variable type would be **String**.

Choosing Variable Names. Names of variable should be meaningful and reflect their function in the program. For integers, don't use single letters; instead use something like **Count** or **Age**. For real numbers, use name like **Interest**, not **I**, or **Energy**, not **E**. Visual Basic places a few restrictions of the names of variables, however. The variable name:

- Must begin with an alphabetic character (not with, say, #, &, $, and so on)
- Cannot contain an embedded period or type-declaration character
- Must be unique within the same scope (more about scope later)
- Must be no longer than 255 characters (no problem there!)

Some legal names are, **MyName**, **Count**, **Result1**, **Result_1**, **Interest**, **Energy**, or even **This_is_a_very_long_name** (legal, but not advisable). The name cannot be the same as Visual Basic's reserved words, such as **Integer**, **Boolean**, **Dim**, **String**, **Single**, **Double**, and so on. Some illegal variable names are **1stResult**, **Dim**, **#ofTrials**, **Valid?**, **Final Result**, and **X/Y**. Why is each illegal? Try to use variable names that make sense in terms of what they represent. Usually it is best to avoid single letters as variable names. Use **Length**, not **L**. Use **Velocity**, not **V**. In addition to using sensible names for variables, it is often recommended to place a three-letter prefix in front of the variable name in order to identify the variable's type.

Type Prefixes for Names. Knowing the type, as well as the variable's name, often facilitates reading code to debug or just understand it. Even programs that you write for yourself often look a bit strange after you haven't looked at them for a long period of time. So it's useful to have all the help you can get. Table 11-4 gives a list of three-letter prefixes that have become customary. They become part of the variable's name.

Table 11-3

Data Types in Visual Basic

Data Type	Range of Value	Variable Declaration Examples
Byte	0 to 255	Dim Checker as Byte
Boolean	True or False	Dim Linear as Boolean
Integer	–32,768 to 32,767	Dim Trial as Integer
Long (integer)	–2,147,483,648 to 2,147,483,647	Dim CustNo as Long
Single	–3.402823E38 to –1.401298E–45 for negative numbers 1.401298E–45 to 3.402823E38 for positive numbers	Dim Speed as Single
Double	Like Single, but with 15 significant figures	Dim Ratio as Double
Currency	±922,337,203,685,477.5808	Dim Assets as Currency
Decimal	Approx ±79 billion	Dim Interest as Decimal
Date	January 1, 100, to December 31, 9999	Dim Birthday as Date
Object	Any	Dim Praxis as Object
String (var. len)	0 to approximately 2 billion	Dim CityName as String
String (fix. len)	0 to approximately the range of a Double	Dim StateName as String
Variant	Same as above	Default
Constant		Const Counter = 25
Static array		Dim MyArray (50) As Integer
Dynamic array		Dim HerArray () As Double

Table 11-4

Conventional Prefixes for Variable Data Types

Prefix	Type	Example
bln	Boolean	blnFinished
byt	Byte	bytSection
cur	Currency	curBottomLine
dte	Date	dteBirth
dbl	Double	dblWaveLength
int	Integer	intSpecies
lng	Long	lngEnrollment
obj	Object	objPhoneBook
sng	Single	sngCalories
str	String	strCity
var	Variant	varAddedValue

Declaring Variables. The declaration of variables is the way that the program gives advance notice to Visual Basic what variables are to be used in the program. The variable declaration gives the name and type of the variable. The syntax for the declaration statement is

```
Dim (variable name) As (type)
```

Here **Dim** is a reserved word that VB recognizes as the beginning of a variable declaration. **As** is a reserved word that is part of the declaration. For type, you use whatever type name is appropriate, such as **Integer**, **String**, **Single**, and so on. Table 11-3 gives some examples of variable declarations for Visual Basic's various variable types. For example:

```
Dim strPerson As String      ' This statement declares to the computer
                             ' program that strPerson is the name
                             ' of a variable of type String

Dim intCounter As Integer    ' This statement declares to the computer
                             ' program that intCounter is the name
                             ' of a variable of type Integer

Dim sngArea As Single        ' This statement declares to the computer
                             ' program that sngArea is the name
                             ' of a variable of type Single
```

The first line of code declares that a variable with the name **strPerson** is of type **String**; As a variable, **strPerson** can have a value. The only type of value that it can store is a value of **String** type (because of the **Dim** statement). Values are assigned to variables. The type of the value must match the type of the variable. Notice that the names in these sample declaration statements include the appropriate type prefixes.

It makes sense that VB needs to know what variables it will encounter. If you host a wedding party, then you need to know that Jabarie is the groom, Kelli is the bride, Latrell is the best man, and DeMya, Tangela, Yolanda, Edna, and Ticha are the bridesmaids. The guests have different names and are of different types.

Scope of Variables. You might consider the scope of a variable as its jurisdiction. When a variable is declared within a procedure, only code in that procedure can assign values to that variable. The variable is known to the procedure in which it is declared. It is not known to other procedures and cannot be accessed or changed in procedures outside the one in which it is declared. Its scope is local to the procedure in which it is declared.

When a variable is declared at the very beginning of a program, in the general declaration section of the program, then it is accessible to all the procedures. Its scope is global. At times it is useful to declare variables globally, but in general it is better programming to declare variables locally wherever possible. You could imagine a circumstance when one procedure unwittingly changes the value of a variable that is subsequently to be used in another procedure. The results might be unpredictable and difficult to track down. Local declaration of variables tends to make finding errors easier.

The scope of the wedding party guest list (see above) is local to the families of the bride and groom. If the list is published in the newspaper, then its scope is global. It is known everywhere.

The Assignment Operator

We have already seen that a variable stores data. How do we get data into a variable for storage and later use? The assignment operator does this for us. The equal sign (=) is the assignment operator. It assigns the value of whatever stands to its right to the variable or property that stands to its left. It does not signify equality between the left and right side.

```
Dim strPerson As String                'This statement declares to the computer
                                       'program that Person is the name of a variable
                                       'of type string

strPerson = "George Washington"        'This statement assigns the string "George
                                       'Washington" to the variable strPerson where
                                       'the string will be stored
```

VB requires quotation marks to enclose the value of a string variable. **"George Washington"** is the value of the string variable **Person**. This example shows how the assignment operator assigns a value to a variable. In the Justify program described earlier, the program used the assignment operator to assign a property to a control, namely a text box named **txtPhrase.**

```
txtPhrase.Alignment = 2                ' VB knows that assigning 2 means to center

txtPhrase.Font.Italic = True           ' The text box font is assigned italic
```

Another example serves to demonstrate that the assignment statement does not imply equality between the left and right side of the "equal" sign, because the "equal" sign in this instance is an assignment operator. Consider the following lines of code.

```
Dim intTestNum As Integer                   ' variable declaration
Dim intCount As Integer                     ' variable declaration
intTestNumt = 25                            ' intTestNum stores the integer 25
intCount = 3                                ' intCount stores the integer 3
intCount = intCount + 1                     ' Reading from right to
                                            ' left: 1 is added to the 3 currently
                                            ' stored in intCount; then this integer
                                            ' is stored in intCount

intTestNum = intCount                       'The 25 stored in intTestNum is
                                            ' replaced by the integer 4, the
                                            ' current value of intCount
```

After the declaration statement in the first two lines, **25** is assigned to the variable **intTestNum**. In the next line, **3** is assigned to the variable **intCount**. Then, **1** is added to the current value of **intCount**, and that total is assigned to the variable **intCount** (replacing the last value of **intCount**). Finally, the value of **intCount** in the third line is assigned to the variable **intTestNum** so that its value is now **4**. At the end, **intTestNum** has the value 4, since the value in **intCount** (**4**) is assigned to the variable **intTestNum**.

Notice that the assignment operator looks like an equal sign. However, it is not, and does not even imply equality. The line of code **intCount = intCount + 1** surely suggests the no equality is implied. A line of code with an assignment statement is best read from left to right: Add **1** to the current value of **intCount** and assign that sum to the variable **intCount**.

The Form

You've already been introduced to the form, but now let us make the introduction more formal. Like controls, the form is an object, and as such has its own set of properties, methods, and events. Some of them you'll recognize from the few controls that you've already encountered. We've seen that the form is the ubiquitous container for controls. Table 11-5 presents an abbreviated list of some of the form's properties, methods, and events. Check the **Object Browser** window for a complete list. Many of the form's properties, methods, and events will be illustrated in this and subsequent chapters.

Table 11-5

Selected Properties, Methods, and Events of the Form

Properties		Methods	Events
BackColor	FontName	Cls	Click
BorderColor	FontSize	EndDoc	DblClick
Caption	FontStrikeThrough	Line	KeyPress
FillColor	FontUnderline	Load	Load
FillStyle	ForeColor	PrintForm	MouseMove
FontBold	Name	Print	
FontItalic	WindowState	Show	

Previous VB programs used some of the entries in this table. The next program demonstrates the **Print**, **Show**, and **EndDoc** methods and some of the **Font** properties.

Printing to the Form

The print method is useful for displaying output from the program by printing it to the form. A form can print the value of a variable on itself with its **Print** method. If we have declared some variables of different types and assigned values,

```
Dim strPerson As String
Dim intAge As Integer
Dim sngWeight

strPerson = "George Washington"
intAge = 25
sngWeight = 165.7
```

then the statements

```
Form1.Show
Form1.Print strPerson
Form1.Print intAge
Form1.Print sngWeight
```

cause the form to print on itself, beginning at the very upper left corner of the screen:

```
George Washington
25
165.7
```

The word **Print** is a method of the form. When you enter the word **Forml** in the code window, a list of properties and methods appear. On the list, the icon to the left of the name of a method is a little green brick; the icon to the left of the name of a property is a hand holding a note. The word **Show** is also a method of the form. In order for the printed text to show on the form, it is necessary to precede the **Form1.Print** statement with the **Form1.Show** statement.

If the program includes the statement **Form1.Cls**, the form will clear itself of everything that has previously been printed on it. **Cls** is short for clear screen. **Print** and **Cls** are methods of the form. The **Show** method of the form allows that which the form prints to be shown on the form. In the preceding examples, we have used the default name of **Forml**.

The name of an object's method is appended to the name of the object with a period. Some methods take an argument (like **Print**), others do not (like **Show**). When you enter the code for an object and its method, Visual Basic gives an automatic list of properties and methods as soon as the period is entered after the name of the object. After you type the first few letters of the property or method, pressing the TAB key or SPACEBAR completes the entry of the property or method name (automatic code completion).

To use the form's properties, it is actually not necessary to use the name of the form. Some other objects have some of the same properties as the form. With these objects, it is necessary to precede the name of the method by a period and the name of the object. The statement:

```
.Show
.Print strPerson
.Print intAge
.Print sngWeight
```

executes the same as the statement with the preceding form name, **Form1**. In other words, the form is the default object of these methods written in this manner.

Printing to the Printer

The **Print** method is also a method of the printer, another object. Recall that a program is made of objects and code. The objects include, for example, your computer's drives, files, monitor, keyboard, forms, and printer. Just as the form prints (to itself), so does the printer print (to its paper). The statements

```
Printer.Print strPerson
Prinert.Print intAge
Printer.Print sngWeight
```

cause the printer to print on a sheet of printer paper beginning at the very upper left corner of the paper:

```
George Washington
25
165.7
```

If the name of the object is omitted, so that the code reads as follows:

```
Print strPerson
Print intAge
Print sngWeight
```

then, because the default object of the **Print** method is the form, the data are printed to the form. When the **Printer** method is used with any object other than the form, it is necessary to include the object's name, followed by a period, followed by **Print**, followed by a space, followed by the print list.

The Print Demo Program

This short program illustrates how to get output from a program by printing the output to the form. It also is our first program that has a variable declaration. Examples illustrate the default format of numbers and dates and also show how the programmer can control the format of numbers and dates. The **Show** method and the **EndDoc** method are illustrated. The program demonstrates the **Now** function for displaying and printing the current date and time.

Starting the Project

- Start Visual Basic by clicking on the **VB** icon on the desktop.
- Select the **Standard EXE** icon in the **New Project** window, and click the **OK** button.
- A **New Project** window opens with a fresh blank form with a title bar that reads **Project1 – Form1 (Form)**.
- Select the form, and in the properties window change the **(Name)** property to **frmPrintDemo**.
- Click **Save**; the file name will be frmPrinDemo. Click **OK**.
- The file name changes to **Project1**; change it to **The Print Property Program**, and click the **OK** button.
- On the form, place three label controls, three text box controls, and two command button controls, as shown in Figure 11-2.
- Set the properties according to Table 11-6.

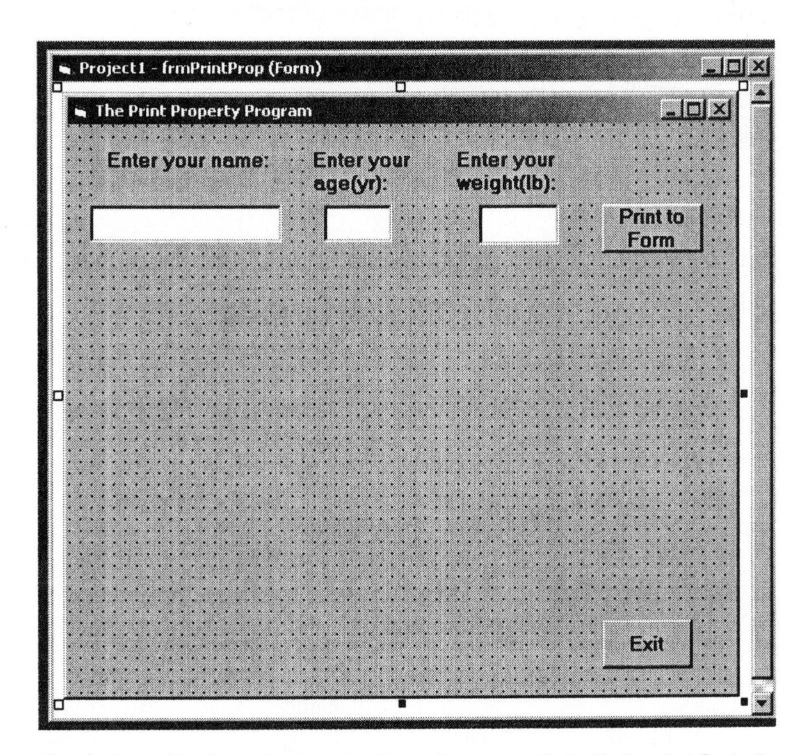

Figure 11- 2. The Form for the Print Demo Program with Its Embedded Controls

Writing the Print Demo Program. Enter the following code in the general declaration section of the code window:

```
Option Explicit
' The Print Demo Program
```

Enter the following code for the **Private Sub cmdPrint_Click()** procedure:

```
Private Sub cmdPrint_Click()
   Show
   Cls
   Print: Print: Print: Print: Print: Print
   FontBold = True
   FontSize = 12
```

```
    FontName = "Times New Roman"
    Print txtName
    Print txtAge
    Print txtWt
    Print txtName; txtAge; txtWt; Tab(60); "Semicolons"
    Print txtName, txtAge, txtWt; Tab(60); "commas"
    Print txtName; vbTab; txtAge; vbTab; txtWt; Tab(60); "vbTabs"
    Print txtName; Spc(3); txtAge; Spc(3); txtWt; Tab(60); "Spaces"
    Print txtName; ", your age is "; txtAge; " yr" _
          ; " and your weight is "; txtWt; " lb" _
          ; " or "; txtWt / 2.2; " kg"
    Print
    Print txtName; ", your age is "; txtAge; " yr" _
          ; " and your weight is "; txtWt; " lb" _
          ; " or "; Format(txtWt / 2.2, "##.###"); " kg"
    Print
    FontName = "Arial"
    Print "Today is "; Now; " or "; Format(Now, "Long Date")
End Sub
```

Enter the following code for the **Private Sub cmdExit_Click()** procedure:

```
Private Sub cmdExit_Click()
    End
End Sub
```

Table 11-6

The Properties Table for the Print Demo Program

Object	Property	Setting
Form1	Name	frmPrintDemo
	Caption	The Print Demo Program
	WindowState	2-Maximized
Label1	Name	lblName
	Caption	Please enter your name
Label2	Name	lblAge
	Caption	Enter your age
Label3	Name	lblWt
	Caption	Enter you weight
Text1	Name	txtName
	Caption	Make it blank
Text2	Name	txtAge
	Caption	Make it blank
Text3	Name	txtWt
	Caption	Make it blank
Command1	Name	cmdPrint
	Caption	Print
Command2	Name	CmdExit
	Caption	Exit

Executing the Print Demo Program. Click the **Start** button to execute the program. The form appears with three text boxes, each under a label prompting the user to action. Two command buttons offer the user two options: **Print** or **Exit**. Enter your name, age, and weight in the appropriate text boxes. Click the **Print** button. Examine the appearance and position of the text and numbers printed to the form, and compare them with the code.

Understanding the Print Demo Program. Executing the program loads the form with its attached controls. The program waits for the user to take action. The labels over the three text boxes suggest what action the user should take. Notice that the labels spell out what the units are of age and weight. It may seem trivial here, but in programs written for scientists and engineers, it is imperative that the units be given clearly. Entering the user's name in the first text box establishes the **Text** property of that text box. At design time, the **(Name)** property of this text box was set to **txtName**, and the **Text** property was made blank. Similar assignments were made for the other two text boxes. After the user establishes the value for **txtName**, **txtAge**, and **txtWt**, she clicks the command button with the caption **Print**. At design time it was named **cmdPrint**.

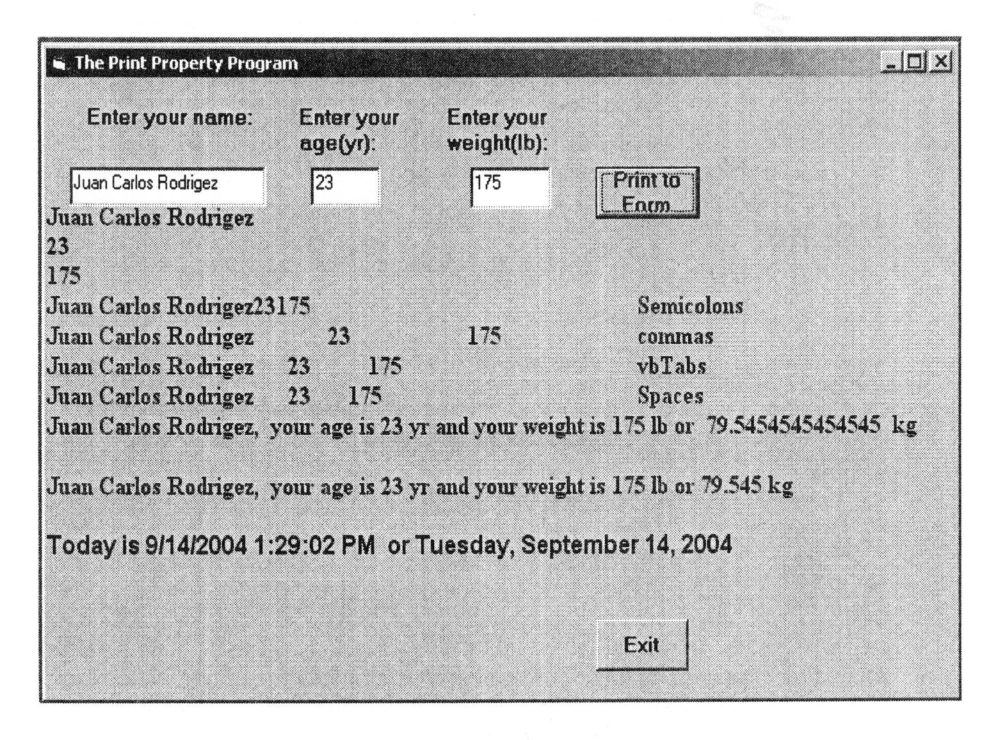

Figure 11-3. The Output of the Print Demo Program Printed to Its Form

Clicking **cmdPrint** causes it to respond to its click event, and the **cmdPrint_Click** procedure executes. The first three lines of the procedure are the **Show** method, the **Cls** method, and the **Print** method. These methods are methods of the form, so it is not necessary to include the name of the form:

```
frmPrintDemo.Show
frmPrintDemo.Cls
frmPrintDemo.Print
```

It is always necessary to use the **Show** method to show what the **Print** method prints to the form. The **Cls** method clears the screen of any text previously printed to the form.

The colon (:) separating the **Print** statements allows several statements to occupy the same line. This practice is not advisable unless the statements are very short, as is the case here. The **Print** statement without an output list prints a blank line. Normally, printing begins in the upper left corner of the form and proceeds downward. We want the printing to begin on the form below the row of text boxes, so we insert five blank lines.

The **Print** method prints to the form all the items on the output list. The argument of **Print** is the output list. The members of the output list may be separated by a semicolon (;), a comma (,), a vbTab constant, a Spc(n) function, or a Tab(n) function. Choosing a variety of list separators (Table 11-7) permits the programmer to display the program's output in a manner that is clear and attractive to the user.

Table 11-7

Separators for the Print Method

Separator	Action
Semicolor (;)	Causes the second item on the list to be printed after the first item in the next available space
Comma (,)	Causes the second item on the list to be printed after the first item, but at the next available print zone
vbTab	Inserts a horizontal tab
Spc(n)	Inserts *n* spaces before printing the next item in the print or printer output list.
Tab(n)	Causes the printing of the next item in the print output list to be printed beginning at the *n*th column number of the screen or printed page.

The first three **Print** statements each have one item in the output list, the value of a text box. When this item is printed, a line feed and carriage return occur so that the printed output of the next **Print** statement occurs on the next line, as show in Figure 11-3.

The next **Print** statement uses semicolons between each item in the output list. The last two items in the output list are the function **Tab(75)** separated from a descriptor, **"Semicolon,"** causing the word **Semicolon** to be printed beginning at the 75th column of the screen. Since the last item in the output list is nothing, a line feed and carriage return is executed by Visual Basic so that the next **Print** statement prints on the next line. In the next two **Print** statements, vbTab and Spc(3) are used as output list separators, again with an appropriate descriptor printed beginning in the 75th column of the screen.

The next **Print** statement has eight items in the output list, all separated by semicolons to give the user readable output. The items are

1. **txtName**
2. **", Your Age is "**
3. **txtAge**
4. **" yr, and your weight is "**
5. **txtWt**
6. **"lb, or "**
7. **txtWt/2.2**
8. **"kg"**

On a single line, seven semicolons would be required to separate these eight items in the output list. However, because the lineis continued twice (with a **space _**), Visual Basic adds a semicolon at the beginning of the second and third lines of the print statement.

Printing and Formatting Numbers

Remember that the text that goes into a text box is of the **Text** type. But the statement **txtWt/2.2** appears to correspond to the division of a **Text** type by a **Decimal** type. Visual Basic, not being a highly typed language, looks at the value of text weight, 133, and assumes that this text representation of a number is really a number, and automatically retypes it as a **Double** and carries out the division. The quotient is a **Double** and is printed in the full glory of its fifteen digits, which doesn't look so great. Fortunately, Visual Basic, being a visual programming language, furnishes a formatting function that enables the programmer to make the number print exactly as desired.

In the next line of our program, the value of the output list item

```
txtWt / 2.2
```

a **Double**, is formatted with the expression

```
Format(txtWt / 2.2, "##.###")
```

In general, the expression for formatting an expression that has a numeric value is

Format(*expression*,"*Formatting characters*")

where the formatting characters and some examples are shown in Table 11-8 and Table 11-9. Named number formats are shown in Table 11-10.

Table 11-8

Printing Custom-Formatted Numbers to the Form

Format	Printed Result
Print Format(3287.412, "#####.###")	3287.412
Print Format(3287.412, "#####.#")	3287.4
Print Format(3287.412, "$####.##")	$3,287.41
Print Format(3287.412, "00000.000")	03,287.412

Table 11-9

Custom Formatting Characters

Symbol	Description
0	Digit placeholder; prints a trailing or lead zero if appropriate
#	Digit placeholder; never prints trailing or leading zeros
.	Decimal placeholder
,	Thousands separator
- + $ ( ) space	Characters that are displayed exactly as typed into the format string

Table 11-10

Printing Named Number Formats to the Form

Named Number Format	Example	Result
Currency	Format (1234.56, “Currency”)	$1,234.56
Fixed	Format (1234.56, “Fixed”)	1234.56
Standard	Format (1234.56, “Standard”)	1,234.56
Percent	Format (0.1234, “Percent)	12.34 %
Scientific	Format (1234.56, “Scientific”)	1.23E03

Printing and Formatting Dates and Time

This is an appropriate time to demonstrate formatting dates and time with a simple example in the program. Some examples of formatting dates and time are shown in Table 11-11.

The statement **Print Now** prints the current date and time to the form according to your computer system's regional settings. To change this, on the Windows **Help** menu, point to **Regional Setting**, and click **Date and Time**. The last few lines of the program show how to print out the current time and how to format it. The statement

```
Print "Today is "; Now; " or "; Format(Now, "Long Date")
```

first prints the current date and time to the form in default format and then in one of several possible named date and time formats, shown in Table 11-12.

Table 11-11

Printing Custom-Formatted Dates and Times to the Form

Format	Result
Print Format(Now, “m/d/yy”)	3/17/04
Print Format(Now, “dddd,mmmm dd,yyyy”)	Saturday, March 17, 2004
Print Format(Now, “d-mmm”)	17-March
Print Format(Now, “hh:mm AM/PM”)	08:23 PM
Print Format(Now, “d-mmmm h:mm”)	17-March 8:23

Table 11-12

Printing Named Date and Time Formats to the Form

Named Date or Time Formats	Example	Result
General Date	Format (Now, “General Date”)	9/15/2004 10:12:36 AM
Long Date	Format (#7/3/31#), “Long Date”)	Friday, July 03,1931
Medium Date	Format (Now, “Medium Date”)	04-May-04
Short Date	Format(Now, “Short Date”)	9/15/2004
Long Time	Format (Now, “Long Time”)	2:03:13 PM
Medium Time	Format (Now, “Medium Time”)	02:03 PM
Short Time	Format (Now, “Short Time”)	14:03

The Printer, its Methods and Properties

When it comes to displaying output, the form and the printer have many properties in common, especially the various font properties. And they both have the **Print** method.

Form1.Print or **Print** prints items in the output list to **Form1**. If the **Name** property of **Form1** has been changed to **frmPrintDemo**, then **frmPrintDemo.Print** prints to **frmPrintDemo**. The name of the form is optional.

Printer.Print prints items in the output list to a piece of paper in the printer. The name of the object, **Printer**, is not optional.

Once printed, the paper can't be erased, so obviously the printer has no **Cls** method.

Nor does the printer have the **Show** method, but it does have a method that is somewhat analogous: the **EndDoc** method. The **Printer.Print** statement prints nothing until the program executes the **Printer.EndDoc** statement. This printer method tells the program that the end of the (printed) document has been reached and it is OK to print all previous **Printer.Print** statements. If you forget to place the **Printer.EndDoc** statement after your **Printer.Print** statements, the program will print the **Printer.Print** statements after you close the program. After all, if you close the program, you must be at the end of the document, or so assumes Visual Basic. If you forget to write **Printer.EndDoc** and instead write **EndDoc**, no printing occurs, and no error is signaled. Printing resumes when you figure out your mistake.

To show how similar the properties and methods of the form and the printer are, we shall edit the Print Demo program so that it prints to the printer instead of to the form. Carry out the following steps:

- Delete **Show**.
- Delete **Cls**.
- Leave the pointer where it is, below the **cmdPrint_Click** declaration.
- Click **Editor_Replace**.
- For **Find**, enter **Print**; for **Replace**, enter **Printer.Print**.
- Click **Replace all**.
- Replace the three occurrences of **Font ...** with **Printer.Font ...**
- Insert **Printer.EndDoc** after the last line, before **End Sub**.

Click the **Start** button to run the program. Enter your name, age, and weight. Click the **Print** command button. The program will print to the printer exactly as it printed to the form.

Option Explicit and Declaring Variables

Visual Basic does not actually require the explicit declaration of variables. However, it is highly advantageous to do so, because the explicit declaration of variables enables Visual Basic to detect errors and to give an error message identifying where the error occurs. This feature of Visual Basic saves a great deal of time in debugging programs.

Programs OptExp1, OptExp2, OptImp1, and OptImp2

The next four programs are almost the same. They serve to clarify the meaning, use, and importance of the **Option Explicit** statement. Each program is highly commented to aid in understanding the slight differences in the programs.

Starting the Project

- Start Visual Basic by clicking on the **VB** icon on the desktop.
- Select the **Standard EXE** icon in the **New Project** window and click the Open button
- A **New Project** window opens with a fresh blank form with a title bar that reads **Project1-Form1(Form)**.
- Click the **Save** icon; the file name will be Form1. Click **Save**.
- The file name changes to **Project1**; change it to **The OptExp1 Program** and click the **Save** button.

Using this form, we will write a program called OptExp1 and then modify it slightly to change it to the OptExp2, OptImp1, and OptImp2 programs. By running these small programs and comparing their behavior, we will get a better idea of the importance and utility of Visual Basic's **Option Explicit** statement.

Double-click the form. This opens the code window with the first and last lines of the **Form_Load** procedure written for you:

```
Private Sub Form_Load()
End Sub
```

The procedure list (left side of the code window) lists just two entries: **general** and **frmOpExp1**. The events list (right side of the code window) lists many events to which the form can respond. The first event listed is **Load**, and that is the one we want. When the **Form_Load** procedure is on the form, alone or with several other procedures, it is the first procedure to execute when the program is run. In the case of the OptExp1 program, it is the only procedure, and the code for the program is placed between the lines **Private Sub Form_Load()** and **End Sub.** Enter the code listed below for the OptExp1 program.

Enter the code for the first program, but to save time in your version, you could omit the comments:

```
Option Explicit                    ' Forces explicit declaration of all variables
' Program OptExp1
Private Sub Form_Load()            ' Declaration of a procedure
   Dim Person As String            ' Explicit declaration of the variable Person
   Person = "Geo Washington"       ' Assignment statement
   Form1.Show                      ' Show is a method of Form1
   Form1.Print Person              ' Person is a variable name; Print is a method
End Sub                            ' End statement
                                   ' This program executes normally.
```

Executing Program OptExp1. Execute the OptExp1 program by clicking the **Start** button on the design window. The OptExp1 program executes normally and prints to the form the string **"Geo Washington"** in the upper left corner of the form.

Program OptExp2. Modify the program so that the code appears as shown below. Make one change: in the last line, change the variable name **Person** to **Persn** so that it is misspelled. Misspelling and typographical error occur frequently in writing the initial version of code.

```
Option Explicit                       ' Forces explicit declaration of all variables
' Program OptExp2
Private Sub Form_Load()               ' Declaration of a procedure
   Dim Person As String               ' Explicit declaration of the variable Person
   Person = "Geo Washington"          ' Assignment statement
   Form1.Show                         ' Show is a method of Form1
   Form1.Print Persn                  ' Persn is a misspelled variable name
End Sub                               ' End statement
' This program does not execute. It generates a compiler error,
' which Visual Basic displays on the screen, over the code window.
```

Executing Program OptExp2. Click the **Start** button on the design window to execute the program. The program does not execute, but instead displays a message box over the code window indicating that a compiler error has occurred (Figure 11-4): **Compiler Error: Variable not defined**. You have used variable name **Persn**, which has not previously been defined. The **Option Explicit** statement requires that any variable used must be previously explicitly defined (delared). Here, the use of the **Option Explicit** statement has detected a misspelled word.

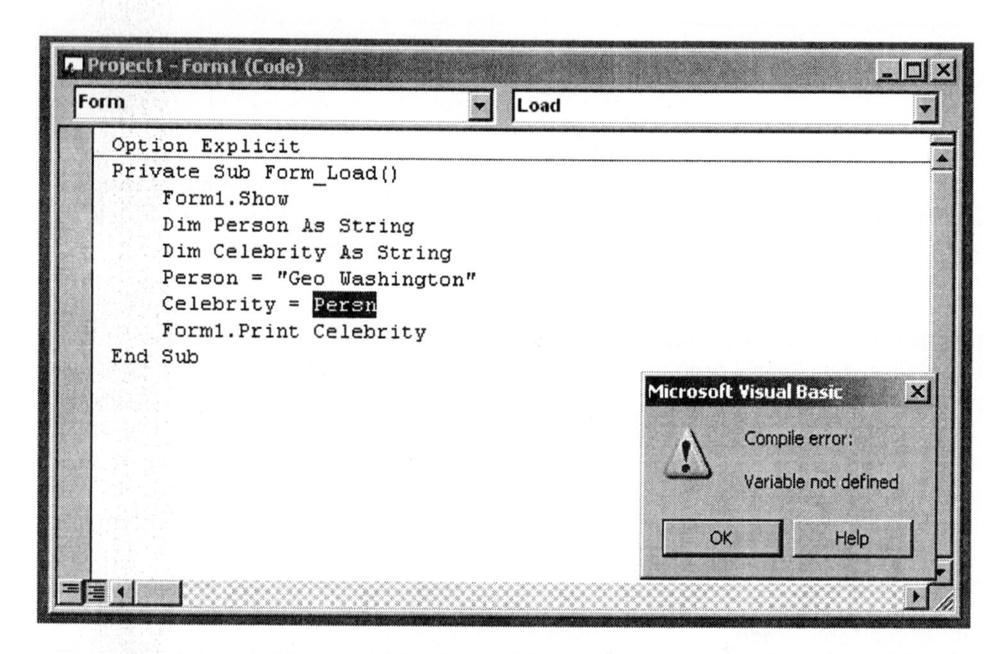

Figure 11-4. A helpful compiler error message made possible by including the **Option Explicit** statement.

Program OptImp1. Modify the program so that the code appears as shown below. Delete the line in which the variable **Person** is explicitly declared. Make one change: in the last line, change the variable name **Prson** to **Person** so that it is spelled correctly.

```
' Program OptImp1
' This program omits the Option Explicit statement
' The variable Person is implicitly declared
Private Sub Form_Load()               ' Declaration of a procedure
                                      ' The variable Person is not declared on this line
   Person = "Geo Washington"          ' Assignment statement & implicit declaration of
   Form1.Show                         ' Show is a method of Form1
   Form1.Print Person                 ' Person the name of an implicitly declared variable
End Sub                               ' End statement
' This program executes normally, and the ouput is
' printed to the form exactly like the OptExp1 program.
```

Executing Program OptImp1. Click the **Start** button on the design window to execute the program. The program has no explicit declaration of the only variable in the program: Person. The string **"Geo Washington"** is assigned to what is clearly a variable, namely **Person**. Visual Basic is a very tolerant language and accepts the implicit declaration of the variable **Person**. The program executes normally, and the output is printed to the form exactly like the OptExp1 program.

Program OptImp2. Now make one change to the OptImp1 program to get to the OptImp2 program. In the last line of the program, change the spelling of **Person** to **Persn**.

```
' Program OptImp2
' This program omits the Option Explicit statement
' The variable Person is implicitly declared
' Program OptImp2
Private Sub Form_Load()             ' Declaration of a procedure
                                    ' The variable Person is not declared on this line
   Person = "Geo Washington"        ' Assignment statement & implicit declaration
   Form1.Show                       ' Show is a method of Form1
   Form1.Print Persn                ' Persn is the misspelled name of
                                    ' an implicity declared variable
End Sub                             ' End statement

'This program executes normally, but nothing is printed to the form
```

Executing Program OptImp2. Now execute the program by clicking the **Start** button. The program executes and displays no error message, but the form is blank. Nothing is printed to it. What happened?

Understanding What Happened. In the last line, the variable name **Persn** generates no error, because it is implicitly declared in this line just as **Person** was implicitly declared in a previous line. However nothing was ever assigned to the variable **Persn**, so the **Print** method in the last line cannot and does not print anything to the screen.

In a program five lines long, the consequences of implicit declaration of variables may seem trivial. In a program of five hundred lines or even fifty lines, the advantage of using the **Option Explicit** statement as the first line of the general declaration section becomes dramatic. Visual Basic not only beeps and displays an error message box but also highlights the exact line where the error occurred. You should place the **Option Explicit** statement as the first line of every program that you write.

Automatic Option Explicit

Without the **Option Explicit** statement, Visual Basic does indeed allow you to make up variables as you go along, but it cannot effectively check for errors, making it difficult for you to debug your program. If Visual Basic does not automatically place the **Option Explicit** statement at the beginning of every new project, you can tell it to do so as follows:

- In the Visual Basic **Tools** menu, click **Options**.
- The **Options** dialog box appears (Figure 11-5), probably with most check boxes selected.
- Select the **Require Variable Declaration** check box, if it is not selected.
- Click **OK**.

From now on, Visual Basic will automatically supply the line of code **Option Explicit** at the very beginning of the program in the general declaration section of the program.

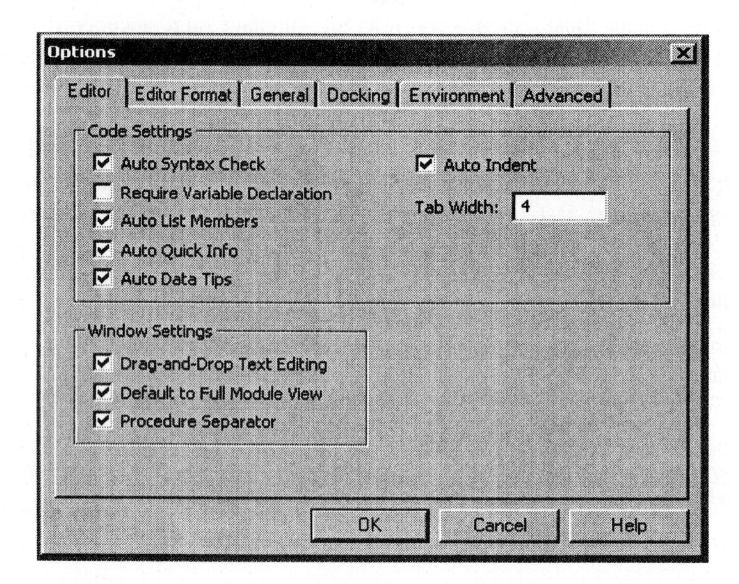

Figure 11-5. The **Options** Dialog Box. Select the **Require Variable Declaration** check box to install the **Option Explicit** statement automatically.

Chapter 12

Visual Basic Operators, Control Structures, and Functions

Chapters 10 and 11 provide a brief introduction to writing a Visual Basic program in the design window, getting input and output, and making text and numbers look pretty. But programming is not always just a matter of input, action, and output. After input, what action should occur may depend on various conditions. If a user wants to divide one number by another number and enters a zero for the divisor, the computer program should recognize this unacceptable condition and warn the user of the mistake.

Beside the obvious arithmetic operators like addition, subtraction, multiplication, and division, Visual Basic has relational and conditional operators with which the programmer can control the path of the program. Relational operators include, for example, symbols for equal to (=), greater than (>), and less then (<). Some examples of conditional operators are AND, OR, and NOT. Used with a decision structure like **If ... Then**, conditional operators provide the programmer with powerful tools for taking control of the program.

This chapter also introduces the **MsgBox** statement, several kinds of built-in functions, and some new controls: the scroll bar control, the list box control, the menu control, and the option button control. To illustrate these new VB features, this chapter include several short VB programs: the Boolean Operator program, the IfThenElse program, the Random Guess program, the pH program, the ColorBox program, the List Box program, the Arithmetic program, and the Functions program.

Relational Operators

Relational operators are also known as comparison operators. Table 12-1 gives the properties of relational operators. The result of an arithmetic operation is a numeric value. The result of a relational operation is a Boolean value, either true or false. Compare the similar syntax of forming an arithmetic expression (Table 12-11) with forming a Boolean expression (Table 12-1). The values for arithmetic expressions can be any number, but the value of a Boolean expression can only be true or false.

Conditional Operators

Conditional operators compare two relational Boolean expressions. For example:

$$(A < B) \text{ AND } (B > C)$$

Each relational expression has a value, so conditional operators compare Boolean values, in this case (false) AND (false), if the values for A, B, and C given in Table 12-2 are used. The result of the conditional comparison has a Boolean value, in this case, false.

Table 12-1

Properties of Relational Operators

Relational Operator	Description	Boolean Expression	Value of Expression for A=10, B=2, C=5, D=3
=	Equal	A = B	False
<>	Not equal	A <> B	True
<	Less than	A < B	False
>	Greater than	A > B	True
<=	Less than or equal	A <= B	False
>=	Greater than or equal	A >= B	False

Table 12-2

Logical Operators (Conditional Operators)

Conditional Operators	Description	Boolean Expression	Value of Expression for A=10, B=2, C=5, D=3
AND	Logical conjunction	(A > B) AND (B > D)	False
OR	Logical disjunction	(A > B) OR (B > D)	True
NOT	Logical negation	NOT (A > B)	False
XOR	Logical exclusion	(A > B) XOR (B < D)	False

Since Boolean expressions have values, it can also be stated that conditional operators compare Boolean values. Both relational operators and conditional operator play a role in decision structures that will be discussed shortly. The value of the resulting Boolean expression depends on the various combinations of Boolean values that Expression1 and Expression2 may have. A truth table presents all the various possibilities (Tables 12-3, 12-4, and 12-5).

Table 12-3

Truth Table for the AND Conditional Operator

If Expression1 Is	AND If Expression2 Is	The Result Is
True	True	True
True	False	False
False	True	False
False	False	False

Table 12-4

Truth Table for the OR Conditional Operator

If Expression1 Is	OR If expression2 Is	Then the Result Is
True	True	True
True	False	True
False	True	True
False	False	False

Table 12-5

Truth Table for the XOR Conditional Operator

If Expression1 Is	XOR If Expression2 Is	The Result Is
True	True	False
True	False	True
False	True	True
False	False	False

If one and only one of the two expressions compared by the XOR operator is true, then the result is true.

The Boolean Operator program demonstrates how relational operators and conditional operators function in a Visual Basic program. Almost all computer programs use relational operators and conditional operators in conjunction with decision structures to control the flow of the program. Their importance cannot be underestimated.

The Boolean Operator Program

The Boolean Operator program illustrates a number of features of relational and conditional operators. Just as Visual Basic can print the value of an integer, real or string, it can print the value of a Boolean expression. If the Boolean expression is false, VB prints **False**; if the Boolean expression is true, VB prints **True**.

Writing the Boolean Operator Program

- Start Visual Basic by clicking the **VB** icon on the Windows desktop.
- In the **New Project** dialog box, select the **Standard EXE** icon in the **New Project** window, and click the **OK** button.
- A **New Project** window opens with a fresh form.
- Click **Save**; the file name will be Form1, so click **OK**.
- The file name changes to Project1; change it to **The Boolean Operator Program**, and click the **OK** button.

Double-click the form to open the code window. In the general declaration section of the program, add the lines:

```
' Option Explicit
' Program BoolOp demonstrates Boolean values, relational
' and conditional operators
```

After the procedure declaration, fill in the following lines to complete the program:

```
Private Sub Form_Load()
   Dim A, B, C, D As Integer
   A = 10: B = 8: C = 6: D = 1

   Show
   Print
   Print " A  B  C  D"
   Print A; B; C; D
   Print
   Print "Relational Operators"
   Print
   Print "B < A", Spc(10), (B < A)
   Print "B > C", Spc(10), (B > C)
   Print "A < B", Spc(10), (A < B)
   Print "D < B", Spc(10), (D < B)
   Print "A < B", Spc(10), (A < B)
   Print
   Print "Relational Operators with Conditional Operators"
   Print
   Print " (B > A  Or  B > C)", (B > A Or B > C)
   Print " (A > B  And  D > B)", (A > B And D > B)
   Print " (B > D  Or  C > A)", (B > D Or C > A)
   Print " (B > D  Or  C > A)", (B > D Or C > A)
End Sub
```

Executing the Boolean Operator Program. After entering the code for the Boolean Operator program, click the **Start** button to execute the program. Figure 12-1 displays the output. Carefully compare the output with the program code in light of the previous discussion of conditional operators. Exit the program, and enter some random different numbers for the variables A, B, C, and D. Predict the corresponding Boolean value of the Boolean expressions in the Boolean Operator program. Run the program again. Check your predictions.

Understanding the Boolean Operator Program. When the **Start** button is clicked, the program executes by immediately loading the form with the **Form_Load** procedure:

```
Private Sub Form_Load()
```

The only code in the program is the code of the form (Form1), so its code executes. The first few lines of the program declare some integers. Notice that three colons separate four assignment statements permitting all four short assignment statements to appear on one line. This is convenient at this point and makes the code compact. Ordinarily, however, it is not good practice to put more than one long statement on a single line.

```
Dim A, B, C, D As Integer
A = 10: B = 8: C = 6: D = 1
```

The **Show** method enables printing values to show up on the form. Whenever you want to print to the form, your must use the **Show** method. The complete syntax is

```
Form1.Show
```

If the name of the form is omitted, it is assumed to be the current form. A series of **Print** statements comes next. A **Print** statement without an argument just prints a blank line to make the output more readable.

```
Print
Print " A  B  C  D"
Print A; B; C; D
Print
Print "Relational Operators"
Print
```

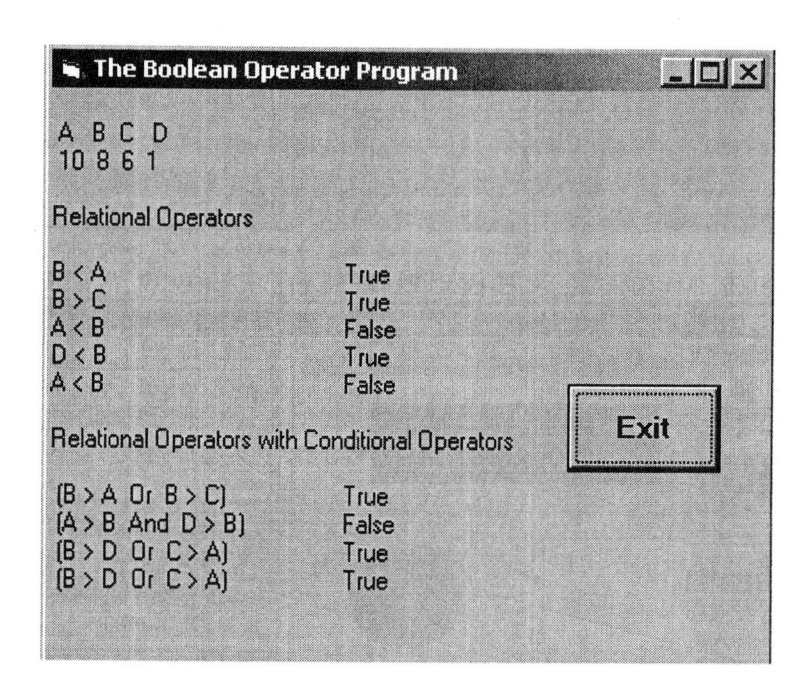

Figure 12-1. Output of the Boolean Operator Program

Each of the next four statements prints four items in its output list. The first is a string, "B < A". Next the **Spc(*n*)** function, which returns *n* spaces before the next item in the list is printed. This lines the output up with the output of the last five **Print** statements. The third item in the output list is a Boolean expression. Boolean expressions can have only two values, true and false. Visual Basic prints out either **True** or **False**, depending on the value of the Boolean expression.

```
Print "B < A", Spc(10), (B < A)
Print "B > C", Spc(10), (B > C)
Print "A < B", Spc(10), (A < B)
Print "D < B", Spc(10), (D < B)
Print "A < B", Spc(10), (A < B)
```

After printing some more blank lines and a label, four more print statements appear, this time with only two arguments. The first is a string, then the value of a complex Boolean expression. These are called complex because they use both relational operators and conditional operators. Complex or simple, a Boolean expression can have only two values, true and false, and these values are printed following the string.

```
Print
Print "Relational Operators with Conditional Operators"
Print
Print " (B > A  Or  B > C)", (B > A Or B > C)
Print " (A > B  And  D > B)", (A > B And D > B)
Print " (B > D  Or  C > A)", (B > D Or C > A)
Print " (B > D  Or  C > A)", (B > D Or C > A)
```

Change some of the expressions and see if you can predict their values. Run the program to check your logic.

Control Structures

Combined with Boolean expressions, Visual Basic's decision statements give the user control structures for controlling the course of the program. Control structures allow the program to carry out actions selectively, repeatedly, conditionally, unconditionally, or sequentially. For example, the **If ... Then** statement tests a condition (the value of a Boolean expression) and then makes a decision concerning what action to do next. The **If...Then...ElseIf...Else** statement extends the control further.

The If ... Then Decision Statement

The **If ... Then** statement (between **Statement1** and **Statement5**) has the following structure:

```
Statement1
If (Boolean Expression) Then
        Statement2
        Statement3
        Statement4
End If
Statement5
```

After **Statement1** is executed, if the Boolean expression has the value **True**, then the block of **Statement2**, **Statement3**, and **Statement4** between the **If** and the **End If** are executed. Then **Statement5** is executed. However, if the Boolean expression has the value **False**, then after **Statement1** is executed, control skips to **Statement5**, which is executed. The program ignores **Statement2**, **Statement3**, and **Statement4**. The following code shows how the **If ... Then** statement can test a condition

```
A =2:  B = 5
If A < B Then
        Form1.Print "Yes, A is less than B"
End If
```

If a program executes this series of statements, the string **"Yes, A is less than B"** is printed on the form. However, if the assignments are reversed so that **A = 5: B = 2**, then control skips the **Form1.Print** statement, and statements following the **End If** are executed.

The If...Then ... Else Decision Statement

The **If ... Then ... Else** statement is similar to the **If .. Then** statement, but it provides the option of branching. It has the following structure:

```
Statement 1
        If (Boolean Expression) Then
                Statement2              ' This is the
                Statement3              ' first block
                Statement4              ' of statements
        Else
                Statement5              ' This is the
                Statement6              ' second block
                Statement7              ' of statements
        End If
Statement8
```

If the Boolean expression has the value **True**, then the statements are executed in the order 1, 2, 3, 4, 8. If the Boolean expression has the value **False**, then the statements are executed in the order 1, 5, 6, 7, 8. If the Boolean expression is true, the first block of the statement is executed, and the second block is ignored. If the Boolean expression is false, then the first block of the statement is ignored, and the second block of statements is executed.

The **If ... Then ... Else** structure can be extended with the **If ... Then ...ElseIf ....Then ElseIf ...** statement. This extended structure is illustrated in the Random Guess program.

```
Statement1
        If (Boolean Expression) Then
                Statement2              ' This is the
                Statement3              ' If block
                Statement4              ' of statements

        ElseIf (Boolean Expression1) Then
                Statement5              ' This is the
                Statement6              ' first ElseIf block
                Statement7              ' of statements

        ElseIf (Boolean Expression2) Then
                Statement8              ' This is the
                Statement9              ' second ElseIf block
                Statement10             ' of statements

        ElseIf (Boolean Expression3)     Then
                Statement11             ' This is the
                Statement12             ' third ElseIf block
                Statement13             ' of statements

        Else (Boolean ExpressionN) Then
                Statement14             ' This final Else
                Statement15             ' expression
                Statement16             ' is optional
Statement17
```

- **Statement14**, **Statement15**, and **Statement16** are executed only if no previous Boolean expression is true.
- If the first **If** statement (after **Statement1**) at the beginning is true, then **Statement2**, **Statement3**, and **Statement4** are executed, but no other statements.
- If the Boolean expression following an **ElseIf** statement is true, the block of statements following it is executed, but none other.

The IfThenElse Program

The IfThenElse program demonstrates the **If ... Then ... Else** decision structure by comparing the values of two variables. This is a short demonstration program that has no controls. Clicking the **Start** button executes the **Form** procedure to which the program code is attached.

Writing The IfThenElse Program

- Start Visual Basic by clicking the **VB** icon on the Windows desktop.
- In the **New Project** dialog box, select the **Standard EXE** icon in the **New Project** window, and click the **OK** button.
- A **New Project** window opens with a fresh form.
- Click **Save**; the file name will be Form1, so click **OK.**
- The file name changes to Project1; change it to **The IfThenElse Program**, and click the **OK** button.

Double-click the form to open the code window. In the general declaration section of the program, add the lines:

```
Option Explicit
' The IfThenElse Program
```

Enter the following code in the **Form_Load** procedure:

```
Private Sub Form_Load()
Dim A As Integer
Dim B As Integer
  A = 2
  B = 5
  Show
  If (A > B) Then
    Print "If statement is True"
  ElseIf (A = B) Then
    Print " First ElseIf is True"
  ElseIf (B < A) Then
    Print " Second ElseIf is True"
  ElseIf (A < B) Then
    Print "Third ElseIf is True"
  Else
    Print "Else gets to print"
  End If
End Sub
```

Executing the program. Click the **Start** button to execute the program. Observe the output that is printed on the form. Run the program with various Boolean expressions to see for yourself how the **If ... Then ... ElseIf ...** statement works.

Understanding the IfThenElse Program. Compare the output with the program code. Switch the values assigned to the variables **A** and **B**. Run the program again, observe the output, and compare with the program code.
Change the code so that **5** is assigned to both **A** and **B**. Run the program again, observe the printed output, and compare it with the program code.

If more than one statement is true, then the first one encountered is executed or printed in this case. If none of the statements is true, then the final **Else** statement executes. This would occur if you make one modification to the program as it is written: in the third **ElseIf** statement, change the **(A < B)** to **(A > B)**. Then none of the **ElseIf** statements is true, and the program prints the final **Else** statement, **"Else gets to print"**.

This control structure gets a little tedious if it contains several **ElseIf** statements. In these situations, it is usually better to use the **Select Case** statement described later. But before that, let's look at the Random Guess program, which not only gives further practice with the **If ... Then ... Else** structure but also introduces some frequently encountered Visual Basic statements and functions.

The Random Guess Program

The Random Guess program puts into practice some of the concepts that have already been introduced, such as the **If ... Then ... Else** decision structure, input to a text box control, and captions for a label control. In addition it provides an introduction to the **Timer** function, the **Rnd** function, the **Randomize** statement, the **MsgBox** statement, and the **Call** statement. First, we'll look at isolated examples of these functions and statements and then utilize them in the Random Guess program.

The Rnd Function

The **Rnd** function generates a random number greater than or equal to zero and less than one. Write this program, attaching it to the form just as you did with the IfThenElse program:

```
Option Explicit
Private Sub Form_Load()
   Show
   Print Rnd
   Print Rnd
   Print Rnd
End Sub
```

When this program is run, it prints three random numbers in the upper left corner of the form. For example:

```
0.7055475
0.5334247
0.5785186
```

If the program is run again, today, tomorrow, or whenever, the identical three random numbers result.

The Randomize Statement

Slip in the **Randomize** statement in the preceding code as follows:

```
Option Explicit
Private Sub Form_Load()
   Randomize                      'add this statement to the code
   Show
```

```
    Print Rnd
    Print Rnd
    Print Rnd
End Sub
```

When this program is run, three random numbers are printed on the form. If you run it again, three different random numbers result. The **Randomize** statement initializes the **Rnd** function with a new seed value generated by the Windows system timer. Consequently every time you run it, the random numbers are different, because the seed generated by the current time of the system clock is different.

The more complete syntax for the **Rnd** function is: **Rnd(*number*)**. The number is the seed number. Both it and the parentheses are optional. You can modify the preceding program to use your own seed number and see what happens.

The Timer Function

The **Timer** function returns the number of seconds that have elapsed since midnight, according to your computer's system clock. Write this program and attach it to the form just as you did with the IfThenElse program described earlier. Note the time on your computer, and run the program.

```
Private Sub Form_Load()
    Dim NumOfSec As Single
    Show
    NumOfSec = Timer
    Form1.Print NumOfSec
End Sub
```

If you ran the program at exactly 9:02 a.m., the program would print on the form the value of **NumOfSec: 32551.35**. The **Timer** function is useful in science and engineering for timing experiments and events.

The MsgBox Statement

The **MsgBox** statement displays a dialog box with one or more buttons that the user must respond to with a click. Users of Windows applications such as Microsoft Word are familiar with the frequently encountered **MsgBox** dialog box. For example, if you try to close your work without saving it, you may get a dialog box like the one shown in Figure 12-2.

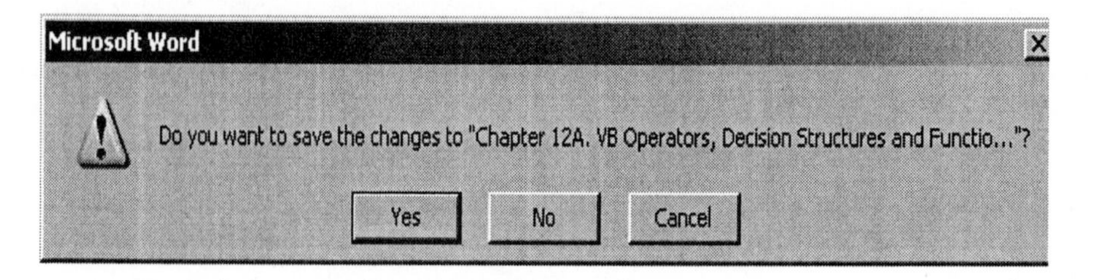

Figure 12-2. A Familiar Message Box in Microsoft Word

When you invoke the **MsgBox** statement in your code, Visual Basic generates the code to display the message box. The code is invisible to us and is hidden deep within VB, so we don't have to learn how to do what is does. It does a lot. It creates on the screen a little box with a title bar on top, and underneath a button that says "OK," and it may create one or two more buttons, if you wish. Between the title bar and the button, it writes a message. In the **MsgBox** statement, commas serve to separate the three parameters. We often use only two of the parameters, the first and third. The second, missing parameter must still be separated from the first and third by two commas. The first parameter is the message that you want the message box to deliver on screen, written in the message box. The third parameter is the name we want to give to the title bar of the message box.

```
MsgBox "Your Box Message", vbConstant, "Your Bar Message"
```

The second parameter (optional, but its delimiting commas are not) generates an icon, described in Table 12-6. Using the constants listed in Table 12-6 is preferable to using the integer values, since the resulting code is more readable.

To get a quick look at the message box, write this program, and attach it to the form just as you did with the IfThenElse program. Notice the line continuation character (a space followed by an underline character) in the **MsgBox** statement line. This combination allows a long line of code to be continued on the next line.

```
Option Explicit
Private Sub Form_Load()
   MsgBox "First string to the Box", vbExclamation, _
   "Second String to the Bar"
  Show
   Print "That's all folks"
End Sub
```

Table 12-6

Message Box Constants

vbConstant	Integer Value	Description
vbOKOnly	0	Displays the **OK** button only
vbOKCancel	1	Displays the **OK** and **Cancel** buttons
vbAbortRetryIgnore	2	Displays the **Abort**, **Retry**, and **Ignore** buttons
vbYesNoCancel	3	Displays the **Yes**, **No**, and **Cancel** buttons
vbYesNo	4	Displays the **Yes** and **No** buttons
vbRetryCancel	5	Displays the **Retry** and **Cancel** buttons
vbCritical	16	Displays the **Critical Message** Icon
vbQuestion	32	Displays the **Warning Query** icon
vbExclamation	48	Displays the **Warning Message** icon
vbInformation	64	Displays the **Information Message** icon
vbDefaultButton1	0	Displays the first button as the default
vbDefaultButton2	256	Displays the second button as the default
vbDefaultButton3	512	Displays the third button as the default
vbDefaultButton4	768	Displays the fourth button as the default

When the message box is onscreen, your computer execution halts and waits for you to click **OK**. After you click **OK**, execution resumes, and the form's **Print** method prints **"That's all folks"** on the form.

Writing the Random Guess Program

- Start Visual Basic by clicking the **VB** icon on the Windows desktop.
- In the **New Project** dialog box, select the **Standard EXE** icon in the **New Project** window, and click the **OK** button.
- A **New Project** window opens with a fresh form.
- Click the form to select it. Then, in the properties window, change the **(Name)** property to **frmRandomGuess**.
- Click **Save**; the file name will be frmRandomGuess, so click **OK**.
- The file name changes to Project1; change it to **The Random Guess Program**, and click the **OK** button.
- Place controls on frmRandomGuess according to Figure 12-3.
- Set the properties of the controls according to Table 12-7.

The Random Guess Program Form

Figure 12-3 displays the user interface for the Random Guess program. When you execute the program, the program generates a random number between 1 and 15. You try to guess the number by entering your guess in a text box and then clicking a button to check whether your guess is correct. Upon the first click of the **Check your guess** button, the program stores the seconds since midnight. You keep guessing and checking until you guess the correct number. At this moment the computer again stores the seconds past midnight and calculates the number of seconds you needed to guess correctly. You are then congratulated. The program congratulates the user with a statement printed on the form, and a message box displays the number of tries required to guess correctly (Figure 12-4).

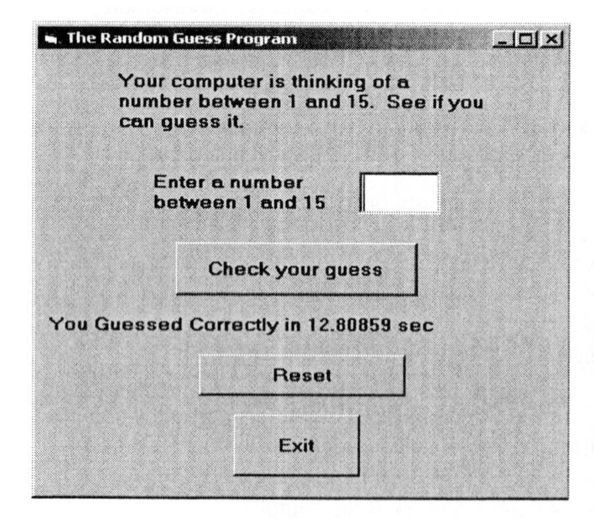

Figure 12-3. The Form for the Random Guess Program

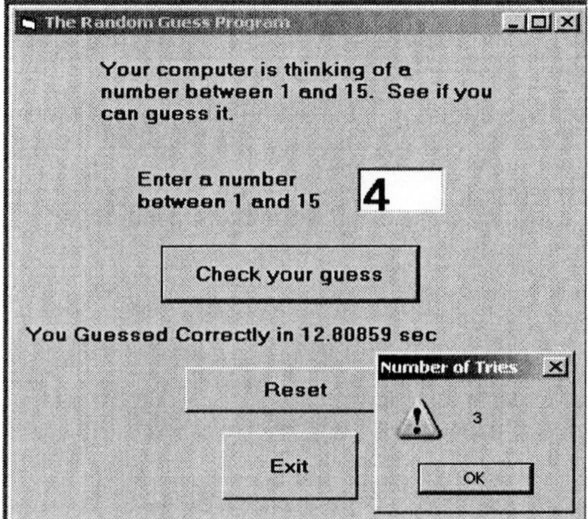

Figure 12-4. The Form for the Random Guess Program with a Message Box

Table 12-7

The Properties Table for the Random Guess Program

Object	Property	Value
Form1	Name	frmRandomGuess
	Caption	The Random Guess Program
Label1	Name	lblInfo
	Caption	Your computer is thinking of a number between 1 and 15. See if you can guess it.
	Font	10 point bold
Label2	Name	lblDirect
	Caption	Enter a number between 1 and 15
	Font	10 point bold
Label3	Name	lblResult
	Caption	Leave it blank
	Font	10 point bold
Text box1	Name	txtInput
	Text	Make it blank
	Font	24 point bold
Command1	Name	cmdCheck
	Caption	Check your guess
	Font	10 point bold
Command2	Name	cmdReset
	Caption	Reset
	Font	10 point bold
Command2	Name	cmdExit
	Caption	Exit
	Font	10 point bold

Entering the Code of the Random Guess Program. In the general declarations section of the program, enter some comments giving the name of the program and its function:

```
Option Explicit
' The Random Guess program demonstrates the
' If...Then...Else control structure, the Timer function,
' the Rnd function, the Randomize statement
' and the MsgBox statement

Dim sngStartTime As Single, sngEndTime As Single
Dim sngYourTime As Single
Dim Tries As Integer, answer As Integer
```

Double-click the form, or select **Form** from the procedure list, and enter:

```
Private Sub Form_Load()
   Tries = 0
   txtInput.Text = "0"
   Randomize
   answer = 14 * Rnd + 1
End Sub
```

Double-click the command button, and enter:

```
Private Sub cmdCheck_Click()

  Tries = Tries + 1
  If Tries = 1 Then sngStartTime = Timer
  If txtInput = answer Then
    sngEndTime = Timer ' Save seconds from midnight to when correct guess made
    sngYourTime = sngEndTime - sngStartTime
    lblResult.Caption = "You Guessed Correctly in " & CStr(sngYourTime) Sec"
    MsgBox Tries, vbExclamation, "Number of Tries"
  ElseIf (txtInput < 1) Or (txtInput > 15) Then
    lblResult.Caption = " Your guess must be between 1 and 15"
  ElseIf (txtInput < answer) Then
    lblResult.Caption = "Your guess is too low"
  ElseIf (txtInput > answer) Then
    lblResult.Caption = "Your guess is too high"
  End If

End Sub
```

Double-click the **Reset** button, and enter:

```
Private Sub cmdReset_Click()
  txtInput.SetFocus
  txtInput.Text = ""
  lblResult.Caption = ""
  Call Form_Load
End Sub
```

Double-click the **Exit** button, and enter:

```
Private Sub cmdExit_Click()
  End
End Sub
```

Executing the Random Guess Program. Click the **Start** button to execute the program. The first label at the top of the user interface tells the user what the game is about. The second label directs the user to enter a number in the text box. The caption on the button prompts the user to check the guess. Depending on the relation of the guess to the correct answer, the label (invisible until now) above the **Exit** button displays one of four messages:

1. "You Guessed Correctly in " & CStr(sngYourTime) & " sec"
2. "Your guess must be between 1 and 15"
3. "Your guess is too low"
4. "Your guess is too high"

The program calculates the number of seconds required to get a correct answer. This value is stored in a variable named **sngYourTime**, which CStr converts to a string.

Understanding the Random Guess Program. Clicking the **Start** button causes the **Form_Load** procedure to execute immediately. Because it is the first procedure to execute, the **Form_Load** procedure often contains a few statements that initialize a few of the program's parameters. In the Random Guess program, the user tries to guess a random number generated by the program, and tries again until the correct number is guessed. The program counts the number of tries, and when the user finally guesses the number correctly, the computer displays the number of tries in a message box and the lapsed time as a **lblResult**. Since the user may ask the computer to reset the number and try again, the **Form_Load** procedure initializes the number of tries at zero, generates a random number, and assigns it to the variable answer.

```
Private Sub Form_Load()
   Tries = 0
   txtInput.Text = "0"
   Randomize
   answer = 14 * Rnd + 1
End Sub
```

Since the **Rnd** function returns a number between 0 and 1, it must be multiplied by 14 and then 1 added to give an upper limit of 15 and a lower limit of 1. Thus when **Rnd** returns 0, the lower limit is 1. When **Rnd** returns 1, the upper limit is 15. The computer "...is thinking of a number between 1 and 15."

When the **Check your guess** button is clicked, 1 is added to the counter **Tries**, which was initialized to have the value 0. On the first guess, when **Tries** equals 1, the current value of the **Timer** function is stored in **sngStartTime**.

Whenever the **cmdCheck** button is clicked, the **If** statement compares the user's guess entered in **txtInput** with the answer, the randomly generated number between 1 and 15. The **cmdCheck** procedure first uses the relational operator = to see if the answer is correct. If it is correct, the current value of the **Timer** function is stored in **sngEndTime**, and the lapsed time (**sngYourTime**) is calculated from the difference between **sngStartTime** and **sngEndTime**. The string **"You Guessed Correctly in" & CStr(sngYourTime) & "sec"** is assigned to the **lblResult**, the **Caption** property, which is displayed on the screen. Because the caption of this label was made blank at design time, nothing appears on the label until the **cmdCheck** button is clicked.

```
If txtInput = answer Then
   sngEndTime = Timer  ' Save seconds from Midnight to when correct guess made
   sngYourTime = sngEndTime - sngStartTime
   lblResult.Caption = "You Guessed Correctly in " & CStr(sngYourTime) & " sec"
   MsgBox Tries, vbExclamation, "Number of Tries"
End if
```

The value of **sngYourTime** is converted to a string and concatenated to the preceding and following strings. The number of guesses required is stored in the current value of **Tries**, which is displayed in a message box.

The first **ElseIf** uses the conditional operator OR to check whether the user's guess is outside the game's boundaries (1 to 15) and, if the guess is outside the boundaries, prints the string **"Your guess must be between 1 and 15".**

```
ElseIf (txtInput < 1) Or (txtInput > 15) Then
   lblResult.Caption = " Your guess must be between 1 and 15"
```

If the guess is wrong, but within the game's boundaries, the procedure uses the two relational operators (< and >) to determine whether the guess is too low or too high and prints this information on the **lblResult** label. With this information, the user can converge on the correct answer in fewer than three or four guesses.

```
ElseIf (txtInput < answer) Then
   lblResult.Caption = "Your guess is too low"
```

```
ElseIf (txtInput > answer) Then
    lblResult.Caption = "Your guess is too high"
```

In order to try a new number, the user clicks the **Reset** button. The **Command_Reset** procedure returns the focus to the text box to signal the user that a number may be entered. It also clears the text box of the last guess that the user entered and calls the **Form_Load** procedure. Remember, the **Form_Load** procedure was automatically executed when the **Start** button was clicked. We don't want to close the program and restart it to get a new number; we just want a new number. We can do that with the **Call** statement. **Call Form_Load** causes the **Form_Load** procedure to execute again, providing a new number to guess. Since loading the form sets **Tries** to zero, the **Timer** function will start with a fresh **sngStartTime** when **cmdCheck** is clicked again. **Call** is an optional keyword and may be omitted.

```
Call ProcedureName
```

and

```
Procedure Name
```

are equivalent ways of calling another procedure and causing it to execute. Try deleting the **Call** part of the call to the **Form_Load** procedure, and you will see that the program runs unchanged.

Invoking the **Timer** function returns the number of seconds since midnight. Invoking the **Timer** function twice and taking the difference between the two values returned gives the lapsed time. The **Exit** button functions as usual.

The Scroll Bar Control

The toolbox contains controls for vertical and horizontal scroll bars. Following is the **Vertical Scroll Bar** icon as it appears in the toolbar:

The **Horizontal Scroll Bar** icon appears as follows:

The scroll bar functions through the use of two of its events: the scroll event and the change event. If we have a vertical scroll bar named **vsbBar**, then the **Scroll** procedure might look like this:

```
Private Sub vsbBar_Scroll()
    vshBar_Change
End Sub
```

This simple short procedure is executed every time the mouse scrolls detectably, a tiny pixel-sized amount. Every time it executes, it calls the **Change** procedure. It is in the **Change** procedure that the programmer writes code for the action desired. The change event occurs when the value of a scroll bar control changes as a result of a change in the position of the scroll bar. Usually the mouse drags the scroll bar to a new position. In outline form, a **Change** procedure might look like this:

```
Private Sub vsbBar_Change()
    statement
    statement
    lot's
    of
    action here
End Sub
```

Let's see how the pH program uses these characteristics of the vertical scroll bar.

The pH Program

All aqueous solutions contain hydrogen ions (H^+) at some concentration, which is relatively small. The concentration is usually expressed in molarity (M), the units of which are moles/liter. The concentration of hydrogen ions in pure water is 1.00 x 10^{-7} M. A solution in which the hydrogen ion concentration is higher than 1.00 x 10^{-7} M is called acidic. A solution in which the hydrogen ion concentration is less than 1.00 x 10^{-7} M is called basic. The chemical properties of many substances in aqueous solution are extremely sensitive to the hydrogen ion concentration. The taste of food, the character of fine wines, and the health of blood all change with tiny changes in hydrogen ion concentration. People who work with food (chefs), with fine wines (vintners), and with blood (health care professionals) have neither the time nor inclination to use terms like *molar hydrogen ion concentration* or numbers expressed in scientific notation. The concept of pH eliminates all these problems with the following definition:

$$\mathrm{pH} = \mathrm{p}K_a + \log[\mathrm{H}^+] \qquad \text{or} \qquad [\mathrm{H}^+] = 10^{=\mathrm{pH}} \tag{12-1}$$

Thus, the pH of pure water is –log [1.00 x 10^{-7}] equals 7. If the pH of an acid is 2, then the hydrogen ion concentration is 1.00 x 10^{-2}. The pH of vinegar is about 2, the pH of saliva is about 6, and the pH of household ammonia is about 11.

In aqueous solutions, water, hydrogen ions, and hydroxyl ions exist in a state of dynamic equilibrium:

$$H_2O \rightleftharpoons H^+ + OH^- \tag{12-2}$$

The equilibrium constant K_w for this equilibrium equals 1.00 x 10^{-14}.

$$K_w = [\mathrm{H}^+][\mathrm{OH}^-] = 1.00x10^{-14} \tag{12-3}$$

Whenever the hydrogen ion concentration is known, so is the hydroxyl ion concentration, since

$$[\mathrm{OH}^-] = \frac{K_w}{[\mathrm{H}^+]} \tag{12-4}$$

Furthermore, if the pH is known, so is the pOH, since taking the –Log of both sides of Equation 12-3 gives

$$\mathrm{p}K_w = \mathrm{pH} + \mathrm{pOH} \tag{12-5}$$

According to Equation 12-5, the sum of the pH and the pOH always equals 14. For the purpose of the pH program, we will approximate that the minimum pH and pOH equal 0 and the maximum pH and pOH equal 14.

The pH program illustrates some of these relationships by featuring the vertical scroll bar (vsb) control. Like pH and pOH, the scroll bar has a **Max** property and a **Min** property. It also has a **Value** property, which depends on the position of the scroll bar. Its value lies between the **Min** and the **Max** properties, which are set at design time.

Starting the Project

- Start Visual Basic by clicking the **VB** icon on the desktop.
- Select the **Standard EXE** icon in the **New Project** window, and click the **OK** button.
- A **New Project** window opens with a fresh form.
- Select the form, and in the properties window, change the **(Name)** property to **frmPH**.
- Click **Save**; the file name will be frmPH, so click **OK**.
- The file name changes to Project1; change it to **The pH Program**, and click the **OK** button.
- Place controls on frmPH according to Figure 12-5.
- Set the properties of the controls according to Table 12-8.

Table 12-8

The Properties Table for the frmPH Form

Object	Property	Value
Form1	Name	frmPH
	Caption	The pH Program
Label1	Name	lblInfo
	Caption	This program demonstrates the vertical scroll bar, Boolean expressions, and properties of the text box and label controls
	BorderStyle	1-Fixed Single
Label2	Name	lblVyBasic
	Caption	Very Basic
	Font	12 Point Sans Serif Bold
	ForeColor	Blue
Label3	Name	lblVyAcidic
	Caption	Very Acidic
	Font	12 Point Sans Serif Bold
	ForeColor	Red
Text box1	Name	txtHeq
	Text	Make it blank
	Locked	True
Text box2	Name	txtPH
	Caption	Make it blank
	Locked	True
Vertical scroll bar1	Name	vsbPH
	Max	1400
	Min	0
Command1	Name	cmdExit
	Caption	E&xit

Entering the Code of the pH Program. Enter the following code in the general declarations section of the program:

```
Option Explicit
' The pH program provides an introduction to the
' scroll bar, the scroll event, and the change event.
' It provides an example of a call to a procedure
' from within another procedure. And it makes
' frequent use of Booleans.
```

Enter the following code in the **cmdExit_Click()** procedure:

```
Private Sub cmdExit_Click()
   End
End Sub
```

Enter the following code in the **Form_Load()** procedure:

```
Private Sub Form_Load()                    ' This procedure runs before any others
   lblVyBasic.Visible = False              ' Make label invisible
   lblVyAcid.Visible = False               ' Make label invisible
   vsbPH.Value = 700                       ' Begin with pH = 7 ( = vshPh.Value/100)
End Sub
```

Enter the following code in the **vsbPH_Change()** procedure:

```
Private Sub vsbPH_Change()                 ' This procedure is called by vsb_Scroll( )
                                           ' for every detectable movement of
                                           ' the vertical scroll bar (see below )

  ' Text box displays: pH = its value at the moment
  txtPH.Text = "pH = " & vsbPH.Value / 100

  ' Convert pH to concentration and express in scientific notation
  txtHeq.Text = "H ion conc. = " & _Format(10 ^ -(vsbPH.Value / 100), "Scientific")

  lblVyBasic.Visible = False           ' These two statements keep the acid and base
  lblVyAcid.Visible = False            ' labels invisible until the next two If...Then
                                       ' blocks make one or the other visible

  If vsbPH.Value < 300 Then            ' Let's say a pH < 3 is very acidic
     lblVyBasic.Visible = False
     lblVyAcid.Visible = True
  End If

  If vsbPH.Value > 1100 Then           ' Let's say a pH > 11 is very basic
     lblVyBasic.Visible = True
     lblVyAcid.Visible = False
  End If

End Sub
```

Enter the following code in the **vsbPH_Scroll()** procedure:

```
Private Sub vsbPH_Scroll()             ' Each measureable scroll movement calls
   vsbPH_Change                        ' procedure vsbPh_Change (see above )
End Sub
```

Executing the pH Program. When the pH program is run, the vertical scroll bar is set to its middle so that the left text box displays an initial pH value of 7.00 and the text box on the right displays the corresponding hydrogen ion concentration of 1.00 E –7. Dragging the vertical scroll bar up makes the system more acidic, and the pH decreases while the hydrogen ion concentration increases. When the pH drops below 3.00 (a rather arbitrary figure), the red label “Very Acidic” becomes visible. Similarly, dragging the vertical scroll bar down makes the system more basic, and the pH increases while the hydrogen ion concentration decreases. When the pH rises above 11.00 (another rather arbitrary figure), the red label “Very Basic” becomes visible.

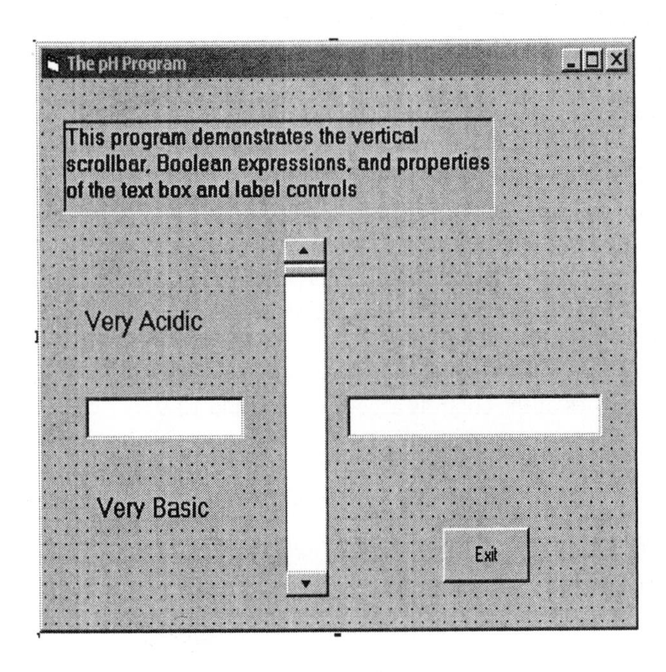

Figure 12-5. The Form for the pH Program

Understanding the Code of the pH Program. At design time, the **Max** property of the vertical scroll bar (**vsbPH**) is set to 1400 and the **Min** property to 0. The pH is calculated by dividing the value property (**vsbPH.Value**) by 100 so that as **vsbPH** ranges from 0 to 1400, the pH ranges from 0.00 to 14.00, a precision of 0.01 pH units.

When the pH program is run, the **Form_Load** procedure runs first and makes the "Very Acidic" and Very Basic" labels invisible. It also sets the initial vertical scroll bar value (**vsbPH.Value**) to equal 700, corresponding to a pH of 7.00.

Every perceptible movement of the vertical scroll bar executes the procedure **vshPH_Scroll( )**, which in turn calls the procedure **vsbPH_Change ( )**.

At any given position, the scroll bar has a value (**vsbPH.Value**), which is converted to pH by dividing by 100. The pH value is assigned to the text box **txtPH.Text**, which displays the value. After calculating the equilibrium concentration of hydrogen ion, it is assigned to **txtHeq.Text** to display its value. Two **If ... Then** statements constantly monitor the pH. If the pH falls below 3.00, the label **lblVyAcid** becomes visible. If the pH rises above 11.00, the label **lblVyBasic** becomes visible. Between a pH of 3.00 and 11.00, both labels are invisible.

If the last procedure (**vsbPH_Scroll**) is omitted, the program still runs, but instead of displaying the continuously changing pH and H ion concentration, the values displayed reflect the last value of **vsbPH.Value**, which is the value when you release the mouse button. You could temporarily comment out this procedure to get a better feel for the way it functions.

The ColorBox Program

In the previous program, a single vertical scroll bar changed the **Text** property of two text boxes. In the ColorBox program, three horizontal scroll bars change the **BackColor** property of a single text box.

Starting the Project

- Start Visual Basic by clicking the **VB** icon on the desktop.
- Select the **Standard EXE** icon in the **New Project** window, and click the **OK** button.
- A **New Project** window opens with a fresh form.
- Select the form, and in the properties window, change the **(Name)** property to **frmColorBox**.
- Click **Save**; the file name will be frmColorBox, so click **OK**.
- The file name changes to Project1; change it to **The ColorBox Program**, and click the **OK** button.
- Place controls on frmColorBox according to Figure 12-6.
- Set the properties of the controls according to Table 12-9.

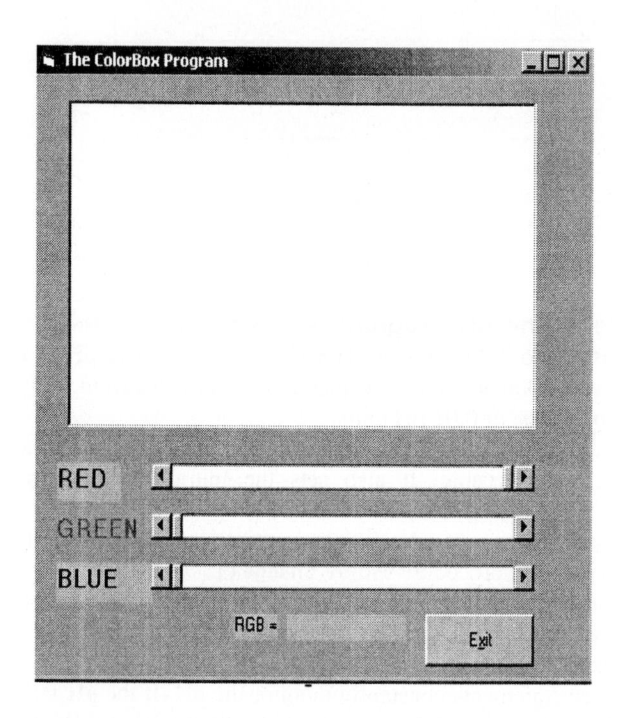

Figure 12-6. The Form for the ColorBox Program

Entering the Code of the ColorBox Program

Enter the following code in the general declarations section:

```
Option Explicit
' the ColorBox program demonstrates the horizontal scroll bar and the RGB (color) function
Dim R As Integer
Dim G As Integer
Dim B As Integer
```

Table 12-9

The Properties Table for the frmColorBox Form

Object	Property	Value
Form1	Name	frmColorBox
	Caption	The ColorBox Program
Label1	Name	lblInfo
	Caption	This program demonstrates the horizontal scroll bar and the RGB (color) function
	BorderStyle	1- Fixed Single
Label2	Name	lblRGB
	Caption	RGB =
Label3	Name	lblRGBNum
	Caption	Make it blank
Label4	Name	lblRed
	Caption	RED
	ForeColor	Red
	Font	Sans Serif Bold 12 Point
Label5	Name	lblGreen
	Caption	GREEN
	ForeColor	Green
	Font	Sans Serif Bold 12 Point
Label6	Name	lblBlue
	Caption	BLUE
	ForeColor	Blue
	Font	Sans Serif Bold 12 Point
Horizontal scroll bar1	Name	hsbRed
	Max	255
Horizontal scroll bar2	Name	hsbGreen
	Max	255
Horizontal scroll bar3	Name	hsbBlue
	Max	255
Command1	Name	cmdExit
	Caption	E&xit

Enter the following code in the **Form_Load( )** procedure:

```
Private Sub Form_Load()                  ' This procedure runs first and immediately
   R = hsbRed.Value                      ' the values for the current settings of
   B = hsbBlue.Value                     ' the horizontal scroll bar are
   G = hsbGreen.Value                    '  assigned to the integer variables R, B, and G
   txtColorBox.BackColor = RGB(R, G, B)          'The RGB function returns a color and
                                                 ' assigns it to the BackColor property
                                                 ' of the text box
   lblRGBNum.Caption = (R & Space(3) & G & Space(3) & B)       'The values of the RGB
                                                               ' parameter are assigned
                                                               ' to the label caption

End Sub
```

Enter the following code in the **hsbBlue_Change( )** procedure:

```
Private Sub hsbBlue_Change()
   B = hsbBlue.Value          ' The current value of the scroll bar is assigned to B
   Form_Load                  ' Running this procedure updates the text box color
End Sub
```

Enter the following code in the **hsbGreen_Change)** procedure:

```
Private Sub hsbGreen_Change()     ' similar to above
   G = hsbGreen.Value
   Form_Load
End Sub
```

Enter the following code in the **hsbRed_Change( )** procedure:

```
Private Sub hsbRed_Change()       ' similar to above
   R = hsbRed.Value
   Form_Load
End Sub
```

Enter the following code in the **hsbRed_Scroll)** procedure:

```
Private Sub hsbRed_Scroll()
   hsbRed_Change
End Sub
```

Enter the following code in the **hsbGreen_Scroll)** procedure:

```
Private Sub hsbGreen_Scroll()
   hsbGreen_Change
End Sub
```

Enter the following code in the **hsbBlue_Scroll)** procedure:

```
Private Sub hsbBlue_Scroll()
   hsbBlue_Change
End Sub
```

Executing the ColorBox Program. When the program is run, the user initially sees a red text box and a label that reads **RGB = 255, 0, 0**. The red slide is way to the right, while the green and blue sliders are way to the left, since these settings were made at design time. Why red? An initial bright color is more dramatic to the user than black or white.

Slide the blue slider to the far right, and the screen turns magenta (RGB = 255, 0, 255). Now slide the green slider to the far right, and the screen turns white (RGB = 255, 255, 255). Next, slide all three sliders to the far left, and the screen turns black. Try a variety of intermediate settings to observe a few the 16,777,216 possible colors.

Understanding the Code of the ColorBox Program. The **RGB** function permits the programmer to establish colors by specifying the proportions of red, green, and blue components in a mixture. The amount of each color is specified by an integer from 0 to 255. **RGB(0, 0, 0)** corresponds to the absence of color or white. **RGB(255, 255, 255)** corresponds to black. You can use other combinations to generate 16,777,216 colors. A few colors are noteworthy: **RGB(255, 0, 255)** is magenta, **RGB(255, 255, 0)** is yellow, and **RGB(0, 255, 255)** is cyan. Any method or property of an object that accepts a color specification expects that specification to represent a number returned by the **RGB** function. For example:

```
txtColorBox.BackColor = RGB(R, G, B)
```

When the ColorBox program is run, the **Form_Load** procedure executes first. It assigns the values of the three scroll bars that were set at design time to the integer variables **R**, **G**, and **B**. These values are **255**, **0**, and **0**, respectively. These three parameters are passed to the **RGB** function, which returns the color assigned to the text box color property, **txtColorBox.BackColor**. Upon running the program, the user sees a red text box, since the color of the text box was set to red at design time. The label (**lblRGBNum**) displays the three RGB numbers, which are **255**, **0**, and **0** initially.

When any of the horizontal scroll bars are moved, the scroll event occurs, and depending on which scroll bar is moved, one of the following three procedures executes:

```
Private Sub hsbRed_Scroll()
  hsbRed_Change
End Sub

Private Sub hsbGreen_Scroll()
  hsbGreen_Change
End Sub

Private Sub hsbBlue_Scroll()
  hsbBlue_Change
End Sub
```

Every infinitesimal scroll movement executes one of these procedures and causes a call to one of the following three procedures:

```
Private Sub hsbGreen_Change()
  Form_Load  ' Updates txtColorBox.BackColor and lblRGBNum.Caption
End Sub

Private Sub hsbRed_Change()
  Form_Load  ' Updates ditto
End Sub

Private Sub hsbRed_Change()
  Form_Load  ' Updates ditto
End Sub
```

When any of the horizontal scroll bars are moved, the value is assigned to **R** or **G** or **B**, depending on which scroll bar is moved. The **Form_Load** procedure is then executed to immediately update the text box color. Since the scroll event constantly updates the colors, the colors appear to change continuously as the scroll bar is scrolled. Try commenting out the three scroll procedures and rerunning the program. Now when you scroll the scroll bar, the color does not change until the mouse button is released and the then-current value of the scroll bar is established.

The List Box Control

The **List Box** icon in the toolbox appears as follows:

As its name suggests, the list box presents to the user a list of items (strings or numbers) from which the user can choose. The programmer establishes the items in the list at run time with the **AddItem** method of the list box. The user cannot add an item to the list box, but can select items that are already listed in the list box. To add a few string items to a list box named **lstCheese**, the programmer's code would appear like this:

```
lstCheese.AddItem "Cheddar"
lstCheese.AddItem "Roquefort"
lstCheese.AddItem "Swiss"
lstCheese.AddItem " Cottage"
```

or more compactly

```
With lstCheese
      .AddItem "Cheddar"
      .AddItem "Roquefort"
      .AddItem "Swiss"
     .AddItem " Cottage"
End With
```

The **With ... End With** statement enables you to execute a series of statements on one specified object (here, the list box named **lstCheese**) without restating the name of the object. This usage saves programming time and execution time, and the code appears more clear and readable.

The list box has a number of useful methods beside **AddItem**, including **RemoveItem**, **Clear**, and **SetFocus**. Some of its important properties are **ListCount**, **ListIndex**, **Sorted**, and **Text**. Several of these will be illustrated in later programs.

The Select Case Decision Statement

The **Select Case** statement is a simpler alternative to using several **If ... Then ... ElseIf** statements. The **Select Case** structure usually takes the following form (assume a previous declaration: **Dim TestVariable As Integer**):

```
Select Case TestVariable
        Case 1
                Execute this block of
                statements if TestVariable =1
        Case 2
                Execute this block of
                statements if TestVariable =2
```

Case 3
Execute this block of statements if TestVariable =3
Else
Execute this block of statements if TestVariable is not equal to any of the above cases

If none of the values of **TestVariable** satisfy the **Select Case** statement, then the block following the **Else** statement is executed. This is similar to the final **Else** in the **If ... Then ... ElseIf ... Else** control structure. Some variations are possible. Compare with the above statement, where:

Case 1 is equivalent to **If TestVariable = 1 Then** ...
Cases 2 and **3** are equivalent to **If TestVariable = 2 OR 3 Then** ...

The List Box Program

The List Box program illustrates some features of the **Select Case** statement and some properties of the list box control. Two version of the List Box program are presented. The first version features the **Text** property of the list box, and the second version features the **ListIndex** property of the list box.

Starting the Project

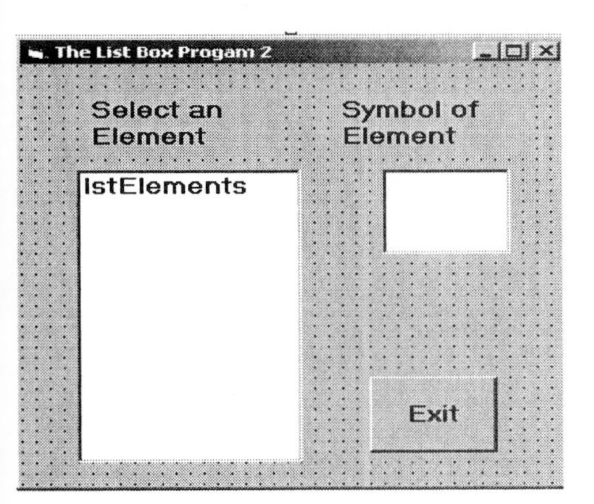

Figure 12-7. The Form for the List Box Program

- Start Visual Basic by clicking the **VB** icon on the desktop.
- Select the **Standard EXE** icon in the **New Project** window, and click the **OK** button.
- A **New Project** window opens with a fresh form.
- Select the form, and in the properties window, change the **(Name)** property to **frmList box**.
- Click **Save**; the file name will be frmSelectList, so click **OK**.
- The file name changes to Project1; change it to **The List Box Program**, and click the **OK** button.
- Place controls on frmList box according to Figure 12-7.
- Set the properties of the controls according to Table 12-10.

Table 12-10

The Properties Table of the List Box Program

Object	Property	Value
Form1	Name	frmList box
	Caption	The List Box Program
Text box1	Name	txtSymbol
	Text	Make it blank
	Font	24 point bold
List box1	Name	lstElements
	Text	Make it blank
	Font	14 point bold
Label1	Name	lblListElements
	Caption	Select an Element
Label2	Name	lblSymbol
	Caption	Symbol of Element
Command1	Name	cmdExit
	Caption	Exit
	Font	10 point bold

Entering the Code of the List Box Program. In the general declarations section of the program, enter some comments giving the name of the program and its function:

```
Option Explicit
' The List Box program illustrates the list box control,
'  its .AddItem, and its click event.
' The Select Case decision structure is also demonstrated.
```

Double-click the form to open the code window and enter:

```
Private Sub Form_Load()
   With lstElements
      .AddItem "Silver"        'add these items to lstElements
      .AddItem "Gold"
      .AddItem "Mercury"
      .AddItem "Potassium"
      .AddItem "Sodium"
      .AddItem "Tin"
      .AddItem "Tungsten"
      .AddItem "Antimony"
      .AddItem "Iron"
      .AddItem "Lead"
   End With
End Sub
```

Double-click the list box labeled **lstElements** to open the code window, and enter:

```
Private Sub lstElements_Click()

  Select Case lstElements.Text
    Case "Silver"
      txtSymbol.Text = "Ag"
    Case "Gold"
      txtSymbol.Text = "Au"
    Case "Mercury"
      txtSymbol.Text = "Hg"
    Case "Potassium"
      txtSymbol.Text = "K"
    Case "Sodium"
      txtSymbol.Text = "Na"
    Case "Tin"
      txtSymbol.Text = "Sn"
    Case "Tungsten"
      txtSymbol.Text = "W"
    Case "Antimony"
      txtSymbol.Text = "Sb"
    Case "Iron"
      txtSymbol.Text = "Fe"
    Case "Lead"
      txtSymbol.Text = "Pb"
  End Select

End Sub
```

Double-click the **Exit** command button to open the code window, and enter:

```
Private Sub cmdExit_Click()
  End
End Sub
```

Executing the List Box Program. Execute the List Box program by clicking the **Start** button. The form shown in Figure 12-7 appears, with a list of ten elements displayed in the list box. Click one of the elements to select it. The text box displays the symbol of the element in 24-point bold font. Select other elements and get their symbols. Click **Exit** to exit the program.

Understanding the Code of the List Box Program. When you click the **Start** button to execute the List Box program, the **Form_Load** procedure executes immediately. The **AddItem** property of the list box, named **lstElements**, adds the names of ten elements to the list box, which displays the list of elements to the users.

Clicking an item in the list box selects it, triggers the click event of the list box, and establishes the value of the **Text** property of the list box at the moment. When the item **"Sodium"** is clicked, the **Text** property of the list box is **"Sodium"**. In this program, the test expression of the **Select Case** statement is **lstElements.Text**. Thus when **"Sodium"** is selected, the string **"Na"** is assigned to the **Text** property of the text box named **txtSymbol**:

```
Case "Sodium"
        txtSymbol.Text = "Na"
```

If the **Sorted** property of the list box had been set to **True** at design time, then all the items in the list would appear in alphabetical order.

The List Box Program, Version 2

The list box maintains an array named List in which the added items are stored. The **ListIndex** property of a list box returns the number or index, that is, the position of each item in the list of the currently selected item in the list. Numbering begins at zero. Clicking an item in the list selects the item and establishes a value for the **ListIndex**.

In the following revised program, the test expression of the **Select Case** statement is **lstElements.ListIndex**. With ten items in the list, the **ListIndex** can have values from 0 through 9. Thus if **"Mercury"** is selected, then **lstElements.ListIndex** has a value equal to 2. The value of 2 causes the string **"Hg"** to be assigned to the **txtSymbol.Text**, which appears in the text box.

For the **Private Sub lstElements_Click()** procedure, substitute the following **Select Case** statement, and run the program with the same form. To the user, the results are identical.

```
Private Sub lstElements_Click()

    Select Case lstElements.ListIndex          ' ListIndex = 0 for first item
        Case 0                                 '... if user selects first item in list
            txtSymbol.Text = "Ag"
        Case 1                                 '... if user selects second item in list
            txtSymbol.Text = "Au"
        Case 2                                 '... and so on
            txtSymbol.Text = "Hg"
        Case 3
            txtSymbol.Text = "K"
        Case 4
            txtSymbol.Text = "Na"
        Case 5
            txtSymbol.Text = "Sn"
        Case 6
            txtSymbol.Text = "W"
        Case 7
            txtSymbol.Text = "Sb"
        Case 8
            txtSymbol.Text = "Fe"
        Case 9
            txtSymbol.Text = "Pb"
    End Select

End Sub
```

This procedure is written as a demonstration of the **ListIndex** property of the list box and to show how the **ListIndex** property can be used to examine the selected item in a list box list. The preferred method is the first procedure, which used the **Text** property rather than the **ListIndex** property. We'll see later that the **ListIndex** property is also a powerful and useful property of the list box.

Arithmetic Operators

Beside the usual arithmetic operators that work with real numbers, Visual Basic supports operators that deal primarily with integers and strings. Table 12-11 lists some of the important operators and some simple illustrative examples.

Operators in Visual Basic use symbols and syntax similar to those used in other programming languages. Addition, subtraction, multiplication, division, and exponentiation operators work with either real numbers or integers or on mixtures of reals and integers. Division of one integer by another integer may yield an integer (12/3) or a real (3/12). The Div and Mod operators are integer operators, although Visual Basic allows the use of real numbers by rounding real numbers to integers before operating.

The Mod operator is used to divide one number by another number and return only the remainder. The numbers are rounded to integers *before* the operation. Some examples:

39 Mod 5 returns 4
28.1 Mod 5.2 returns 3
11 Mod 2 returns 1
12 Mod 2 returns 0

As the last two examples suggest, the Mod operator can be used to determine whether a number is odd or even.

Table 12- 11

Arithmetic Operators

Arithmetic Operators	Description	Arithmetic Expression	Value of Expression for A=11, B=2.1, C=5, D=3.3
+	Addition	A + D	14.3
–	Subtraction	A – D	7.7
*	Multiplication	A*D	36.3
/	Division	A/C	2.2
^	Exponentiation	A^C	161051
Integer Operators			
\	Div (integer division)	A\C	2
Mod	Modulus (integer remainder)	A Mod C	1
String Operator			
&	Concatenation	A&D	113.3

The \ operator is used to divide one number by another and return an integer result. The numbers are rounded to integers *before* the operation. Some examples:

39\5 returns 7
28.1\5.2 returns 5

11\2 returns 5
12\2 returns 6

The concatenation operator (&) forces concatenation of two string expressions. In Table 12-11, the example of 113.3 for the concatenation of 11 & 3.3 considers 11 and 3.3 to be the text representations of the values. It is more usual to concatenate two strings:

"oxy" & "moron" returns "oxymoron"
"Laugh" & "able" returns Laughable
"per" & "manganate" returns "permanganate"
"1.27" & "9.43" returns "1.279.43"

A string is just a list of characters. Words are strings. In Visual Basic (and other languages) "1.27" is a string, but 1.27 is a number.

The Arithmetic Program

The Arithmetic program provides an introduction to arithmetic operators for addition (+), subtraction (–), multiplication (*), division (/), and exponentiation (^). The program also includes the modulus (or remainder) operator (Mod), the \ operator (Div), and the concatenation operator (&).

The Arithmetic program also demonstrates interactive input, using text boxes for both input by the user and output by the program. The text box is useful not only for the input and output of text (string data types) but also for the input and output of numbers or any other data type. Since numbers entered into a text box are entered as their text (string) representation, it may be necessary to convert a string to a number. Visual Basic often does this automatically, but the programmer can take control with conversion functions (Table 12-14) that carry out the necessary conversion. Finally, the Arithmetic program demonstrates how to create a menu control with the Menu Editor.

Starting the Project

- Start Visual Basic by clicking the **VB** icon on the desktop.
- Select the **Standard EXE** icon in the **New Project** window, and click the **OK** button.
- A **New Project** window opens with a fresh form.
- Select the form, and in the properties window, change the (**Name**) property to **frmArithmetic**.
- Click **Save**; the file name will be frmArithmetic, so click **OK**.
- The file name changes to Project1; change it to **The Arithmetic Program**, and click the **OK** button.
- Place controls on frmArithmetic according to Figure 12-8.
- Set the properties of the controls according to Table 12-12.

Table 12-12

The Properties Table of the Arithmetic Program

Object	Property	Value
Form1	Name	frmArithmetic
	Caption	The Arithmetic Program
Text box1	Name	txtNumber1
	Text	Make it blank
Text box2	Name	txtNumber2
	Text	Make it blank
Text box3	Name	txtAnswer
	Text	Make it blank
Label1	Name	lblInfo
	Caption	This program demonstrates arithmetic operations +, -, *, / , \ , Mod, concatenation, and exponentiation
	Alignment	2-Center
	BorderStyle	1-Fixed Single
Label2	Name	lblIn1
	Caption	Enter the first of two numbers
Label3	Name	lblIn2
	Caption	Enter the second of two numbers
Label4	Name	lblDirections
	Caption	Select the desired operation from the menu
	Font	Bold
Label5	Name	lblAnswer
	Caption	For number 1 (operation) number 2
	Font	Bold
Label6	Name	lblResult
	Caption	the answer is ->
	Font	Bold
Command1	Name	cmdExit
	Caption	Exit
Command2	Name	cmdClear
	Caption	Clear Numbers

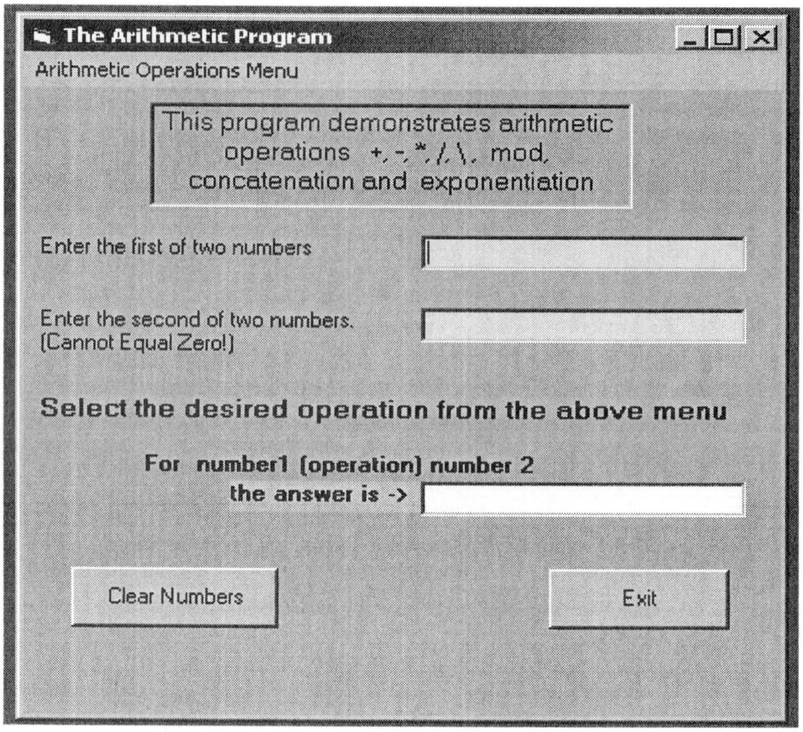

Figure 12-8. The Form for the Arithmetic Program

Creating a Menu Control. You have probably encountered menus in programs that you have used. Menus consist of a list of commands that appears after you click the menu's name. Clicking a menu command usually executes an underlying procedure for which code has been written. This is really not very different from the operation of a command button, with which you are already familiar. Clicking a command button executes a procedure; clicking a menu command also executes a procedure.

Getting the Menu Editor. Visual Basic provides an easy-to-use Menu Editor for creating and editing a menu control.

- Open the Arithmetic program for design mode.
- Click frmArithmetic to select it.
- On the VB **Tools** menu, click **Menu Editor**.

Before you create the menu for the Arithmetic program, the Menu Editor appears as shown in Figure 12-9. Figure 12-10 shows the form for the Arithmetic program with the completed menu when its name, **Arithmetic Operations Menu**, has been clicked. On the **Arithmetic Operations Menu** are ten commands, beginning with **Addition** and ending with **Exit**.

1. **Addition**
2. **Subtraction**

3. **Multiplication**
4. **Division**
5. **Mod**
6. **Div**
7. **Concatenation**
8. **Exponentiation**
9. **A Separator Bar**
10. **Exit**

Using the Menu Editor. Use the following steps to create the **Arithmetic Operations Menu** with the Menu Editor dialog box (Figure 12-9):

- In the **Caption** box, type &**Arithmetic &Operations Menu** for the menu caption.
- In the **Name** box, type **mnuOperations** as the name.
- Click **Next**.
- In the group of four arrows, click the arrow pointing to the right ➔.
- In the **Caption** box, type **&Addition** for the caption.
- In the Name box, type **mnuAdd** as the name.
- Click **Next**.
- In the **Caption** box, type **&Subtraction** as the caption.
- In the **Name** box, type **mnuSub** as the name.

Table 12-13

The Menu Table of the Arithmetic Program

Caption	Name
&Arithmetic Operations Menu	mnuOperations
&Addition	mnuAdd
&Subtraction	mnuSub
&Multiplication	mnuMult
&Division	mnuDivide
M&od	mnuMod
Di&v	mnuDiv
&Concatenation	mnuConcat
&Exponentiation	mnuExpo
-	mnuSepBar1
E&xit	mnuExit

Continue creating the menu commands in this manner with the captions and names listed in Table 12-13. The **A Separation Bar** command between the **Exit** command and the **Exponentiation** command adds clarity and aesthetic appearance, but serves no function. Entering a minus sign (-) as the caption for a command creates a separation bar between two commands.

- When completed, the Menu Editor appears to the user as shown in Figure 12-9.
- Click the **OK** button of the Menu Editor when you are finished.

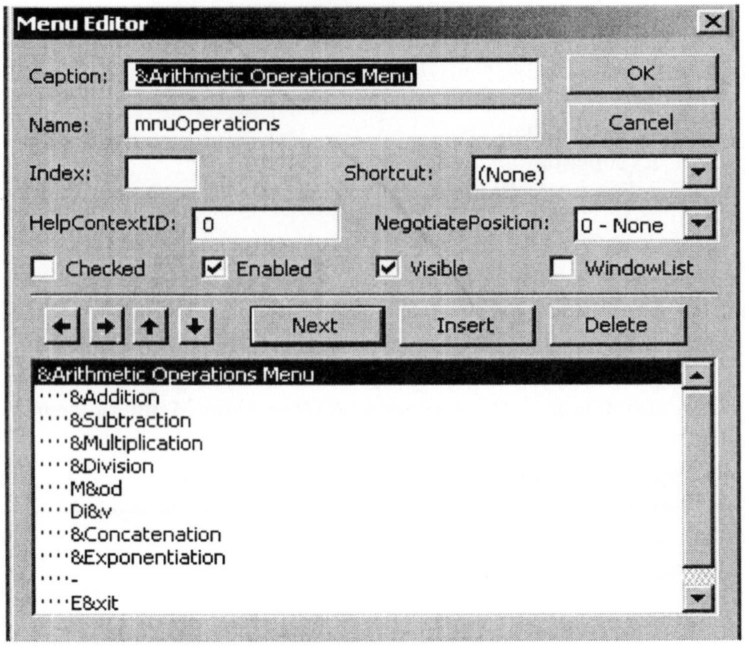

Menu Editor

Caption: &Arithmetic Operations Menu
Name: mnuOperations
Index:
Shortcut: (None)
HelpContextID: 0
NegotiatePosition: 0 - None
Checked Enabled Visible WindowList
OK Cancel Next Insert Delete

&Arithmetic Operations Menu
····&Addition
····&Subtraction
····&Multiplication
····&Division
····M&od
····Di&v
····&Concatenation
····&Exponentiation
····-
····E&xit

Figure 12-9. The Menu Editor Dialog Box Ready for the Arithmetic Program

The Arithmetic Program

Arithmetic Operations Menu

Addition
Subtraction
Multiplication
Division
Mod
Div
Concatenation
Exponentiation
Exit

gram demonstrates arithmetic erations +, -, *, /, \, mod, tenation and exponentiation

mbers

numbers.

Select the desired operation from the above menu

For number1 (operation) number 2
the answer is ->

Clear Numbers Exit

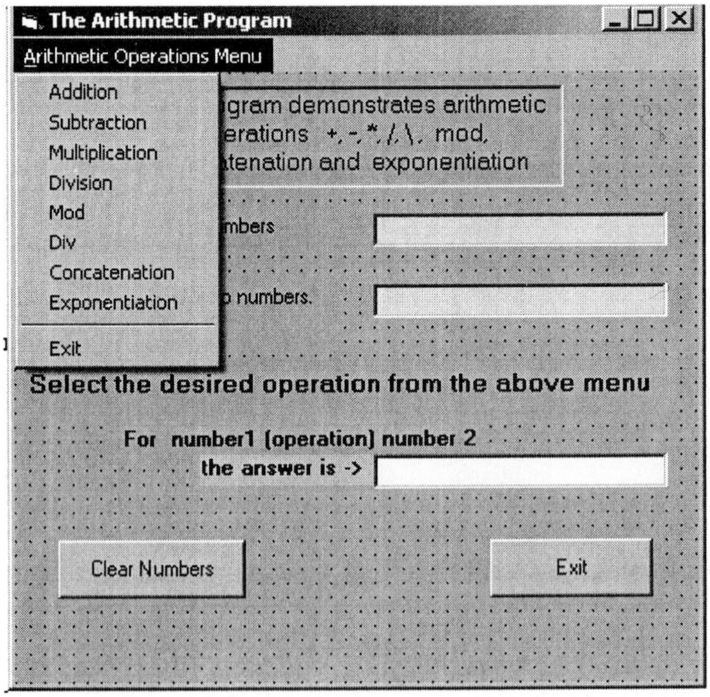

Figure 12-10. Menu for the Arithmetic Program

Entering the Code of the Arithmetic Program

Enter the following code in the general declaration section of the program:

```
' The Arithmetic program
' This program provides an introduction to the menu control and
' to miscellaneous arithmetic operators: +, -, /, *, and so on
Option Explicit
```

Enter the following code in the **cmdExit_Click( )** procedure:

```
Private Sub cmdExit_Click()
   End
End Sub
```

Enter the following code in the **mnuExit_Click( )** procedure:

```
Private Sub mnuExit_Click()
   End
End Sub
```

Enter the following code in the **mnuAdd_Click( )** procedure. The **VarType** Function is discussed in a later section. Note that these **Print** statements are comments for the time being.

```
Private Sub mnuAdd_Click()
   txtAnswer = CSng(txtNumber1) + CSng(txtNumber2)
   ' Print VarType(txtNumber1)
   ' Print VarType(CSng(txtNumber1))
End Sub
```

Enter the following code in the **mnuConcat_Click( )** procedure:

```
Private Sub mnuConcat_Click()
   'The & operator (concatenation operator) joins two string together
   txtAnswer = txtNumber1 & txtNumber2
End Sub
```

Enter the following code in the **mnuDiv_Click( )** procedure:

```
Private Sub mnuDiv_Click()
    If txtNumber2 = 0 Then
      MsgBox "Your divisor cannot equal zero!", vbExclamation, "Try again"
   Else
      txtAnswer = txtNumber1 \ txtNumber2
   End If
End Sub
```

Enter the following code in the **mnuDivide_Click( )** procedure:

```
Private Sub mnuDivide_Click()

On Error GoTo ErrorHandler
   txtAnswer = txtNumber1 / txtNumber2
   Exit Sub

ErrorHandler:
```

```
    MsgBox Str(Err.Number) & ": " & Err.Description, vbExclamation, "Enter Again"
End Sub
```

Enter the following code in the **mnuExpo_Click( )** procedure:

```
Private Sub mnuExpo_Click()
   txtAnswer = txtNumber1 ^ txtNumber2
End Sub
```

Enter the following code in the **mnuExit_Click( )** procedure:

```
Private Sub mnuExit_Click()
   End
End Sub
```

Enter the following code in the **mnuMod_Click( )** procedure:

```
Private Sub mnuMod_Click()
    If txtNumber2 = 0 Then
      Beep
      txtAnswer = "The divisor cannot equal zero!"
   Else
      txtAnswer = txtNumber1 Mod txtNumber2
   End If
End Sub
```

Enter the following code in the **mnuMult_Click( )** procedure:

```
Private Sub mnuMult_Click()
   txtAnswer = txtNumber1 * txtNumber2
End Sub
```

Enter the following code in the **mnuSub_Click( )** procedure:

```
Private Sub mnuSub_Click()
   txtAnswer = txtNumber1 - txtNumber2
End Sub
```

Enter the following code in the **mnuClear_Click( )** procedure:

```
Private Sub cmdClear_Click()
   txtNumber1.Text = ""
   txtNumber2.Text = ""
   txtAnswer.Text = ""
   txtNumber1.SetFocus
End Sub
```

Executing the Arithmetic Program. Execute the Arithmetic program by clicking the **Start** button. Enter decimal numbers in the text boxes indicated by the program. Click the **Arithmetic Operations Menu**. Then click in turn each of the commands for the operations indicated. The program used three different methods for handling division by zero, which is a danger for the Mod, Div, and Division procedures. Type a **0** for the second number, and observe the results for these three procedures. Next, type the letter **o**, and notice that two procedures cause the program to crash, and it must be restarted to try again. The **Division** procedure survives use of the letter **o** for the second number.

As an experiment, enter two words in the **txtNumberl** and **txtNumber2** text boxes. Then, select the **Concatenation** command on the menu. The program concatenates the two words and displays the result in the **txtAnswer** box. Selecting any other menu item will, of course, result in a run-time error.

Understanding the Code of the Arithmetic Program. The Arithmetic program introduces several new concepts in addition to the simple arithmetic operators. Use of the Menu Editor dialog box facilitates the creation of custom menus. The program uses a type conversion function (**Cstr**) to convert a number to a string. The **On Error** statement is used and compared with other methods of handling errors.

Creating a Custom Menu. For the moment, let's focus on the first two properties of the menu (Figure 12-9), the **Caption** and the **Name** properties. These properties are exactly analogous to the same properties of the command button. The **Caption** property is the name that appears on the menu command that is visible to the user. The **Name** property is the name of the menu command that is used in the underlying code. The menu commands act just like a bunch of closely spaced command buttons. Compare:

Caption	Name	Control
Exit	**cmdExit**	Command button
Exit	**mnuExit**	Menu command

Clicking a command in a menu or a command button causes the underlying procedure code to execute. You can exit the Arithmetic program either by clicking the **Exit** command in the **Arithmetic Operations Menu** or by clicking the **Exit** command button to execute the underlying procedures.

```
Private Sub mnuExit_Click()
  End
End Sub
```

or

```
Private Sub cmdExit_Click()
  End
End Sub
```

The entire program consists of short procedures written for each menu item. The code for the **mnuAdd_Click ( )** procedure is typical.

Except for the **Clear**, **Exit**, and **Add** procedures, the result of an operation between two numbers is give by a statement of the following form:

```
txtAnswer = txtNumber1 (operator) txtNumber2
```

The default property of the text box controls the **Text** property, so it is only necessary to enter, for example, **txtNumber1** instead of the complete expression **txtNumber1.Text.** Text boxes accept user input as text, the default property of a text box, even if the input consists of numerical characters. Visual Basic then performs the desired numerical operation and assigns the result to **txtAnswer**, a third text box. The result assigned to **txtAnswer** is still of type **Text**.

Conversion Functions

In a more extended program, it may be necessary to convert the text representation of a number to, for example, an **Integer**, **Single**, or **Double** type. In fact, it is necessary to convert the input text numbers to numerical numbers before adding them in the **Add** procedure:

```
txtAnswer = CDbl(txtNumber1) + CDbl(txtNumber2)
```

Without conversion, this statement concatenates the two numbers. Both the + operator and the & operator concatenate strings. The + operator adds numbers, but not their string representations. The & operator cannot add numbers, but it can concatenate both strings and numbers. Table 12-14 lists several common conversion functions.

The VarType Function

Use **Print VarType(*Expression*)** to find the type of a variable or expression if you're not sure what the type is. Unlike Pascal, Visual Basic is not a highly typed language and is forgiving of seeming type mismatches. If you run into a type mismatch error, you can learn the type of a variable with the **VarType** function. The argument of the **VarType** function is the name of the variable whose type you want to know. Since **VarType** is a function, it returns a value, namely an integer value corresponding to the variable's type as shown in Table 12-15. The type should be clear from the name of the constant.

```
Private Sub mnuAdd_Click()
   txtAnswer = CDbl(txtNumber1) + CSng(txtNumber2)
   ' Print VarType(txtNumber1)
   ' Print VarType(CSng(txtNumber1))
End Sub
```

In your Arithmetic program, remove the comment apostrophes from the two statements with the **VarType** functions, and run the program. These statements will print on the form (in the upper right corner) an **8** and below it a **4**. From Table 12-15, it is seen that **8** indicates that the variable type is a **String** and that the **4** indicates that the variable type is a **Single**.

Table 12-14

Type Conversion Functions

Function	Return Type	Function	Return Type
Cbool(*Expression*)	Boolean	CInt(*Expression*)	Integer
Cbyte(*Expression*)	Byte	CLng(*Expression*)	Long
CCur(*Expression*)	Currency	CSng(*Expression*)	Single
CDate(*Expression*)	Date	CStr(*Expression*)	String
CDbl(*Expression*)	Double	CVarl(*Expression*)	Variant
CDec(*Expression*)	Decimal		

Table 12-15

The VarType Function and Its Return Values

Constant	Value	Constant	Value
vbEmpty	0	vbString	8
vbNull	1	vbObject	9
vbInteger	2	vbError	10
vbLong	3	vbBoolean	11
vbSingle	4	vbVariant	12
vbDouble	5	vbDataObject	13
vbCurrency	6	vbDecimal	14
vbDate	7	vbByte	17

Handling Errors

The **On Error GoTo** statement enables an error-handling routine marked **YourLineLabel** (followed by a colon) identifying a brief section of code. The syntax for the **On Error GoTo** statement is shown in the pseudocode below:

```
YourProcedure
    On Error GoTo YourLineLabel
            The code for
            YourProcedure
            goes here
    Exit Sub
YourLineLabel:                ' The colon is required

    Your short error handling
    code ... usually a MsgBox ...
    goes here

End Sub
```

The **LineLabel** is a descriptive string or a number. A frequently used string for the line label is **ErrorHandler**, but you could use **MyError**, **YourError**, **OurErrorHandler**, or whatever seems appropriate. It must be followed by a colon. The use of the **On Error GoTo** statement protects the subroutine (**YourProcedure**) from fatal errors. In the Arithmetic program, the **On Error GoTo** statement protects against the user entering a zero in the divisor. Division by zero is a fatal run-time error: the program crashes. With the **On Error GoTo** statement, the name and number of the error can be given to the user in a message box. Clicking **OK** in the message box returns execution to the procedure, and the user can reenter a nonzero number. Compare the following code in the Arithmetic program with the previously described syntax:

```
        Private Sub mnuDivide_Click()

            On Error GoTo ErrorHandler
                    txtAnswer = txtNumber1 / txtNumber2
            Exit Sub

        ErrorHandler:            ' The colon is required
            MsgBox Str(Err.Number) & ": " & Err.Description, vbExclamation, _
                                                    "Enter Again"

    End Sub
```

Here is boilerplate for using the **On Error GoTo** statement:

```
    Private Sub YourProcedure_Click()

    On Error GoTo ErrorHandler

            Block of one or
            more statements
            in YourProcedure.

    Exit Sub

    ErrorHandler:        ' The colon is required

            Block of one or
            more statements
            in the error-handling
            routine, but usually just a message box

    End Sub
```

When an error occurs, the **Exit Sub** statement shifts execution from **YourProcedure** to the **ErrorHandler.** A **MsgBox** statement is often a simple, satisfactory method of communicating the problem to the user:

```
MsgBox Str(Err.Number) & ": " & Err.Description, vbExclamation, "Enter Again"
```

Let us dissect the **MsgBox** statement from left to right. Associated with the **On Error GoTo** statement is the **Err** object, which has two properties of interest to us: the **Err.Number** property and the **Err.Description** property. The **Number** is the error number, which is an integer. The **Description** of the error is, not surprisingly, a string that gives a terse description of the nature of the error.

The first parameter of the **MsgBox** statement is a string, which appears in the message box itself. We create this string by converting **Err.Number** to a string, concatenating it to the string ": " and concatenating these two strings to the string value of **Err.Description**. A comma separates this first string from the Visual Basic constant **vbExclamation**, which places an exclamation point icon on the message box. A second comma separates **vbExclamation** from the last string, **"Enter Again"**, which appears on the message box's title bar.

The **Mod** and **Div** procedures in this program use difference approaches to catch the error and inform the user. The **Mod** procedure checks for zero, and if the check is true, the program beeps and places a message in the **txtAnswer** box. If no zero is found, then the **Mod** operation is carried out as usual. Unlike the **On Error GoTo ...ErrorHandler** procedure, the **If ... Then ...Else** only checks for zero input.

```
Private Sub mnuMod_Click()

If txtNumber2 = 0 Then
    Beep
    txtAnswer = "The Divisor cannot equal zero!"
  Else
    txtAnswer = txtNumber1 Mod txtNumber2

  End If

End Sub
```

The **Div** procedure is similar, except that upon detecting a zero, it calls a message box into play with appropriate messages. It also only detects zero input, and it is not as effective as the error handler for detecting a variety of input errors and giving the user a second chance to enter acceptable data.

The Option Button Control

In the toolbox, the **Option Button** icon looks like this:

Unlike most controls, the option button control is nearly always used in a group, from which the user can select only one. The selected option button displays a black dot in its center, while an unselected option button displays a blank center. Usually, a group of option buttons is placed on the form, but they can be placed on other controls, such as a frame control or a picture box control. More than one group can be utilized by placing one group, for example, on one frame and a second group on another frame. Only one option button control in a given group can be selected at a time.

The most important property of the option button is its **Value** property, which is of type **Boolean**. If an option button is selected, then its **Value** property is **True**. If an option button is unselected, then its **Value** property is **False**.

The Functions Program

Besides demonstrating several Visual Basic mathematical functions (Table 12-16), the Functions program introduces the option button control, discussed in the previous section.

Earlier, the Random Guess program introduced the **MsgBox** statement. The Functions program introduces the **MsgBox** function, which is very similar to the **MsgBox** statement. This program introduces some of Visual Basic's built-in mathematical functions.

Table 12-16

Some Selected Visual Basic Built-in Mathematical Functions

Function	Description
Abs(x)	Returns the absolute value of *x*
Atn(x)	Returns the arctangent of *x*
Cos(x)	Returns the cosine of *x*
Exp(x)	Returns e^x
Fix(x)	Returns the integer part of *x*
Hex(x)	Returns the hex equivalent of *x*
Int(x)	Returns the greatest integer
Log(x)	Returns the log of *x* to the base e
Rnd(x)	Returns a random number between 0 and 1 for *x* as a seed
Sgn(x)	Returns the sign of *x*
Sqr(x)	Returns the square root of *x*
Tan(x)	Returns the tangent of *x* in radians

Starting the Project

- Start Visual Basic by clicking the **VB** icon on the desktop.
- Select the **Standard EXE** icon in the **New Project** window, and click the **OK** button.
- A **New Project** window opens with a fresh form.
- Select the form, and in the properties window, change the **(Name)** property to **frmFunctions**.
- Click **Save**; the file name will be frmFunctions, so click **OK**.
- The file name changes to Project1; change it to **The Functions Program**, and click the **OK** button.
- Place controls on frmFunctions according to Figure 12-11.
- Set the properties of the controls according to Table 12-17.

Table 12-17

The Properties Table for the Functions Program

Object	Property	Value
Label1	Name	lblInfo
	Caption	This is a program for demonstrating some of Visual Basic's built-in functions, the **MsgBox** function and the option button control
	Border Style	1-Fixed Single
Label2	Name	lblDirections
	Caption	Enter a decimal number between –50 and +50
Label3	Name	lblX
	Caption	= X
Label4	Name	lblFunName
	Caption	Make it blank
	Font	18 point bold
Label5	Name	lblResult
	Caption	Make it blank
	Font	18 point bold
Text box1	Name	txtInput
	Text	Make it blank
Option1	Text	optAbs
	Caption	Abs(x)
Option2	Text	optCos
	Caption	Cos(x)
Option3	Text	optExp
	Caption	Exp(x)
Option4	Text	optInt
	Caption	Int(x)
Option5	Text	optLog
	Caption	Log(x)
Option6	Text	optSqrt
	Caption	Sqr(x)
Option7	Text	optClear
	Caption	Clear Entry

Entering the Code of the Functions Program

Enter the following code in the general declaration section of the program:

```
' Option Explicit
' The Functions program demonstrates option buttons,
' functions, and the MsgBox function.
```

Enter the following code in the **optAbs_Click()** procedure:

```
Private Sub optAbs_Click()
If optAbs.Value = True Then
     lblFunName.Caption = "Abs(x)"
     lblResult.Caption = Abs(txtInput.Text)
  End If
End Sub
```

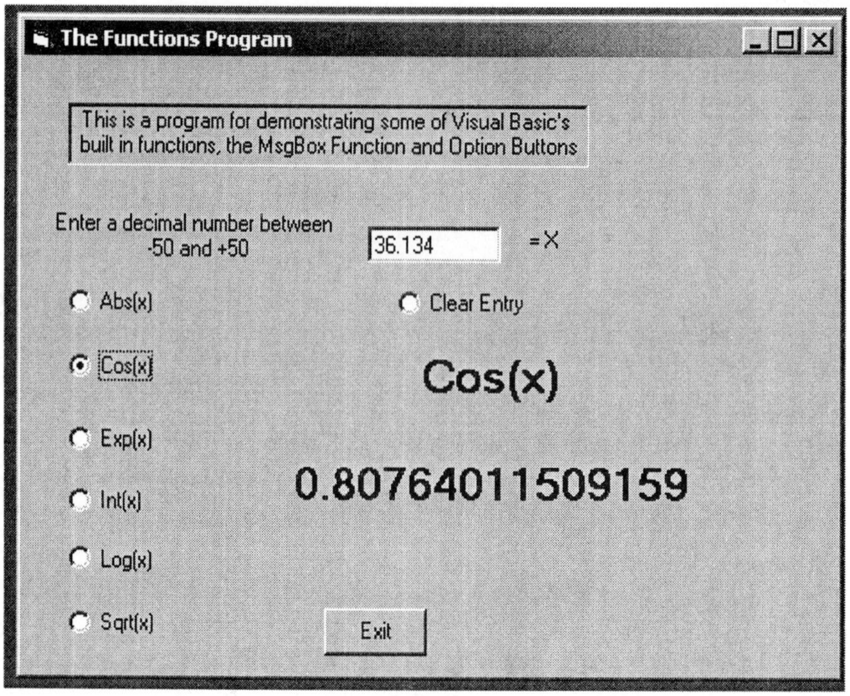

Figure 12-11. The Form for the Functions Program

Enter the following code in the **optClear_Click()** procedure:

```
Private Sub optClear_Click()
If optClear.Value = True Then
      lblFunName.Caption = ""
      lblResult.Caption = ""
      txtInput.Text = ""
      txtInput.SetFocus
   End If
End Sub
```

Enter the following code in the **optCos_Click()** procedure:

```
Private Sub optCos_Click()
   Const pi = 3.14159265359
   Dim Radians As Single
   If optCos.Value = True Then
      'Convert degrees to radians
      Radians = (txtInput.Text * pi / 180)
      lblFunName.Caption = "Cos(x)"
      lblResult.Caption = Cos(Radians)
   End If
End Sub
```

Enter the following code in the **optExp_Click()** procedure:

```
Private Sub optExp_Click()
If optExp.Value = True Then
      If (txtInput.Text < -50) Or (txtInput.Text > 50) Then
```

```
        MsgBox "Your number must be between -50 and +50", _
        vbExclamation, "Error"
      Else
        lblFunName.Caption = "Exp(x)"
        lblResult.Caption = Exp(txtInput.Text)
      End If
   End If
End Sub
```

Enter the following code in the **optInt_Click()** procedure:

```
Private Sub optInt_Click()
If optInt.Value = True Then
      lblFunName.Caption = "Int(x)"
      lblResult.Caption = Int(txtInput.Text)
   End If
End Sub
```

Enter the following code in the **optLog_Click()** procedure:

```
Private Sub optLog_Click()
   If optLog.Value = True Then
      If txtInput.Text < 0 Then
        MsgBox "Your number must be greater than zero", _
        vbExclamation, "Error"
      Else
        lblFunName.Caption = "Log(x)"
        lblResult.Caption = Log(txtInput.Text)
      End If
   End If
End Sub
```

Enter the following code in the **optSqr_Click()** procedure. Notice that the **MsgBox** statement is broken into two lines. The first line ends with the line continuation character (a space followed by an underline character). A single statement can be broken into multiple lines, as many as you feel necessary to make the code more readable.

```
Private Sub optSqr_Click()
   If optSqr.Value = True Then
      If txtInput.Text <= 0 Then
        MsgBox "Your number must be greater than zero", _
        vbExclamation, "Error"
      Else
        lblFunName.Caption = "Sqrt(x)"
        lblResult.Caption = Sqr(txtInput.Text)
      End If
   End If
End Sub
```

Enter the following code in the **cmdExit_Click()** procedure:

```
Private Sub cmdExit_Click()
   Dim UserResponse As Integer, Buttons As Integer
   Buttons = vbYesNo + vbQuestion
   UserResponse = MsgBox("Do you really want to quit?", Buttons, "Quit?")
   If UserResponse = vbYes Then End
End Sub
```

Executing the Functions Program. Execute the Functions program by clicking the **Start** button. The form with its controls appears like Figure 12-11, except the text box is clear as are the two label boxes for output named **lblFunName** and **lblResult**. Enter a decimal number in the text box where prompted. Click an option button, and the name of the function and the result of the function's calculation is displayed in 18-point bold Arial font. Click the **Clear** option button. Try out all the remaining options. The program displays a **MsgBox** statement to alert the user to an error if the number entered is zero or negative (Exp and Log). Upon clicking the **Exit** button, the user is asked, **Do you really want to quit?** (Figure 12-12). Depending upon the user's response, the program either continues or ends. The exit procedure demonstrates the **MsgBox** function. Elsewhere, the program uses the **MsgBox** statement, which was described in detail earlier in the Random Guess program.

Understanding the Code of the Functions Program. A function is a block of code that accepts one or more values and returns a single value, which is assigned to a variable.

```
VariableName = FunctionName (argument)

Answer = Sqr(12.34)
```

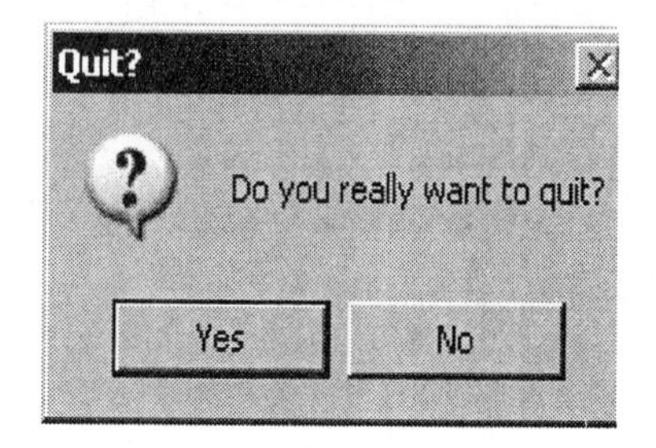

Figure 12-12. The Functions Program Message Box for Quitting the Program

The value **3.5128** is returned by the **Sqr(60)** function and argument and is assigned to the variable named **Answer**. In this program, the value returned by the function is assigned to the **Text** property of a text box, from which the user can read the result. With the built-in functions demonstrated in this program, the function's block of code is invisible to the user, but is part of Visual Basic. In a later program, we will learn how to write the code for a user-defined function. Actually, we have been using Visual Basic functions all along. In the Color Properties program, we used the **RGB** function to return the value of a color and assign it to the color property of a shape.

```
shpCircle.FillColor = RGB(255, 0, 0)
```

Some of the other functions you have already encountered are listed in Table 12-18.

Table 12-18

Previously Used Functions

Function Example	Returns	Where Previouly Used
RGB(0, 255, 0)	green	Color Properties Program
Format(Name,"#.##"}	number as a formatted string	Print Demo Program
Now	the current date and time	Print Demo Program
Tab(n)	*n* tabs	Print Demo Program
VarType (*expression*)	type of expression	Arithmetic Program
Spc(n)	*n* spaces	Boolean Operator Program
CDbl(*expression*)	a double (converts)	Arithmetic Program
VarType(*expression*)	an integer (see Table 12-15)	Arithmetic Program
Timer	time elapsed since midnight	Random Guess Program
Rnd	a random number	Random Guess Program
CSng (*expression*)	a single (converts)	Arithmetic Program

The user enters a number, decimal or integer, positive or negative, in a text box named **txtNumber**. This number is the **Text** property of the text box: **txtNumber.Text**. Clicking an option button causes the appropriate function to use **txtNumber.Text** as its argument. For example, **Abs(txtNumber.Text)** returns the absolute value of **txtNumber.Text** that the user enters and assigns it to **Answer**.

In all six functions illustrated in this program, the value of **Answer** is passed as a caption parameter of the label control named **lblResult**. With two of the function (**Log** and **Sqr**), it does so only after establishing that the value that the user input is not negative. If **txtNumber.Text** equals zero or is negative, a **MsgBox** statement informs the user that the value must be positive. The name of the function is passed to the **Caption** property of the label control named **lblFunName**.

The MsgBox Function. All of the message boxes in the program, except one in the **cmdExit** procedure, are **MsgBox** statements. (The **MsgBox** statement was discussed earlier in the Random Guess program.) The **cmdExit** procedure uses a **MsgBox** function. When the user clicks the **cmdExit** button, execution of the command buttons's code displays a message box with:

- A **Yes** button and a **No** button
- A question mark
- A title (**The Functions Program**)
- A message (**Do you really want to quit?**)

The Appearance of the Message Box. Clicking the **Exit** button causes execution of the **cmdExit** procedure. The first statement declares two integer variables. The second statement assigns the sum of two VB constants to the variable **Buttons.** When the third statement is executed, the **MsgBox** appears on the screen, and execution halts and waits for the user to respond by clicking either the **Yes** button or the **No** button. If the **Yes** button is clicked, then **End** is executed and the program exits.

```
Private Sub cmdExit_Click()
   Dim UserResponse As Integer, Buttons As Integer
   Buttons = vbYesNo + vbQuestion
   UserResponse = MsgBox("Do you really want quit?", Buttons, "Quit?")
   If UserResponse = vbYes Then End
End Sub
```

The middle parameter, **Buttons**, determines what the user sees on the message box. The valid integer values for these message box constants are listed in Table 12-19. For example, if the value of **Buttons** equals **4**, then the message box displays the **Yes** button and the **No** button. If the value of **Buttons** equals **32**, then the message box displays a **Warning Query** icon. If the value of **Buttons** equals **32** + **4** or **36**, then the message Box displays a **Warning Query** icon, a **Yes** button, and a **No** button. However, the code is more readable if the sum of the VB constants is used:

```
Buttons = vbYesNo + vbQuestion
```

Try changing the code by substituting **36** for **vbYesNo + vbQuestion**, and notice that the program runs the same. Experiment with some other values and see what happens.

The Values Returned by the Message Box Function. The **MsgBox** function returns a value, as all functions do, but the only values that the **MsgBox** function can return are those listed in Table 12-19. (See also Table 12-6.) These values tell the program what button the user clicked, and that, in turn, allows the program to take alternative actions. In this program, if the **MsgBox** function returns the value equal to **6**, then the program exits, otherwise nothing happens. Again, the code is more readable if the constant **vbYes** is used rather than its value of **6**. Change the code to a **6** instead of **vbYes**, and notice that the program runs the same. Experiment with some other values and see what happens

Using the VB constant instead of the equivalent number makes the code more readable. In this program the **MsgBox** function is followed by the control statement:

```
If UserResponse = vbYes Then End
```

which is more readable than the statement

```
If UserResponse = 6 Then End
```

Notice that when the **If ... Then** statement is short and simple, it may be written on one line. However, if a block of code lies between the **If** and the **End**, it is necessary to use the usual long form.

```
If UserResponse = vbYes Then
    Block of statements
End If
```

Table 12-19

Return Values of the MsgBox Function

The MsgBox function returns a value depending on which button the user clicks

Button Constant	Value Returned
vbOK	1
vbCancel	2
vbAbort	3
vbRetry	4
vbIgnore	5
vbYes	6
vbNo	7

Chapter

13

Visual Basic Loops without Arrays

The loop control structures give a programming language the capacity for iteration. Iteration, the ability to carry out tedious calculations rapidly, relentlessly, and repetitively, without complaints, without boredom, and without errors, is what makes a computer such a powerful tool for calculation and communication. The loop structures described in this section are the **For ... Next** loop, the **Do ... While** loop, and the **Do ... Until** loop control structures. This chapter also introduces ASCII characters, the KeyPress event, the Object Browser, the timer control, and user-defined functions. To illustrate these VB features, the chapter includes several VB programs:

The Loops program
The HKL program
The KeyPress program
The Rotational Energy program
The Nested Loops program
The ASCII Print program
The Timer A to Z program
The Vapor Pressure program

Unconditional Looping

The For ... Next Control Statement. The **For ... Next** control statement loops a predetermined number of times, set by the programmer, using a counter, which is incremented positively or negatively by the programmer. The **For ... Next** control statement has the following structure:

```
For (counter start) To (counter end) Step
    Statement block1
Next
Statement block 2
```

The ***counter start*** to ***counter end*** values could be:

1 To 50
1 To 100 Step 2
50 to 0 Step –1
99 to 0 Step –3
and so on

All statements in **Statement block1** are executed repeatedly until the counter end value is reached. Then execution goes on to **Statement block 2**.

- **Step** can be positive or negative. If omitted, the default value is 1.
- VB sets the counter equal to the start value (usually 1).
- VB executes the loop if **(*counter start*)** < **(*counter end*)**.

- VB increments the counter by the step value (usually 1).
- VB repeats comparing the counter value with the counter end value.

Conditional Looping

The conditional loop structure loops depending on the outcome of a test of a conditional value or expression. Notice that if the condition never becomes false, the loop is infinite; it never stops.

The Do ... While Statement. The **Do ... While** control statement has the following structure:

```
Do ... While (condition)
   Statement(s)
Loop
```

First, the condition is tested. If the condition is true, the statement block is executed. When the condition becomes false, execution moves out of the loop to succeeding statements. If the condition is false on the first test, no statement in the loop is ever executed.

An alternative structure is:

```
Do
   Statements
While (condition)
```

This control structure executes at least one statement before testing the condition. As long as the condition is true, execution continues. Thus, the preceding loop structure loops as long as the condition is true.

The Do ... Until Statement. The next two variations loop until the condition is false. Otherwise, they are quite analogous. The **Do ... Until** statement has the following structure:

```
Do Until (condition)
   Statement(s)
Loop
```

The program checks the condition first, then executes if the condition is true. If the condition is false at the beginning, then no statement is ever executed, and control leaves the loop structure. This structure loops zero or more times. An alternative structure is:

```
Do
   Statement(s)
Loop Until (condition)
```

At least one statement is executed before the condition is tested. As long as the condition remains false, the statements within the loop structure are executed. This structure loops one or more times.

With any of the conditional loops, infinite looping is possible, and then you must take drastic action to stop execution, which may result in loss of programmed code. Consequently, it is wise to save your work before you test run a newly written conditional control statement so you do not lose what you have recently programmed.

The Loops Program

The Loops program is a short program written to demonstrate the **For ... Next**, the **Do ... While**, and the **Do ... Until** loop structures. It also provides review and practice in printing to the form and formatting numerical output.

Starting the Project

- Start Visual Basic by clicking the **VB** icon on the desktop.
- Select the **Standard EXE** icon in the **New Project** window, and click the **OK** button.
- A **New Project** window opens with a fresh form.
- Select the form, and in the properties window, change the **(Name)** property to **frmLoops**.
- Click **Save**; the file name will be frmLoops, so click **OK**.
- The file name changes to Project1; change it to **The Loops Program**, and click the **OK** button.
- Place a single command button on frmLoops.
- Set the properties of the controls according to Table 13-1.

Table 13-1

The Properties Table for the Loops Program

Object	Property	Value
Form1	Name	frmLoopDemo
	Caption	The Loops Program
	Windows State	2-Maximized
Command1	Name	CmdExit
	Caption	E&xit

Entering the Code of the Loops Program. Enter the following code in the general declarations section of the program:

```
Option Explicit
'The Loops program demonstrates the
' For ... Next, the Do ... While, and the Do ... Until structures
```

Enter the following code in the **cmdExit_Click( )** procedure:

```
Private Sub cmdExit_Click()
  End
End Sub
```

Enter the following code in the **Form_Load()** procedure:

```
Private Sub Form_Load()
   Dim n As Integer
   Show
   Print
   Print " A For ... Next statement did this:"
   Print
   For n = 1 To 10
      Print Spc(5); " For n = "; n; _
      ", the square of n = "; n ^ 2
   Next

   Print
   Print " A Do ... While statement did this:"
   Print
   n = 0
   Print Spc(5); "n"; Spc(5); "8 ^ n"
   Do While n < 10     'Loops if condition is True
      n = n + 1
      Print Spc(5); n; Spc(5); 8 ^ n
   Loop
   Print
   Print "A Do ... Until statement did this:"
   Print
   n = 0
   Print Spc(5); "n"; Tab; "Sqr Root n"
   Do Until n = 10     'Loops if condition is False
      n = n + 1
      Print Spc(5); n; Tab; _
      Format(Sqr(n), "#.00000")
   Loop

End Sub
```

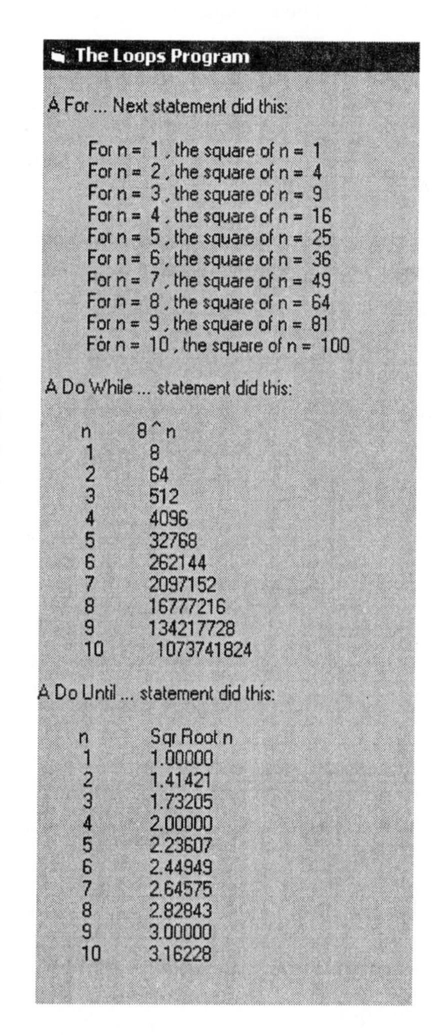

Figure 13-1. The Output of the Loops Program

Executing the Loops Program. Click the **Start** button to execute the Loops program. Compare the output on the form with the code that produced it. Click the **Exit** button to end the program. The output of the program, printed on the form, is shown in Figure13-1.

Understanding the Code of the Loops Program. Besides the **cmdExit_Click()** procedure, this demonstration program has only the **Form_Load** procedure, which executes immediately after the user clicks the **Start** button. The variable *n* is dimensioned as an integer. Then the **Show** statement and the **Print** statements follow. Both the **Show** and the **Print** statements are methods of the Form1 object. It is not necessary to print the complete statement, **frmLoopDemo.Show** or **frmLoopDemo.Print**. The **Show** statement is always required for the printing to show on the form. The first **Print** statement prints a blank line on the form. Next an informative string is printed for the benefit of the user, followed by another blank line.

```
Private Sub Form_Load()
    Dim n As Integer
    Show
    Print
    Print " A For ... Next statement did this:"
     Print
```

Within this procedure, three loops execute. This first is a **For ... Next** loop.

```
    For n = 1 To 10
        Print Spc(5); " For n = "; n; _
        ", the square of n = "; n ^ 2
    Next
```

The **Print** statement is continued on a second line instead of one long line:

```
Print Spc(5); " For n = "; n; ", the square of n = "; n ^ 2
```

To continue a long line of code, type an underline character followed by a space, and press ENTER. The **Print** method here has five items in the output list, which it prints on one line.

1. **Spc(5)** inserts five spaces before the next argument is printed.
2. The semicolon causes an item in the output list to follow the last character of the previously printed item in the output list.
3. The string **" For n = "** is printed.
4. The current value of the **For ... Next** counter *n* is printed.
5. The string **", the square of n = "** is printed.
6. The current value of the square of *n* is printed.

Upon entering the **For ... Next** loop, the initial value of *n* is 1. After printing the preceding five arguments of the **Print** statement, the value of *n* is incremented by one, so that its value equals two. The five arguments are printed again on the next line. Incrementation of *n* continues until *n* has the value 10, at which point execution leaves the **For ... Next** loop and proceeds to the next statement of the program, which in this program is a **Do ... While** statement. The first three statements inform the user what is happening:

```
Print
Print " A Do ... While statement did this:"
Print
```

Then the counter *n* is initialized with the value equal to 0, and a string is printed that gives a label to the columns of numbers that will follow:

```
n = 0
Print Spc(5); "n"; Spc(5); "8 ^ n"
```

Next the loop is entered, and the condition is checked. If *n* is less than 10, then execution continues through the loop. The counter *n* is incremented by 1, so that its value on the first pass through the loop equals $0 + 1 = 1$. Next, five spaces are printed, followed by the current value of *n*, then five more spaces, followed by 8 raised to the power of the current value of *n*. Execution then loops back for another test of the value of *n*. When the value of $n = 10$, one more line is printed, but when execution loops back for another test of the value of *n*, since now *n* is not less than 10, execution leaves the loop and continues on to the next statement of the program.

```
Do While n < 10    'Loops if condition is True
  n = n + 1
  Print Spc(5); n; Spc(5); 8 ^ n
Loop
```

When the program exits the previous loop, it moves on to a **Do ... Until** loop. The first three statements inform the user what is happening:

```
Print
Print "A Do ... Until statement did this:"
Print
```

Then *n* is initialized to have a value of 0, and a **Print** statement prints column labels for the number that will follow.

```
n = 0
Print Spc(5); "n"; Tab; "Sqr Root n"
Do Until n = 10    'Loops if condition is False
  n = n + 1
  Print Spc(5); n; Tab; _
  Format(Sqr(n), "#.00000")
Loop
```

Upon entering the loop for the first time, the program checks to see if *n* equals 10. Since *n* at this point was initialized to equal 0, the condition is false, and the loop is entered. The counter *n* is incremented from 0 to 1. Then a long line is continued by splitting it with an underline and a space. The **Print** method has four arguments:

1. **Spc(5)** prints five spaces on a line.
2. The current value of *n* is printed.
3. A tab is inserted.
4. The square root of *n* is printed and is custom formatted.

The **#.00000** forces zeros to be printed, if there are any zeros. Change the format to **#.#####** and run the program again to see how this change affects the format of the numbers.

The Nested Loops Program

This program illustrates the use of two nested loops without using an array. In the next chapter, three programs use nested loops with arrays: the Bubble Sort program, the TotoLoto program, and the DesLandres program.

Starting the Project

- Start Visual Basic by clicking the **VB** icon on the desktop.
- Select the **Standard EXE** icon in the **New Project** window, and click the **OK** button.
- A **New Project** window opens with a fresh form.
- Select the form, and in the properties window, change the **(Name)** property to **frmNested**.
- Click **Save**; the file name will be frmNested, so click **OK**.
- The file name changes to Project1; change it to **The Nested Loops Program**, and click the **OK** button.

Entering the Code for the Nested Loops Program. Double-click the form. This action opens the code window with the first and last line of the **Form_Load** procedure.

```
Option Explicit

Private Sub Form_Load()
  Dim R As Integer
  Dim C As Integer
  Dim N As Integer
  Show
  For R = 1 To 4
    For C = 10 To 12
      N = R + C              ' Do a simple calculation
      Print R; C; N          ' Print results to form
    Next C
  Next R
End Sub
```

On the **View** menu, click **Object** to get the form back. Scroll the **Events List** box to **DblClick** to get the first and last lines of the **Form_DblClick** procedure. Enter the following code:

```
Private Sub Form_DblClick()
  End
End Sub
```

Executing the Nested Loops Program. Click the VB **Start** button. The result of executing the program is show in Figure 13-2. Double-click anywhere in the form to exit the program.

Understanding the Nested Loops Program. Examine the program loop statements and compare them with the printed output in Figure 13-2. During the first pass through the outer loop, the value of *R* is fixed at 1, while the value of *C* in the inner loops changes from 10 to 11 and then to 12. After each increment of *C*, the current value of *C* is added to 1 (the initial value of *R*), which is assigned to *N*. At this point, the statement **Print R; C; N** prints the current value of *R*, which equals 1; the current value of *C*, which equals 10; and *N*, the sum of *R* and *C*, which equals 11. These values are printed in a row on the first line of the output.

In the second line of the output, the value of *R* in the outer loop remains at 1, while the value of *C* is incremented from 10 to 11. Again, these current values of *R* and *C* are added and assigned to *N*. All three are printed on the form in the second line.

After *C* is incremented from 11 to 12, the inner loop is satisfied, so *R* is now incremented from 1 to 2, and the inner loop again loops three times as *C* repeats its journey for 10 to 11 to 12.

The inner loop loops three time for each of the three times that the outer loop loops, giving rise to twelve lines of output.

Instead of using a command button to exit the program, this program uses the double-click (**DblClick**) event of the form to exit the program.

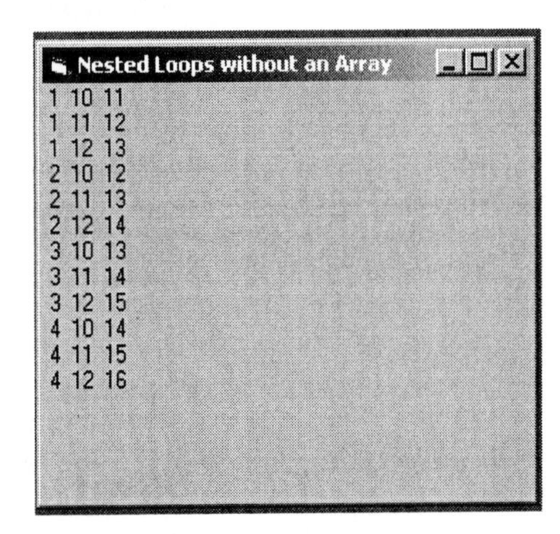

Figure 13-2. Output of the Nested Loops Program

The HKL Program

This program illustrates thee nested loops without arrays. From the length of the edge of a cubic unit cell *a*, the spacing between diffracting planes d_{hkl} and the sine *S* of the diffraction angles are calculated. Two programs in the next chapter (the Bubble Sort program and the TotoLoto program) use nested loops with arrays.

Starting the Project

- Start Visual Basic by clicking the **VB** icon on the desktop.
- Select the **Standard EXE** icon in the **New Project** window, and click the **OK** button.
- A **New Project** window opens with a fresh form.
- Select the form, and in the properties window, change the **(Name)** property to **frmHKL**.
- Click **Save**; the file name will be frmHKL, so click **OK**.
- The file name changes to Project1; change it to **The HLK Program**, and click the **OK** button.

Entering the Code of the HKL Program. Enter the following code in the general declarations section of the program:

```
Option Explicit
' The HKL program demonstrates the used of nested loops without arrays.
' The HKL program calculates d spacing between planes of a crystal from the unit
' cell dimension, the wavelength, and the Miller indices of the plane.
```

Double-click the form to get the first and last statements of the **Form_Load** procedure, and enter the following code:

```
Private Sub Form_Load()
  Dim h As Integer, k As Integer, l As Integer
  Dim d As Single, S As Single
  Const a = 4.56                      ' The unit cell dimension (angstroms)
  Const Lambda = 1.5418               ' The X-ray wavelength (angstroms)
  Show
  Print "h", "k", "l", "d", "Sin(Theta)"
  Print
  For h = 1 To 3
    For k = 0 To 2
      For l = 0 To 2
        d = a / Sqr(h ^ 2 + k ^ 2 + l ^ 2)      ' calculate spacing between planes
        S = Lambda / (2 * d)                    ' calculate sines of the diffraction angles
        Print h, k, l, Format(d, "#.###0"), Format(S, "0.####0")
      Next l
    Next k
  Next h
End Sub
```

On the **View** menu, click **Object** to get the form back. Scroll the **Events List** box to **DblClick** to get the first and last lines of the **Form_DblClick** procedure. Enter the following code:

```
Private Sub Form_DblClick()
  End
End Sub
```

Executing the HKL Program. Click the **Start** button. The result of executing the program is shown in Figure 13-3. Double-click anywhere in the form to exit the program.

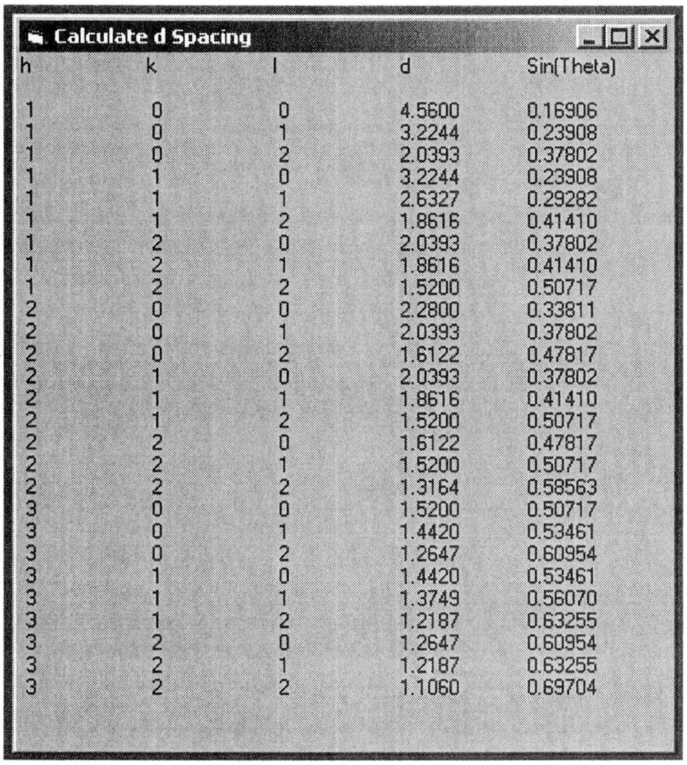

Calculate d Spacing

h	k	l	d	Sin(Theta)
1	0	0	4.5600	0.16906
1	0	1	3.2244	0.23908
1	0	2	2.0393	0.37802
1	1	0	3.2244	0.23908
1	1	1	2.6327	0.29282
1	1	2	1.8616	0.41410
1	2	0	2.0393	0.37802
1	2	1	1.8616	0.41410
1	2	2	1.5200	0.50717
2	0	0	2.2800	0.33811
2	0	1	2.0393	0.37802
2	0	2	1.6122	0.47817
2	1	0	2.0393	0.37802
2	1	1	1.8616	0.41410
2	1	2	1.5200	0.50717
2	2	0	1.6122	0.47817
2	2	1	1.5200	0.50717
2	2	2	1.3164	0.58563
3	0	0	1.5200	0.50717
3	0	1	1.4420	0.53461
3	0	2	1.2647	0.60954
3	1	0	1.4420	0.53461
3	1	1	1.3749	0.56070
3	1	2	1.2187	0.63255
3	2	0	1.2647	0.60954
3	2	1	1.2187	0.63255
3	2	2	1.1060	0.69704

Figure 13-3. Output of the HKL Program

Understanding the HKL Program

The Code. The indices h, k, and ℓ are declared as integers. The spacing d between diffraction planes and the sin of the diffraction angle S are declared as the **Single** type. The unit cell dimension and the X-ray wavelength are declared as constants. **Show** is a method of the form required to print to the form. The labels for the output are printed on the form.

The next three loops function similarly to the two nested loops described in the previous program. In the outer loop, 1 is assigned to h; in the first inner loops, 0 is assigned to k, and then the most inner loop is satisfied after 0, 1, and 2 are assigned to ℓ.

The value of h is incremented from 1 to 2, and the inner loops run again. When h reaches its last value (3 in this example), the two inner loops execute and all the counters are satisfied. The innermost loop prints the current values of h, k, ℓ, d, and S.

The Science. The details of X-ray powder diffraction are reviewed in Chapter 1, Excel Example 1. In this Visual Basic exercise, the cell spacing $d_{hk\ell}$ is calculated for a crystal with the cell edge a equal to 4.56 angstroms for a nested loop of h, k, and ℓ values with the use of Equation 1-1.

$$d_{hk\ell} = \frac{a}{\sqrt{h^2 + k^2 + \ell^2}}$$

Simultaneously, the sines of the diffraction angles are calculated with Equation 1-2.

$$d_{hkl} = \frac{\lambda}{2Sin\theta}$$

This calculation is the reverse of the determination of a from measurements of diffraction angles. Such a calculation is frequently used to check whether the observed diffraction angles are consistent with the measurement cell dimension a.

The ASCII Character Codes

When you press a key on the computer's keyboard, the text editor of Visual Basic or any application puts a character code corresponding to that character into a text file. The character codes, integers from 0 to 256, are standardized and called the ASCII code. ASCII is an acronym for American Standard Code for Information Interchange. The printable characters on an English language keyboard range from ASCII 32 to ASCII 126 and are shown in Table 13-2.

Table 13-2

The ASCII Codes for Keyboard Characters

Code	Char	Code	Char	Code	Char	Code	Char	Code	Char
32	Spc	51	3	70	F	89	Y	108	l
33	!	52	4	71	G	90	Z	109	m
34	“	53	5	72	H	91	[	110	n
35	#	54	6	73	I	92	\	111	o
36	$	55	7	74	J	93	]	112	p
37	%	56	8	75	K	94	^	113	q
38	&	57	9	76	L	95	_	114	r
39	‘	58	:	77	M	96	`	115	s
40	(	59	;	78	N	97	a	116	t
41	)	60	<	79	O	98	b	117	u
42	*	61	=	80	P	99	c	118	v
43	+	62	>	81	Q	100	d	119	w
44	,	63	?	82	R	101	e	120	x
45	-	64	@	83	S	102	f	121	y
46	.	65	A	84	T	103	g	122	z
47	/	66	B	85	U	104	h	123	{
48	0	67	C	86	V	105	I	124	|
49	1	68	D	87	W	106	j	125	}
50	2	69	E	88	X	107	k	126	~

The ASCII Print Program

The ASCII Print program provides familiarization with the ASCII codes and gives further practice in looping with relational and conditional operators in a **For ... Next** loop structure. The ASCII Print program demonstrates Boolean values, expressions, and assignments.

Starting the Project

- Start Visual Basic by clicking the **VB** icon on the desktop.
- Select the **Standard EXE** icon in the **New Project** window, and click the **OK** button.
- A **New Project** window opens with a fresh form.
- Select the form, and in the properties window, change the **(Name)** property to **frmAsciiPrint**.
- Click **Save**; the file name will be frmAsciiPrint, so click **OK**.
- The file name changes to Project1; change it to **The ASCII Print Program**, and click the **OK** button.
- Place a single command button control down in the lower right corner of frmAsciiPrint.
- Set the properties of the controls according to Table 13-3.

Table 13-3

The Properties Table for the ASCII Print Program

Object	Property	Value
Form1	Name	frmAsciiPrint
	Caption	The ASCII Print Program
	WindowState	2-Maximized
Command1	Name	cmdExit
	Caption	Exit

Entering the Code of the ASCII Print Program. Enter the following code in the general declarations section:

```
Option Explicit
' The ASCII Print program demonstrates Boolean values,
' expressions, and assignments. It also demonstrate the
' use of relational and conditional operators in a
' For ... Next loop structure.
```

Enter the following code in the **cmdExit_Click()** procedure:

```
Private Sub cmdExit_Click()
  End
End Sub
```

Enter the following code in the **Form_Load** Procedure:

```
Private Sub Form_Load()

  Dim AsciiNum As Integer
  Dim Condition1 As Boolean
  Dim Condition2 As Boolean
  frmAsciiNum.Show
  frmAsciiNum.Print
  For AsciiNum = 97 To 122
    frmAsciiNum.Print " "; Chr(AsciiNum);
  Next
  frmAsciiNum.Print
  For AsciiNum = 65 To 90
    frmAsciiNum.Print " "; Chr(AsciiNum);
  Next
  frmAsciiNum.Print
  For AsciiNum = 0 To 127
```

```
        Condition1 = AsciiNum > 32 And AsciiNum < 65
        Condition2 = AsciiNum > 122 And AsciiNum < 122
        If Condition1 = True Or Condition2 = True Then
            frmAsciiNum.Print " "; Chr(AsciiNum);
        End If
    Next
    frmAsciiNum.Print
    For AsciiNum = 65 To 90
        frmAsciiNum.Print _
        AsciiNum; vbTab; Chr(AsciiNum); vbTab; _
        AsciiNum + 32; vbTab; Chr(AsciiNum + 32); vbTab; _
        AsciiNum - 32; vbTab; Chr(AsciiNum - 32); vbTab
    Next
    Print """Axe"""                         ' printing quotation marks
    Print Chr(34); "Axe"; Chr(34)           ' printing quotation marks
End Sub
```

Executing the ASCII Print Program. Click the **Start** button to run the program. The form loads, and the program executes, displaying the printable ASCII characters from char number 33 to 122, first in rows and then in columns. The output is shown in Figure 13-4. Click the **Exit** button to end the program.

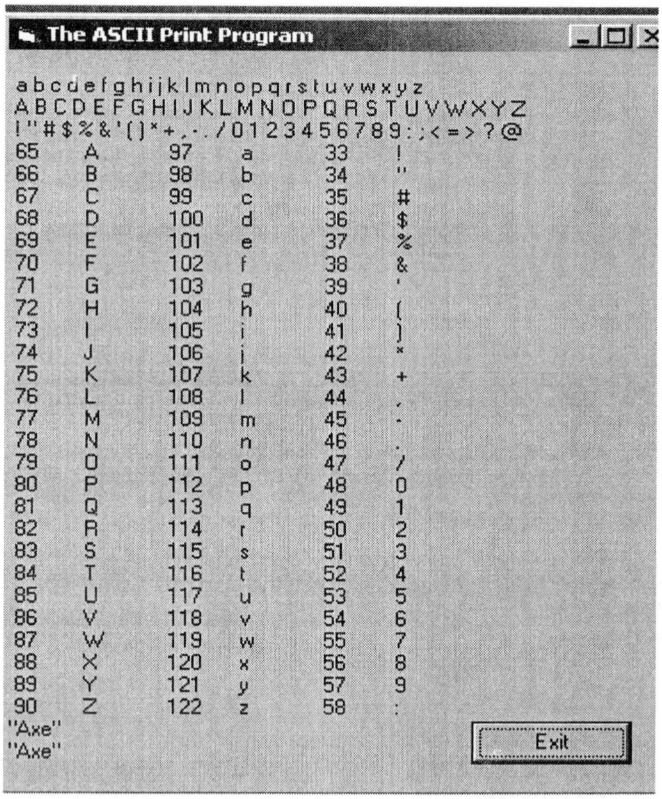

```
The ASCII Print Program
abcdefghijklmnopqrstuvwxyz
ABCDEFGHIJKLMNOPQRSTUVWXYZ
!"#$%&'()*+,-./0123456789:;<=>?@
65  A   97  a   33  !
66  B   98  b   34  "
67  C   99  c   35  #
68  D   100 d   36  $
69  E   101 e   37  %
70  F   102 f   38  &
71  G   103 g   39  '
72  H   104 h   40  (
73  I   105 i   41  )
74  J   106 j   42  *
75  K   107 k   43  +
76  L   108 l   44  ,
77  M   109 m   45  -
78  N   110 n   46  .
79  O   111 o   47  /
80  P   112 p   48  0
81  Q   113 q   49  1
82  R   114 r   50  2
83  S   115 s   51  3
84  T   116 t   52  4
85  U   117 u   53  5
86  V   118 v   54  6
87  W   119 w   55  7
88  X   120 x   56  8
89  Y   121 y   57  9
90  Z   122 z   58  :
"Axe"
"Axe"
                                Exit
```

Figure 13-4. Output of the ASCII Print Program

Understanding the Code of the ASCII Print Program. Besides the **cmdExit_Click()** procedure, the program has just one procedure, the **Form_Load** procedure, which runs immediately after the user clicks the **Start** button. This procedure has four **For ... Next** loops.

The first procedure prints in a row on the form the ASCII numbers from 97 to 122, which are the lowercase letters. This loop generates the first line of output printed on the form. The second **For ... Next** loop prints in a row on the form the ASCII numbers from 65 to 90, which are the uppercase letters. This loop prints the second line of output printed on the form. The remaining characters lie between ASCII numbers 65 and 32, so we define a Boolean variable named **Condition.**

Condition = AsciiNum > 32 And AsciiNum < 65

The **If ... Then** statement within the third **For ... Next** loop tests the value of the Boolean variable **Condition.** If **Condition** equals **True,** then the character corresponding to that ASCII number is printed. This loop prints the third line of output printed on the form.

The fourth and last **For ... Next** loop prints six items on a single line for each pass through the loop (excluding the **vbTabs** for spacing) for ASCII numbers 65 to 90. The six items for each pass through the fourth and last **For ... Next** loop are as follows:

AsciiNum	(This equals 65 on the first pass through the loop.)
AsciiNum + 32	(This equals 97 for the first pass through the loop.)
Chr (AsciiNum + 32	(This equals "a" for the first pass through the loop.)
AsciiNum – 32	(This equals 33 for the first pass through the loop.)
Chr (AsciiNum – 32	(This equals ! for the first pass through the loop.)

Next, the loop increments the counter (**AsciiNum**) by 1 and prints the six items again on a new line with their new values. Notice that the first three **For ... Next** loops all end with a semicolon (;), but the fourth **For ... Next** loop ends with no punctuation mark at all. Experiment by ending one of the loops with a comma (,), a semicolon (;), or nothing ( ). Observe the changes in the printing format.

Further Comments on ASCII characters. Notice that the statement

```
Print Chr (90)
```

gives the same result as

```
Print "Z"
```

and that

```
Print Chr (65); Chr(120); Char(101)
```

gives the same result as

```
Print "Axe"
```

and that result is **Axe**. To print quotation marks in addition to the string they enclose requires some trickery. The statement

```
Print Chr(34); "Axe"; Chr(34)
```

results in printing "**Axe**". The same result can be obtained with the following statement:

```
Print """Axe"""
```

Three quotation marks precede **Axe** and three quotation marks follow **Axe** with no spaces between.

Nonprinting ASCII Characters. The first 32 ASCII characters are nonprintable, but **Chr** values of 0, 8, 9, 10, 12, and 13 are useful at times. These are shown in Table 13-4, along with the corresponding Visual Basic constants.

Table 13-4

Nonprinting ASCII Characters

Chr (*n*)	Constant	Description
Chr (0)	vbNull	Null character
Chr (8)	vbBack	Backspace
Chr (9)	vbTab	Tab
Chr (10)	vbLf	Line feed
Chr (11)	vbVericalTab	Vertical tab
Chr (12)	vbFormFeed	Form feed
Chr (13)	vbCr	Carriage return

ASCII Char Numbers 128 to 255: The Second Half. In the second half, character numbers from 128 through 161 are nonprinting. **Chr** number 161 is the space. To get a quick look at the remaining printable characters, modify the last **For ... Next** loop so that it looks like this:

```
For AsciiNum = 161 To 191
  frmAsciiNum.Print _
  AsciiNum; vbTab; Chr(AsciiNum); vbTab; _
  AsciiNum + 32; vbTab; Chr(AsciiNum + 32); vbTab; _
  AsciiNum + 64; vbTab; Chr(AsciiNum + 64); vbTab _
Next
```

Now run the program again, and the characters with ASCII numbers from 161 to 255 will be displayed.

The KeyPress Event. The syntax for the KeyPress event is

```
Private Sub object_KeyPress (keyascii As Integer)
```

The event occurs when the user presses a key in an object that has the keyPress event. The event belongs to the object that has the focus. The object is usually a text box, list box, or combo box. The **keyascii** parameter returns an integer corresponding to an ASCII key code. For example, if the key for the letter *a* is pressed, then **keyascii** has the value 97, according to Table 13-2.

As an example, suppose you have a text box into which you want the user to enter a serial number, and you want to prevent the user from accidentally entering a letter. If the text box were named **txtSerialNo**, then the following code would do what is desired:

```
Private Sub txtSerialNo_KeyPress (Keyscii As Integer)
        If Keyascii < Asc("0") OR Keyascii >Asc("9") Then Keyascii = 0
        The remaining code goes here
End Sub
```

Keyascii equal to 0 corresponds to no key pressed, so nothing happens, and nothing is entered. The next example shows how to force capitalization of all letters entered into a text box named **txtCity**:

```
Private Sub txtCity_KeyPress (Keyascii As Integer)
        Char = Chr (Keyascii)
        Keyascii = Asc(Ucase (Char))
        The remaining coded goes here
End Sub
```

The KeyPress Program

To try out the KeyPress event for a text box, start a new project, drop a text box on the form, and change its name property to **txtTestKeys**. Double-click the text box to open the code window with the following procedure declaration, ready-made for you.

Starting the Project

- Start Visual Basic by clicking the **VB** icon on the desktop.
- Select the **Standard EXE** icon in the **New Project** window, and click the **OK** button.
- A **New Project** window opens with a fresh form.
- Select the form, and in the properties window, change the **(Name)** property to **frmKeyPress**.
- Click **Save**; the file name will be frmKeyPress, so click **OK**.
- The file name changes to Project1; change it to **The KeyPress Program**, and click the **OK** button.
- Place a text box control and a command button on the form (Figure 13-5).
- Set the properties of the controls according to Table 13-5.

Table 13-5

Properties Table for the KeyPress Program

Object	Property	Value
Form1	Name	frmKeyPress
	Caption	The KeyPress Program
Text box1	Name	txtTestKeys
	Text	Make it blank
	TabIndex	0
Command1	Name	cmdExit
	Caption	E&xit

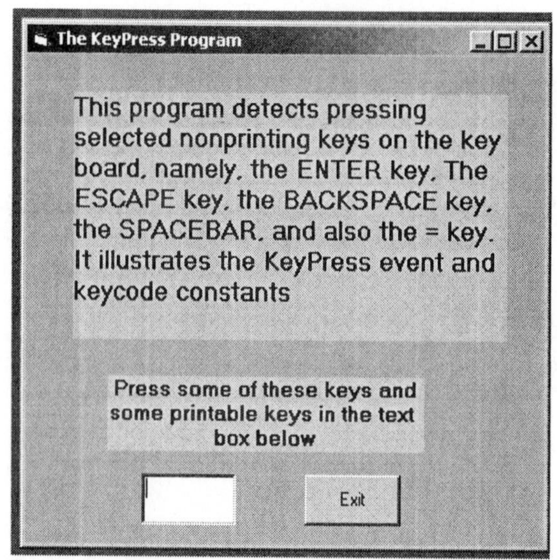

Form 13-5. The Form for the KeyPress Program

Entering the Code. Enter the following code in the declaration section:

```
Option Explicit
```

Double-click the **cmdExit** button, and enter the following code:

```
Private Sub cmdExit_Click()
   End
End Sub
```

Double-click the text box, and scroll through the **Events List** to **KeyPress**. Enter the following code:

```
Private Sub txtTestKeys_KeyPress(Keyascii As Integer)
   If Keyascii = vbKeyReturn Then
         MsgBox "You Pressed the Enter Key", , "Testing Keys"
   ElseIf Keyascii = vbKeyEscape Then
         MsgBox "You Pressed the Escape Key", , "Testing Keys"
   ElseIf Keyascii = vbKeyBack Then
         MsgBox "You Pressed the Backspace", , "Testing Keys"
   ElseIf Keyascii = vbKeySpace Then
         MsgBox "You Pressed the Spacebar", , "Testing Keys"
   ElseIf Keyascii = 61 Then
         MsgBox "You Pressed the equal (=) key", , "Testing Keys"
   End If
   txtTestKeys.Text = ""
End Sub
```

Press the **Start** button to run the program. Enter a couple of letters and numbers from the keyboard. Then press the ENTER key. A message box appears with the message: **You Pressed the Enter Key**. Try the ESC key, the BACKSPACE key, the SPACEBAR, and the = key to see how the KeyPress event detects the pressed key.

The Object Browser

In the preceding procedure, the ASCII number was used for the equal (=) sign, but key code constants were used to represent the ASCII number for the other keys tested. The key code constants form more readable code than integers.

Visual Basic offers a large number of key code constants, which are easily found in the **Object Browser** window (Figure 10-7). In the **Object Browser** window, scroll down the left side (**Classes**) until you can select **KeyCode Constants**. On the right side (**Members of KeyCode Constants**), scroll until you see **vbKeyEscape**, and select it. At the very bottom of the window, this constant is identified as **Esc Key**.

The **Object Browser** window is useful for looking up the properties and methods of the objects that you use in your program. In the **Object Browser** window, you will see an object that you have used several times, the label control. Click it, and you will be able to browse through its methods and properties as they are listed on the right side of the **Object Browser** window.

The Timer Control

The Visual Basic timer control is really an interval timer. In fact, it has few properties, and only two of them are of interest to us: **Enabled** and **Interval**. When the timer control is enabled, it precipitates its timer event after an interval of time. **Enabled** is a Boolean that has values of **True** or **False**. **Interval** is a period of time expressed in milliseconds: 1500 ms equals 1.5 s. The timer event occurs repeatedly at the specified interval until the value of **Enabled** becomes equal to **False**. In the toolbox, the **Timer Control** icon looks like a tiny clock:

The timer control is placed on the form as usual. Unlike most other controls, it cannot be resized, and while it is visible on the screen during design time, it is invisible when the program is running. For this reason, you can place it anywhere on the form that is convenient.

The default name of the timer control is **Timer1**, but suppose we set the name property at design time to **tmrMyTimer** and set the value of **Interval** to 500. If we double-click the timer, Visual Basic declares the following procedure in the code window:

```
Public Sub tmrMyTimer_Timer ()
   Statement
   Statement
   Lots of
   Action
End Sub
```

When the **Enabled** property becomes true, the preceding procedure is executed. For example, the following statement would make **Enabled** true:

```
MyTimer.Enabled = True
```

When this statement is executed, the **MyTimer_Time()** procedure executes repeatedly at a period of **Interval** seconds. With these characteristics of the timer control in mind, let's write a program that uses the timer control.

The Timer A to Z Program

Earlier, in Chapter 12, the Random Guess program introduced the **Timer** function. The Timer A to Z program demonstrates the timer control. We'll also have a little fun with disappearing and appearing labels, colored labels, and with sound effects.

Starting the Project

- Start Visual Basic by clicking the **VB** icon on the desktop.
- Select the **Standard EXE** icon in the **New Project** window, and click the **OK** button.
- A **New Project** window opens with a fresh form.
- Select the form, and in the properties window, change the **(Name)** property to **frmAZ.**
- Click **Save**; the file name will be frmAZ, so click **OK.**
- The file name changes to Project1; change it to **The Timer A to Z Program**, and click the **OK** button.
- Place controls on frmAZ according to Figure 13-6.
- Set the properties of the controls according to Table 13-6.

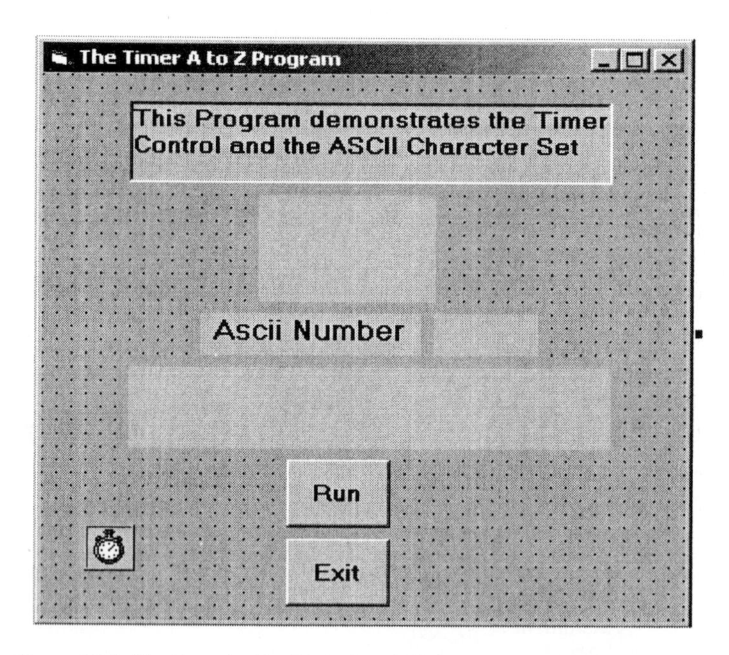

Figure 13-6. The Form for the Timer A to Z Program

Entering the Code of the Timer A to Z Program. Enter the following code in the general declarations section:

```
' The Timer A to Z program
' This program demonstrates the timer control by displaying successive
' letters of the alphabet every 300 milliseconds
Option Explicit
Dim AsciiNum As Integer        ' A global declaration for use in two procedures
```

Enter the following code in the **cmdExit_Click()** procedure:

```
Private Sub cmdExit_Click()            ' To exit the program
   End
End Sub
```

Enter the following code in the **cmdRun_Click()** procedure:

```
Private Sub cmdRun_Click()           ' This click
   tmrTimer.Enabled = True           ' starts the timer,
   lblInfo.Visible = False           ' makes the top label control disappear,
   lblAsciiNo.Visible = True         ' makes the ASCIINo label control appear,
   lblAscii.Visible = True           ' and makes the ASCII label control appear
End Sub
```

Table 13-6

Properties Table for the Timer A to Z Program

Object	Property	Value
Form1	Name	frmAZ
	Caption	The Timer A to Z Program
Label1	Name	This program demonstrates the timer control and the ASCII character set
	Caption	1- Fixed Single
	BorderStyle	10 point bold
	Font	lblAZ
Label2	Name	24 point bold
	Font	Red
	ForeColor	Make it blank
	Caption	lblAscii
Label3	Name	ASCII number:
	Caption	2-Center
	Alignment	Blue (choose from palette)
	ForeColor	False
	Visible	12 point bold
	Font	lblAsciiNo
Label4	Name	2-Center
	Alignment	12 point bold
	Font	lblTime
Label5	Name	2-Center
	Alignment	14 point bold
	Font	Dark green (choose from palette)
	Color	tmrTimer
Timer1	Name	300
	Interval	cmdRun
Command1	Name	&Run
	Caption	cmdExit
Command2	Name	E&xit
	Caption	

Enter the following code in the **Form_Load()** procedure:

```
Private Sub Form_Load()             ' This is the first procedure to run
  AsciiNum = 64                     ' Initialize the ASCII number one less than that of "A"
  tmrTimer.Enabled = False          ' Begin the program with the timer disabled
End Sub
```

Enter the following code in the **tmrTimer_Timer()** procedure:

```
Public Sub tmrTimer_Timer()         ' This procedure runs after cmdRun is clicked,
  Dim Char As String                ' which enables the timer control
  AsciiNum = AsciiNum + 1           ' AsciiNum now equals 65 (for "A") after first beep
  Char = Chr(AsciiNum)              ' Char is the string representation of the Ascii number
  lblAZ.Caption = Char              ' The label's Caption property displays the letter
  lblAsciiNo.Caption = AsciiNum     ' The ASCII number is also displayed
  lblTime.Visible = True            ' Display the Time  label
  lblTime.Caption = "The Current Time Is " & Format(Now, "long time")
  If AsciiNum > 90 Then             ' When we reach the letter "Z" ...
    Beep
    tmrTimer.Enabled = False        '... the timer is turned off
    lblAZ.Caption = "That's All!"   '... and a message is displayed
    lblAsciiNo.Visible = False      ' Then the ASCII number is erased,
    lblAscii.Visible = False        '  the ASCII label is erased,
    lblTime.Visible = False         '  and the time label is erased
    Form_Load                       ' This call to Form_Load reinitializes the program
  End If

End Sub
```

Executing the Timer A to Z Program. Click the **Run** button. The top label (**lblInfo**) disappears. A red letter **A** appears on the form followed every 600 milliseconds by another letter of the alphabet. At the bottom of the form, the current time is displayed, and the seconds change. After **Z** appears on the form, the message **"That's All"** appears, and the labels, the ASCII number, and the current time display disappear.

Understanding the Code of the Timer A to Z Program. The procedures in the Timer A to Z program execute in the following order:

```
Private Sub Form_Load()             ' This is the first procedure to run
Private Sub cmdRun_Click()          ' This click starts the timer
Public Sub tmrTimer_Timer()         ' This procedure runs after cmdRun is clicked,
Private Sub cmdExit_Click()         ' To exit the program
```

When the program is run, the first procedure to execute is **Form_Load**. This procedure turns the timer off before the user clicks the **Run** button. It also initializes the value of the integer variable **AsciiNum** to 64. During the course of the program, its value will be incremented by 1 whenever the timer event occurs. At design time the timer **Interval** property was set to 300 milliseconds.

When the **Run** button is clicked, the big label (**lblInfo**) at the top is made invisible, and the ASCII number of the character and its label are made visible. At this point the timer is enabled.

The timer control is visible at design time, but invisible to the user at run time. It has just two properties of interest: the **Interval** property and the **Enabled** property. Both may be set at design time or changed at run time. The timer control's only event is the timer event, so the only procedure for the timer control is the event procedure. The default name for the timer control is **Timer1**. In this program, the timer control is named **tmrTimer**, so its procedure declaration and code looks like this:

```
Public Sub tmrTimer_Timer()
  Statement
  Statement
  Statement
End Sub
```

Every time that the timer control's interval elapses, Visual Basic generates the timer control's timer event, and all the statements in the procedure are executed.

As the timer begins its first 300-millisecond interval, **AsciiNum** is incremented from 64 to 65, and the **Chr(AsciiNum)** function changes this to an "A" and assigns that string value to the string variable **Char**. The caption property of the label, **lblAZ.Text**, is set to **Char** to display the "A". At design time the caption's **Font** property was set to 24 point bold red. The ASCII number itself is also displayed, in blue.

The timer control acts like a control loop in that it executes a block of code every 300 milliseconds (the interval set at design time), until some statement intervenes to disable the timer. The last statement in the timer's block is an **If ... Then** statement that checks how high **AsciiNum** has increased. When **AsciiNum** exceeds 90, the timer control is disabled.

After "Z" is reached (ASCII number equal 90), the next time that the interval lapses, the ASCII number exceeds 90, so **tmrTimer** is disabled, **lblAscii**, and **lblAsciiNo** are made invisible, and the message **"That's All!"** is displayed. The **Form_Load** procedure is called to reinitialize the program. Clicking **Run** repeats the performance.

Named Date and Time Formats

In Chapter 11, the Print Demo program demonstrated custom formatting of numbers, date, and time. In the Timer A to Z program, printing the time on the form resulted in printing the time of the computer's clock at the instant that the program was executed. In the Timer A to Z program, the statement for displaying the time is contained in the **tmrTimer_Timer** procedure.

```
Public Sub tmrTimer_Timer()
   Statement
   lblTime.Caption = "The Current Time Is " & Format(Now, "long time")
   Statement
   Statement
End Sub
```

Every time the specified interval elapses, the current time at the moment the timer event is triggered is displayed as the **lblTime** caption. Since the **Interval** property was set to 300 milliseconds, the time changes more than once per second, and the number of actual seconds displayed is rounded off. To display the time exactly every 1 second, the **Interval** property must be set to 1000. Try changing the interval to various values, and run the program. Try changing the named date and time format from **"Long Time"** to some of the other possibilities listed in Table 13-7.

Table 13-7

Named Date and Time Formats

Named format	Description
General Date	Shows the day, date, and time; for example, 7/18/01 3:45:25 PM
Long Date	Shows the day and date; for example, Wednesday, July 18, 2001
Medium Date	Uses the dd-mmm-yy format; for example, 18-Jul-01
Short Date	Uses the short date format; for example, 7/18/01
Long Time	Displays hours, minutes, and seconds; for example, 1:4:56 PM
Medium Time	Displays hours and minutes; for example, 9:15 AM
Short Time	Displays hours and minutes; for example, 3:45

Looping with the List Box

In Chapter 12, the List Box program demonstrated loading the list box with a fixed number of items with the **Form_Load** procedure. The list box is also a natural choice for displaying output that consists of a list of items. The **AddItem** method within a **For ... Next** loop is an appropriate choice for filling the list box with calculated output. Some simple equations from physics provide something to practice with.

```
For J = 1 To 12
    v = J * (J + 1) * B    ' Rotational energy: Units of E are 1/cm
    E = 100 * v * h * c    'Convert E(1/cm) to v(Joules)

    ' For each set of J, v, and E
    ' add a line with AddItem to a list box named lstOutPutE

    lstOutPutE.AddItem J & vbTab & v & vab & E
Next J
```

The Rotational Energy program calculates a list of values of *v* and *E* and places them in a single list box with some additional formatting.

The Rotational Energy Program

The science underlying the Rotational Energy program was reviewed in Excel Example 10 in Chapter 3. This exercise is a Visual Basic version of Excel Example 10.

Starting the Project

- Start Visual Basic by clicking the **VB** icon on the desktop.
- Select the **Standard EXE** icon in the **New Project** window, and click the **OK** button.
- A **New Project** window opens with a fresh form.
- Select the form, and in the properties window, change the **(Name)** property to **frmRotEnergy**.
- Click **Save**; the file name will be frmRotEnergy, so click **OK**.
- The file name changes to Project1; change it to **The Rotational Energy Program**, and click the **OK** button.
- Place controls on frmRotEnergy according to Figure 13-7.
- Set the properties of the controls according to Table 13-8.

Entering the Code of the Rotational Energy Program. Enter the following code in the general declarations section:

```
Option Explicit
  ' The Rotational Energy program calculates the rotational
  ' energies of a diatomic molecule in wavenumber and
  ' Joules when its rotational constant in wavenumbers is
  ' input. Calculations are for the lowest 12 quantum numbers.
  Const h = 6.626186E-34    ' Planck constant, J s
  Const c = 299792600       ' Speed of light, m/s
```

Enter the following code in the **cmdExit_Click()** procedure:

```
Private Sub cmdExit_Click()
  End
End Sub
```

Table 13-8

The Properties Table for the Rotational Energy Program

Object	Property	Value
Form1	Name	frmRotEnergy
	Caption	The Rotational Energy Program
Label1	Name	lblInfo
	Caption	The Rotational Energy program accepts input from a text box and sends output to a list box
	BorderStyle	1-Fixed Single
Label2	Name	lblInputB
	Caption	Enter B in wavenumbers (1/cm)
Text box1	Name	txtInputB
	Text	Make it blank
	TabIndex	0
List box1	Name	lstOutPutE
Command1	Name	cmdRun
	Caption	&Run
Command2	Name	cmdExit
	Caption	E&xit
Command3	Name	cmdClear
	Caption	Clear B

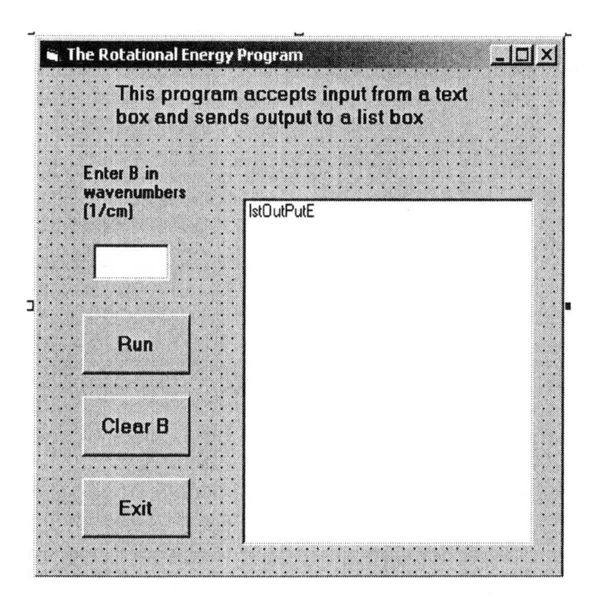

Figure 13-7. The Form for the Rotational Energy Program

Enter the following code in the **cmdClear_Click()** procedure:

```
Private Sub cmdClear_Click()
   txtInputB.Text = ""
   txtInputB.SetFocus
End Sub
```

Enter the following code in the **cmdRun_Click()** procedure:

```
Private Sub cmdRun_Click()

  Dim v As Double            ' Energy in wavenumbers
  Dim E As Double            ' Energy in Joules
  Dim J As Integer           ' Rotational quantum number
  Dim B As Double            ' The rotational constant, 1/cm

  lstOutPutE.Clear           'Clears the list box of any previous output

  On Error GoTo ErrorHandler

    ' Assign the value of B entered in the text box
    ' to the variable B
    B = txtInputB.Text

    ' Insert an empty line in lstOutputE list box for looks
    lstOutPutE.AddItem " "

    ' Print a label for the contents of the list box
    lstOutPutE.AddItem " J" & vbTab & "v(1/cm)" & vbTab & "E(joules)"

    ' Insert an empty line in lstOutputE list box for looks
    lstOutPutE.AddItem " "

    ' Calculate E and v for each rotational level J
    For J = 1 To 12
        v = J * (J + 1) * B          ' Rotational energy: Units of E are 1/cm
        E = 100 * v * h * c          'Convert E(1/cm) to v(Joules)

        ' For each set of J, v, and E
        ' add a line with AddItem to the list box named lstOutPutE

        lstOutPutE.AddItem "  " & J & vbTab & _
                    Format(v, "####.##") & vbTab & _
                    Format(E, "#.###E-00")
    Next J
    Exit Sub

 ErrorHandler:
      MsgBox Str(Err.Number) & ":" & Err.Description, vbExclamation, "Enter Again"
 End Sub
```

Executing the Rotational Energy Program. Click the **Start** button to execute the Rotational Energy program.

- In the text box, enter a value for the rotational energy constant **B** from Table 3-2. For testing, try the value for the HCl molecule: 10.591 cm^{-1}. Click **Run**.
- The list box displays a table of J, v, and E, printed on the form.
- Clear the text box, enter another value for **B** from Table 3-2, and click **Run** again.

Understanding the Code of the Rotational Energy Program. The general declarations section of the program has the usual **Option Explicit** statement to force declaration of variables. Since a listing of the program has the name of the form but not the caption of the form, this is a good place to place the name of the program as a comment. Widely used constants, such as *h* and *c*, are declared here, but could be declared locally. The scope of constants declared in the general declaration section is global. That is, they are known and available to all procedures. Constants (and variables) declared in a private procedure are known only to that procedure.

Execution of the program results in loading the form and its controls. After execution, we want the text box to have focus so the user can immediately enter a new value of **B** and run the program again if desired. The focus of each control depends on the order in which you placed the controls on the form at design time. By setting the **TabIndex** property of **txtInputB** to equal 0 at design time, **txtInputB** will have focus, contain a blinking insertion point, and be ready for the input of a new value of **B**, the rotational constant. Regardless of the order in which you placed the controls on the screen, **txtInputB** will have focus, and we don't care what the order of the tab index is for the other controls.

Commenting. Clicking the **Run** button executes the **cmdRun_Click()** procedure. The variables are declared, and the units of the physical quantities are explained with comments. Comments clarifying the units are especially important in calculations in science and engineering. Comments explaining individual statements increase greatly in importance as time goes by.

On Error GoTo. The **cmdRun_Click()** procedure is placed between the **On Error GoTo Error (Label)** statement and the **Exit Sub** statement. The label chosen here is **ErrorHandler**, which is often used. But you could use for the label **ErrorTrap**, **ErrorInfo**, **YourError**, or something else that conveys the meaning.

```
        On Error GoTo ErrorHandler

        The cmd Run_Click ( ) procedure is placed here

        Exit Sub

ErrorHandler:

        MsgBox Str(Err.Number) & ":" & Err.Description, vbExclamation, "Enter Again"

EndSub
```

statement: **B = txtInptB.Text**

This bit of code protects the program from crashing (run-time error) caused by the user entering faulty data for the value of the variable **B**. These include clicking **Run** before entering any datum and typographical errors such as typing a letter instead of a number. It does not protect against the accidental entry of a comma instead of a period. If the user enters 10,5909 instead of the intended 10.5909, the program will ignore the comma and read the value as 105909.

- The **lstOutPutE.Clear** statement clears the list box of any previous output from a possible preceding run of the program.
- The value of **txtInput.Text** input by the user is assigned to **B** with the statement **B = txtInptB.Text.**

The next three statements in the **Else** block serve to improve the appearance and readability of the output that will appear in the list box.

The AddItem Method. The **AddItem** method of the list box does just that: it adds an item to the list. The argument of **AddItem** may be a string or a variable or a combination of both. The argument is not, however, optional. To add an empty line, the argument must be an empty string, "", or a space, " ".

- **LstOutPutE.AddItem** "" inserts an empty line.
- **LstOutPutE.AddItem " J" & vbTab & "v(1/cm)" & vbTab & "E(joules)"**. The item added is a string, which is a label in the list box for the columns of numerical output to come next. The & is the concatenation operator, and **vbTab** is a Visual Basic string constant. It inserts a horizontal tab.
- **LstOutPutE.AddItem** "" inserts an empty line.

A **For ... Next** loop calculates the value of the rotational energy V in wavenumbers (cm^{-1}) from the statement:

```
v = J * (J + 1) * B
```

In this program V (vee) is used instead of the conventional scientific symbol $\tilde{\nu}$ (Greek nu bar). The value of ν wavenumbers (cm^{-1}) is converted to E (joules) in the next statement:

```
E = 100 * v * h * c
```

$$E(J) = cm^{-1} \times 100\, cm\, m^{-1} \times J\, s \times m\, s^{-1}$$

After each cycle of the loop for the counter J, the **lstOutPutE.AddItem** adds the current values of J, v, and E to a single line in the list box, with formatting to control the appearance of the output. The user can run the program again by clicking the **Clear B** button, entering a new value of **B**, and clicking the **Run** button. A new set of 12 energies will be written to the list box.

Programmers writing scientific programs encounter numbers ranging from small integers to decimal numbers to very large or very small numbers that require formatting in scientific notation (exponential notation). Visual Basic furnishes a variety of methods for accommodating just about any desired format. The Rotational Energy program illustrates the typical problem. The line

```
lstOutPutE.AddItem "  " & J & vbTab & v & vbTab & E
```

adds three numeric items to the list box, J, v, and E, which are separated by tabs. The list box displays the three column of number in their default format.

The display of the integer J is satisfactory, but the display of v and E shows too many decimal places. The Visual Basic **Format** function converts a number into a string and allows you to control the number of digits and the location of the decimal point and other formatting symbols. Custom formatting numbers and dates was described earlier with the Print Demo program. With custom formatting, the desired format is achieved:

```
lstOutPutE.AddItem "  " & J & vbTab & _
        Format(v, "####.##") & vbTab & _
        Format(E, "#.###E-00")
```

In addition to custom formatting, Visual Basic provides for "semicustom" formatting with the use of named number formats and named date and time formats, both described earlier in the Print Demo program (Tables 11-10 and 11-12).

User-Defined Functions

In Chapter 12, the Functions program illustrated many of Visual Basic's built-in functions, such as **Abs(x)**, **Cos(x)**, or **Exp(x)**. Recall that a function is a procedure that returns a value. The value may be of any type. Often one or more parameters are passed to the function, but that is optional. The usual form of a function procedure is:

```
Public Function name (parameter list)
   Statement
   Statement
   Statement
   name = expression
End Function
```

The name of the function can be any legal variable name. The two occurrences of **name** in the function procedure must be identical. The parameter list consists of the parameters (separated by commas) passed to the function when it is called. As with the **Sub** procedure, the declaration may be **Public** or **Private**. Any number of statements and data and control structures can follow the function, but the very last statement must be an assignment statement to the name of the function. As an example, let's write a function for calculating the volume of a parallelepiped, the volume of which is given by Volume = length x width x height.

```
Public Function Volume (L As Single, W As Single, H As Single)
   V = L*W*H
   Volume = V
End Function
```

The call to the function is simply the assignment of the function (with parameters) to a variable:

```
V = Volume(23.2, 15.3, 8.7)
```

Alternatively, the following call works:

```
len = 23.2
wid = 15.8
hgt = 8.7
V = Volume (len,wid,hgt)
```

The names in the call parameter list do not have to match the names in the declaration parameter list. It is their position and value that matters.

In the Color Properties program, in Chapter 10, we have already encountered a built-in function with three parameters: the **RGB(R, G, B)** function. This function returns a color depending on the three integers **R**, **G**, and **B**, each of which ranges from 0 to 255. The built-in function **Cos(x)** accepts a single parameter, the angle in radians, and returns the cosine of the angle.

The Vapor Pressure Program

The thermodynamics of liquid-vapor equilibrium is reviewed in Excel Example 4 in Chapter 2. A table of Antoine constants is also found in Chapter 2, Excel Example 4 (Table 2-2).

Starting the Project

- Start Visual Basic by clicking the **VB** icon on the desktop.
- Select the **Standard EXE** icon in the **New Project** window, and click the **OK** button.
- A **New Project** window opens with a fresh form.
- Select the form, and in the properties window, change the (**Name**) property to **frmPvap**.
- Click **Save**; the file name will be frmPvap, so click **OK**.
- The file name changes to Project1; change it to **The Vapor Pressure Program**, and click the **OK** button.
- Place controls on frmPvap according to Figure 13-8.
- Set the properties of the controls according to Table 13-9.

Table 13-9

The Properties Table for the frmPvap Form

Object	Property	Value
Form1	Name	frmPvap
	Caption	The Vapor Pressure Program
Label1	Name	lblInfo
	Caption	The Vapor Pressure program demonstrates the user-defined function
	BorderStyle	1-Fixed Single
	Font	Sans serif, bold,10 point
Label2	Name	lblDirections
	Caption	Run or enter your A, B, and C
	Font	Sans serif, bold, 8 point
Label3	Name	lblA
	Caption	A
Label4	Name	lblB
	Caption	B
Label5	Name	lblC
	Caption	C
Text box1	Name	txtA
	Text	7.11714
Text box2	Name	txtB
	Caption	120.595
Text box3	Name	txtC
	Caption	229.664
Command1	Name	cmdRun
	Caption	&Run
Command2	Name	cmdExit
	Caption	E&xit

Entering the Code of the Vapor Pressure Program. Enter the following code in the general declarations section of the program:

```
Option Explicit
' The Vapor Pressure program
'This program demonstrates the user-defined function
' The user inputs three Antoine constants to three text boxes
' The text box.Text default values are for the vapor pressure of acetone
' The program prints the output to the form
```

To enter the code for the function, do the following:

- On the **Tools** menu, click **Add Procedure** to open the **Add Procedure** dialog box (Figure 13-9).

- In the **Name** box, type **VapPressure** as the name.
- In the **Type** section, select **Function**.
- In the **Scope** section, select **Public**.
- Click **OK**.

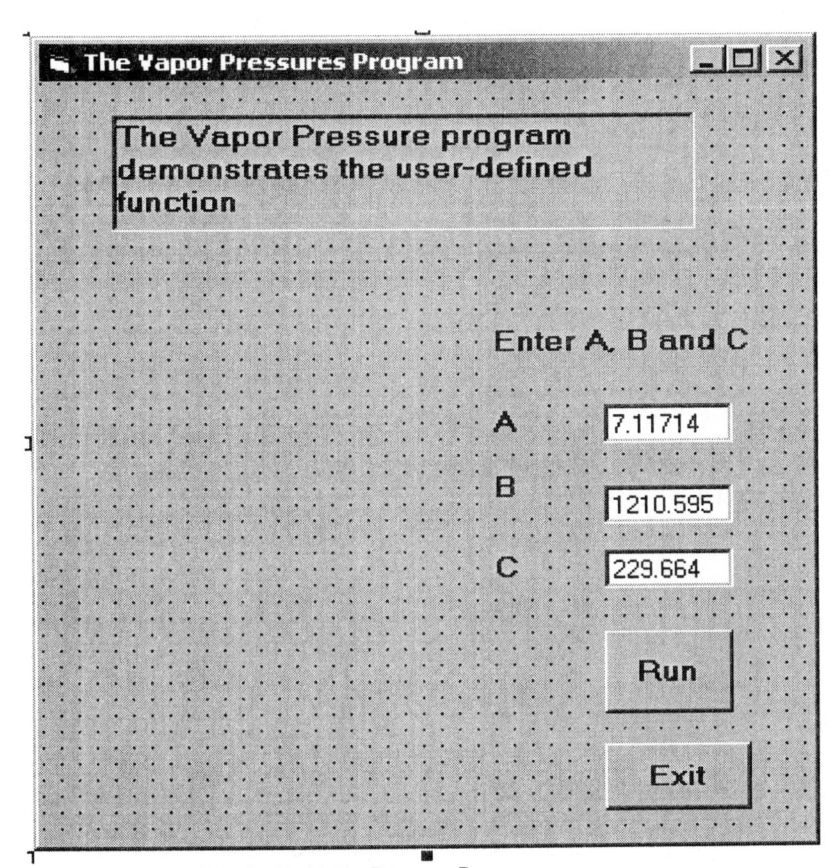

Figure 13.8. The Form for the Vapor Pressure Program

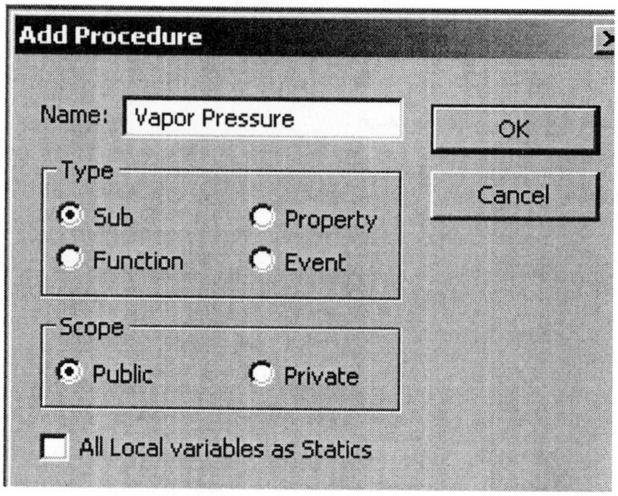

Figure 13-9. The Add Procedure Dialog Box

Enter the following code for the function:

```
Public Function VapPressure(t As Double)
        Dim A As Double            ' These are the constants
        Dim B As Double            ' in the Antoine vapor pressure equation
        Dim C As Double            ' for acetone

        A = txtA                   ' The user can change the default values
        B = txtB                   ' in the text boxes
        C = txtC                   ' for A, B, and C

        VapPressure = 10 ^ (A - B / (t + C))   ' Antoine vapor pressure equation

    End Function
```

Enter the following code in the **cmdExit_Click()** procedure:

```
Private Sub cmdExit_Click()
  End
End Sub
```

Enter the following code in the **cmdRun_Click()** procedure:

```
Public Sub cmdRun_Click()

  Dim VP As Double                                  ' Vapor pressure im mm Hg
  Dim SpaceNo As Integer, DataNo As Integer         ' For ... Next loop counters
  Dim deg As Double                                 ' Temperature in deg C

  frmPvap.Cls                                       ' Clear the form of any
                                                    ' previous output

  For SpaceNo = 1 To 6                              ' Insert 6 spaces before
    frmPvap.Print                                   ' printing output
  Next
```

```
    deg = 0
    frmPvap.Print "          Vapor Pressures"                    ' Title for output
    frmPvap.Print "     t/deg C"; "     Vapor Pressure/mm     ' Title for output
    frmPvap.Print                                              ' blank line
    For DataNo = 1 To 13
      ' Calculate every 5 deg from 5 up
      deg = deg + 5

          ' Call the VapPressure function in a Print statement, and print a table of
          ' temperatures and vapor pressures

      frmPvap.Print vbTab & deg & vbTab & Format(VapPressure(deg), "##.#")
    Next

End Sub
```

Proof your work and save it.

Executing the Vapor Pressure Program. Click **Run**, and the program prints the vapor pressure of acetone from 5 to 65 deg C (Figure 13-10). Or enter different Antoine constants for another substance, and then click **Run**. Some selected constants are listed in Excel Example 4. See Table 2-2.

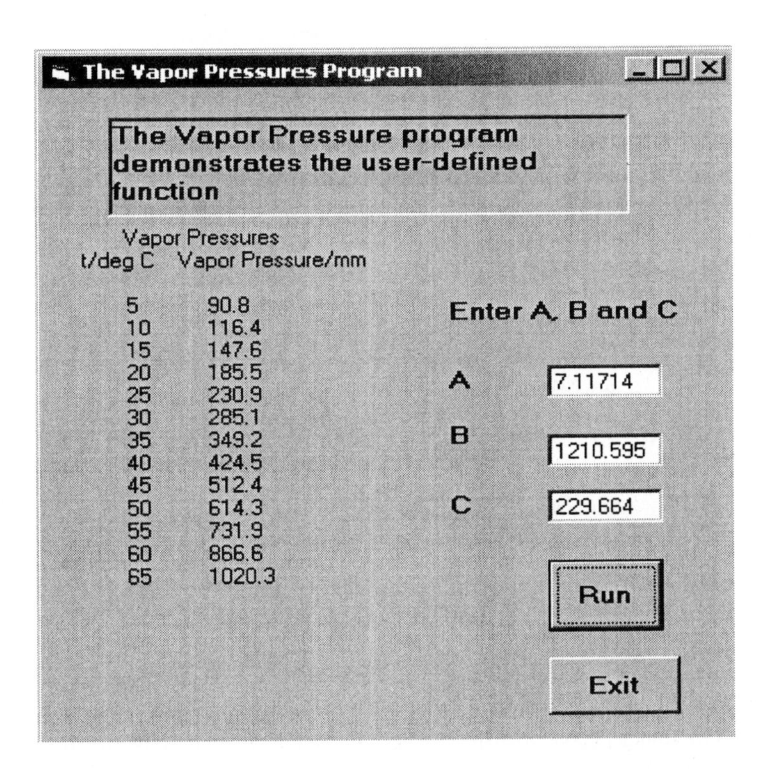

Figure 13-10. Output of the Vapor Pressure Program Printed on the Form

Understanding the Code of the Vapor Pressure Program. The Antoine equation (Equation 2-22) for vapor pressure is of the form:

$$\log P = A - \frac{B}{t+C}$$

We wish to calculate the vapor pressures from the Antoine equation, so the function we wish to program is

$$P = 10^{\left[A - \frac{B}{t+C}\right]}.$$

The VB form of a user-defined function is

```
Public Function NameOfFunction ( )
        Dim ....                              ' Dim declarations are optional
        Statement1                            ' As many statement blocks as you like
        Statemnts2
        .
        .
        NameOfFunction = an expression        ' The assignment of a value to the
End Function                                  ' function is a mandatory
                                              ' last statement
```

The last statement in a function must be a **NameOfFunction** that is identical to the **NameOfFunction** in the function header. The main difference between a procedure and a function is that a function always returns a value, while a procedure does not. Passing one or more parameters to either a function or a procedure is optional. In our Vapor Pressure program, the **NameOfFunction** is **VapPressure**, and one parameter is passed to the function, the temperature **(t As Double)**. The syntax for a function call is

	X **= NameOfFunction**
Or	***X*** **= NameOfFunction (*expression1*, *expression2*,...)**
for example,	***X*** **= NameOfFunction (13.43, Sin(.876))**

In our Vapor Pressure program, the call is of the form:

X **= VapPressure (*deg*)**

where ***deg*** is the incremented temperature in a loop ranging from 5 to 65. Notice that the parameter passed (***deg***) can have a different variable name than that in the function parameter declaration (**t As Double**), but the type must match.

We call the function and assign the value returned by the function to the **Print** property of the form:

```
frmPvap.Print VapPressure (deg)
```

We actually want to print to the form both the temperature and pressure, separate them by tabs, and format the numbers so that the complete call is

```
frmPvap.Print vbTab & deg & vbTab & Format(VapPressure(deg), "##.#")
```

The **frmPvap.Print** statement prints the value returned by the call 13 times as the temperature is incremented 5 degrees for each cycle of the loop.

Before entering the loop, the program inserts six blank lines, so the printing begins below the label (**lblInfo**) and then clears the form of any previously printed output if different values of **A**, **B**, and **C** were entered for more than one run. Recall that objects have properties and methods. One of the form's methods is **Cls**, a method for clearing the form. The form's name property is **frmPvap**, so the statement that uses the **Cls** method for clearing the form is

```
frmPvap.Cls
```

Notice that when you enter this code, at the instant that you enter the period after **frmPvap**, a small window appears on your screen, showing the properties and methods of the object just named (**frmPvap**). If you scroll to and select **Cls** and then touch the SPACEBAR, Visual Basic automatically places the **Cls** after the period. Notice also that when you select **Cls**, VB gives a short description of what **Cls** is. This is handy when you can't remember the exact syntax of the procedure or method. See Figures 10-13 and 10-14.

Chapter 14

Visual Basic Loops with Arrays

An *array* is a list (or table) of variables of the same type. You can use a one-dimensional array (a list) to store a list of strings or numbers or dates, as long as the data type is the same. The items in a list are the *elements* of the array. A two-dimensional array is a table, and you can use it to store a table of strings or numbers or dates, as long as the data type is the same.

First we shall look at one-dimensional arrays, their characteristics, and their use in programs. Later on we shall look at some examples of two-dimensional arrays. This chapter also introduces the combo box and compares it with the text box and list box. Several programs illustrate the topics of this chapter: the Array of Pressures program, the Text Box to List Box program, the Combo Box to Array program, and the 2-D Array program are demonstration programs. The TotoLoto program picks a set of numbers for playing the lottery, and the Bubble Sort program illustrates a well-known sorting method.

The IR Spectrum program, the Heat Capacity program, and the DesLandres program illustrate some scientific calculations from spectroscopy and thermodynamics. The Trapezoid Integration program calculates the area under a curve defined by a set of points. The Visual Basic programs in this chapter are:

The Array of Pressures program
The Heat Capacity program
The Combo Box to Array program
The 2-D Array program
The Bubble Sort program
The IR Spectrum program
The Text Box to List Box program
The Trapezoid Integration program
The TotoLoto program
The DesLandres program

Arrays

The elements of an array have the same name as the array itself but are characterized by an integer index that gives the position of the element in the array.

A simple variable type can store a single value. An array is a structured data type that can hold more than one value. An array is characterized by its size and type, as shown in the sample declaration of an array designated **Pressure(*n*)** where ***n*** is an integer.

```
Dim Pressure (50) As Single
```

The name of the array is **Pressure**, and it may store 51 numerical values (of the **Single** type) of pressures. Unless otherwise specified, the index begins at zero and runs to the declared maximum number of elements. It is not necessary to use the default value of zero for the first index value, as shown by the following legal declarations:

```
Dim Pressure (1 To 50) As Single
Dim Temperature (-100 To 300) As Single
```

Values are assigned to each element of the array just as a value is assigned to any variable. For example:

```
Pressure (14) = 761.24
```

Usually the elements of an array are filled with a **For ... Next** statement, the counter of which is the index of the array.

```
Dim P (50) As Single
Dim Pinit As Single
Dim J As Integer
Pinit = 770.00
For J = 1 to 5
Pressure(J) = Pinit – 10
Next J
```

The values stored in the array **Pressure(50)** are the same as if the following assignment statements were used:

```
Pressure(1) = 760
Pressure(2) = 750
Pressure(3) = 740
Pressure(4) = 730
Pressure(5) = 720
```

As already mentioned, a one-dimensional array is also called a list. The list of a list box (which was discussed earlier) or a combo box (which will be discussed later) is an indexed one-dimensional array. One-dimensional arrays are frequently encountered in all programming languages and are extremely useful.

The Array of Pressures Program

The Array of Pressures program is designed to illustrate the use of arrays in numerical calculations. It shows how to declare an array and how to load an array, that is, how to assign values to its elements. The program then gives an example of operating on the elements of one array and assigning the new values to the elements of another array: the list of the list box. The program gives further practice in formatting numbers. Subsequent programs in this book will make frequent use of arrays.

Starting the Project

- Start Visual Basic by clicking the **VB** icon on the desktop.
- Select the **Standard EXE** icon in the **New Project** window, and click the **OK** button.
- A **New Project** window opens with a fresh form.

- Select the form, and in the properties window, change the **(Name)** property to **frmArrayDemo**.
- Click **Save**; the file name will be frmArrayDemo, so click **OK**.
- The file name changes to Project1; change it to **The Array of Pressures Program**, and click the **OK** button.
- Place controls on frmArrayDemo according to Figure 14-1, and size them accordingly.
- Set the properties of the controls according to Table 14-1.

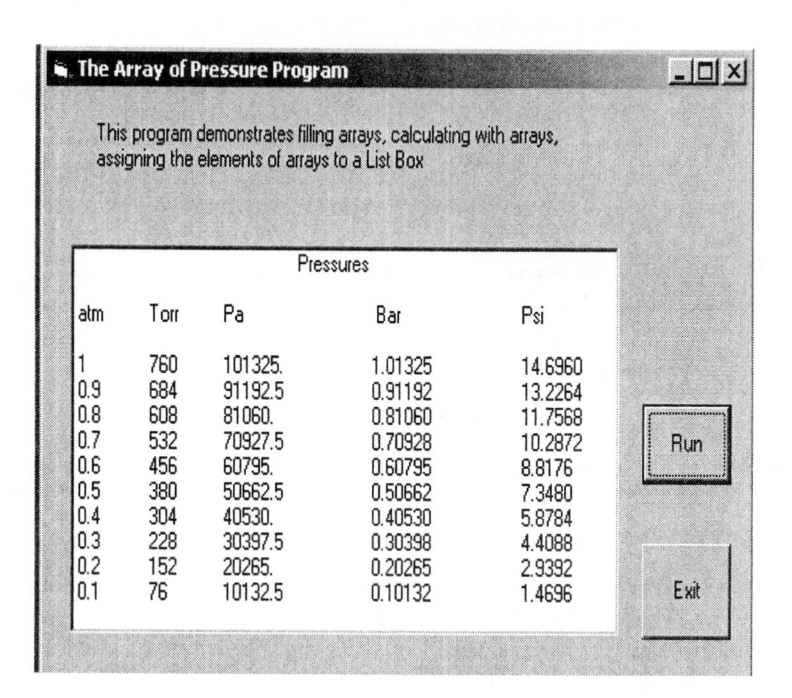

Table 14-1

The Properties Table for the Array of Pressures Program Form

Object	Property	Value
Form1	Name	frmArrayDemo
	Caption	The Array of Pressures Program
Label1	Name	lblInfo
	Caption	This program demonstrates filling arrays, calculating with arrays, assigning the elements of arrays to a list box
	BorderStyle	1-Fixed Single
Command1	Name	CmdRun
	Caption	&Run
Command2	Name	cmdExit
	Caption	E&xit
List box1	Name	LstP

Entering the Code of the Array of Pressures Program. Enter the following code in the general declaration section:

```
' The Array of Pressures program
' This program demonstrates the use of arrays in numerical calculations, filling
' arrays, assigning array elements to a list box, and formatting numbers.
Option Explicit
Const PtoTorr = 760                ' Convert atm to Torr
Const PtoPa = 101325               ' Convert atm to Pascal
Const PtoBar = 1.01325             ' Convert atm to Bar
Const PtoPsi = 14.696              ' Convert atm to pounds/sq in
```

Enter the following code in the **cmdExit_Click()** procedure:

```
Private Sub cmdExit_Click()
  End
End Sub
```

Enter the following code in the **cmdRun_Click()** procedure:

```
Private Sub cmdRun_Click()
  Dim J As Integer
  Dim P(30) As Single
  Dim Torr(30) As Single
  Dim Pa(30) As Single
  Dim Pbar(30) As Single
  Dim Ppsi(30) As Single
  AddColumnTitles                  ' Assigns title and column labels to the list box
  For J = 0 To 9
    P(J) = 1 - J * 0.1             ' Fills P(J) with P's 1.0, 0.9, 0.8, and so on
    Torr(J) = P(J) * PtoTorr       ' Fills Torr(J) with P's in Torr
    Pa(J) = P(J) * PtoPa           ' Fills Pa(J) with P's in Pascals
    Pbar(J) = P(J) * PtoBar        ' Fills Pbar(J) with P's in Bar
    Ppsi(J) = P(J) * PtoPsi        ' Fills Ppsi(J) with P's in lb/sq in

  ' Now assign all these pressures to a list box for display
  lstP.AddItem P(J) & vbTab & Torr(J) _
  & vbTab & Format(Pa(J), "##.#####") & vbTab _
  & vbTab & Format(Pbar(J), "0.00000") & _
  vbTab & vbTab _
  & Format(Ppsi(J), "##.0000")
  Next
End Sub
```

Enter the following code in the **AddColumnTitles** procedure. To do this, on the **Tools** menu, click **AddProcedure** to open the **Add Procedure** dialog box. In the **Name** box, type **AddColumnTitles** as the name. In the **Type** section, select **Sub**; and in the **Scope** section, select **Public**. Click **OK**, and the **AddColumnTitles** procedure declaration appears in your code window.

```
Public Sub AddColumnTitles()       ' Assigns title and column labels to the list box
  lstP.AddItem vbTab & vbTab & vbTab & "Pressures"
  lstP.AddItem ""
  lstP.AddItem "atm" & vbTab _
      & "Torr" & vbTab & "Pa" & vbTab _
      & vbTab & " Bar" & vbTab & vbTab & "Psi"
  lstP.AddItem ""
End Sub
```

Executing the Array of Pressures Program

- Click the **Start** button to execute the Array of Pressures program.
- Click **Run**.
- Under a title and column labels, the program prints four lists of pressures in units of atm, Torr, Pascal, Bar, and lbs/in^2, as shown in Figure 14-1.

Understanding the Code of the Array of Pressures Program

Running the program opens the form with an information label, a **Run** button, and an **Exit** button displayed to the user. Conversion factors for converting atm to Torr, Pascal, Bar, and lbs/in^2 are declared in the general declarations section of the program:

```
Option Explicit
Const PtoTorr = 760              ' Convert atm to Torr
Const PtoPa = 101325             ' Convert atm to Pascal
Const PtoBar = 1.01325           ' Convert atm to Bar
Const PtoPsi = 14.696            ' Convert atm to pounds/sq in
```

Clicking the **Run** button causes execution of the **cmdRun_Click** procedure.

- The loop counter *J* and five arrays are declared.
- The **AddColumnTitles** procedure is called. With the **AddItem** method of the list box, this procedure writes the title **"Pressures"** on one line, and below a blank line, it writes the column headings **atm, Torr, Pa, Bar,** and **Psi.**
- A single **For ... Next** loop does the array calculations:
  - *P*(*J*) is filled with pressures in atm from 1.00 down to 0.1.
  - *P*(*J*) is converted to Torr and stored in Torr(*J*).
  - *P*(*J*) is converted to Pascal and stored in Ps(*J*).
  - *P*(*J*) is converted to Bar and stored in Bar(*J*).
  - *P*(*J*) is converted to lbs/in^2 and stored in Psi(*J*).

The last line of the loop assigns each value of pressure to a list box with the **AddItem** method of the list box. The items in the **AddItem** list are separated by tabs so that the beginning of each number lines up with the column labels, all of which are also separated by tabs.

The IR Spectrum Program

The quantum mechanics of rotational energies was reviewed in Excel Example 12 in Chapter 3. In this Visual Basic program, we revisit Excel Example 12. Figure 3-14 shows the energy level diagram of HCl and the origin of the infrared (IR) spectrum.

Starting the Project

- Start Visual Basic by clicking the **VB** icon on the desktop.
- Select the **Standard EXE** icon in the **New Project** window, and click the **OK** button.
- A **New Project** window opens with a fresh form.
- Select the form, and in the properties window, change the **(Name)** property to **frmIRSpec**.
- Click **Save**; the file name will be frmIRSpec, so click **OK**.
- The file name changes to Project1; change it to **The IR Spectrum Program**, and click the **OK** button.
- Place controls on frmIRSpec according to Figure 14-2.
- Set the properties of the controls according to Table 14-2.

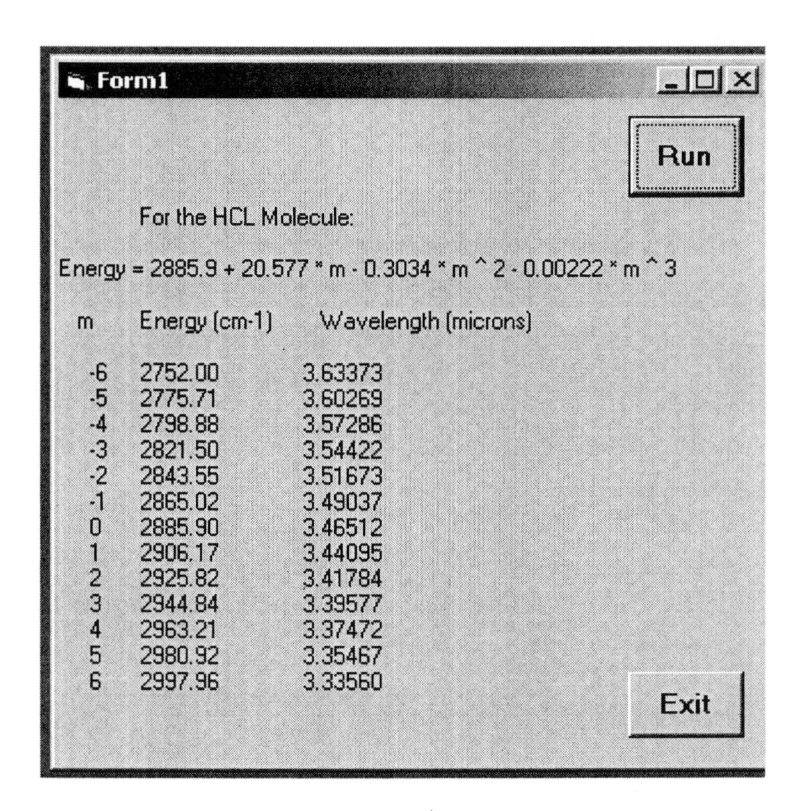

Figure 14-2. The Form for the IR Spectrum Program

Table 14-2

The Properties Table for the IR Spectrum Program Form

Object	Property	Value
Form1	Name	frmIRSpec
	Caption	The IR Spectrum Program
Command1	Name	CmdRun
	Caption	&Run
Command2	Name	cmdExit
	Caption	E&xit

Entering the Code of the IR Spectrum Program. Enter the following code in the general declarations section:

```
Option Explicit
' The IR Spectrum program
' This program demonstrates the use of arrays in numerical calculations.
' It give further practice with functions and formatting numbers.
 Dim EnergyAry(50) As Double                ' Declare array of energies
 Dim LengthAry(50) As Double                ' Declare array of wavelengths
```

Enter the following code in the **cmdExit_Click()** procedure:

```
Private Sub cmdExit_Click()
   End
End Sub
```

Enter the following code in the **WriteLabels()** procedure:

```
Public Sub WriteLabels()
   'This procedure writes a title and labels above the printed output
   Dim Index As Integer
   For Index = 1 To 4                                    ' This inserts 4 blank lines
      frmIRSpec.Print
   Next
   frmIRSpec.Print vbTab & "For the HCL Molecule:"
   frmIRSpec.Print
   frmIRSpec.Print _
   "Energy = 2885.9 + 20.577 * m - 0.3034 * m ^ 2 - 0.00222 * m ^ 3"
   frmIRSpec.Print
   ' Print title lines over the output columns
   frmIRSpec.Print Space(3) & "m" _
          & vbTab & "Energy (cm-1)" _
          & Space(8) & "Wavelength (microns)"
   frmIRSpec.Print              ' Insert a blank line
End Sub
```

Enter the following code in the **cmdRun_Click()** procedure:

```
Private Sub cmdRun_Click()
   Dim Index As Integer                          ' Declare a loop counter
   For Index = 0 To 12                           ' Load an array with energies
      EnergyAry(Index) = SpecLine(Index - 6)     ' Call the SpecLine function
   Next
   'Convert Energy in cm-1 to Wavelength in microns)
   For Index = 0 To 12
      LengthAry(Index) = 10000 / EnergyAry(Index)
   Next
   WriteLabels                                   ' Prints a title and labels about printed output
   For Index = 0 To 12                           ' Prints the output
      frmIRSpec.Print Space(5) & Index - 6 _
      & vbTab & Format(EnergyAry(Index), "#####.00") & vbTab _
      & vbTab & Format(LengthAry(Index), "#.00000")
   Next

End Sub

'This function calculates the transition energies
'for the infrared absorption spectrum of the HCl molecule
Public Function SpecLine(m As Integer)
   SpecLine = 2885.9 + 20.577 * m - 0.3034 * m ^ 2 - 0.00222 * m ^ 3
End Function
```

Executing the IR Spectrum Program

- Click the **Start** button on the VB menu bar to execute the IR Spectrum program.
- Click **Run**.
- The program prints the values of the index and the elements of the two arrays (**EnergyAry** and **LengthAry**) on the form.
- Click **Exit** to end the program.

Understanding the Code of the IR Spectrum Program. The object of the program is:

- To fill the elements of the **EnergyAry** array with the energies of the HCL IR spectrum calculated with a simple user-defined function.
- To convert the energies in cm^{-1} stored in **EnergyAry** to wavelengths in microns and store them in the array named **LengthAry**.
- To print to the form three columns, the values of the parameter *m* and the elements of **EnergyAry** and **LengthAry**, along with column titles.

The **SpecLine** function calculates the energies in the IR absorption spectrum of the hydrogen chloride molecule. The parameter *m* equals zero at the center of the spectrum and ranges from negative to positive integers. To calculate the absorption line at the band center and six lines on each side, *m* must range from –6 to +6. At the same time, the index for the array **EnergyAry** must make room to store these thirteen values by ranging from 0 to 12. The energies of the absorption line in cm^{-1} is given by Equation 3-53 in Chapter 3.

$$\tilde{v} = 2885.9 + 20.577m - 0.3034m^2 - 0.00222m^3$$

This is the basis for the function named **SpecLine**.

```
Public Function SpecLine(m As Integer)
        SpecLine = 2885.9 + 20.577 * m - 0.3034 * m ^ 2 - 0.00222 * m ^ 3
End Function
```

The first **For ... Next** loop fills each element of the array **EnergyAry** with the energies calculated with function **SpecLine**.

```
For Index = 0 To 12                              ' Load an array with energies
  EnergyAry(Index) = SpecLine(Index - 6)     ' Call the SpecLine function
Next
```

Notice that the call to the function passes the parameter (**Index - 6**), so that on the first call **Index** equals 0, (**Index - 6**) equals –6, and this is the first value of *m* for which an energy is calculated and assigned to the array element **EnergyAry(0)**.

The second **For ... Next** loop converts each element of **EnergyAry** in cm^{-1} to microns and assigns the value to **LengthAry**. A micron equals 10^{-6} meter, and a meter equals 100 cm.

$$E\,micron = \frac{(10^6\,micron\,meter^{-1})(10^2\,meter\,cm^{-1})}{E\,cm^{-1}} = \frac{10000}{E\,cm^{-1}} \tag{14-1}$$

```
'Convert energy in cm-1 to wavelength in microns)
For Index = 0 To 12
  LengthAry(Index) = 10000 / EnergyAry(Index)
Next
```

A call to the **WriteLabels** procedure prints the title and column labels before printing the output to the form.

The last **For ... Next** loop prints the value of m (which equal **Index –6**), the energy in cm^{-1}, and the wavelength in microns. The columns of values of the array elements are separated by **vbTab**, a string constant that inserts a horizontal tab. Another **vbTab**, placed before the index values, causes the printing to begin one tab to the right of the first column. Format functions present the output numbers with the proper precision.

The Heat Capacity Program

The statistical thermodynamics treatment of the energy and heat capacity was reviewed in Excel Example 14 and Excel Example 15 in Chapter 4. In this Visual Basic exercise we calculate the heat capacity and internal energy of a diatomic molecule.

The total energy E of any molecule equals the sum of the external (translational) and internal (rotational, vibrational, and electronic) contributions. Since the electronic contribution is zero for most ordinary molecules having no unpaired electrons, the total energy is given by

$$E(\text{tot}) = E(\text{trans}) + E(\text{rot}) + E(\text{vib}) \tag{14-2}$$

A diatomic molecule has three translational degrees of freedom, two rotational degrees of freedom, and one vibrational degree of freedom. Each translational and rotational degree of freedom contributes $RT/2$ to the total energy for a total of $5RT/2$. A complete vibrational degree of freedom contributes a maximum of RT to the total energy, but at moderate temperature, the contribution is less by a factor given by the Einstein function for energy

$$E(\text{tot}) = \frac{5RT}{2} + RT\frac{xe^x}{e^x - 1} \tag{14-3}$$

where

$$x = \frac{hc\tilde{v}}{kT} \tag{14-4}$$

and h is Planck's constant, c is the speed of light, k is the Boltzmann constant, $\tilde{v}$ is the fundamental vibration frequency, R is the gas constant, and T is the temperature (K).

$h = 6.6262^{-34}$ J s^{-1}
$k = 1.3807^{-23}$ J K^{-1}
$c = 299792500$ m s^{-1}
$R = 8.3144$ J K^{-1} mol^{-1}

The heat capacity at constant volume is defined as

$$C_v = \left(\frac{\partial E_v}{\partial T}\right)_V \tag{14-5}$$

so differentiation of Equation 14-5 gives

$$C_V = \frac{5R}{2} + R\frac{x^2 e^x}{(e^x - 1)^2} \tag{14-6}$$

where $\frac{x^2 e^x}{(e^x - 1)^2}$ is another Einstein function. Both Einstein functions vary from 0 to 1 as x varies from small values (high T, small $\tilde{V}$) to large values (low T, large $\tilde{V}$).

Equations 14-3 and 14-6 are the equations for the total energy and total heat capacity of a diatomic molecule. In the following program these two equations are declared as Visual Basic functions. With a temperature range and intervals, and a vibration frequency selected by the user, the program calculates the total energy and total heat capacities over the desired temperature range.

Starting the Project

- Start Visual Basic by clicking the **VB** icon on the desktop.
- Select the **Standard EXE** icon in the **New Project** window, and click the **OK** button.
- A **New Project** window opens with a fresh form.
- Select the form, and in the properties window, change the (**Name**) property to **frmHeatCap**.
- Click **Save**; the file name will be **frmHeatCap**, so click **OK**.
- The file name changes to Project1; change it to **The Heat Capacity Program**, and click the **OK** button.
- Place controls on frmHeatCap according to Figure 14-3.
- Set the properties of the controls according to Table 14-3.

Entering the Code of the Heat Capacity Program. Enter the following code in the general declarations section:

```
Option Explicit
'This program calculates the energies and heat capacities of a diatomic molecule
' from a fundamental vibration frequency and upper and lower temperature limits.

Const h As Double = 6.6262E-34              ' Planck's constant   J/s
Const k As Double = 1.3807E-23              ' Boltzmann constant  J/K/molecule
Const c As Long = 299792500                 ' Speed of light    m/s
Const R As Double = 8.3144                  ' Gas constant      J/K mole
Dim Tlow As Double, Thigh As Double         ' Lowest and highest temperature K
Dim v As Double                             ' Fund. vib. freq. in wavenumbers (1/cm)
Dim Nint As Integer                         ' Number of temperature intervals
Dim AryE(20) As Double, AryC(20) As Double, AryT(20) As Double
' These are arrays of energy, heat capacity, and temperature
```

Enter the following code in the **cmdExit_Click()** procedure:

```
Private Sub cmdExit_Click()
  End
End Sub
```

Enter the following code in the **cmdPrint_Click()** procedure:

```
Private Sub cmdPrint_Click()        ' This prints to the printer parallel arrays
  Dim J As Integer                  ' of temperature, energy, and heat capacity
  Printer.Print
  Printer.Print
```

```
   Printer.Print "      T/K      Energy J/mole     Cv J/K mole"
   Printer.Print
   For J = 1 To Nint + 1
      Printer.Print Tab(6); Format(AryT(J), "####.0"); _
              Tab(18); Format(AryE(J), "######.0"); _
              Tab(32); Format(AryC(J), "###.00")
   Next
   Printer.EndDoc               'Tells printer we're at end of document and it's OK to print
End Sub
```

Enter the following code in the **cmdRun_Click()** procedure:

```
Private Sub cmdRun_Click()

   Dim Tint As Double, T As Double, v As Double, x As Double
   Dim J As Integer
   frmHeatCap.Cls
   Thigh = txtThigh.Text                 ' Get from user via text box
   Tlow = txtTLow.Text                   ' Get from user via text box
   Nint = txtTint.Text                   ' Get from user via text box
   v = txtNuBar.Text                     ' Get from user via text box

   Tint = (Thigh - Tlow) / Nint          ' The temperature interval
                                         ' between calculations
   ' Calculate all the energies and Cvs and load them into arrays
   lblHeader.Visible = True              ' was invisible before running
   T = Tlow                              ' Set initial temper
   For J = 1 To Nint + 1
      x = 100 * h * c * v / (k * T)      ' x is used in the Cv Einstein function
      AryT(J) = T                        ' Loads array with temperatures
      AryE(J) = Etotal(T, x)             ' Loads array with energies
      AryC(J) = CvTotal(x)               ' Loads array with heat capacities
      T = T + Tint
   Next

   ' Print all the temperatures, energies, and Cvs to the form (frmHeatCap)
   For J = 1 To 14
      frmHeatCap.Print                   '  Insert 14 blank lines before printing to form
   Next

   For J = 1 To Nint + 1
      frmHeatCap.Print Tab(6); Format(AryT(J), "####.0"); _
              Tab(18); Format(AryE(J), "######.0"); _
              Tab(32); Format(AryC(J), "###.00")
      T = T + Tint
   Next

End Sub
```

To enter the code for the **Etotal** function, do the following:

- On the **Tools** menu, click **AddProcedure** to open the **Add Procedure** dialog box (Figure 13-9).
- In the **Name** box, type **Etotal** as the name.
- In the **Type** section, select **Function**.
- In the **Scope** section, select **Public**.
- Click **OK**, and enter the following code for the function:

```
Public Function Etotal(T, x)
   Dim Etran As Double, Erot As Double, Evib As Double

   Etran = 1.5 * R * T                   ' translational energy
   Erot = R * T                          ' rotational energy
   Evib = R * T * x / (Exp(x) - 1)       ' Evib = RT times first Einstein function
   Etotal = Etran + Erot + Evib          ' total energy
End Function
```

To enter the code for the second functions, repeat the preceding steps; in the Name box, type **CvTotal** as the name.

```
Public Function CvTotal( x)
   Dim Cvtran As Double, Cvrot As Double, Cvvib As Double
   Cvtran = 1.5 * R                                   ' translational heat capacity
   Cvrot = R                                          ' rotational heat capacity
   Cvvib = R * x ^ 2 * Exp(x) / (Exp(x) - 1) ^ 2      ' R times sec. Einstein function
   CvTotal = Cvtran + Cvrot + Cvvib                   ' total heat capacity
End Function
```

Table 14-3

The Properties Table for the Heat Capacity Program Form

Object	Property	Value
Form1	Name	frmHeatCap
	Caption	The Heat Capacity Program
Label1	Name	lblInfo
	Caption	This program demonstrates the use of user-defined functions and the use of parallel arrays in repetitive numerical calculations. (Change the default input values as you wish.)
	BorderStyle	1- Fixed Single
Label2	Name	lblHeader
	Caption	T/k Energy J/mole Cv J/K/mode
	Visible	False
Label3	Name	lblTint
	Caption	Enter the number of temperature intervals to use
Label4	Name	lblThigh
	Caption	Enter the highest temperature to use, in T/K
Label5	Name	lblTlow
	Caption	Enter the lowest temperature to use, in T/K
Label6	Name	lblNuBar
	Caption	Enter the fundamental vibrational frequency, in 1/cm
Text box1	Name	txtNuBar
	Text	2
Text box2	Name	TxtTint
	TabIndex	7
Text box3	Name	txtTlow
	Caption	300
Text box4	Name	txtThigh
	Caption	1000
Command1	Name	CmdRun
	Caption	&Run
Command2	Name	cmdPrint
	Caption	Print to the Printer
Command3	Name	cmdExit
	Caption	E&xit

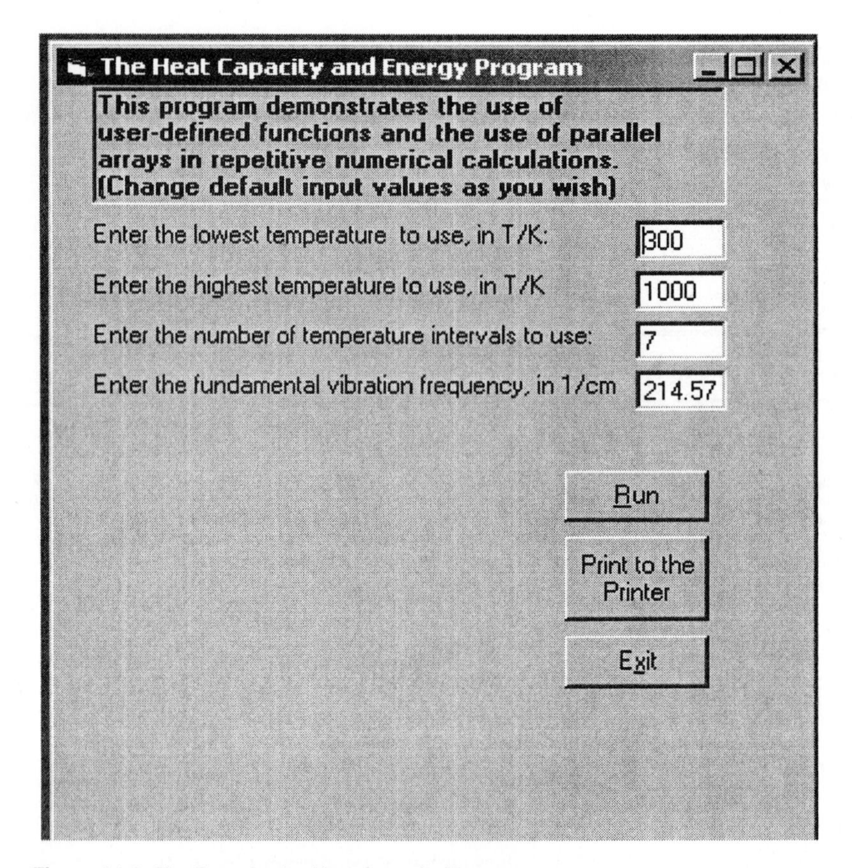

Figure 14-3. The Form for the Heat Capacity Program

Executing the Heat Capacity Program. Execute the program by clicking the **Start** button, which opens the form with four text boxes for the user to input the high and low temperatures, the number of temperature intervals, and the fundamental vibrational frequency.

Clicking the **Run** button causes the program to calculate the energies and heat capacities of a mole of a diatomic molecule from a low temperature (input by the user) to a high temperature (input by the user), at **Nint** equal intervals (input) by user. The user also inputs the fundamental vibrational frequency $\tilde{V}$ for the diatomic molecule of interest (Table 7-4). These parameters are input in four text boxes. The **Text** property of these text boxes could have been set to blank at design time; instead some default parameters were set: **Tlow** = 300 K, **Thigh** = 1000 K, **Nint** = 7, and the fundamental vibrational frequency = 214.57 cm^{-1}, the value for the iodine molecule, $^{127}I_2$. Table 14-4 lists a few other fundamental vibrational frequencies. After running the program, the parameters may be reentered, and the **Run** button clicked again.

Notice that labels on the form by each text box for input not only prompt the user but also state clearly the physical units that are required. Since all the output data are stored in arrays, the user has the option of printing the data to the printer after having seen it displayed on the form.

Table 14-4

Some Fundamental Vibrational Frequencies (cm^{-1})

Diatomic molecule	Symbol	$\tilde{\nu}$ (cm^{-1})
Chlorine	$^{35}Cl_2$	564.9
Carbon monoxide	$^{12}C^{16}O$	2170.21
Hydrogen	$^{1}H_2$	4395.2
Deuterium	$^{2}D_2$	3118.4
Hydrogen chloride	$^{1}H^{35}Cl$	2989.74
Iodine	$^{127}I_2$	214.57
Nitrogen	$^{14}N_2$	2359.61
Oxygen	$^{16}O_2$	1580.361
Phosphorus	$^{31}P_2$	780.43

Understanding the Code of the Heat Capacity Program. It is quite natural to define Visual Basic functions for the total energy (the **Etotal** function) and for the total heat capacity (**CvTotal**). The values of T and x are passed to the **Etotal** function (Equation 14-3); the value of x is passed to the **CvTotal** function (Equation 14-6). The user enters the desired temperature limits and the fundamental vibrational frequency in text boxes on the form. The results are printed on the form. Arrays store the results, so that the data may be printed on the printer if the user desires.

After entering the temperature and molecular data in the text boxes, the user clicks the **Run** button, and after the usual variable declarations, the **Cls** method clears the form of any previous run data. Using the **Thigh**, **Tlow**, and number of temperature intervals (**Nint**), the program calculates the temperature interval (**Tint**).

The program then assigns **Tlow** to the first **T** in the **For ... Next** loop. In each cycle of the loop, the program calculates **x** (Equation 14-4) and passes it and the current **T** to the **Etotal** and **CvTotal** functions. The temperature is incremented and the cycle repeated until the loop is satisfied and the arrays are loaded.

Before printing the output to the form, the program inserts 14 blank lines so that the output printed on the form begins below the label (**lblInfo**). The **Format** functions and the **Tab** functions make the output clear to the user. Having seen the output, the user has the option of printing the results.

Boxes Again: The Combo Box, the Text Box, and the List Box

The combo box derives its name from the fact that it may be considered a combination of a text box and a list box, both of which we've used in previous programs. The text box was used in every program so far that required interactive input from the user, and the list box was used to display lists of output data.

The List Box program (Chapter 12) utilizes both a text box and a list box. The main difference between a text box and a list box is that the user can input an item from the keyboard into a text box, but not into a list box. The list box can hold a large number of items, which are added to the list box with the **AddItem** method of the list box when the program executes. The user can select an item in a list box by clicking the item, and then that selection can generate further action. Both the list box and the combo box have the **Sorted** property, but the text box does not. All three boxes have the **Text** property.

You might want to review some of the characteristics of the list box used in the List Box program. The next short program (the Text Box to List Box program) serves as a review of the text box and list box controls. This program also illustrates the relationship between a combo box, a text box, and a list box and will serve as preparation for the Combo Box to Array program.

The Text Box to List Box Program

This program shows how the user can repeatedly enter single items into a text box and then assign them to the list elements of a list box. In the program after this one (the Combo Box to Array program), we will see how the combo box takes care of both the input and assignments.

Starting the Project

- Start Visual Basic by clicking the **VB** icon on the desktop.
- Select the **Standard EXE** icon in the **New Project** window, and click the **OK** button.
- A **New Project** window opens with a fresh form.
- Select the form, and in the properties window, change the **(Name)** property to **frmTextList**.
- Click **Save**; the file name will be frmTextList, so click **OK**.
- The file name changes to Project1; change it to **The Text Box to List Box Program**, and click the **OK** button.
- Place controls on frmTextList according to Figure 14-4.
- Set the properties of the controls according to Table 14-5.
- Resize and relocate the controls so that the form and its controls appear similar to Figure 14-4.

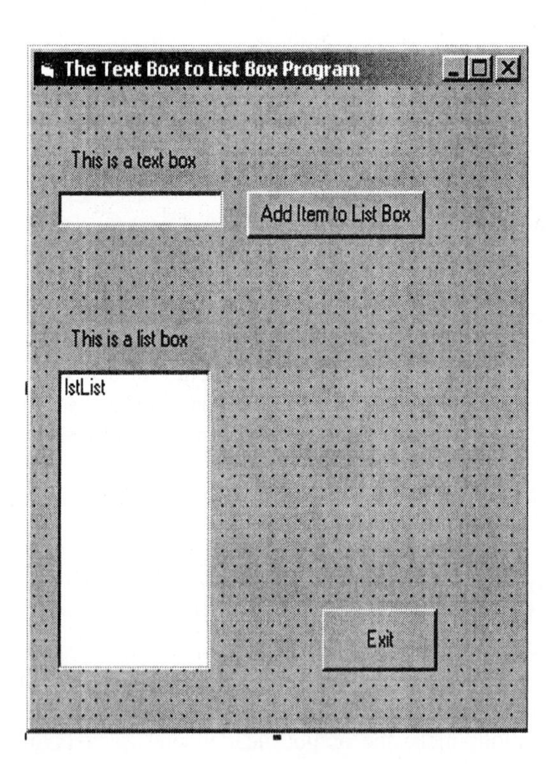

Figure 14-4. The Form for the Text Box to List Box Program

Table 14-5

The Properties Table for the Text Box to List Box Program Form

Object	Property	Value
Form1	Name	frmTextList
	Caption	The Text Box to List Box Program
Label1	Name	lblText box
	Caption	This is a text box
Label2	Name	lblList box
	Caption	This is a list box
Text box1	Name	txtInput
	Text	Make it blank
	Tab Index	0
List box1	Name	lstList
	Sorted	True
Command1	Name	cmdAddItem
	Caption	Add Item to List Box
Command2	Name	cmdExit
	Caption	E&xit

Entering the Code of the Text Box to List Box Program. Enter the following in the general declarations section of the program:

```
Option Explicit
' The Text Box to List Box program
' This program demonstrate how a text box
' and a list box combine to do what a
' combo box does by itself
```

Enter the following code in the **cmdAddItem_Click()** procedure:

```
Private Sub cmdAddItem_Click()
   lstList.AddItem txtInput.Text
   txtInput.Text = ""
   txtInput.SetFocus
End Sub
```

Enter the following code in the **lstList_Click()** procedure:

```
Private Sub lstList_Click()
   MsgBox "You selected # " & lstList.ListIndex + 1 _
   & ":  " & lstList.Text, , "Select from list:"
End Sub
```

Enter the following code in the **cmdExit_Click()** procedure:

```
Private Sub cmdExit_Click()
   End
End Sub
```

Executing the Text Box to List Box Program

- Click the **Start** button on the VB menu bar to execute the Text Box to List Box program.
- Type the string **"dog"** in the text box, and then click the **Add Item to List Box** button.
- Repeat with the strings **"cat"**, **"pony"**, **"horse"**, **"beaver"**, **"zebra"**, **"monkey"**.

The items in the list are automatically alphabetized. Notice that if the number of items added to the combo box exceeds the length of the list portion of the combo box, Visual Basic automatically adds a scroll bar so that the user can scroll up and down the list.

Now click one of the items in the list. A message box appears, on the header bar of which is printed **"Select from list:"** If the third item in your list is **dog** and you selected it, then the message box reads: **"You selected # 3: dog."** Note that the list index of the third item is two.

Understanding the Text Box to List Box Program

This short program demonstrates how the user can input an item into a text box and then assign that item to a list box. The item typed into the text box is the **Text** property of the text box. The **AddItem** method of the list box adds the item to the list, which is an intrinsic part of the list box. The first item added receives the **ListIndex** value of 0. You can consider the list of the list box to be a one-dimensional array with elements having **ListIndex** values as their indices. The essence of the transaction lies in the statement:

```
lstList.AddItem txtInput.Text
```

In the properties listed in Table 14-5, the sorted property of the list box was set to true, so the items in the list box are sorted alphabetically. If you enter a series of numbers, the numbers are sorted only according to their first digit.

The **Private Sub lstList_Click()** procedure invokes the click event of the list box. Clicking an item in the list selects it and causes the subsequent statements to be executed. Here only one statement is executed, the **MsgBox** statement, which has three parameters.

1. The **MsgBox** message, which consists of four concatenated parts:
   a. **"You selected # "** concatenated with
   b. **lstList.ListIndex +1** concatenated with
   c. ": " concatenated with
   d. **lstList.Text**
2. No parameter is given between the two commas. You could put in a constant like **vbExclamation**, if you like.
3. The **MsgBox** header bar message, **"Select from List:"**

In the next program we will see by comparison with this program that the combo box is a combination of the text box and the list box utilized in this program.

The Combo Box Control

The combo box combines many of the features of the text box and of the list box. Like the text box, it is a convenient control for allowing the user to input data. All three boxes display data well. Unlike the text box and the list box, the combo box comes in more than one flavor; three in fact. They have many, but not all, methods and properties in common. Table 14-6 provides a brief comparison of the text box, the list box, and the combo box.

Table 14-6

Comparing Selected Properties and Methods of the Text Box, the List Box, and the Combo Box

Property or Method*	Text Box	List Box	Combo Box
AddItem*	No	Yes	Yes
List(n)	No	Yes	Yes
ListCount	No	Yes	Yes
ListIndex	No	Yes	Yes
Locked	Yes	No	Yes
Multiline	Yes	No	No
RemoveItem*	No	Yes	Yes
Sorted	No	Yes	Yes
Text	Yes	Yes	Yes

The combo box is available in three styles: 0–drop-down combo box, 1–simple combo box, and 2–drop-down list combo box, all shown in Figure 14-5 The program has loaded all three combo boxes with the same list of items, visible in the center box (simple combo box). Clicking the detached down arrow on the drop-down combo box (**Style 0**) or the drop-down list combo box (**Style 2**) causes the list to drop down so that all three boxes look nearly identical. With the simple combo box (**Style 1**), the list is always visible. If the list is too long to fit, Visual Basic automatically adds a scroll bar to the right side of the simple combo box. You can resize its length and width, but you cannot resize the length of the other two combo boxes. With all three boxes, the user can select items that have been added to the list.

With the drop-down combo box and the simple combo box, the user can add items to the list. With the drop-down list combo box, the user cannot add to the list. For this reason the drop-down list combo box is like a list box and is sometimes called a drop-down list box.

The Combo Box to Array Program

This program allows the user to input data into the list of a combo box and then to assign the elements of the list to the elements of an array. The program does some simple mathematical operations on the elements of the array and prints the results on the form. We want to write a program that meets several requirements for inputting data.

1. Data input should be interactive by the user.
2. The user should be able to delete the last item input.
3. The user should be able to select and delete an item anywhere in the list.
4. The user should be able to clear the data already input and begin again.
5. The data in the list should be readily accessible to other parts of the program.

This program illustrates how to enter data into a combo box and how to read that data into an array. The program also demonstrates several properties and methods associated with the combo box. The previous program (the Text Box to List Box program) used a text box to input data and a list box to select data after the data were assigned to the list box from the text box. With a combo box, you can both input data and subsequently select the data.

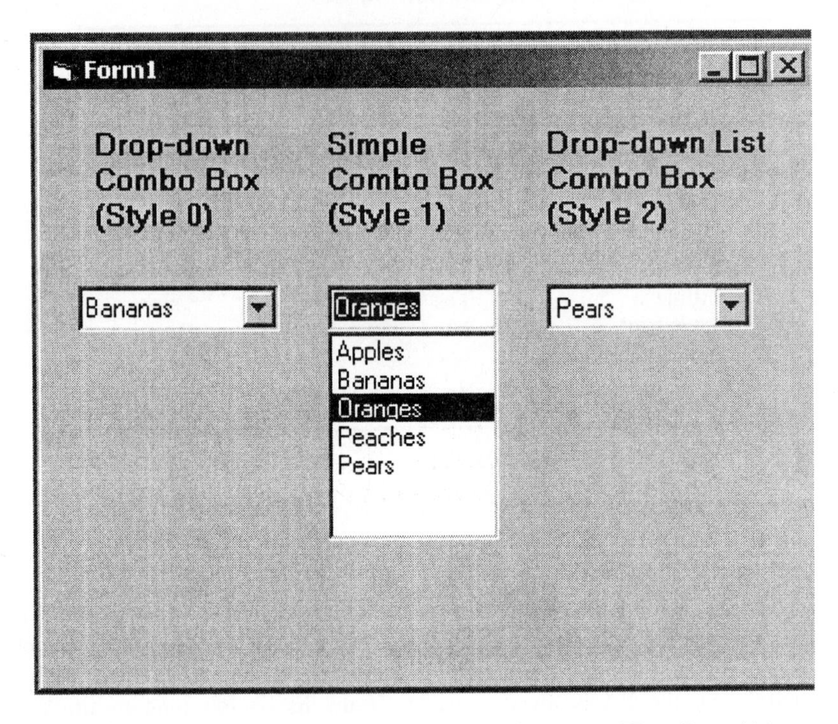

Figure 14-5. Three Styles of Combo Box: Drop-Down, Simple, and Drop-Down List

Starting the Project

- Start Visual Basic by clicking the **VB** icon on the desktop.
- Select the **Standard EXE** icon in the **New Project** window, and click the **OK** button.
- A **New Project** window opens with a fresh form.
- Select the form, and in the properties window, change the **(Name)** property to **frmBoxToArray**.
- Click **Save**; the file name will be frmBoxToArray, so click **OK**.
- The file name changes to Project1; change it to **The Combo Box to Array Program**, and click the **OK** button.
- Place controls on frmBoxToArray according to Figure 14-6.
- Set the properties of the controls according to Table 14-7.
- Resize and relocate the controls so that the form and its controls appear similar to Figure 14-6.

When you place the combo box control on the form and stretch it out a bit to size it, you will notice the top line of the combo box is separated from the larger lower portion by a thin line or bar. You can look at the top line as a text box for adding items one at a time and the lower part of the combo box as a list box for displaying the added item. Many of the methods and properties of the list box and the combo box are the same (**AddItem**, **RemoveItem**, **ListCount**, **ListIndex**, and so on). All three boxes (text, list, and combo) have the **Text** property.

Table 14-7

The Properties Table for the Combo Box to Array Program

Object	Property	Value
Form1	Name	frmBoxToArray
	Caption	The Combo Box to Array Program
Label1	Name	lblInfo
	Caption	This program demonstrates how to enter numeric data into a combo box and transfer to an array for further processing. The numbers are squared and printed on the form.
	BorderStyle	1-Fixed Single
Combo box1	Name	cboInputBox
	Sorted	False
	Style	1-Simple Combo
Command1	Name	cmdAdd
	Caption	< - Add a number here
Command2	Name	cmdDelete
	Caption	Delete last entry
Command3	Name	cmdDelSel
	Caption	Delete selected item
Command4	Name	cmdClear
	Caption	Clear all entries
Command5	Name	cmdExit
	Caption	E&xit
Command6	Name	cmdPrint
	Caption	Print squares to this form

Entering the Code of the Combo Box to Array Program. Enter the following code in the general declarations section:

```
' The Combo Box to Array program
' This program demonstrates how to enter numerical
' data into a combo box and assign those data into
' an array for further processing. Several methods
' and properties of the combo box are illustrated.

Option Explicit
```

Enter the following code in the **cmdAdd_Click()** procedure:

```
Private Sub cmdAdd_Click()
   cboInputBox.AddItem cboInputBox.Text     ' Adds the item entered by user
   cboInputBox.Text = ""                    ' Clears text portion of combo box
   cboInputBox.SetFocus                     ' Returns focus to text portion
End Sub
```

Enter the following code in the **cmdClear_Click()** procedure:

```
Private Sub cmdClear_Click()
   cboInputBox.Clear                  ' Clears the combo box of all entries
   frmBoxToArray.Cls                  ' Clears the form of all entries
   cboInputBox.SetFocus
End Sub
```

Enter the following code in the **cmdDelete_Click()** procedure:

```
Private Sub cmdDelete_Click()
   Dim Count As Integer
   Count = cboInputBox.ListCount      ' The value of ListCount is assigned
                                      ' to Count (ListCount is always one more
                                      ' than the largest ListIndex value
   If Count = 0 Then
      MsgBox "The last item is already deleted", 0, "The BoxToArray Program"
   Else
      cboInputBox.RemoveItem (Count - 1)      ' Removes last item from the list
   End If
   cboInputBox.SetFocus
End Sub
```

Enter the following code in the **cmdExit_Click()** procedure:

```
Private Sub cmdExit_Click()
    End
End Sub
```

Enter the following code in the **cmdPrint_Click()** procedure:

```
Private Sub cmdPrint_Click()
   Dim Counter As Integer, Max As Integer
   Dim dblAry(20) As Double

   frmBoxToArray.Cls                  ' Clears the form
   For Counter = 0 To 7
      frmBoxToArray.Print             ' Insert blank lines
   Next

   Max = cboInputBox.ListCount
   ' Note: ListCount is always one greater
   ' than the largest ListIndex value

   For Counter = 0 To Max - 1
   ' Convert strings to doubles and assign to dblAry
      dblAry(Counter) = CDbl(cboInputBox.List(Counter))
      dblAry(Counter) = dblAry(Counter) ^ 2    ' Squares each element of dblAry
      frmBoxToArray.Print dblAry(Counter)      ' Prints elements of dblAry
   Next

End Sub
```

Enter the following code in the **cmdDelSel_Click()** procedure:

```
 Private Sub cmdDelSel_Click()      ' This is executed when an item _
                                          ' in the list is clicked
   If cboInputBox.ListIndex = -1 Then
      Beep
      MsgBox "You must select an item to delete it", _
                 vbExclamation, "Deleting an item"
   Else
      cboInputBox.RemoveItem cboInputBox.ListIndex   'Removes selected item
   End If
   cboInputBox.SetFocus
End Sub
```

Executing the Combo Box to Array Program

- Click the **Start** button on the VB menu bar to execute the Combo Box to Array program.
- Maximize the program form.
- Enter a decimal or integer number. Then click the **Add a number here** button.
- Enter another number. Then click the **Add a number here** button.
- Enter another number. Then click the **Add a number here** button.
- Try entering a few numbers in scientific notation such as **2.34E4** and **8.954E-5**.
- Similarly, enter several more numbers.
- Click the **Delete last entry** button.
- Click the **Add a number here** button.
- Click a number in the middle of the list to select it.
- Click the **Delete selected item** button
- Click the **Print squares to this form** button. The square of the remaining numbers is printed on the form opposite the listed numbers.
- Click the **Clear all entries** button.
- Click the **Exit** button.

Notice that if the number of items added to the combo box exceeds the length of the list portion of the combo box, Visual Basic automatically adds a scroll bar so that the user can scroll up and down the list.

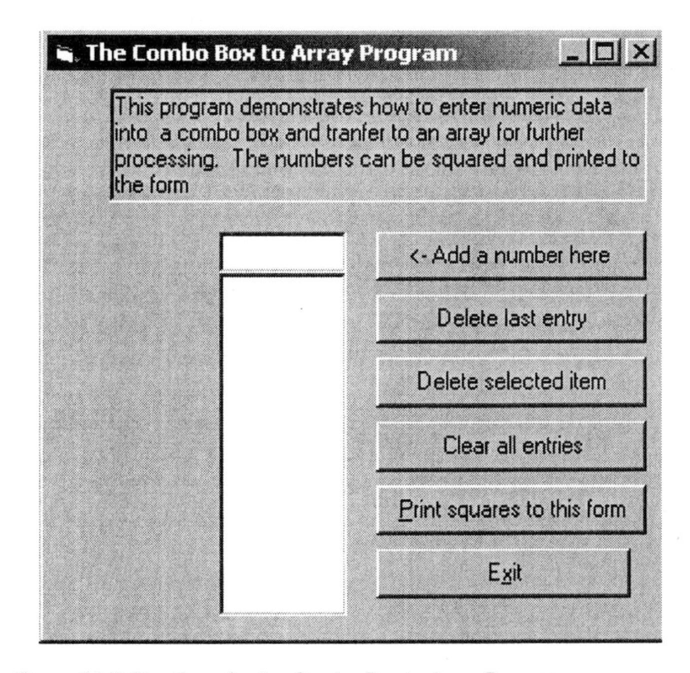

Figure 14-6. The Form for the Combo Box to Array Program

Understanding the Code of the Combo Box to Array Program. Running the program displays the form with its controls, and the focus is on the input part of the combo box, which is waiting for the user to enter an item, that is, a number. After the user enters a number, clicking the **cmdAdd** button causes the **cmdAdd_Click()** procedure to execute. The procedure adds the item entered by the user at the top of the combo box. **AddItem** is a method of the combo box. This method takes an argument, namely, the item to be added. The item to be added is the text that the user entered. This text is the **Text** property of the combo box, at the moment **cboInputBox.Text.** The user builds up a list of items in the combo box by repeatedly entering an item and clicking the **cmdAdd** button.

```
Private Sub cmdAdd_Click()
    cboInputBox.AddItem cboInputBox.Text
    cboInputBox.Text = ""
    cboInputBox.SetFocus
End Sub
```

Assigning the null string ("") to the **Text** property of the combo box clears the text portion of the combo box. At this time, the **SetFocus** method returns focus to the text portion of the combo box.

The number of items in the list equals **ListCount**, a property of the combo box. The position of an item in the list equals **ListIndex**, also a property of the combo box.

Recall that a one-dimensional array is a list. A list is made up of elements, and each element has an integer index beginning with zero and increasing upwards. You can regard the list of a list box and the list of a combo box as one-dimensional arrays. Just as a value can be assigned to the element of an array, so can an item be added to a list in a list box or combo box. The number of items in the list is one more than the value of the list index, since the list index for the first item is zero, for the second item is one, and so on, just like any one-dimensional array.

Clicking on the **cmdClear** button causes the **cmdClear_Click()** procedure to respond to the click event and to execute. **Clear** is a combo box method that clears the combo box of all entries. **Cls** is a form method that clears the form of any text previously printed on the form. Neither method takes an argument. After you clear the combo box, focus is returned to it.

```
Private Sub cmdClear_Click()
  cboInputBox.Clear
  frmBoxToArray.Cls
  cboInputBox.SetFocus
End Sub
```

The **cmdDelete_Click()** procedure deletes the last item entered by the user into the list. It uses two combo box methods, the **ListCount** method and the **RemoveItem** method. The **ListCount** method returns an integer whose value is always one more than the largest **ListIndex** value, that is, the index value of the last item in the list. It does not take an argument.

```
Sub cmdDelete_Click()

   Dim Count As Integer
   Count = cboInputBox.ListCount

   If Count = 0 Then
      MsgBox "The last item is already deleted", 0, "The BoxToArray Program"
   Else
      cboInputBox.RemoveItem (Count - 1)  ' Removes last item from the list
   End If

   cboInputBox.SetFocus
End Sub
```

The **cmdPrint_Click()** procedure squares numbers that the user has input to the combo box list and prints them on the form. The number of items in the list (**Max**) equals **ListCount**.

```
Max = cboInputBox.ListCount
```

The indices of the list range from zero for the first item in the list to **Max** –1 for the last item in the list.

Since all the numbers entered into the list are entered as text (strings), they must be converted to numbers (doubles) and then assigned to an array of doubles:

```
For Counter = 0 To Max - 1
    dblAry(Counter) = CDbl(cboInputBox.List(Counter))
```

In the next statement of the **For ... Next** loop, each element of the array **dblAry** is squared and assigned back to array **dblAry**.

```
dblAry(Counter) = dblAry(Counter) ^ 2
```

In the last statement of the **For ... Next** loop, the elements of the array are printed on the form. A few blank lines were previously inserted so the squares printed on the form line up with the numbers in the combo box list.

```
frmBoxToArray.Print dblAry(Counter)
```

The **cmdDelSel_Click()** procedure deletes an item from the list. Clicking an item in the list selects the item and establishes the value for the **ListIndex** for the selected item. If no item in the list is selected, Visual Basic assigns a value of –1 to the **ListIndex** property of the combo box. The **RemoveItem** method has an argument, which is the **ListIndex** of the item in the list to be removed.

```
Private Sub cmdDelSel_Click()
    If cboInputBox.ListIndex = -1 Then
    MsgBox "You must select an item to delete it", _
                    vbExclamation, "Deleting an item"
     Else
     cboInputBox.RemoveItem cboInputBox.ListIndex
     End If
End Sub
```

After you enter and edit the list, clicking the **Print** button executes the **Private Sub cmdPrint_Click()** procedure. After declaring two integers and one array variable:

```
Dim Counter As Integer, Max As Integer
Dim dblAry(20) As Double
```

the form is cleared, and the first **For ... Next** loop inserts seven blank lines so that the printing of the output begins in a suitable position on the form.

```
frmBoxToArray.Cls
For Counter = 0 To 7
   frmBoxToArray.Print        'Insert blank lines
Next
```

The second **For ... Next** loop converts the text representations of the numbers entered into the combo box into doubles and assigns them to the array **dblAry**. The numbers in the array **dblAry** are then squared and assigned back to the array **dblAry**. Finally, the **Print** method of the form frmBoxToArray prints on the form the numbers stored in the array **dblAry**.

```
    Max = cboInputBox.ListCount
    ' Note: ListCount is always one greater
    ' than the largest ListIndex value

    For Counter = 0 To Max - 1
    ' Convert strings to doubles and assign to dblAry
       dblAry(Counter) = CDbl(cboInputBox.List(Counter))
       dblAry(Counter) = dblAry(Counter) ^ 2    ' Squares each element of dblAry
       frmBoxToArray.Print dblAry(Counter)      ' Print elements of dblAry
    Next

End Sub
```

The next procedure exits the program:

```
Private Sub cmdExit_Click()
    End
End Sub
```

The Trapezoid Integration Program

One of the fundamental problems that gave rise to the development of integral calculus was that of finding the area under a curve. If $f(x)$ defines a curve, then the area under the curve between points a and b is given by the definite integral:

$$Area = \int_a^b f(x)dx \tag{14-7}$$

In Figure 14-7 a curved line is drawn through five x,y points. If the function that represents the curve is known, then the definite integral can provide the area. If only the coordinates of the five points in known, the approximate area can still be found.

The area under the curve between a pair of points approximates the area of the subtended trapezoid. The height of the rectangle within the trapezoid is drawn so that the area of the rectangle is approximately equal to that of the trapezoid. The width of each gray rectangle equals to the difference between adjacent x values. The height of each rectangle equals the average of the high and low y values. The trapezoid rule says that the area of this rectangle approximates closely the area of the corresponding trapezoid. Notice that N pairs of points gives rise to $N - 1$ areas.

$$\begin{aligned}
&\text{Area 1} = (X_2 - X_1)\,(Y_2 - Y_1)/2 \\
&\text{Area 2} = (X_3 - X_2)\,(Y_3 - Y_2)/2 \\
&\text{Area 3} = (X_4 - X_3)\,(Y_4 - Y_3)/2 \\
&\text{Area 4} = (X_5 - X_4)\,(Y_5 - Y_4)/2 \\
&\text{Total Area} = \text{Area 1} + \text{Area 2} + \text{Area 3} + \text{Area 4}
\end{aligned} \tag{14-8}$$

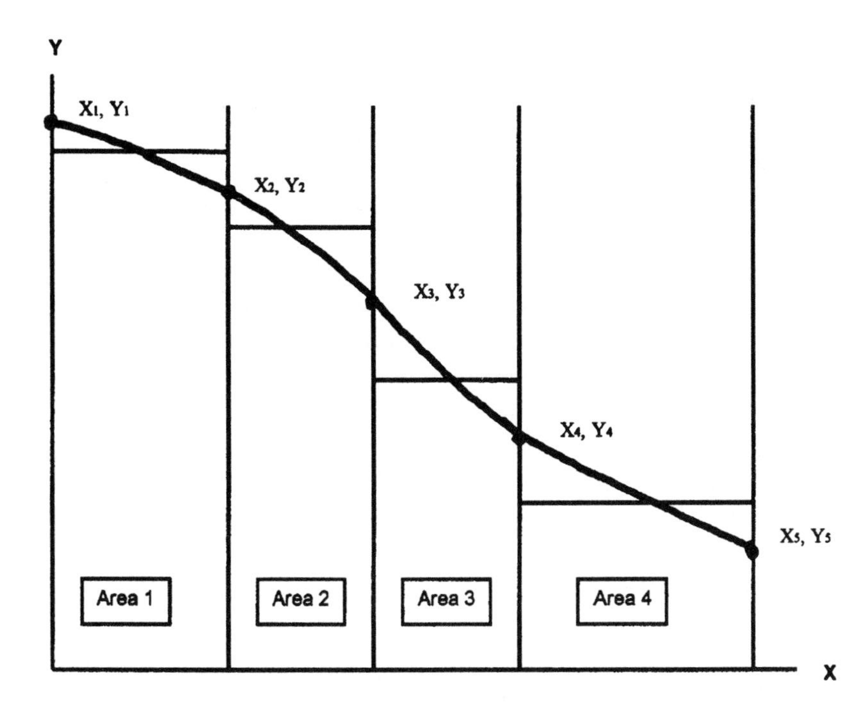

Figure 14-7. The Trapezoid Rule for Numerical Integration

Visual Basic accomplishes the calculation of the area under the curve in Figure 14-7 with a single statement in a **For ... Next** loop:

```
Area = 0
For I = 0 To N -1
        Area = Area + (X (I + 1) - X (I)) * (Y (I + 1) + Y (I)) / 2
Next
```

The Trapezoid Integration program implements this algorithm to find the area under a curve defined by N discrete points. The height of a rectangle is the average of the Y_i values of the X_i, Y_i points between which it is drawn. The X_i values need not be equal.

Experimental work in science and engineering often involves the measurement of a dependent variable (y) and independent variable (x). Multiple measurements give rise to lists (arrays) of y_i and x_i values. In the previous program, a combo box was used to allow the user to input values that were assigned to an array. In the Integration program, two combo boxes are used to input y_i and x_i values into two parallel arrays. These values are then used to calculate the area under a curve defined by the set of y_i,x_i pairs with the aforementioned algorithm. Later we'll see how to write a file of y_i and x_i values that the Trapezoid Integration program can read.

Starting the Project

- Start Visual Basic by clicking the **VB** icon on the desktop.
- Select the **Standard EXE** icon in the **New Project** window, and click the **OK** button.
- A **New Project** window opens with a fresh form.

- Select the form, and in the properties window, change the **(Name)** property to **frmIntegrate**.
- Click **Save**; the file name will be frmIntegrate, so click **OK**.
- The file name changes to Project1; change it to **The Trapezoid Integration Program**, and click the **OK** button.
- Place controls on frmIntegrate according to Figure 14-8.
- Set the properties of the controls according to Table 14-8.

Table 14-8

The Properties Table for the Trapezoid Integration Program

Object	Property	Value
Form1	Name	frmIntegrate
	Caption	The Trapezoid Integration Program
Combo box1	Name	cboYinput
	Style	1 – Simple Combo
Combo box2	Name	cboXinput
	Style	1 – Simple Combo
Command1	Name	cmdDelete
	Caption	Delete Last Pair
Command2	Name	cmdDelSel
	Caption	Delete Selected Pair
Command3	Name	cmdClear
	Caption	Clear All Entries
Command4	Name	cmdExit
	Caption	Exit
Command5	Name	cmdIntegrate
	Caption	Integrate
Label1	Name	lblInfo
	Caption	Enter Xi and Yi in Pairs
	Font	10 Point Bold
Label2	Name	lblYin
	Font	10 Point Bold
	Caption	<- Enter Yi
Label3	Name	lblXin
	Caption	Enter Xi ->
	Font	10 pt Bold
Label4	Name	lblTitle
	Caption	The Trapezoid Integration Program
	Font	10 pt Bold
Label5	Name	lblDirections
	Caption	Enter Xi, then Tab, then Yi, then Press Enter
	Font	10 pt Bold
Label6	Name	lblArea
	Caption	Make blank
	Font	10 pt Bold
	Border Style	1-Fixed Single

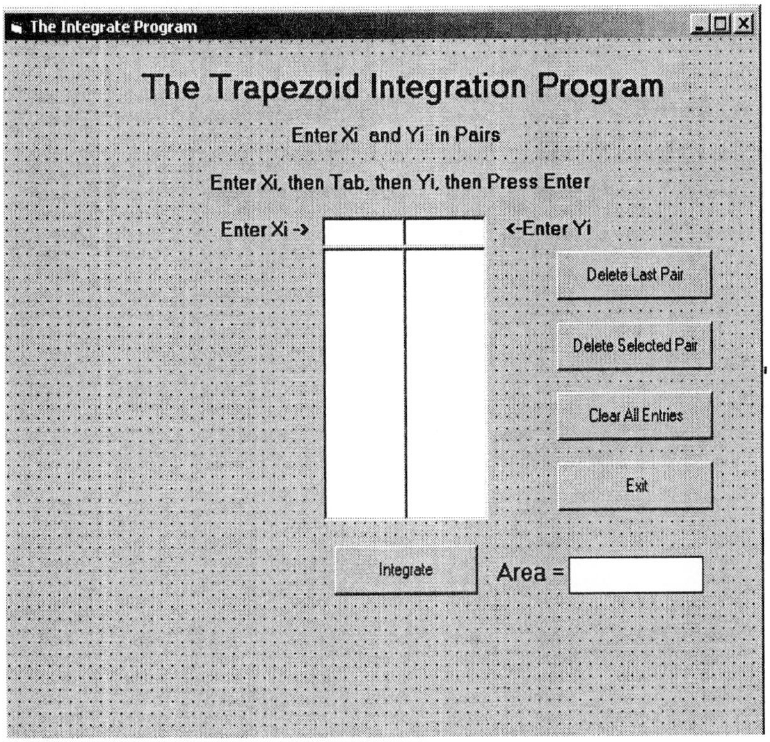

The Integrate Program

The Trapezoid Integration Program

Enter Xi and Yi in Pairs

Enter Xi, then Tab, then Yi, then Press Enter

Enter Xi -> <-Enter Yi

Delete Last Pair

Delete Selected Pair

Clear All Entries

Exit

Integrate

Area =

Figure 14-8. The Form for the Trapezoid Integration Program

Entering the Code of the Trapezoid Integration Program. Enter the following code in the general declarations section:

```
Option Explicit
' The Trapezoid Integration program
' This program inputs x,y pairs with
' two combo boxes into parallel arrays.
' The area under a curve drawn through the points
' is calculated by numerical integration
' (the trapezoid rule)
```

Enter the following code in the **cboYinput_KeyPress** procedure:

```
Private Sub cboYinput_KeyPress(Keyascii As Integer)

If Keyascii = vbKeyReturn Then          ' Press the ENTER key to add an Xi, Yi pair

   'Check if Yi was input
   If cboYinput.Text = "" Then
      MsgBox "You must enter both Xi and Yi ", vbInformation, "Input Error"
      cboXinput.Text = ""
      cboYinput.Text = ""
      cboXinput.SetFocus
      Exit Sub
   End If

   ' Check if Xi was input
   If cboXinput.Text = "" Then
      MsgBox "You must enter both Xi and Yi ", vbInformation, "Input Error"
      cboXinput.Text = ""
      cboYinput.Text = ""
```

```
      cboXinput.SetFocus
      Exit Sub
   End If

   ' Add Xi, Yi pair if both were input
   cboXinput.AddItem cboXinput.Text
   cboYinput.AddItem cboYinput.Text
   cboXinput.Text = ""
   cboYinput.Text = ""
   cboXinput.SetFocus

End If     ' Keyascii

End Sub
```

Enter the following code in the **cmdClear_Click()** procedure:

```
Private Sub cmdClear_Click()
   cboXinput.Clear
   cboYinput.Clear
   cboXinput.SetFocus
End Sub
```

Enter the following code in the **cmdDelete_Click()** procedure:

```
Private Sub cmdDelete_Click()
   Dim Index As Integer
   Index = cboXinput.ListCount
      If Index = 0 Then
         Beep
         MsgBox "The Last item is already deleted", 0, _
         "The XY Input Program"
      Else
         cboXinput.RemoveItem (Index - 1)
         cboYinput.RemoveItem (Index - 1)
      End If
      cboXinput.SetFocus
End Sub
```

Enter the following code in the **cmdDelSel_Click()** procedure:

```
Private Sub cmdDelSel_Click()
   If (cboXinput.ListIndex = -1) Or (cboYinput.ListIndex = -1) Then
      Beep
      MsgBox "You must select both items to delete the pair", _
      vbExclamation, "Deleting an item"
   Else
      cboXinput.RemoveItem cboXinput.ListIndex
      cboYinput.RemoveItem cboYinput.ListIndex
   End If
   cboXinput.SetFocus
End Sub
```

Enter the following code in the **cmdExit_Click()** procedure:

```
Private Sub cmdExit_Click()
   End
End Sub
```

Enter the following code in the **cmdIntegrate_Click()** procedure:

```
Private Sub cmdIntegrate_Click()
   Dim I As Integer
   Dim sngX(20) As Single
   Dim sngY(20) As Single
   Dim sngArea As Single

   For I = 0 To cboXinput.ListCount - 1
      sngX(I) = CSng(cboXinput.List(I))          ' Converts Xi to single and assigns to array
      sngY(I) = CSng(cboYinput.List(I))          ' Converts Yi to single and assigns to array
   Next

   sngArea = 0
   For I = 0 To cboXinput.ListCount - 2
      sngArea = sngArea + (sngX(I + 1) - sngX(I)) * (sngY(I + 1) + sngY(I)) / 2
   Next

   lblArea.Caption = Format(sngArea, "#####.000")
   lblArea.BackColor = RGB(255, 125, 125)
   lblArea.ForeColor = RGB(55, 50, 255)
End Sub
```

Executing the Trapezoid Integration Program. Click the **Start** button to run the program. The form appears with two combo boxes, side by side. Before entering any test data, click the **Integrate** button. The label control at the lower right should show **0.00** (in blue) against a red-orange background. Click the command button labeled **Clear All Entries.**

Table 14-9

Test Data for the Trapezoid Integration Program

$y = x^2 + 2x + 4$

x	*y*
0	4.00
0.1	4.21
0.2	4.44
0.3	4.69
0.4	4.96
0.5	5.25
0.6	5.56
0.7	5.89
0.8	6.24
0.9	6.61
1.0	7.00

To check the program, run some test data for the function

$$y = x^2 + 2x + 4 \tag{14-9}$$

To test the program, integrate the pairs of points from $x = 0.2$ to $x = 0.6$. Type **0.2** in the text portion of the left combo box, press the TAB key, and type **4.44** in the text portion of the right combo box. Press the ENTER key. Repeat until the last pair is entered. Click the **Integrate** button. The two label controls should display **Area = 1.990**. Compare with the hand-calculated definite integral of the function:

$$\int_{0.2}^{0.6} (x^2 + 2x + 4)dx = 1.989 \tag{14-10}$$

The agreement is satisfactory.

Understanding the Code of the Trapezoid Integration Program. Compared to the Combo Box to Array program, the Trapezoid Integration program uses several procedures essentially unchanged: the **cmdClear_Click** procedure, the **cmdDelete_Click** procedure, the **cmdDelSel_Click** procedure, and the **cmdExit_Click** procedure. However the code for adding the items to the combo box list has been changed. In the previous program, the **cmdAdd_Click** procedure was used to add items to the single combo box, but in the Trapezoid Integration program, the KeyPress event of one of the combo boxes is used to add X_i to one combo box and simultaneously to add Y_i to the second combo box.

In the Trapezoid Integration program, the keypress event is used to detect the user pressing the ENTER key. If the ENTER key is pressed, the code of the **cboYinput_keypress (Keyascii as Integer)** procedure executes. After detecting that the ENTER key was pressed, the first **If ... Then** statement checks if a Y_i value was input. The second **If ... Then** statement checks if an X_i value was input. If both were input to the two combo boxes, then everything is OK, and the values are added to the respective lists of the two combo boxes. The KeyPress event was introduced in Chapter 13.

To prepare for the entry of the next X_i, Y_i pair, the text portions of the two combo boxes are cleared (assigned the empty string, ""), and focus is returned to the combo box for the next X_i value (**cboXimput**).

Before pressing the ENTER key to add an X_i, Y_i pair to the combo box, the user has the opportunity to edit the entries already added by using the buttons labeled **Delete Last Pair, Delete Selected Pair**, or **Clear All Entries**. These procedures are essentially the same as those described in the Combo Box to Array program.

The 2-D Array Program

In addition to one-dimensional arrays, Visual Basic allows multidimensional arrays. A one-dimensional array is a list, and a two-dimensional (2-D) array is a matrix or table. Three- and higher-dimensional arrays are permitted but are not frequently encountered. The declaration of a two-dimensional array is similar to the declaration of a one-dimensional array:

```
Dim MyTable (20,25) As Single
Dim MyDisplay (1 To 10, 1 To 10) As Integer
Dim MatrixYX (5, 1 To 15) as String
```

The first index determines the number of rows and the second index determines the number of columns in the table. The index ranges from 0 to n, unless otherwise specified.

Just as with one-dimensional arrays, loops are used to fill the elements of a table. With a two-dimensional array, a pair of nested loops does the job. To fill a table with the product of the indices:

```
Dim J As Integer, K As Integer
Dim MyDisplay (10,10) As Integer
For J = 1 To 10
    For K = 1 To 10
        MyDisplay(J,K) = J*K
    Next K
Next J
```

This code fills the elements of **MyDisplay** as follows:

1	2	3	4	5
2	4	6	8	10
3	6	9	12	15
4	8	12	16	20
5	10	15	20	25

The 2-D Array program demonstrates declaring two-dimensional arrays, using loops to fill them, calculating with arrays, and printing values of the elements on the form and on the printer.

Starting the Project

- Start Visual Basic by clicking the **VB** icon on the desktop.
- Select the **Standard EXE** icon in the **New Project** window, and click the **OK** button.
- A **New Project** window opens with a fresh form.
- Select the form, and in the properties window, change the **(Name)** property to **frm2dArray**.
- Click **Save**; the file name will be frm2dArray, so click **OK**.
- The file name changes to Project1; change it to **The 2-D Array Program**, and click the **OK** button.
- Place controls on frm2dArray according to Figure 14-9.
- Set the properties of the controls according to Table 14-10.

Table 14-10

The Properties Table for the 2-D Array Program

Object	Property	Value
Form1	Name	frm2dArray
	Caption	The 2-D Array Program
Command1	Name	cmdExit
	Caption	Exit

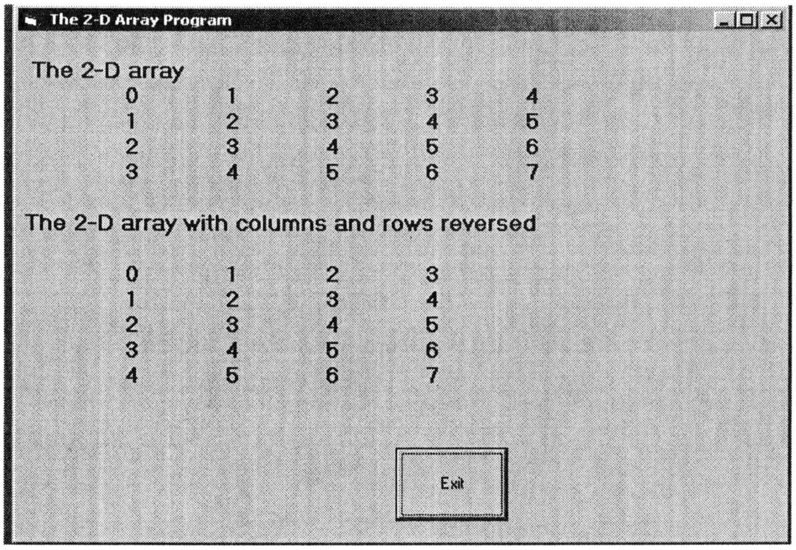

Figure 14-9. The Form for the 2-D Array Program

Entering the Code of the 2-D Array Program

Enter the following code in the general declarations section:

```
Option Explicit
' The 2-D Array program demonstrate how to do an iterative
' calculation, load the results into a 2-D array, and
' print the elements of the array as a table
```

Enter the following code in the **Form_Load()** procedure:

```
Private Sub Form_Load()
   Dim R As Integer
   Dim C As Integer
   Dim N As Integer
   Dim TableN(0 To 10, 0 To 10) As Integer
   ' TableN is a two-dimensional array, that is, a table

   'Load the array
   For R = 0 To 3
      For C = 0 To 4
         N = R + C                ' do a simple calculation
         TableN(R, C) = N         ' load N into an array element
      Next C
   Next R

   frm2dArray.Show
   frm2dArray.Print
   frm2dArray.Print "  The 2-D array"

   For R = 0 To 3                 ' Print the array
      For C = 0 To 4
         frm2dArray.Print vbTab; TableN(R, C);
      Next C
      frm2dArray.Print
   Next R

   frm2dArray.Print
   frm2dArray.Print " The 2-D array with columns and rows reversed"
   frm2dArray.Print

   For C = 0 To 4
      For R = 0 To 3
         frm2dArray.Print vbTab; TableN(R, C);
      Next R
      frm2dArray.Print
   Next C
End Sub
```

Enter the following code in the **cmdExit_Click()** procedure:

```
Private Sub cmdExit_Click()
   End
End Sub
```

Executing the 2-D Array Program

- Click the **Start** button to execute the program.
- The program prints a two-dimensional array on the form, and then reverses the row and column values and prints the array again.
- Study and compare the code of the two **For ... Next** loops that generate these two arrays.
- Click the **Exit** button to end the program.

Understanding the Code of the 2-D Array Program. The program runs immediately since its code is contained in the **Form_Load** procedure. The first **For ... Next** loop loads the array:

```
For R = 0 To 3
    For C = 0 To 4
      N = R + C
      TableN(R, C) = N
    Next C
Next R
```

Since *R* in the outer loop runs from 0 to 3, there will be four rows. The first (zeroth) row begins with $C = 0$ and runs to $C = 4$ for a total of five columns.

After printing a title line for the array, the second **For ... Next** loop prints out the array as it was filled with the first **For ... Next** loop. A tab is inserted between each column element printed. The **frm2dArray.Print** statement acts like a carriage return.

```
For R = 0 To 3
    For C = 0 To 4
      frm2dArray.Print vbTab; TableN(R, C);
    Next C
    frm2dArray.Print
Next R
```

This loop prints the following output on the screen:

```
0    1    2    3    4
1    2    3    4    5
2    3    4    5    6
3    4    5    6    7
```

After printing a title line for the next array, the third **For ... Next** loop prints out the array with the column and row values reversed. Notice that the number of rows is always determined by the range of the outer loop, whether you name it *R* or *C* or *J* or *K* or ....

```
Private Sub Form_Load()
  Dim R As Integer, C As Integer
  Dim MyDisplay(10, 10) As Integer
  Show
  For R = 0 To 4
    For C = 0 To 3
      MyDisplay(R, C) = R + C
      Print vbTab; MyDisplay(R, C);
    Next C
    Print  'prints nothing, but generates a carriage return
  Next R
End Sub
```

This procedure prints the following output on the screen:

```
0    1    2    3
1    2    3    4
2    3    4    5
3    4    5    6
4    5    6    7
```

The **Print** statement is necessary for printing, but not for loading the array. At the end of each row the **Print** statement prints nothing, but it does generate a carriage return, so that the second row begins underneath the first row.

The TotoLoto Program

The TotoLoto program enables the user to generate a set of six different random numbers from 1 to 50 for use in playing the lottery. The user can generate repeated sets of different numbers by repeatedly click a control button.

The TotoLoto program provides more practice with the **For ... Next** loop with nested arrays. In addition, this program introduces the **Control Array**. Recall that all the elements of an array have the same name and type. Usually we think of arrays of integers, singles, characters, or strings. Visual Basic allows for the **Control Array** data structure, in which all the elements of the array are controls of the same name and type. In this program it is convenient to use an array of label controls.

Starting the Project

- Start Visual Basic by clicking the **VB** icon on the desktop.
- Select the **Standard EXE** icon in the **New Project** window, and click the **OK** button.
- A **New Project** window opens with a fresh form.
- Select the form, and in the properties window, change the (**Name**) property to **frmTotoLoto**.
- Click **Save**; the file name will be frmTotoLoto, so click **OK**.
- The file name changes to Project1; change it to **The TotoLoto Program**, and click the **OK** button.
- Place controls on frmTotoLoto according to Figure 14-10.
- Set the properties of the controls according to Table 14-11.

Table 14-11

The Properties Table for the TotoLogo Program

Object	Property	Value
Form1	Name	frmTotoLoto
	Caption	The TotoLoto Program
Command1	Name	cmdPick
	Caption	Pick Your Numbers
Command2	Name	cmdExit
	Caption	Exit
Label1	Name	Label
Label2	Name	Label
Label3	Name	Label
Label4	Name	Label
Label5	Name	Label
Label6	Name	Label

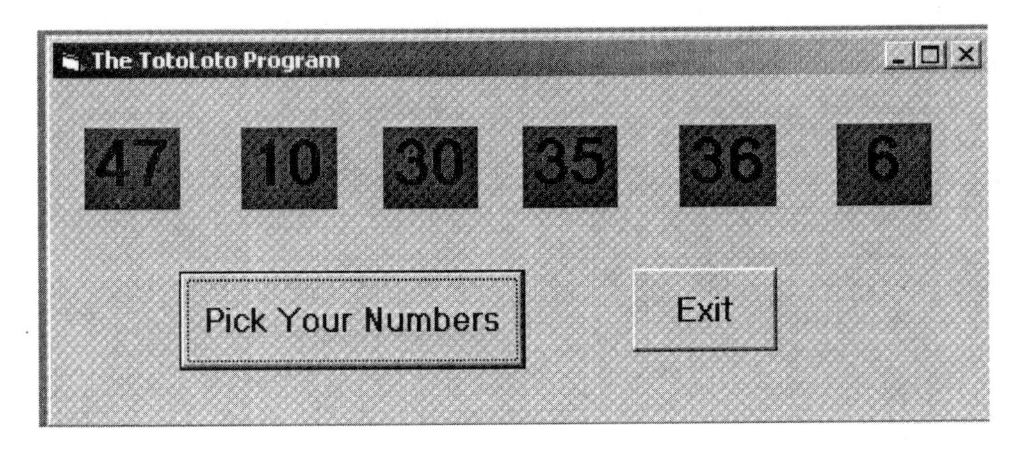

Figure 14-10. The Form for the TotoLoto Program

A glance at Table 14-11 reveals that the form contains two command buttons and six label controls. Place the first label control on the form, and in the properties window set the **Name** property to **Label**. Place the second label control on the form, and in the properties window set the **Name** property to **Label**, the same name as for the first label control. Because the name of the second control is the same as the name of the first label control, Visual Basic responds with a message box (Figure 14-11) informing you that **You already have a control named 'Label.' Do you want to create a control array?** Click **Yes**, and continue to add label controls to the form. After each label control is added, change the **Name** property to **Label**. You now have a Control array. Each element of the control is a Label Control with the name Label. The elements of the array are Label(0), Label(1), Label(2) ....Label(5). Having a Control Array facilitates clear and simple code.

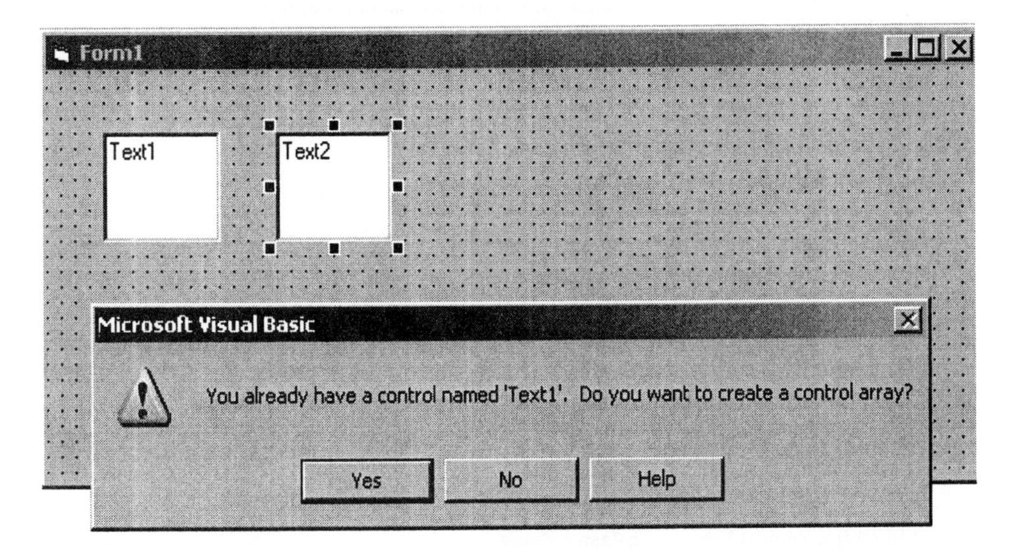

Figure 14-11. The Control Array Dialog Box. **Note:** the **BorderStyle** property in Figure 14-11 has been set to **1-Fixed single** to show the label more clearly. In Figure 14-10, the **BorderStyle** property has been set at design time to **0-none**, which is more dramatic when color is used. The **BackColor** property is set at run time to red in the **cmdPick_Click()** procedure.

Entering the Code of the TotoLoto Program. Double-click the form, and enter the following statements in the general declarations section:

```
Option Explicit
Dim Num(10) As Integer
Dim GoodPass As Boolean
```

Double-click the form, and fill in the code for the **Form_Load()** procedure:

```
Private Sub Form_Load()    ' Initializes all six label controls
   Dim I As Integer
   For I = 0 To 5
      Label(I).Visible = False
      Label(I).Alignment = 2 'Center justify
      Label(I).Font.Bold = True
      Label(I).Font.Size = 24
   Next I
End Sub
```

Double-click the **cmdPick** command button, and fill in the code for the **cmdPick_Click()** procedure:

```
Private Sub cmdPick_Click()
' Note: The default property of the label control is Caption
   Dim J As Integer, K As Integer

   Randomize
   Do                              ' Fill array Num(J) with random numbers
      For J = 0 To 5
      Num(J) = Round(50 * Rnd)
   Next
   CheckForEqualNumbers                  ' No numbers in Num(J) can be the same
   Loop Until GoodPass = True      ' Loop until all numbers
                                   ' in Num(J) are different

   If GoodPass = True Then
      For J = 0 To 5
         Label(J) = Num(J)         ' Label(J) is same as Label(J).Caption
         Label(J).Visible = True
         Label(J).BackColor = RGB(255, 0, 0)
      Next
   End If

End Sub
```

On the **Tools** menu, click **AddProcedure** to open the **Add Procedure** dialog box. In the **Name** box, type **CheckForEqualNumbers** as the name. In the **Type** section, select **Sub**; and in the **Scope** section, select **Public**. Click **OK**. Fill in the code for the **CheckForEqualNumbers** procedure.

```
Public Sub CheckForEqualNumbers()         ' Check if any numbers
                                                  ' in Num(J) are the same
   Dim J As Integer, K As Integer
   GoodPass = True
   For J = 0 To 5                         ' compare each with the next one
      For K = J + 1 To 6
         If Num(J) = Num(K) Then GoodPass = False
         If (Num(J) Or Num(K)) = 0 Then GoodPass = False
      Next K
   Next J
End Sub
```

Double-click the **Exit** button, and fill in the code of the **cmdExit_Click()** procedure.

```
Private Sub cmdExit_Click()
   End
End Sub
```

Executing the TotoLoto Program. In the Visual Basic Editor, click the **Start** button. When the program executes, only the two command buttons are visible. Click the command button with the caption **Pick Your Numbers**. The six label controls become visible, and each displays a random number different from any other. Click again if you believe this set is not the winning set. Repeat until you believe you have a winning set of lottery numbers. Then click the **Exit** button, and forget about playing the lottery. The odds are astronomically large against winning.

Understanding the TotoLoto Program. Playing the California State Lottery or any lottery is not to be recommended. But those who do, pick six different, random (and hopefully, lucky) numbers. The TotoLoto program picks six different random numbers, but not necessarily lucky numbers.

When the program is first executed, the array of label controls is initialized with a **For ... Next** loop that sets the **Visible** property value of each label to **False** and the **Alignment** property value to **2**, or center justified. The program waits for the user to click the command button with the caption **Pick Your Number,** which precipitates the **cmdPick_Click** procedure.

After a **Randomize** statement, a **For ... Next** loop fills the array **Num(J)** with six random numbers that are greater than 1 and less than 50. It is possible that not all the numbers are different, so the program checks for equal numbers by calling the **ChckForEqualNumbers()** procedure. The program will refill the array **Num(J)** until all the numbers are different.

The **ChckForEqualNumbers()** procedure assumes that the first pass through the array **Num(J)** has no equal numbers (**GoodPass = True**). But just to make sure, it passes through the array **Num(J)**, comparing each number with the number subsequent to it. If it finds that two numbers are the same, it sets **GoodPass** to **False**. This procedure is inside the **Do ... Until** loop statement of the **cmdPick_Click** procedure.

When **GoodPass** finally becomes **True**, then the last **For ... Next** loop in the **cmdPick_Click** procedure assigns the elements of the array **Num(J)** to the **Label** property of each element of the control array, makes the labels visible, and gives them a red color.

The Bubble Sort Program

The Bubble Sort program generates a list of ten random numbers and prints them on the form. It then sorts those ten random numbers in ascending order and prints the sorted list on the form. This program demonstrates the manipulation of the contents of an array with **For ... Next** loops, both nested and unnested.

Starting the Project

- Start Visual Basic by clicking the **VB** icon on the desktop.
- Select the **Standard EXE** icon in the **New Project** window, and click the **OK** button.
- A **New Project** window opens with a fresh form.
- Select the form, and in the properties window, change the **(Name)** property to **frmBubble**.
- Click **Save**; the file name will be frmBubble, so click **OK**.
- The file name changes to Project1; change it to **The Bubble Sort Program**, and click the **OK** button.
- Place two command buttons on frmBubble according to Figure 14-12.
- Set the properties of the controls according to Table 14-12.

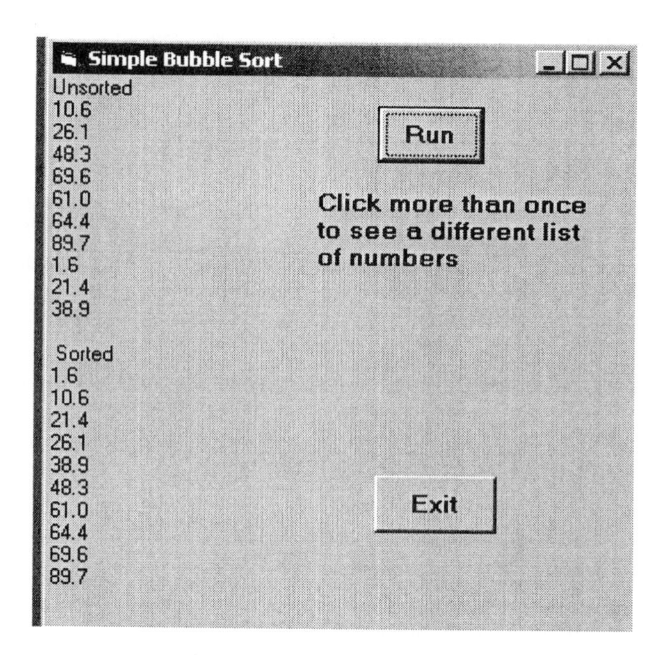

Figure 14-12. The Form for the Bubble Sort Program

Table 14-12

The Properties Table for the Bubble Sort Program

Object	Property	Value
Form1	Name	frmBubble
	Caption	The Bubble Sort Program
Command1	Name	cmdRun
	Caption	Run
Command2	Name	cmdExit
	Caption	Exit

Entering the Code of the Bubble Sort Program. Enter the following code in the general declarations section of the program:

```
Option Explicit
' The Bubble Sort program
Dim List(50) As Single
Const N = 10
```

Double-click the **cmdExit** button, and enter the following code:

```
Private Sub cmdExit_Click()
   End
End Sub
```

Double-click the **cmdRun** button, and enter the following code:

```
Private Sub cmdRun_Click()
frmBubble.Cls
Dim I As Integer, J As Integer
Randomize
For I = 1 To N
   List(I) = 100 * Rnd
Next I
Show
Print "Unsorted"
For I = 1 To N
   Print Format(List(I), "##.0")
Next I
SortTheNumbers
End Sub
```

On the **Tools** menu, click **AddProcedure** to open the **Add Procedure** dialog box. In the **Name** box, type **SortTheNumbers** as the name. In the **Type** section, select **Sub**; and in the **Scope** section, select **Public**. Click **OK**, and enter the following code:

```
Private Sub SortTheNumbers()
Dim Temp As Single
Dim I As Integer, J As Integer
For I = 1 To N - 1
   For J = I + 1 To N
      If List(I) > List(J) Then
         Temp = List(I)
         List(I) = List(J)
         List(J) = Temp
      End If
   Next J
Next I
Print
Print " Sorted"
For I = 1 To N
   Print Format(List(I), "##.0")
Next I
End Sub
```

Executing the Bubble Sort Program. Click the **Start** button to execute the program and place the form on the screen. Click the **Run** button. An unsorted column of random numbers appears. Below it appears the same list of numbers, but sorted from low to high. Click the **Run** button again. A new unsorted list of numbers appears and below it the same list, sorted. Continue to click the **Run** button for more lists. To quit the program, click the **Exit** button.

Understanding the Bubble Sort Program. The Bubble Sort program consists of two procedures in addition to the exit procedure. The first procedure, **cmdRun_Click**, carries out three actions.

1. It generates ten random numbers and stores them in the array **List(50)**.
2. It prints the unsorted list of numbers on the form.
3. It calls the second procedure named **SortTheNumbers**.

The **SortTheNumbers** procedure uses two nested **For ... Next** loops to sort the numbers in ascending order. The numbers either "bubble up" or "bubble down" depending on whether the < or > operator is used in the conditional statement: **If List(I) > List(J) Then....**

```
For I = 1 To N - 1
   For J = I + 1 To N
      If List(I) > List(J) Then
         Temp = List(I)
         List(I) = List(J)
         List(J) = Temp
      End If
   Next J
Next I
```

Suppose we have an unsorted list of eight numbers so that the outer loop passes seven times ($N - 1 = 7$). Compare this loop with the last loop in the previous program (the TotoLoto program). The loops are quite similar, but in the TotoLoto program, the conditional check is for equality.

Table 14-13

The Sorting Progress with the Bubble Sort Program

Unsorted	44	80	53	47	4	9	39	46
After $I = 1$	4	80	53	47	44	9	39	46
After $I = 2$	4	9	80	53	47	44	39	46
After $I = 3$	4	9	39	80	53	47	44	46
After $I = 4$	4	9	39	44	80	53	47	46
After $I = 5$	4	9	39	44	46	80	53	47
After $I = 6$	4	9	39	44	46	47	80	53
After $I = 7$	4	9	39	44	46	47	53	80

The DesLandres Program

The quantum theory underlying the DesLandres table is reviewed in Excel Example 13, Chapter 3. The algorithm is contained in Equations 3-54, 3-55, 3-56, and 3-57. The observed difference between the two electronic states is $\tilde{\upsilon}_{00}$, which equals 39699.01 cm^{-1} for the transition between v" = 0 in the B electronic state and the v' = 0 vibrational level in the A electronic state (Figure 3-16). The $v' = 0$ to $v'' = 0$ transition corresponds to the center of the spectrum. The DesLandres table is a table of the energies of the general emission spectral lines $\tilde{\upsilon}$ between the B and A states given by Equation 3-57 in Chapter 3.

$$\tilde{\upsilon} = 39699.01 + 1094.8v' - 7.25\, v'^{2} - 1329.38v'' + 6.98\, v''^{2} \qquad (14\text{-}11)$$

Starting the Project

- Start Visual Basic by clicking the **VB** icon on the desktop.
- Select the **Standard EXE** icon in the **New Project** window, and click the **OK** button.
- A **New Project** window opens with a fresh form.

- Select the form, and in the properties window, change the **(Name)** property to **frmDeslandres**.
- Click **Save**; the file name will be frmDeslandres, so click **OK**.
- The file name changes to Project1; change it to **The DesLandres Program**, and click the **OK** button.
- Place controls on frmDeslandres according to Figure 14-13.
- Set the properties of the controls according to Table 14-14.

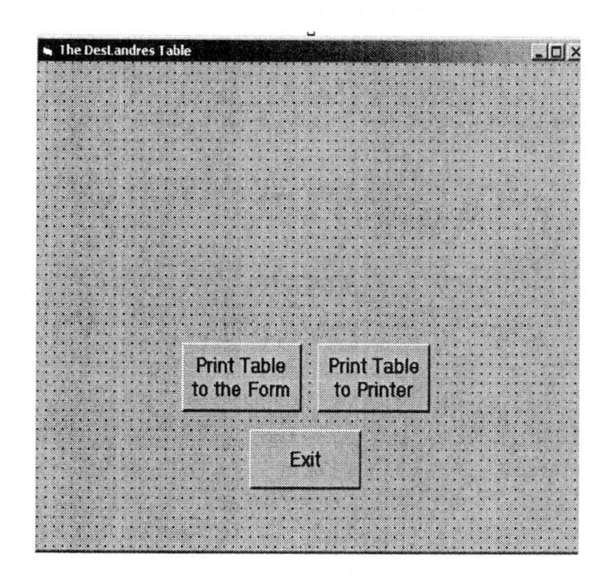

Figure 14-13. The Form for the DesLandres Program

Table 14-14

The Properties Table for the DesLandres Program

Object	Property	Value
Form1	Name	frmDeslandres
	Caption	The Deslandres Table
Command1	Name	cmdPrinter
	Caption	Print Table to Printer
	Font	12 point bold
Command2	Name	cmdPrintToForm
	Caption	Print Table to Form
	Fond	12 point bold
Command3	Name	cmdExit
	Caption	Exit
	Fond	12 point bold

Entering the Code of the DesLandres Program. Enter the following code in the general declarations section:

```
Option Explicit
' The DesLandres program demonstrates how to handle a two-
' dimensional array. The DesLandres table is a table of
' electronic transition for a diatomic molecule.
' The user may print the table to the form, the printer, or a file.
Dim Vp As Integer
Dim Vpp As Integer
Dim DesLandresTable(0 To 10, 0 To 10) As Single
```

Enter the following code in the **cmdExit_Click()** procedure:

```
Private Sub cmdExit_Click()
   End
End Sub
```

Enter the following code in the **cmdPrinter_Click()** procedure:

```
Private Sub cmdPrinter_Click()                    ' Prints the table to the printer
   TitleforPrinter
   For Vp = 0 To 6
      For Vpp = 0 To 6
         Printer.Print ; vbTab; DesLandresTable(Vp, Vpp);
      Next Vpp
      Printer.Print
   Next Vp
   Printer.EndDoc                                 'Tells printer that it's OK to print
End Sub
```

Enter the following code in the **_Load()** procedure:

```
Private Sub Form_Load()
   Dim Upper As Single
   Dim Lower As Single
   Dim WaveNumber As Single
   Dim Center As Single
   Center = 39699.01                              ' Energy above ground electronic state
   ' Load the 2-D array
   For Vp = 0 To 6                                ' Upper electronic state quantum numbers
      For Vpp = 0 To 6                            ' Lower electronic state quantum numbers
         Upper = 1094.8 * Vp - 7.25 * Vp ^ 2      ' Upper state energy
         Lower = 1329.38 * Vpp - 6.98 * Vpp ^ 2   ' Lower state energy
         WaveNumber = Center + (Upper - Lower)    'Transition energy
         DesLandresTable(Vp, Vpp) = WaveNumber    ' Fill the table
      Next Vpp
   Next Vp
End Sub
```

Enter the following code in the **cmdPrintToForm_Click()** procedure

```
Private Sub cmdPrintToForm_Click()                'Prints the table to the form `
   frmDeslandres.Show
   TitleForForm                                   'Prints the title to the form
   For Vp = 0 To 6
      For Vpp = 0 To 6
         frmDeslandres.Print vbTab; DesLandresTable(Vp, Vpp);
      Next Vpp
      frmDeslandres.Print                  ' Prints nothing, but generates a carriage return
   Next Vp
End Sub
```

On the **Tools** menu, click **AddProcedure** to open the **Add Procedure** dialog box. In the **Name** box, type **TitleForForm** as the name. In the **Type** section, select **Sub**; and in the **Scope** section, select **Public**. Click **OK.** Enter the following code:

```
Public Sub TitleForForm()
For Vp = 0 To 5
      frmDeslandres.Print     ' Insert 5 blank lines before printing output
   Next Vp
   frmDeslandres.FontSize = 10
   frmDeslandres.FontBold = True
   frmDeslandres.Print Space(50) & "A DesLandres Table for PN Molecule"
   frmDeslandres.FontBold = False
   frmDeslandres.FontSize = 6
   frmDeslandres.Print
End Sub
```

On the **Tools** menu, click **AddProcedure** to open the **Add Procedure** dialog box. In the **Name** box, type **TitleForPrinter** as the name. In the **Type** section, select **Sub**; and in the **Scope** section, select **Public**. Click **OK.** Enter the following code:

```
Public Sub TitleforPrinter()
Printer.Print
   Printer.Print
   Printer.FontSize = 12
   Printer.FontBold = True
   Printer.Print Space(30) & "A DesLandres Table for PN Molecule"
   Printer.FontBold = False
   Printer.FontSize = 10
   Printer.Print
End Sub
```

Executing the DesLandres Program. Click the VB **Start** button to execute the DesLandres program. The form appears with three command button controls as shown in Figure 14-13. Click the button labeled **Print Table to Printer**, which results in the following table being printed by the printer (Table14-15).

Table 14-15

The DesLandres Table for the PN Molecule as Printed by the Printer

39699.01	38376.61	37068.17	35773.69	34493.17	33226.61	31974.0
40786.56	39464.16	38155.72	36861.24	35580.72	34314.16	33061.5
41859.61	40537.21	39228.77	37934.29	36653.77	35387.21	34134.6
42918.16	41595.76	40287.32	38992.84	37712.32	36445.76	35193.1
43962.21	42639.81	41331.37	40036.89	38756.37	37489.81	36237.2
44991.76	43669.36	42360.92	41066.44	39785.92	38519.36	37266.7
46006.81	44684.41	43375.97	42081.49	40800.97	39534.41	38281.8

Click the **Print Table to the Form** button and the program prints the Deslandres table to the form (Figure 14-15). Click **Exit** to end the program.

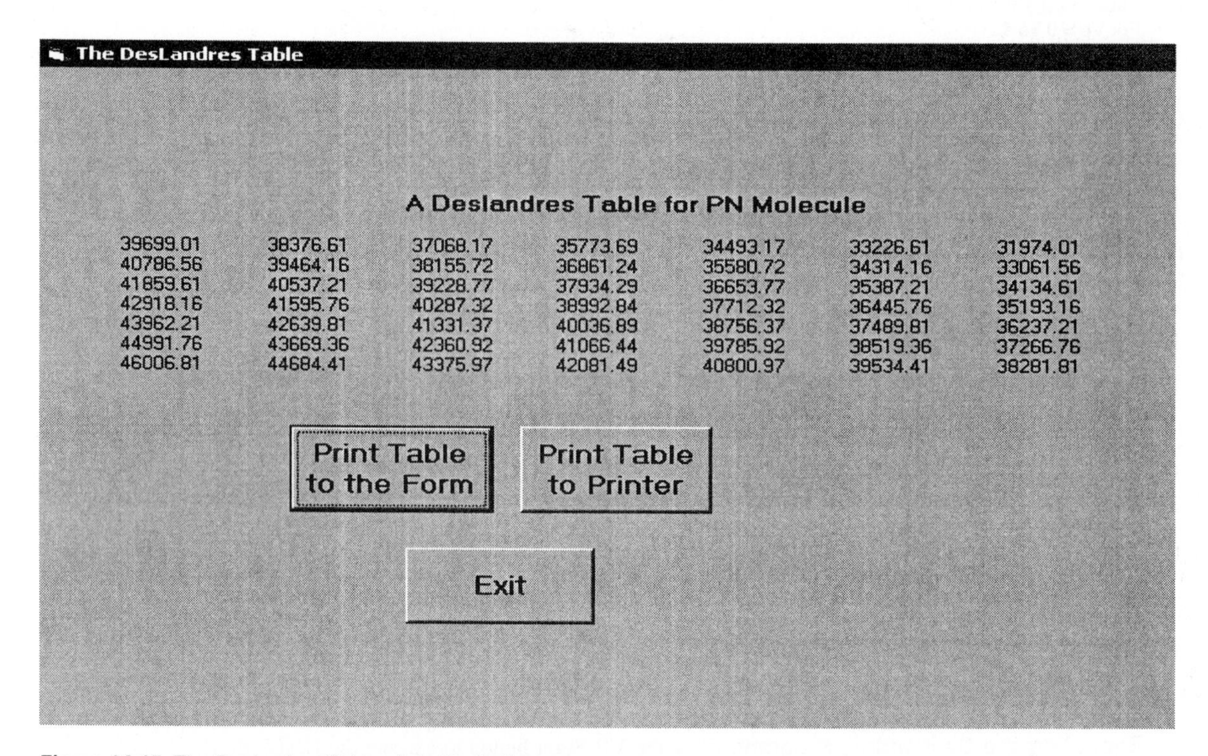

Figure 14-15. The Deslandres Table printed to the form

Understanding the Code of the DesLandres Program

After the **Option Explicit** statement, a few comments serve to remind the programmer what the purpose of the program is. Three variables are declared in the general declarations section of the program.

```
Dim Vp As Integer
Dim Vpp As Integer
Dim DesLandresTable(0 To 10, 0 To 10) As Single
```

Vp and **Vpp** are the two counters for the two-dimensional array named **DesLandresTable** and correspond to the upper level and lower level quantum numbers v' and v'', respectively.

Clicking the Visual Basic **Start** button causes the **Form_Load** procedure to execute. The **Form_Load** procedure begins with declaration of the variables in Equations 3-54, 3-55, and 3-56.

```
Dim Upper As Single
Dim Lower As Single
Dim WaveNumber As Single
Dim Center As Single
```

and assigns a value to the variable **Center**:

```
Center = 39699.01
```

Then the transition energies are calculated and assigned their place in the **DesLandresTable** array:

```
For Vp = 0 To 6
  For Vpp = 0 To 6
    Upper = 1094.8 * Vp - 7.25 * Vp ^ 2                ' Equation 3-54
    Lower = 1329.38 * Vpp - 6.98 * Vpp ^ 2             ' Equation 3-55
    WaveNumber = Center + (Upper - Lower)              ' Equation 3-56
      DesLandresTable(Vp, Vpp) = WaveNumber            ' Fill the table
  Next Vpp
Next Vp
```

The program has now filled the **DesLandresTable** array with values and waits for the user to click a command button. Clicking the **Print Table to the Form** button causes the **cmdPrintToForm_Click** procedure to execute and to print the DesLandres table on the form (Figure 14-15).

```
Private Sub cmdPrintToForm_Click()          'Prints the table to the form  `
  frmDeslandres.Show
  TitleForForm                               'Prints the title to the form
  For Vp = 0 To 6
    For Vpp = 0 To 6
      frmDeslandres.Print vbTab; DesLandresTable(Vp, Vpp);
    Next Vpp
    frmDeslandres.Print                      ' Prints nothing
  Next Vp
End Sub
```

Notice the statement near then end: **frmDeslandres.Print**. This statement prints nothing, but serves to generate a carriage return so that a new row can be printed. Try removing it to see what happens.

The **PrintToForm** procedure calls the **TitleForForm()** procedure to print a title on the form in its larger, darker font.

```
Public Sub TitleForForm()
  For Vp = 0 To 5
    frmDeslandres.Print                      ' Insert 5 blank lines before printing output
  Next Vp
  frmDeslandres.FontSize = 10
   frmDeslandres.FontBold = True
   frmDeslandres.Print Space(50) & "A DesLandres Table for PN Molecule"
   frmDeslandres.FontBold = False
   frmDeslandres.FontSize = 6
   frmDeslandres.Print
End Sub
```

Clicking the **Print Table to Printer** button prints the table (Figure 14-15 on the printer. This procedure for printing on the printer is similar to the procedure for printing on the form. This procedure calls the **TitleForPrint** procedure, which is similar to the **TitleForForm()** procedure. The Exit button exits the program.

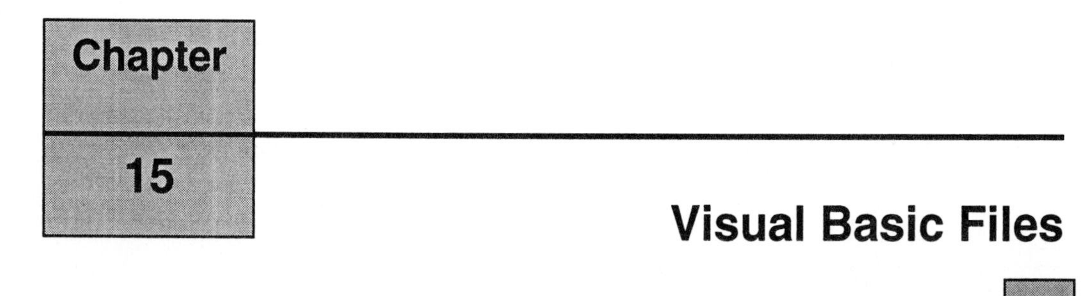

Visual Basic Files

You are already familiar with the concept of files as collections of information, data, or programs stored on floppy, hard, or compact disks. The programs you buy are distributed as files on disks. You save your word processing documents as files, and your spreadsheet workbooks as files. You have stored your Visual Basic programs as files. We will use text files as an introduction to Visual Basic files. In the short programs that we've written until now, output was made available to the user by printing it on the form (**Form.Print**) or printing it with the printer (**Printer.Print**). By sending the output to a file, you can save it for later use. Using files in Visual Basic involves three steps:

1. Opening the file.
2. Writing to or reading from the file.
3. Closing the file.

This chapter contains the following Visual Basic programs:

The Write to File program
The Write to File 2 program
The Write to File 3 program
The Write to a File from a List program
The Write to a File from a Table program
The Read a File of Names program
The Read X_iY_i File program
The Show the Boxes program
The CD Color program
The CD Printer program
The CD Fonts program
The CD Open File program

Opening, Closing, and Deleting a File

The syntax for the statement for opening a file is

Open *pathname* For *mode* As #*n*

The ***pathname*** is a string that gives the name of the file and must include the complete path to the file. Quotation marks must enclose a pathname, since is it a string. For example:

"c:\MyDocuments\MyFile.txt"

Since a program may have more than one file open at the same time, each file must be given a number from 1 to 255. The number sign (#) is optional, but it is usually included. The three most common modes are **Output**, **Append**, and **Input**.

Opening a File for Output. Opening a file for output prepares your Visual Basic program to write to the file. If the file has not yet been created, then the **Open** statement creates the file with the name and path given in the pathname. If the file already exits, then the **Open** statement erases the file and then re-creates the file with the same name and path given in the pathname. For example:

Open "c:\MyFile.Txt" For Output As # 1

Once a file is open, data contained in an expression list can be written to the open file. Normally, the **Write #*n*** command follows the **Open For Output** statement with the syntax described as follows:

Write #1, *expression list*

The ***expression list*** is a list of all the items that you want to write to the file: strings, numbers, and so on, each separated from another by a comma (the list separator).

Opening a File for Append. Opening a file with an **Open For Append** statement prepares your program to append data to the data in an existing file. If the file to which material is to be appended does not exit, the **Append** statement creates the file with the name and path given in the pathname. In this case, the data is appended to an empty file. Subsequent execution of this **Append** statement appends data to the now existing data in that file.

Open "c:\MyFile.Txt" For Append As # 1

Normally, the **Write #*n*** command follows the **Open For Append** statement:

Write #1, *expression list*

Opening a File for Input. Opening a file for with an **Open For Input** statement prepares your program to read data from an existing file. The syntax for opening an existing file for input is as follows:

Open "c:\MyFile.Txt" For Input As # 1

If the file from which data is to be read does not exist, Visual Basic issues a run-time error. Normally, the **Input #*n*** command follows the **Open For Input** statement:

Input #1, *expression list*

Closing Files. After writing to or reading from a file, the file is closed with the **Close** command. The syntax for closing one or more files is

Close #1

Or, if several files are open:

Close #1, #2, #5

To close all files, the command is:

Close

Deleting Files. To delete a file from within a program, use the **Kill** command:

```
Kill "c:\Myfile.txt"
```

Writing to a File

With the **Write #*n*** command you can write constants, numbers, strings, and variables to a file. This syntax for the **Write #*n*** command is

Write #*n*, *expression list*

The ***n*** is the file number, usually **1**, and this file must previously have been opened for output. The ***expression list*** is a list of values that can be written, each separated by a comma, the list separator. For example:

```
Age = 52
Weight = 168.7
Write #1, "George", Age, Weight
Write #1, "Testing 1, 2, 3"
```

If, previously, **52** was assigned to the variable **Age**, and **168.7** was assigned to the variable **Weight**, then the **Write #1** command would write the following two lines:

```
"George", 52, 168.7
"Testing 1, 2, 3"
```

When the **Write #1** command finds no list separator (comma) after the last item in the first line (**Weight**), it inserts a carriage return and line feed. If a comma were present (**Weight,**), then the consecutive **Write #1** statements would write all these items on a single line:

```
"George", 52, 168.7,"Testing 1, 2, 3"
```

Let's try out a short, simple program, the Write to File program, to experiment with the **Open**, **Write**, and **Close** commands. Open a new program in Visual Basic.

Writing to Files: Sample Programs. The following programs in this chapter serve to illustrate some of the features of writing from a program to an external file:

The Write to File program
The Write to File 2 program
The Write to File 3 program
The Write to a File from a List program (the IR Spectrum program revisited)
The Write to a File from a Table program (the DesLandres program revisited)

The first three of the preceding programs stand on their own, but the last two are file-writing procedures that are added to previously written programs. A later section in this chapter contains some programs that illustrate reading an external file by a program.

The Write to File Program

In this short, simple demonstration program, some of the items in the list of expressions that are written to the file are values of a numeric or string variable. A simple calculation within the program provides a numeric value to be written to the file. The file that is created is saved so that its contents can be examined with a word processor or Microsoft Notepad. The effect of the comma as a list separator is demonstrated with the **Write** statement.

Entering the Code of the Write to File Program

- Start Visual Basic by clicking the **VB** icon on the Windows desktop.
- In the **New Project** dialog box, click the **Standard EXE** icon, and click the **OK** button.
- A **New Project** window opens with a fresh form.
- Click the form to select it. Then, in the properties window, change the **(Name)** property to **frmWriteFile.**
- Click **Save**; the file name will be frmWriteFile, so click **OK.**
- The file name changes to Project1; change it to **The Write to File Program**, and click the **OK** button.
- No controls need be placed on frmWriteFile.
- In the properties window:
    1. Change the **(Name)** property of Form1 to **frmWriteFile.**
    2. Change the **Caption** property of Form1 to **The Write to File Program.**

Enter the following code in the general declarations section:

```
Option Explicit
' The Write to File program writes a few strings
' and numbers as text to a file named C:\MyFile.txt
' You can examine the file with Microsoft Notepad
```

Enter the following code in the **Form_Load** procedure:

```
Private Sub Form_Load()
  Dim J As Integer
  Dim T As Single
  T = 3 * 5 / 7                              ' Do a little calculation
  Open "c:\MyFile.txt" For Output As #1      ' Open a file
  Write #1, "Experiments with Write #1 "     ' Write a title
  Write #1,                                  ' Write a blank line
  Write #1, 12                               ' Write a number
  Write #1, "Red"                            ' Write a string
  Write #1, "Apples"                         ' Write another string
  Write #1,                                  ' Write a blank line
  Write #1, 12,                              ' Write a number + comma
  Write #1, "Red",                           ' Write a string + comma
  Write #1, "Apples"                         ' Write another string
  Write #1,                                  ' Write a blank line
  Write #1, 12,"Red","Apples"                ' Write an output list
  Write #1,                                  ' Write a blank line
  Write #1, T                                ' Write an unformatted value
  Write #1,                                  ' Write a blank line
```

```
    Write #1, Format(T,"##.###")              ' Write a value
    Write #1,                                 ' Write a blank line
    Write #1, Format(Now,"Short Date")        ' Write a formatted value
    Close #1                                  ' Close a file
    ' So you can see something happened
   frmWriteFile.Show

    frmWriteFile.Print "The program has executed. "
    frmWriteFile.Print "Examine the output in MS Notepad"
    frmWriteFile.Print "The path is: c:\MyFile.txt"
  End Sub
```

Executing the Write to File Program. Click the **VB Start** button to execute the program. The file is quietly and quickly written. You wouldn't know that anything happened if the program hadn't written to the form the message **The program has executed** (Figure 15-1). If you had given the file name a:\MyFile.txt, then you would hear drive A operate momentarily.

Viewing the Output of the Write to File Program

Examine the output of the Write to File program in Microsoft Notepad.

- Click the Windows **Start** button.
- On the **All Programs** menu, point to **Accessories**, and click **Notepad**.
- After Notepad opens, click the **Open** icon,
- The file you have just created is in the file directory list as MyFile.txt. Double-click **MyFile.txt** to open it.

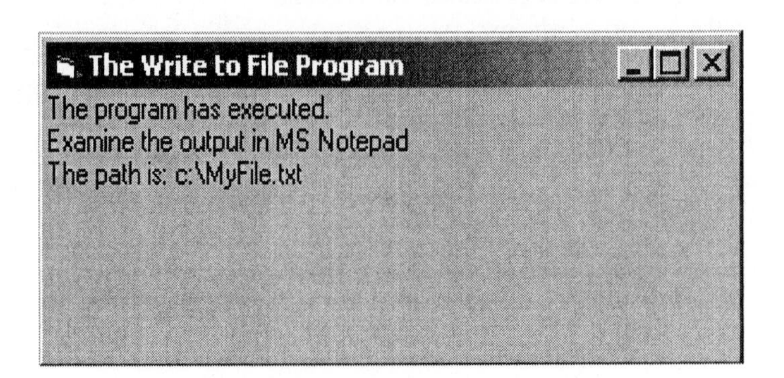

The output of the program, written to the file MyFile, opens in Microsoft Notepad and appears as follows:

```
"Experiments with Write #1 "

12
"Red"
"Apples"

12,"Red","Apples"

12,"Red","Apples"
```

2.142857

"2.14"

"9/22/2004"

Notice that strings are written with quotation marks. You can play with this program by making a few changes and observing the results. Instead of writing to MyFile.txt in the drive C root directory, try writing to a subdirectory or to a blank disk in drive A.

Understanding the Write to File Program. Some characteristics of Visual Basic's **Write #*n*** command are evident.

- Unlike the **Form.Print** and **Printer.Print** statements, the **Write #*n*** command writes all strings between quotation marks. In the preceding output listing, notice that numbers are not enclosed in quotation marks.
- The **Write #1** command followed by an output list adds a carriage return and line feed to the end of each line.
- The **Write #1** statement alone generates a blank line. The trailing comma (the list separator) is necessary.
- The **Write** command writes integers and numbers (single in this case) as expected in their default format. They can, of course, be formatted with the **Format** function as shown in the program.
- The **Write** command writes dates and times the same way that the **Print** method prints dates and times to the form or printer. The usual formatting function can be used to change the default format.

Compare the output that appears in Notepad with the code of the Write to File program. Check whether the characteristics listed earlier are consistent with the observed output.

The Write to File 2 Program

For more practice in writing to a file, let us write a program that carries out some simple calculations in a **For ... Next** loop and then writes pairs of data on each output line. This situation arises frequently in science and engineering when dealing with dependent and independent variables that occur in experimental work.

Beginning the Project

- Start Visual Basic by clicking the **VB** icon on the Windows desktop.
- In the **New Project** dialog box, click the **Standard EXE** icon, and click the **OK** button.
- A **New Project** window opens with a fresh form.
- Click the form to select it.
- Click **Save**; the file name will be Form, so click **OK**.
- The file name changes to Project1; change it to **The Write to File 2 Program**, and click the **OK** button.
- No controls need be placed on Form1.
- In the properties window, change the caption to **The Write to File 2 Program**.

Entering the Code for the Write to File 2 Program

```
Option Explicit
' The Write to File 2 program
Private Sub Form_Load()
   Dim J As Integer, K As Integer, L As Integer
   Dim T As Single
   K = 10
   Open "c:\MyFile2.txt" For Output As #1
   For J = 1 To 8
      T = (J * K) ^ 1.5
                  Write #1, J, T ' This writes values of J and T on the same line
   Next
   Close #1
   Form1.Show
   Form1.Print "The program has executed. "
   Form1.Print "Examine the output in Microsoft Notepad"
End Sub
```

Executing the Write to File 2 Program. Click the **VB Start** button to execute the program. The message on the form is similar to Figure 15-1

Viewing the Output of the Write to File 2 Program. Examine the output of the Write to File 2 program in Microsoft Notepad.

- Click the Windows **Start** button.
- On the **All Programs** menu, point to **Accessories**, and click **Notepad**.
- After Notepad opens, click the **Open** icon,
- The file you have just created is in the file directory list as MyFile2.txt. Double-click **MyFile2.txt** to open it.

The output of MyFile2 appears in the Notepad text box:

```
1,31.62278
2,89.44272
3,164.3168
4,252.9822
5,353.5534
6,464.758
7,585.662
8,715.5417
```

The Write to File 3 Program

This short program creates a short temperature conversion table. It uses a `For ... Next` loop to calculate the temperature values, which are then written to a file that can be read with Microsoft Notepad or some other word processor. The program also writes a text heading to the file.

Beginning the Project

- Start Visual Basic by clicking the **VB** icon on the Windows desktop.
- In the **New Project** dialog box, click the **Standard EXE** icon, and click the **OK** button.
- A **New Project** window opens with a fresh form.
- Click the form to select it.

- Click **Save**; the file name will be Form1, so click **OK**.
- The file name changes to Project1; change it to **The Write to File 3 Program**, and click the **OK** button.
- No controls need be placed on Form1.
- In the properties window, change the caption to **The Write to File 3 Program.**

Entering the Code of the Write to File 3 Program

```
Option Explicit
' The Write to File 3 program
' This program creates a short
' table of temperature conversions
' and writes the number to a text file

Private Sub Form_Load()
   Dim sngC As Single, sngF As Single
   Open "c:\MyFile3.txt" For Output As #1
   Write #1, Tab(5), "C deg", Tab(15), "F deg"
   Write #1,
   For sngC = 0 To 100 Step 10
      sngF = 9 * sngC / 5 + 12
      Write #1, Tab(5), sngC, Tab(15), sngF
   Next
   Write #1,
   Close #1
   Form1.Show
   Form1.Print "The program has executed. "
   Form1.Print "Examine the file output in Microsoft Notepad"
   Form1.Print "The path to the file is: c:\MyFile3"
End Sub
```

Executing the Write to File 3 Program. Click the VB **Start** button to execute the Write to File 3 program. The message on the form is similar to Figure 15-1. In the same way, examine the file output with Microsoft Notepad, which appears as follows.

```
,"C deg", ,"F deg"

,0,    ,12
,10,   ,30
,20,   ,48
,30,   ,66
,40,   ,84
,50,   ,102
,60,   ,120
,70,   ,138
,80,   ,156
,90,   ,174
,100,  ,192
```

Understanding the Write to File 3 Program. After the variables are dimensioned, the file named MyFile3 is opened, and a title line for the output that follows is written. After **Write #1** writes a blank line, a **For ... Next** loop calculates temperatures in Fahrenheit (**sngF**) from temperatures in Celsius (**sngC**) and prints each pair on a line. The commas separating the items in the output list keep the writing on the same line. Not until the end of the output list is reached, indicated by the absence of a comma, does Visual Basic insert a line feed and carriage return, for writing the next line. After you close the file, the program prints three lines of text to the form to ensure the user that the program ran and that the file was written.

The Write to a File from a List Program

For this demonstration of writing to a file, we will write a short file-writing procedure and copy it into a previously written program (the IR Spectrum program). To save time, let's revisit the IR Spectrum program and add a procedure that writes from an array of output data to a file. The IR Spectrum program generates two lists, one of the spectral line wavelengths and the other of the spectral line energies. The following procedure copies the data in the two lists to a file named IRSpec.txt which can be read as usual with Microsoft Notepad (or any other word processor).

Starting the Project

- Open the IR Spectrum Program (Chapter 14).
- Drop a command button on the form of the IR Spectrum program between the **Run** and **Exit** buttons already in place.
- In the Visual Basic properties table, change the name of the new command button to **cmdWriteFile**.
- In the properties table, change the **Caption** property to **Write to a file**.
- Double-click cmdWriteFile, and add the following code:

```
Private Sub cmdWriteFile_Click()
' Add this procedure to the IR Spectrum program
   Dim Index As Integer
   Open "c:\IRSpec.txt" For Output As #1
   Write #1, " The Infrared Spectrum of HCl"
   For Index = 0 To 12                    'Index runs from -6 to +6
      Write #1, Index – 6 & vbTab _
      & vbTab & Format(EnergyAry(Index), "#####.00") & vbTab _
      & vbTab & Format(LengthAry(Index), "#.00000")
   Next
   Close #1
   cmdWriteFile.FontSize = 18
   cmdWriteFile.Caption = "OK!"
End Sub
```

Executing the Write to a File from a List Program. This program is the IR Spectrum program with the file-writing procedure. Click the **Start** button as usual. The form appears as shown in Figure 15-2, with three command buttons, labeled **Run**, **Exit**, and **Write to a File**. Click the **Run** button, and the output appears printed to the screen as shown in Figure 15-2. When you click the **Run** button of the IR Spectrum program, the wavelengths and energies are printed to the form, as shown in Figure 15-2. When you click the **Write to a File** button, the caption on the button changes from **Write to a File** to **OK!** (Figure 15-2). This lets the user know that the file writing did indeed take place OK. As usual, examine the output with Microsoft WordPad. This time, open the file named IRSpec.txt, and the file shows the following list of data.

```
" The Infrared Spectrum of HCl"
"-6        2752.00        3.63373"
"-5        2775.71        3.60269"
"-4        2798.88        3.57286"
"-3        2821.50        3.54422"
"-2        2843.55        3.51673"
"-1        2865.02        3.49037"
"0         2885.90        3.46512"
"1         2906.17        3.44095"
"2         2925.82        3.41784"
"3         2944.84        3.39577"
"4         2963.21        3.37472"
"5         2980.92        3.35467"
"6         2997.96        3.33560"
```

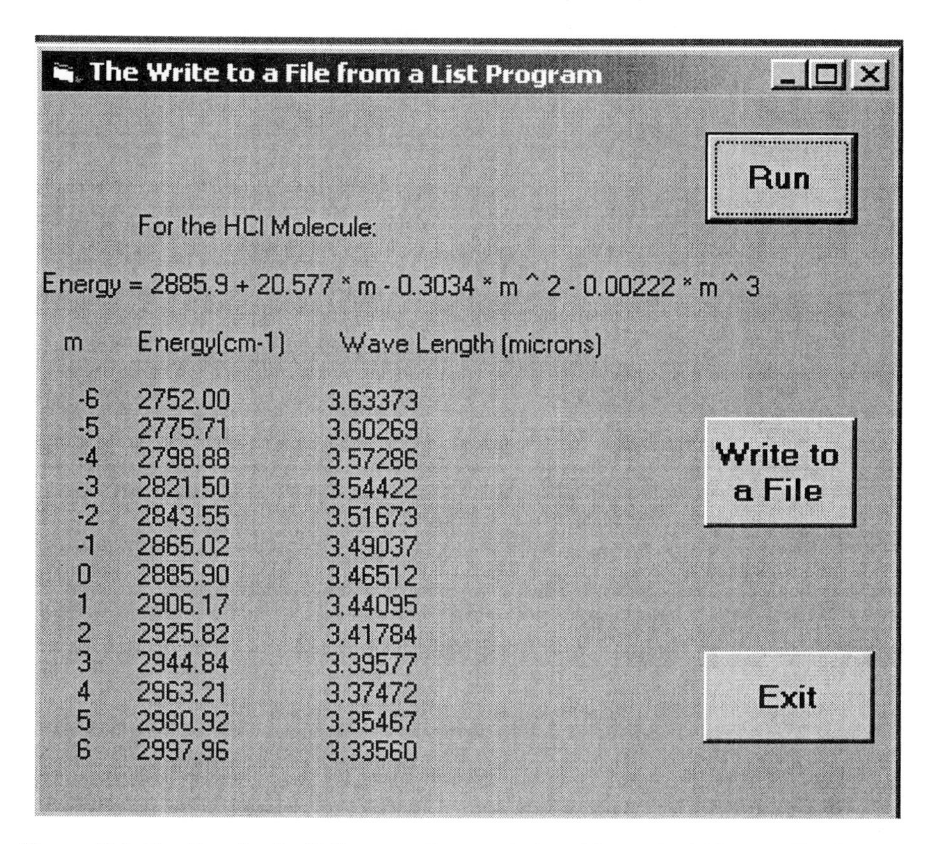

Figure 15-2. The Form for the IR Spectrum Program before Clicking the **Write to a File** Button

Understanding the IR Spectrum Program with a File Writing Procedure. The first third of the **cmdWriteFile** procedure takes care of formalities: **Index** is declared to be a variable of type **Integer**, the file is opened, and the **Write #1** command writes a title from the program to the file.

```
Private Sub cmdWriteFile_Click()
    ' Add this procedure to the IR Spectrum program
        Dim Index As Integer
        Open "c:\IRSpec.txt" For Output As #1
        Write #1, " The Infrared Spectrum of HCl"
```

The middle of the procedure is nearly identical to the last **For ... Next** loop in the **cmdRun** procedure of the original IR Spectrum program, except that the **Write #1** command replaces the **frmIRSpec.Print** command.

```
For Index = 0 To 12
  Write #1, Index - 6 _
  & vbTab & Format(EnergyAry(Index), "#####.00") & vbTab _
  & vbTab & Format(LengthAry(Index), "#.00000")
Next
```

The last third of the procedure closes the file and then changes the caption of the **cmdWriteFile** command button. At design time, the caption was **Write to a File**. When the **cmdWriteFile** procedure executes, its last two statements increase the font size to 14 and change the caption to **OK!** This caption reassures the user that the file was written OK.

```
Close #1
  cmdWriteFile.FontSize = 14
  cmdWriteFile.Caption = "OK!"
End Sub
```

The Write to a File from a Table Program

With the addition of a single procedure, the DesLandres program can send the DesLandres table to a file that can be saved on a hard disk or floppy disk. It can then be opened in Microsoft WordPad (or any word processor), read, and printed if desired.

Starting the Project. Add the following procedure to the DesLandres program from Chapter 14:

- Open the DesLandres Table program (Chapter 14).
- Drop a command button on the form of the Deslandres Table program to the right of the **Print Table to Printer** button (Figure 15-3).
- In the Visual Basic properties Window, change the name of the new command button to **cmdWriteFile**.
- In the Visual Basic properties window, make the **Caption** property **Write Table to a file.**
- Double-click the **Write Table to a File** button, and add the following code:

```
Private Sub cmdWriteFile_Click()
  'This procedure is optional for writing to a file
  Open "c:\Deslandres.txt" For Output As #1
  Write #1, "DesLandres Table for PN"
  Write #1,

  For Vp = 0 To 6
    For Vpp = 0 To 6
      Write #1, vbTab; DesLandresTable(Vp, Vpp);
    Next Vpp
    Write #1,
  Next Vp

  Close #1
End Sub
```

Executing the DesLandres Program with cmdWriteFile. Click the **Start** button to execute the program. The form appears (Figure 15-3), offering the user the choice of clicking four buttons. Clicking the **Print Table to the Form** button, the **Print Table to the Printer** button, or the **Exit** button gives results identical to the results of the DesLandres program described in Chapter 14. The program prints the DesLandres table to the form or prints it to the printer or exits the program.

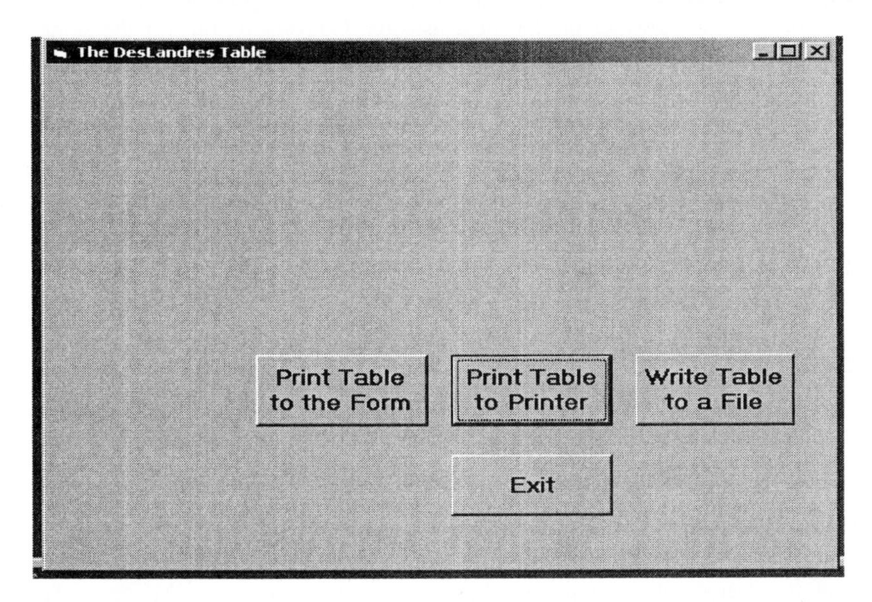

Figure 15-3. The Form for the DesLandres Program with the **Write Table to a File** Button

Clicking the **Write Table to a File** button writes the DesLandres table to the file DesLandres.txt. It appears similar to Table 14-15 when viewed in Microsoft Notepad or printed from Microsoft Notepad as seen below. The extraneous apostrophes and commas can be removed

```
"DesLandres Table for PN"

"  ",39699.01,"  ",38376.61,"  ",37068.17,"  ",35773.69,"  ",34493.17,"  ",33226.61,"  ",31974.01,
"  ",40786.56,"  ",39464.16,"  ",38155.72,"  ",36861.24,"  ",35580.72,"  ",34314.16,"  ",33061.56,
"  ",41859.61,"  ",40537.21,"  ",39228.77,"  ",37934.29,"  ",36653.77,"  ",35387.21,"  ",34134.61,
"  ",42918.16,"  ",41595.76,"  ",40287.32,"  ",38992.84,"  ",37712.32,"  ",36445.76,"  ",35193.16,
"  ",43962.21,"  ",42639.81,"  ",41331.37,"  ",40036.89,"  ",38756.37,"  ",37489.81,"  ",36237.21,
"  ",44991.76,"  ",43669.36,"  ",42360.92,"  ",41066.44,"  ",39785.92,"  ",38519.36,"  ",37266.76,
"  ",46006.81,"  ",44684.41,"  ",43375.97,"  ",42081.49,"  ",40800.97,"  ",39534.41,"  ",38281.81,
```

Understanding the DesLandres Program with cmdWriteFile. Addition of a single new procedure to the DesLandres program described in Chapter 14 permits the user to write the DesLandres table, a two-dimensional array, to a file. The **cmdWriteFile_Click** procedure is very similar to the already existing **cmdPrintoToForm** procedure. The **For ... Next** loops in these two procedures differ only in the substitution of the statement:

```
Write #1, vbTab; DesLandresTable(Vp, Vpp);
```

for the statement:

```
frmDeslandres.Print vbTab; DesLandresTable(Vp, Vpp);
```

Reading from Files

Now that we've written data to a file, the next step is to read data from a file. When a program reads data from a file, it assigns the data to a variable or object property in the program. Usually, a file is a long list of items of the same type, so it is quite natural to read from a file and assign the items to the elements of an array or the list of a list box. Conversely, but for the same reason, programs usually write data from an array to a file. Later, we will involve arrays and files, but for now we will keep the file demonstration programs simpler.

Reading from Files: Sample Programs. The following programs in this chapter serve to illustrate some of the features of reading from a file:

The Read a File of Names program
The Read X_iY_i File program

The Read a File of Names Program

This program reads the items in the file named c:\NameFile.txt and then assigns these items to the list of a list box.

Creating a File for the Program to Read. If we write a program that reads a file, then we must have a file that can be read, so let us create one. A text file consisting of a list of names is suitable and can be created with a simple word processing program such as Microsoft WordPad. Open WordPad, and enter a list of names in random order, with no spaces preceding the names, and no punctuation marks following the names.

Gilbert
Dalton
Thompson
Becquerel
Bohr
Einstein
Rutherford
Curie
Heisenberg
Fermi
Hahn
Meitner

Save this text file with the name c:\NameFile.txt.

Starting the Project. Start Visual Basic by clicking the **VB** icon on the Windows desktop.

- In the **New Project** dialog box, click the **Standard EXE** icon, and click the **OK** button.
- A **New Project** window opens with a fresh form.
- Click the form to select it. Then, in the properties window, change the (**Name**) property to **frmReadFile**.
- Click **Save**; the file name will be frmReadFile, so click **OK**.
- The file name changes to Project1; change it to **The Read a File of Names Program**, and click the **OK** button.
- Place controls on frmReadFile according to Figure 15-4.
- Set the properties of the controls according to Table 15-1.

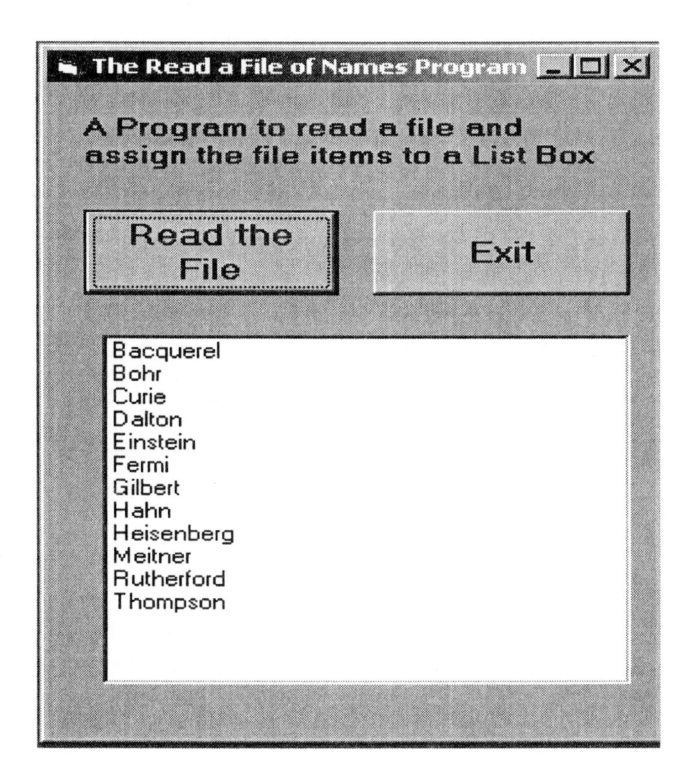

Figure 15-4. The Form for the Read a File of Names Program

Table 15-1

The Properties Table for the Read a File of Names Program

Object	Property	Setting
Form1	Name	frmReadFile
	Caption	The Read a File of Names Program
Label1	Name	lblInfo
	Caption	A program to read from a file and assign items to a list box
Command1	Name	cmdRead
	Caption	Read the File
Command2	Name	cmdExit
	Caption	Exit
List box1	Name	lstFromFile
	Sorted	True

Entering the Code of the Read a File of Names Program. Enter the following code in the general declarations section.

```
Option Explicit
' The Read a File of Names program reads the items in the
' file named c:\NameFile.txt and then
' assigns these items to the list of a list box
```

Enter the following code for the **Private Sub cmdExit_Click()** procedure:

```
Private Sub cmdExit_Click()
   End
End Sub
```

Enter the following code for the **Private Sub cmdReadFile_Click()** procedure.

```
Private Sub cmdReadFile_Click()
   Dim strScientist As String
   Open "c:\NameFile.txt" For Input As #1
   Do Until EOF(1) = True
     Input #1, strScientist
     lstNameList.AddItem strScientist
  Loop
  Close #1
End Sub
```

Executing the Read a File of Names Program. Click the VB **Start** button to execute the program. The form appears with a large empty list box and two command controls. Click the button labeled **Read the File**. Compare the items in the list box with the output of c:\MyFile displayed in Notepad. The list in the list box is sorted alphabetically, while the original list was not sorted. Click **Exit** to end the program.

Understanding the Read a File of Names Program. Since the user of a file-reading program ordinarily does not know how many lines a file holds, Visual Basic provides the **EOF** function to tell the program when the end of a file has been reached. The syntax for the **EOF** function is

```
EOF (n)
```

where ***n*** is the file number. It is not optional to use **#*n***. The **EOF** function returns the Boolean value **False** until the end of the file has been reached. When the end of the file is reached, the **EOF** function returns the Boolean value **True**. For example:

```
Do Until EOF(1) = True
   Input #1, strScientist
   lstNameList.AddItem strScientist
Loop
```

This **Do ... Until** loop does the file reading. The file is opened before it and closed after it. In this sample file-reading program, the lines in the file are added to a list box until the end of the file is reached. When the end of the file is reached, the **EOF** function returns **True**, and execution drops out of the loop.

The default setting for the **Sorted** property is **False**. We set it to **True**, so that our file of names would appear sorted alphabetically. Change the **Sorted** property to **False**, and run the program again.

The Read X_iY_i File Program

Reading Data from a File of X_i, Y_i Pairs. In the experimental laboratory, one of the most common situations that scientists and engineers encounter is the measurement of dependent and independent variables In mathematics, $y = f(x)$ expresses the dependency of y on the independent variable x. In a laboratory experiment, x represents a measured parameter, and the object of the experiment is to determine the dependency of y upon x. Usually the experiment is varied by changing the value of x and measuring the effect upon y, the dependent variable. A series of experiment generates parallel lists of x_i and y_i, and usually these pairs are read into a computer file for further data processing.

The Read X_iY_i File program reads the items in the file named c:\MyXYpairs.txt. It prints the pairs to the screen and adds them to a list box for display.

Creating an X_iY_i File to Read. We can simulate a set of x_i and y_i pairs by using Microsoft WordPad to enter some pairs of numbers separated by commas. For example:

```
1.2,3.5
2.3,4.6
3.5,7.01
4.01,9.03
5.24,11.83
6.54,13.023
7.98,15.32
8.03,16.34
9.32,19.23
10.11,20.45
35.22, 55.66
```

Enter these data, and save this WordPad file with the name MyXYpairs.txt. The purpose of the Read X_iY_i File program is to read the file of numbers listed above and display them to the user. The program demonstrates two ways of displaying the data read from the file. One way is by printing the data to the form; the other way is by adding the item in the file to a list box for display.

In practice, further calculations could be done on the numbers read from the file. In this case the numbers would normally be assigned to parallel one-dimensional arrays and further calculation done before displaying the final results in a list box or printing them to a form.

Starting the Project

- Start Visual Basic by clicking the **VB** icon on the Windows desktop.
- In the **New Project** dialog box, click the **Standard EXE** icon, and click the **OK** button.
- A **New Project** window opens with a fresh form.
- Click the form to select it. Then, in the properties window, change the **(Name)** property to **frmReadFile**.
- Click **Save**; the file name will be frmReadFile, so click **OK**.
- The file name changes to Project1; change it to **The Read X_iY_i File Program**, and click the **OK** button.
- Place controls on frmReadFile according to Figure 15-5.
- Set the properties of the controls according to Table 15-2.

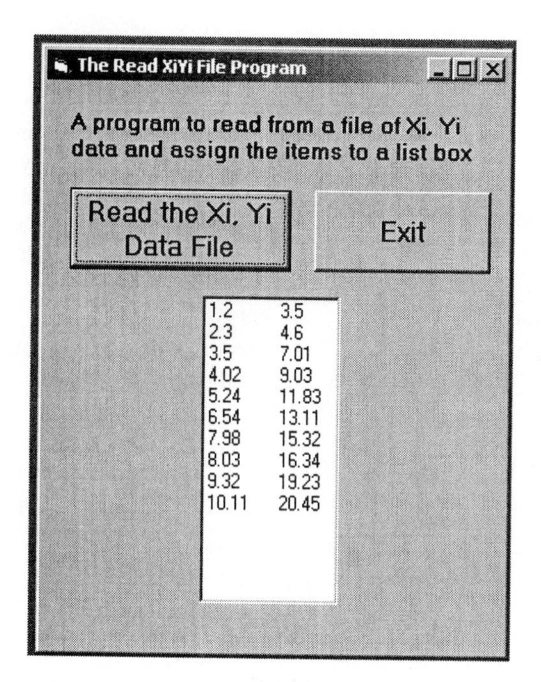

Figure 15-5. The Form for the Read X_iY_i File Program

Table 15-2

The Properties Table for the Read X_iY_i File Program

Object	Property	Setting
Form1	Name	frmReadFile
	Caption	The Read X_iY_i File Program
Label1	Name	lblInfo
	Caption	A program to read from a file of Xi, Yi data and assign the items to a list box
Command1	Name	cmdRead
	Caption	Read the Xi, Yi Data File
Command2	Name	cmdExit
	Caption	Exit
List box1	Name	lstFromFile

Entering the Code of the Read X_iY_i File Program. Enter the following code in the general declarations section:

```
Option Explicit
' The Read XiYi File program reads the items in the
' file named c:\MyXYpairs.txt. It prints the pairs to
' the screen and adds them to a list box for display
```

Enter the following code in the **cmdExit_Click()** procedure:

```
Private Sub cmdExit_Click()
   End
End Sub
```

Enter the following code in the **cmdReadFile_Click()** procedure:

```
Private Sub cmdReadFile_Click()
      Dim strLeft As String, strRight As String
      Open "c:\MyXYpairs.txt" For Input As #1
      Do Until EOF(1) = True
            Input #1, strLeft, strRight
            frmReadFile.Print strLeft, strRight
            lstDataPairs.AddItem strLeft & _
                     vbTab & strRight
      Loop
      Close #1
End Sub
```

Executing the Read X_iY_i File Program. With the Read X_iY_i File program open, click the **Run** button to execute the program. The form appears with an information label, a blank list box, an **Exit** button, and a button labeled **Read the Xi, Yi Data File**. Click the **Read the Xi, Yi Data File** button. The program reads the file named MyXiYiPairs.txt and displays the contents by listing in a list box as shown in Figure 15-5.

Understanding the Read X_iY_i File Program. The code attached to the **Read the Xi, Yi Data File** button, begins with a statement declaring the string variables **strLeft** and **strRight**. These variables correspond to the right and left members of the x_i, y_i data pairs listed in the file named MyXiYiPairs.txt.
As previously pointed out, the three statements required for reading a file are

1. **Open "c:\MyXYpairs.txt" For Input As #1**
2. **Input #1, strLeft, strRight**
3. **Close #1**

The **Input #1** statement is enclosed in a **Do . . . Until** loop control structure. With each iteration of the loop, the **Input #1** statement reads an x_i, y_i data pair from the file named MyXiYi.txt. The iterations continue until no more x_i, y_i data pairs are to be found; that is, until the end of file (**EOF**) is encountered. **EOF(1)** refers to the end of file #1. When **EOF(1)** equals **True** (the end of the file is encountered), the loop is satisfied and iterations cease. Execution passes to the **Close #1** statement, which closes the file, and the subroutine is ended. Notice the difference between the statements:

```
frmReadFile.Print strLeft(Index), strRight(Index)
```

and

```
lstDataPairs.AddItem strLeft(Index) & _
                vbTab & strRight(Index)
```

The **Print** method prints the items in the print list to the form. The items in the print list are separated from each other by a comma (the list separator). The items may be a mixture of different types.

The **AddItem** method adds a single item to a line in the list of the list box. The item must be a string or string expression. In order to place a pair of numbers on a single line, the numbers must be of the text type (the text representation of a number) so that the two numbers can be concatenated and thus placed on the same line. Placing the **vbTab** constant between the two numbers separates the numbers in each pair and gives rise to two columns of numbers as the list box is filled. The output printed to the form and the output listed in the list box appear essentially identical.

The Common Dialog Boxes Control

In working your way through some of the previous Visual Basic programs, you have encountered a number of Visual Basic's dialog boxes: the **Add Procedure** dialog box, the **Control Array** dialog box, the **Options** dialog box, the **Object Browser** dialog box, and others. These dialog boxes are specifically Visual Basic dialog boxes. In other Windows applications, you have also used dialog boxes for printing, saving, and opening files and other tasks. You may have noticed that dialog boxes that are common to a variety of Windows application appear nearly identical to the user, and for this reason are called common dialog boxes (Figure 15-6). The common dialog boxes control allows the Visual Basic programmer to use these common dialog boxes in Visual Basic programs.

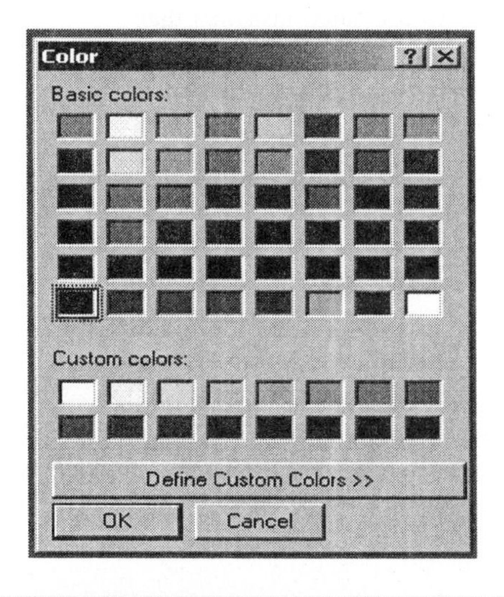

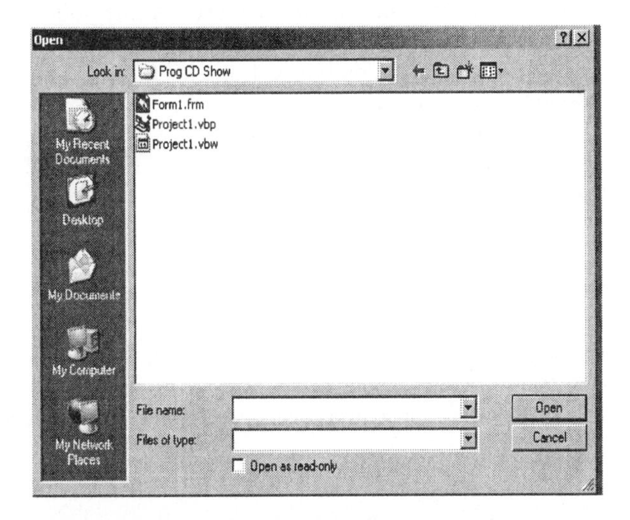

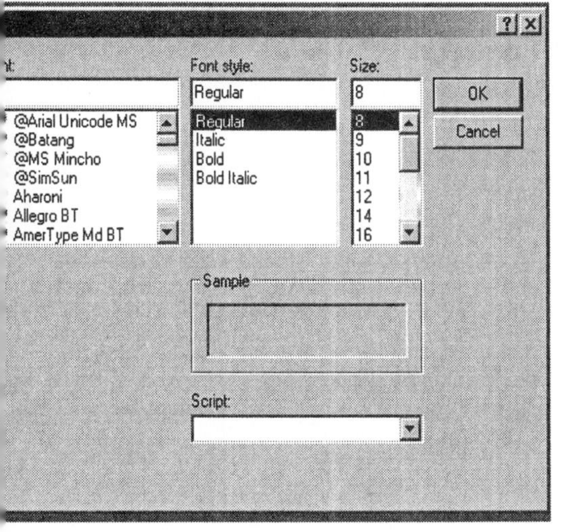

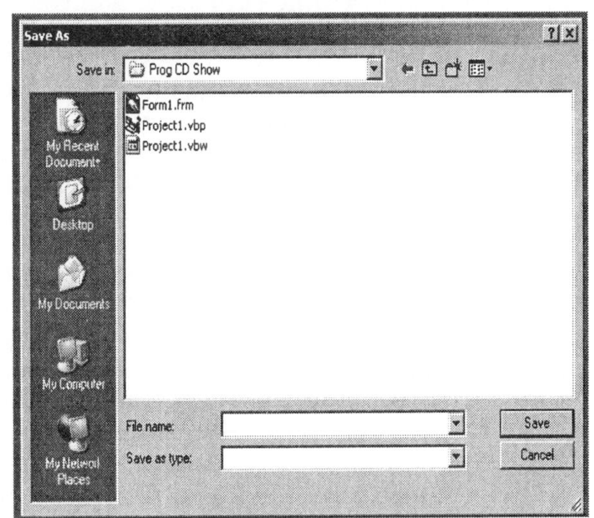

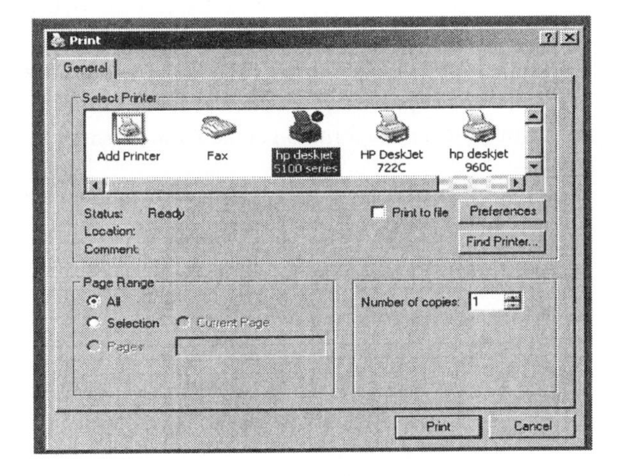

Left: The **Color** Dialog Box

Right: The **Open** Dialog Box

ter Left:: The **Font** Dialog Box

ter Right: The **Save As** Dialog Box

om Right: The **Print** Dialog Box

ure 15-6. The Five Common Dialog Boxes Described in this Chapter. The Show the Boxes program displays these common og boxes.

The common dialog boxes permit the user to carry out a wide variety of unrelated tasks, so they could be described in nearly any chapter on Visual Basic. This chapter is on Visual Basic files. Two of the common dialog boxes deal with files, so it is appropriate to treat the common dialog boxes control in this chapter.

The insignificant appearing little **Common Dialog Boxes** icon packs a powerful punch, as it contains all of the common dialog boxes. It is visible on the form at design time but (like the timer control) is invisible at run time.

The Show the Boxes Program

This program simply serves to show what each of the dialog boxes looks like. The common dialog boxes control object contains the various common dialog boxes. Five similar methods of the common dialog boxes control object display each of the five common dialog boxes (Table 15-3).

Table 15-3

The Show... Methods of the Common Dialog Boxes Control

Method	Common Dialog Box Displayed
ShowColor	The **Color** Dialog Box
ShowFont	The **Font** Dialog Box
ShowOpen	The **Open** Dialog Box
ShowPrinter	The **Print** Dialog Box
ShowSave	The **Save** Dialog Box

Starting the Project

- Start Visual Basic by clicking the **VB** icon on the Windows desktop.
- In the **New Project** dialog box, click the **Standard EXE** icon, and click the **OK** button.
- A **New Project** window opens with a fresh form.
- Double-click the form to get the Form_Load procedure in the code window.

The **Common Dialog Boxes** icon is displayed in the toolbox as:

If the **Common Dialog Boxes** icon is not in your toolbox:

- Right click the toolbox.
- Click **Components** on the shortcut menu to open the **Components** dialog box (Figure 15-7).
- Scroll to **Microsoft Wilndows Common Controls**.
- Select its check box on the left, and then click **OK**.

The **Common Dialog Boxes** icon appears in the toolbox. Placed on the form, it appears in design mode as shown in Figure 15-8.

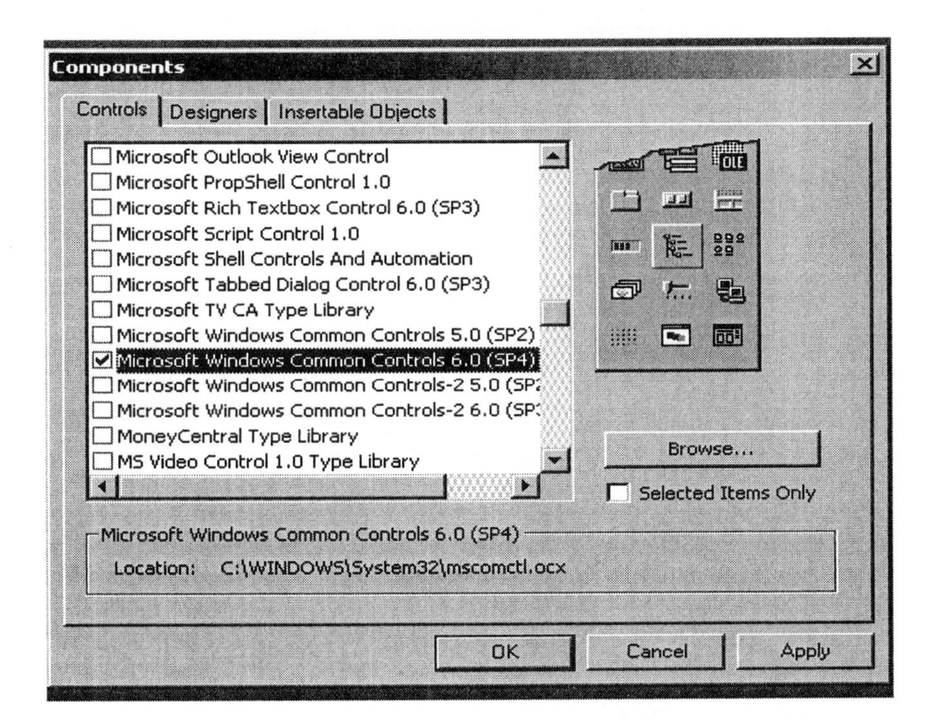

Figure 15-7. The **Components** Dialog Box for Adding the **Common Dialog Boxes** Icon to the Toolbox

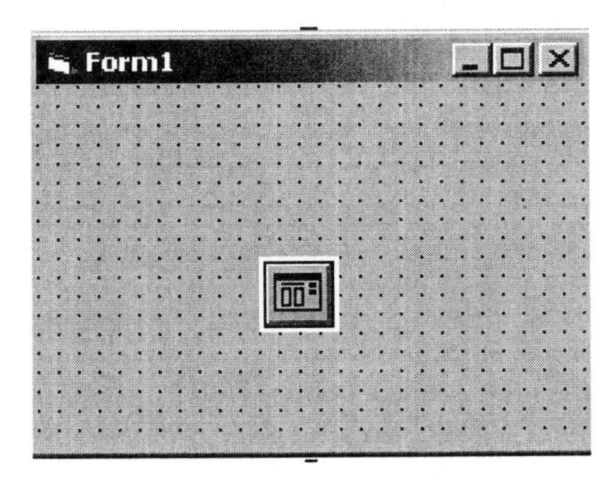

Figure 15-8. The **Common Dialog Boxes** Icon as It Appears in Design Mode on the Form

Entering the Code for the Show the Boxes Program. With the form selected in the procedure window, select **Click in the Event Window**, and enter the following code in these two procedures:

```
Option Explicit
' The Show the Boxes program
' A program to show the six common dialog boxes

Private Sub Form_Click()
   End
End Sub

Private Sub Form_Load()
   CommonDialog1.ShowColor
   CommonDialog1.ShowPrinter
   CommonDialog1.ShowOpen
   CommonDialog1.ShowSave
   CommonDialog1.Flags = cdlCFBoth
   CommonDialog1.ShowFont
End Sub
```

Running the Show the Boxes Program. Click the **Start** button. The **Form_Load** procedure executes immediately, displaying the **Color** dialog box. Close the **Color** dialog box by clicking **Cancel**. The program shows the next dialog box, the **Print** dialog box. Cancel it, and the next one shows. Continue until just the form shows. Notice that the **Common Dialog Boxes** icon is not visible on the form except in design mode. It is invisible in run mode. Click anywhere in the form to exit the program.

Understanding the Show the Boxes Program. Like all the other controls in the toolbox, the common dialog boxes control is just a control. Like the other controls, it has methods and properties, although it responds to no events. **ShowColor**, **ShowPrinter**, **ShowOpen**, **ShowSave**, and **ShowFont** are all methods of the common dialog boxes control. This program uses one property of the common dialog boxes: **Flags**. It is necessary to set the **Flags** property to one of the following values before showing the **Font** dialog box:

- **cdlCFScreenFonts**
- **cdlCFPrinterFonts**
- **cdlCFBoth**

When the program shows the **Font** dialog box, examine the list box with the names of the fonts. Select one by clicking it, and notice how the **Font** dialog box displays a sample of the font. Try changing the size and the other font properties while the **Font** dialog box is showing. This is the same dialog box that you recognize in your Windows word processing applications (Word, WordPerfect, and others). Visual Basic makes it easy for you to incorporate professional appearing dialog boxes in your Visual Basic applications.

The next three short programs illustrate more properties of the common dialog boxes control and show how dialog boxes may be used in an application.

The Common Dialog Color Program

The Common Dialog Color program demonstrates the **Color** dialog box by assigning the **Color** property of the common dialog boxes to the **FillColor** property of a shape control. The user can conveniently select the desired color from the **Color** dialog box. The user can cancel a selection and return to the **Color** dialog box to select an alternative color.

Starting the Project

- Start Visual Basic by clicking the **VB** icon on the Windows desktop.
- In the **New Project** dialog box, click the **Standard EXE** icon, and click the **OK** button.
- A **New Project** window opens with a fresh form.
- Select the form, and in the properties window, change the (**Name**) property to **frmColor**.
- Click **Save**; the file name will be frmColor, so click **OK**.
- The file name changes to Project1; change it to **The Common Dialog Color Program**, and click the **OK** button.
- Place controls on frmColor according to Figure 15-9.
- Set the properties of the controls according to Table 15-4.

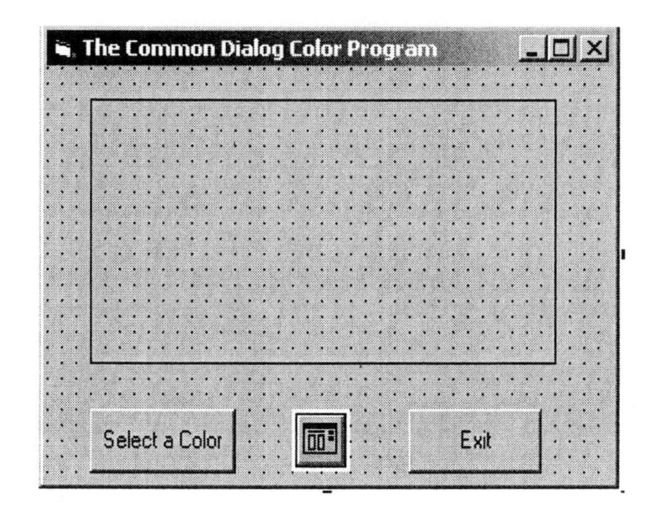

Figure 15-9. The Form for the Common Dialog Color Program

Table 15-4

The Properties Table for the Common Dialog Color Program

Object	Property	Value
Form1	Name	frmColor
	Caption	The Common Dialog Color Program
	WindowState	2-Maximized
Shape1	Name	shpRect
Command1	Name	cmdSelect
	Caption	Select a Color
Command2	Name	cmdExit
	Caption	Exit

Entering the Code of the Common Dialog Color Program. Enter the following code in the general declarations section:

```
Option Explicit
' The CD Color program
' A program to demonstrate the Color dialog box
Private Sub cmdSelect_Click()
   CommonDialog1.CancelError = True
   On Error GoTo CancelMessage
   CommonDialog1.Flags = cdlCCRGBInit
   CommonDialog1.ShowColor
   shpRect.FillColor = CommonDialog1.Color
   shpRect.FillStyle = vbFSSolid
   Exit Sub
CancelMessage:
   MsgBox "You cancelled your selection", _
       vbInformation, "Cancelled"
   Exit Sub
End Sub

Private Sub cmdExit_Click()
   End
End Sub
```

Executing the Common Dialog Color Program. Click the **Start** button to execute the program. Click the **Select a Color** button to open the **Color** dialog box. Click the **Select a Color** button to select a few colors, pressing **OK** after each selection. After selecting a color, click the **Cancel** button, and observe the message box. After clicking the **OK** button of the message box, you are still in the **Color** dialog box and are free to make an alternative selection.

In the **Color** dialog box, click the **Define Custom Colors** button, and try a few custom color selections. Click the **Exit** button to end the program.

Understanding the Common Dialog Color Program. The code associated with the **Color** dialog box and selecting a color is attached to the **cmdSelect** command button. In order to use the **Cancel** button of the **Color** dialog box, it is necessary to set its **CancelError** property to equal **True**. Then, when the **Cancel** button is clicked, Visual Basic generates an error, which is trapped by the **On Error GoTo CancelMessage** statement. This action displays the message box, which displays to the user the message: **You cancelled your selection**.

The **Flags** property of the Color **dialog box** is set to **cdlCCRGBInt**, which determines the default color selection. Change this constant to **cdlCCFullOpen**, and run the program again. This time, the **Color** dialog box opens with the custom color part open as well. Change this constant to **cdlCCPreventFullOpen**, and custom colors are no longer available. You can find all the constants in the Object Browser listed in the color constants class.

Selecting a color in the **Color** dialog box by clicking it establishes the **Color** property of the common dialog boxes control. This color is assigned to the **FillColor** property of **shpRect** with the assignment statement:

```
shpRect.FillColor = CommonDialog1.Color
```

The **Color** property of one control (the common dialog boxes **Color** property) is assigned to the **FillColor** property of another control.

The Common Dialog Printer Program

The CD Printer program uses the same **On Error GoTo** construction as the CD Color program to allow the user to use the **Cancel** button. The user enters a phrase in the text box. Clicking the **Print** button allows the user to print the phrase. No properties are set with printer constants. The Object Browser lists many printer constants that are sometimes useful. Click the **Object Browser** icon, and under **Classes**, scroll down to **Printer Constants** and click to see the large number of printer constants.

Starting the Project

- Start Visual Basic by clicking the **VB** icon on the Windows desktop.
- In the **New Project** dialog box, click the **Standard EXE** icon, and click the **OK** button.
- A **New Project** window opens with a fresh form.
- Select the form, and in the properties window, change the **(Name)** property to **frmCDPrinter**.
- Click **Save**; the file name will be frmCDPrinter, so click **OK**.
- The file name change to Project1; change it to **The Common Dialog Printer Program**, and click the **OK** button.
- Place controls on frmCDPrinter according to Figure 15-10.
- Set the properties of the controls according to Table 15-5.

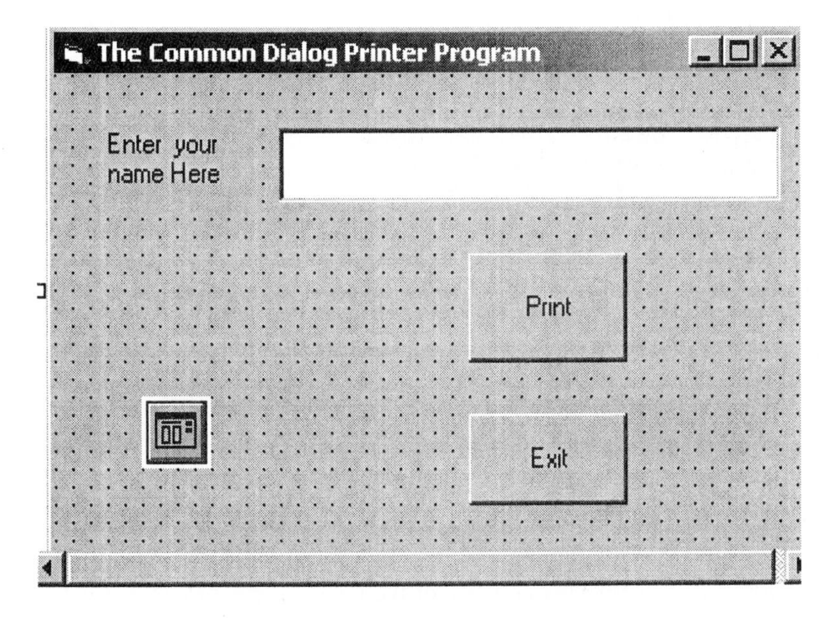

Figure 15-10. The Form for the Common Dialog Printer Program

Table 15-5

The Properties Table for the Common Dialog Printer Program

Object	Property	Value
Form1	Name	frmCDPrinter
	Caption	The Common Dialog Printer Program
	WindowState	2-Maximized
Text1	Name	txtPhrase
	Multiline	True
	Text	Make it blank
Label1	Name	Label1
	Caption	Enter your Name here
Command1	Name	cndPrint
	Caption	Print
Command2	Name	cmdExit
	Caption	Exit

Entering the Code of the Common Dialog Printer Program. Enter the following code in the general declarations section:

```
Option Explicit
' The CD Printer program
' A program to demonstrate the Print dialog box
Private Sub cmdPrint_Click()
   CommonDialog1.CancelError = True
   On Error GoTo CancelMessage
   CommonDialog1.ShowPrinter
   Printer.Print txtPhrase.Text
   Printer.EndDoc
   Exit Sub
CancelMessage:
   MsgBox "You cancelled printing", _
       vbInformation, "Cancelled"
   Exit Sub
End Sub

Private Sub cmdExit_Click()
   End
End Sub
```

Executing the Common Dialog Printer Program. Click the **Start** button to execute the program. Enter a phrase or your name in the text box. You can enter your address on more than one line, if you wish, since the **Multiline** property of the text box is set to **True**. Click the **Print** button, and the **Print** dialog box icon appears over the form. This is the same **Print** dialog box that appears on most Windows applications that feature printing. At this point, you can click **OK**, and your name will be printed. Or you can change the print range or number of copies or click **Properties** and proceed to change the printer properties. You can change the orientation of printing from its default setting (portrait) to landscape. If you have FAX software installed on your computer, you can FAX the phrase you entered.

Understanding the Common Dialog Printer Program. After you enter your name or a phrase in the text box, clicking the command button labeled **Print** cause the **cmdPrint_Click()** procedure to execute. The error trapping in the CD Printer program is the same as in the CD Color program. The phrase that the user enters into the text box is the **Text** property of the text box (**txtPhrase.Text**). The statement:

```
CommonDialog1.ShowPrinter
```

executes, displaying the **Print** dialog box. Execution halts, waiting for action on the part of the user. The user can use the features offered by the **Print** dialog box. Clicking **OK** causes the printer to print the phrase with the statement:

```
Printer.Print txtPhrase.Text
```

Notice that the **Print** dialog box does not actually print anything; it does not send data to the printer. It is still necessary to use the **Printer.Print** statement to send the output list to the printer. In this program the output list is simple **txtPhrase.txt**. As usual, the **EndDoc** method tells the printer that it's OK to begin printing.

The Common Dialog Fonts Program

The CD Fonts program uses the same **On Error GoTo** construction as the CD Color program to allow the user to use the **Cancel** button.

Starting the Project

- Start Visual Basic by clicking the **VB** icon on the Windows desktop.
- In the **New Project** dialog box, click the **Standard EXE** icon, and click the **OK** button.
- A **New Project** window opens with a fresh form.
- Select the form, and in the properties window, change the **(Name)** property to **frmCDFonts**.
- Click **Save**; the file name will be frmCDFonts, so click **OK**.
- The file name changes to Project1; change it to **The Common Dialog Fonts Program,** and click the **OK** button.
- Place controls on frmCDFonts according to Figure 15-11.
- Set the properties of the controls according to Table 15-6.

Entering the Code of the Common Dialog Fonts Program

Enter the following code in the general declarations section:

```
Option Explicit
' The CD Fonts program
' A program to demonstrate the Font dialog box
Private Sub cmdSelect_Click()
   CommonDialog1.CancelError = True
   On Error GoTo CancelMessage

   CommonDialog1.Flags = cdlCFBoth Or cdlCFEffects
   CommonDialog1.ShowFont
```

```
        With txtPhrase.Font
            .Bold = CommonDialog1.FontBold
            .Italic = CommonDialog1.FontItalic
            .Name = CommonDialog1.FontName
            .Size = CommonDialog1.FontSize
            .Strikethrough = CommonDialog1.FontStrikethru
            .Underline = CommonDialog1.FontUnderline
        End With
        txtPhrase.SetFocus
        Exit Sub

    CancelMessage:
        MsgBox "You didn't choose a font", _
            vbInformation, "Cancelled"
    End Sub

    Private Sub cmdExit_Click()
        End
    End Sub
```

Table 15-6

The Properties Table for the Common Dialog Fonts Program

Object	Property	Value
Form1	Name	frmCDFonts
	Caption	The Common Dialog Fonts Program
	WindowState	2-Maximized
Text1	Name	txtPhrase
	Text	Make it blank
	Multililne	True
Label1	Name	Label1
	Caption	Enter a phrase
Command1	Name	cmdSelect
	Caption	Select a Font
Command2	Name	cmdExit
	Caption	Exit

Executing the Common Dialog Fonts Program. Click the **Start** button to execute the program. You can enter a phrase in the text box first, and then select a font, or reverse the order. Any time you select a font, you can click the **Cancel** button of the **Font** dialog box, and then choose an alternative selection.

Understanding the Common Dialog Fonts Program. The error trapping with the **On Error GoTo** statement is the same in this program as it is in the CD Color program. On the first statement after **On Error GoTo**, two flags are set. You can set more than one flag for a dialog box using the **Or** operator. For example:

```
CommonDialog1.Flags = cdlCFBoth Or cdlCFEffects
```

With the **Font** dialog box, selecting one of its properties (**FontBold**, **FontItalic**, **FontUnderline**, or another property) establishes the value of those properties, which the program assigns to the **Font** properties of the text box named **txtPhrase**.

Note. You must set the **Flags** property to **cdlCFScreenFonts**, **cdlCFPrinterFonts**, or **cdlCFBoth** before displaying the **Font** dialog box. Otherwise, the error **No Fonts Exist** occurs.

cdlCFScreenFonts: With this constant, the **Font** dialog box lists only the screen fonts supported by the system.
cdlCFPrinterFonts: With this constant, the **Font** dialog box lists only the printer fonts supported by the system.
cdlCFPrinterFonts: With this constant, the **Font** dialog box lists both the screen fonts and the printer fonts supported by the system.

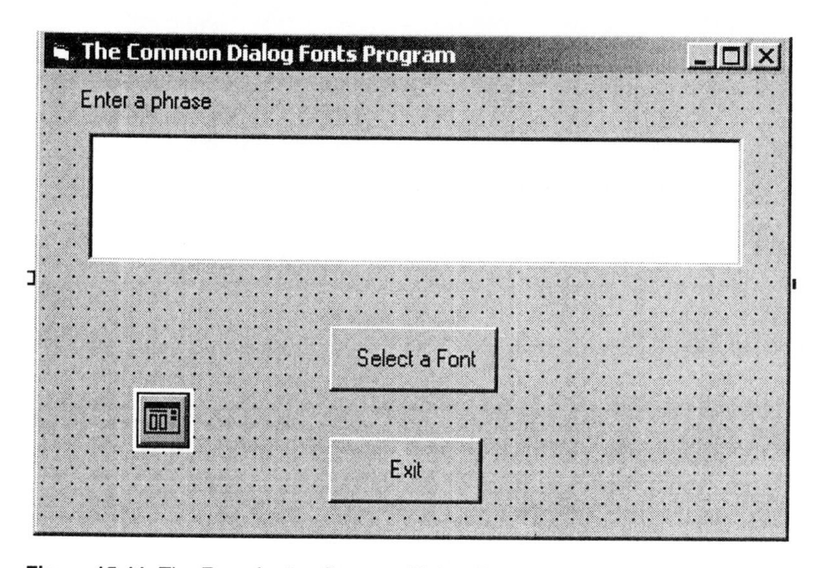

Figure 15-11. The Form for the Common Dialog Fonts Program

The Common Dialog Open File Program

Starting the Project

- Start Visual Basic by clicking the **VB** icon on the Windows desktop.
- In the **New Project** dialog box, click the **Standard EXE** icon, and click the **OK** button.
- A **New Project** window opens with a fresh form.
- Select the form, and in the properties window, change the **(Name)** property to **frmCDOpen**.
- Click **Save**; the file name will be frmCDOpen, so click **OK**.
- The file name changes to Project1; change it to **The Common Dialog Open File Program**, and click the **OK** button.
- Place controls on frmCDOpen according to Figure 15-12.
- Set the properties of the controls according to Table 15-7.

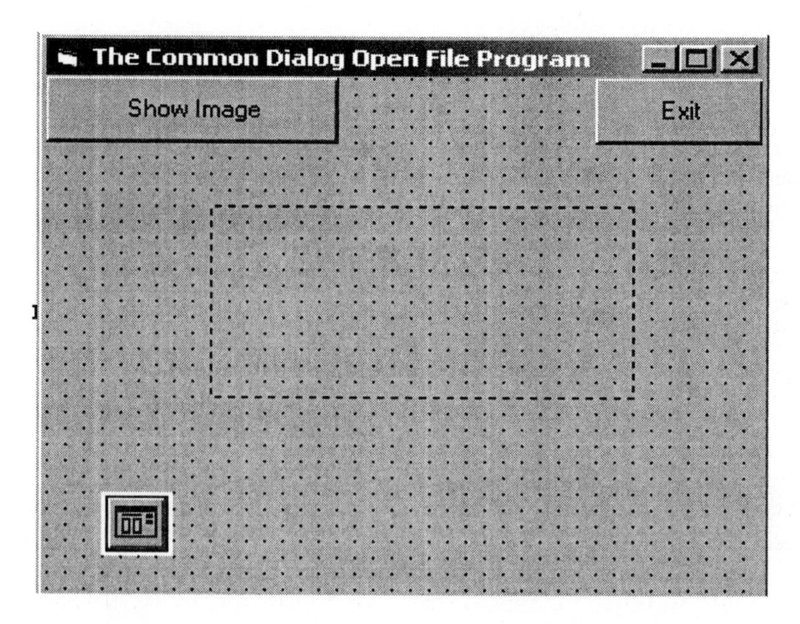

Figure 15-12. The Form for the Common Dialog Open File Program

Table 15-7

The Properties Table for the Common Dialog Open File Program

Object	Property	Value
Form1	Name	frmCDOpen
	Caption	The Common Dialog Open File Program
	WindowState	2-Maximized
ImageBox1	Name	Image1
CommonDialog1	Name	CommonDialog1
Command1	Name	cmdShowImage
	Caption	Show Image
	Font	10 Point Bold
Command2	Name	cmdExit
	Caption	Exit
	Font	10 Point Bold

Entering the Code of the Common Dialog Open File Program

```
Option Explicit
' The CD Open File program
' Opens a *.bmp file that the user selects

Private Sub cmdExit_Click()
   End
End Sub
```

```
Private Sub cmdShowImage_Click()
   Dim MyFile As String
   CommonDialog1.CancelError = True

   On Error GoTo CancelMessage
   'Set Filters
   CommonDialog1.Filter = "(*.BMP)|*.BMP|(*.JPG)|*.JPG"
   'Set Filter Index for Default Filter
   CommonDialog1.FilterIndex = 1
   CommonDialog1.ShowOpen
   MyFile = CommonDialog1.FileName
   Image1.Picture = LoadPicture(MyFile)
      Exit Sub
CancelMessage:
   MsgBox "You didn't choose a file", _
       vbInformation, "Cancelled"
End Sub
```

Finding Some *.BMP FILES to Open and Display. If you don't have any *.BMP files or *.JPG files of your own to open to look at, you can find many *.BMP files tucked away in Windows. Click the **Start** button at the lower left corner of the Windows desktop, then click **Search** (Windows XP) or **Find** (Windows 98) and search drive C for files of type *.BMP. In addition to hundreds of *.BMP files, you'll find some head and shoulder photos of some young men and women who probably participated in producing Windows XP.

It's handy to have a little collection of *.BMP files. After your search gives you a screen full of BMP files, right-click the icon of a file, click **Copy**, and copy and paste the BMP file into a convenient directory of your choice. Name the directory something like MyBMPFiles.

Windows XP has a file named MyPictures that contains some sample *.JPG files that are suitable for use with the CD Open File program. The photo CDs furnished by photo processors are in the *.JPG format.

Executing the Common Dialog Open File Program. Click the **Start** button to execute the program. Click the **Show Image** button on the form. The **Open** dialog box appears (Figure 15-13).

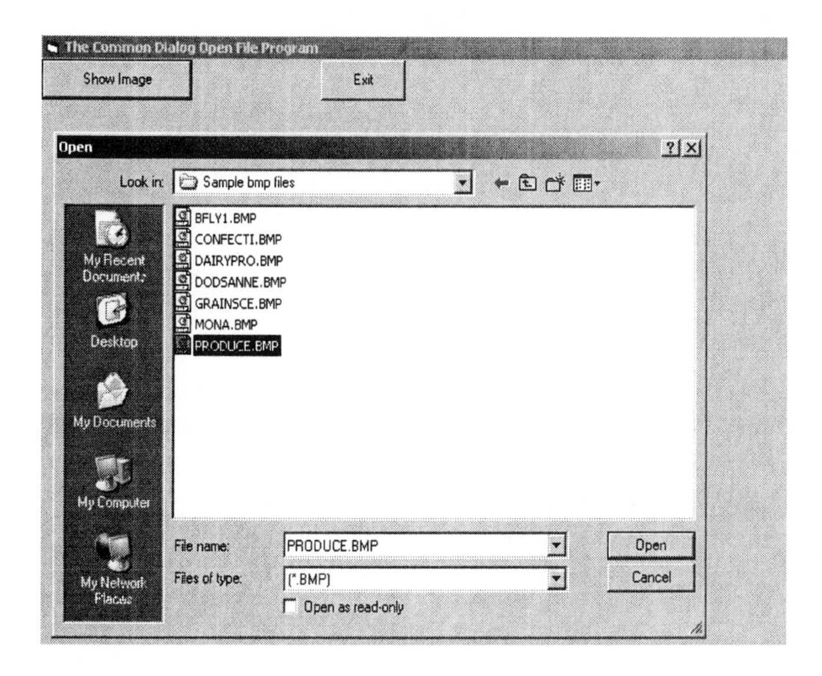

Figure 15-13. Common Dialog Open File Program. Select a BMP file, and click **Open.**

At the top of the **Common Dialog Open form (Figure 15-13)** is a drop-down box labeled **Look in**. Scroll this until you reach a folder that contains either JPG or BMP file as shown in Figure 15-13. Select a BMP file, and click **Open** to see the image displayed (for example, the image shown in Figure 15-14).

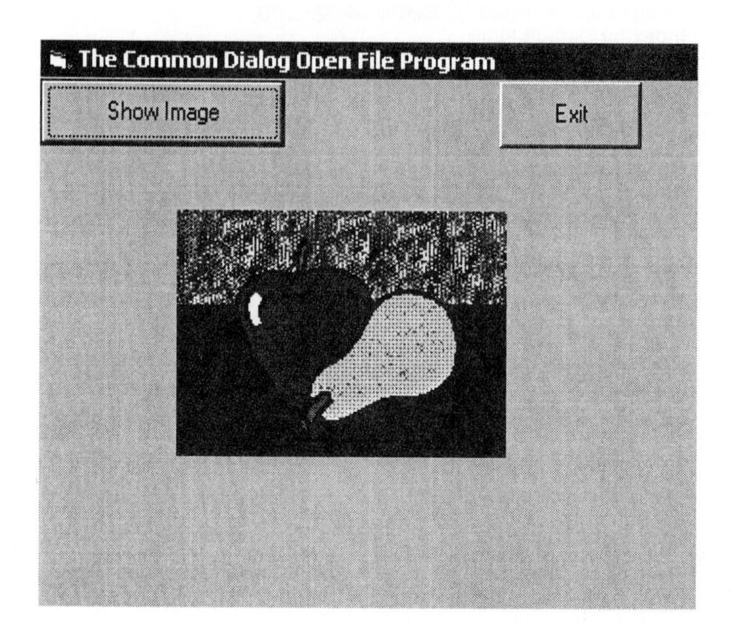

Figure 15- 14. An Image Opened and Displayed in the Common Dialog Open File Program

Understanding the Common Dialog Open File Program. The CD Open File program uses the same **On Error GoTo** construction as the CD Color program to allow the user to use the **Cancel** button. File names are variables of type **String**, so the first line of the program declares MyFile to be a string variable. MyFile is the file that we want to open.

```
Dim MyFile As String
```

The next two statements begin the usual error trapping structure used in the previous common dialog boxes program:

```
CommonDialog1.CancelError = True
On Error GoTo CancelMessage
```

Since we want only graphic images that we can assign to an image box control, we want to filter out files of types other than BMP and JPG. The next statement assigns a filter to the **Filter** property of the common dialog boxes control:

```
CommonDialog1.Filter = "(*.BMP)|*.BMP|(*.JPG)|*.JPG"
```

The right side of the statement is the filter. It has its own peculiar syntax, all of which is enclosed in quotation marks: "   ". Each file that is selectable (not filtered out) consists of two parts: a descriptor and a filter.

(*.BMP) and **(*.JPG)** are descriptors

***.BPM** and ***.JPG** are filters

Each descriptor | filter pair must be separated by a pipe symbol (|). The descriptor can be any text. Thus the brackets are optional in the descriptor. The following three examples are legitimate descriptor pairs:

(All Files)|*.* **Text | *.TXT** **(Document) | *.DOC**

The next statement sets the **FilterIndex** for the default file type. This is the file type that will appear in the **File name** box at the bottom of the Open dialog box.

```
CommonDialog1.FilterIndex = 1
```

In this statement, **1** is the value assigned to the **FilterIndex** property of the common dialog boxes control. It refers to the first filter, which in this program is ***.BMP**.

When the CD Open File program is first run, the form appears with control buttons on it. Click the **Show Image** button to open the **Open** dialog box. The **File name** box at the bottom is blank. The **Files of type** box under it displays (***.BMP**). If you click the down arrow of the **Files of type** box, the drop-down list now shows both (***.BMP**) and (***.JPG**). When you select a program, its name appears in the **File name** box.

In any common dialog boxes program, there is always a statement like the next one, a statement in which a property of the common dialog boxes control is assigned to a variable. In this program, the property is the (selected) file name, and the variable is MyFile.

```
MyFile = CommonDialog1.FileName
```

This statement assigns the selected file to the variable **MyFile**.

After you have selected and viewed a couple of graphic images, exit the program, and change the filter index from 1 to 2, and run the program again, You will find that (***.JPG**) shows up first in the **Files of type** box.

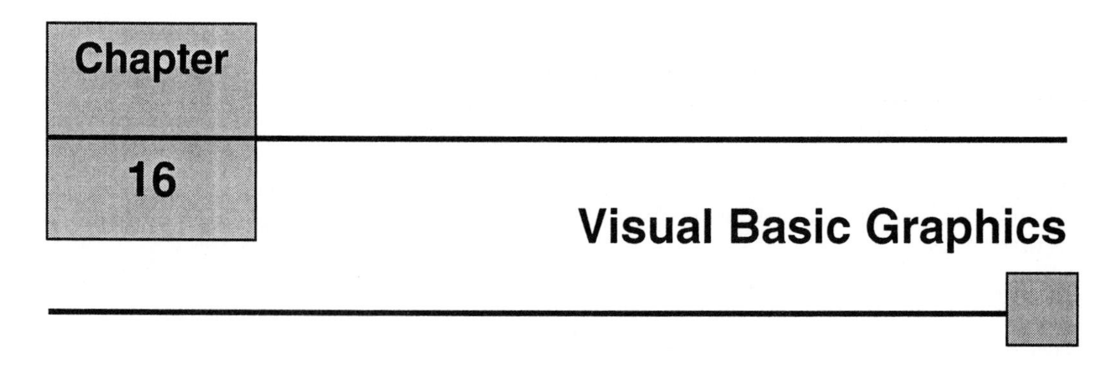

Visual Basic Graphics

This chapter gradually develops some of the concepts needed to work with Visual Basic's graphics offerings. Several controls and methods are specialized to work with graphics. Visual Basic has several coordinate systems within which drawings and images can be located, and these are discussed in detail.

By means of its various graphic objects, their properties, and their methods, Visual Basic offers programmers many options for creating graphic shapes and displaying graphic images. The form itself is as well suited for displaying graphic images as it is for displaying text. We have frequently used the form to display numerical and text output to the user. Just as a text box attached to a form is a control that specializes in displaying text, the image box and picture box are controls that specialize in displaying graphic images. We have already used the shape control, which can create various geometric shapes (Table 10-6) on the form. Similarly, the line control can draw straight lines on the form.

Objects and Controls for Drawing and Graphics

Some specific Visual Basic objects and controls are designed for use with graphics. These are:

- Form
- Image box control
- Picture box control
- Shape control
- Line control

Graphic images can be displayed on the form, the image box control, and the picture box control. All three of these objects have the **Picture** property, which allows a graphic image to be loaded onto the object. The syntax is quite simple:

```
Form1.Picture = file name
ImageBox.Picture = file name
PictureBox.Picture = file name
```

The acceptable file name has an extension of the type .bmp, .jpg, .gif, .dib, .ico, .cur, .wmf, or .emf. For example, these files include bitmaps, icons, and files from paint programs and digital photographs (Table 16-1). In some short example programs to follow, the **Picture** property will be demonstrated.

Table 16-1

Graphic Formats Supported by the Form, Picture Box Control, and Image Control

Format	Extension	Description
Bitmap	.bmp	An image formed as a pattern of pixels
Icon	.ico	A 32 x 32 pixel bitmap
Cursor	.cur	Similar to the icon format, but contains a hot spot to define *x* and *y* positions
GIF	.gif	Graphic Interchange Format; common on the Internet
JPEG	.jpg	Joint Photographic Experts Group; common on the Internet
Metafile	.wmf/.emf	An image as code lines and shapes for Microsoft Windows

The Image Control. The image control is used to display graphics in any of the formats listed in Table 16-1. Images are loaded into an image control with its **Picture** property. Clicking its **Picture** property in the properties window allows the programmer to select the path and name of the desired image.

The image control has a unique property, the **Stretch** property. The **Stretch** property allows the programmer to stretch pictures in an image control to fit the control's size. No other control has this property, including the picture box control.

The Image Program

This program uses an image from an ordinary Kodak Photo CD of the type that you receive when you develop a roll of 35 mm color negative film and receive the negatives, color prints, and a CD. The snapshot used here was taken in Berlin and given the name Berlin00010.

Starting the Project.

- Start Visual Basic by clicking the **VB** icon on the desktop.
- Select the **Standard EXE** icon in the **New Project** window, and click the **OK** button.
- A **New Project** window opens with a fresh form.
- Select the form, and in the properties window, change the (**Name**) property to **frmImageBerlin**.
- Click **Save**; the file name will be frmImageBerlin, so click **OK**.
- The file name changes to Project1; change it to **The Image Program**, and click the **OK** button.
- Place controls on frmImageBerlin according to Figure 16-1.
- Set the properties of the controls according to Table 16-2.

Entering the Code of the Image Program. The program has a form but no controls, so no properties table is needed. In the code window, enter the following two procedures, both of which are form events:

```
Option Explicit

Private Sub Form_DblClick()
   End
End Sub

Private Sub Form_Load()
   Image2.Picture = Image1.Picture
   Image3.Picture = Image1.Picture
End Sub
```

Table 16-2

The Properties Table for the Image Program

Object	Property	Value
Form1	Name	frmImage Berlin
	Caption	The Image Program
Image1	Name	Image1
	Border Style	1 – Fixed Single
	Picture	C:\Kodak Pictures\Berlin00010, as an example; use your own graphics file of a type shown in Table 16-1.
	Stretch	True
Image2	Name	Image2
	Border Style	1 – Fixed Single
	Stretch	True
Image3	Name	Image3
	Border Style	1 – Fixed Single
	Stretch	True

Executing the Image Program. Click the **Start** button to run the program. Double-click anywhere in the form to exit the program.

Figure 16-1. The Form for the Image Program

The Picture Box Control and the Line Control. The picture box control bears many similarities to the image control. The picture box control is used to display graphics in any of the formats listed in Table 16-1. Images are loaded into a picture box control with its **Picture** property. Clicking its **Picture** property in the properties window allows the programmer to select the path and name of the desired image.

Although the picture box control does not have the **Stretch** property, it is a more "powerful" control than the image control because it has more properties and methods, especially those that relate to graphics.

The line control, not to be confused with the **Line** method, which is discussed later, is another lightweight control, like the image control. It is used to draw straight lines on a form, picture box control, or frame control.

The Picture Box Program

The image used in the Picture Box program is also an ordinary snapshot taken in Alaska of Mt. Denali (Mt. McKinley) and stored on an ordinary Kodak Photo CD. Its format is JPG. You could use any digital camera or photo CD snapshot, although a mountain or craggy face would come close to the example in the Picture Box program.

Starting the Project

- Start Visual Basic by clicking the **VB** icon on the desktop.
- Select the **Standard EXE** icon in the **New Project** window, and click the **OK** button.
- A **New Project** window opens with a fresh form.
- Select the form, and in the properties window, change the **(Name)** property to **frmDenali**.
- Click **Save**; the file name will be frmDenali, so click **OK**.
- The file name changes to Project1; change it to **The Picture Box Program**, and click the **OK** button.
- Place controls on frmDenali according to Figure 16-2.
- Set the properties of the controls according to Table 16-3.

Entering the Code of the Picture Box Program. Enter the following procedure in the code window:

```
Option Explicit

Private Sub Exit_Click()
   End
End Sub

Private Sub cmdHide_Click()
   Line1.Visible = False
   Line2.Visible = False
   Line3.Visible = False
   Line4.Visible = False
   Line5.Visible = False
   Line6.Visible = False
   Line7.Visible = False
   Line8.Visible = False
   Line9.Visible = False
   Line10.Visible = False
   Line11.Visible = False
   Line12.Visible = False
End Sub

Private Sub cmdView_Click()
```

```
    Line1.Visible = True
    Line2.Visible = True
    Line3.Visible = True
    Line4.Visible = True
    Line5.Visible = True
    Line6.Visible = True
    Line7.Visible = True
    Line8.Visible = True
    Line9.Visible = True
    Line10.Visible = True
    Line11.Visible = True
    Line12.Visible = True
End Sub
```

Table 16-3

The Properties Table for the Picture Box Program

Object	Property	Value
Form1	Name	frmDenali
	Caption	The Picture Box Program
Picture1	Name	Picture1
	Border Style	1 – Fixed Single
	Picture	C:\Kodak Pictures\Danali0008 as an example; or use your own graphics file of any type shown in Table 16-1.
Line1 and so on	Name	Line1 through Line*n*, as many as you need. The programmer puts a series of lines on the image that is already on the picture box control. Each line segment is a separate line control. Click and drag from where you want the line segment to begin to where you want the line segment to end. The picture box in Figure 16-2 has eleven line controls to mark the route on the picture of the mountain.
Command1	Name	cmdView
	Caption	View Route
Command2	Name	cmdHide
	Caption	Hide Route
Command2	Name	cmdExit
	Caption	Exit

The **Picture** property of the picture box control loads the image from a JPG file. Then the programmer puts a series of lines on the image that is already on the picture box control. Each line segment is a separate line control.

Executing the Picture Box Program. Click the **Start** button to execute the program. The form (Figure 16-2) displays the picture box with its picture of Mt. Denali and the line segments, each of which is a separate line control. Clicking the **Hide Route** button causes the lines marking the route to disappear. Clicking the **View Route** button causes the lines marking the route to reappear. Click back and forth between these two buttons. Click **Exit** to end the program.

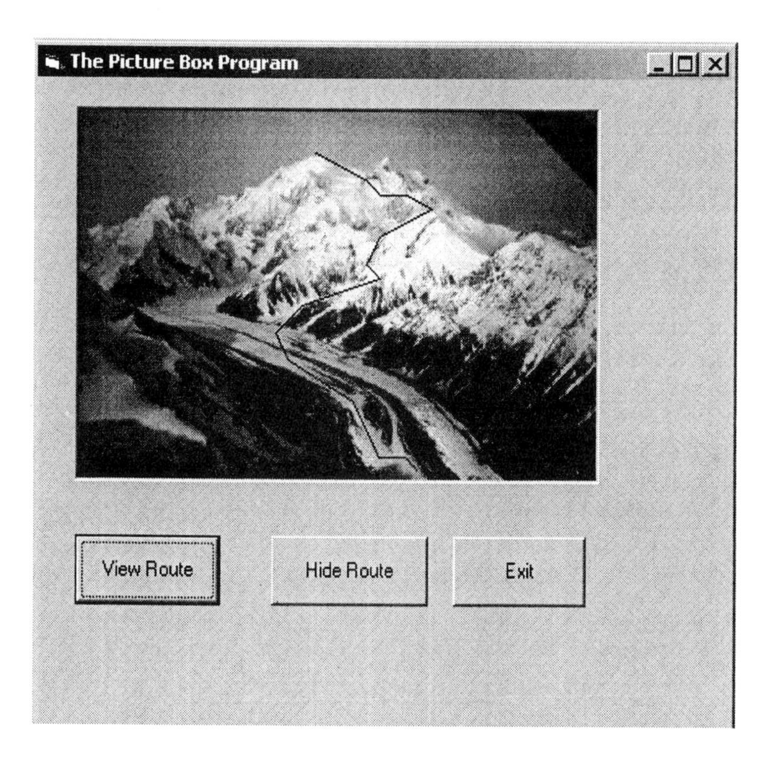

Figure 16-2. The Form for the Picture Box Program Showing Mt. Denali

Understanding the Picture Box Program. When the program is run, the form is loaded, and the **Picture** property of the picture box control is assigned the graphic image whose name and path you specified in the **Picture** property of the picture box control. At design time, the **Visible** property of all fourteen line controls was left in its default value of **True**.

Clicking the **Hide Route** button executes the **cmdHide_Click** procedure, which sets the **Visible** property of all fourteen line controls to **False**.

Clicking the **View Route** button executes the **cmdView_Click** procedure, which sets the **Visible** property of all fourteen line controls to **True**.

Examination of the code for this program suggests that there has got to be a better way to handle a large group of controls that avoids such a long, tedious list. In fact, there is: the control array.

The Control Array

The fourteen line controls of the procedures in Picture Box program do have the appearance of a list, but they do not yet have the data structure of a list, that is, of a one-dimensional array. Visual Basic, however, does provide a data structure called a control array, which is an array of controls. The control array was described in the TotoLoto program of Chapter 14. Each element of a control array is a control, and the values (properties) of those elements can be set by the programmer as desired.

Changing a Group of Controls to a Control Array

- Open the Picture Box program.
- Click the line control at the summit of Mt. Denali with the default name Line1.
- Change the name Line1 to **linRt** (for line route).
- Click Line2, and change its name to the same name as the previous line control, **linRt**.
- A dialog box appears with the message **You already have a control named 'linRt.' Do you want to create a control array?** Click the **Yes** button.
- Click Line3, and change its name to **linRt**; change the names of all the other line controls to **linRt**.

With the program still open, click each line control from the summit to the base camp and observe the **Index** property of each control. The index equals 0 for the first control, 1 for the second control, and so on. You have created an array named **linRt**, the elements of which are **linRt(0)**, **linRt(1)**, and so on. Each element is a line control. The code for a control array can be written just like for any array.

Modifying the Code of the Picture Box Program. Open the Picture Box program, and delete all the statements within the **cmdHide_Click** and **cmdView_Click** procedures. Substitute the following code within each of these procedures in order to use the control array structure. Execute the program as you did before modifying it. The external results are the same, but internally, you know that the code is more compact and more elegant.

```
Option Explicit
' The Picture Box program demonstrates the control array

Private Sub cmdExit_Click()
  End
End Sub

Private Sub cmdHide_Click()
  Dim J As Integer
  For J = 0 To 13
    linRt(J).Visible = False
  Next
  End Sub
  Private Sub cmdView_Click()
  Dim J As Integer
  For J = 0 To 13
    linRt(J).Visible = True
  Next
End Sub
```

Properties for Drawing and Graphics

The form and the picture box control have many properties with which the programmer can control the appearance of the image or drawing (Table 16-4). The image control, the shape control, and the line control have fewer properties, methods, and events, and for this reason are referred to as "lightweight controls." However, they also use less system resources and consequently run faster. The image control is used only for displaying pictures. You can not only display images on the form and picture box control, but you can draw lines and circles and draw on them with pixels. Many of the properties listed in Table 16-4 are demonstrated later in this chapter.

Table 16-4

Common Object Properties for Drawing or Appearance

Property	Form	Picture Box Control	Image Control	Shape Control	Line Control
AutoRedraw	Yes	Yes	No	No	No
BackColor	Yes	Yes	No	Yes	No
BorderColor	Yes	Yes	No	Yes	Yes
BorderStyle	Yes	Yes	Yes	Yes	Yes
BorderWidth	Yes	Yes	No	Yes	Yes
DrawMode	Yes	Yes	No	Yes	Yes
DrawStyle	Yes	Yes	No	No	No
DrawWidth	Yes	Yes	No	No	No
FillColor	Yes	Yes	No	Yes	No
FillStyle	Yes	Yes	No	Yes	No
ForeColor	Yes	Yes	No	No	No
Picture	Yes	Yes	Yes	No	No
Shape	No	No	No	Yes	No
Stretch	No	No	Yes	No	No

The Coordinate System

In describing the Visual Basic coordinate system, it is useful to think of the screen as a container for the form and of the form as a container for other objects, for example the picture box control. All these objects have their own independent default coordinate systems and their own position in their container object. The default origin of all these coordinate systems is at the upper left corner of the object.

As shown in Figure 16-3, the units of *X* increase from left to right; the units of *Y* increase from top to bottom. The units of measure along these axes are called the scale. Since **Scale** is a property of each object (that can have a coordinate system), each object can have its own scale, that is, its own coordinate system. The default scale unit is the twip. The user may choose other scale modes listed in Table 16-5. However the screen always has the unit of the twip.

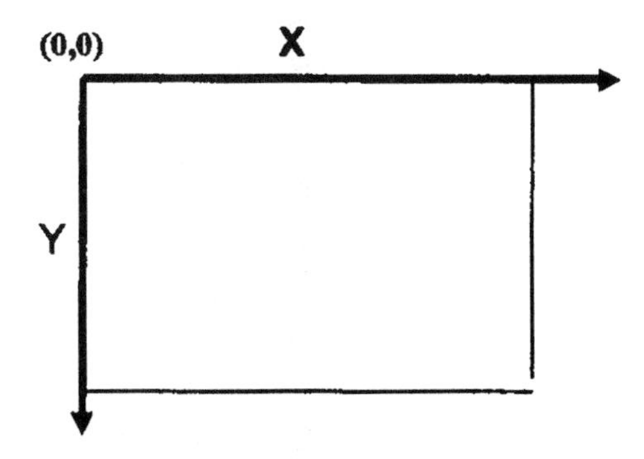

Figure 16-3. The Visual Basic Coordinate System

Notice that the origin of the screen, the form, and the picture box are all (0,0). The origin of the screen is always in the upper left corner of the screen. The origin of the form is always in the upper left corner of the form and can be located anywhere on the screen. In Figure 16-4, the origin of the form is near the upper left corner of the screen. The origin of the picture box can be anywhere on the form. In Figure 16-4, the origin of the picture box is near the lower left corner of the form.

Table 16-5

The Scale Mode Property

Constant	Setting	Description
vbUser	0	User defined
vbTwips	1	Twip (1440 twips per inch) (This is the default for all objects.)
vbPoints	2	Point (72 points per inch)
vbPixels	3	Pixel (Smallest unit of monitor or printer resolution
vbCharacters	4	Character (horizontal 120 twips per character) (vertical 240 twips per character)
vbInches	5	Inch
vbMillimeters	6	Millimeter
vbCentimeters	7	Centimeter

The scale mode identifies the scale unit (twips, pixels, cm, or other measurement) currently in use. The **ScaleMode** property sets a value establishing the unit in use with the syntax:

```
object.ScaleMode = Setting
```

The values of the setting are listed in Table 16-5. If you are not sure what your current **ScaleMode** setting is, you can display the current setting of an object with the syntax:

```
Print object.ScaleMode
```

Table 16-6

Properties Relating to the Coordinate System

Property	Form	Picture Box Control	Image Control
ScaleMode	Yes	Yes	No
ScaleLeft	Yes	Yes	No
ScaleTop	Yes	Yes	No
ScaleHeight	Yes	Yes	No
ScaleWidth	Yes	Yes	No
CurrentX	Yes	Yes	No
CurrentY	Yes	Yes	No
Left	Yes	Yes	Yes
Top	Yes	Yes	Yes

The Coordinate Demo Program

The Coordinate Demo program shows the relationship between the coordinate systems of a picture box contained on a form contained on a screen. It should also be noted that the screen also has a coordinate system, but its units are always pixels, and the origin is always in the upper left corner of the screen. It has only two properties related to its coordinate system: **Height** and **Width**, which depend on the size and resolution of the computer's associated monitor. Your monitor may have different values, but the meaning of the property values is the same. The output to the form is shown in Figure 16-4.

- Notice that the origin of the screen, the form, and the picture box are all (0,0). The origin of the screen is always in the upper left corner of the screen. The origin of the form is always in the upper left corner of the form and can be located anywhere on the screen. In Figure 16-4, the origin of the form is near the upper left corner of the screen. The origin of the picture box can be anywhere on the form. In Figure 16-4, the origin of the picture box is near the lower left corner of the form.
- Notice that the **Height**, **Width**, **Left**, and **Top** properties relate to the position of an object, not its size.
- Notice that the **ScaleLeft** and **ScaleTop** properties give the coordinates of the upper left corner.
- Notice that **ScaleHeight** and **ScaleWidth** properties give the number of scale units vertically and horizontally across the object (its size, not its position).

Starting the Project

- Start Visual Basic by clicking the **VB** icon on the desktop.
- Select the **Standard EXE** icon in the **New Project** window, and click the **OK** button.
- A **New Project** window opens with a fresh form.
- Click **Save**; the file name will be Form1, so click **OK**.
- The file name changes to Project1; click the **OK** button.
- Place a picture box control on Form1 according to Figure 16-4.
- Leave the properties of the controls with their default values.

To keep it simple, no properties table is set up for this program, and so the **Picture** property of the picture box control is empty (the default). If you want to put one of your pictures in Picture1, then set the **Picture** property of Picture1 to the path to your picture as shown in Figure 16-4.

To generate Figure 16-4, open a new Visual Basic program, and place a picture box near the bottom of the form as shown in Figure 16-4. Double-click the form to open the code window. On the **Procedure List** of the code window, select **Click** to get the **Form_Click** procedure, and enter the following code:

```
Option Explicit
' The Coordinate Demo program

Private Sub Form_Click()
   End
End Sub

Private Sub Form_Load()
   Show
   Form1.Print "Screen.Width: "; Screen.Width
   Form1.Print "Screen.Height: "; Screen.Height
   Form1.Print
   Form1.Print "Form1.Left: "; Form1.Left
   Form1.Print "Form1.Top: "; Form1.Top
   Form1.Print "Form1.Height: "; Form1.Height
```

```
    Form1.Print "Form1.Width:  "; Form1.Width
    Form1.Print "Form1.ScaleLeft: "; Form1.ScaleLeft
    Form1.Print "Form1.ScaleTop: "; Form1.ScaleTop
    Form1.Print "Form1.ScaleHeight: "; Form1.ScaleHeight
    Form1.Print "Form1.ScaleWidth: "; Form1.ScaleWidth
    Form1.Print
    Form1.Print "Picture1.Left: "; Picture1.Left
    Form1.Print "Picture1.Top: "; Picture1.Top
    Form1.Print "Picture1.Height:  "; Picture1.Height
    Form1.Print "Picture1.Width:  "; Picture1.Width
    Form1.Print "Picture1.ScaleLeft: "; Picture1.ScaleLeft
    Form1.Print "Picture1.ScaleTop: "; Picture1.ScaleTop
    Form1.Print "Picture1.ScaleHeight: "; Picture1.ScaleHeight
    Form1.Print "Picture1.ScaleWidth: "; Picture1.ScaleWidth

End Sub
```

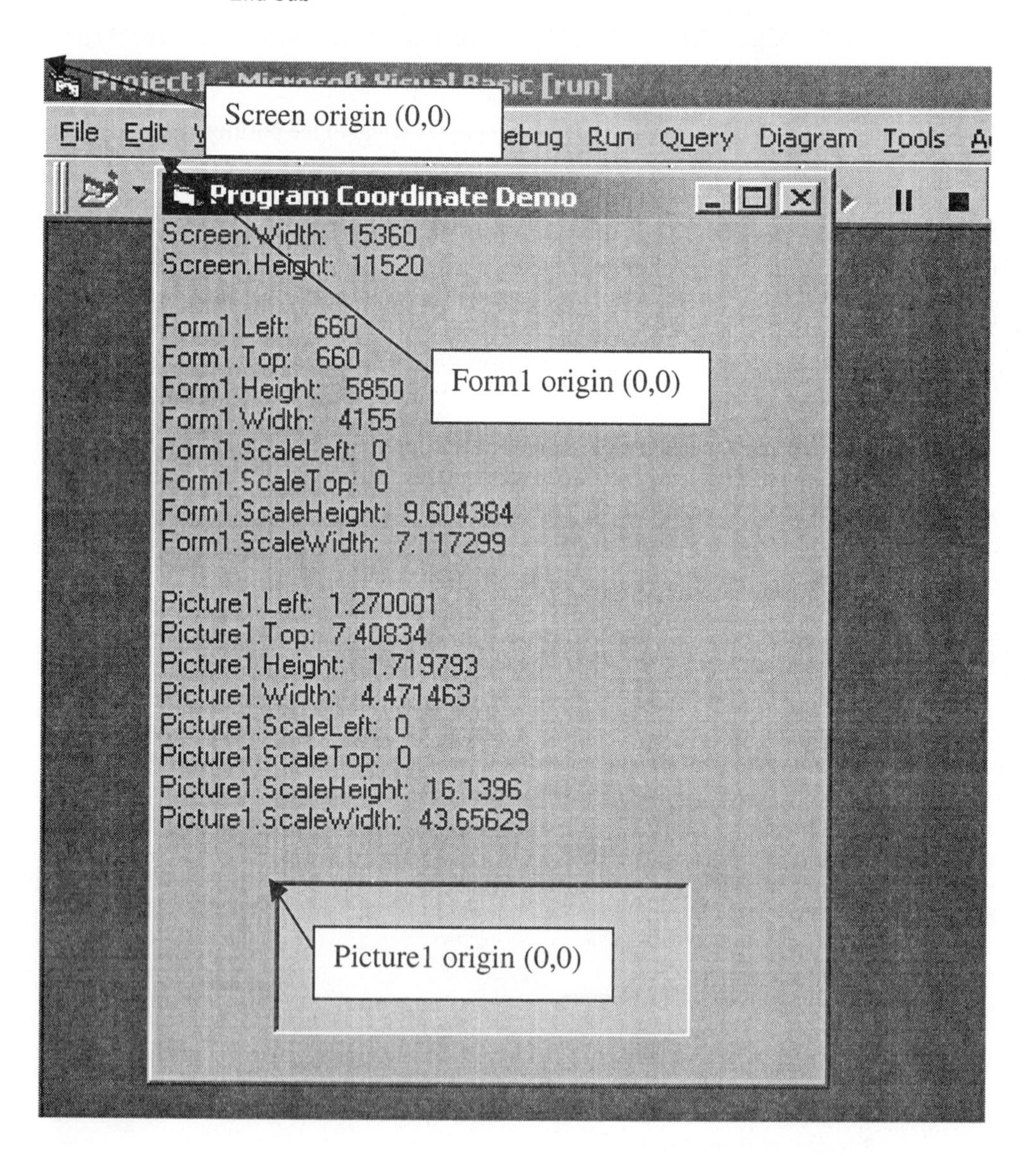

Figure 16-4. Form for the Coordinate Demo Program

Methods for Drawing and Graphics

In addition to the **Picture** property of the form, image control, and picture box control described earlier, several methods are of great utility for placing graphic images on an object. Some of these are summarized in Table 16-7.

Table 16-7

Common Methods for Graphics

Object Method	Object			
	Form	Image Box	Picture Box	Printer
Line	Yes	No	Yes	Yes
Circle	Yes	No	Yes	Yes
PSet	Yes	No	Yes	Yes
Cls	Yes	No	Yes	No
Scale	Yes	No	Yes	Yes
Show	Yes	No	Yes	No
EndDoc	No	No	No	Yes

Notice that the toolbox contains a line control, not to be confused with the **Line** method, which is a method of the form, picture box, and printer but not of the image box. These methods will be demonstrated in some demonstration programs to follow. The line control was used in the Picture Box program, and the **Cls** method has been used in several previous programs.

The Line and Box Methods. With the **Line** method, the programmer can place a straight line on an object, such as a form or picture box. A slight variation permits drawing a box, filled or unfilled with a specified color. The syntax for the **Line** method is

```
Form1.Line (X1,Y1) – (X2,Y2),[Color], [B],[F]
```

This statement draws a line on Form1 from the beginning of the line at the initial form coordinates (**X1,Y1**) to the end of the line at final form coordinates (**X2,Y2**). The form is the default object, so if its name is omitted, the **Line** method draws the line or box to the form. The statement has three optional parameters:

- The parameter **[Color]** is optional. If included, it can be set with the **RGB** function or the **QBColor(*n*)** function (Table 16-8)
- If **[B]** is included, then a box is drawn with (**X1,Y1**) and (**X2,Y2**) specifying the upper left and lower right corners of the box.
- The parameter **[F]** can be included only if **[B]** is included. If **[F]** is included, the box is filled with the current **FillColor** and **FillStyle** parameters. The Line and Box program illustrates these parameters.

Table 16-8

QBColor Function Syntax: QB(Color)

Number	Color	Number	Color
0	Black	8	Gray
1	Blue	9	Light blue
2	Green	10	Light green
3	Cyan	11	Light cyan
4	Red	12	Light red
5	Magenta	13	Light magenta
6	Yellow	14	Light yellow
7	White	15	Bright white

DrawWidth and DrawStyle. The **DrawWidth** and **DrawStyle** properties can be used to set the width and style of the line. These properties are of those of the object on which the line is drawn. Thus, the syntax is

```
object.DrawWidth [= size]
```

and

```
object.DrawStyle [= setting]
```

To make the code more readable, Visual Basic furnishes some constants for the various **DrawStyle** settings (Table 16-9).

Table 16-9

Settings and Constants for the DrawStyle Property

DrawStyle	Setting	Constant
Solid (the default)	0	vbSolid
Dash	1	vbDash
Dot	2	vbDot
Dash-Dot	3	vbDashDot
DashDotDot	4	vbDashDotDot
Transparent	5	vbInvisible
Inside Solid	6	vbInsideSolid

The Line and Box Program

The Line and Box program demonstrates the **Line** method; the **Box** method is just a variation on the **Line** method. It also demonstrates the following properties: **CurrentX**, **CurrentY**, **ScaleHeight**, **ScaleWidth**, **ScaleTop**, **ScaleBottom**, and **ScaleMode**.

Starting the Project. To generate Figure 16-5, open a new Visual Basic program. This demonstration program has no controls, so no properties table is needed.

Entering the Code of the Line and Box Program. Double-click the form to get the code window. On the **Procedure List** of the code window, select **Click** to get the **Form_Click** procedure, and enter the following code:

```
Option Explicit
' The Line and Box program
Private Sub Form_Load()

   Show
   DrawWidth = 2

   ' Draw a line
   Line (500, 250)-(2500, 250), QBColor(12)
   Print "Last X and Y are: "; CurrentX; ","; CurrentY

   ' Draw a box under the line
   Line (500, 500)-Step(2000, 1000), vbBlue, B
   Print "Last X and Y are: "; CurrentX; ","; Round(CurrentY)

   ' Draw a triangle under the line
   Line (500, 1750)-(500, 2500), RGB(0, 170, 0)
   Line (500, 2500)-(2500, 2500), RGB(0, 170, 0)
   Line (500, 1750)-(2500, 2500), RGB(0, 170, 0)

   Print "Last X and Y are: "; CurrentX; ","; CurrentY
   Print
   Print vbTab; "ScaleLeft is:  "; ScaleLeft
   Print vbTab; "ScaleTop is:  "; ScaleTop
   Print vbTab; "ScaleHeight is:  "; ScaleHeight
   Print vbTab; "ScaleWidth is:  "; ScaleWidth
   Print vbTab; "ScaleMode is:  "; ScaleMode

   ' Use the last point referenced to draw lines
   Line (500, 4000)-(500, 5000)
   Line -(1000, 5000)
   Line -(3000, 6000)
   Line -(3000, 4000)
   Line -(500, 4000)
End Sub

Private Sub Form_Click()
   End
End Sub
```

After you run the program, a single click in the form exits the program.

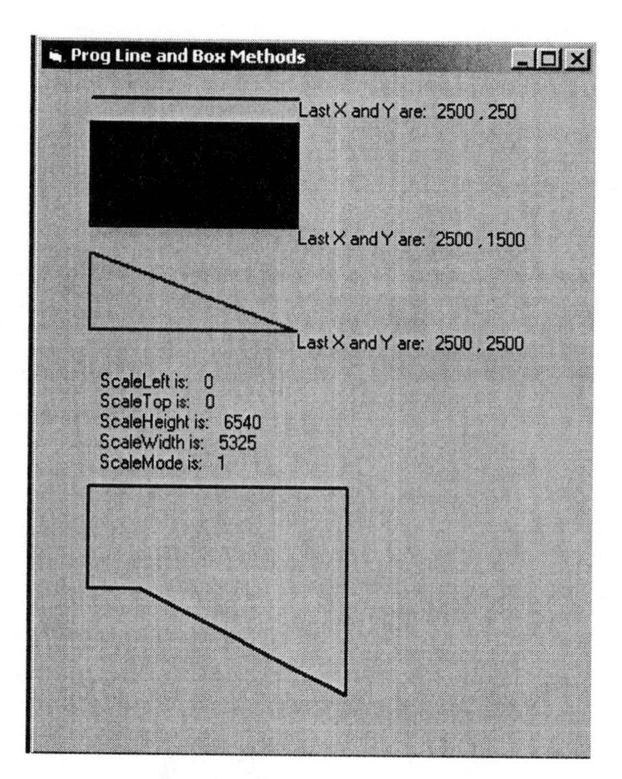

Figure 16-5. The Form for the Line and Box Program

Understanding the Line and Box Program. The first action by the program is to draw a straight line from *x*1 equals 500 and *y*1 equals 250 to *x*2 equals 2500 and *y*2 equals 250. The units of the coordinate system are twips. The default origin (0,0) lies at the upper left corner of the screen. The color of the line is specified with the **QBColor** function. After the program draws the line, the invisible pointer is at the end of the line where *x*2 equals 2500 and *y*2 equals 250. A print statement begins at this point. The top of the **L** in **Last X** lies at this point.

```
' Draw a line
Line (500, 250)-(2500, 250), QBColor(12)
Print "Last X and Y are: "; CurrentX; ","; CurrentY
```

The second action of the program is to draw a box. The last parameters in the **Line** method statement are a **B** and an **F**. The **B** causes a box to be drawn by specifying the upper left corner and the lower right corner of the box. Attaching an **F** to the **B** causes the box to be filled with the specified color, specified in this statement by the **QBColor** function.

```
'Draw a box under the line
Line (500, 500)-Step(2000, 1000), vbBlue, BF
Print "Last X and Y are: "; CurrentX; ","; Round(CurrentY)
```

The next action of the program is to draw a triangle that begins at a point exactly below where the first line began, at *x*1 equals 500. The next two lines begin where the previous line ends. The last line ends where the first begins, completing a triangle.

```
' Draw a triangle under the line
Line (500, 1750)-(500, 2500), RGB(0, 170, 0)
Line (500, 2500)-(2500, 2500), RGB(0, 170, 0)
Line (500, 1750)-(2500, 2500), RGB(0, 170, 0)
Print "Last X and Y are: "; CurrentX; ","; CurrentY
```

The third action is to summarize the graphics parameters that characterize the size (in units of twips) of the form on which all the graphics and drawing are taking place. Comparing these values with the **CurrentX** and **CurrentY** values and the drawing parameters gives you a better idea of the relationship of the figure to the form coordinates.

```
Print vbTab; "ScaleLeft is:  "; ScaleLeft
Print vbTab; "ScaleTop is:  "; ScaleTop
Print vbTab; "ScaleHeight is:  "; ScaleHeight
Print vbTab; "ScaleWidth is:  "; ScaleWidth
Print vbTab; "ScaleMode is:  "; ScaleMode
```

Finally, we illustrate drawing a line with the last point referenced: the **CurrentX** and **CurrentY** at the end of a line or figure. If the beginning coordinates of a line are omitted ($x1$ and $y1$), then the line is drawn from the last point referenced to the specified coordinates of the end of the line. Here a line is drawn, and then several lines are drawn to the last point referenced until a closed figure is drawn.

```
' Use the last point referenced to draw lines
Line (500, 4000)-(500, 5000)
Line -(1000, 5000)
Line -(3000, 6000)
Line -(3000, 4000)
Line -(500, 4000)
```

The Circle and Scale Methods

With the **Circle** method, Visual Basic draws a circle, ellipse, or arc on a form or picture box or to the printer. It responds to **FillColor**, **FillStyle**, **DrawStyle**, and **DrawMode** properties. The **Scale** method provides the programmer with a simple method for customized the scale of the form, picture box, or printer. Recall that the default scale value is **ScaleMode 3**, with pixels as the unit and the origin at the upper right corner.

Understanding the Scale Method. The **Scale** method is a wonderfully simple method that makes it easy for the programmer to customize her coordinate system. The syntax for the **Scale** method is

***object*.Scale (*x*1,*y*1) – (*x*2,*y*2)**

The ***object*** is the form, the picture box, the printer, or any other object that has the **Scale** method.

The (***x*1,*y*1**) values are the minimum values of the x-axis and y-axis (top left – see Figure 16-3).
The (***x*2,*y*2**) values are the maximum values of the x-axis and y-axis.(bottom right)
The (***x*1,*y*1**) and (***x*2,*y*2**) values are single precision values. For readability by the user, the programmer should usually select integer values.

Understanding the Circle Method. With the **Circle** method, the program can draw a circle, an ellipse, or an arc on an object such as a form or picture box. Table 16-10 lists the meaning of the parts of the **Circle** method syntax. The syntax for the **Circle** method is

***object*.Circle [Step] (x,y), *radius*, [*color*, *start angle*, *end angle*, *aspect*]**

The quantities in brackets are optional.

Table 16-10

Meaning of the Parts of the Circle Method Syntax

Part	Description
object	The default is the form. Picture box and printer also have the **Circle** method.
Step	Optional. **Step** specifies the circle center coordinates relative to the current *x* and *y*.
(x,y)	Required. Coordinates for the center of the circle (in the current **ScaleMode**).
radius	Required. Value for the radius of the circle (in the current **ScaleMode**).
color	Optional. Use the **RGB** function or the **QBColor** function.
start angle, end angle	Optional for drawing arcs. These specify the beginning and end of the arc in radians.
aspect	Optional aspect ratio of the circle (ellipse). The default is 1 for a perfect circle.

The Graph Circle Program

This program draws three circles, which form the figure's head (filled with yellow) and two eyes (filled with black). It draws an ellipse for a nose and an arc for a smile, both red. The scale of the three figures the program draws to the form is set with the **Scale** method, so the origin of the drawing is at the center of the form. The minimum (–10) and maximum (+10) of the drawing axes are set with the **Scale** method. The Graph Circle program first draws a blue circle of radius 10 with its center at the origin (0,0), then a black ellipse.

Entering the Code for the Graph Circle Program. The program has a form but no controls, so no properties table is needed. Open a new Visual Basic program. Double-click the form to get the code window. On the **Procedure List** of the code window, select **Click** to get the **Form_Click** procedure. Enter the following code:

```
Option Explicit
' The Graph Circle program
Private Sub Form_Click()
   End
End Sub

Private Sub Form_Load()
   Const PI = 3.14159265
   Show
   Scale (-10, 10)-(10, -10)
   FillStyle = 0                          'Fill figures with solid color
   FillColor = QBColor(14)                'Fill the next figure yellow
   Circle (0, 0), 6, vbBlack, , , 1       'Draw a black circle (the head)
   FillColor = QBColor(12)                'Fill the next figure red
   Circle (0, -0.8), 1.5, vbBlack, , , 3  'Draw an ellipse (the nose)
   FillColor = QBColor(9)                 'Fill the next figure blue
   Circle (-3, 1), 0.8, vbBlack, , , 1    'Draw a black circle (the left eye)
   Circle (3, 1), 0.8, vbBlack, , , 1     'Draw a black circle (the right eye)
   DrawWidth = 6                          'Draw (arc) with a heavy line
                                          'Draw a smile
   Circle (0, -1), 3, vbRed, 225 * PI / 180, 315 * PI / 180
End Sub
```

Executing the Program. To run the Graph Circle program, click the VB **Start** button, and the program prints three circles, one ellipse, and one arc to the screen. The large circle (the head) is centered on a square form, 20 units high and 20 units wide.

After you run the program, a single click in the form will exit the program.

Figure 16-6. The Form for the Graph Circle Program

Understanding the Graph Circle Program

Units and Scaling. In the preceding program listing, the statement

Scale (-10, 10)-(10, -10)

sets the coordinate system.. The x-axis units range in value from –10 on the left to +10 on the right. The y-axis units range in value from +10 at the top to –10 at the bottom. The corresponding Cartesian coordinate system chosen by the programmer is shown to the right, although the form is square. The origin lies at the center of a square form.

When you print graphics to a page, this same **Scale** method places the origin at the center of the page, which is an 8 ½ x 11 piece of paper, as shown in Figure 16-7.

The **Scale** method of setting the coordinate system is exactly equivalent to using the **ScaleTop**, **ScaleLeft**, **ScaleWidth**, and **ScaleHeight** properties together.

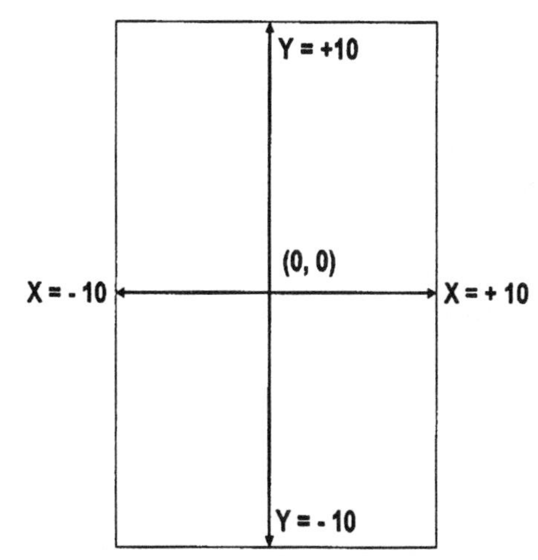

Figure 16-7. The Coordinate System Set with the **Scale (-10,10)-(10,-10)** Statement

For Example:

```
Scale (-2, 3)-(6, -1)
```

is equivalent to:

```
ScaleLeft = -2
ScaleTop = 3
ScaleHeight = -4
ScaleWidth = 8
```

The only surprise here is the negative value for the **ScaleHeight** property. Recall that the origin of the default coordinate system is located at the upper left corner of the form. This means that, while *x* must increase from left to right as it usually does, *y* has no place to go except down. Thus increasing values of *y* correspond to points further down the form. In order to have *y* increase in an upward manner, it is necessary to give **ScaleHeight** a negative value.

FillColor and FillStyle. The syntax of the **Circle** method was given earlier. It should be noted that the **FillColor** method fills the current color over any other objects present. For this reason the order of drawing and filling is important. As an experiment, cut and paste the two lines:

```
FillColor = QBColor(14)          'Fill the next figure yellow
Circle (0, 0), 6, vbBlack, , , 1 'Draw a black circle (the head)
```

just in front of the last line of the program. The result is a yellow head with a red smile. The eyes and nose are filled over with yellow. Table 16-11 lists the **FillStyle** parameters.

***object*.Circle [Step] (*x,y*), *radius*, [*color, start angle, end angle, aspect*]**

Drawing Arcs. Drawing an arc consists of drawing just part of a circle defined by the angles subtending the arc. The angles must be expressed in radians. Visual Basic sets the zero of angle along the positive x-axis. This convention places 90 degrees ($\pi/2$ radians) along the positive y-axis, 180 degrees (π radians) along the negative x-axis, and 270 degrees ($3\pi/2$ radians) along the negative y-axis. As an experiment, change the start and stop angles in the program to 45 – 135, 135 – 225 and 315 – 45, to see the effect.

Table 16-11

The FillStyle Property

FillStyle Value	Effect
0	Solid fill
1	No fill – transparent
2	Horizontal lines
3	Vertical lines
4	Diagonal lines – ////
5	Diagonal lines - \\\\\
6	Crosshatch
7	Diagonal crosshatch

The Resize Event. The form resize event occurs when the form is first displayed, just like the form load event. However the form resize event also occurs when the form is maximized, minimized, or restored or when its size is changed.

The Resize the Form Program

This short program utilizes the form resize event to redraw a centered circle whenever the state of the form is changed.

Entering the Code for the Resize the Form Program

```
Option Explicit
' The Resize the Form program
Private Sub Form_Click()
   End
End Sub

Private Sub Form_Resize()
Dim x As Integer, y As Integer
   Form1.Cls
   Form1.Scale (-10, 10)-(10, -10)
   Form1.DrawWidth = 3
   Form1.Circle (0, 0), Form1.ScaleWidth / 4, RGB(255, 0, 0)
End Sub
```

Understanding the Resize the Form Program. When the program is executed, the form is loaded, triggering the resize event, because the state of the form is changed. The **Form_Resize** procedure executes and draws a circle centered on the form as its state was when the program first executed. The radius of the circle is one quarter of whatever the **ScaleWidth** property of the form was when the program was first executed.

If you maximize the form, the **Form_Resize** procedure executes once again, but now the size of the form is larger, and its **ScaleWidth** property is larger. The circle drawn by the **Circle** method of the form is larger, corresponding to a new radius equal to one quarter of the new, larger **ScaleWidth** value.

Try resizing the form by moving the pointer to an edge of the form until it changes to a double arrow. Then drag the double arrow to increase or decrease the dimension of the form. Try on the other edges. Notice that changing the height of the form does not change the size of the circle. This is because the radius of the circle is always specified in terms of the horizontal units.

Comment out the **Form1.Cls** statement, and experiment with the size of the form again. The need for the **Cls** statement becomes apparent, although the effect of commenting out the **Cls** statement is interesting.

Graphing with the PSet Method

The PSet method permits the programmer to turn on (set) a single pixel (point) at a specified coordinate on an object. PSet, then, means Pixel Set or Point Set. The syntax for the PSet method is:

object.*PSet* (*x*,*y*), [*color*]

The object is usually the form or picture box control. The *x,y* coordinates of the point do not have to be in **ScaleMode 3** (pixels). The ***color*** parameter is optional and can be set in a number of ways: by using color constants, the **QBColor** function, or the **RGB** function. The next three programs illustrate the use of **PSet** in graphing mathematical functions.

The *r2R2* Program

The first function chosen for graphing with the **PSet** method has the form

$$y = 4x^2e^{-2x}$$

This function has the form of the probability distribution function for the 1s orbital of a hydrogen atom, which will be reviewed below. In graphing a mathematical function, whether on a piece of paper or on a computer screen, one needs to get an idea of the range of the function in order to scale the axes and decide on a range of the independent variable x. Examination of this function shows that y is always positive and approaches 0 for large or small x. In between, it reaches a maximum, shown by differentiation, at x equals exactly 1. In addition to plotting the function, we shall draw the coordinate system on which the function is plotted. This involves drawing the x-axis and y-axis and their tick marks.

Writing the *r2R2* Program. The program has a form but no controls, so no properties table is needed. Open a new program, double-click the form to get the code window, and enter the following code:

```
Option Explicit

Private Sub Form_Load()
' The r2R2 program
   Dim I As Integer
   Dim x As Single
   Scale (-1, 1)-(5, 0)
   DrawWidth = 2

   ' Plot the points
   Show
   For x = 0 To 5 Step 0.01
      PSet (x, 4 * x ^ 2 * Exp(-2 * x)), vbRed
   Next x

   ' Draw the axes and ticks
   Line (0, 0)-(0, 1)              ' Draw y-axis
   DrawWidth = 4
   Line (-2, 0)-(5, 0)             ' Draw x-axis
   CurrentX = - 2                  ' Put cursor at
   CurrentY = 0                    ' left end of x-axis
   For I = -1 To 4                 ' Draw ticks on x-axis
      CurrentX = CurrentX + 1
      Line (CurrentX, 0.05)- (CurrentX, 0)
   Next
End Sub

Private Sub Form_Click()
   End
End Sub
```

Executing the *r2R2* Program. Click the VB **Start** button to execute the program. The graphic output appears on the form (Figure 16-8). Exit the program by clicking anywhere in the form.

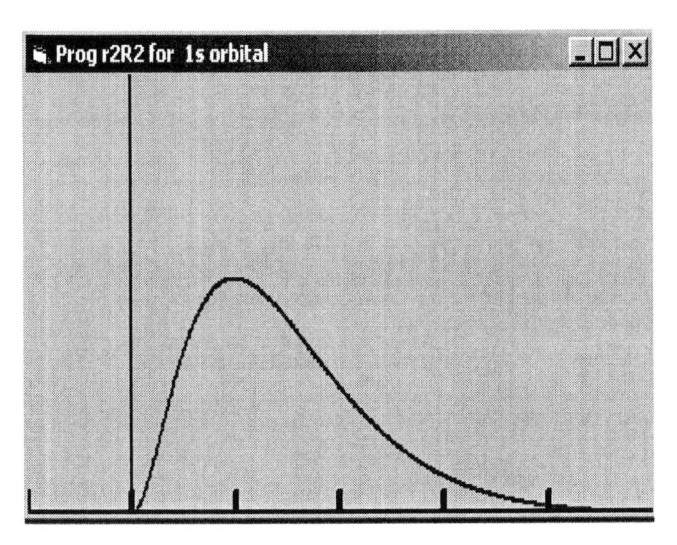

Figure 16-8. The Form of the *r2R2* Program with Graphic Output

Understanding the *r2R2* Program

The Code. After establishing the variable, the scale of the plot is set with the **Scale** method. The syntax for the **Scale** method is

object.*Scale* (*x1*,*y1*) – (*x2*,*y2*)

where ***x1*,*y1*** are the coordinates of the upper left corner of the object, and ***x2*,*y2*** are the coordinates of the lower right corner of the object. In our program, the **Scale** method statement becomes:

Scale (-1, 1) – (5,0)

This shows that the y values may range from 0 to 1, and the x values may range from –1 to 5, even though our plot only ranges from x equals 0 to 5.

The **PSet** method within a single **For ... Next** loop graphs the function by setting a point whose y coordinate is the value of the function at x. The way the pixel appears depends on the value of the **DrawWidth** property on which the pixel is set. The appearance of the lines drawn as axes also depends on the **DrawWidth** property.

The next statements draw the x-axis and y-axis. Then a **For ... Next** loop draws tick marks along the x-axis, according the scale set earlier.

The Science. Some properties of wave functions were reviewed in Excel Example 7, Chapter 3. The radial wave function for the ground state of a hydrogen atom is given by

$$R_{1s}(r) = \left(\frac{1}{a_0}\right)^{3/2} 2e^{\frac{-r}{a_0}} \tag{16-1}$$

where a_0 is the Bohr radius, which equals 52.9177249 pm. This equation says that the magnitude of the wave function is independent of direction, that is, of θ and φ, and falls off inverse exponentially as r increases in any direction. The square of a wave function is interpreted as the probability density or the probability per unit volume (per unit length in one dimension) of finding the electron at some position r. The probability of finding the electron between r and $r + dr$ (regardless of direction) is given by

$$r^2 R_{1s}^2(r) = \left(\frac{1}{a_0}\right)^{3} 4r^2 e^{\frac{-2r}{a_0}} \tag{16-2}$$

which can be rearranged to give

$$y = r^2 R_{1s}^2(r) = \left(\frac{4}{a_0}\right)\frac{r^2}{a_0^2} e^{\frac{-2r}{a_0}} = \left(\frac{4}{a_0}\right) x^2 e^{-2x} \tag{16-3}$$

where $x = r/a_0$.

It is most convenient to plot in units of a_0^{-1}, which means that we plot y / a_0^{-1} (dimensionless) versus $4x^2 e^{-2x}$ (dimensionless). Plotting in units of a_0^{-1} shows clearly that the maximum probability of finding the electron in a hydrogen atom lies at $r = a_0$ but that there is a finite probability of finding the electron closer or further from the nucleus. To find the position of the maximum, differentiate the function, set it equal to zero, and solve for x: $x = a_0$. The Bohr theory of the hydrogen atom has the electron in orbit about the central nucleus at a fixed distance of a_0. Quantum mechanics has the electron in an orbit of varying radius, and one can only have knowledge of the probability of finding the electron at any particular distance r from the nucleus. In quantum mechanics, the term *orbit* has evolved into *orbital*. An orbital is not only the region in space where there is a finite probability of finding the electron, but the word *orbital* is also used as a label for the wave function itself, or its square or r^2 times its square. Thus, we have graphed an orbital.

The $2p_x$ Orbital Program

The angular wave functions were discussed in some detail in Excel Example 8, Chapter 3. The angular wave function for the $2p_x$ orbital is given by

$$p_x = \Psi_{2p_x} = \Theta(\theta)\Phi(\phi) = \left(\frac{3}{8\pi}\right)^{1/2} \sin\theta\cos\phi \tag{16-4}$$

In the XY plane, according to the coordinate system shown in Figure 3-5, θ equals 90 degrees and sinθ equals 1, so a graph of this angular wave function in the XY plane reduces to a plot of

$$y = \left(\frac{3}{8\pi}\right)^{1/2} \cos\varphi \qquad (16\text{-}5)$$

To visualize the directional characteristics of this wave function, the constant term can be omitted from the plot. If P_x is not squared, the plot consists of two circles, tangent at the origin.

Entering the Code for the P_x Orbital Program. The program has a form, but no controls so no properties table is needed. Open a new Visual Basic program, double-click the form to get the code window, and enter the following code:

```
Option Explicit
Private Sub Form_Load()
  ' The Px Orbital program
  Dim X As Single, Y As Single
  Dim Px As Single, Phi As Single
  Dim Pi As Single
  Pi = 4 * Atn(1)
  Scale (-2, 2)-(2, -2)
  Show
  DrawWidth = 2

  For Phi = 0 To 2 * Pi Step 0.01
    Px = (Abs(Cos(Phi))) ^ 2              ' Length of Px squared
    X = Px * Cos(Phi)                      ' its component along X
    Y = Px * Sin(Phi)                      ' its component along Y
    PSet (X, Y), QBColor(13)               ' Plot the point
  Next Phi

  Form1.FontBold = True
  Form1.FontSize = 14
  Line (-2, 0)-(1.8, 0): Print " X"       ' Draw line and label it X
  Line (0, -2)-(0, 2): Print " Y"         ' Draw line and label it Y
End Sub
```

Executing the Program. Click the **Start** button. The typical P_x orbital graph appears on the screen (Figure 16-9), which is really the square of the P_x orbital. Go back to the program and delete the ^2, which squares the P_x orbital. Run the program again. This time, the graph consists of two tangent circles, the graph of the P_x orbital.

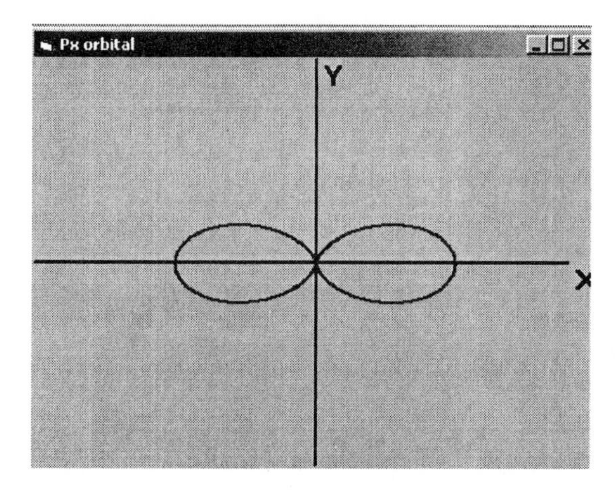

Figure 16-9. The Form for the P_x Orbital Program

The Graph Vib Program: Vibrational Wave Functions

Vibrational wave functions were discussed in Chapter 3, Excel Example 11. The Graph Vib program graphs the ground state wave function for the vibrational quantum number $v = 0$, and the next three excited states, $v = 1$, $v = 2$, and $v = 3$. The parameters for constructing wave functions are given in Table 3-3 in Excel Example 11 and were used for the wave functions plotted in this program. All four wave functions are plotted on the same form but in different colors, so that their shapes can be compared. The click event of the form is used to display the square of the wave function. When the program is run, the four wave functions are displayed. Clicking anywhere on the form changes the display to the square of the wave function. Clicking again returns the display to the original form. Continued clicking causes the display to go back and forth between these two displays. A double-click event ends the program.

Entering the Code for the Graph Vib Program. The program has a form but no controls, so no properties table is needed. Open a new Visual Basic program, double-click the screen to get the code window, and enter the following code:

```
Option Explicit
' The Graph Vib program
Dim n As Integer
Dim t As Integer

Private Sub Form_Click()
  Increment                         ' call to the Increment procedure
  Cls
  Dim X As Single, Y As Single      ' coordinates of a plotted point
  Scale (-5, 1)-(5, -1)
  DrawWidth = 4
  For X = -5 To 5 Step 0.01
    Y = Vib0(X) ^ n                 ' For n, see the Increment procedure
    PSet (X, Y)                     ' Plot black for v = 0

    Y = Vib1(X) ^ n
    PSet (X, Y), RGB(255, 255, 0)   ' Plot yellow for v = 1
    Y = Vib2(X) ^ n
    PSet (X, Y), RGB(0, 0, 255)     ' Plot blue for v = 2
    Y = Vib3(X) ^ n
    PSet (X, Y), RGB(255, 0, 0)     ' Plot red for v = 3
  Next X
  Line (-5, 0)-(5, 0)               ' Draw x-axis
  Line (0, 1)-(0, -1)               ' Draw y-axis' v = 1
  End Sub

Public Function Vib0(X As Single)                   ' Ground state: v = 0
  Vib0 = Exp(-(X ^ 2 / 2))
End Function

Public Function Vib1(X As Single)                   ' v = 1
  Vib1 = 0.7071 * 2 * X * Exp(-(X ^ 2 / 2))
End Function

Public Function Vib2(X As Single)                   ' v = 2
  Vib2 = 0.3535 * (4 * X ^ 2 - 2) * Exp(-(X ^ 2 / 2))
End Function

Public Function Vib3(X As Single)                   ' v = 3
  Vib3 = 0.1443 * (8 * X ^ 3 - 12 * X) * Exp(-(X ^ 2 / 2))
End Function

Private Sub Form_DblClick()
  End
End Sub

Private Sub Form_Load()
```

```
    WindowState = vbMaximized
    t = 0   ' The number of times the user has clicked the form
    Show
    Print Spc(15); "Click anywhere to see wave functions"
    Print Spc(15); "Click again to see their squares"
    Print Spc(15); "Click again, and so on"
    Print Spc(15); "Double-click to end the program"
End Sub

Public Sub Increment()          'called each time user clicks the form
    t = t + 1                   ' t is the number of times the user clicked the form
    If t Mod 2 = 1 Then n = 1   ' If t is uneven, then n = 1
    If t Mod 2 = 0 Then n = 2   ' If t is even, then n = 2 (square the functions)
End Sub
```

Executing the Graph Vib Program. Click the **Start** button to execute the Graph Vib program. Four color graphs appear on the full form (Figure 16-10, bottom graphs). Click again, and the squares of the vibrational wave functions appear (Figure 16-10, top graphs). Click again to return to the original graphs of the vibrational wave functions. Further clicks cause alternating displays of the functions and their squares. Double-clicking exits the program.

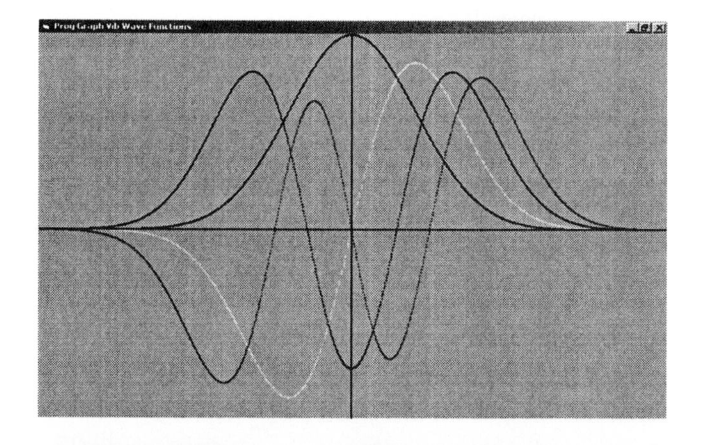

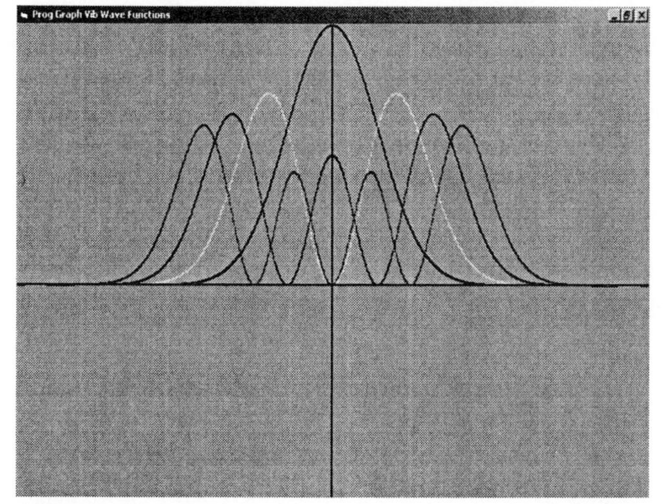

Figure 16-10. Vibrational Wave Functions (Upper) and Their Squares (Bottom)

Understanding the Graph Vib Program. When the program is executed, the **Form_Load** procedure executes first. It maximizes the screen, initializes the global variable *t*, and prints to the form some directions useful to the user. The program keeps track of the number of times *t* that the user clicks the form by incrementing the variable *t* by one for each click.

The purpose here is not to use the total number of clicks, but to determine whether *t* is an even number or an odd number. In the **Increment** procedure, *t* is incremented one for each click. If *t* is odd, then the integer variable *n* is assigned the value 1. If *n* is even, then the integer variable *n* is assigned the value 2. If *t* is odd, then the wave functions are displayed. If *t* is even, the squares of the wave functions are displayed.

When the form is clicked, the **Increment** procedure is called immediately by the **Form_Click** event. This establishes the value of *n* (1 or 2). The **Scale** function sets the user-defined coordinate system for the form.

The **For ... Next** loop graphs the four wave functions. Since the range of *x* is the same for all four functions, the statement **x = -5 To 5 Step 0.01** appears only once. Then each function is called, and the current value of *x* is passed to it. A Visual Basic **Function** procedure is written for each vibrational wave function. The parameters for the wave functions were taken from Table 3-3, Excel Example 11, Chapter 3. Finally, the user exits the program by double-clicking the form. Double-clicking the form executes the **Form_DblClick** event, which calls the **End** statement.

The MouseMove Event. The **MouseMove** event occurs continuously when the user moves the mouse across the form, so it is a natural event to use for executing the **PSet** method to set point after point as the mouse moves, generating a continuous sequence of points. The sequence of points is not quite continuous but depends on how fast the mouse moves and the speed of the computer operating environment. A sample sketch drawn with the Sketch program is shown in Figure 16-11. The syntax for the **MouseMove** event is:

```
Private Sub Form_MouseMove (button As Integer, shift As Integer, x As Single, y As Single)
```

The arguments for the **MouseMove** event are shown in Table 16-12. Its use is illustrated in the Sketch program.

Table 16-12

The Arguments of the MouseMove Event

Mouse Button and Shift Constants

Constant (Button)	Value	Description
vbLeftButton	1	Left button is pressed.
vbRightButton	2	Right button is pressed.
vbMiddleButton	4	Middle button is pressed.
Constant (Shift)	**Value**	**Description**
vbShiftMask	1	SHIFT key is pressed.
vbCtrlMask	2	CTRL key is pressed.
vbAltMask	4	ALT key is pressed.

The Sketch Program

This is a barebones program that illustrates some of the features of the **MouseMove** event. The program has a form but no controls, so no properties table is needed.

Entering the Code for the Sketch Program. Open a new Visual Basic program, double-click the screen to get the code window, and enter the following code:

```
Option Explicit
' The Sketch program
' Enables the user to sketch on the form with the mouse

Private Sub Form_MouseMove(Button As Integer, Shift As Integer, _
                    X As Single, Y As Single)
'  Draws a series of pixels as the mouse is dragged.
 DrawWidth = 2                  ' sets the dot size
   If Button Then               ' While button is pressed.
      PSet (X, Y)    ' Draws a point
   End If
End Sub
```

Executing the Sketch Program. Click the VB **Start** button to execute the program. Drag the mouse across the form without clicking any mouse button. Click a mouse button, and drag slowly, then quickly. The slow line is continuous; the fast line is a series of points. Simple sketches can be made with a mouse, but a drawing tablet, substituting for a mouse, facilitates creating more complex drawing such as that in Figure 16-11.

Figure 16-11. A Drawing Made on the Form with the Sketch Program

Understanding the Sketch Program. Whenever the mouse moves, the **MouseMove** procedure is called and executed. When the **MouseMove** procedure is executed, the **PSet** method (of the form) is executed, and a point is set on the form. The **MouseMove** procedure detects whether any button, or combination of buttons, is down. Right-click, and drag the mouse. Click both buttons, and drag the mouse. Both actions sketch out a line on the form.

Close the program. Open the program, and for the statement **If Button Then**, substitute **If vbLeftButton Then**. Run the program again, and you'll find that only the left button works. The **Shift** parameter in the procedure's parameter list functions like the **Button** parameter, but applies to the SHIFT, CTRL, and ALT keys. Try editing the **If Button Then** statement again, and check the results. Try these variations:

```
If Button Then

If vbLeftButton Then

If (Shift) and (Not Button) Then
```

The button constants and the shift constants are listed in the Object Browser and in Table 16-12.

The Random Points Program

The Random Points program demonstrates the **PSet** method and the use of the **ScaleWidth** and **ScaleHeight** properties.

Entering the Code for the Random Points Program. The program has a form but no controls, so no properties table is needed.

```
Option Explicit
' The Random Points program
Sub Form_Click()
  Dim XPos As Integer, YPos As Integer
  Dim Point As Integer, Color As Integer
  DrawWidth = 2                              ' Set the point size
  For Point = 1 To 20000
    XPos = (Rnd * ScaleWidth)                ' horizontal position.
    YPos = (Rnd * ScaleHeight)               ' vertical position.
    PSet (XPos, YPos), RGB(Rnd * 255, Rnd * 255, Rnd * 255)
   Next
   MsgBox "Finished!"
   End
End Sub
```

Executing the Random Points Program. Click the **Start** button to load the program. Click the screen anywhere to run the program. The screen rapidly fills points in random positions with random colors (Figure 16-12).

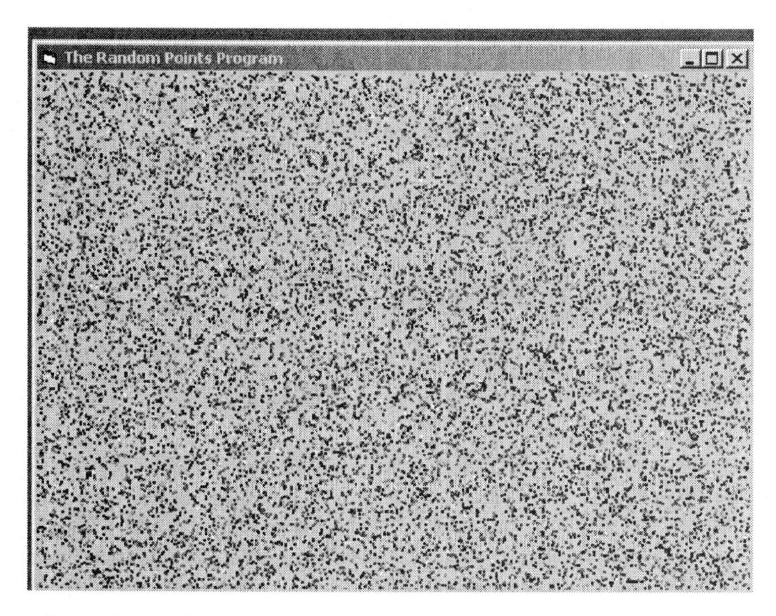

Figure 16-12. The Form of the Random Points Program at the End of a Run

Understanding the Random Points Program. Recall that the **Rnd** function returns a value between 0 and 1. The statement:

```
XPos = (Rnd * ScaleWidth)
```

returns a value for **XPos** that is somewhere between zero and the width of the form. The next statement returns a value for **YPos** that is somewhere between zero and the height of the form. Thus the statement:

```
PSet (XPos, YPos), RGB(Rnd * 255, Rnd * 255, Rnd * 255)
```

sets a point at a random position within the boundaries of the form's coordinate system. The color values of the **RGB** function parameter range from 0 to 255, so multiplying by the **Rnd** function gives all possible random colors.

When the **For ... Next** loop is satisfied, a message box appears on the screen advising the user that the program has ended. Clicking **OK** terminates the program.

Printing Graphic Methods to the Printer

The printer is an object with its own properties and methods that will be described and demonstrated in the following sections and programs.

The Circle to Printer Program

This program is similar to the Graph Circle program, which prints graphics to the form. The name of the form may be omitted when using methods and properties of the form, but when the printer is the object, it is necessary to include **Printer** as the name of the object, as shown in Table 16-13.

Table 16-13

Comparing the Actions of Some Methods of the Form and of the Printer

Form Object		Printer Object	
Scale( )	Sets scale on form	Printer.Scale( )	Sets scale of printer (paper)
Circle( )	Draws circle on form	Printer.Circle ( )	Draws circle with printer on paper
Print " "	Prints string on form	Printer.Print " "	Prints string with printer on paper

Entering the Code for the Circle to Printer Program. Open a new Visual Basic program. Double-click the form to get the code window. On the **Procedure List** of the code window, select **Click** to get the **Form_Click** procedure, and enter the following code:

```
Option Explicit
' The Circle to Printer program
Private Sub Form_Click()
   End
End Sub

Private Sub Form_Load()
   Printer.Scale (-10, 10)-(10, -10)
   Printer.DrawWidth = 8
   Printer.Circle (0, 0), 2, vbRed, , , 3        'A vertical ellipse
   Printer.Circle (0, 3), 3, vbBlack, , , 1      'A circle
   Printer.Circle (0, -2), 1.5, vbBlue, , , 0.3  'A horizontal ellipse
   'Printer.EndDoc                                ' Start the printer
End Sub
```

Executing the Circle to Printer Program. To run the Circle to Printer program, turn on the printer and click the VB **Start** button. The program runs, printing a circle and two ellipses to the printer (Figure 16-12). The center of the middle figure lies at (0, 0), the middle of the 8 ½ x 11 paper. After you run the program, a single click in the form will exit the program. Compare the syntax when the form is the default object and when the printer is the assigned object.

Understanding the Circle to Printer Program

The Coordinate System. In the preceding program listings, the statement:

```
Printer.Scale (-10, 10)-(10, -10)
```

sets the coordinate system. The origin (0,0) is at the intersection of the two axes. The x-axis units range in value from –10 on the left to +10 on the right. The y-axis units range in value from +10 at the top to –10 at the bottom. The corresponding Cartesian coordinate system chosen by the programmer is shown in Figure 16-7. The origin in this case lies at the center of the page, which is an 8 ½ x 11 piece of paper. The **Scale** method of setting the coordinate system is exactly equivalent to using the **ScaleTop**, **ScaleLeft**, **ScaleWidth**, and **ScaleHeight** properties together.

Actually, nothing is printed until the **Printer.EndDoc** method is executed. Incidentally, if you place a **Printer.EndDoc** statement in the middle of the program, say, after the last **Circle** method statement, the circles will be printed on one page and the text on a second page (a final **Printer.EndDoc** is still required).

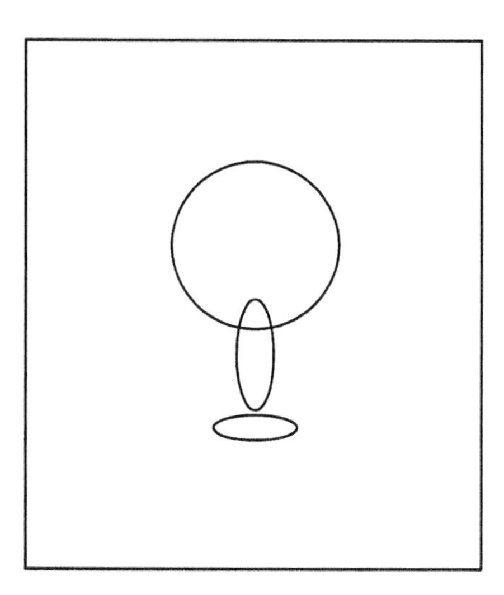

Figure 16-13. The Output of the Circle to Printer Program, Printed to the Printer

The PSet to Printer Program

Previous programs have illustrated how the **PSet** method of the form is useful in graphing mathematical functions. The **PSet** method is also a method of the printer, and so printing **PSet** to the printer is almost the same as printing **PSet** to the form. With the form, the name of the form may be omitted, but with the printer, its name, **Printer**, must precede the method.

With the printer, the PSet to Printer program prints a graph of the function e^{-x^2}, a function similar to the normal distribution function. The program has a form but no controls, so no properties table is needed.

Entering the Code for the PSet to Printer Program

Open a new Visual Basic program; double-click the screen to get the code window. On the **Procedure List** of the code window, select **Click** to get the **Form_Click** procedure. Enter the following code:

```
Option Explicit
' The PSet to Printer program
Private Sub Form_Load()
' Graph Exp(-x^2)
   Dim x As Single
   Printer.Scale (-4, 2)-(4, 0)
   Printer.DrawWidth = 5
   For x = -3 To 3 Step 0.01
      Printer.PSet (x, Exp(-x ^ 2))
   Next x
   Printer.Line (0, 0)-(0, 1.2)
   Printer.Line (-3.5, 0)-(3.5, 0)
   Printer.EndDoc
End Sub
```

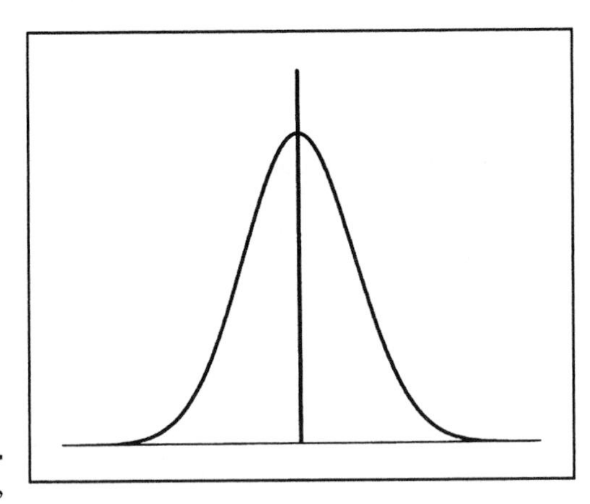

Figure 16-14. The Printed Output of the PSet to Printer Program

Executing the PSet to Printer Program. Turn on your printer. Click the VB **Start** button, and the program causes the printer to print a graph of the function e^{-x^2} (Figure 16-14).

Understanding the PSet to Printer Program

This statement **Printer.Scale (-4, 2)-(4, 0)** sets up the Cartesian coordinates system, with x ranging from –4 to +4 and y ranging from 0 to +2. The **For ... Next** loop does the printing while x ranges from – 3 to + 3, so the graph does not fill the entire page from right to left.

The NoForm Program

Because the NoForm program has no form, no output can be printed to the form. The NoForm program has no controls, because controls require a form to which the controls may be attached. Consequently, no output can be printed to a control, such as a list box or label control. In order to see some printed output, we'll use a message box.

So why write such a crude and limited program at this point, the last program in the chapter on Visual Basic? The answer: to help you understand the nature of Visual Basic for Applications, the subject of the next chapter.

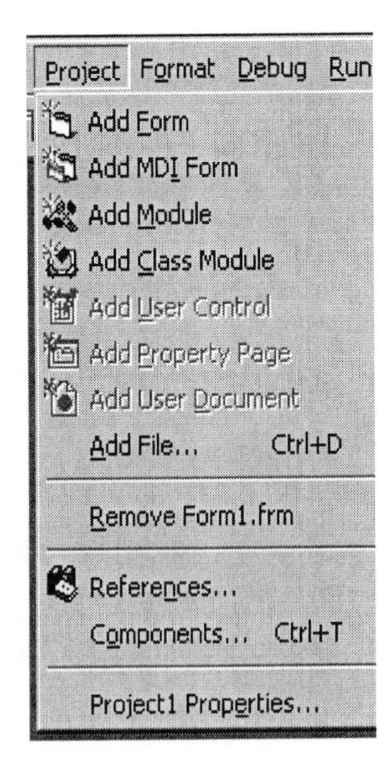

Figure 16-15. The Project Menu

Beginning the NoForm Program Project. Open a new Visual Basic program. As usual, Visual Basic responds with the Visual Basic design window (Figure 10-3), containing the **Project Explorer** window, the properties window, the form, and the form's code window.

Removing the Form and Adding a Module. Because the form is created automatically in Visual Basic, we will remove it. But then the form's code window will also disappear, so where will we write the code for our formless program? We will add a module and write the code on it. Here's how:

- In the **Project Explorer** window (Figure 10-3), click once on any plus sign to expand the list, and then click once on the line **Form1 (Form1)** to select it.
- On the **Project** menu on the Visual Basic menu bar, click **Remove Form1.frm** (Figure 16-15), and the form disappears.
- On the **Project** menu on the Visual Basic menu bar, click **Add Module** (Figure 16-15) to open the **Add Module** dialog box (Figure 16-16). Click Open to get the Window with the title bar **Project1 – Module1 (Code)** (Figure 16-17).
- The insertion point is blinking in the upper left corner, waiting for you to enter the code for the program. Figure 16-17 shows the design window after the code has been added to **Project1 – Module1 (Code)**.

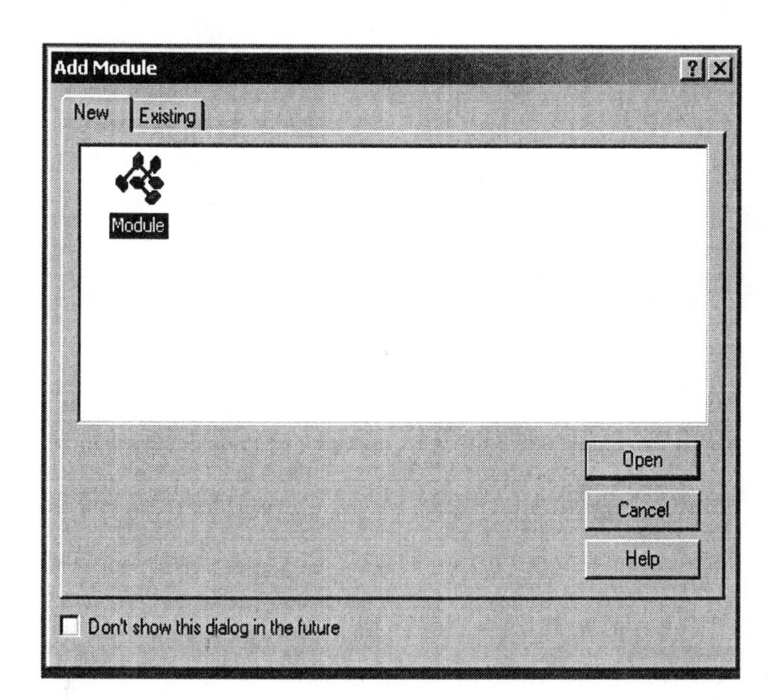

Figure 16-16. The Add Module Dialog Box

Entering the Code for the NoForm Program. The project has no form and no controls, so no properties table is needed.

- You can enter the code as written in Figure 16-17.
- Alternatively, on the **Tools** menu in the menu bar, click **Add Procedure**.
  1. In the **Name** box, type **Main** as the name.
  2. In the **Type** section, select **Sub**; in the **Scope** section, select **Public**. Click **OK**.
  3. The first and last lines of the procedure are written for you.
- Enter one line of code between the first and last lines, as shown in Figure 16-17:

```
MsgBox "This program has a module, but no form"
```

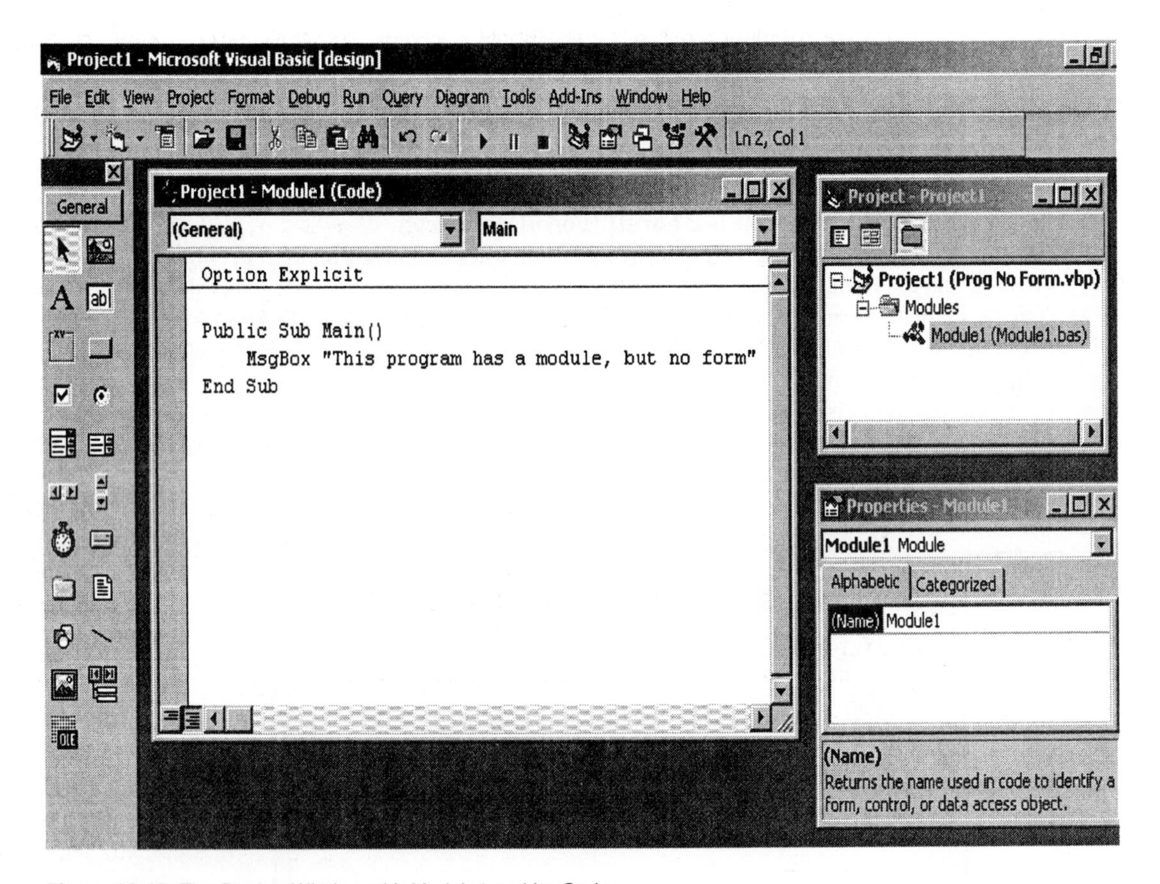

Figure 16-17. The Design Window with Module1 and Its Code

Executing the Program. Click the VB **Start** button to run the program. The program responds by displaying a message box with the message **This program has a module, but no form** (Figure 16-18). With no form, no message can be written to the form. And with no form, the program can have no controls on which to write. So a message box provides a way of getting the text message to the user.

Examine the **Project Explorer** window (Figure 16-17, upper right). In the expanded list, under **Modules**, it shows the line **Module1 (Module1.bas)**. This is the module on which you just wrote the code. In a program with a form (Figure 10-3), the Project Explorer shows, under Project1, the forms, and then Form1 (Form1). The item in parentheses is the **(Name)** property of the object, which you could change, if you wish, in the properties window.

In the next chapter on Visual Basic for Applications, we'll use this Visual Basic example to understand the structure of macros and macro procedures.

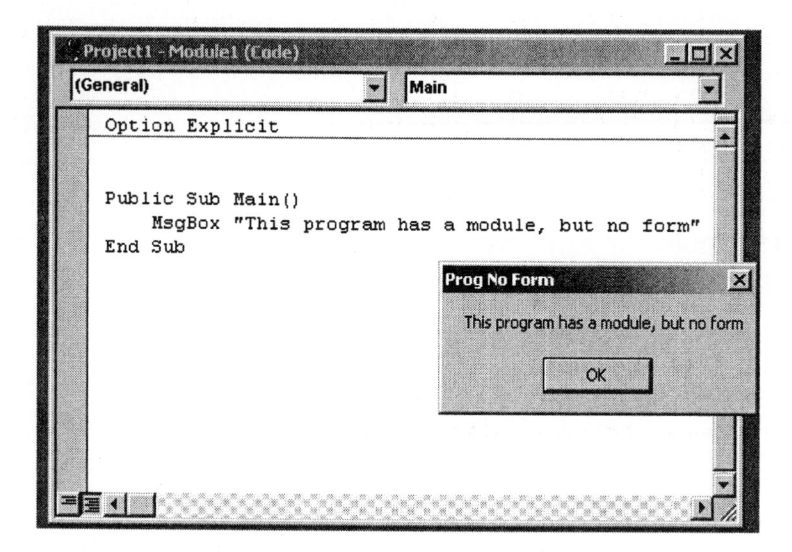

Figure 16-18. Output of the NoForm Program

Part III. Visual Basic for Applications

Chapter 17. Visual Basic for Applications: Microsoft Word
Chapter 18. Visual Basic for Applications: Microsoft Excel

Chapter 17

Visual Basic for Applications: Microsoft Word

This chapter and the next illustrate some applications of Visual Basic for Applications (VBA) to Microsoft Word and to Microsoft Excel. It is useful to separate the more numerical aspects of Microsoft Excel from Microsoft Word. VBA for Word provides good preparation for VBA for Excel.

The Relationship between VB and VBA

Visual Basic (VB) is a programming language for creating stand-alone programs that run in a Microsoft Windows environment. Visual Basic for Applications, on the other hand, is an accessory that is built in to many Microsoft applications, such as Excel, Word, PowerPoint, Access, and Outlook. A number of non-Microsoft applications are licensed to include VBA: AutoCAD, Corel WordPerfect, and Visio, for example. The VB in VBA is almost as complete and powerful as VB. That part of the language that is common to VB and VBA is nearly identical syntactically. The integrated development environment (IDE), the editor, and the windows are virtually identical. Consequently, if you have gone through the previous chapters on Visual Basic, you will find that most of the features of Visual Basic for Applications are familiar. VBA has fewer controls in its toolbox than VB. You will recall that the VB design window opens by default with a form; the VBA design window opens without a form. However, the similarities between VB and VBA exceed their differences.

You have already seen the important role that the form plays in Visual Basic. In VB, the form is the graphical interface between the user and the program and its controls (command buttons, list boxes, labels, and others). However, VBA is subservient to its host document, and a VBA program frequently does not even include a form or control at all.

In addition, VBA is closely related to macros. We will now explore this relationship.

Macros

A *macro* is a sequence of an application's commands that the user has collected and stored in a file with a name. At some time or another, you have probably written a macro for Word or Excel. A macro is created by "recording" with a macro recorder a series of keyboard entries that correspond to some simple task that is frequently used. For example, your Microsoft Word application has some default font with which you routinely write your documents. Perhaps in some Word documents you frequently want to change a selection of words from the default font to Arial 10-point bold. Using a macro for this task, you can change the font name and size and make it bold with as little as a keystroke or two. You might write another macro to undo this change.

The user may not realize it, but the macro recorder stores the sequence of commands as a Visual Basic procedure, or perhaps more exactly, a VBA procedure. Any Microsoft program that features macros and a macro recorder also includes a Visual Basic Editor. Thus, a VBA procedure written by the macro recorder may be edited to fine-tune and personalize the macro. Indeed, the user may use the Visual Basic Editor to create powerful VBA procedures from scratch. It should be clear at this point that a macro is a VBA procedure. The two terms are interchangeable. The Macros/VBA procedures included in this chapter are listed below. All use the same host program, MyTestDoc.

The Arial10B macro	(Host Microsoft Word file: MyTestDoc)
The Times12 macro	(Host Microsoft Word file: MyTestDoc)
The Table3_5 macro	(Host Microsoft Word file: MyTestDoc)
The CountStuff macro	(Host Microsoft Word file: MyTestDoc)
The DateName macro (VBA procedure)	(Host Microsoft Word file: MyTestDoc)

Word VBA Example 1

The Arial10B Macro

Recording the Arial10B Macro

A macro requires a host program, so the first step in writing a macro is to open the host file.

Step I. Create the host document.

- Open a new document in Word.
- Enter the following text, which will serve as a reminder of the name of the host document and the name of the macro to be written

  This is a Word document for hosting and testing a recorded macro.
  I will save this document with the name MyTestDoc.
  I will record a macro that changes selected text to Arial 10-point bold.
  I will name the first macro Arial10B.
- Save the host document with the name MyTestDoc.

Step II. Record the Macro.

At this point you are ready to prepare for recording the keystrokes and mouse clicks required to carry out the desired task. If the macro contains several keystrokes, you may want to make a list so you won't forget any keystrokes. If you really mess up, you can delete the macro and begin again.

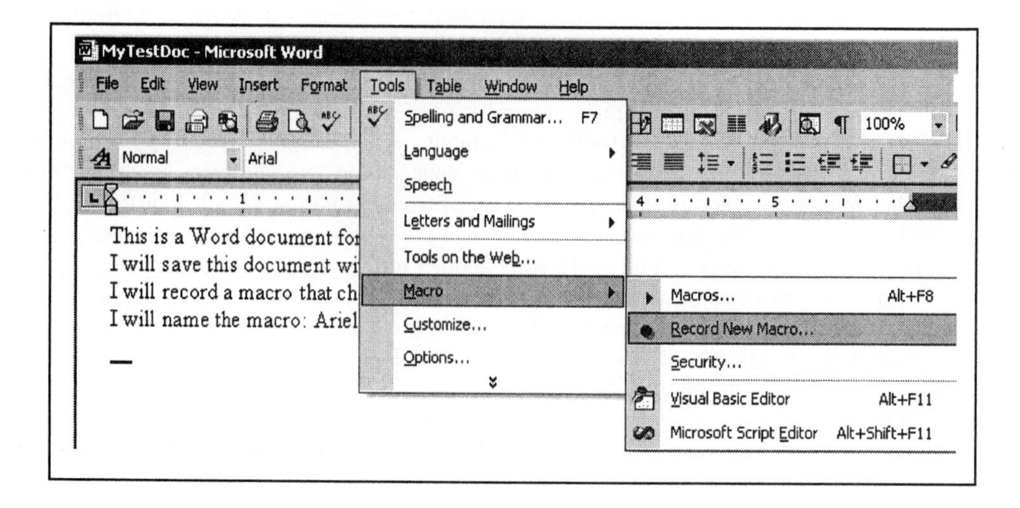

Figure 17-1. The **Tools** Menu and the **Macro** Submenu with **Record New Macro** Selected

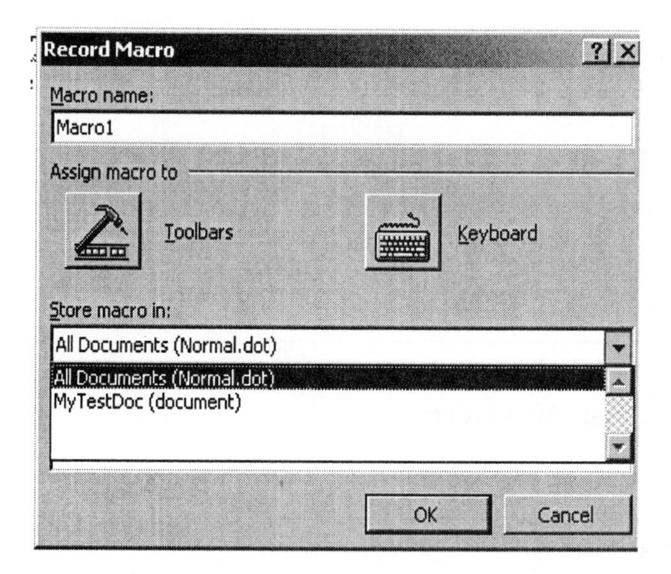

Figure 17-2. The **Record Macro** Dialog Box

- With the host document open, on the **Tools** menu, point to **Macro**, and click **Record New Macro** (Figure 17-1). Clicking **Record New Macro** causes the **Record Macro** dialog box (Figure 17-2) to appear.
- In the **Record Macro** dialog box:

1. In the **Macro name** box, type **Arial10B** as the name. Choose a macro name that describes the macro sufficiently well that you will recognize it some time in the future.
2. For now, skip the assignment icons, which will be discussed later.
3. The **Store macro in** box allows you two choices for the scope of your macro:
    a. For global scope, select **All Documents (Normal.dot)**.

b. For local scope, select **MyTestDoc (document)**.
c. Select **MyTestDoc (document)** so that your macro can be used in only this particular document and in none of your other Word documents.

4. In the **Description** box, add your description of your macro's function to whatever the **Record Macro** dialog box has already written. (In Figure 17-2, the **Store macros in** box hides the **Description** box.)
5. Click **OK**. the **Record Macro** dialog box disappears, Word's tiny **Stop Recording** toolbar (Figure 17-3) appears, and the pointer changes to a tiny tape recorder.

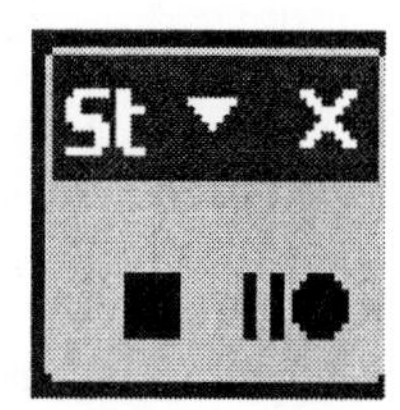

Figure 17-3. Stop Recording Toolbar

At this point you are ready to record. Don't click the **Stop Recording** toolbar yet. Every keystroke, command, and click that you enter is recorded, mistakes and all. (So take care not to make any typos.) While the **Stop Recording** toolbar is visible, enter the following keystrokes or clicks:

- On the **Format** menu on the Word menu bar, click **Font**, and then click **Arial** (scroll if necessary).
- In the **Font style** box, click **Bold**.
- In the **Size** box, click **10**, and then click **OK**.
- To stop recording, click the square button (■) on the **Stop Recording** toolbar (Figure 17-3). It will then disappear.
- Save your document (MyTestDoc), and close it.

Running the Arial10B Macro. Open the document MyTestDoc that you have used to record the macro named Arial10B. Your document will open with a Microsoft message box (Figure 17-4) warning that **Macros may contain viruses**. Since you just created the Arial10B macro, you don't need to worry that it contains viruses, so you may safely click the **Enable Macros** button and then close the message box. This warning will appear every time that the MyTestDoc document is opened. If, when you created the Arial10B macro, you had selected **All Documents (Normal.dot)** to make the scope global, this message box would appear with all documents that you have in your Word application. To avoid this, it is usually advantageous to assign local scope to your macros.

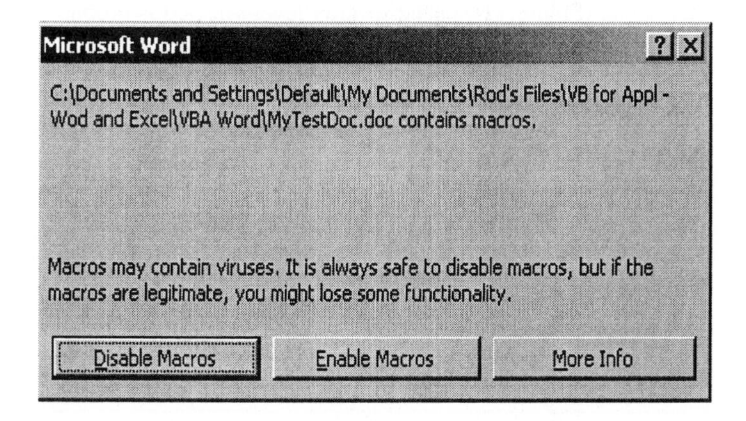

Figure 17-4. Virus Warning Message Box for Disabling and Enabling Macros

In the host document (MyTestDoc), select the text you want to change to Arial 10-point bold (click and drag over the text to select it).

- On the **Tools** menu, point to **Macro**, and click **Macros**. The **Macros** dialog box (Figure 17- 5) appears. The **Macros** dialog box lists all the macros available to MyTestDoc if the **Macros in** box shows "MyTestDoc."
- Select the macro named **Arial10B**, and click **Run**.

The selected text changes to Arial 10-point bold. In a document in which you regularly use two kinds of fonts, it is useful to have a second macro to change back to the default font. We will write it shortly.

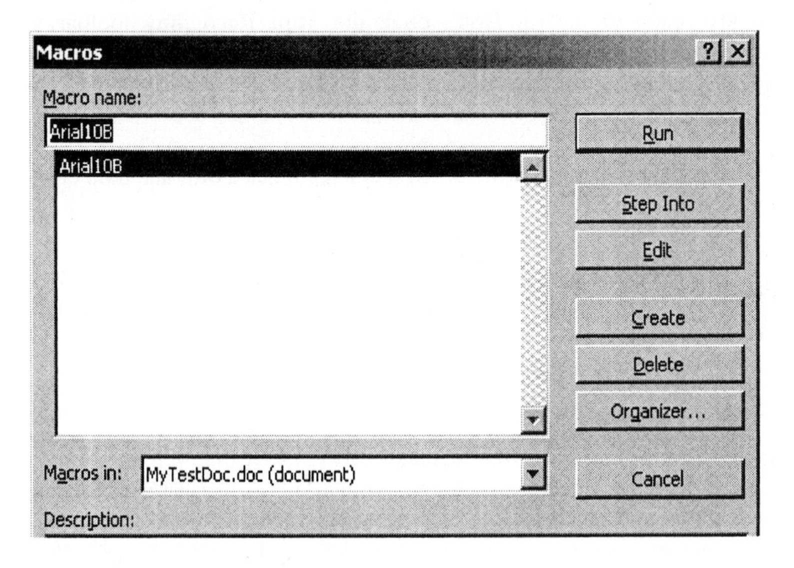

Figure 17-5. The **Macros** Dialog Box

Enabling and Disabling Macros. When you attempt to run a macro, Microsoft VBA may display a dialog box (Figure 17-6) announcing that **The macros in this project are disabled**, and your macro won't run. In that case, on the **Tools** menu, point to **Macro**, and click **Security** to open the **Security** dialog box (Figure 17-7). The security level selected is probably **High**. Change it to **Medium**, click **OK**, save your document, reopen it, and try again to run the macro. Now it should function OK.

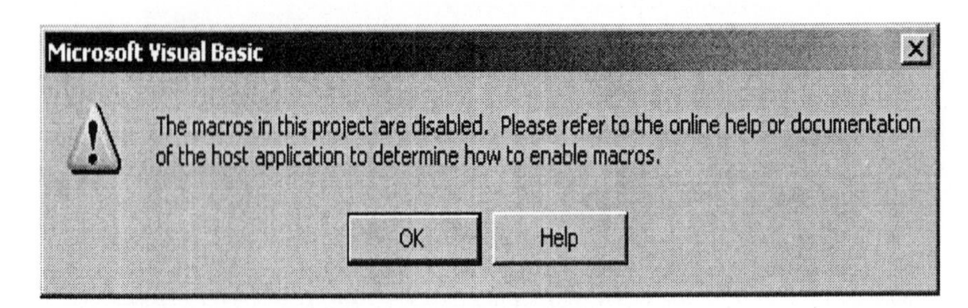

Figure 17-6. The Microsoft VBA Disabled Macros Dialog Box

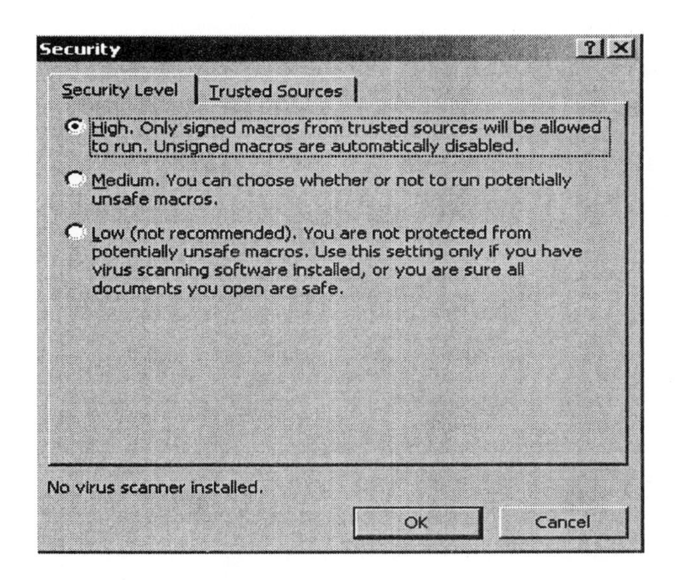

Figure 17-6. The Microsoft VBA **Disabled Macros** Dialog Box

Word VBA Example 2

The Times12 Macro

Recording a Second Word Macro: Times12

For practice, record a second macro that reverses the action of the macro just recorded (the Arial10B macro). This next Word macro changes any selected text to Times New Roman 12-point font. Open the Word document named MyTestDoc, which will also serve as the host document for the Times12 macro.

- On the **Tools** menu, point to **Macro**, and click **Record New Macro** to open the **Record Macro** dialog box (Figure 17-2).
- In the **Macro name** box, type **Times12** as the name.
- In the **Store macro in** box, scroll to **MyTestDoc**.
- In the **Assign macro to** section, click the **Keyboard** icon. The **Customize Keyboard** dialog box appears (Figure 17-8), and in the **Commands** box (top right), it reads **Project NewMacros.Times12**.
  1. In the **Specify keyboard sequence** section, under **Press new shortcut key**, press **Alt +T**, and **Alt +T** appears in the one-line text box.
  2. In the **Save changes in** box, scroll to **MyTestDoc**.
  3. Click the **Assign** button, and **Alt+T** appears in the **Current keys** box.
  4. Click the **Close** button.
- The **Record Macro** dialog box returns (Figure 17-2). In the **Store macro in** box, scroll to **MyTestDoc**, and then click **OK**. (This keeps the macro local in scope.)
- The dialog box disappears, and the tiny **Stop Recording** toolbar appears (Figure 17-3).

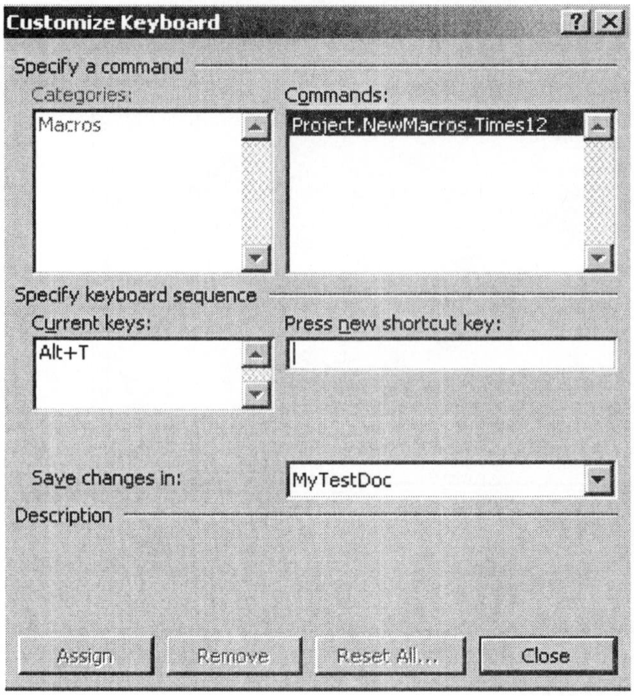

Customize Keyboard
Specify a command
Categories:
Macros
Commands:
Project.NewMacros.Times12
Specify keyboard sequence
Current keys:
Alt+T
Press new shortcut key:
Save changes in:
MyTestDoc
Description
Assign
Remove
Reset All...
Close

Figure 17-8. The **Customize Keyboard** Dialog Box

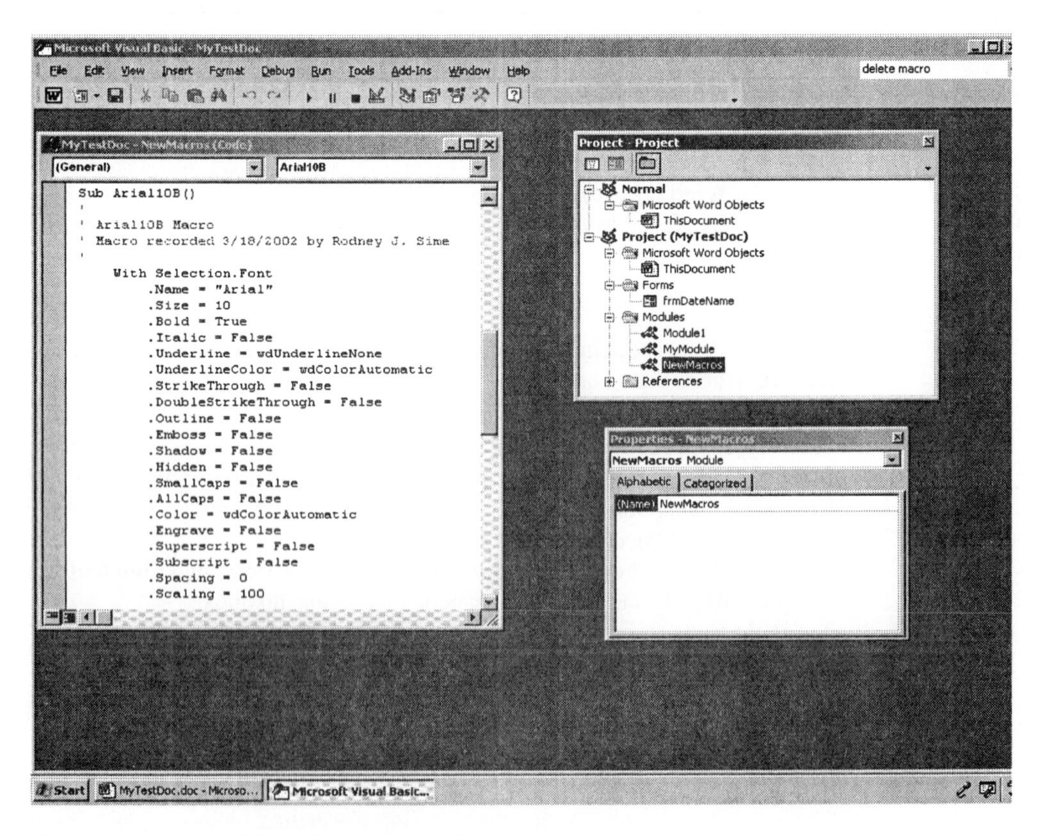

Figure 17-9. The Visual Basic for Applications Design Window

Running the Times12 Macro. Run the Times12 macro exactly as you did the Arial10B macro. Experiment with the two macros. Open the Word document named MyTestDoc. Select a phrase or sentence. Run the Arial10B macro to change the font to Arial 10-point bold. Now select all or part of that sentence, and run the Times12 macro by pressing ALT+T. Compare the ease of running a macro from the **Macro** dialog box (on the **Tools** menu, point to **Macro**, and click **Macros**) and with the shortcut key (ALT+T, in this example).

Using the shortcut key is clearly more convenient. However, if you assign all your macros to shortcut keys, you will quickly run out of available shortcut keys. Also, if you don't make a written record of the nature of the macro associated with a shortcut key, you may forget which shortcut key is for what macro. To extend the number of shortcut keys, you can combine keys, as in ALT+CTRL+T.

The Relationship between Macros and Visual Basic for Applications

What is happening when you record a macro for your application? Your application is automatically writing a Visual Basic procedure with the name you gave to the macro. The name of the procedure is the name that you gave in the **Record Macro** dialog box (Figure 17-2). When you run the macro, you are essentially calling the procedure by that name. Thus, a macro is a VBA procedure consisting of a number of statements written in the Visual Basic language and collected in a named file. Normally the VBA procedure is invisible to the casual macro user.

To see the VBA procedure corresponding to the macros you have created, it is necessary to open the VBA editor, which incidentally, looks like and behaves like the Visual Basic Editor you are familiar with from the previous chapters on Visual Basic. To open the VBA editor for the procedures you have just written:

- On the **Tools** menu, point to **Macro**, and click **Visual Basic Editor**, and the VBA design window appears (Figure 17-9. If the project and properties windows are missing, on the **View** menu, click **Project Explorer**, or on the **View** menu, click **Properties**.

  Or ...
- On the **Tools** menu, point to **Macro**, and click **Macros** to open the **Macros** dialog box (Figure 17-5). A list box displays the names of the macros.
- Select either the Arial10B macro or the Times12 macro.
- Click **Edit**, and the Visual Basic for Applications integrated development environment (or design window) appears (Figure 17-9). It is labeled **Microsoft Visual Basic – MyTestDoc.** The code window in the Visual Basic for Applications editor is labeled **MyTestDoc – NewMacros (Code)**.

The VBA Editor. The VBA editor looks just like the Visual Basic Editor ... in fact, the title bar simply states **Microsoft Visual Basic**. As with Visual Basic in Chapters 10 through 16, the editor window contains three child windows. First there's the code window for which the title bar now says **MyTestDoc – New Macros (Code)**. The next window is the **Project Explorer** window, for which the title bar is **Project – Project**. The third window is the properties window with the title bar **Properties – New Macros**.

The VBA Project Explorer. The **Project Explorer** window is similar to the Visual Basic project and the Windows Explorer. Clicking a box with a negative sign collapses the subheading. Clicking a box with a positive sign displays the subheadings. The **Project Explorer** shows that one document is open, and it is listed as **Project (MyTestDoc)**. If you also have other Microsoft Word documents open, they will be listed in the project window as **Project (*the name you gave it*)**. Under **Project (MyTestDoc)** are listed the modules associated with this document. There is only one module, which has the name **NewMacros**.

The VBA Properties Window. This window shows the properties of the selected object. The selected object (darkened in the **Project Explorer** window) is a module with the default name. The properties window shows that this module has only one property, and that is the **(Name)** property. You could rename this module something more to your liking, if you wish.

Global versus Local Macros. All the local macros associated with the MyTestDoc document are stored in the module named **NewMacros**. All global macros are stored in **Normal** (at the top of the **Project Explorer** window.) The two Word macros that we have recorded so far (Arial10B and Times12) have been local because we chose to store them with MyTestDoc in the **Record Macro** dialog box. If we had chosen to store our macro with the **All Documents (Normal.dot)** scope, then the macro would have been stored in the **NewMacros** module (top of Figure 17-9) under **Normal**.

With Figure 17-9 on your screen, close the code window (with the title bar **MyTestDoc – NewMacros (Code)**). Next, select the **NewMacros** line in the **Project (MyTestDoc)** section of the **Project – Project** window, which is highlighted in Figure 17-9. The code window opens

The Project Explorer lists all the forms (if any) and modules (if any) in your VB or VBA project. A project is the collection of all the files used in constructing your VB application or VBA macro. For the most part, you don't have to deal very much with the **Project Explorer** window. That's the job for Visual Basic or Visual Basic for Applications, both of which need to keep track of all the files in your project so that they are there the next time you open your application.

The VBA Code Window. The code for the macros stored in **NewMacros** is written in the code window, the title bar of which is **MyTestDoc – NewMacros (Code)**. Figure 17-9 shows this window displaying the code for the Arial10B macro. The VBA code for your macro is stored on the module named **NewMacros** and is displayed in the code window with the title bar **MyTestDoc – New Macros (Code)**. The code window displays listings for two procedures. The first procedure is named **Sub Arial10B()**, and the second procedure is named **Sub Times12().**

```
Sub Arial10B()
'
' Arial10B macro
' Macro recorded 3/18/2004
   With Selection.Font
      .Name = "Arial"
      .Size = 10
      .Bold = True

          etc

End Sub
```

Only the **Name**, **Size**, and **Bold** properties are shown in these two code listings.

```
Sub Times12()
'
' Times12 macro
' Macro recorded 3/18/2004 by John Dalton
'
   With Selection.Font
      .Name = "Times New Roman"
      .Size = 12
      .Bold = False

          etc

End Sub
```

The macro recorder automatically added the comments regarding the macro's name and creation date. The programmer can add to the descriptive comment in the **Description** box of the **Macros** dialog box at the time of creation.

Notice that the name of the procedure is the name given to the recorded macro. A macro is a VBA procedure. Each procedure (Arial10B and Times12) consists of three different assignment statements. (Ignore the duplicate procedures.) The first procedure assigns the name of the font, **Arial**, to the **Name** property of the **Font** property of the object **Selection**.

Compare Figure 17-9 (this chapter) with Figure 16-16 (previous chapter). The project window for the Visual Basic NoForm program (Figure 16-16) is similar to the project window for the VBA macros stored in **NewModules** (Figure 17-9). The usual macro is a Visual Basic program with a module, but no form. The usual Visual Basic program has a form, but no module. However, we have just seen that a Visual Basic program can be written without a form (the NoForm program, Chapter 16), and we will later write macros that do have forms.

Word VBA Example 3

The Table3_5 Macro

Recording a Third Word Macro: Table3_5

When you record a macro in Word, you usually select some text—a word, sentence, or paragraph—and then enter some keystroke that performs some action such as formatting or inserting text. When VBA automatically writes the procedure for the macro, it creates the **Selection** object, and then changes its properties according to your keystrokes, mouse clicks, and commands.

Text is not the only object that you can modify or create with a macro. A macro can record any sequence of keystrokes that you use in Word. For example, you can record a macro that creates a table, perhaps of a generic type, that you use many times over in the course of recording data or writing reports.

To try this kind of macro, open the previous Word document named MyTestDoc. Go to the end of the document, and press the ENTER key once to enter a blank line. The blank line isn't necessary, but it is often useful to separate the last line of text from a following table.

- On the **Tools** menu, point to **Macro**, and click **Record New Macro**. The **Record Macro** dialog box (Figure 17-2) appears.

1. In the **Macro name** box, type **Table3_5** as the name.
2. Skip the assignment icons; in the **Store macro in** box, select **MyTestDoc**.
3. In the **Description** box, type **Creates a 3 x 5 table** to add a description of the macro's function.
4. Click **OK**. The **Record Macro** dialog box disappears, Word's **Stop Recording** toolbar(17-3) appears, and the pointer changes to a tiny tape recorder.

At this point you are ready to record. Use the TAB key or the up and down arrow keys to place the pointer as directed. Enter the following keystrokes and mouse clicks:

- On the **Table** menu, point to **Insert**, and click **Table**, and the **Tables** dialog box appears.
- Set the number of columns to **3** and the number of rows to **5**, and click **OK**.
- Place the pointer in column 1 row 1, and type **Run** as the text.
- Place the pointer in column 2 row 1, and type **Concentration** as the text.
- Place the pointer in column 3 row 1, and type **Temperature** as the text.
- Place the pointer in column 1 row 2, and type **1**.
- Place the pointer in column 1 row 3, and type **2**.
- Place the pointer in column 1 row 4, and type **3**.
- Place the pointer in column 1 row 5, and type **4**.
- Place the pointer outside the table by pressing the down arrow key.
- Press the ENTER key once.
- To stop recording, click the **Stop Recording** toolbar.

Table 17-1, which follows, will have formed as you record the macro.

Table 17-1

A Table Created with Macro Table3_5

Run	Concentration	Temperature
1		
2		
3		
4		

Running the Table3_5 Macro:

- On the **Tools** menu, point to **Macro**, and click **Macros**. The **Macros** dialog box appears. Scroll in the **Macros in** box to **MyTestDoc**.
- Select the macro named **Table3_5**, and click **Run**.

You can replicate the table readily. If such a table were to be used routinely, you could assign it to a keystroke shortcut and generate the table with a single keystroke.

In VBA editor, examine the procedure for the Table3_5 macro. Try to identify objects, their properties, and their methods. Recording a macro for a desired procedure is one of the ways to learn how objects, their properties, and their methods are used in an application.

Open your document named MyTestDoc. On the **Tools** menu, point to **Macro**, and click **Visual Basic Editor**. Scroll from the top to the bottom of the code window, which is titled **MyTestDoc – NewMacros (Code)**. You will find the Arial10B, Times12, and Table3_5 macros keeping each other company.

The VBA Code Generated by Recording the Table3_5 Macro

```
Sub Table3_5()
'
' Table3_5 Macro
' Macro recorded 9/17/2002
'
  ActiveDocument.Tables.Add Range:=Selection.Range, NumRows:=5, NumColumns:= _
    3, DefaultTableBehavior:=wdWord9TableBehavior, AutoFitBehavior:= _
    wdAutoFitFixed
  With Selection.Tables(1)
    If .Style <> "Table Grid" Then
      .Style = "Table Grid"
    End If
    .ApplyStyleHeadingRows = True
    .ApplyStyleLastRow = True
    .ApplyStyleFirstColumn = True
    .ApplyStyleLastColumn = True
  End With
  Selection.TypeText Text:="Run"
  Selection.MoveRight Unit:=wdCell
  Selection.TypeText Text:="Concentration"
  Selection.MoveRight Unit:=wdCell
  Selection.TypeText Text:="Temperature"
  Selection.MoveDown Unit:=wdLine, Count:=1
  Selection.MoveLeft Unit:=wdCharacter, Count:=2
  Selection.TypeText Text:="1"
  Selection.MoveDown Unit:=wdLine, Count:=1
  Selection.TypeText Text:="2"
  Selection.MoveDown Unit:=wdLine, Count:=1
  Selection.TypeText Text:="3"
  Selection.MoveDown Unit:=wdLine, Count:=1
  Selection.TypeText Text:="4"
  Selection.MoveDown Unit:=wdLine, Count:=1
  Selection.TypeParagraph
End Sub
```

Objects, Properties, Methods, and Events

Just as in VB, a VBA procedure is made up of objects that have properties and methods and that respond to events. Many of the objects in VBA are familiar to you from the previous chapters on VB. The form, modules, and controls were all used in Visual Basic and are also used in VBA. A few selected objects in a Word application are applications, words, sentences, paragraphs, fonts, and tables. A few selected properties of applications are **ActiveDocument**, **Count**, **Name**, **FullName**, and **Selection**.

VBA handles properties in exactly the same way as Visual Basic. The property of an object can be changed by assigning a new value to it. In the reverse direction, the value of the property of an object can be assigned to a variable. Table 17-2 displays a comparison of Visual Basic and VBA. Their handling of methods and events is equally similar.

Table 17-2

Objects and Properties in Visual Basic and VBA

Visual Basic Examples	VBA Examples
Examples of Assignments: Object.Property = Value	
txtPhrase.Font.Size = 14 lblResult.Caption = "Your guess is too low"	Selection.Font.Name = "Arial" ActiveDocument.Range.Sentences(2).Italic = True
Variable = Object.Property	
R = hsbRed.Value Thigh = txtThigh.Text	CountNum = ActiveDocument.Words.Count MsgBox "The name of this document is " & ActiveDocument.Name

In shifting our attention from Visual Basic to VBA, we encounter good news and bad news. The good news, presented in Table 17-2, is that Visual Basic and VBA are virtually identical syntactically in dealing with objects and their properties. The bad news is that the number, complexity, and ambiguity of objects in VBA for Word (and Excel) are somewhat greater than in Visual Basic. For example, the VBA keyword **Selection** can refer both to an object or a property. The VBA keyword **Range** can refer to an object, a property, or a method.

Fortunately, help is available from a number of sources on ways to deal with objects, properties, and methods while working in VBA.

1. The **Object Browser** (Figure17-10). In VBA, on the **View** menu, click **Object Browser**, or click its icon, or press F2. You can now browse through the objects and their properties, methods, and events.
2. The **Automatic Code Completion** list (Figure 10-12 and 10-13). As with Visual Basic, entering an object in VBA followed by a period generates an **Automatic Code Completion** list that displays the properties, methods, and events of that object.
3. Microsoft Visual Basic Help. Click **Help** on the menu bar, and then enter the name of the object, property, or method with which you need help. The help page often gives links you can click on such as **Example, See Also,** and **Applies to**.
4. Recording a macro. If you know what you want to do but don't know the syntactically correct names for the objects, procedures, and methods that you'll probably need, record a macro that does or nearly does what you want it to do. Then go to the VBA editor and edit it to fit your needs. Many of the sophisticated VBA programs that you see in the literature were written this way.

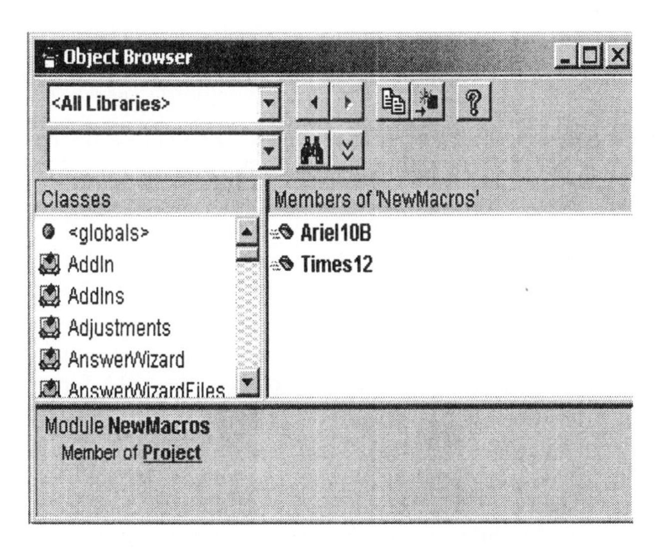

Figure 17-10. The Visual Basic **Object Browser** Window Showing Two **NewMacros**

Writing a Word VBA Procedure without a Form

Writing a procedure directly into the VBA editor is not very different from writing a Visual Basic program as first described in Chapter 10. Most Visual Basic programs use a form as a user interface; in fact, the Visual Basic Editor opens with a form ready for use. With VBA, the host program can be considered to be the user interface, although VBA can also use forms with embedded controls. When VBA has no form or controls, it is necessary to insert a module on which to write the code for the procedure.

To access VBA in any application, on the **Tools** menu, point to **Macro**, and the drop-down menu that appears includes a command at the bottom labeled **Visual Basic Editor** (Figure 17-1). As an alternate method for accessing the **Visual Basic Editor**, press ALT+F11. Click the icon or press ALT+F11, and the **Visual Basic Editor** window opens.

Why is VBA grouped with the macro tools? It is because a VBA procedure is a macro in an application. VBA not only has an intimate relationship with its host application, but even more specifically with the application's macros. In the VBA editor:

- Macros created with the host application's macro recorder can be edited in the VBA editor.
- Macros can be created from scratch in the VBA editor.
- A macro is a VBA procedure.
- All macros can be played (in macro terminology).
- All procedures can be run (in VBA terminology).

The last two statements are synonymous.

The VBA editor in the design window (Figure 17-9) differs only slightly from the Visual Basic Editor with which you are by now familiar if you have worked your way through the Visual Basic chapters (Chapters 10–16).

- When the Visual Basic Editor opens, it automatically opens with a form (default name: Form1), with the **Project Explorer** window and the properties window, and the toolbox.
- When the VBA editor opens, it opens with the **Project Explorer** window and the properties window, exactly as in VB, but with no form.

If you need a form in VBA, then after opening the VBA editor for your document, on the **Insert** menu, just click **UserForm**. A form appears, along with the toolbox containing the VBA controls. The **Project Explorer** window and the properties window remain open. If you select one of these windows (by clicking it), the toolbox disappears, but the form remains. Clicking the form brings back the toolbox. Other than the disappearing toolbox, the VBA forms and VBA windows appear and behave just as they did in Visual Basic. If you use a form, you click the form, then add controls, and write code for each control just like in Visual Basic. We will use the form in VBA shortly.

If you need a module in VBA, then after opening the VBA editor for your document, on the **Insert** menu, just click **Module**. A module appears, and you can begin writing your VBA code.

VB and VBA not only have the same kind of forms, windows, and editor, but they also have many of the same controls with most of the same properties. However, VBA for different applications is not identical. For example, Microsoft Word and Microsoft Excel have different objects. Excel has cells, but Word does not. Word has sentences, but Excel does not. Both have fonts, and the font properties are similar. But be reassured. If you have gone through the previous Visual Basic programs, you will feel very much at home with Visual Basic for Applications.

Word VBA Example 4

The VBA CountStuff Procedure, aka the CountStuff Macro

This time we'll write a VBA procedure directly on a module without using the macro recorder. It is still a macro, and the name you give to the VBA procedure will be listed in the **Macros** dialog box. A VBA procedure without a form is written on a module. At this point you already have a module named **NewMacros**, and it has three macros (VBA procedures) written on it: Arial10B, Times12, and Table3_5.

So in which module should you place your new Word procedure? You can write the new macro in the module named **NewMacros**, where your first three macros are stored. Or you can insert a new module and write your fourth macro in it. You have two choices, and it doesn't make any difference—it's a matter of personal organization. Since you haven't created a module before, let's do that and name it MyModule.

Programming the CountStuff Macro in Visual Basic for Applications

As an example, we shall write a procedure that counts the number of words, sentences, and paragraphs in a Word document. Again, we'll use the document named MyTestDoc as the host document for the CountStuff macro.

- Open the document MyTestDoc.
- On the **Tools** menu, point to **Macro**, and click **Visual Basic Editor** (Figure 17-9).

- On the **Insert** menu, click **Module**, and a blank module appears with the name **MyTestDoc – Module1 (Code)**.
- If the **Project Explorer** window or the properties window is not visible, on the **View** menu, click **Project Explorer** or **Properties**.
- In the **Project Explorer** window, find the module you just created, and select it.
- In the properties window, change the **(Name)** property of your module to **MyModule**.
- Click in the code window.
- On the **Insert** menu, click **Procedure**, which opens the **Add Procedure** dialog box (Figure 13-9).
- In the **Add Procedure** dialog box, in the **Name** box, type **CountStuff** for the name; in the **Type** section, select **Sub**; and in the **Scope** section, select **Public**. Then click **OK**.

Just like Visual Basic, VBA inserts the procedure declaration and the **End** statement:

```
Public Sub CountStuff()

End Sub
```

In between these two statements, enter the procedure code so that the procedure appears as follows:

```
Public Sub CountStuff()
    Dim CountNum As Integer
    CountNum = ActiveDocument.Words.Count
    MsgBox "The number of words in this document is " & CountNum
    CountNum = ActiveDocument.Paragraphs.Count
    MsgBox "The number of paragraphs in this document is " & CountNum
    CountNum = ActiveDocument.Sentences.Count
    MsgBox "The number of sentences in this document is " & CountNum
End Sub
```

Running the CountStuff Macro

- On the **Tools** menu, point to **Macro**, and click **Macros**. The **Macros** dialog box appears, which lists all the available macros if the **Macros in** box shows **All active templates and documents**.
- Select the macro named **CountStuff**, and click **Run**.
- A message box appears with the message **The number of words in this document is 49**, this latter number depending on the words in your document. Click **OK**.
- A second message box appears with the message **The number of paragraphs in this document is 4**, this latter number depending on how many times you click the ENTER key after the final period. Click **OK**.
- A third message box appears with the message **The number of sentences in this document is 4**, this latter number depending on the sentences in your document.

Examination of Figure 17-11 shows the presence of two modules attached to **Project (MyTestDoc)**: the default module, **NewMacros**, and the module you inserted, **MyModule**.

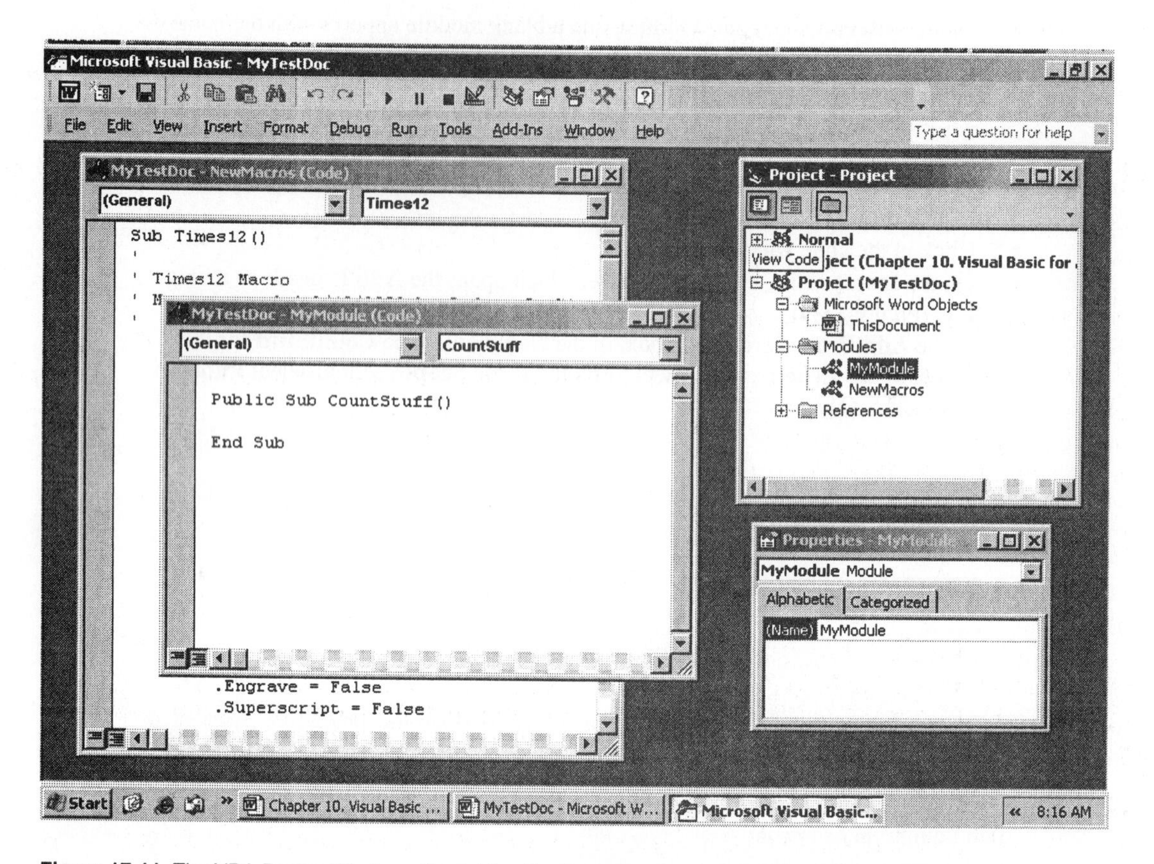

Figure 17-11. The VBA Design Window *after* Adding the Procedure but *before* Filling in the Code

Understanding the CountStuff Macro. The VBA CountStuff procedure begins with the declaration of an integer that will be used several times:

```
Dim CountNum As Integer
```

This integer declaration is followed by an assignment statement:

```
CountNum = ActiveDocument.Words.Count
```

ActiveDocument is an object that contains the collection **Words. Count** is a property of the collection **Words.** Its value equals the number of **Word** objects in the collection **Words.**

Similarly, a **Paragraph** object is any string delimited by a carriage return (ENTER key). **Paragraphs** is a collection of **Paragraph** objects.

A **Word** object begins with a space and ends with a space or punctuation mark. **Words** is a collection of **Word** objects.

A **Sentence** object begins with a space and an uppercase letter and ends with a punctuation mark. **Sentences** is a collection of **Sentence** objects. The actual number of each property that the macro reports depends on the way Word defines paragraphs, sentences, and words. The number of Microsoft Word objects, collections, properties, and methods is very large and complex. Fortunately, Visual Basic takes care of the organization for us in these simple macros and VBA procedures.

Word VBA Example 5

The DateName Macro and VBA Procedure

Writing a Word VBA Procedure with a Form: The DateName Macro

So far, none of the Word macros that we've written or recorded have included a form, but the next macro, which is also the last macro in this chapter, will contain a form. Why include a form with a VBA procedure? If your VBA procedure has a form, you can place controls on it, add code to the controls, and greatly increase the power of your VBA procedure (macro).

Again, open the Word document MyTestDoc. We will use it as the host document to write a VBA procedure with a form.

- On the **Tools** menu, point to **Macro**, and click **Visual Basic Editor** to open the VBA design window. The design window may display the **NewModule** and **MyModule** forms from our previous work. Ignore them, or close them if you wish.
- On the **Insert** menu, click **UserForm**, and a blank form should appear in the design window. Its title bar reads **MyTestDoc – UserForm1 (UserForm)**.
- If the toolbox is not visible, click the form, and the toolbox appears. Or click its icon in the VBA toolbar.
- If the **Project Explorer** window is not visible, on the **View** menu, click **Project Explorer**, and it appears. Under **Project (MyTestDoc)**, a line for **Forms** is now displayed, and under it is listed the new form, **UserForm1**.
- If the properties window is not visible, on the **View** menu, click **Properties**, and it appears.
- In the properties window, change the **(Name)** property of the form from **Form1** to **frmDateName**. The name of the form in the properties window changes to **frmDateName**, and so does the name of the form listed in the project window.

Figure 17-12 shows the appearance of the VBA design window, which is very similar to the Visual Basic design window (Figure 10-3).

The **Project Explorer** window is in the upper left. It lists the project (**MyTestDoc**), the forms (**frmDateName**), and the modules (**MyModule** and **NewMacros**).

Below it is the properties window. If the form has been selected, it displays the **(Name)** property of the form, **frmDateName**.

If the form has been selected, the toolbox is visible. If it is not visible, click the form (to select it), and the toolbox becomes visible.

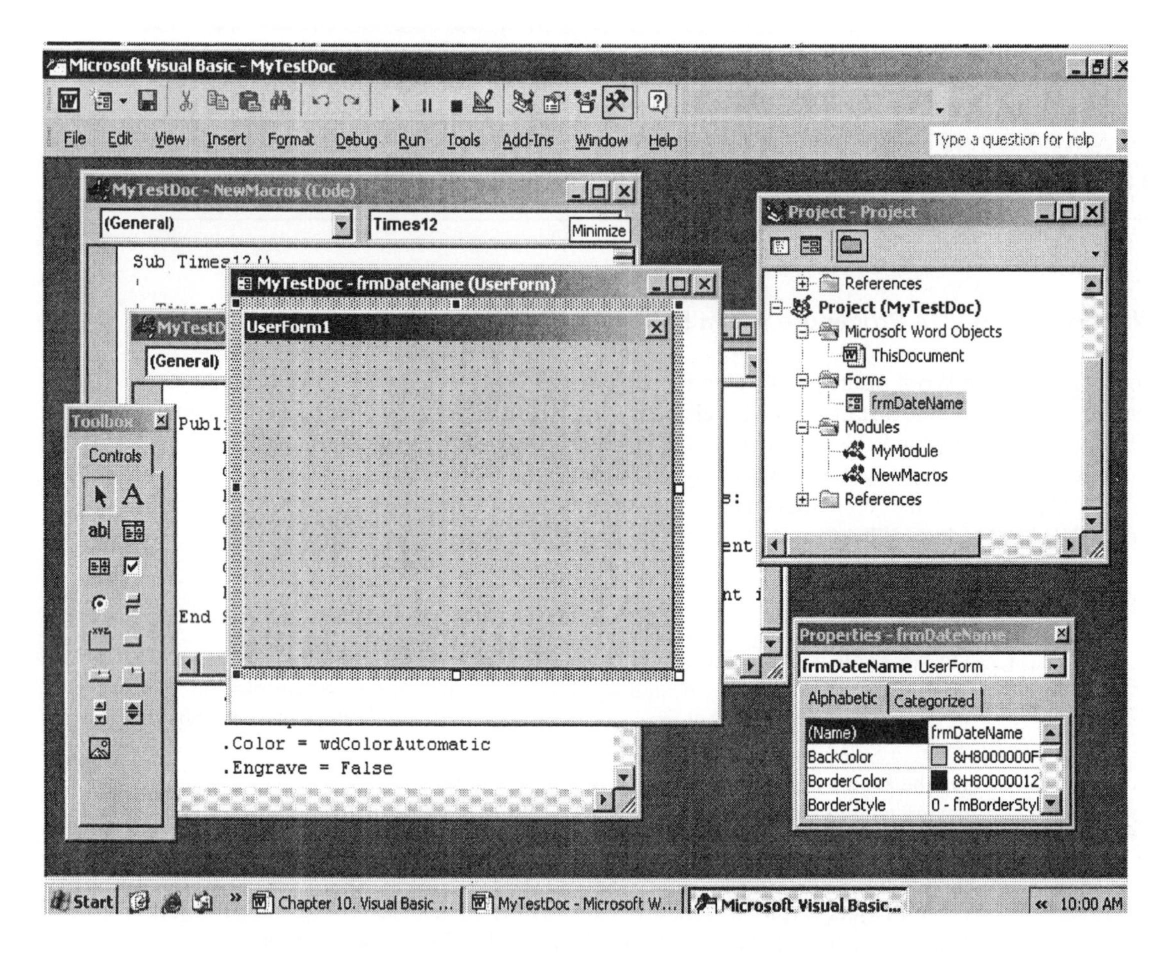

Figure 17-12. The VBA Design Window for Creating the DateName Macro. The form has been inserted, and the **(Name)** property of the form has changed from **Form1** (the default name) to **frmDateName**.

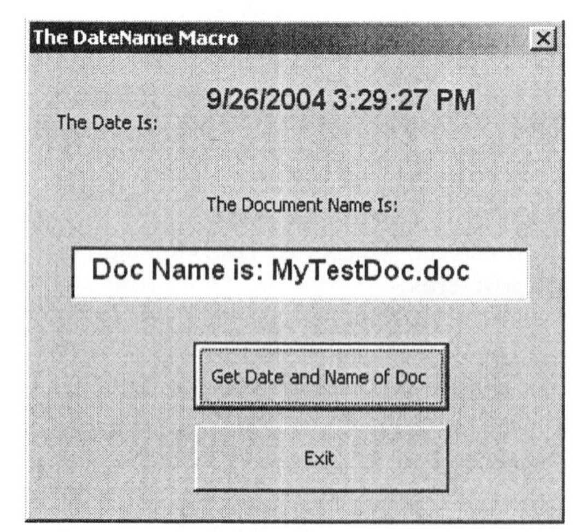

Figure 17-13. The **frmDateName** Form with Its Controls and Caption: **The DateName Macro**

Setting the Properties of the Form and Controls. Place the controls listed in Table 17-3 on the form. Adjust their positions and size so that they appear approximately as shown in Figure 17-13. Click the form, and set its properties in the properties window according to Table 17-3. Click each control, and set the properties of each according to Table 17-3.

After setting the properties of the form and controls, the form should like very much like Figure 17-13. Also, notice that the **Project Explorer** window shows that the **frmDateName** form has been added to the project.

Table 17-3

Property Settings for frmDateName

Object	Property	Setting
UserForm1	Name	frmDateName
	Caption	The DateName Macro
Label1	Name	lblDateInfo
	Caption	The Date Is:
Label2	Name	lblDate
	Caption	Make it blank
Label3	Name	lblName
	Caption	The Document Name Is:
Text box1	Name	txtName
Command1	Name	cmdGetInfo
	Caption	Get Date and Name of Doc
Command2	Name	cmdExit
	Caption	Exit

Entering the Code. Double-click one of the command buttons to open the code window (Figure 17-14). Its identifier on the title bar is **MyTestDoc – frmDateName (Code)**. The VBA code window has a procedure list box on the top left and an events list box on the top right, exactly the same as Visual Basic. Scroll the procedure list box and select **cmdGetInfo**, and scroll the events list box to select **Click**. Enter the following code:

```
Private Sub cmdGetInfo_Click()
  lblDate = Now
  TextBox1.Text = "Doc Name Is: " & ActiveDocument.Name
End Sub
```

Now scroll the procedure list box and select **cmdExit**, and scroll the events list box to select **Click**. Enter the following code:

```
Private Sub cmdExit_Click()
  End
End Sub
```

Click **Save**.

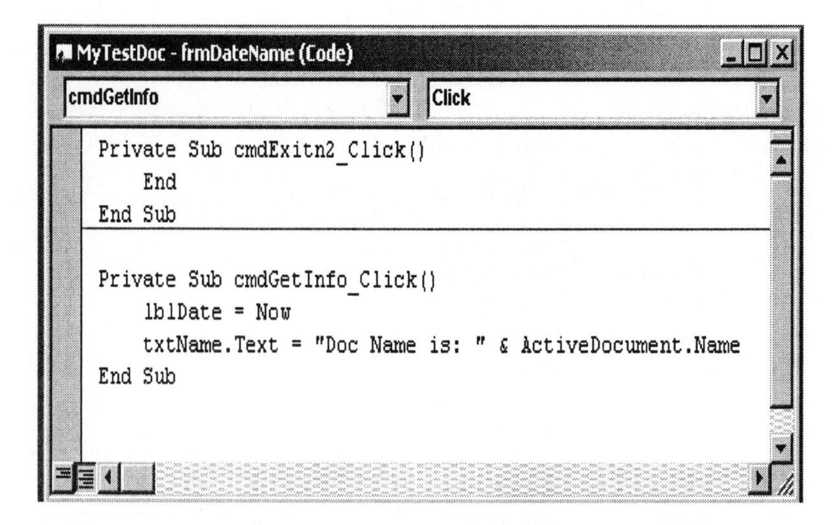

Figure 17-14. The Code Window, Titled **MyTestDoc – frmDateName (Code)**

Running a VBA Procedure from the VBA Design Window. In VBA, the standard toolbar displays three icons: ▶ = ■ , which can also be accessed by clicking the **Run** command on the menu bar. The **Run** command displays a drop-down menu that identifies the three icons:

▶ **Run Macro**	This runs the procedure
= **Break**	This interrupts execution
■ **Reset**	This resets your code to the beginning

Clicking the VBA editor **Run Macro** button runs the VBA program and places the form on the application's window. Clicking the **Get Date and Name of Doc** button executes that control's code. The program displays the date and time in a label box and displays the name of the document in a text box. This procedure does not change or perturb the host document in any way. It simply provides information.

Review. All the macros that we have recorded so far were run from the **Macros** dialog box, by opening a document and, on the **Tools** menu, pointing to **Macro,** and clicking **Macros**. This gave us a list of macros we had recorded. We selected the desired macro by clicking it, and ran it by clicking **Run**.

With the last example, we used the VBA editor to insert a form, add controls, and write procedural code. It was necessary to run it from within the VBA editor by clicking the **Run** button on the toolbar. If, on the **Tools** menu, we had pointed to **Macro** and clicked **Macros**, the macro list would not have included any information about the procedure we just created with the VBA editor.

Running a VBA Procedure from the Macros Dialog Box. To run a VBA program without entering the VBA editor, it is necessary that the program include a module as well as a form. A module is a container for procedures and declarations; in other words, it is just another separate page on which you can write some code. It is separate from the code window where you write the controls' procedures.

To add a module so that you can run your VBA program without the editor, use the following steps:

- Open the Word document named MyTestDoc.
- On the **Tools** menu, point to **Macro**, and click **Visual Basic Editor** to get back to the VBA editor.
- On the **Insert** menu, click **Module**, and **Module1 (Code)** appears in the design window. In the project window, **Module1** appears listed under **Project (MyTestDoc)**.
- On the **Insert** menu, click **Procedure**. In the dialog box that appears, in the **Name** box, type **ShowDateName** as the name. In the **Type** section, select **Sub**; in the **Scope** section, select **Public**. Then click **OK**.
- Enter the following code in the module:

```
Sub ShowDateName()
    frmDateName.Show
End Sub
```

The code window with the title bar **MyTestDoc – Module1 (Code)** and the project window at this point are shown in Figure 17-15. The code window shows the **ShowDateName()** procedure.

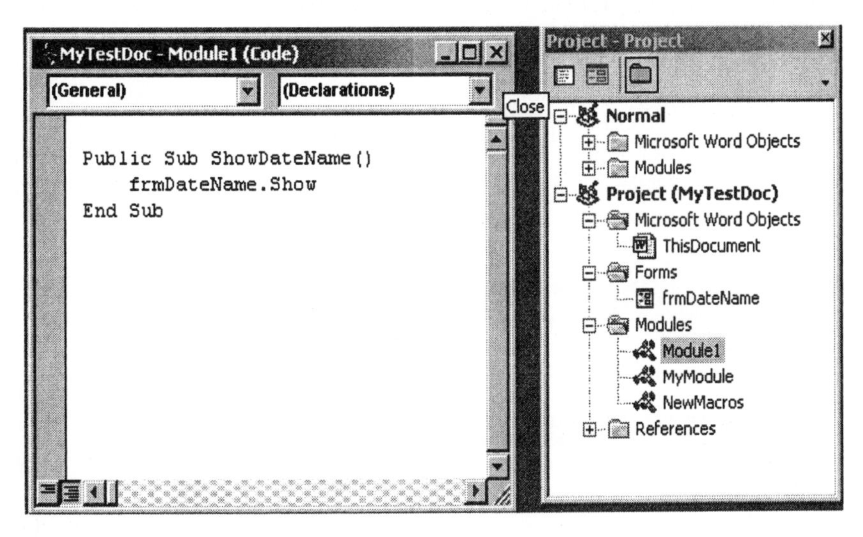

Figure 17-15. The Code Window and the Project Window for the ShowDateName Macro

Review the Project. The project window shows all the open Word projects, of which there is one: **Project (MyTestDoc)**. This project includes the document itself (**This Document**) and the attached forms and modules. Our project has only one form, which we named **frmDateName**.

Our project has three modules. The last one listed, but the first one we used, is the default module **NewMacros**. In it, we recorded and stored our first three macros: Arial10B, Times12, and Table3_5. Next we wrote the CountStuff macro in the Visual Basic Editor. We inserted a macro of our own and named it MyMacro. Finally, we inserted another module and left it with the default name **Module1**. With the Visual Basic Editor we wrote a short procedure name **ShowDateName()** This macro is listed in the **Macros** dialog box with all our other macros. When it is run, it shows the form named **frmDateName** and its controls, which can then be run to get the current date and time and the name of the host document.

Running the Procedure ShowDateName. The short procedure **Sub ShowDateName()** written on **Module1** has no other function than to run the VBA program that gives us the date and name of the host application. We saw that we can run this VBA program with the **Run** button from within the VBA editor. Now you can run the VBA program from the **Macro** dialog box.

- Open the host Word document named MyTestDoc.
- On the **Tools** menu, point to **Macro**, and click **Macros** to open the **Macros** dialog box (Figure 17-16).
- Select the macro named **ShowDateName**.
- Click **Run**, and the **frmDateName** form appears on the application.
- On the form, click the **Get Date and Name of Doc** button.

Now anytime that you want to run this VBA procedure, you just open the **Macros** dialog box, select the macro named **ShowDateName**, and click **Run**.

Review the DateName Macro (VBA Procedure). First, we opened Word and named the host document MyTestDoc. Next we opened the VBA editor by, on the **Tools** menu, pointing to **Macro**, and clicking **Visual Basic Editor**. The **Project Explorer** window (Figure 17-15) responded with the line **Project (MyTestDoc)**.

In the VBA editor, we inserted a form by, on the **Insert** menu, clicking **UserForm**. In the properties window, we changed the name of **Userform1** to **frmDateName**. The **Project Explorer** window responded to this action by adding **frmDateName** to the forms list of the project (Figure 17-12). Then we added controls to the form: three label controls, one text box control, and two command button controls. Next we added code to the two command button controls. We could run this VBA program from the VBA editor with the **Run** button, but we wanted to be able to run it from the **Macros** dialog box. For this reason, we inserted a module by, on the **Insert** menu, clicking **Module** in the VBA editor. The **Project Explorer** window responded to this action by adding **Module1** to the modules list of the project (Figure 17-15). We then wrote the code to show the form so that the program could be run from the **Macros** dialog box.

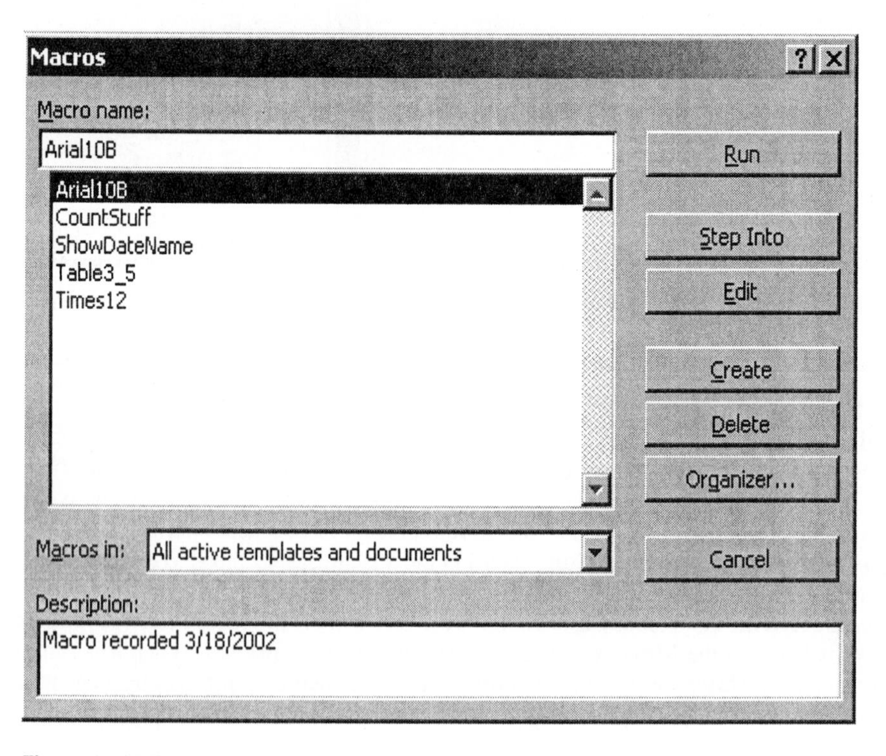

Figure 17-16. The **Macros** Dialog Box Listing Five Macros

Chapter 18

Visual Basic for Applications: Microsoft Excel

Excel and Word: Macros and VBA

Fortunately, the methods for recording macros and writing VBA procedures are the same for Excel as they are for Word. Everything that we've covered so far applies to Excel as well as to Word. Since one application is a word processor and the other a spreadsheet program, the two applications differ greatly in the kinds of objects that they support. For example, Word supports the paragraph object, Excel does not. Excel supports the cell object, Word does not. As in the previous chapter on VBA for Word, we shall begin by recording a few macros and examining them in the VBA editor. Then we'll use the VBA editor to write some VBA procedures without the use of a form. Finally, we'll write some VBA procedure using a form. Some of these VBA procedures will use Excel applications from Part 1 as the host program.

Writing Excel VBA Procedures without a Form

Excel VBA Example 1: The SwimHeader macro	(Host program: Alcatraz.xls)
Excel VBA Example 2: The AutoGraph macro	(Host program: DataToPlot.xls)
Excel VBA Example 3: The CheckNames procedure	(Host program: CheckStuff.xls)
Excel VBA Example 4: The ColorBlock procedure	(Host program: AddColor.xls)
Excel VBA Example 5: The ChessBoard procedure	(Host program: Chess.xls)
Excel VBA Example 6: The DesLandresTable procedure	(Host program: DesLandres.xls)

Writing Excel VBA Procedures with a Form

Excel VBA Example 7: The FirstForm procedure	(Host program: Host for Form.xls)
Excel VBA Example 8: The NumberStats procedure	(Host program: DataForStats.xls
Excel VBA Example 9: The SilverEntropy procedure	(Host program: Cp of Silver.xls)
Excel VBA Example 10: The ABCUser procedure	(Host program: ConsecReact.xls)
Excel VBA Example 11: The ABCScroll procedure	(Host program: ConsecReact.xls)
Excel VBA Example 12: The Eutectic procedure	(Host program: Eutectic Host.xls)

Recording Excel Macros. Just as we did with Microsoft Word, we shall begin our study of Microsoft Excel macros by recording them. Viewing recorded Excel macros with the VBA editor reveals some of the objects, properties, and methods of Excel's macros. We shall then proceed to writing Excel macros (VBA procedures) directly in the VBA editor.

Excel VBA Example 1

The SwimHeader Macro

This Excel macro involves no calculations but instead uses text for information that is written on the worksheet. By recording such an Excel macro, running it, and examining it in the VBA editor, we can learn something about Excel objects and how they compare with Word objects.

Creating the SwimHeader Macro

- For a host program, open a blank workbook in Excel, and save it with the name **Alcatraz.**
- Select cell B1 in an Alcatraz worksheet.
- On the **Tools** menu, point to **Macro**, and click **Record New Macro** to open the **Record Macro** dialog box.
- In the **Macro** name box, type **SwimHeader** as the name. In the **Store Macro in** box, select **This Workbook.**
- Select cell B2, and type **University of Alcatraz** as the label.
- Select cell B3, and type **A Study in Tidal Swimming** as the label.
- Select cell B4, and type **Principle Investigator: T. Head Warden** as the label.
- Select cell B5, and type **Swimmer's Name** as the label.
- Select the range of cells D5:E5.
- In the **Format** menu, click **Cells**; in the **Format Cells** dialog box, click the **Patterns** tab and select **25%**, and then click **OK.**
- Select Cell A7, and type **Date** as the label.
- Select Cell B7, and type **Tide** as the label.
- Select Cell C7, and type **Start** as the label.
- Select Cell D7, and type **Finish** as the label.
- Select Cell E7, and type **Time** as the label.
- Select Cell F7, and click the **Stop Recording** toolbar.

Running the SwimHeader Macro. The act of recording writes the text in the SwimHeader macro to Sheet1 of the Alcatraz workbook. Activate Sheet2 by clicking its tab at the bottom of the sheet page. On the **Tools** menu, point to **Macro,** and click **Macros** to open the **Macros** dialog box (Figure 17-5). Select **SwimHeader** (the only macro name listed at this time), and click **Run**. The text already written to Sheet1 is now written to Sheet2. The SwimHeader macro will write its text to any sheet in the Alcatraz workbook. The results of running the SwimHeader macro are shown in Figure 18-1.

	A	B	C	D	E
1					
2		University of Alcatrax			
3		A Study in Tidal Swimming			
4		Principle Investigator: T. Head Warden			
5		Swimmer's Name			
6					
7	Date	Tide	Start	Finish	Time
8					
9					

Figure 18-1. The Result of the Excel SwimHeader Macro Entered on Sheet1 of the Alcatraz.xls Workbook and Recorded with the Macro Recorder

In the VBA editor you can examine the listing of the SwimHeader macro, shown below. With the Alcatraz workbook open, on the **Tools** menu, point to **Macro**, and click **Visual Basic Editor**. The VBA editor (Figure 18-2) appears with a listing of the macros that you and VBA created.

```
Sub SwimHeader()
'
' The SwimHeader macro
' Macro recorded 3/26/2004 by John Doe

'
    Range("B2").Select
    ActiveCell.FormulaR1C1 = "University of Alcatraz"
    Range("B3").Select
    ActiveCell.FormulaR1C1 = "A Study in Tidal Swimming"
    Range("B4").Select
    ActiveCell.FormulaR1C1 = "Principle Investigator: T. Head Warden"
    Range("B5").Select
    ActiveCell.FormulaR1C1 = "Swimmer's Name"
    Range("A7").Select
    ActiveCell.FormulaR1C1 = "Date"
    Range("B7").Select
    ActiveCell.FormulaR1C1 = "Tide"
    Range("C7").Select
    ActiveCell.FormulaR1C1 = "Start"
    Range("D7").Select
    ActiveCell.FormulaR1C1 = "Finish"
    Range("E7").Select
    ActiveCell.FormulaR1C1 = "Time"
    Range("D5:E5").Select
    With Selection.Interior
        .ColorIndex = 0
        .Pattern = xlGray25
        .PatternColorIndex = xlAutomatic
    End With
End Sub
```

Observations on the SwimHeader Macro. With the Alcatraz workbook open, on the **Tools** menu, point to **Macro**, and click **Visual Basic Editor** (Figure 18-2). The code window's title bar is **Microsoft Visual Basic – Alcatraz.xls – [Module1 (Code)]**. It lists the code for the SwimHeader macro generated by the macro recorder.

The Project Explorer's title bar is **Project – VBAProject** and is similar to the Project Explorer for VBA in Word. The open VBA project is listed as **VBAProject (Alcatraz.xls)** , and under it are **Microsoft Excel Objects** (**Sheet1** through **Sheet3** and **ThisWorkbook**) and **Modules**

(**Module1**). At the moment, **Module1** is selected, so it is the object for which properties are listed in the properties window, whose title bar reads **Properties – Module1**. A module has only one property, its **(Name)** property. Its default name is **Module1**, and if you wish, you can change its name in the properties window. Although the objects are different, the organization of projects, modules, and forms in Excel VBA is similar to their organization in Word VBA.

It was suggested in Chapter 17 that one of the ways to learn about objects, properties, and methods while working in VBA is to record macros. The SwimHeader macro illustrates some of Excel's objects. The first line of the listing shows a **Range** object and the **Select** method of the **Range** object. (With the macro recorder on, we selected cell D2.) The string **"University of Alcatraz"** was assigned to the **ActiveCell** object, as we entered that string while the macro recorder was on. Later we selected two cells and changed the properties of their interior to a 25% pattern. The VBA recorder responded with the **With ... End With** block.

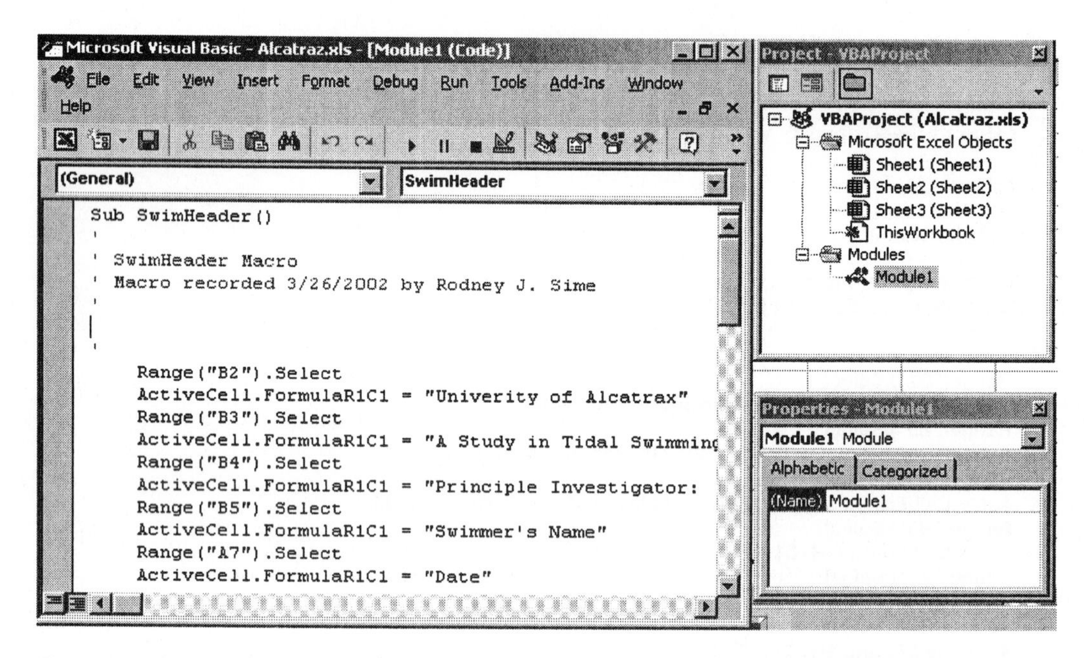

Figure 18-2. The VBA Editor's Code, Project, and Properties Windows for the SwimHeader Macro

Absolute and Relative Cell References. You can run the SwimHeader macro to produce the header text in any worksheet in the Alcatraz workbook. The header will occupy the same position regardless of where the pointer is located on that sheet when the macro is run, because the cell references are absolute.

The bottom half of the **Stop Recording** toolbar has two buttons. The button on the left is the **Stop Record** button. The button on the right is the **Relative Reference** button. In its "out," or unclicked, position, the cell references in the macro code are absolute, as shown in the preceding listing. In the "in," or clicked, position, the cell references are relative to the cell selected at the time the recording was begun, which is normally the pointer position. If the keystrokes listed earlier were entered with relative references, then the header begins wherever the pointer is located.

The SwimHeaderRelative Macro for Excel. Duplicate the SwimHeader macro recording with the **Relative Reference** button clicked to the "in" position, and give it a different name, say, **SwimHeaderRelative**. The macro's listing generated by the macro recorder follows:

```
Sub SwimHeaderRelative()
'
' The SwimHeaderRelative macro
' Macro recorded 3/26/2004 by W. Gates
'

'
    ActiveCell.FormulaR1C1 = "University of Alcatraz"
    ActiveCell.Offset(1, 0).Range("A1").Select
    ActiveCell.FormulaR1C1 = "A Study in Tidal Swimming"
    ActiveCell.Offset(1, 0).Range("A1").Select
    ActiveCell.FormulaR1C1 = "Principle Investigator: T. Head Warden"
    ActiveCell.Offset(1, 0).Range("A1").Select
    ActiveCell.FormulaR1C1 = "Swimmer's Name"
    ActiveCell.Offset(2, -1).Range("A1").Select
    ActiveCell.FormulaR1C1 = "Date"
    ActiveCell.Offset(0, 1).Range("A1").Select
    ActiveCell.FormulaR1C1 = "Tide"
    ActiveCell.Offset(0, 1).Range("A1").Select
    ActiveCell.FormulaR1C1 = "Start"
    ActiveCell.Offset(0, 1).Range("A1").Select
    ActiveCell.FormulaR1C1 = "Finish"
    ActiveCell.Offset(0, 1).Range("A1").Select
    ActiveCell.FormulaR1C1 = "Time"
    ActiveCell.Offset(-2, -1).Range("A1:B1").Select
    With Selection.Interior
        .ColorIndex = 0
        .Pattern = xlGray25
        .PatternColorIndex = xlAutomatic
    End With
End Sub
```

You can run the SwimHeaderRelative macro to produce the header text in any worksheet in the Alcatraz workbook beginning in any cell location you desire, except for a cell in column A. Run the macro with the pointer located in a column other than column A. The header will begin at the position where the pointer is located when the macro is run, because the cell references are relative. Now run the macro with the pointer located in a cell in column A. The result is a Microsoft Visual Basic dialog box with the message **Run-time error**. Can you guess why this happens? For a clue, click the **Debug** button on the dialog box. Visual Basic highlights the line of code:

```
ActiveCell.Offset(2, -1).Range("A1").Select
```

Relative to the initial position of the pointer in column A, text entry is offset by 2 rows (that's OK) and by –1 columns (that's not OK, because there's no column to the left of column A).

Referencing Objects. The SwimHeader and SwimHeaderRelative macros display some new and unfamiliar objects, methods, and properties. The Excel application is itself an object. It contains a collection of workbooks. A workbook contains a collection of worksheets. **Excel** is a container object. **Workbooks** is a container object. An object expression consists of all the necessary containers joined together by a dot separator (.). For example:

```
Excel. ActiveWorkBook.WorkSheets("Sheet3").Range ("A3").Select
```

Excel is a container for **ActiveWorkBook. ActiveWorkBook** is a container for **WorkSheets. Sheet3** is a member of the **WorkSheets** collection. **Range("A3")** refers to the single cell A3. **Select** is a method of the **Range** object. What does this statement do? It is as though Sheet3 were open in Excel and you clicked in cell A3 to select it.

In practice it is often not necessary to reference the entire hierarchy of container names. If Sheet3 of your Excel workbook were open, then the following statement in your macro would accomplish the same thing:

```
Range ("A3").Select
```

On the other hand, if you wanted to run this statement from a different notebook than Alcatraz, you would need a preceding statement:

```
Excel.Workbooks("Alcatraz").Activate
Excel. ActiveWorkBook.WorkSheets("Sheet3").Range ("A3").Select
```

Excel VBA Example 2

The AutoGraph Macro

With Microsoft Word, we saw that we could create a complex macro to produce a table with labeled columns and rows. Before creating an Excel macro with the VBA editor, we shall record a macro named AutoGraph that will automatically produced an XY (scatter) graph from x_i,y_i data entered into the A and B columns of an Excel worksheet. The AutoGraph macro in Microsoft Excel is analogous to the Table3_5 macro in Microsoft Word.

Recording the Excel AutoGraph Macro. Producing a macro always involves two steps. The first step is the preparation of the host application. The second step is the recording of the macro (or writing a VBA procedure).

Preparing the Host Document

- Open an Excel workbook, and save it as **DataToPlot**.
- On Sheet1, in the range A1:A6, type **0**, **1**, **2**, **3**, **4**, and **5** as the values.
- On Sheet1, in the range B1:B6, type **0**, **1**, **4**, **9**, **16**, and **25** as the values.
- Click and drag from cell A1 to cell B20 to select this range.

Preparing to Record the Macro

- On the **Tools** menu, point to **Macro**, and click **Record New Macro** to open the **Record Macro** dialog box.
- In the **Macro name** box, type **AutoGraph** as the name.
- In the **Store macro in** box, select **This WorkBook**.
- In the **Description** box, add your description of your macro's function, something like: **This macro automatically prepares an XY scatter graph of XY pairs.**
- Click **OK**; the **Record Macro** dialog box disappears, VBA's tiny **Stop Recording** toolbar appears, and the pointer changes to a tiny tape recorder.

At this point you are ready to record. While the **Stop Recording** toolbar is visible, enter the following mouse clicks and keystrokes.

Recording the Macro

- On the Excel toolbar, click the **Chart Wizard** icon.
- In the **Chart Wizard – Step 1 of 4 – Chart Type** dialog box, select **XY (Scatter)**, and then click the second icon (data points connected with smooth line). Then click **Next**.

- In the **Chart Wizard – Step 2 of 4 – Chart Source Data** dialog box, just click **Next.**
- In the **Chart Wizard – Step 3 of 4 – Chart Options** dialog box, in the **Chart title** box, type **My Title** for the title; in the **Value (X) axis** box, type **My X** as the label; and in the **Value (Y) axis** box, type **My Y** as the label. Then click **Next.**
- In the **Chart Wizard – Step 4 of 4 – Chart Location** dialog box, select **As object in Sheet 1**. Then click **Finish.**
- To stop recording, click the **Stop Recording** toolbar, which disappears.

The data you selected will have been graphed on an XY (scatter) chart embedded in Sheet1 (Figure 18-3). Edit the titles and axis labels by clicking twice slowly (not double-clicking) the title or a label. Then edit it as you would in any word processor.

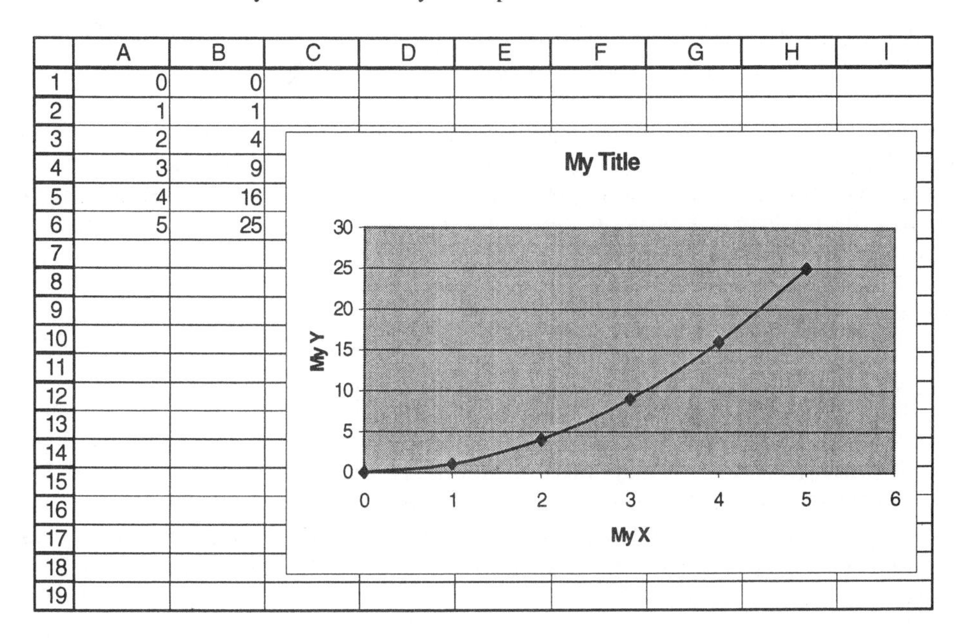

Figure 18-3. Result of Executing the AutoGraph Macro on the Data in Sheet1 of the DataToPlot.xls Worksheet

Executing the Excel AutoGraph Macro. Test your macro by activating a new sheet in this workbook (click the tab for Sheet2). Enter some suitable data in columns A and B. You can enter up to 20 *x,y* data pairs, if you originally selected the range A1:B20 while recording. Then, on the **Tools** menu, point to **Macro**, and click **Macros**. Select the **AutoGraph** macro, and then click **Run**.

Understanding the Excel AutoGraph Macro. To examine the code generated by the Excel macro recorder, on the **Tools** menu, point to **Macro**, and click **Macros**, and select the **AutoGraph** macro. Click **Edit**, and the following listing is displayed:

```
Sub AutoGraph()
'
' The AutoGraph macro
' Macro recorded 3/27/2004 by Sir Isaac Newton
' This macro automatically prepares an XY (scatter) graph
'

'
    Charts.Add
    ActiveChart.ChartType = xlXYScatterSmooth
```

```
    ActiveChart.SetSourceData Source:=Sheets("Sheet1").Range("A1:B20"), PlotBy _
        :=xlColumns
    ActiveChart.Location Where:=xlLocationAsObject, Name:="Sheet1"
    With ActiveChart
        .HasTitle = True
        .ChartTitle.Characters.Text = "My Title"
        .Axes(xlCategory, xlPrimary).HasTitle = True
        .Axes(xlCategory, xlPrimary).AxisTitle.Characters.Text = "My X"
        .Axes(xlValue, xlPrimary).HasTitle = True
        .Axes(xlValue, xlPrimary).AxisTitle.Characters.Text = "My Y"
    End With
End Sub
```

- The first line assigns the type of chart (**xlXYScatterSmooth**) to the **ChartType** property of the **ActiveChart**. This is done in Step 1 of the Chart Wizard.
- The second line assigns the **Range("A1:B20")** to the **SetSourceData Source** property of the **ActiveChart.**
- The next few lines assign the title, x-axis labels, and y-axis labels the appropriate property of the **ActiveChart**. These assignments are done within a **With ... End With** block.

The **ActiveChart** is a container for **ChartTitle** and **Axes**. Their properties can easily be edited to refine the macro. With a few more keystrokes, the macro could do a linear regression and chart the results as described in Partr 1. Remember, any of the keystrokes, command, and mouse clicks that you apply to an Excel workbook sheet can be recorded. You can always examine what you have recorded in the Visual Basic Editor to learn how Visual Basic has identified its objects, their properties, and their methods.

Writing Excel VBA Procedures without a Form. It will be useful to write some simple VBA procedure to get more experience with referencing objects in a VBA procedure. Writing Excel VBA procedures in the Visual Basic Editor is no different than writing Word VBA procedures in the Visual Basic Editor. In both Excel and Word, the VBA procedures may or may not have a form. When they do not have a form, the name of the VBA procedure appears in the **Macros** dialog box, and the VBA procedure can be run from there. If the VBA procedure has a form, then it is necessary to add a module and write a short procedure to show the form. This was done earlier with the Word VBA DateName macro, which has a form. The short procedure's name appears in the **Macros** dialog box along with other macros, regardless of how they were created (as a macro recording or in the Visual Basic Editor).

Excel VBA Example 3

The CheckNames Procedure

The Excel VBA CheckNames Procedure. Our first Excel VBA procedure, named CheckNames, demonstrates some features of referencing objects and their methods. This VBA procedure has no form, so it can be run from the **Macros** dialog box. As in writing VBA procedures with Microsoft Word, we need a host application for the VBA procedure. It will be an empty worksheet in a workbook named CheckStuff, just to remind us that it is the host program for the VBA procedure named CheckNames.

- Open a new workbook, and name it **CheckStuff.**
- With Sheet1 the active sheet, point to **Format**, select **Sheet,** then **Renme** and rename Sheet1 as **Run1.**

- On the **Tools** menu, point to **Macro**, and click **Visual Basic Editor**.
- In the VBA editor, on the **Insert** menu, click **Module**.
- In the VBA editor, on the **Insert** menu, click **Procedure**, and in the **Name** box, type **CheckNames** as the name.
- Enter the following code for the CheckNames procedure::

```
Public Sub CheckNames()

    ' 1. Assign values to cells, and use different object references
    Excel.ThisWorkbook.ActiveSheet.Range("A1").Value = "Hello there!"      'Longest
    ThisWorkbook.ActiveSheet.Range("A2").Value = "What's up?"              'Less long
    ActiveSheet.Range("A3").Value = "Nice day"                             'Short
    Range("A4").Value = "Indeed!!!."                                       'Shortest
    Range("A5").Value = 6.02E+24
    Range("A6").Value = 1.23
    Range("A6").Select

    ' 2. Let VBA describe itself
    MsgBox Application.Name & " is application name"
    MsgBox Application.ActiveWorkbook.Name & " is name of workbook "
    MsgBox Application.ActiveWorkbook.FullName & " is full name of workbook "
    MsgBox Application.ActiveSheet.Name & ": name of active sheet"
    MsgBox ActiveWorkbook.Worksheets.Count & " = number of sheets in workbook"

    ' 3. Assign cell values to MsgBox
    MsgBox Application.ActiveCell.Value & " is the value of the selected cell"
    MsgBox Sheets("Run1").Range("A5").Value & " is the value of what's in cell A5"

    ' 4. From Sheet1 to Sheet3 and back
    Worksheets("Sheet3").Activate
    ActiveSheet.Range("A1").Value = "Now we're in Sheet3"
    MsgBox "Let's return to the sheet named Run1"
    Worksheets("Run1").Activate
    MsgBox "Have a " & Range("A3").Value
    MsgBox "That's all"

End Sub
```

Organization of the CheckNames Procedure. The VBA CheckNames procedure is arranged in four parts for easy readability.

1. In the first part of CheckNames, various strings and numbers are assigned to the **Value** property of a range. The long, complete expressions are compared with the minimal expressions.
2. In the second part, various properties of workbooks, worksheets, and cells are assigned to a message box. This assignment is the opposite direction of the assignments in the first part of the VBA CheckNames procedure.
3. In the third part, the **Value** property of various ranges is assigned to a message box.
4. In the fourth part, Sheet3 is activated, a string is assigned to its cell A1 and to a message box, and then the sheet assignment goes back to the sheet we named Run1 (its default name was Sheet1).

These two kinds of assignments, going in opposite directions between a worksheet and a VBA procedure, lie at the heart of Visual Basic for Applications. Examine the object references in the VBA CheckNames procedure, especially those labeled with the comments longest, less long, short, and shortest.

A Compact Version of the CheckNames Macro. In the VBA CheckNamesShort procedure (which follows), the object references have been trimmed as short as possible, but the VBA CheckNames procedure and the VBA CheckNamesShort procedure function identically. The code is clearly more compact.

```
Public Sub CheckNamesShort()

  ' 1. Assign value to cells, and use different object references
  Range("A1").Value = "Hello there!"
  Range("A2").Value = "What's up?"
  Range("A3").Value = "Nice day"
  Range("A4").Value = "Indeed!!!"
  Range("A5").Value = 6.02E+24
  Range("A6").Value = 1.23
  Range("A6").Select

  ' 2. Let VBA describe itself
  MsgBox Application.Name & " is application name"
  MsgBox ActiveWorkbook.Name & " is name of workbook "
  MsgBox Application.ActiveWorkbook.FullName & " is full name of workbook "
  MsgBox ActiveSheet.Name & ": name of active sheet"
  MsgBox Worksheets.Count & " = number of sheets in workbook"

  ' 3. Assign cell values to MsgBox
  MsgBox ActiveCell.Value & " is the value of the selected cell"
  MsgBox Range("A5").Value & " is the value of what's in cell A5"

  ' 4. From Run1 to Sheet3 and back
  Worksheets("Sheet3").Activate
  Range("A1").Value = "Now we're in Sheet3"
  MsgBox "Let's return to the sheet named Run1"
  Worksheets("Run1").Activate
  MsgBox "Have a " & Range("A3").Value
  MsgBox "That's all"
End Sub
```

Running the CheckNames and CheckNamesShort Macros. To run the macros, first be sure that Run1 (originally Sheet1) is the active sheet of the CheckStuff.xls workbook. On the **Tools** menu, point to **Macro**, and click **Macros**. Select the **CheckNames** macro, and click **Run.** In the range A1:A6, text is written, and a message box appears on the sheet named Run1 (Figure 18-4). Click **OK** on the successive message boxes that appear on the screen. These message boxes give properties of various objects referenced in the VBA procedure. Notice that the VBA procedure switches to Sheet3 and then back to Run1 (originally Sheet1). Select Sheet3, and the text written in cell A1 is still there.

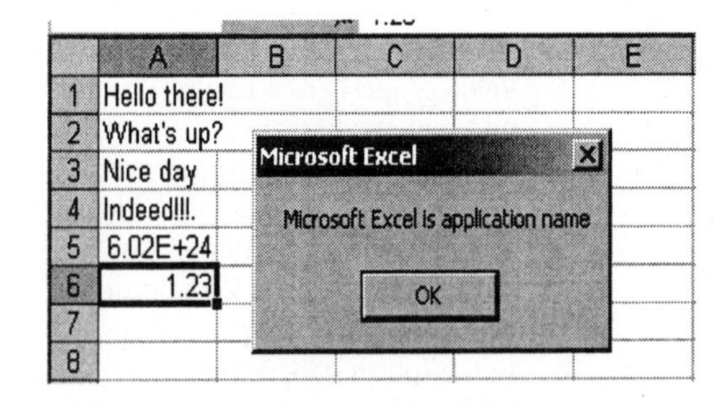

Figure 18-4. The Result of Running the Excel VBA CheckNames Procedure with the Host Program CheckStuff.xls

Click **OK**, and another message box appears. This time the **Name** property of the active workbook is assigned to **MsgBox**.

Click **OK**, and the **FullName** property of the active workbook is assigned to **MsgBox**.

Click **OK**, and the **Name** property of the active sheet is assigned to **MsgBox**. That name is Run1 if you changed the worksheet name in the beginning. Otherwise, the name returned will be Sheet1.

Click **OK**, and the **Count** property of the **Worksheets** collection is assigned to **MsgBox**. It is an integer and normally equals 3.

Click **OK** on the next message box, and the string values assigned in the first part of the VBA CheckNames macro are now assigned to **MsgBox**.

After clicking **OK** on the message box that displays the value of "what's in cell A5," the **Activate** method of the **Worksheets** collection activates Sheet3. The string "Now we're in Sheet3" is assigned to the **Value** property of **Range("A1")**.

A message box displays the message **Let's return to the sheet named Run1**. Clicking **OK** causes execution of the next statement, in which the **Activate** method of the **Worksheets** collection actives sheet Run1, and we're back where we started.

Click **OK**, and the string "Have a " is concatenated to "nice day" so the message box displays the message **Have a nice day**.

Understanding the Excel VBA CheckNames Procedure. The VBA CheckNamesShort macro shows how the object references can be shortened to give more compact code. Most of our Excel VBA procedures use only short object references, and that is usually to a range. When the **Range** object is referenced in the "shortest form" (shown below), VBA assumes that the range is contained in the **ActiveSheet**, that the **ActiveSheet** is contained in **ThisWorkbook**, and that **ThisWorkbook** is contained in **Excel**.

```
Excel.ThisWorkbook.ActiveSheet.Range("A1").Value = "Hello there!"      ' Longest
      ThisWorkbook.ActiveSheet.Range("A2").Value = "What's up?"        ' Less long
                   ActiveSheet.Range("A3").Value = "Nice day"          ' Short
                               Range("A4").Value = "Indeed!!!."        ' Shortest
```

The references to **Excel, ThisWorkbook**, and **ActiveSheet** are unnecessary because they are all implied when not specifically stated.

Summary of the CheckNames and CheckNamesShort Macros. Some properties of Excel objects are demonstrated in the CheckNames macro: **Value, Name, FullName**, and **Count.** One method of Excel objects is demonstrated: **Activate.** These macros demonstrate referencing some common objects in Microsoft Excel, such as the active workbook, the active sheet, and the range.

Two kinds of assignment statements are demonstrated.

1. An assignment to an Excel object from the VBA procedure
2. An assignment to a VBA object (a message box in this example) from an Excel object

The Essence of Excel VBA Macros. These two kinds of assignment form the essence of Excel VBA macros. They permit exchange of information and data in either direction between a host Excel program and its macro(s). In the VBA **CheckNames** procedure, a value is assigned to a macro message box from the host program. We'll see later that a macro with a form can also receive data from a macro. Data from an Excel worksheet can be written to a form or to controls on the form such as a text box or label, just as we did in previous chapters with Visual Basic.

More on the Ranges. With macros and VBA procedures written for Microsoft Word, the most common object is the **Selection** object. This is used to change the font type or size for a selected block of text. Such macros are also useful with Microsoft Excel, but with Microsoft Excel the most common object is the **Range**. The **Range** may consist of a single cell, or it may cover a block of cells in more than one row and column. A **Range** may also consist of cells in a single column or a single row.

The **Range** object is returned by the **Range** property of an object, which is almost always a worksheet, and that is usually the active sheet or a named sheet like Sheet1.

```
ActiveSheet.Range(cell1,[cell2])
```

The identifier of the cell is in R1C1 notation, examples of which follow:

A single cell	**Range("B4")**
In a column	**Range("C4:C8)**
In a row	**Range("D5:H5")**
A block	**Range("D3:E7")**
Entire row	**Range("B:B")**
Entire column	**Range("6:6")**

The quotation marks enclosing the R1C1 notation are required.

Excel VBA Example 4

The ColorBlock Procedure

Our next Excel VBA procedure, named ColorBlock, demonstrates the use of a named range. In addition, it demonstrates how to change the **Font** property of the active cell. And finally, it demonstrates changing the **Interior** property of a cell. Here, a macro makes assignments to the properties of a cell in an Excel worksheet. This VBA procedure has no form, so it can be run from the **Macros** dialog box. As always we prepare a host application and then write the VBA procedure.

Writing the Excel VBA ColorBlock Procedure

- Open a new workbook, and name it **AddColor**.
- On the **Tools** menu, point to **Macro**, and click **Visual Basic Editor**.
- In the VBA editor, on the **Insert** menu, click **Module**.
- In the VBA editor, on the **Insert** menu, click **Procedure**, and in the **Name** box, type **ColorBlock** as the name.
- Enter the following code for the ColorBlock procedure:

```
Public Sub ColorBlock()
  Dim RedBlock As String
  RedBlock = "B3: D4"
  Range(RedBlock).Select
  With ActiveCell
    .Font.Bold = True
    .Font.Size = 14
    .FormulaR1C1 = "This is a red block"
    .Font.Name = "New Times Roman"
```

```
    End With
    With Selection.Interior
        .ColorIndex = 0
        .Pattern = xlGray25
        .Color = RGB(255, 100, 100)
        .PatternColorIndex = xlAutomatic
    End With
    ActiveSheet.Range("A1").Select
End Sub
```

Running the ColorBlock Macro

- Click the **Excel** icon at the left end of the toolbar to return to the host document (AddColor.xls).
- On the **Tools** menu, point to **Macro**, and click **Macros**; select the **ColorBlock** macro, and click **Run**.
- A red block appears in the range B3:D4. It contains the message **This is a red block**, as shown in Figure 18-5.

Figure 18-5. Result of Running the VBA ColorBlock Procedure with Sheet1 in the AddColor.xls Workbook

Understanding the VBA ColorBlock Procedure. Anything that a programmer can do to enhance the readability of code is worthwhile. A name for a range that says something about the nature of the range usually makes the code using that range more readable. It is good programming practice to name ranges, especially if a large number of different ranges are specified.

To name a range, the name must first be declared as a string. Then the value of the range is assigned to the name.

```
Dim RedBlock As String
RedBlock = "B3:D4"
```

The string **RedBlock** can be used anywhere that the string **"B3:D4"** would normally be used.

The VBA ColorBlock procedure also shows how the font properties of the active cell can change within a VBA procedure. The second **With ... End With** block was suggested by an earlier recording (the SwimHeader macro). Notice that as you enter each period (.) after **With ActiveCell** and after **With Selection.Interior**, VBA's automatic code completion furnishes a list of properties that can be referenced.

The last statement in the VBA ColorBlock procedure:

```
ActiveSheet.Range("A1").Select
```

simply serves to get the selected cell off the red block and place it out of the way at cell A1.

The Cells Property of the ActiveSheet. The syntax for the Cells property of the active sheet is

***Object*.Cells(*RowIndex*,*ColumnIndex*)**

The ***Object*** can be either the active sheet or a range that contains the cell of interest. **Cells** is a property of both the **ActiveSheet** object and of the **Range** object. The **Cells** property returns only a single cell, not a block of cells, or a block of cells contained in a single row or single column.

The ***RowIndex*** is an integer that equals the row number of the active sheet in which the cell of interest is located. The ***ColumnIndex*** is an integer corresponding to the column index of the row in which the cell of interest is located.

The following statement assigns the string "**I did this with the Cells property**" to the **Value** property of the cell located at row 10 and column 2 (column B):

```
ActiveSheet.Cells(10, 2).Value = " I did this with the Cells property"
```

It appears at first glance that the **Cells** property is more awkward to use in referencing a block of cells than the **Range** property. However, the **Cells** property turns out to be very powerful when the block size is variable, because ***RowIndex*** and ***ColumnIndex*** can be integer variables that can be used in a **For ... Next** loop. If **J** is previously declared to be an integer variable, then the block of code:

```
For I = J To 6
   Cells (11, I).Value = "Hello"
Next
```

writes the string "**Hello**" in Row 11 from column 1 to 6 (A to F). In a similar fashion, the block:

```
For J = 12 To 18
   Cells (I, 3).Value = "Bye"
Next
```

writes the string "**Bye**" in column 3 (column C) from row 12 to row 18.

The statement **Cells (1, 1).Select** selects the cell A1. In these examples, VBA assumes that **Cells** is a property of the active sheet. The next two VBA procedures illustrate the use of the **Cells** property.

Excel VBA Example 5

The ChessBoard Procedure

The Excel VBA ChessBoard Procedure. Open a new workbook, and save it with the name **Chess**. It will serve at the host document for the VBA ChessBoard procedure. The host document has no initial entries and serves only to give access to VBA and to provide a place for the output of the VBA procedure. The VBA ChessBoard procedure demonstrates the use of the **For ... Next** control structure with the **Cells** property of the active sheet.

- On the **Tools** menu, point to **Macro**, and click **Visual Basic Editor**.
- In the VBA editor, on the **Insert** menu, click **Module**.
- In the VBA editor, on the **Insert** menu, click **Procedure**, and in the **Name** box, type **ChessBoard** as the name.
- Enter the following code into the ChessBoard procedure:

```
Public Sub ChessBoard()
  Dim Rw As Integer, Cm As Integer
  For Rw = 1 To 24
    For Cm = 1 To 12
      ' Color the odd rows and odd column squares
      If (Rw Mod 2 = 1) And (Cm Mod 2 = 1) Then
        Sheet2.Cells(Rw, Cm).Interior.Color = vbBlack
      End If
      ' Color the even rows and even column squares
      If (Rw Mod 2 = 0) And (Cm Mod 2 = 0) Then
        Sheet2.Cells(Rw, Cm).Interior.Color = vbBlack
      End If
    Next Cm
  Next Rw
End Sub
```

Running the Excel VBA ChessBoard Procedure

- Click the **Excel** icon at the left end of the toolbar to return to the host document (Chess.xls).
- On the **Tools** menu, point to **Macro**, and click **Macros**; select the **ChessBoard** macro, and click **Run**.

The visible region of the worksheet, shown in Figure 18-6, appears with alternate cells colored black.

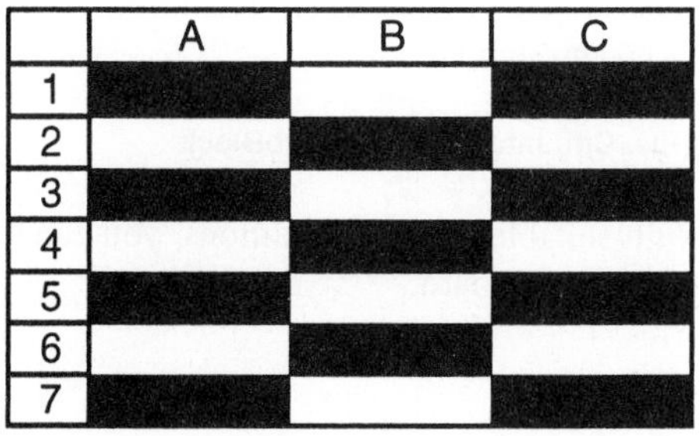

	A	B	C
1			
2			
3			
4			
5			
6			
7			

Figure 18-6. The Result of Running the VBA ChessBoard Procedure with the Chess.xls Workbook

Understanding the Excel VBA ChessBoard Procedure. To make a chessboard on a worksheet, it is necessary to blacken alternate squares on the worksheet. More exactly the ***RowIndex*** and ***ColumnIndex*** of the **Cells** property of the active worksheet must be either all even or all odd for each cell that is to be black. The **Mod** function is convenient for testing whether a number is even or odd.

Since a chessboard is a two dimensional table of black and white squares (cells), it makes sense to fill in the table with two nested **For ... Next** loops just as we did with Visual Basic. The VBA ChessBoard procedure fills the visible spreadsheet (A1:L25) with alternating black and white cells.

If you want just a chessboard, then run both **For ... Next** indices from 1 to 8. Then to make the cells square rather than rectangular, click the **Select All** button, which lies at the upper left corner of the worksheet. This action darkens (selects) the entire worksheet, including the row numbers and column letters. Then click the bottom edge of any row number and drag it down until that cell is square. When you release, all the cells of the worksheet assume a square configuration (Figure 18-7).

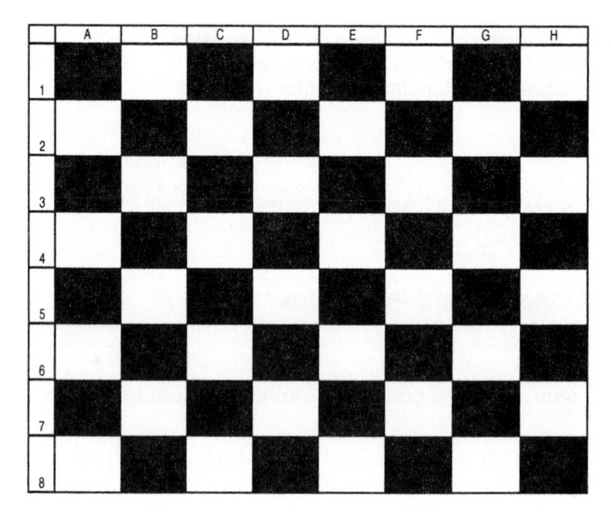

Figure 18-7. The Result of Running the VBA ChessBoard Procedure with the Chess.xls Workbook and Dragging to Change the Row Height Property of All Selected Cells

The syntax of the nested **For ... Next** loops is the same as described originally in Chapter 13 on Visual Basic. The Excel VBA code is the assignment of color (**vbBlack**) to a cell in Sheet2 whose ***RowIndex*** equals **Rw** and whose ***ColumnIndex*** equals **Cm** for each pass through the loop. **Rw** ranges from 1 to 25, and **Cm** ranges from 1 to 12 (A through L).

```
Sheet2.Cells(Rw, Cm).Interior.Color = vbBlack
```

If you add two more **For ... Next** loops with suitable control conditions, you can assign a red color (**vbRed**) to the remaining cells and create a checkerboard.

Excel VBA Example 6

The DesLandresTable Procedure

The Excel VBA DesLandresTable Procedure. The VBA DesLandresTable procedure demonstrates the use of the **For ... Next** control structure with the **Cells** property of the active sheet to carry out numerical calculation and place the results in a two-dimensional table on the active sheet. The host program has no initial entries; it serves only to access VBA and to provide a worksheet on which to place the output of the VBA DesLandresTable procedure.

Writing the Excel VBA DesLandresTable Procedure

- Open a new workbook, and save it with the name **DesLandres**. It will serve as the host document for the VBA DesLandresTable procedure. It will have no initial entries.
- On the **Tools** menu, point to **Macro**, and click **Visual Basic Editor**.
- In the VBA editor, on the **Insert** menu, click **Module**.
- In the VBA editor, on the **Insert** menu, click **Procedure**; in the **Name** box, type **DesLandresTable** as the name.

Enter the following code into the module:

```
Public Sub DesLandresTable()
  ' Data from Herzberg, "Spectra of Diatomic Molecules"
  ' Van Nostrand, Princeton, NJ, 1950, pages 36-41

  Dim Vp As Integer, Vpp As Integer
  Dim Upper As Single, Lower As Single
  Dim WaveNumber As Single, Center As Single
  Center = 39699.01
  For Vp = 0 To 6                          ' Upper electronic state quantum numbers
    For Vpp = 0 To 6                       ' Lower electronic state quantum numbers
      Upper = 1094.8 * Vp - 7.25 * Vp ^ 2                ' Upper state energy
      Lower = 1329.38 * Vpp - 6.98 * Vpp ^ 2             ' Lower state energy
      WaveNumber = Center + (Upper - Lower)                      'Transition energy
      Sheet1.Cells(Vp + 1, Vpp + 1).Value = WaveNumber ' Fill the table
    Next Vpp
  Next Vp

End Sub
```

Executing the Excel VBA DesLandresTable Procedure. The host sheet is a blank Sheet1 of the DesLandres.xls workbook. With Sheet1 the active sheet, on the **Tools** menu, point to **Macro**, and click **Macros**. In the **Macros** dialog box, select the **DesLandresTable** macro, and click **Run**. The result is shown in Figure 18-8.

	A	B	C	D	E	F	G	H
1	39699.01	38376.61	37068.17	35773.69	34493.17	33226.61	31974.01	
2	40786.56	39464.16	38155.72	36861.24	35580.72	34314.16	33061.56	
3	41859.61	40537.21	39228.77	37934.29	36653.77	35387.21	34134.61	
4	42918.16	41595.76	40287.32	38992.84	37712.32	36445.76	35193.16	
5	43962.21	42639.81	41331.37	40036.89	38756.37	37489.81	36237.21	
6	44991.76	43669.36	42360.92	41066.44	39785.92	38519.36	37266.76	
7	46006.81	44684.41	43375.97	42081.49	40800.97	39534.41	38281.81	
8								
9								
10								

Figure 18-8. A DesLandres Table Calculated with the VBA DesLandresTable Program in the DesLandres.xls Workbook

Understanding the Excel VBA DesLandresTable Procedure. Both the underlying science and the Visual Basic code for the DesLandres table are given in Chapter 14, in the DesLandres program. It is instructive to compare the VBA code for this procedure with the Visual Basic code for the DesLandres table presented in Chapter 14:

```
Dim DesLandresTable(0 To 10, 0 To 10) As Single
For Vp = 0 To 6                                    ' Upper electronic state quantum numbers
  For Vpp = 0 To 6                                 ' Lower electronic state quantum numbers
    Upper = 1094.8 * Vp - 7.25 * Vp ^ 2            ' Upper state energy
    Lower = 1329.38 * Vpp - 6.98 * Vpp ^ 2         ' Lower state energy
    WaveNumber = Center + (Upper - Lower)          ' Transition energy
    DesLandresTable(Vp, Vpp) = WaveNumber          ' Fill the table
  Next Vpp
Next Vp
```

Only the last line inside the nested **For ... Next** loops differs between Visual Basic and Excel VBA. In Visual Basic, the value of wavenumbers is assigned to its position in the previously declared two-dimensional table, **DesLandresTable**. In Excel VBA , the value of wavenumbers is assigned to its cell in the worksheet. Since the vibrational quantum numbers **Vp** and **Vpp** range from 0 on up, it is necessary to add 1 to their values, because ***RowIndex*** and ***ColumnIndex*** range from 1 on up.

Writing Excel VBA Procedures with a Form. A dialog box is a form with some controls on it that allow the user to interact with an application. A VBA procedure with a form provides the user with a dialog box with which to interact with the underlying host application, an Excel worksheet. You may think of a dialog box not only as holding a dialog with the user but also as having a dialog with the host application: the dialog box exchanges data with the application. The exchange may take place in both directions. That is, the dialog box may accept data from the user and assign that data to the application. Or the application may assign data to the dialog box for the user to see or act upon. The Excel VBA **NumberStats** procedure demonstrates these exchanges of information. But first let's write an even simpler VBA procedure that has a form, the FirstForm procedure. A form is added to the procedure using the **Insert User Form** command on the **Insert** menu.

Excel VBA Example 7

The FirstForm Procedure

The Excel VBA FirstForm Procedure. This short procedure doesn't do anything except serve as an introduction to the use of a form with VBA. It doesn't interact with the user. It doesn't accept data from its host application. It doesn't assign data to the host application. We'll see later that VBA applications with a form can do all of these things, but for now we'll keep things simple.

Preparing the Host Program

- Open a new Excel workbook.
- Beginning at cell A1, type **This is the host Excel program** as the label.
- Save it with the name **Host for Form.**
- On the **Tools** menu, point to **Macro**, and click **Visual Basic Editor.**
- On the **Insert** menu, click **User Form**, and add the three controls listed in Table 18-1.
- Set the properties of the controls according to Table 18-1.
- Notice that the UserForm1 form is now added to the object list in the project window (Figure 18-9).

Table 18-1

The Properties Table for the FirstForm Procedure

Object	Property	Setting
UserForm1	Name	frmFirstForm
	Caption	VBA Demonstration Form
Label1	Name	lblInfo
	Caption	Make it blank
	Font	Bold 16 point
Command1	Name	cmdRun
	Caption	Click Me
	Font	Bold 10 Point
Command2	Name	cmdOK
	Caption	OK
	Font	Bold 10 Point

After setting the properties of the controls, double-click the **cmdRun** button, and enter the following code:

```
Private Sub cmdInfo_Click()
   Dim Phrase As String
   Phrase = "Hello! Welcome to VBA with forms"
   lblInfo.Caption = Phrase
End Sub
```

Double-click the **cmdOK** button, and enter the following code:

```
Private Sub cmdOK_Click()
   End
End Sub
```

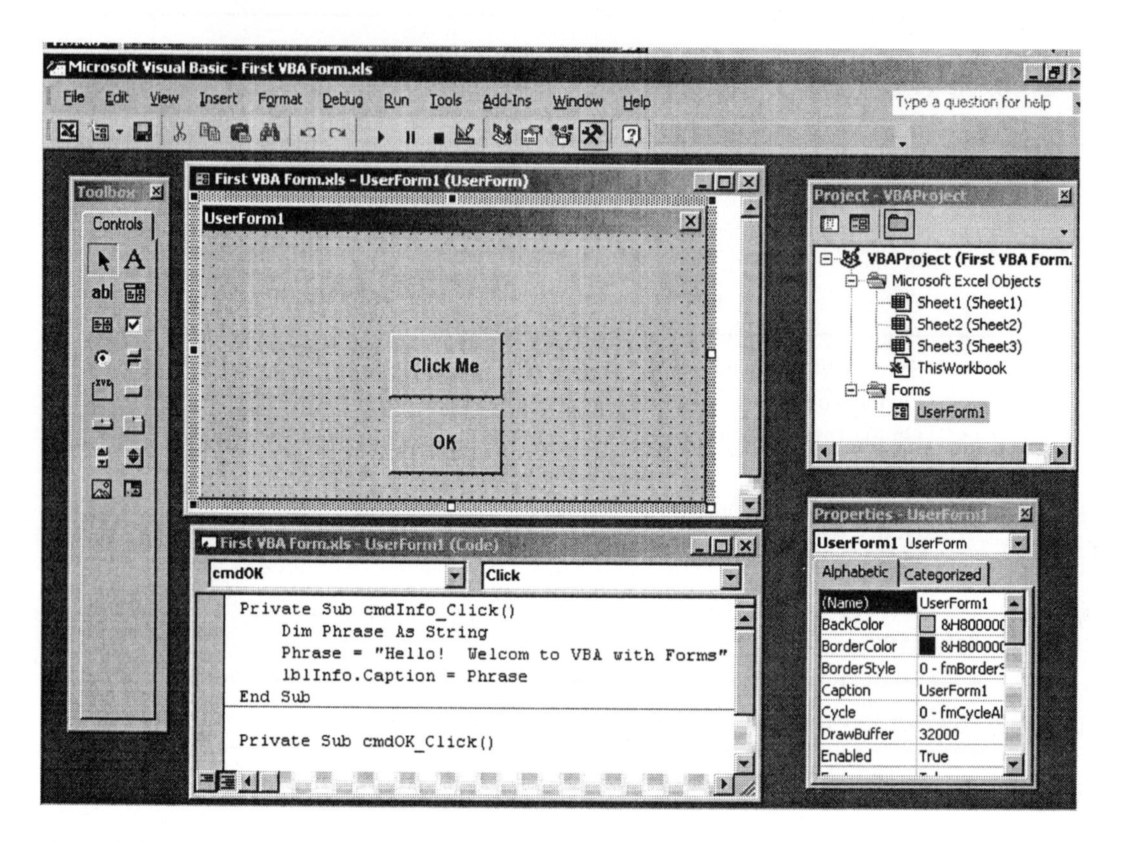

Figure 18-9. The Visual Basic Editor Design Window for Host for Form.xls.

So far you have run all VBA procedures from the VBA editor. Now we shall compare running a VBA procedure from the VBA editor and from the **Macros** dialog box.

Running the VBA Procedure from the VBA Editor

- Click the **Start** button on the VBA editor, and the form appears.
- Click the **Click Me** button, and a message appears in the label (Figure 18-10).
- Click the **OK** button to end the program.

To run the procedure from the **Macros** dialog box, it is necessary to write a short procedure to show the form as we've done previously.

Running the VBA Procedure from the Macros Dialog Box

- With the VBA editor open, on the **Insert** menu, click **Module.** Notice that the project window now shows **Module1** in addition to **frmFirstForm.**
- On the **Insert** menu, click **Procedure**, and in the **Name** box, type **DisplayMyForm** as the name.
- Enter the following code for the DisplayMyForm procedure in Module1 as follows:

```
Public Sub DisplayMyForm()
   frmFirstForm.Show
End Sub
```

- Click the **Excel** icon on the upper left corner of the VBA editor to get the **Host for Form** workbook.
- On the **Tools** menu, point to **Macro**, click **Macros**, and select the **DisplayMyForm** macro.
- Click **Run** in the **Macros** dialog box, and the same results show as when the VBA procedure was run from the VBA editor (Figure 18-10).

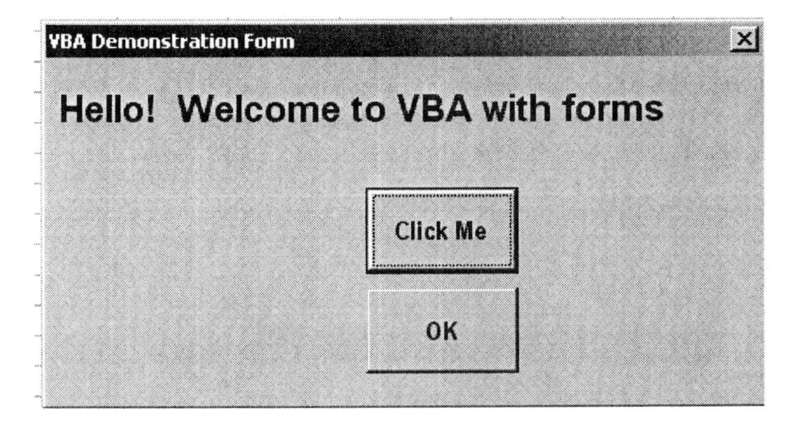

Figure 18-10. The VBA Demonstration Form, frmFirstForm

Review and Preview

A VBA project must have a host application. Out host application is Excel, and the name of the host workbook is **Host for Form**. The default name for a VBA form is UserForm1, and we gave it the name frmFirstForm. We also added a caption to the form: VBA Demonstration Form. To show the form so that we could run its underlying code, we inserted a module and left it with its default name, Module1. The code in Module1 serves one purpose: to show the form to the user. It does this with the **frmFirstForm.Show** statement. The name of the procedure in Module1 that executes this statement is DisplayMyForm. That is the name that appears in the **Macros** dialog box This is exactly what we did with the Word VBA DateName procedure at the end of the previous chapter: **frmDateName.Show**.

Summary of VBA Procedures with a Form

The next several Excel VBA projects are structured in a similar manner, with a host application, a form, and a module to show the form. The nomenclature for the various objects in the forthcoming Excel VBA projects is listed in Table 18-2.

Table 18-2

Nomenclature for Excel VBA Projects

Excel Host Name	Form Name	Form Caption	Module Name	Macro Name
Host for Form.xls	frmFirstForm	VBA Demonstration Form	Module1	DisplayMyForm
DataForStats.xls	frmStatFun	Calculate Statistics for a List	Module1	DoStatistics
Cp of Silver	frmIntegrate	Calculate Entropy of Silver	Module1	GetEntropy
ConsecReact.xls	frmABCUser	The ABCUser Program	RunABCUser	RunABCUser
ConsecReact.xls	frmABCScroll	The ABCScroll Program	RunABCScroll	RunABCScroll
Eutectic Host.xls	frmEutectic	The Eutectic Program	Module1	CalcEutectic

Excel VBA Example 8

The NumberStats Procedure

Suppose you frequently use Excel to list some repetitive data for which you need the same set of statistical analyses. These analyses might include the average, the standard deviation, the minimum, the maximum, and the range. The Excel VBA procedure described here does these operations with a couple of mouse clicks.

Preparing the Host Program

- Open an Excel worksheet, and name it **DataForStats**. Enter a few decimal numbers in cells from A1 to A5, for example, **2.25**, **2.21**, **2.19**, **2.32**, and **2.22**.
- On the **Tools** menu, point to **Macro**, and click **Visual Basic Editor**.
- On the **Insert** menu, click **Form**, and add the controls shown in Figure 18-11.
- Set the properties of the controls according to Table 18-3.

After setting the properties of the controls (Table 18-3), double-click the **cmdRun** button, and enter the following code:

```
Private Sub cmdInfo_Click()
   Dim MyRange As Range
   Set MyRange = Worksheets("Sheet1").Range("A1:A20")
   txtAve = Excel.WorksheetFunction.Average(MyRange)
   txtStDv = Excel.WorksheetFunction.StDev(MyRange)
   txtMin = Excel.WorksheetFunction.Min(MyRange)
   txtMax = Excel.WorksheetFunction.Max(MyRange)
   txtRange = Abs(txtMin - txtMax)
End Sub
```

Next, double-click the **cmdExit** button, and add the remaining code.

```
Private Sub cmdExit_Click()
   End
End Sub
```

Running the VBA NumberStats Procedure from the VBA Editor. A VBA procedure with a form must be run from the VBA editor until a procedure has been written on a module that shows the form (see the next section). The Visual Basic Editor with the controls on the form, the code on the module, and the project and properties windows is shown in Figure 18-9. Click the Visual Basic Editor **Start** button. The properties window, the code window, and the project window disappear, leaving the form with its controls and inviting the user to action. **Note:** to run a procedure from the VBA editor, either the code window or the form must be selected. If the project window or the properties window happen to be selected, clicking **Run** will not execute the VBA procedure.

Click the **Get Statistics** button. If a series of numbers are enter anywhere in the range A1:A20, the five statistical values will be displayed in the text box controls of the form (Figure 18-11).

Table 18-3

The Properties Table for the VBA NumberStats Procedure

Object	Property	Setting
Form1	Name	frmStatFun
	Caption	Calculate Statistics for a List
Label1	Name	lblInfo
	Caption	Statistical functions calculate data in the range A1:A20
	Font	Arial 16 point bold
For the remaining labels, make the Font Arial 10 point bold and use default names		
Label2	Caption	Average
Label3	Caption	Standard Deviation
Label4	Caption	Minimum
Label5	Caption	Maximum
Label6	Caption	Range
Text box1	Name	txtAve
Text box2	Name	txtStDv
Text box3	Name	txtMin
Text box4	Name	txtMax
Text box5	Name	txtRange
Command1	Name	cmdInfo
	Caption	Get Statistics
Command2	Name	cmdExit
	Caption	Exit

Running the VBA NumberStats Procedure from the Macros Dialog Box. As with Word VBA procedures that have a form, it is necessary to insert a module that has code on it for a procedure to show the form. The name of the procedure will appear on the **Macros** dialog box.

- With the Visual Basic Editor open, on the **Insert** menu, click **Module**.
- On the **Insert** menu, click **Procedure**, and in the **Name** box, type **DoStatistics** as the name.
- Enter a single statement in the **frmStatFun.Show** procedure:

```
Public Sub DoStatistics()
  frmStatFun.Show
End Sub
```

Now, anytime that the Excel workbook named Data For Stats is open, the **Macros** dialog box, which you open by, on the **Tools** menu, pointing to **Macro** and clicking **Macros**, will list DoStatistics. Select **DoStatistics** and click **Run** to show the form. Click the **Get Statistics** button on the form to run the VBA procedure (Figure 18-11).

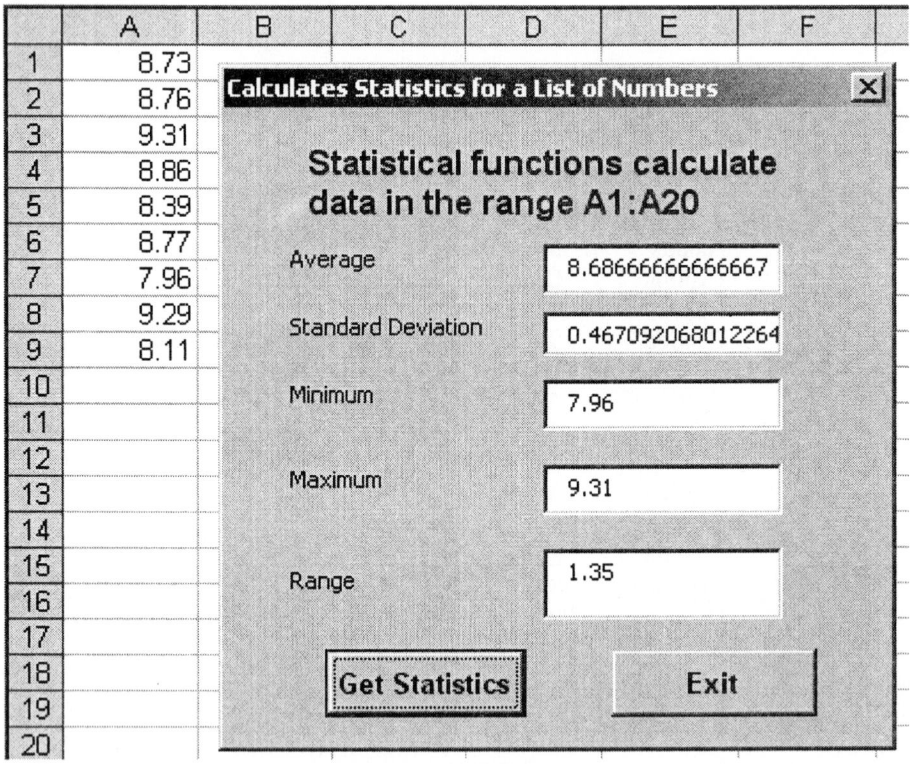

Figure 18-11. Data in Sheet1 with the Statistical Functions on the Form

Understanding the Excel VBA NumberStats Procedure. When the **Run** button is clicked, the first statement declares the variable **MyRange** as a **Range** object:

```
Dim MyRange As Range
```

In the next statement, **MyRange** is set to the range A1:A20 in Sheet1:

```
Set MyRange = Worksheets("Sheet1").Range("A1:A20")
```

Worksheets() is a collection of worksheets. Instead of **Worksheets("Sheet1")**, you could use **Worksheets(1)**, an entirely equivalent expression. You can look at the collection **Worksheets()** as an array of worksheets. Sheet1 is a container for the **Range** object. The **WorksheetFunction** is used as a container for Excel worksheet functions.

```
txtAve = Excel.WorksheetFunction.Average(MyRange)
```

In this statement, a value (**Average(MyRange)**) is assigned to the text value of the text box control on the form. In this way, all Excel functions become available for use in Visual Basic.

The form is often called a custom dialog box—custom, because you, the programmer (not Microsoft, Inc.), created it; dialog, because it offers a two-way exchange of information between the user and the dialog box. One, the user clicks. Two, the dialog box displays information.

Figure 18-12 shows the design window displaying the code window, the frmStatFun form, the project window, the properties window, and Module1. Module1 serves to show the form, and its procedure name appears in the **Macros** dialog box. At the moment it is the selected object

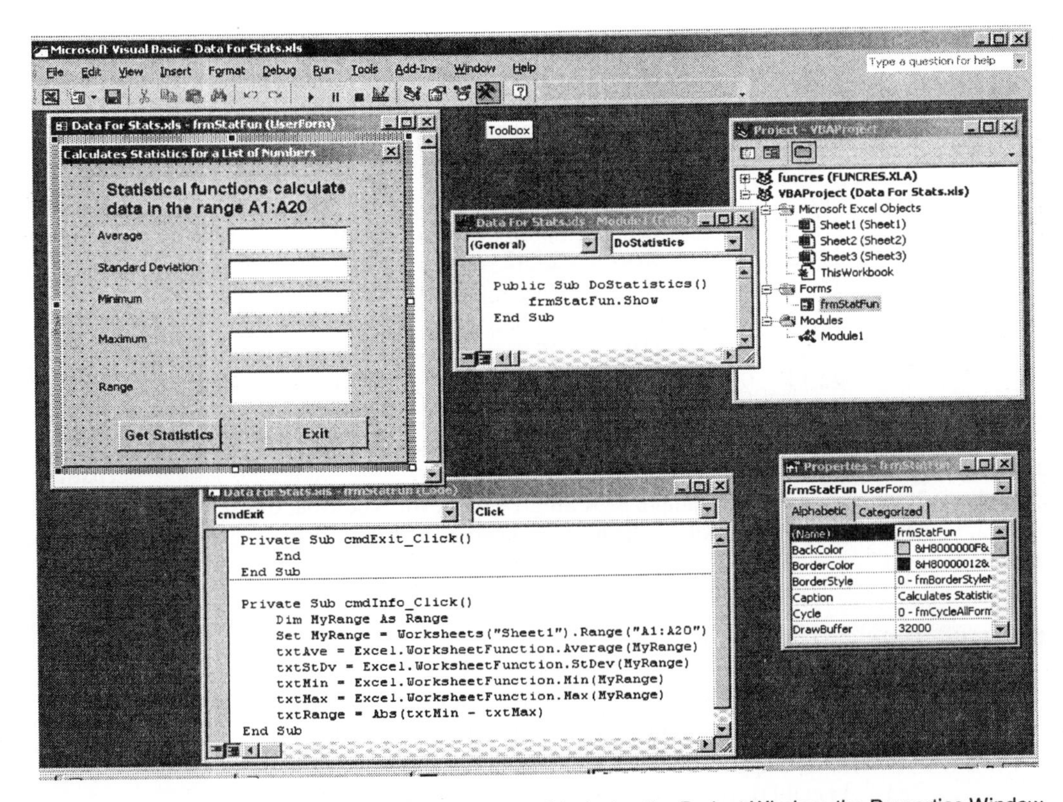

Figure 18-12. The Visual Basic Editor Design Window Displaying the Project Window, the Properties Window, the Module1 (DoStatistics), and the frmStatFun Form and its Code Window. The host application is the Excel workbook named DataForStats.xls.

Excel VBA Example 9

The SilverEntropy Procedure

For this VBA project, we will use an Excel workbook created in Chapter 2, Excel Example 3. This workbook is a simple XY (scatter) plot of the heat capacity (C_p) of silver versus the log of temperature (ln*T*). The object of this VBA project is to calculate the area under the heat capacity curve. The form on which you do this is shown in Figure 18-13.

The heat capacity C_p of most crystalline substances decreases as the temperature decreases. If the substance undergoes no phase transitions as the temperature is lowered, a plot C_p versus T follows a smooth curve, and C_p approaches zero as the temperature approaches absolute zero. For theoretical reasons to be explained later, it is more useful to plot C_p versus ln*T*. Some experimental value of C_p and T are listed in Table 18-4.

For the host application of this project, we shall enter the data in Table 18-4 into an Excel worksheet, calculate ln*T* for each *T*, and use Excel to prepare a plot C_p versus ln*T*.

For the VBA procedure, in the VBA editor, we shall write a procedure that will calculate the area under the curve of C_p versus ln*T*. From this area, the VBA procedure will calculate the absolute entropy of silver at 300 K. The final result is shown in Figure 18-13.

Table 18-4

The Heat Capacity of Silver

C_p/(J/K mol)	*T*/K	C_p/(J/K mol)	*T*/K
0.67	15	23.61	170
4.77	30	24.09	190
11.65	50	24.42	210
16.33	70	24.73	230
19.13	90	25.03	250
20.96	110	25.31	270
22.13	130	25.44	290
22.97	150	25.50	300

Preparing the Host Application. If you saved Excel Example 3 (A Single Point-Plot: The Heat Capacity of Silver) in Chapter 2, you can use that workbook as the host application. If you did not save it, then you can create it as follows.

Entering the Data

- Open a new workbook, and save it with the name **Cp of Silver**. It will serve as the host document for the VBA SilverEntropy procedure.
- On the host application, enter a title and column labels:
  1. In cell D1, type **The Heat Capacity of Silver** as the title.
  2. In cell A3, type **T/K** as the label.
  3. In cell B3, type **lnT** as the label.
  4. In cell C3, type **Cp/(J/K mol)** as the label.
- Enter the temperature data from Table 18-4 in the range A:5:A20.
- Enter the C_p data from Table 18-4 in the range C5:C20.

- In cell B5, type **=LN(A5)** as the formula.
- Select cell B5, and drag the fill handle in the lower right corner down to cell B20. This selection calculates the ln*T* values.

Creating the Chart. The chart is actually optional. But if you are calculating the area under a curve defined by a number of *x*,*y* pairs, it's useful to see the chart.

- Click and drag from cell B5 to cell C20.
- Click the **Chart Wizard** icon and prepare the XY (scatter) graph shown in Figure 18-14. The graph in Figure 18-14 was edited slightly to improve its appearance, and the entropy calculation results are shown.

Writing the VBA Code in the Editor

- On the **Tools** menu, point to **Macro**, and click **Visual Basic Editor**.
- In the VBA editor, on the **Insert** menu, click **Form**.
- Click the **Toolbox** icon, and add the controls listed in Table 18-5 and shown in Figure 18-13.
- Set the properties of the controls according to Table 18-5.

Table 18-5

The Properties Table for the VBA SilverEntropy Procedure

Object	Property	Setting
Form1	Name	frmIntegrate
	Caption	Calculate Entropy of Silver
Label1	Caption	From the area under a curve, this VBA procedure calculates the entropy of silver
	Font	Bold 10 point
Text box1	Name	txtEntropy
Command1	Name	cmdIntegrate
	Caption	Integrate
Command2	Name	cmdExit
	Caption	Exit

Double-click the **Integrate** button to open the code window. Enter the following code. (Much of this procedure can be copied from the **Private Sub cmdIntegrate_Click()** procedure in the Trapezoid Integration program described in Chapter 14.)

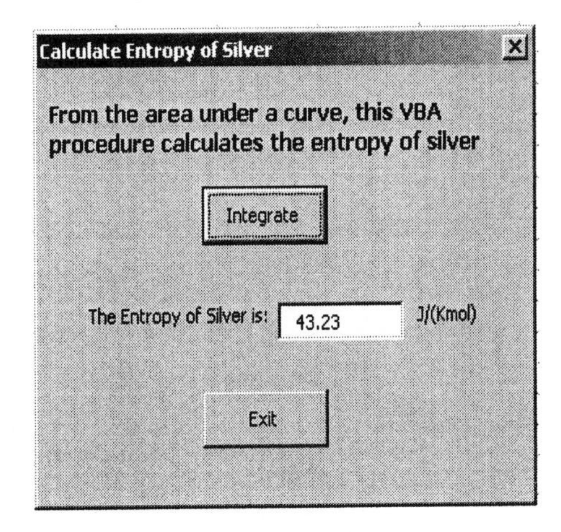

Figure 18-13. The frmIntegrate Form for the VBA SilverEntropy Procedure

```
Private Sub cmdIntegrate_Click()
   Dim N As Integer, I As Integer
   Dim X(50) As Single
   Dim Y(50) As Single
   Dim Area As Single
   Dim Entropy As Single

   For I = 5 To 20                                   ' Data are in cell rows 5 to 20
      X(I - 4) = Cells(I, 2).Value                   ' and columns 2 and 3
      Y(I - 4) = Cells(I, 3).Value                   ' Assign cell values to
   Next                                              ' Array X(50) and Y(50)

   N = 16                                            ' Number of x,y pairs
   Area = 0                                          ' Initialize Area
   For I = 1 To N - 1                                ' N points subtend N - 1 segments
      Area = Area + (X(I + 1) - X(I)) * (Y(I + 1) + Y(I)) / 2
   Next

   Range("A21:I22").Font.Bold = True
   Range("A21:i22").Font.Size = 12
   Cells(21, 5).Value = "Area Under Curve = " & Format(Area, "###.00") & " J/K/mol"
   Entropy = Area + Cells(5, 3) / 3
   Cells(22, 5).Value = "Entropy = " & Format(Entropy, "###.00") & " J/K/mol"
   txtEntropy = Format(Entropy, "###.00")
End Sub
```

Running the Excel VBA SilverEntropy Procedure from the VBA Editor

With the Excel VBA SilverEntropy procedure open in the VBA editor, click the **Start** button in the VBA editor toolbar. Either the form or the code window must be selected in order for the procedure to run. If the project window or the properties window is selected, clicking the **Start** button will not run the procedure. Examine the project window, and note how frmIntegrate is a member of the Forms collection of the Cp of Silver project.

Running the procedure causes the entropy of silver to appear in a text box with the **(Name)** property **txtEntropy** on frmIntegrate (Figure 18-13). It is also written underneath the chart on the worksheet (Figure 18- 14).

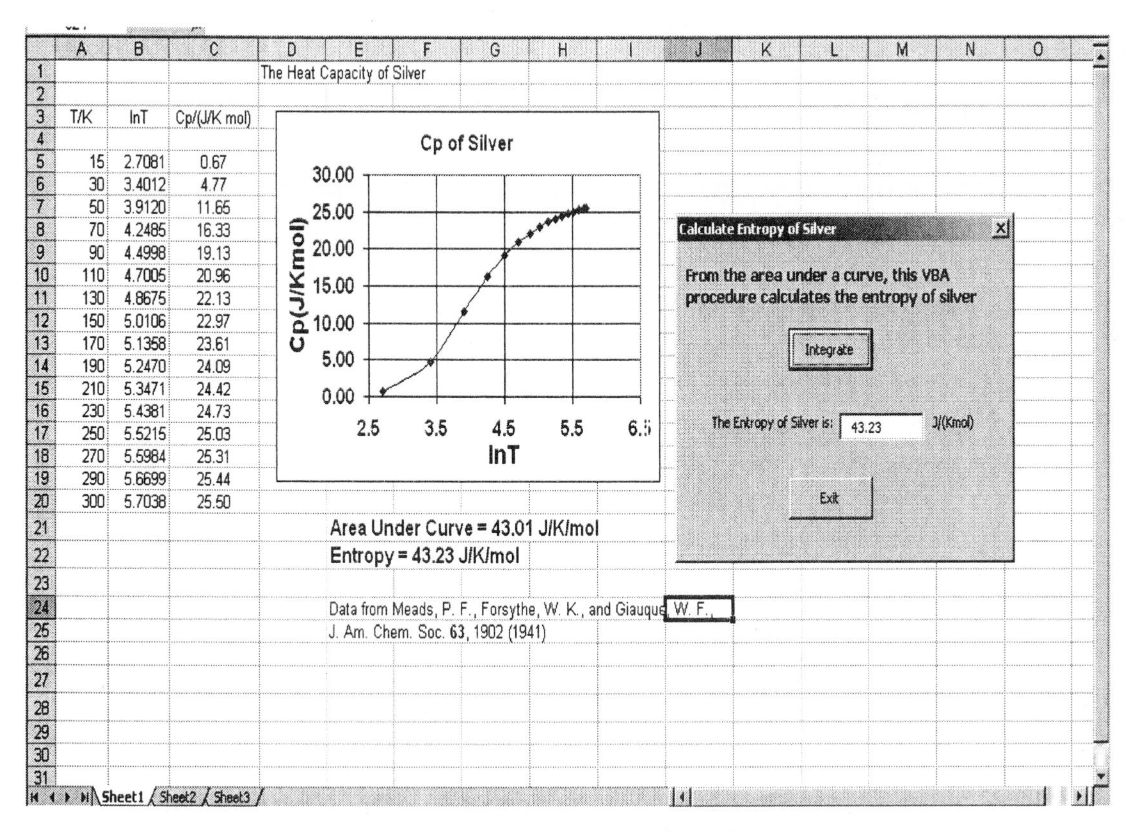

Figure 18-14. The Host Program Cp of Silver. The form of the VBA SilverEntropy procedure displays the entropy of silver. Running the VBA procedure also prints the area under the C_p curve and the total entropy.

Running the Excel VBA SilverEntropy Procedure from the Macros Dialog Box

- With the VBA editor open, on the **Insert** menu, click **Module**. Notice that the project window (Figure 18-15) now shows Module1 in addition to frmIntegrate.
- On the **Insert menu,** click **Procedure**, and in the name box, type **GetEntropy** as the name.
- Enter the following code for the GetEntropy procedure in Module1:

```
Public Sub GetEntropy()
  frmIntegrate.Show
End Sub
```

- Click the **Excel** icon on the upper left corner of the VBA editor to get the Cp of Silver worksheet.
- On the **Tools** menu, point to **Macro**, and click **Macros**, and select the **GetEntropy** macro.
- Click **Run** in the **Macros** dialog box, and the same results show as when the VBA procedure was run from the VBA editor (Figure 18-14).

At this point, the VBA editor shows the code window in the center of the workspace in Figure 18-15. The upper left displays frmIntegrate. The **Project Explorer** window in the upper right displays the file organization, including the host program, Cp of Silver, and Module1, which holds the short GetEntropy procedure for running the **cmdIntegrate** procedure from the **Macros** dialog box.

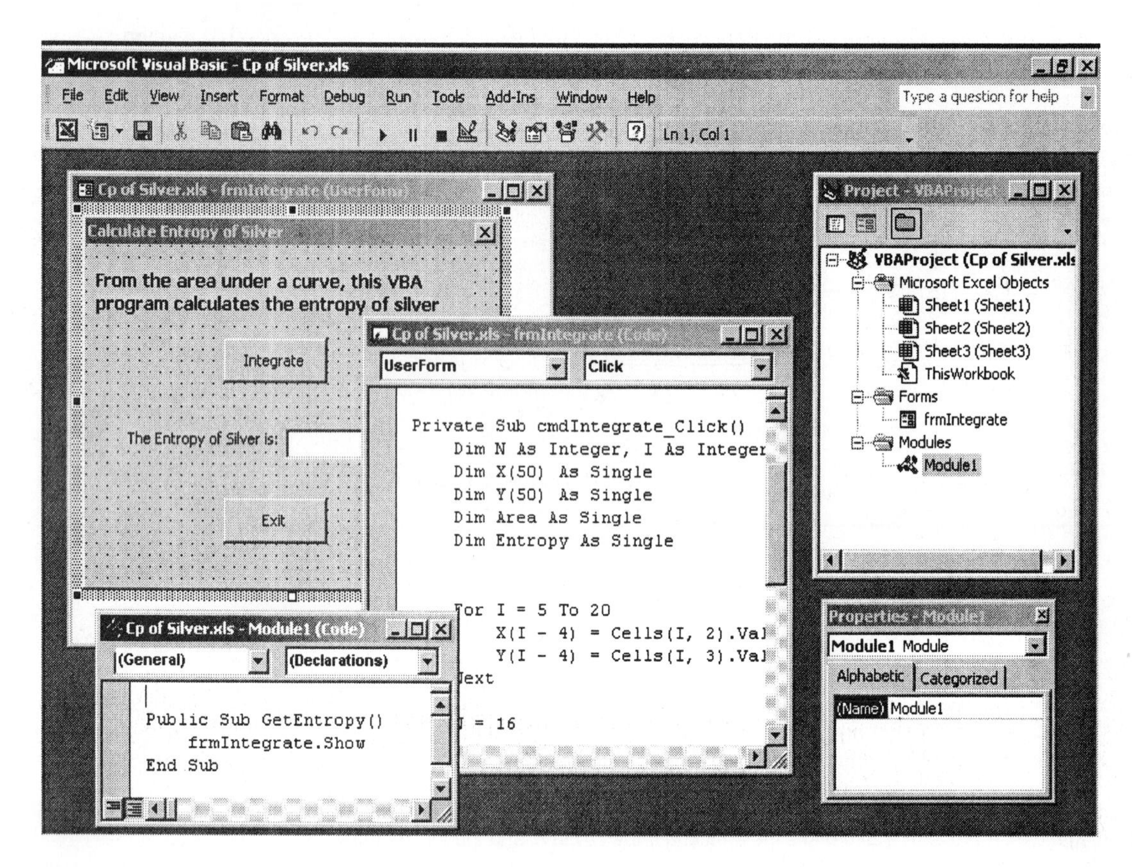

Figure 18-15. The VBA Editor Displaying frmIntegrate, the frmIntegrate Code Window, the Properties Window, and the Project Window, Which Has Added frmIntegrate to the List of Objects

Understanding the Excel VBA SilverEntropy Procedure. The algorithm for finding the area under a curve was described in detail in Chapter 14 in the Trapezoid Integration program. It is useful to compare the Visual Basic code with the Excel VBA code. The Excel VBA code is above; the Visual Basic code is below.

Visual Basic code for the Trapezoid Integration program from Chapter 14.

```
Private Sub cmdIntegrate_Click()
   Dim I As Integer
   Dim sngX(20) As Single
   Dim sngY(20) As Single
   Dim sngArea As Single

   For I = 0 To cboXinput.ListCount - 1
      sngX(I) = CSng(cboXinput.List(I))      ' Converts Xi to single and assigns to array
      sngY(I) = CSng(cboYinput.List(I))      ' Converts Yi to single and assigns to array
   Next

   sngArea = 0
   For I = 0 To cboXinput.ListCount - 2
      sngArea = sngArea + (sngX(I + 1) - sngX(I)) * (sngY(I + 1) + sngY(I)) / 2
   Next

   lblArea.Caption = Format(sngArea, "#####.000")
   lblArea.BackColor = RGB(255, 125, 125)
   lblArea.ForeColor = RGB(55, 50, 255)
End Sub
```

In the Visual Basic Trapezoid Integration program, input was by means of two parallel combo boxes. With the Excel VBA SilverEntropy procedure, the user input is directly into the worksheet. In both programs the raw input data are assigned to arrays for further data processing to get the area under a curve defined by several pairs of points in the XY plane. Slight variations on this macro could be used for any problem requiring the area under a curve defined by several points (*x*,*y* pairs).

The Science. First a quick review of the three laws of thermodynamics.

The first law:

$$dE = \delta q + \delta w$$

where dE is the increase in the internal energy of a system that absorbs δq Joules of heat and has δw Joules of work done on it.

The second law:

$$dS = \frac{\delta q_{rev}}{T}$$

where dS is the increase in entropy of a system that reversibly absorbs δq Joules of heat at a temperature T Kelvin.

The third law:

$$S_0 = 0$$

At absolute zero, the entropy of all pure, perfectly crystalline substances equals zero. It follows that

$$S_T = \int_0^T \frac{\delta q_{rev}}{T}$$

In terms of the heat capacity, the heat absorbed at constant pressure is

$$\delta q_{rev} = C_P\, dT$$

so that

$$S_T = \int_0^T \frac{C_P\, dt}{T} = \int_0^T C_P\, d\ln T$$

A plot of C_P against $\ln T$ from 0 to T equals the area under the curve, which equals S_T.

Zero Kelvin is not attainable, but it is possible to measure heat capacities to within a few units of absolute zero. In the data used in this example, the lowest temperature attained is 15 K. Fortunately, it is possible to make a close approximation of this small entropy, between 0 K and the lowest measured temperature. The Debye extrapolation shows that the entropy for this region between 0 K and T_{lowest} is equal to one third of the value of the heat capacity at the lowest attained temperature.

$$S_{T_{lowest}} = \frac{C_P(atT_{lowest})}{3}$$

The lowest temperature at which heat capacities are actually measured is T_{lowest}.

$$S_T = S_{T_{lowest}} + Area$$

where the *Area* is the area under the curve between T_{lowest} and T.

In this exercise, the data is entered into an Excel worksheet in columns B and C, beginning in row 5. The data are plotted, although this is not necessary to run the Excel VBA procedure to calculate the entropy. It just looks good and reminds us of what we're doing. The program calculates the Debye extrapolation and adds it to the area under the curve to get the total entropy of silver at 300 K.

Excel VBA Example 10

The ABCUser Procedure

In Chapter 6, Excel Example 19 introduces the Excel Consecutive Reactions ABC.xls workbook. The consecutive reaction studies are

$$A \xrightarrow{k_1} B \xrightarrow{k_2} C$$

In an Excel graph, we saw how the concentrations of A, B, and C varies with time, given a pair of rate constants. In this Excel VBA project, we shall use this previously written Excel workbook as the host application for the Excel VBA procedure that we are going to write. This VBA procedure has a form, that is, a dialog box, with which the user can change the value of the k_1 rate constant. Because the shape of all three graphs on Chart1 depends on the relative values of k_1 and k_2, the user can instantly see the dramatic effect of changing k_1.

Starting the Project

- Open the previous Excel workbook named Consecutive Reactions ABC, and save it with the name **ConsecReact.xls**.
- With the ConsecReact workbook open, on the **Tools** menu, point to **Macro**, and click **Visual Basic Editor**.
- On the **Insert** menu, click **UserForm**.
- Select the form, and in the properties window, change the **(Name)** property to **frmABCUser**, and click **Save**.
- Add the four controls to the form shown in Figure 18-16. Set their properties according to Table 18-6.

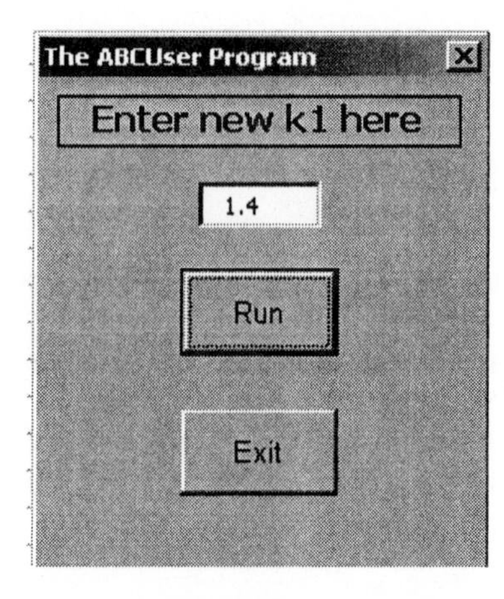

Figure 18-16. The ABCUser Program Dialog Box (frmABCUser)

Table 18-6

The Properties Table for the ABCUser Procedure

Object	Property	Value
Form1	Name	frmABCUser
	Caption	The ABCUser Program
Label1	Name	lblInput
	Caption	Enter new k1 here
	Font	12 Point Bold
	Border Style	1
Text box1	Name	txtInput
	Text	0.15
Command1	Name	cmdRun
	Caption	Run
Command2	Name	cmdExit
	Caption	Exit

Entering the Code of the Excel VBA ABCUser Program. In the VBA editor, open the code window (on the **View** menu, click **Code**). In the general declarations section, enter the following code. The **Option Explicit** statement is probably there already.

```
Option Explicit
'This VBA Program allows the user to input one rate constant (k1) after another
'to an existing Excel Workbook that displays the grapic change in concentrations
'of A, B and C for a consecutive reaction: A -> B -> C.
```

Double-click the **Exit** control (or scroll to it and select it), and enter the following code:

```
Private Sub cmdExit_Click()
   txtInput = 1.4                                        ' Resets k1 to 1.4
   End
End Sub
```

Double-click the **Run** control (or scroll to it and select it), and enter the following code:

```
Private Sub cmdRun_Click()
   Dim strValue As String
   txtInput.SetFocus
   strValue = txtInput          'The text in the text box is
                                'assigned to the variable strValue
   If strValue <> 1.4 Then
      Sheet1.Range("D4").NumberFormat = "[Red];#.00" 'k1 red if not default
      txtInput.ForeColor = vbRed
   Else
      Sheet1.Range("D4").NumberFormat = "[Black]"   'k1 black if default
      txtInput.ForeColor = vbBlack
   End If                               ' ie, if k1 = 1.4
   Sheet1.Range("D4").Value = CSng(strValue)     ' This changes the value on
                                        ' at cell D4 to the value
                                        ' that the user input to
                                        ' the InputBox Control
End Sub
```

Proofread your entries, correct any obvious errors, and click the **Save** button.

Executing the Excel VBA ABCUser Program. The graphic display of the host program may be either a chart object floating on the sheet or the chart object alone. It makes no difference to VBA, but for now let's locate the chart in Sheet1 if it's not already there. If the chart is not in Sheet1, then the first tab at the bottom of the workbook is labeled Chart1, the second tab is Sheet1, the next is Sheet2, and so on. To relocate the chart as an object in Sheet1:

- Click the **Chart1** tab to activate the chart.
- On the **Chart** menu in the Excel menu bar, click **Location**, and select **As object in Sheet1**. Then click **OK**.
- Drag the chart high up on the middle of Sheet1, approximately as shown in Figure 18-17.

At this point, we still must execute the program in the VBA editor. In Excel, on the **Tools** menu, point to **Macro**, and click **Visual Basic Editor**.

- If the form is not visible in the VBA editor, click the **Project Exploror** icon, then double-click **frmABCUser**. If **frmABCUser** is visible along with other objects, click **frmABCUser** to select it
- Click the VBA editor **Start** button. This opens Sheet1 with the chart displayed. The VBA form is displayed over the chart.
- Drag the form out of the way to the lower left or lower right side of the window as shown in Figure 18-17.
- In the text box, select the default, **1.4**, and change it to, say, **0.4**.
- Click **Run**, and see how the shapes of all three graphs on the chart change.
- Select the **1.4** value for k_1, and change it to **0.8**.
- Click **Run**, and see how the shapes of all three graphs on the chart change.

Experiment with a variety of values for k_1. When k_1 is large (compared to k_2), component A disappears rapidly so that the concentration of B piles up. When k_1 is small, the disappearance of A is slow, and the concentration of B does not reach as high a maximum value. If k_1 is very small, then B is used up as fast as it is formed, and its concentration may be barely visible on the chart. Figure 18-18 shows the charts for six different value of k_1.

Consecutive rate processes are quite common in nature. They occur not only in chemistry (consecutive chemical reactions), but in physics (radioactive decay), biology (population ecology), engineering (hydrology), and economics (modeling).

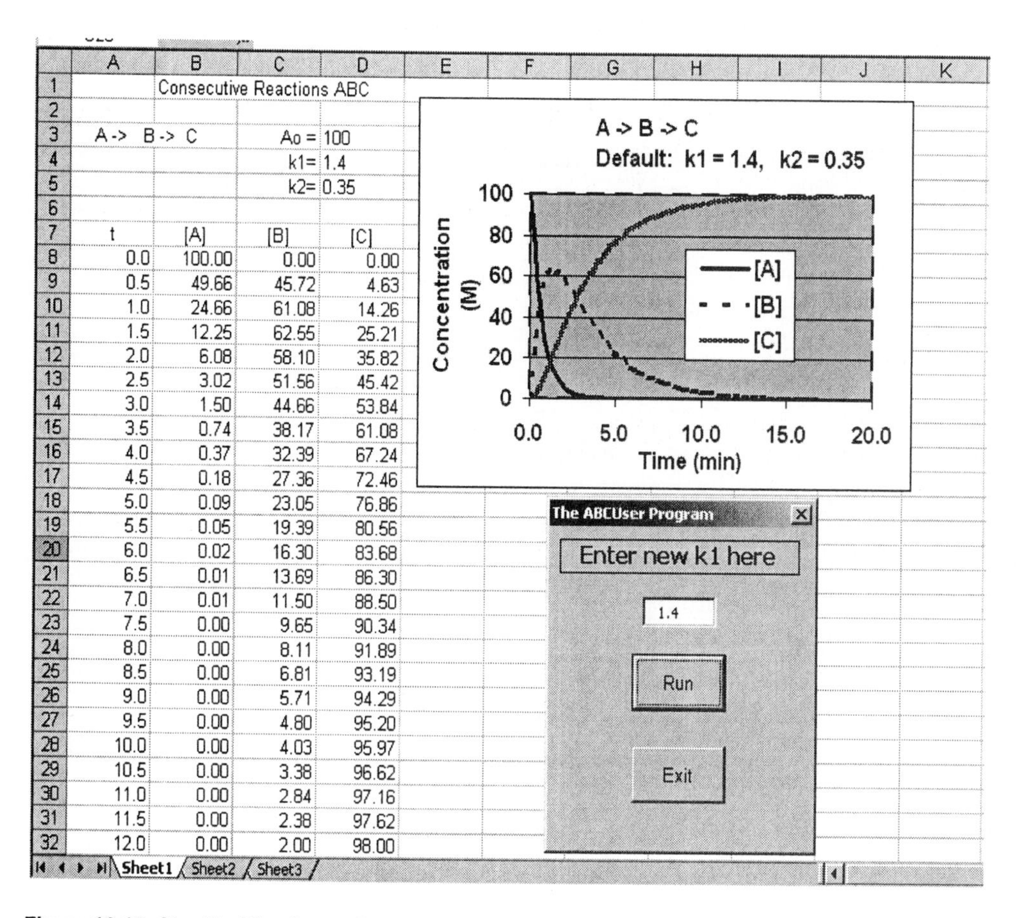

Consecutive Reactions ABC

A-> B-> C

Ao = 100
k1= 1.4
k2= 0.35

t	[A]	[B]	[C]
0.0	100.00	0.00	0.00
0.5	49.66	45.72	4.63
1.0	24.66	61.08	14.26
1.5	12.25	62.55	25.21
2.0	6.08	58.10	35.82
2.5	3.02	51.56	45.42
3.0	1.50	44.66	53.84
3.5	0.74	38.17	61.08
4.0	0.37	32.39	67.24
4.5	0.18	27.36	72.46
5.0	0.09	23.05	76.86
5.5	0.05	19.39	80.56
6.0	0.02	16.30	83.68
6.5	0.01	13.69	86.30
7.0	0.01	11.50	88.50
7.5	0.00	9.65	90.34
8.0	0.00	8.11	91.89
8.5	0.00	6.81	93.19
9.0	0.00	5.71	94.29
9.5	0.00	4.80	95.20
10.0	0.00	4.03	95.97
10.5	0.00	3.38	96.62
11.0	0.00	2.84	97.16
11.5	0.00	2.38	97.62
12.0	0.00	2.00	98.00

Figure 18-17. Sheet1 of the ConsecReacts.xls Workbook Showing the Chart in the Upper Middle Section of the Worksheet

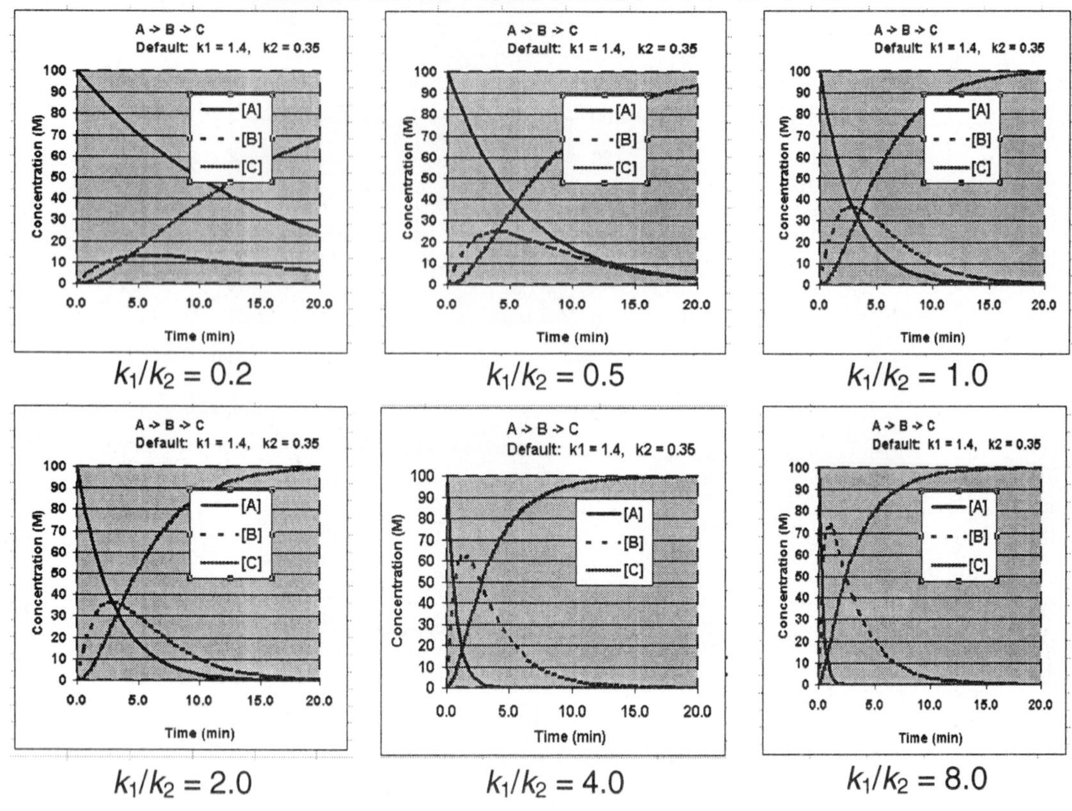

Figure 18.18. Six Different Values of k_1 Give Six Different Snapshots of the Concentrations of *A*, *B*, and *C* Changing with Time

Understanding the Code of the VBA ABCUser Program. The **Option Explicit** statement requires explicit declaration of variables. It is always recommended that a comment at the beginning of the program serve as a reminder of the program's purpose.

```
Option Explicit
'This VBA Program allows the user to input one rate constant (k1) after another
'to an existing Excel Workbook that displays the grapic change in concentrations
'of A, B and C for a consecutive reaction: A -> B -> C.
```

Clicking the **Exit** button exits the program.

```
Private Sub cmdExit_Click()
   txtInput = 1.4                        ' Resets k1 to 1.4   End
End Sub
```

When you run the VBA ABCUser program after opening the host program, the frmABCUser form appears over the host program. The form displays a command button control captioned **Run**, a label, a small text box, and a command button control captioned **Exit**. The text box displays **1.4** for k_1 because the **txtK1.Text** property was set to 1.4 at design time. This is the original value of k_1 in the host program.

The label prompts the user to change the rate constant to another value. When the user clicks the **Run** button, the **cmdRun_Click** procedure executes.

```
Private Sub cmdRun_Click()
   Dim strValue As String
   txtInput.SetFocus
   strValue = txtInput          'The text in the text box is
                                'assigned to the variable strValue
   If strValue <> 1.4 Then
      Sheet1.Range("D4").NumberFormat = "[Red];#.00" 'k1 red if not default
      txtInput.ForeColor = vbRed
   Else
      Sheet1.Range("D4").NumberFormat = "[Black]"   'k1 black if default
      txtInput.ForeColor = vbBlack
   End If                                   ' ie, if k1 = 0.15
   Sheet1.Range("D4").Value = CSng(strValue)     ' This changes the value on
                                    ' at cell D4 to the value
                                    ' that the user input to
                                    ' the InputBox Control
End Sub
```

The value the user enters in the text box is **txtInput**. It is assigned to the variable **strValue**. If the user enters nothing, then the assignment to **strValue** is 1.4, the default value. The program checks to see if **strtValue** is the default value of 1.4, and if it is, the color property of cell D4 is set to black and the color of txtInput is set to black,also. If the user changes 1.4 to some other number, then that new number is assigned to **strValue**, the color property of cell D4 is set to **Red and the color of txtInput is set to red, also**. The **Text** property of a text box is always a string value.

After completing these checks with the **If ... Then ... Else** structure, the program assigns what the user has entered for the new k_1 to cell D4.

```
Sheet1.Range("D4").Value = CDbl(strValue)        ' This changes the value of
                                                 ' cell D4 to the value
                                                 ' that the user input to
                                                 ' the text box control
   txtInput = 1.4                                ' Resets k1 to 1.4

   End Sub
```

The change in k_1 in the host program precipitates a change in its graphic presentation of the $A \xrightarrow{k_1} B \xrightarrow{k_2} C$ consecutive reactions. The last line in the procedure resets the text box to the default value of 1.4.

To summarize, if **strValue** is not equal to 1.4, then the color property of cell D4 is set to **Red**, and the **NumberFormat** property is set to **0.##**. If **strValue** equals 0.15, then the color property of cell D4 is set to **Red**, and the **NumberFormat** property is set to **0.##**. Notice that the program has not yet changed the value property of cell D4.

The next statement summarizes the whole purpose of the program, which is to control the value of a cell in the host program from the outside, that is, from a VBA macro. This statement takes the datum (**strValue**) input by the user, converts it to a double, and assigns it to the value property of cell D4 in the host program

Running the VBA ABCUser Program outside the VBA Editor. To create a module so that you can run your VBA program without the editor, use the following sequence:

- Open the host application, ConsecReact.xls.
- On the **Tools** menu, point to **Macro**, and click **Visual Basic Editor**.
- On the **Insert** menu, click **Module**, and change its (**Name**) property in the properties window to **RunABCUser**.

- Enter the following code in the module:

```
Option Explicit
Sub RunABCUser()
    frmABCUser.Show
End Sub
```

Save your work. Now anytime you want to run this VBA program:

- Open the host program (ConsecReact.xls).
- On the **Tools** menu, point to **Macro**, and click **Macros**.
- Select the macro with the name **RunABCUser**.
- Click **Run**; the VBA frmABCUser form appears.
- Change k_1 as you wish, and click **Run** to see the results.
- Repeat until finished, and then click **Exit**.

Excel VBA Example 11

The ABCScroll Procedure

The previous project (the Excel VBA ABCUser program) allows the user to control a single parameter in the host application, the ConsecReact.xls workbook. Changing the value of k_1 and examining the results gives the user insight into the dependent character of consecutive rate processes.

The previous project suggests that the use of a scroll bar control instead of a text box control to change the value of k_1 should provide an even more dramatic presentation of the $A \xrightarrow{k_1} B \xrightarrow{k_2} C$ reaction sequence. Changing k_1 with a scroll bar control allows for an almost continuously varying ratio of k_1/k_2. This control results in a dramatic display of the plots of concentrations [A], [B], and [C] that vary simultaneously with time as the user changes the rate constant ratio.

After completing the ABCUser project, the host program ConsecReact has a form (frmABCUser) and a module (RunABCUser). To the host program ConsecReact, we shall now add another form, frmABCScroll, and another module, RunABCScroll.

Starting the Project

- Open the previous Excel workbook with the name ConsecReact.xls.
- On the **Chart** menu, click **Location**, and select **As new sheet**. This location will provide a large presentation of the concentration versus time plots for a dramatic display of their changing shapes as the user moves the scroll bar.
- On the **Tools** menu, point to **Macro**, and click **Visual Basic Editor**.
- On the **Insert** menu, click **UserForm**.
- Select the form, and in the properties window, change the **(Name)** property to **frmABCScroll**, and click **Save**.
- Add the seven controls to the form shown in Figure 18-19: three label controls, one horizontal scroll bar control, two text box controls, and one command button control.
- Set their properties according to Table 18-7.

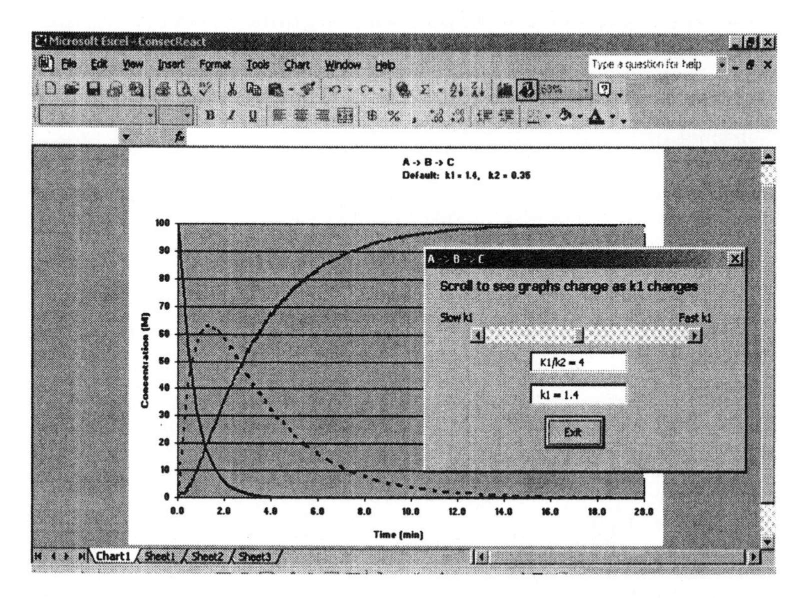

Figure 18-19. The Dialog Box for the VBA ABCScroll Program (frmABCScroll) and the Chart of the Host Program (ConsecReact.xls). The chart is located as a new sheet.

Table 18-7

The Properties Table for the ABCScroll Procedure

Object	Property	Value
Form1	Name	frmABCScroll
	Caption	Dynamic View of Consecutive Reactions
Label1	Name	lblInfo
	Caption	Scroll to see graphs change as k1 changes
	BorderStyle	1- Fixed Single
	Font	10 Point Bold
Label2	Name	lblSlow
	Caption	Slow k1
Label3	Name	lblFast
	Caption	Fast k1
Text box1	Name	txtRatio
	Caption	Make it blank
	Locked	True
Horizontal scroll bar1	Name	hsbK1
	Max	2000
	Min	15
Command1	Name	cmdExit
	Caption	Exit

Entering the Code of the Excel VBA ABCScroll Program. Select the frmABCScroll form from the procedure list. Double-click the **Exit** button, and enter its code:

```
Private Sub cmdExit_Click()
   Sheet1.Range("D4") = 1.4
   End
End Sub
```

Select the frmABCScroll form from the procedure list. Scroll the procedure list box (top left) to **hsbK1**. Then scroll the events list box (top right) to the **Change** event. Enter the code for the **Change** event:

```
Private Sub hsbK1_Change()
   Dim k1 As Single, k2 As Single
   Dim Ratio As Single
   k1 = hsbK1.Value / 1000
   k2 = Sheet1.Range("D5")
   Ratio = k1 / k2
   Sheet1.Range("D4") = k1
   txtK1.Text = "k1 = " & FormatNumber(k1, 2)
   txtRatio.Text = "k1/k2 = " & FormatNumber(Ratio, 2)
End Sub
```

Select the frmABCScroll form from the procedure list. Scroll the procedure list box (top left) to **hsbK1**, if it's not there already. Then scroll the events list box (top right) to the **Scroll** event. Enter the code for the **Scroll** event. This procedure calls the **Change** event procedure whenever the **hsbK1** control is scrolled.

```
Private Sub hsbK1_Scroll()
   hsbK1_Change
End Sub
```

Select the frmABCScroll form from the procedure list . Scroll the procedure list box (top left) to **UserForm**. Then scroll the events list box (top right) to the **Activate** event. Enter the code for the **Activate** event:

```
Private Sub UserForm_Activate()
   hsbK1.Value = 1400
   Sheet1.Range("D4") = hsbK1.Value / 1000
   txtK1.Text = "k1 = " & Sheet1.Range("D4")
   txtRatio = "K1/k2 = " & 1.4 / Sheet1.Range("D5")
End Sub
```

Executing the Excel VBA ABCScroll Program from the VBA Editor

- Open the ConsecReact.xls workbook.
- On the **Tools** menu, point to **Macro**, and click **Visual Basic Editor**.
- Click the **Project Explorer** icon to open its window.
- In the **Project Explorer** window, double-click **frmABCScroll**.
- Click the **Run** button in the VBA editor, or press F5.
- Scroll the horizontal slide bar, and observe that the graphs change their shape.

When the user moves the horizontal scroll bar, the program displays a continuous range of concentrations of A, B, and C caused by the continuously varying ratio k_1/k_2. All of the plots shown in Figure 18-18 are displayed, as well as all intermediate plots. With the ability to go back and forth between various k_1/k_2 ratios, the user can more easily understand the nature of chemical kinetics and radioactive decay sequences.

Executing the Excel VBA ABCScroll Program from the VBA Macros Dialog Box. To create a module so that you can run your VBA program without the editor, use the following sequence:

- Open the host application, ConsecReact.xls.
- On the **Tools** menu, point to **Macro**, and click **Visual Basic Editor**.
- On the **Insert** menu, click **Module**, and in the properties window, change its **(Name)** property to **RunABCScroll**.
- Enter the following code in the module:

```
Option Explicit
Sub RunABCScroll()
    frmABCScroll.Show
End Sub
```

Save your work. Now anytime you want to run this VBA program:

- Open the host program (ConsecReact.xls).
- On the **Tools** menu, point to **Macro**, and click **Macros**.
- Select the macro with the name **RunABCScroll**.
- Click **Run**; the VBA frmABCScroll form appears.
- Adjust the scroll bar to display the graphical interaction with changes in k_1.
- Repeat until finished, and then click **Exit**.

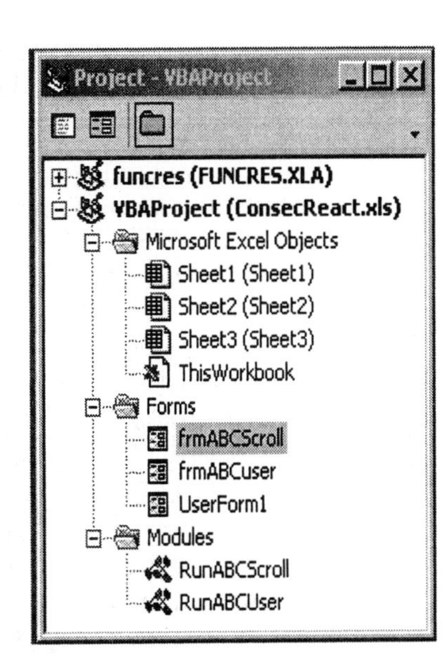

Figure 18-20. The Project Window for the ConsecReact.xls Project. The project window shows two forms and two modules.

The ConsecReact.xls project, now completed, contains two forms and two modules, as displayed in the project window (Figure 18-20). If all the windows contained in the design window are open, the design window is too crowded and confusing to be useful, as Figure 18-21 shows. Close all the windows, forms, and modules that you don't need for your current task. To display a module or form, double-click its icon in the project window. To view the code window of a form, double-click the form or select the form and, on the **View** menu, click **Code**.

Understanding the Excel VBA ABCScroll Program. When the VBA program is run, the first procedure to execute is

```
Private Sub UserForm_Activate()
   hsbK1.Value = 1400
   Sheet1.Range("D4") = hsbK1.Value / 1000
   txtK1.Text = "k1 = " & Sheet1.Range("D4")
   txtRatio = "K1/k2 = " & 1.4 / Sheet1.Range("D5")
End Sub
```

The **Activate** event of the **UserForm** causes this procedure to execute before any other procedure. This procedure initializes the settings so that the first view that the user has of the horizontal scroll bar is at the default setting for k_1. At design time minimum and maximum settings for the scroll bar were set to 36 and 3000, respectively, corresponding to k_1 values ranging from 0.036 to 3.0. The initial k_1 value chosen in the first two lines equals 1400/1000 or 1.4, a middling value. This value is also displayed in a text box (**txtK1**); the ratio k_1/k_2 is calculated and displayed in a text box (**txtRatio**).

Now nothing happens until the user moves the horizontal scroll bar. Every infinitesimal movement of the scroll bar precipitates the **Scroll** event for the scroll bar, causing execution of:

```
Private Sub hsbK1_Scroll()
```

When this procedure executes, it carries out just one action: it calls the **hsbK1_Change()** procedure.

```
Private Sub hsbK1_Change()
   Dim k1 As Single, k2 As Single
   Dim Ratio As Single
   k1 = hsbK1.Value / 1000
   k2 = Sheet1.Range("D5")
   Ratio = k1 / k2
   Sheet1.Range("D4") = k1
   txtK1.Text = "k1 = " & FormatNumber(k1, 2)
   txtRatio.Text = "k1/k2 = " & FormatNumber(Ratio, 2)
End Sub
```

This procedure takes the current value of **hsbK1**, converts it to the value of k_1, assigns this value to its proper place in the worksheet (cell D5). and displays the current k_1 in the text box (**txtK1**). It calculates the current value of the ration of k_1/ k_2 and displays that value in the text box (**txtRatio**).

The user exits the program by clicking the **Exit** button. Before ending the program, this procedure reinitializes the value of k_1 to equal 1.4 in the worksheet so the graphs will have their default values when the workbook is next opened.

```
Private Sub cmdExit_Click()
   Sheet1.Range("D4") = 1.4
   End
End Sub
```

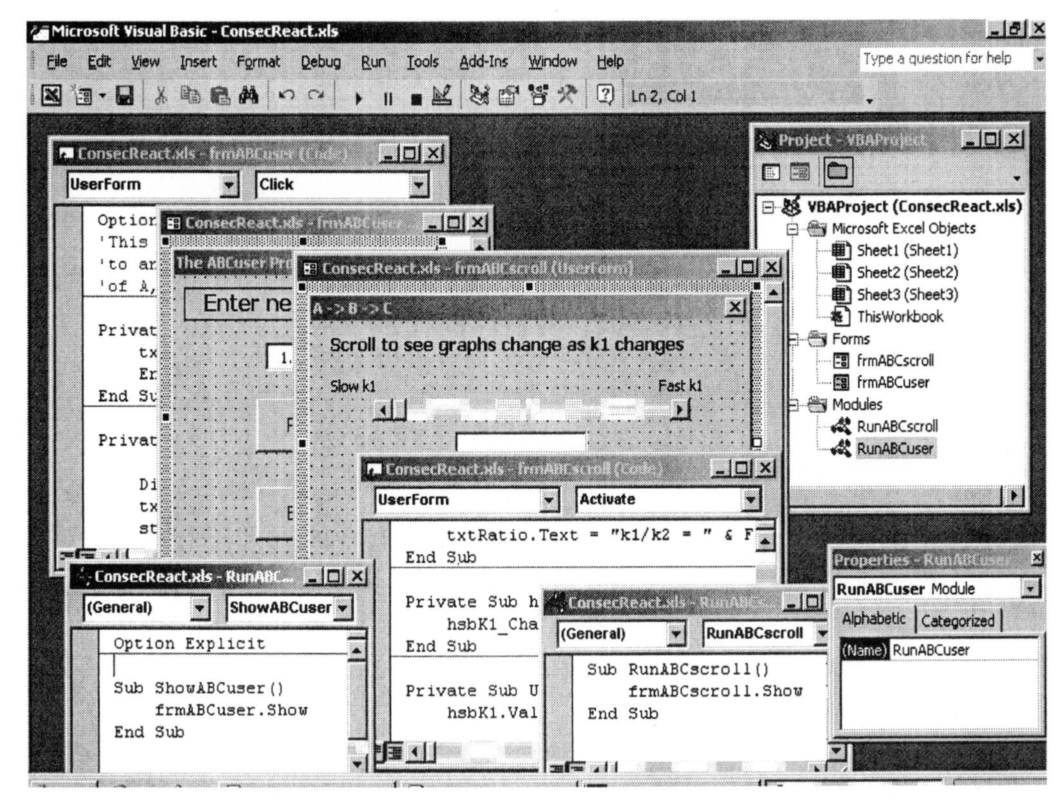

Figure 18-21. The Design Window for the ConsecReact.xls Project. The design window is displaying everything. The two forms are in the center. Their two code windows are to their left above and to their right below. In the upper right corner is the project window, which keeps track of all these forms and modules. At the lower right is the properties window. To the immediate left of the properties window is the module with the name **RunABCScroll**. At the lower left corner is the module with the name **RunABCUser**. The Design Window displayed here is, of course, too complete. Just close all the window you don't want to see.

Results. The plots in Figure 18-18 shows the concentrations [A], [B], and [C] changing with time as the reaction sequence $A \xrightarrow{k_1} B \xrightarrow{k_2} C$ progresses for a variety of k_1/k_2 ratios. With the horizontal scroll bar, the program users can observe a virtually continuous transition of k_1/k_2 ratios from small (upper left, Figure 18-18) to larger (lower right, Figure 18-18). When the k_1/k_2 ratio is small, k_2 is large, and the second step is faster than the first step.

If the k_1/k_2 ratio is very small, then the concentration [B] is low and virtually constant; that is, $d[\mathrm{B}]/dt \cong 0$. Investigate the concentration changes and rate with the horizontal control bar at smaller and smaller k_1/k_2 ratios. A ratio is quickly reached at which the concentration of [B] is virtually constant and small. As shown in Equation 6-3 in Excel Example 19 in Chapter 6, the concentration of B is given by

$$[B] = [A_0]\frac{k_1}{k_2 - k_1}(e^{-k_1 t} - e^{-k_2 t})$$

With a little algebraic rearrangement, the exponential term can be written

$$e^{-k_1 t}(1 - e^{(k_1 - k_2)t})$$

If $k_1 \cong 0$ and k_2 is large, then the value of this term approaches unity, so the concentration of B is given by

$$[B] = [A_0]\frac{k_1}{k_2}$$

This is the steady state approximation. It is the basis for understanding many consecutive rate processes in complex chemical reactions and in physics, especially radioactive decay.

Excel VBA Example 12

The Eutectic Procedure

Most of the macros and VBA procedures that we've written so far have been relatively small. The host program does most of the heavy lifting, while the macro procedure plays a subservient roll. Sometimes it is convenient to have the host program play the supporting roll, while the VBA procedure does the hard work. In this next project, essentially all the code is written in Visual Basic for Applications, and the host program in Excel does all the graphing. The entire program could be written in Visual Basic, including the graphing, but Excel's graphing capabilities are outstanding, so why not use them?

The word *eutectic* comes from the Greek *eu*, meaning well or good, and *tectic*, meaning melting. Your auto's antifreeze mixture is a eutectic mixture of water and ethylene glycol. It freezes at a lower temperature than either pure glycol or water. Ordinary solder is a mixture of tin and lead. It melts at a lower temperature than either pure tin or pure lead. Many binary metal systems form simple eutectic mixtures. The melting point curve of a binary mixture forming an ideal solution provides an interesting graph, and the calculation provides some practice with VBA functions and decision structures.

The Excel VBA Eutectic program generates a list of 99 *x,y* pairs in columns A and B of a worksheet and prepares an X,Y (scatter) graph of the data. The *x,y* pairs are T, X_A pairs, that is, the freezing temperatures and mole fractions of a mixtures of two substances that form an ideal solution. The lowest freezing point of the mixture is called the eutectic temperature.

Creating the Host Excel Program, Eutectic Host.xls. Open a new Excel workbook. Since VBA does all the work for this project, it really isn't necessary to enter anything before saving it with a title. Save the Excel worksheet with the name **Eutectic Host**.

Writing the VBA Program

- With Eutectic Host.xls open and Sheet1 the active sheet, on the **Tools** menu, point to **Macro**, and click **Visual Basic Editor**.
- On the **Insert** menu, click **UserForm**.
- Select the form, and in the properties window, change the **(Name)** property to **frmEutectic**, and click **Save**.
- Insert controls in the form as shown in Figure 18-22. Set the properties of the controls according to Table 18-8. The initial text box values furnished are for the naphthalene (A), benzene (B) system. Try other values listed in Table 18-9.

Table 18-8

The Properties Table for the Eutectic Procedure

Object	Property	Value
Form1	Name	frmEutectic
	Caption	The Eutectic Program
Label1	Name	lblInfo
	Caption	This program demonstrates how a simple VBA program can utilize Excel's powerful graphics
	Font	Arial Bold 10
Label2	Name	lblFPA
	Caption	Enter the FP of pure A (K)
	Font	Arial Bold 8
Label3	Name	lblHeatA
	Caption	Enter the heat of fusion of pure A (kJ/mol)
	Font	Arial Bold 8
Label4	Name	lblFPB
	Caption	Enter the FP of pure B (K)
	Font	Arial Bold 8
Label5	Name	lblHeatB
	Caption	Enter the heat of fusion of pure B (kJ/mol)
	Font	Arial Bold 8
Text box1	Name	txtFPA
	Text	353.4
Text box2	Name	txtHeatA
	Text	19.1
Text box3	Name	txtFPB
	Text	278.6
Text box4	Name	txtHeatB
	Text	9.90
Command1	Name	cmdRun
	Caption	Run
Command2	Name	cmdExit
	Caption	Exit

Table 18-9

Selected Melting Points (K) and Heats of Fusion (J/mol)

Metals			**Organic Compounds**		
Element	T_{MP}	ΔH^{o}_{fus}	Compound	T_{MP}	ΔH^{o}_{fus}
Cd	594.258	6192	Diphenylamine	326.06	4269
Pb	600.652	4799	p-Nitroaniline	420.65	5041
Sn	505.1181	7322	p-Dibromobenzene	360.9	4836
Tl	576.7	4142	p-Dichlorobenzene	326.2	4366
Zn	692.73	7322	Diphenyl	343.2	4441

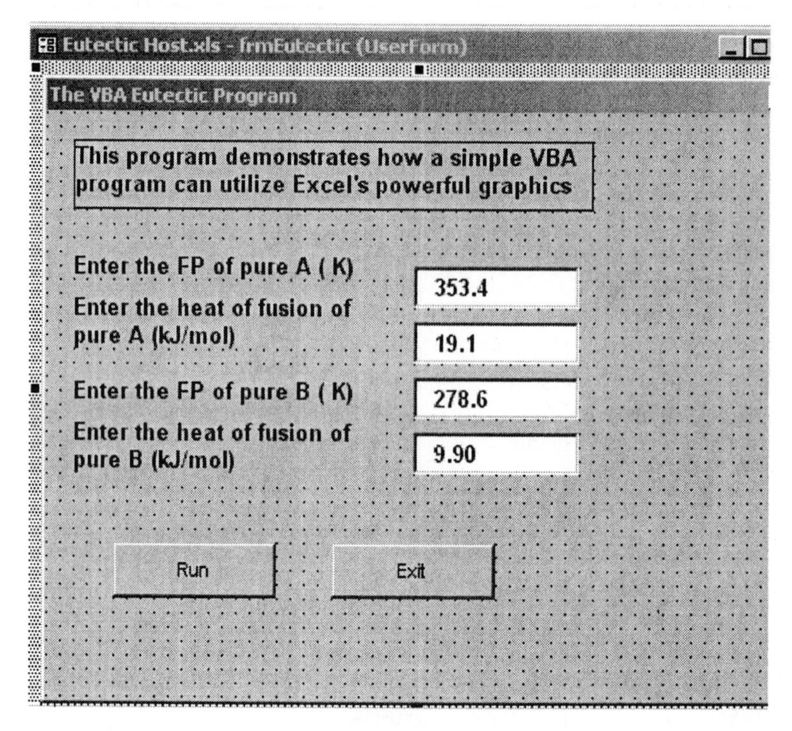

Figure 18-22. The Form for the VBA Eutectic Program

Entering the Code for the VBA Eutectic Program. In the VBA editor, open the code window (on the **View** menu, click **Code**). In the general declarations section, enter the following code. The **Option Explicit** statement is probably there already.

```
Option Explicit
'A VBA program to calculate the freezing point diagram for a binary mixture

Const R As Double = 0.00831434                          ' Gas constant, kJ/deg/mode
Dim T As Double, ToA As Double, ToB As Double          ' MP, MP pure A, MP pure B
Dim Xa As Double, Xb As Double                          ' mole fraction A and B in mixture
Dim DelHa As Double, DelHb As Double                    ' heat of fusion of A and B
```

Double-click the **cmdExit** control (or scroll to it and select it), and enter the following code:

```
Private Sub cmdExit_Click()
  End
End Sub
```

Double-click the **cmdRun** control (or scroll to it and select it), and enter the following code:

```
Private Sub cmdRun_Click()
  Dim I As Integer
  Dim TL As Double, TR As Double          ' TL is FP at Xa < Xa at Eutectic mixture
                                          ' TR is FP at Xa > Xa at Eutectic mixture
```

```
        ToA = txtFPA.Text                ' The user enters the next four data
        ToB = txtFPB.Text
        DelHa = txtHeatA.Text
        DelHb = txtHeatB.Text

        Xa = 0                           ' initialize Xa = 0 at beginning
        For I = 1 To 99
          Xa = Xa + 0.01                 ' increment Xa
          TL = LeftTemp(Xa)              ' calculate FP
          TR = RightTemp(Xa)             ' calculate FP

          If TL > TR Then                ' compare calculated temperature
            T = TL                       ' assign T
          Else
            T = TR                       ' assign T
          End If

          Cells(I, 1).Value = Xa         ' assign mole fractions to worksheet col A
          Cells(I, 2).Value = T          ' assign temperatures to worksheet col B
        Next
        BigChart                         ' This is a call to a chart procedure (see below)
    End Sub
```

In the VBA editor, on the **Insert** menu, click **Procedure** to open the **Add Procedure** dialog box; then:

- In the **Name** box, type **LeftTemp** as the name.
- In the **Type** section, select **Function.**
- In the **Scope** section, select **Public.**
- Click **OK**, and enter the following code for the first function:

```
' Calculate temperature at which A freezes out of solution
Public Function LeftTemp(Xa As Double)
     LeftTemp = 1 / ((1 / ToA) - (R * Log(Xa) / DelHa))
End Function
```

In the VBA editor, on the **Insert** menu, click **Procedure** to open the **Add Procedure** dialog box; then:

- In the **Name** box, type **RightTemp** as the name.
- In the **Type** section, select **Function.**
- In the **Scope** section, select **Public.**
- Click **OK**, and enter the following code for the second function:

```
' calculate temperature at which B freezes out of solution
Public Function RightTemp(Xa As Double)
     RightTemp = 1 / ((1 / ToB) - (R * Log(1 - Xa) / DelHb))
End Function
```

Since we know that we are going to graph the XY pairs with Excel's Chart Wizard, we might as well add the graphics procedure that we recorded earlier, the AutoGraph macro. Just go back to the AutoGraph macro, and copy and paste it into our VBA program at this point.

A few changes are noted with the comment **This is edited**. The procedure's name was changed, the chart location was changed, and the title and labels were changed. You can copy and paste AutoGraph and edit it, or just enter the following code:

```
Sub BigChart()                                                          ' This is edited
'
'
' This is an edited version of the AutoGraph macro                      ' This is edited
'
   Charts.Add
   ActiveChart.ChartType = xlXYScatterSmoothNoMarkers
   ActiveChart.SetSourceData Source:=Sheets("Sheet1").Range("A1:B100"), PlotBy:= _
      xlColumns
  ActiveChart.Location Where:=xlLocationAsObject, Name:="Sheet1"        ' This is new
   With ActiveChart
      .HasTitle = True
      .ChartTitle.Characters.Text = "Freezing Point Diagram"            ' This is edited
      .Axes(xlCategory, xlPrimary).HasTitle = True
      .Axes(xlCategory, xlPrimary).AxisTitle.Characters.Text = "Xa"     ' This is edited
      .Axes(xlValue, xlPrimary).HasTitle = True
      .Axes(xlValue, xlPrimary).AxisTitle.Characters.Text = "T (K)"     ' This is edited
   End With
End Sub
```

Running the VBA Eutectic Program from the VBA Editor

- Open the host program, Eutectic Host.xls.
- On the **Tools** menu, point to **Macro**, and click **Visual Basic Editor**.
- In the VBA editor, click the **Start** button.

The Visual Basic Editor closes, and Sheet1 of the Eutectic Host.xls workbook opens. The dialog box (frmEutectic, with the title bar Freezing Point Diagram) floats over Sheet1 with the default values of freezing points of the pure components and their heats of fusion.

- Change the text box entries if you wish, or run the default values.
- Click **Run** on the VBA Eutectic program form (frmEutectic).

The worksheet now looks like Figure 18-23. On Sheet1, the range A1:B99 fills with numbers corresponding to the *x,y* pairs of temperature and composition. Simultaneously, the chart appears displaying a graph of the freezing point versus composition, with a title and labeled axes. Clicking the **Exit** button causes Sheet1 and the dialog box to disappear. Clicking the **Excel** icon in the upper left corner of the Visual Basic Editor reopens the host workbook.

Editing the Chart. The chart needs a bit of editing to display its information more suitably, as shown in Figure 18-23.

- Right-click the plot area, and change its color to white.
- Right-click any y-axis number, then on **Format Axis.**
- Click the **Scale** tab; in the **Minimum** box, type **260** as the value, and in the **Maximum** box, type **360** as the value; click **OK**.
- Right click any x-axis number, then on the **Format Axis.**,.
- Click the **Scale** tab; in the **Maximum** box, type **1.0** as the value; then click **OK**.
- Right-Click anywhere on the graphed line, then on the **Format Data Series.**
  1. Click the **Patterns** tab, and then scroll the **Weight** box and select a fairly heavy line. Click **OK**.
  2. Click the Legend box (**Series 1** then press the DELETE key.
- The chart now appears approximately like Figure 18-24.

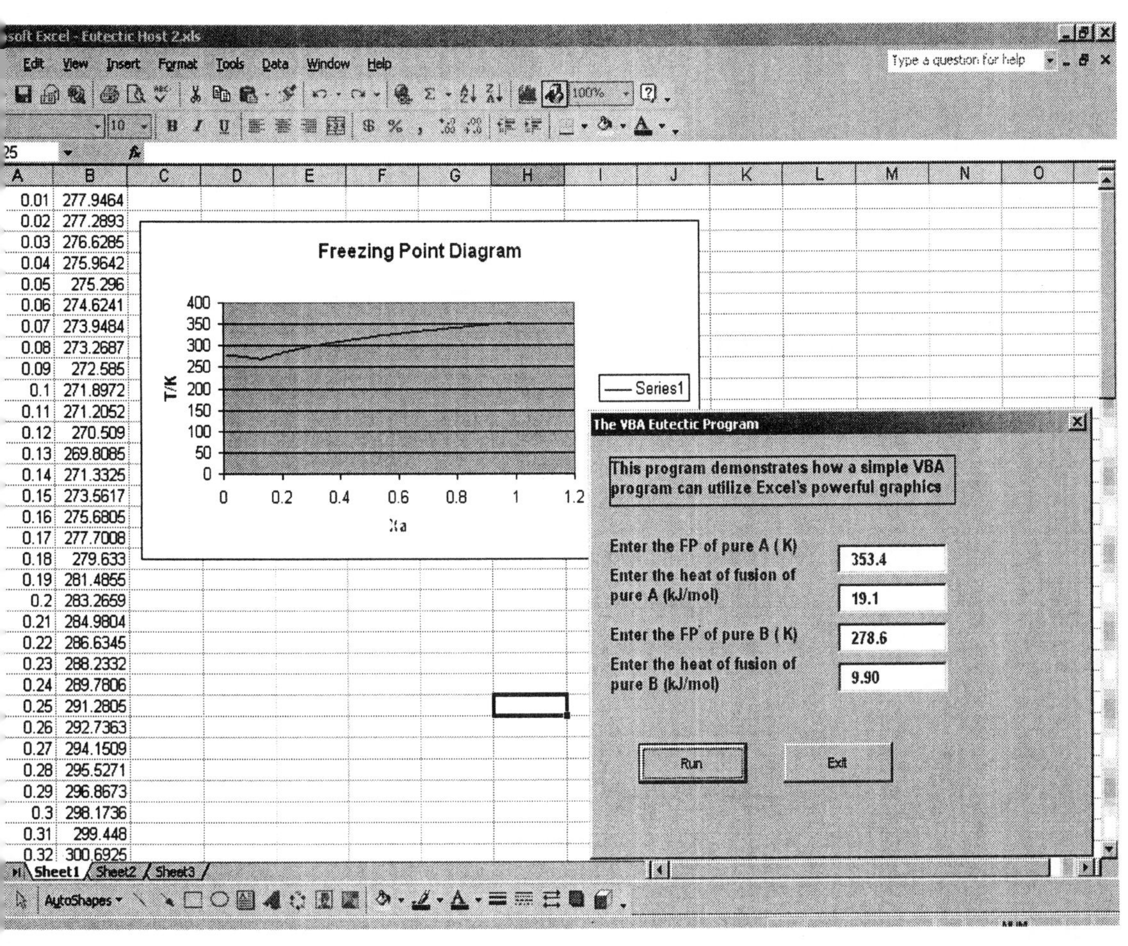

e 18-23. Running the Macro for This Blank Worksheet Puts the Dialog Box on the Sheet. Clicking the **Run** button on the box calculates the freezing point temperatures (column B) and their corresponding compositions (column A) and taneously generates the chart with the title Freezing Point Diagram. Editing the chart produces the chart shown in Figure . The user may change the freezing points and heats of fusion from the default values shown in the figure.

Running the VBA Eutectic Program from the Macros Dialog Box. To create a module so that you can run your VBA program without the editor, use the following sequence:

- Open the host application, Eutectic Host.xls
- On the **Tools** menu, point to **Macro**, and click **Visual Basic Editor**.
- On the **Insert** menu, click **Module**.
- Enter the following code in the module:

```
Option Explicit
Sub CalcEutectic()
    frmEutectic.Show
End Sub
```

Save your work. Now anytime you want to run this VBA program:

- Open the host program (Eutectic Host.xls).
- On the **Tools** menu, point to **Macro**, and click **Macros**.
- Select the macro with the name **CalcEutectic**.
- Click **Run**; the VBA frmEutectic form appears.
- Click **Run** on the VBA Eutectic program form (frmEutectic).

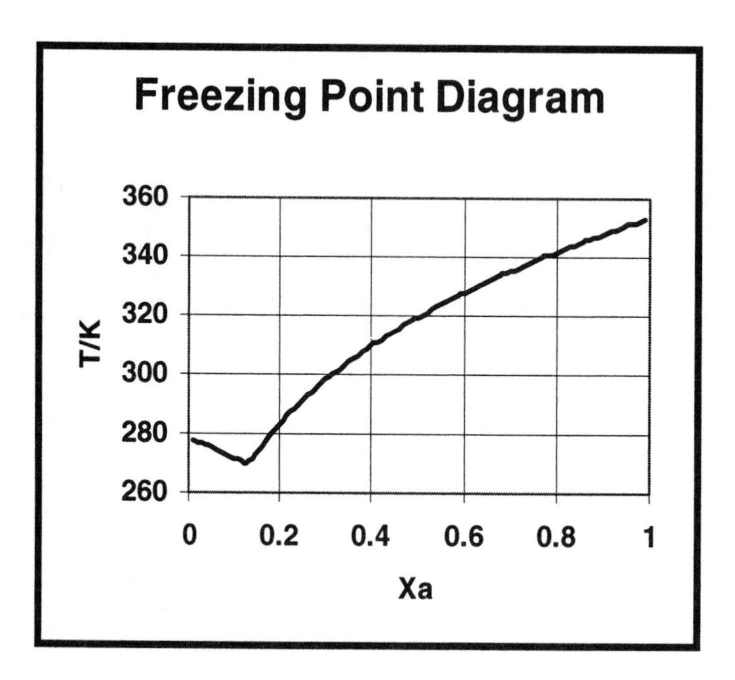

Figure 18-24. The Eutectic Freezing Point Diagram for Benzene and Naphthalene Calculated and Plotted with the VBA Eutectic Program

Understanding the VBA Eutectic Program

The Science: Solid-Liquid Phase Equilibrium. When a solid substance $A(s)$ or a liquid substance $A(\ell)$ goes into solution with another substance, it undergoes a change of state that may be represented by

$$A(s) \quad = \quad A(\ell) \quad = \quad A(soln) \tag{18-1}$$

$$\leftarrow \quad \Delta H_A^0\text{(fusion)} \quad \rightarrow\leftarrow \quad \Delta H_A^0\text{(mixing)} \quad \rightarrow$$

The same relationships may be written for the second component of the solution B. For an ideal solution, ΔH_A^0 (mixing) equals zero for both components. When a solution cools, eventually a solid begins to form. This observation can be described in two ways: one of the components begins to freeze or one of the components precipitates out of the saturated solution. The observed phenomenon is the same, only the words (the semantics) are different. The thermodynamics of ideal solutions is straightforward and provides the relationship between the composition of the solution (X_A) and is melting point (T). For component A:

$$\ln X_A = \frac{\Delta H_A^0(fus)}{R}\left(\frac{1}{T_A^0} - \frac{1}{T}\right) \tag{18-2}$$

where X_A is the mole fraction of A at the freezing temperature, T_A^0 is the freezing point of pure A, T is the freezing point of the solution. R is the gas constant, and ΔH_A^0 (fusion) is the enthalpy of fusion of A. An exactly equivalent equation may be written for the other component, B.

$$\ln X_B = \frac{\Delta H_B^0(fus)}{R}\left(\frac{1}{T_B^0} - \frac{1}{T}\right) \tag{18-3}$$

The dependence of temperature on composition may be expressed explicitly. For an ideal solution between two components, the temperature at which the two phases are in equilibrium is given by

$$T_A = \left(\frac{1}{T_A^0} - \frac{R\ln X_A}{\Delta H_A(fus)}\right)^{-1} \tag{18-4}$$

and

$$T_B = \left(\frac{1}{T_B^0} - \frac{R\ln(1 - X_A)}{\Delta H_B(fus)}\right)^{-1} \tag{18-5}$$

When X_A equals one, $T = T^0{}_A$, and when X_A equals zero, $T = T^0{}_B$. At the eutectic, the solution is saturated with respect to both components, and the saturation temperatures are equal. From a slightly different point of view, the freezing points of both components have been lowering to the same temperature. In either case, a plot of T versus X_B begins at the freezing point of pure A, drops to the eutectic temperature, and then rises to the freezing point of pure A.

The Calculation: Solid-Liquid Phase Equilibrium. To construct a solid–liquid phase diagram for a two-component system, we need five constants: the freezing points of the two pure components, their two heats of fusion, and the gas constant R. As an example, we will use the benzene-naphthalene system. Our strategy with a worksheet is to form three data series. The first should list the composition of naphthalene from 0.00 to 1.00 in steps of 0.01. The second two series will be the freezing points of, first, solutions of benzene in naphthalene, and, second, solutions of naphthalene in benzene. These will be calculated with Equations 18-4 and 18-5 over the entire composition range.

The VBA Project. In this VBA project, Excel does the charting, but the VBA program does all the calculation. The form (frmEutectic) is the dialog box for the convenience of the user. The code attached to the **Run** button does all the calculations. The previously recorded AutoGraph macro, renamed and slightly edited, graphs the temperature versus composition curves.

The Code. The general declaration section of the program contains a short description of the VBA program, the declaration of global variables, and the gas constant *R*. Note that units are provided.

The **cmdRun** procedure begins with the declaration of several variables, after which **Xa** is set to zero to initialize the following calculations, which are carried out in increments of **Xa** equal to 0.01. **TL** and **TR** are equilibrium temperatures to the left of the eutectic (lowest) temperature. **TL** lies in the *B* rich (*A* lean) composition region; **TR** lies in the *A* rich (B lean) composition region. The statements:

```
TL = LeftTemp(Xa)
TR = RightTemp(Xa)
```

call the **LeftTemp(Xa)** and **RightTemp(Xa)** functions, which are inserted toward the end of the procedure.

The **If ... Then ... Else** control statement checks whether the calculated temperature **T** is **TL** or **TR**. The current **Xa** and **T** are then assigned to cells in rows A and B of Sheet1 of the host application, the Eutectic Host.xls workbook.

When the **For ... Next** loop is satisfied, the program calls the BigChart procedure, which automatically charts the data in a new sheet (as opposed to a chart embedded in Sheet1). The chart as automatically created has unsuitable x-axis and y-axis scales, which is frequently the case with Excel's Chart Wizard. However, it is a simple matter to adjust the scales, resulting in the graph shown in Figure 18-20.

Part IV. Mathcad and Mathematica

Chapter 19. Mathcad
Chapter 20. Mathematica

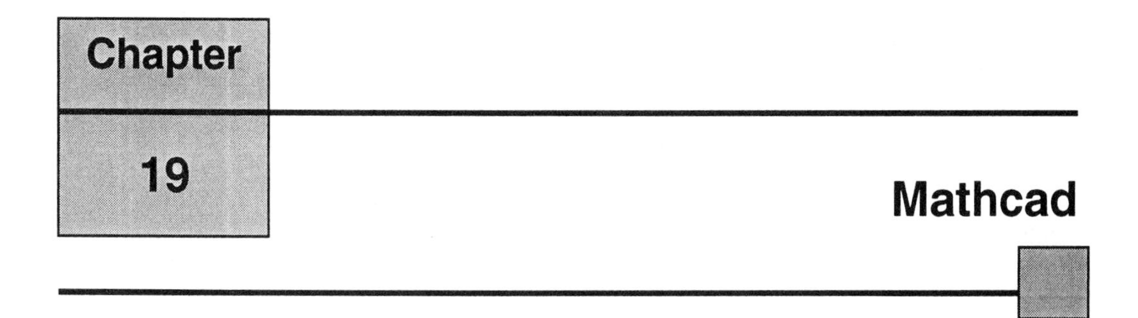

Introduction: the Mathcad Worksheet

Mathcad is an application for working with numerical calculations, equations, graphs, and text. With Mathcad, the equations you type appear on the screen the way you would type them or see them in a book. You can carry out a wide variety of calculations and mathematical manipulations with the equations and expressions that you type. And then you can seamlessly shift to writing text, and back to calculations, as often as you wish. Thus the printed Mathcad page looks like a finished report. For final editing and presentation, you can save a Mathcad file (which has the extension .mcd) as a text file (with the extension .rtf) and open it or insert it into a word processor such as Word or WordPerfect. The usual file, edit, cut, and paste features are all available.

The newly opened Mathcad workspace appears as shown in Figure 19-1. At the moment, all of Mathcad's eleven toolbars are visible on the screen. To display the toolbars, on the **View** menu, **Toolbars** has been clicked, resulting in a submenu of toolbars. All have been selected to display the entire set of toolbars.

At the top of the window is the usual menu bar. The toolbar just under the menu bar is the standard toolbar, and just below the standard toolbar is the formatting toolbar. Many of the icons on these two toolbars are familiar to users of Windows applications.

A fresh Mathcad worksheet shows a pointer in the form of a red cross (+). If you just start writing when the red-cross pointer is visible, the program expects you to work with numerical calculations. If you enter a number or a letter, the red cross disappears, and the number or letter is enclosed in a box. For example, if you enter the number 3, the red cross disappears resulting in

3|

Now if you enter **3 /7=,** the result is

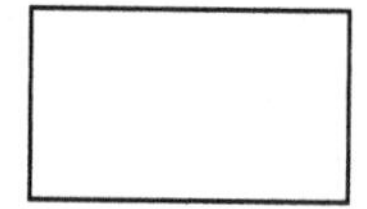

Clicking outside the math region defined by the box results in a return of the red cross. The box outline and the ∎ (placeholder) to disappear.

To enter text, when Mathcad displays the red cross, on the **Insert** menu, click **Text Region**, or press the double-quotation mark (") key. These actions result in a small text region enclosed in a box, and if you type some text into the box, it expands to contain the text in 10 Arial, by default:

This is some text in a text region

Clicking outside the text region causes the box to disappear and the red-cross pointer to reappear.

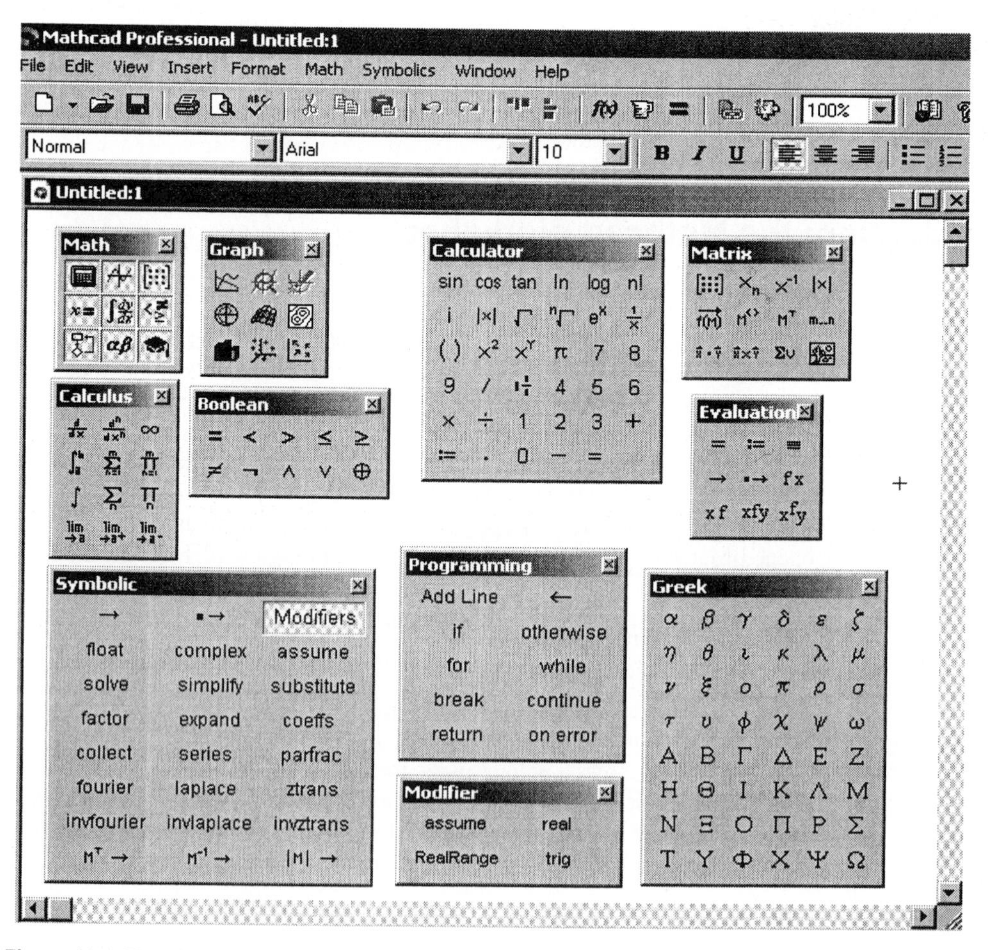

Figure 19-1. The Mathcad Worksheet with its Toolbars

Cursors and Calculations. To enter text, first enter the quotation mark character ("). The quotation mark won't show on the screen, but the red-cross pointer (for numbers) changes to the text pointer, a vertical red line. To resume entering math material, just click outside the text region, and the red-cross pointer reappears. You can place comments as text anywhere you wish, alternating text and math regions as often as you like.

Assigning a Value to a Variable. Click anywhere in the Mathcad worksheet so that the red-cross math pointer is visible. To assign the value **3** to the variable named *Width*, type **Width:3.** When the colon(:) is typed, Mathcad adds the equal sign (=) to generate automatically the assignment operator, which is the colon and equal sign together (:=). The following appears on screen:

Width := 3

Repeat to add the next statement.

Height := 6

Type **Area:Width*Height** next. Mathcad produces the following statement and automatically changes the multiplication operator (*) to a dot (·).

Area := Width·Height

The right side has a value, which has been assigned to the variable named **Area**. To see the value of the variable, it is necessary to type its name followed by an equal sign:

Area = 18

Notice that Mathcad distinguishes between uppercase and lowercase. Typing **area** instead of **Area** causes Mathcad to inform you that **This variable or function is not defined above**.

Mathcad uses the usual symbols for the arithmetic operations: addition (+), subtraction (–), multiplication (*) and division (/). Mathcad automatically changes the symbol for multiplication as seen previously, and for division as seen in the following.

The Mathcad Editing Lines. Type **Ratio1:Area/Height + 5**. Watch the editing lines carefully as you enter each part of the expression. The blue editing lines are the horizontal editing line and the vertical insertion line. The editing lines change as your entries progress:

Ratio1|
Ratio1 := ■|
Ratio1 := Area|
Ratio1 := Area
______■|

$$\text{Ratio1} := \frac{\text{Area}}{\text{Height} + 5}$$

If you misspell **Height** as **Heifht**, place the insertion point between the *f* and *h* and click. The horizontal edit line underlines the entire word, and the insertion line falls between the *f* and the *h*. The usual BACKSPACE and DELETE keys may then be used to edit the word. To evaluate, type **Ratio1** = and the following line appears.

$$\text{Ratio1} = 1.636$$

Typing **Ratio2:area/length [space]+5** produces the following on the Mathcad screen:

$$\text{Ratio2} := \frac{\text{Area}}{\text{Height}} + 5$$

Typing **Ratio2=** gives the result.

$$\text{Ratio2} = 8$$

Again, watch how the editing lines position themselves as you make further entries.

Positioning Math Regions and Making Assignments to Variables. At the red-cross math pointer, type **VolBox:a*b*c** and press the ENTER key. Mathcad responds with:

$$\text{VolBox} := a \cdot b \cdot c$$

When you move the pointer outside of the *VolBox,* the *a* turns red, and if you click it, you get a message box that says **This variable or function is not defined above.**

Mathcad cannot assign a value to *VolBox* because no value has been assigned to *a* (or *b* or *c*). Now assign values to *a, b,* and *c* before making the *VolBox* assignment statement:

$$a := 3$$
$$b := 5$$
$$c := 5$$
$$\text{VolBox} := a \cdot b \cdot c$$

Now that *a, b,* and *c* are defined before the *VolBox* assignment, none of the variables turn red. To see the value of *VolBox*, type **VolBox =**, and **75** appears.

$$VolBox = 75$$

Elementary Calculations. This section introduces some features of Mathcad for simple calculations and iterative calculations. Mathcad has numerous built-in functions, and also it permits the user to utilize user-defined functions. Users will find it convenient to use lists in many calculations. Open a new worksheet, and try out some of Mathcad's features.

Range Variables and Iteration. In a new space on the notebook page, click the page to get a red-cross pointer. Type **Height:6,7;12.** Mathcad furnishes the equal sign after the colon, and the semicolon changes to two dots. The right side indicates that 6 is assigned to the variable *Height*, and then this value is increment by one until it reaches seven. The two dots are the Mathcad range variable operator. The range variable operator continues until the height reaches 12. *Height* is now a range variable.

Next, type **Width*Height=** and Mathcad instantly responds by showing a table of values. (**Width:= 3** was previously entered.) The table display option depends on your Mathcad original settings: from left to right matrix, table with column/row labels and table without column/row labels. To change the display option, click on a table (or matrix). On the **Format** menu, click **Result.** On the **Result Format** dialog box, select the **Matrix display style** box.

$$\mathrm{Width}\cdot\mathrm{Height} = \begin{pmatrix} 18 \\ 21 \\ 24 \\ 27 \\ 30 \\ 33 \\ 36 \end{pmatrix}$$

Width·Height =

	0
0	18
1	21
2	24
3	27
4	30
5	33
6	36

Width·Height =

18
21
24
27
30
33
36

For the left example below, click the **View** menu, click **Toolbars,** then click **Calculator,** and then click the square root icon. Inside the square root sign type **Width*Height=**.

The center example below was calculated by entering **Height/5*Width=**. The up and down arrows can also be used instead of the SPACEBAR to determine the level of the next operation.

The right example below was calculated by entering **Height/5[SPACEBAR]*Width =**.

$\sqrt{\mathrm{Width}\cdot\mathrm{Height}} =$

4.243
4.583
4.899
5.196
5.477
5.745
6

$\sqrt{\mathrm{Width}\cdot\mathrm{Height}} =$

4.243
4.583
4.899
5.196
5.477
5.745
6

$\sqrt{\mathrm{Width}\cdot\mathrm{Height}} =$

4.243
4.583
4.899
5.196
5.477
5.745
6

Mathcad's Built-in Functions. To access a large number of Mathcad functions, click the ***f(x)*** icon on the toolbar (Figure 19-1) or, on the **Insert** menu, click **Function** on the menu bar to open the **Insert Function** dialog box (Figure 19-2). Some common functions are found by, on the **View** menu, pointing to **Toolbars**, and clicking **Calculator** (Figure 19-1). To try out a Mathcad built-in function, let's calculate the cosine of 30 degrees. On the **View** menu, point to **Toolbars**, and click **Calculator**, and then click **cos**. This action gives:

$$\cos(\blacksquare)$$

The argument of the cosine function is the angle in radians. Consequently, we must convert 30 degrees to radians by multiplying by $\pi/180$. To get a Greek letter, enter its English equivalent, and then press CRTL+G. For example, pressing **p** then CTRL+G gives π. (Pressing CTRL+G again gets you out of Greek letter mode.) You can also get Greek letters by, on the **View** menu, pointing to **Toolbars**, and clicking **Greek**. This action opens the **Greek** toolbar with the entire Greek alphabet on display, useful when the English equivalent is not obvious.

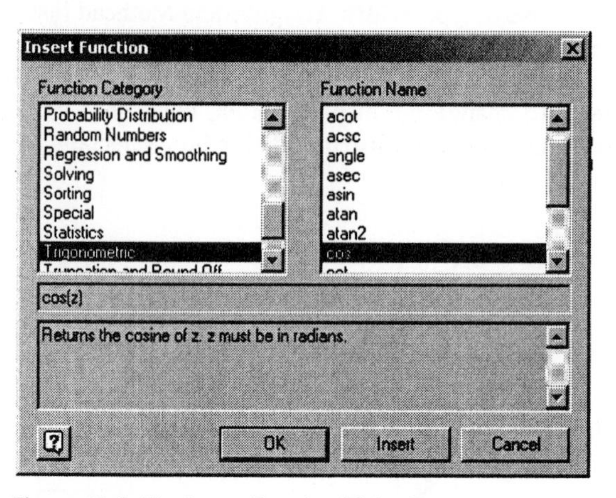

Figure 19-2. The Insert Function Dialog Box

In the cos function placeholder, (■), type **30*pCTRLg/180=**. The result is

$$\cos\left(30\cdot\frac{\pi}{180}\right) = 0.866025$$

Pressing CTRL+G toggles back and forth between English and Greek representations of the selected letter. Selecting each letter in the first line of assignments and toggling with CTRL+G results in the second line of assignments.

$$a := 1 \quad b := 2 \quad c := 3 \quad d := 4 \quad D := 5$$

$$\alpha := 1 \quad \beta := 2 \quad \chi := 3 \quad \delta := 4 \quad \Delta := 5$$

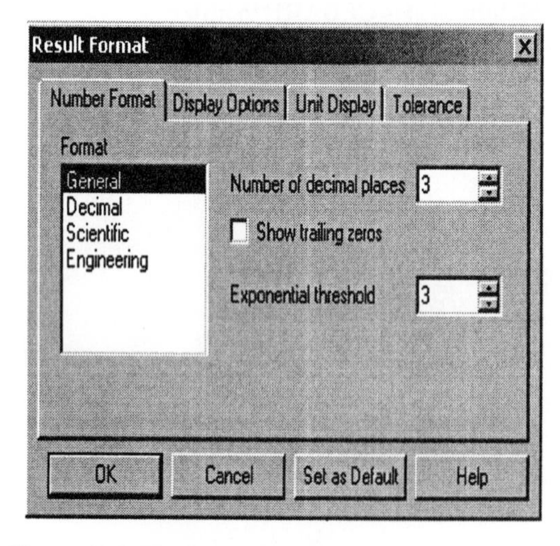

Figure 19-3. The Result Format Dialog Box

To get the desired number of decimal places, double-click the number, and in the **Result Format** dialog box (Figure 19-3), select the desired number of decimal places.

User-Defined Functions. With the usual algebraic notation, we would write, for example, a function like $f(x) = 3x^2 + 4$. To define this function with Mathcad's notation, type **f(x):** and Mathcad responds with **f(x) :=**. Continue by entering **3*x [SHIFT ^2]+4**. Try another on the same line. The result is as follows:

$$f(x) := 3\cdot x^2 + 4 \qquad g(y) := y^2 + 3$$

The name of the function is *f*, and the argument of the variable is *x*. The name of the second function is *g*, and the name of its argument is *y*. To evaluate the function with a particular value of the argument, rewrite the function definition with the particular value substituted for the argument followed by the equal sign:

$$f(5.5) = 94.75 \qquad g(5.5) = 33.25$$

To evaluate the function over a range of argument, define the range variable over the desired range. Then rewrite the function and substitute the range variable name for the argument as shown in the following examples:

$$i := -2,-1..5 \qquad x := -2,-1..5$$

f(i) =	g(i) =	g(x) =
16	7	7
7	4	4
4	3	3
7	4	4
16	7	7
31	12	12
52	19	19
79	28	28

Next, define a function of two variables.

$$V(r,h) := \pi \cdot r^2 h$$

To evaluate this function for *r* equals 3 and *h* equals 4, type **V(3,4)=**, and the result is:

$$V(3,4) = 113.097$$

Now do it again with different values for *r* and *h*.

$$V(4.5,7.33) = 466.314$$

To perform iteration with this function, change one of the variables (for example, *r*) to a range variable.

$$r := 1,1.1..2$$

The result is shown to the right.

V(r,4) =
12.566
15.205
18.096
21.237
24.63
28.274
32.17
36.317
40.715
45.365

Creating Lists of Raw Data: Subscripted Variable. In the laboratory, experiments are generally repeated, resulting in lists of data. With Mathcad, it is usually convenient to enter lists of data into a Mathcad vector, a list or one-dimensional array. As an example, suppose you have a list of six data. Type **i:1;6** (you could click the **m..n** icon on the **Matrix** toolbar instead of entering a semicolon). This entry creates a range variable.

$$i := 1,6..$$

Click the $\mathbf{X_n}$ icon on the **Matrix** toolbar, and two placeholders appear on the screen.

Type $\boldsymbol{x}$ in the upper left placeholder, and type **i** in the lower right placeholder. Then type a colon followed by the first number in the list, **3.4**.

$$x_i := 3.4$$

Next, type a comma after the **3.4**. The comma does not appear, but the box changes to:

$x_i :=$
3.4
■

Place the second number in the list in the placeholder below the **3.4**, and enter a comma after it. Placing the comma generates a third placeholder box, into which you can place the third item in the list. Continue until the sixth item has been placed in the list, but enter no comma after the last item in the list. The box stretches as each new item is added. After you add the last item, click outside the list. The outer line box containing the definition of the subscripted variable and the table of values disappears. Create a second list, y_i. Notice that the range variable i is followed by the assignment operator :=. The declarations of the subscripted variables x_i and y_i are also followed by the assignment operator.

Next, do some operations with the two lists x_i and y_i. Just enter the expression above each list delineated by boxes. The two columns on the left are lists of raw data created by the user. The three columns on the right represent calculations done with the raw data. Notice that to evaluate the indicated operation the expression is followed by an equal sign (not by the assignment operator).

Calculating with Lists: Simple Examples

Use the x_i and y_i lists for some sample calculations. The format chosen for the table is without columns/rows.

$i := 1..6$

$x_i :=$	$y_i :=$	$x_i \cdot y_i =$	$\sqrt{x_i} =$	$(x_i)^2 =$
3.4	11	37.4	1.844	11.56
4.5	22	99	2.121	20.25
6.7	33	221.1	2.588	44.89
8	44	352	2.828	64
5	55	275	2.236	25
6.5	66	429	2.55	42.25

The Mathcad worksheet data are shown above. Mathcad "remembers" or "knows" everything in the math regions above and to the left of the present position of the math pointer. For this reason, you will notice that the data presented on a Mathcad page tend to be slightly out of line and tilt down and to the right. Above, x_i and y_i appear on the same level, but you will notice that the next three evaluation equal signs lie a bit lower. Click the $x_i y_i$ to select this math region. Click the boundary edge to get the little black hand, and drag the region a little above the x_i to the left. The list of data disappears until you move the region back below the level of the x_i. In some cases, Mathcad's **Align** command, found on the **Format** menu, can be helpful.

Mathcad Worksheet 1
Calculating with Lists: X-Ray Powder Diffraction

The calculation of X-ray powder diffraction data provides an example of typical laboratory calculations. No graphing is involved, just repetitive calculations arranged in tabular form. Even if you are not familiar with the nature of X-ray diffraction calculations, doing this exercise illustrates some frequently encountered spreadsheet features.

Excel Example 1 in Chapter 1 provides a brief review of X-ray powder diffraction, and a brief summary is provided here. You can think of a cubic crystal as a solid bounded by six orthogonal planes separated by a distance a_0. Further, imagine all kinds of planes that could be drawn through the edges and sides of the cube (or unit cell). Some of these planes form families that are parallel to each other and are separated by a distance *d*. After a little consideration, you might notice that the spacing of the planes must be less than the spacing of the unit cell a_0. The families of planes are characterized by some integer *h, k,* and ℓ, called Miller indices.

Measurement of the diffraction angle θ from various families of planes permits calculation of the separation between planes d_{hkl} and the size of the unit cell itself, a_0. The relevant equations are Equation 1-1, Chapter 1, Excel Example 1

$$d_{hk\ell} = \frac{\lambda}{2\mathrm{Sin}\theta}$$

and Equation 1-2, Chapter 1, Excel Example 1

$$a_0 = d_{hk\ell}\sqrt{h^2 + k^2 + \ell^2}$$

The measurement of diffraction angles θ allows calculation of interplanar spacing d and the length of the cubic unit cell a.

- Enter a suitable title, and change its font to 14 point bold.
- Type **λ:1.5418** to declare the value of the wavelength λ.
- Type **i:0;5** to declare the range variable i.
- Define the subscripted variable θ_i and fill in the experimental values. Click the $\mathbf{X_n}$ icon on the **Matrix** toolbar to get two placeholders: ■▪. Fill in the upper placeholder with θ and the lower one with i. Type **21.811**, **25.368**, and so on.
- Change θ_i in degrees to Θ_i in radians. For the left side the equation, click the $\mathbf{X_n}$ icon on the **Matrix** toolbar, and fill in the upper placeholder with □ and the lower one with i. Then type a colon (**:**) and again click the $\mathbf{X_n}$ icon on the **Matrix** toolbar. Fill the upper placeholder with θ and the lower one with i. Then type ***π/180**. Several lines below this equation, enter **□** $\mathbf{\square_i}$ **=** to display a vertical vector filled with the diffraction angles in radians.
- In the same way, calculate d_i from λ and Θ_i and display with $\mathbf{d_i}$ **=**.
- In the same way that you defined the subscripted variable θ_I, define the subscripted variables h_i, k_i, and ℓ_i, and fill them in.
- Calculate a_i from d_i and their Miller indices h_i, k_i, and ℓ_i. Use the $\mathbf{X_n}$ icon on the **Matrix** toolbar to enter $\mathbf{a_i := d_i *}$ and then click the $\sqrt{\ }$ icon on the **Calculator** toolbar. Enter $\mathbf{(h_i)\hat{}2 +}$ and so on. The square root symbol expands as more terms are added.
- Display a_i with $\mathbf{a_i}$ **=**.
- Calculate the mean of a_i with *mean*(a). The name of the vector whose elements are a_i is a.

Figure 19-4 shows the Mathcad worksheet for the X-ray diffraction calculation. Notice how every variable that is calculated from a previously defined variable is calculated below and to the right of it. If you don't do this, Mathcad will give an "undefined variable" error message at the point of infraction. Just move your entry a little lower and a little more to the right. The subscripted variables h_i, k_i, and ℓ_i are allowed high up on the page because they use no previously defined variable. Notice also that Mathcad distinguishes between uppercase and lowercase. Here, θ_i and Θ_i are two different variable names. So do *theta, Theta, THETA,* and *theTa* each represent a different variable. Trigonometric functions must be in lowercase, for example, *sin* (θ). The *mean* function must be in lowercase.

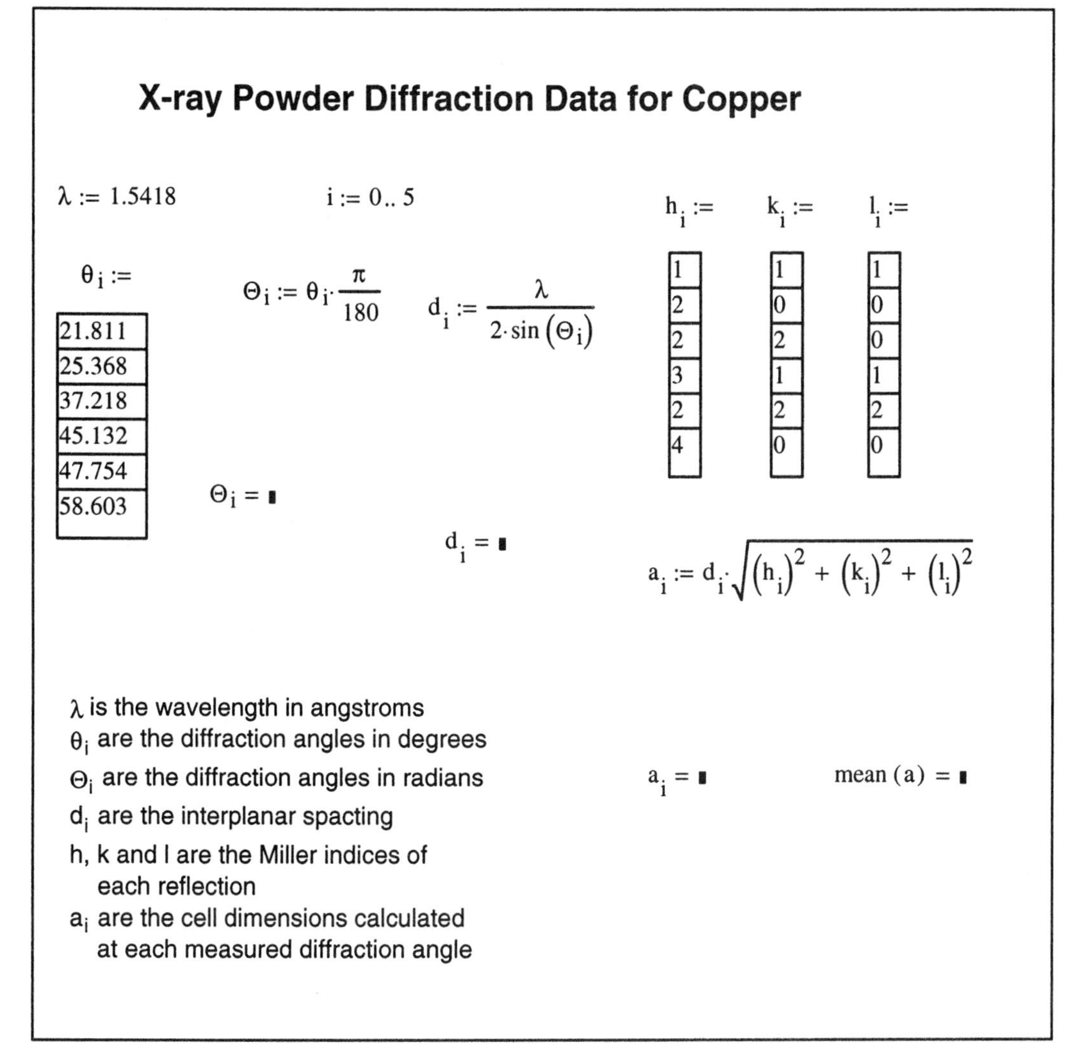

Figure 19-4. Mathcad Worksheet 1 for X-Ray Diffraction

Whenever you use a vector (list) or array, be sure that every element contains a datum. Mathcad assigns 0 to any element to which a value has not been assigned. For this reason, the range variable i is assigned the range from 0..5 and not zero to six. This is important here because the *mean(A)* function returns the arithmetic mean of the element of *array A* (or *vector A*). We defined a_i as a subscripted variable, but Mathcad is perfectly willing to consider it to be an array named a with elements a_i. Mathcad won't accept a_i as an argument for *mean(A),* but it will accept a, the name of a vector with elements a_i. If the range 1..5 were chosen, Mathcad would assign the value 0 to a_0, and *mean(a)* would find the mean of the five calculated values plus zero, which would be wrong. In general it is probably good practice with Mathcad to begin a range with zero, not one.

Two-Dimensional Plots

Two-dimensional plots are simple to create with Mathcad. More than one plot can be placed on the same grid, and more than one graph can be place on a single worksheet.

The vapor pressures of benzene and chloroform, plotted in Excel Example 5, Chapter 2, provide a comparison with the same plot created with Mathcad.

Mathcad Worksheet 2
Plotting with Lists: The Vapor Pressures of Benzene and Chloroform

Plotting a Table of *x,y* Points. In presenting graphical data to your reader, you sometimes face the choice of placing two plots on one graphical figure or presenting two separate plots. Two related plots on the same figure often can make vivid and easy comparisons. If the ranges of the variable in the two plots are reasonably comparable, it is usually feasible to combine the two plots on one graph. While this double plot of points is straightforward in Mathcad, it requires some not-so-straightforward tricks to accomplish with a spreadsheet, as demonstrated in Chapter 2, Excel Example 5. The data for Mathcad Worksheet 2 is from Excel Example 5.

Entering the Data for the Vapor Pressure Plots

- Enter the range variables shown at the top of Figure 19-5 by typing **i:1;7**. The screen will display **i := 1..7**. Then press the TAB key twice, and finally enter **j:1;11**.
- Alternatively, you could click the **m..n** icon in the **Matrix** toolbar (Figure 19-1) instead of typing a semicolon.
- To define the subscripted variable Pb_i, click the **Matrix** toolbar, and then click $\mathbf{X_n}$; a pair of placeholders, $\blacksquare_{\blacksquare}$, appear on the screen.
- In the first placeholder, type **Pb**. Use the right arrow key ($\longrightarrow$) to move the pointer to the second placeholder, and type **i**. Again, use the right arrow key ($\longrightarrow$) to move the pointer to the right , and type a colon (:). A placeholder appears.
- Type the first value of Pb_i from Figure 19-5 into this placeholder, and then type a comma (,). Next, enter the remaining Pb_i values, each separated by a comma.
- Repeat this procedure for tb_i, Pc_j, and tc_j. to create four tables of temperatures and vapor pressure.

Creating Two Vapor Pressure Plots on the Same Graph

- To graph the data in the two vapor pressure data sets, on the **Insert** menu, point to **Graph**, and click **X-Y Plot**, or click the **X-Y Plot** icon on the **Graph** toolbar.
- In the placeholder at the middle-left side of the blank square graph space, click the $\mathbf{X_n}$ icon of the **Matrix** toolbar, and type $\mathbf{Pb_i}$ followed by a comma. The comma creates another placeholder. Click the $\mathbf{X_n}$ icon of the **Matrix** toolbar, and type $\mathbf{Pc_j}$.
- In the placeholder at the middle-bottom side of the blank square graph space, click the $\mathbf{X_n}$ icon of the **Matrix** toolbar, and type $\mathbf{tb_i}$ followed by a comma. Then click the $\mathbf{X_n}$ icon of the **Matrix** toolbar, and type $\mathbf{tc_j}$.
- The graph should look approximately like the one in the bottom left of Figure 19-5.

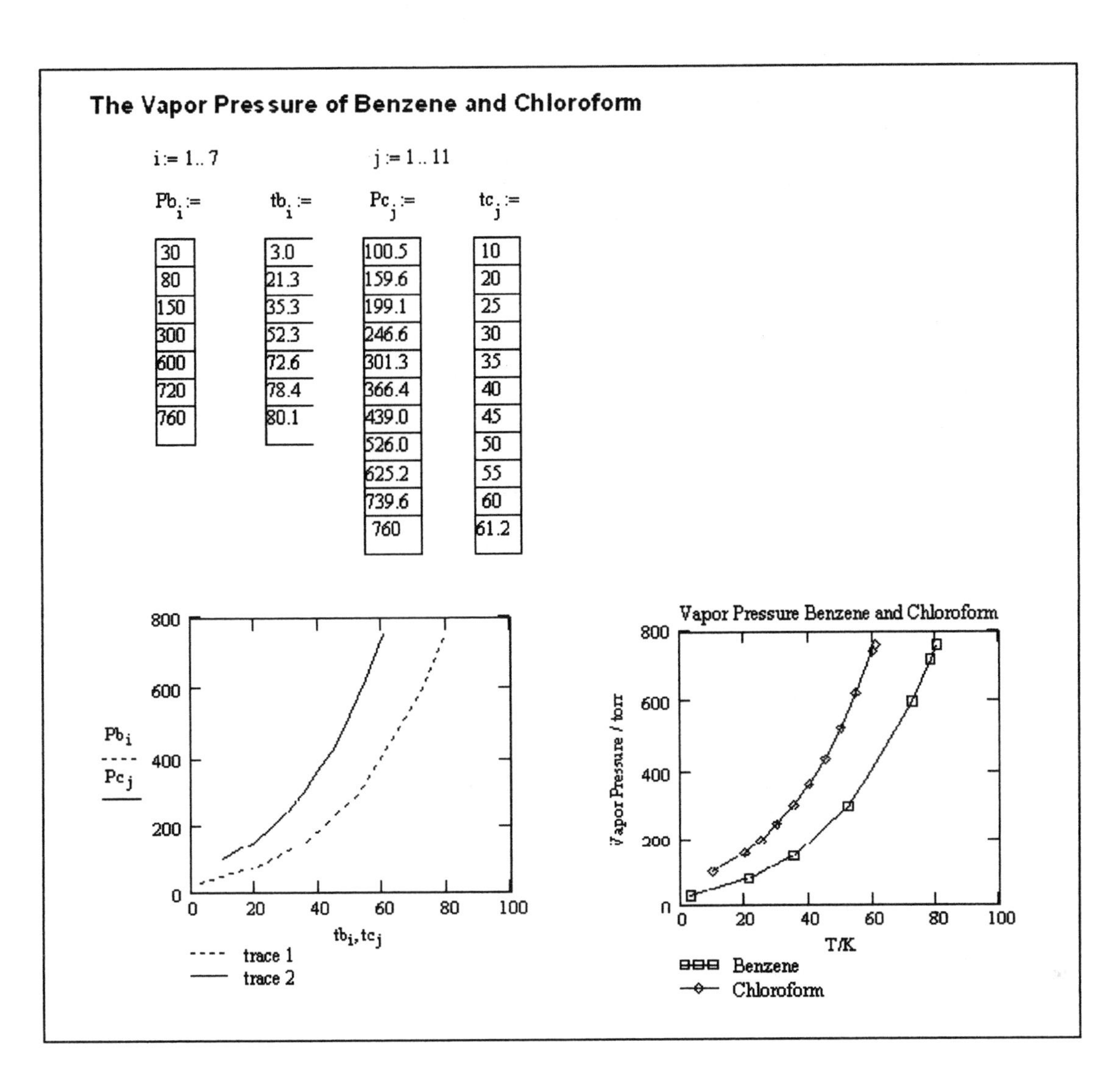

Figure 19-5. Mathcad Worksheet 2 for Vapor Pressure of Benzene and Chloroform

Editing the Vapor Pressure Graph

- Click anywhere inside the graph area, and then, on the **Format** menu, click **Graph** and select **X-Y Plot**. The **Formatting Currently Selected X-Y Plot** dialog box appears. Click the **Trace** tab to edit the two plots (Figure 19-6). In this dialog box, the various plots on the graph are called traces.
- Click **Trace 1**.
  1. Change **Trace 1** to **Benzene**.
  2. Change **Symbol** from **none** to **box**.
  3. Change the line from dotted to solid, and change the color from red to black.
- Click **Trace 2**.
  1. Change **Trace 2** to **Chloroform**.
  2. Change **Symbol** from **none** to **Diamond**.
  3. Change the line from dotted to solid, and change the color from blue to black.

For the final editing, we will add titles and axis labels and hide the legends.

- Click anywhere on the graph.
- On the **Format** menu, click **Graph**, and select **X-Y Plot.**
- Click the **Traces** tab.
  1. Select the **Hide Arguments** check box.
  2. Clear the **Hide Legend** check box.
- Click the **Labels** tab.
  1. In the **Title** box, type **Vapor Pressure of Benzene and Chloroform** as the title, and select the **Show Title** check box.
  2. In the X-Axis box, type **T/K** as the label.
  3. In the Y-Axis box, type **Vapor Pressure/torr** as the label.
- The graph should now look like the one on the bottom right of Figure 19-5.

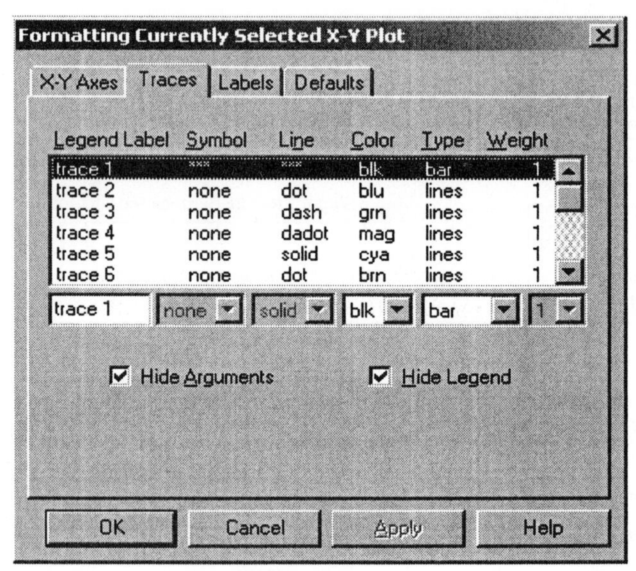

Figure 19-6. The **Formatting Currently Selected X-Y Plot** Dialog Box

Mathcad Worksheet 3
Calculating with Subscripted Variables: A Linear Regression Line

In experimental work, scientists and engineers frequently make repeated measurements of some dependent variable, say y, as the value of some independent variable, say x, is varied. Often, the variable y depends in a linear manner on the variable x. A plot of y versus x reveals such linearity as a straight line, the slope and intercept of which often permit calculation of important properties.

The thermodynamics of equilibria provides such an example. Here we will treat the data for the equilibrium between solid silver, gaseous oxygen, and solid silver oxide.

$$2Ag(s) + \frac{1}{2} O_2(g) = Ag_2O(s)$$

$$K = P_{O_2}^{-1/2} \qquad (19\text{-}1)$$

Table 19-1

Experimental Data for the Silver Oxygen Equilibrium

$K_{eq} = P_{O_2}^{-1/2}$

T/K	K_{eq}
423.0	2.04
449.0	1.34
456.3	1.12
461.4	1.03
464.4	0.98
473.0	0.85

The experimental data consists of the equilibrium constants K (the dependent variable) and the temperatures T (the independent variable) in Kelvin, as displayed in Table 19-1.

The Linear Regression Calculation and Plot with Subscripted Variables. It is convenient to supply Mathcad with data as a list of subscripted variables, which is really a vector or one-dimensional array. Try the following steps to duplicate the math regions of Figure 19-7. The number of elements to fill into the subscripted variable is defined above the definition of the subscripted variable.

Entering the Data

- Type **i:0;5** and the screen shows **i:=0..5**. Alternatively, you could click the appropriate icons in the **Calculator** toolbar (Figure 19-1).
- To define the subscripted variable T_i, click the **Matrix** toolbar, then click $\mathbf{X_n}$, and what appears on the screen is a pair of placeholders: $\blacksquare_{\blacksquare}$. In the first placeholder, type **T**. Use the right arrow key ($\rightarrow$) to move the pointer to the second placeholder, type **i**, then move the pointer to the right again, and type a colon (**:**). Another placeholder appears. Type the first value of T_i from Table 19-1 into this placeholder; and then type a comma (**,**) . Next, type the remaining T_i values, each separated by a comma. Repeat this procedure for K_i.
- In the same way, on a line a bit lower than the K_i entry, type $\mathbf{y_i{:}ln(K_i)}$.
- To display the list of $\ln(K_i)$, type $\mathbf{y_i{=}}$.
- In the same way, on a line a bit lower than the T_i entry, type $\mathbf{x_i{:}1/T_i}$.
- To display the list of reciprocal temperatures, type $\mathbf{x_i{=}}$.

The Mathcad worksheet now contains the four tables shown in the upper half of Figure 19-7. Now you can get the linear regression values for the slope and intercept.

Calculating the Slope and Intercept of the Linear Regression Line. It is not necessary to plot the data to get the slope and intercept of the linear regression line. Mathcad provides several regression functions.

Table 19-2

Mathcad Regression Functions

Regression Function	Description
Intercept(**vx, vy**)	Returns the y-intercept (a scalar) of the linear regression line
Slop (**vx, vy**)	Returns the slope (a scalar) of the linear regression line
Line(**vx, vy**)	Returns the slope and intercept of the linear regression line
Stderr(**vx, vy**)	Returns the standard error (a list) of the elements **vx** on the elements of **vy** of the linear regression line

Note: **vx** and **vx** are the vector representations of the lists or data. Vectors will be treated in more detail in the next section.

In this exercise we will use the slope and intercept functions, not only because we are interested in their values but also because they are necessary for plotting the linear regression line.

- In a space below the table of T_i values, type **m:slope(x,y)**.
- On the same line, but a tiny bit lower, type **m=**.

- Below these two entries, type **b:intercept(x,y)**.
- On the same line, but a tiny bit lower, type **b=**.

The slope and intercept of the linear regression line have now been calculated. Next we will plot the data and regression line.

Plotting the Experimental Data and the Regression Line on the Same Graph. We have previously seen how to plot data consisting of two lists (parallel arrays) when we plotted the vapor pressures of benzene and chloroform. Again, we must make two plots on the same graph, but the plots are quite different. First, we plot the two parallel arrays y_i and x_i. This is a plot of the experimental ln*K* and experimental *1/T*. We edit this plot so that the graph displays the data pair markers, but we hide the line joining the points. Then we plot the calculated linear regression line and edit it so that we display the line but hide the calculated data pair markers.

To Graph the Data Markers Representing the Two Lists of x_i and y_i Values

- On the **Insert** menu, click **Graph**, and select **X-Y Plot**.
- In the graph's placeholder at the middle-bottom side of the blank square graph space, enter $\mathbf{x}_i$. Do this by clicking the $\mathbf{X_n}$ icon in the **Matrix** toolbar (Figure 19-1) and filling in its placeholders with an **x** and an **i**.
- In the graph's placeholder at the middle-left side of the blank square graph space, enter $\mathbf{y}_i$ as above. These actions result in a line graph of the experimental data, but with no markers displayed.
- To edit this first graph, select the graph, and then, on the **Format** menu, click **Graph**, and select **X-Y Plot** (or double-click the graph).
  1. Click the **Traces** tab.
  2. Select **Trace 1**.
  3. Change **Symbols** from **none** to **box**.
  4. Change **Type** from **lines** to **points**.
  5. Change **Label** from **Trace 1** to **Expt**.

To Graph the Linear Regression Line

- Click the y_i in the graph's middle-left placeholder, and adjust the edit lines to include the entire x_i.
- Type a comma; this entry generates a new placeholder.
- In this placeholder, type $\mathbf{m*x_i +b}$.
- Click anywhere outside the graph area, and the linear regression line appears. This graph is a plot of the y_{calc} points versus x_i. The y_{calc} points are given by the $mx_i + b$ values at each x_i.
- To edit this second graph, select the graph and then on the **Format** menu, click **Graph**, and select **X-Y Plot**.
  1 Select the **Traces** tab.
  2 Change **Legend** from **Trace 2** to **Regression**.
  3 Change the line from dotted to solid.
  4 Select both the **Hide Arguments** and **Hide Legend** check boxes.
  5 Click the **Labels** tab, enter **The Silver Oxide Equilibrium** as the title, and select the **Show Titles** check box.
  6 For the axis labels, type **(1/T)/(1/T)** for the x-axis and **lnK** for the y-axis.
  7 Select both the **X-Axis** and **Y-Axis** check boxes.

The graph now appears as it does in Figure 19-7.

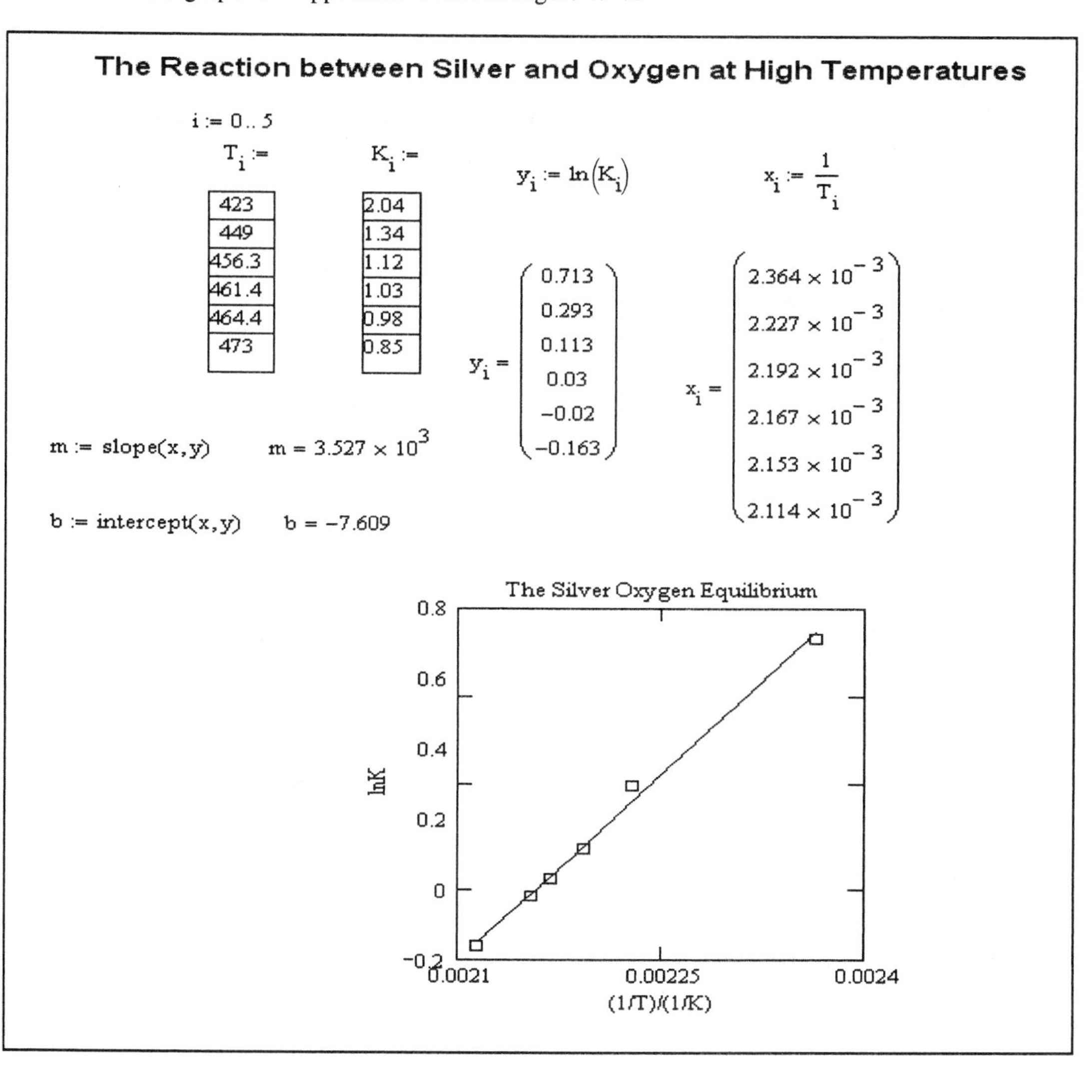

Figure 19-7. Mathcad Worksheet 3 for a Linear Regression Line Calculated with Subscripted Variables

The Linear Regression Calculation and Plot with Vectors

Vector representation of data is straightforward with Mathcad. With experimental data consisting of a set of dependent and independent variables, it is quite natural to use parallel arrays (parallel vectors) for the calculations.

Defining Vectors. Instead of subscripted variables (for example, x_i and y_i), vectors may be used for ordinary data manipulation. Two vectors A and B are equivalent to two subscripted variables or two parallel arrays. A vector is defined in Mathcad by clicking the **Matrix** icon (shown to the right) on the **Matrix** toolbar (Figure 19-1).

To define a vector named *A*, type **A:**, which results in *A* :=. Then click the **Matrix** icon (shown above) to open the **Insert Matrix** dialog box (Figure 19-8). Set **Rows** to **5** and **Columns** to **1**, and click **OK** or **Insert**. This results in a column vector with five empty placeholders.

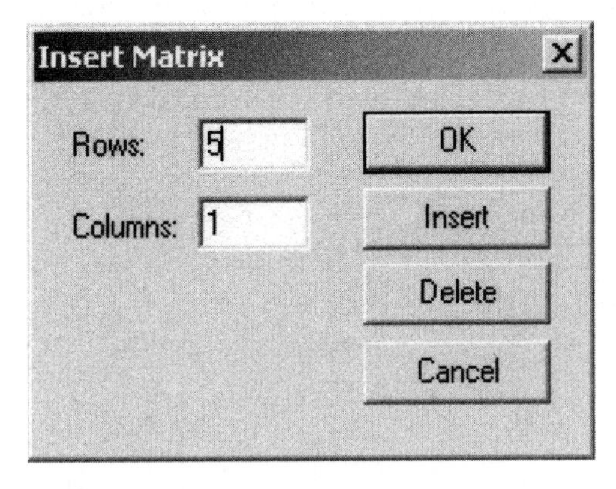

Figure 19-8. The Insert Matrix Dialog Box

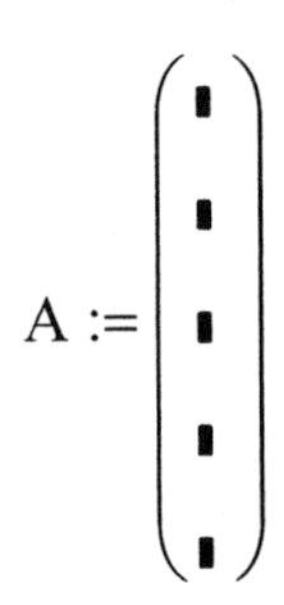

Click the uppermost placeholder and type **2**. Then press the TAB key and type the next number. Continue until the five placeholders are filled. Similarly, create a vector named *B*.

$$A := \begin{pmatrix} 2 \\ 4 \\ 6 \\ 8 \\ 10 \end{pmatrix} \quad B := \begin{pmatrix} 1 \\ 3 \\ 5 \\ 7 \\ 9 \end{pmatrix}$$

Arithmetic Operators and Vectors. Most of the arithmetic operators and functions in the **Calculator** toolbar (Figure 19-1) operate on each element of the vector, as shown in the following examples:

$$5 \cdot A = \begin{pmatrix} 10 \\ 20 \\ 30 \\ 40 \\ 50 \end{pmatrix} \quad \frac{1}{B} = \begin{pmatrix} 1 \\ 0.333 \\ 0.2 \\ 0.143 \\ 0.111 \end{pmatrix} \quad A^2 = \begin{pmatrix} 4 \\ 16 \\ 36 \\ 64 \\ 100 \end{pmatrix} \quad \sqrt{B} = \begin{pmatrix} 1 \\ 1.732 \\ 2.236 \\ 2.646 \\ 3 \end{pmatrix} \quad A + B = \begin{pmatrix} 3 \\ 7 \\ 11 \\ 15 \\ 19 \end{pmatrix}$$

However, multiplication results in the dot (scalar) product. The scalar value of the dot product is evaluated as $\sum a_i b_i$.

$$A \cdot B = 190$$

The Vectorize Operator. It is frequently necessary to multiply each element in a vector A by the corresponding element in another vector B, resulting in a new vector whose elements are $a_i b_i$. Mathcad's vectorize operator allows such an operation. The vectorize operator is displayed as an arrow over the vectors. It is accessible on the **Matrix** toolbar as:

$$\overrightarrow{f(M)}$$

To multiply each element of the vector A by each corresponding element of the vector B:

- Type **A*B**.
- Mathcad responds with $A \cdot B$.
- With the editing line enclosing both A and B, click the vectorize operator on the **Matrix** toolbar.
- Mathcad responds with $\overrightarrow{(A \cdot B)}$
- Type an equal sign (=) and Mathcad responds with:

$$\overrightarrow{(A \cdot B)} = \begin{pmatrix} 2 \\ 12 \\ 30 \\ 56 \\ 90 \end{pmatrix}$$

The vectorize operator is frequently useful when calculating with parallel arrays.

Mathcad Worksheet 4
Calculating with Vectors: A Linear Regression Line

The next example illustrates the use of vectors in data deduction and graphing. The vectorize operator will play a typical role in this calculation. For the data, we will use the heat capacity of silver at very low temperatures, below 5 K (Table 19-3). The theory of the heat capacity of metals at low temperatures fits an equation of the form

$$\frac{C_V}{T} = \alpha T^2 + \gamma \qquad (19\text{-}2)$$

When $\frac{C_V}{T}$ is plotted against T^2, the slope of the line equals α, and the intercept equals γ.

Table 19-3

The Heat Capacity of Silver below 5 K

C_V (J/K mol)	T/K
0.00106	1.35
0.00262	2
0.00657	3
0.01270	4
0.02130	5

Assigning the Data to Vectors. To use the data in Table 19-3, vectors T and C_V are defined.

- Type **T:** and then press CTRL+M or click the **Matrix** icon on the **Matrix** toolbar (Figure 19-1) to open the **Insert Matrix** dialog box (Figure 19-8).
- Set **Rows** to **5** and **Columns** to **1**; click **OK.**
- Type the five temperatures from Table 19-3 in the empty placeholders.
- In the same way, create a vector named C_V.

From the T and C_V vectors just defined (shown at the top of Figure 19-9), we need to define two vectors that hold the values of T_i^2 and $\frac{C_{V\,i}}{T_i}$. First define a vector R (for *reciprocal*) that contains the values $\frac{1}{T_i}$

- Type **R: 1/T** to define the vector R.
- Type **R=** to see the elements of R. This step is optional.
- Type **X:TSHIFT^2** to define a vector containing the squares of T.
- Type **X=** to see the elements of X. This step is optional.
- Type **Y:Cv*RSPACEBAR**. The SPACEBAR puts the editing lines under *Cv*R*.
- While *Cv*R|* is underlined, click the **Matrix** toolbar, and then click the vectorize icon: $\overrightarrow{f(M)}$
- Type **Y=** to see the elements of Y. This step is optional.

Calculating the Slope and Intercept of the Linear Regression Line

- In a space below the definitions of the vectors X and Y, type **m:slope(X,Y)**.
- On the same line, type **b:intercept(X,Y)**.
- Below these two entries, type **b=** and to its right type **m=**.
- Below these four entries, type **R2:corr(X,Y)**.
- Below this, type **R^2 is a measure of the "goodness of fit."**

The slope and intercept of the linear regression line have now been calculated. Next we will plot the data and regression line.

Plotting the Experimental Data and the Regression Line on the Same Graph. Creating these plots on the same graph is similar to the preceding steps used with subscripted variables. The vector *X* contains the values of T_i^2 , and the vector *Y* contains the values of Cv_i/T_i.

Graphing the Data Markers Representing the Two Vectors *X* and *Y*

- On the **Insert** menu, click **Graph**, and select **X-Y Plot**.
- In the graph's placeholder at the middle-bottom side of the blank square graph space, type **X**.
- In the graph's placeholder at the middle-left side of the blank square graph space, type **Y**. These actions result in a line graph of the experimental data, but with no markers displayed.
- To edit this first graph, select the graph and then, on the **Format** menu, click **Graph**, and select **X-Y Plot**.
  1. Click the **Traces** tab.
  2. Select **Trace 1**.
  3. Change **Symbols** from **none** to **box**.
  4. Change **Type** from **lines** to **points**.
  5. Change **Label** from **Tracel** to **Expt**.

Graphing the Linear Regression Line

- Click the *Y* in the graph's middle-left placeholder.
- Type a comma (,); this entry generates a new placeholder.
- In this placeholder, type **m*X +b**.
- Click anywhere outside the graph area, and the linear regression line appears. It is a plot of the values of *Cv/T* calculated by linear regression versus the experimental value of T^2.
- To edit this second graph, select the graph, and then, on the **Format** menu, click **Graph**, and select **X-Y Plot**.
  1. Select the **Traces** tab.
  2. Change **Legendl** from **Trace 2** to **Regression**.
  3. Change the lines from dotted to solid.
  4. Select both the **Hide Arguments** and **Hide Legend** check boxes.
  5. Click the **Labels** tab, and type **The Low Temperature Cv of Silver** as the title, and select the **Show Titles** check box.
  6. For the axis labels, type **T^2** for the x-axis and **Cv/T** for the y-axis.
  7. Select both the **X-Axis** and **Y-Axis** check boxes. The graph now appears as it does in Figure 19-9.

Plotting Functions

In the next examples we'll look at how Mathcad handles some familiar plots. In a previous section we learned about built-in functions and how to define our own functions. The plot function previously used to plot lists also applies to plotting functions.

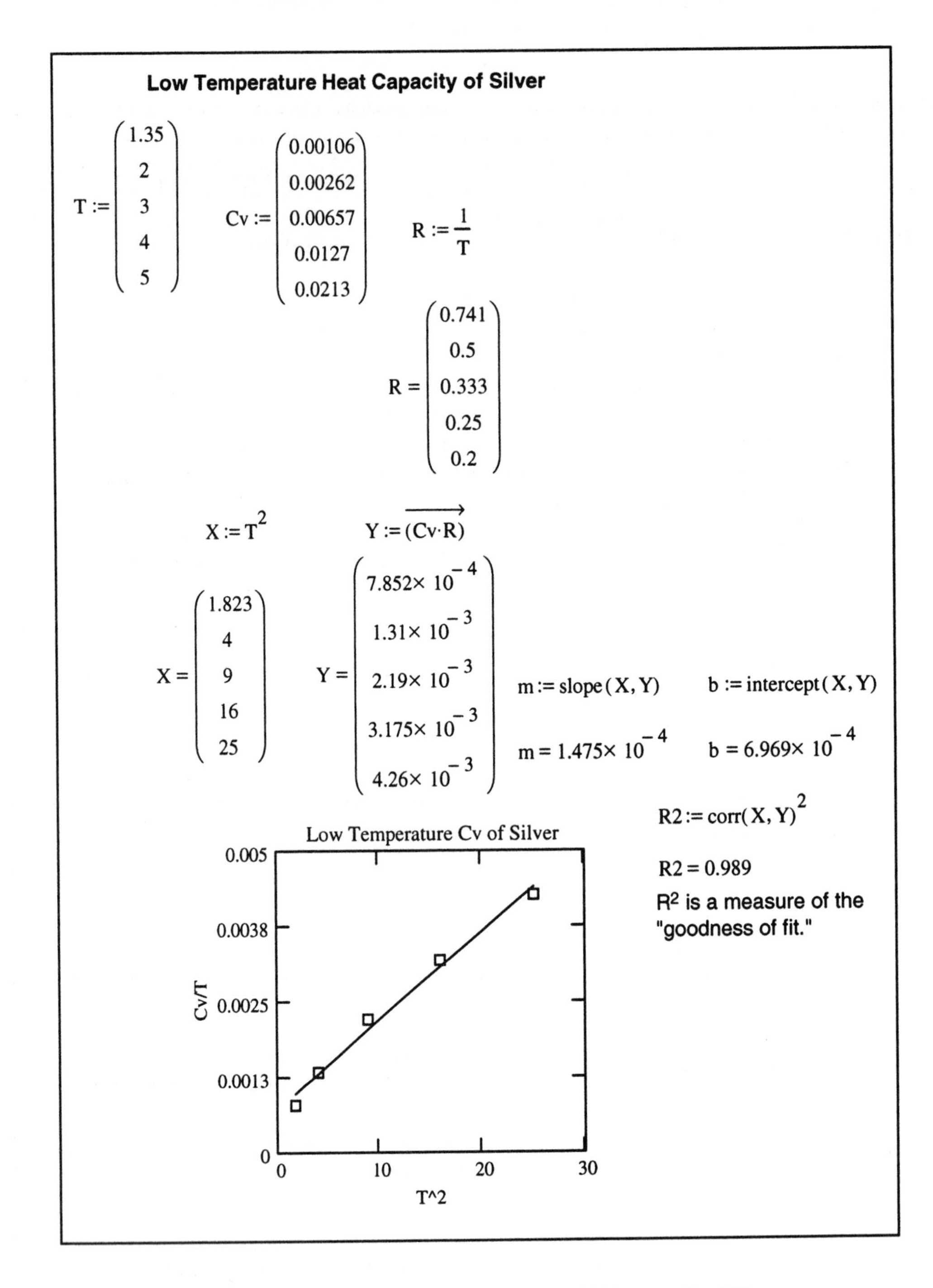

Figure 19-9. Mathcad Worksheet 4 for the Linear Regression with Vectors: *Cv* of Silver

Mathcad Worksheet 5
Plotting Two Functions

In Mathcad Worksheet 2, we learned how to graph two point plots on the same graph. In Worksheet 5, we will learn some ways to graph two functions on the same graph.

Figure 19-10 shows two graphs of plots of two functions on one graph. In one graph, the two functions have two different independent variables, *x* and *y*. In the second graph, the two function have the same independent variable. Use Figure 19-10 as a guide for placing the range and variable definition on the worksheet.

- Define the range variables *x* and *y*.
- Define the two functions *f(x)* and *g(y)* .
- On the **Insert** menu, click **Graph**, and select **X-Y Plot**. In the placeholder at the middle-bottom side of the blank square graph space, type **x,y**.
- In the placeholder at the middle-left side of the blank square graph space, type **f(x),g(y)**.
- Click anywhere outside the graph area, and the first graph in Figure 19-10 appears.
- Double-click anywhere inside the graph area to open the **Formatting Currently Selected X-Y Plot** dialog box (Figure 19-6).
- On the **Format** menu, click **Graph**. Select **X-Y Plot**, and click the **Traces** tab. Format the graph as you wish.

In Figure 19-10, the **Color** of **Trace 1** and **Trace 2** was changed to black. The **Weight** of **Trace 1** was changed to **2**, and the **Weight** of **Trace 2** was changed to **3**. In this same manner, create plots of *p(z)* and *q(z)*.

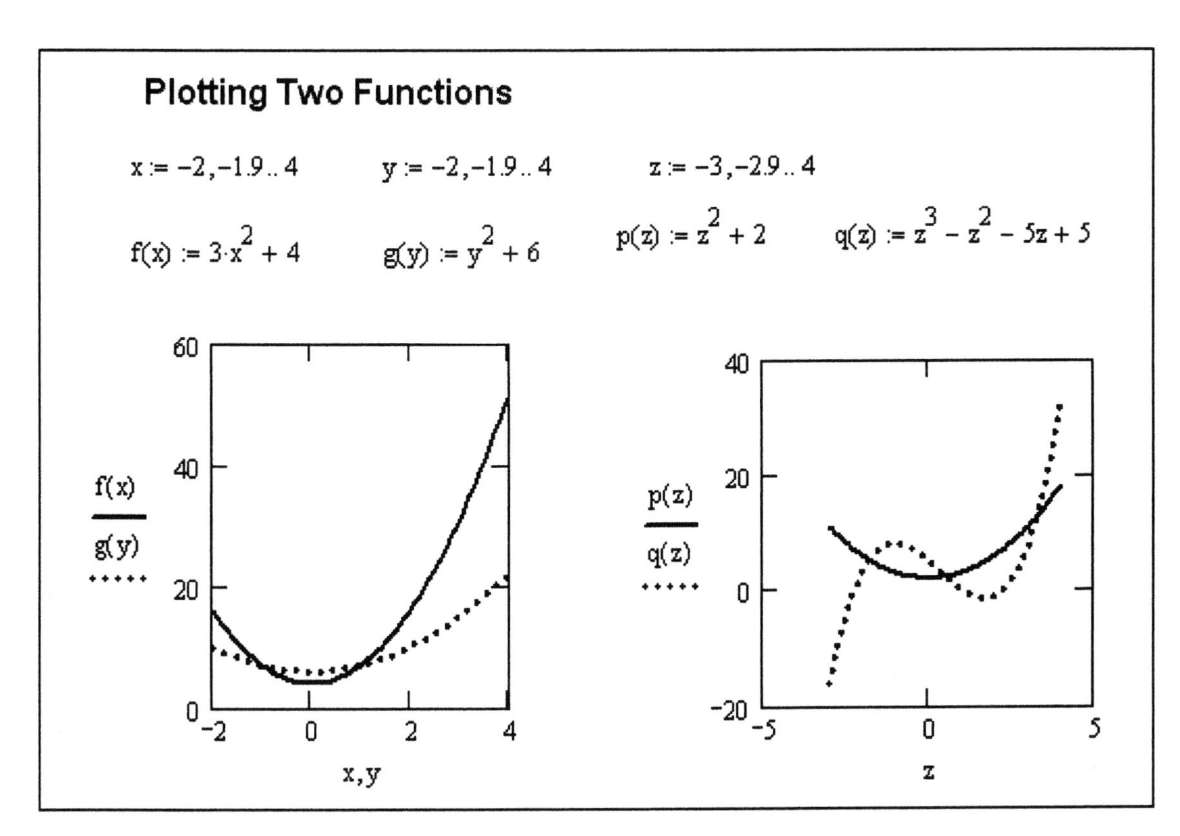

Figure 19-10. Mathcad Worksheet 5: Plotting Two Functions on One Graph

Mathcad Worksheet 6
Plotting a One-Dimensional Velocity Distribution Function for N_2

Since gases are isotropic, that is, they behave the same properties in all directions, it follows that their velocity distribution in one dimension should reflect their behavior in three dimensions. Because velocity has magnitude and direction, one would expect that for every velocity vector in one direction, there is another identical vector in the opposite direction, so that the average of the pair, and all other pairs, is zero. A plot of the one-dimensional velocity distribution should reveal these considerations. The kinetic theory of velocity and speed distribution functions is reviewed in Excel Example 18, Chapter 5.

The Maxwell-Boltzmann distribution of molecular velocities is given by Equation 5-11, Excel Example 18, Chapter 5.

$$f(u) = \sqrt{\frac{M}{2\pi RT}} e^{\frac{-Mu^2}{2RT}}$$

The left side of the equation represents $\frac{dN/N}{du_x}$, where dN/N is the fraction of molecules having velocities between u_x and du_x. The plotting procedure is straightforward:

- Define the variables M, R, T_L, and T_H, and document their units. To enter T_L, type **T.L**; everything after the period is subscripted.
- Type **u:= -1500,-1450..1500** to define the range variable.
- Define the functions $f(u_x)$ and $g(u_x)$ just as written above, but at T_L and T_H.
- Plot the function (Figure 19-11):
  1. On the **Insert** menu, click **Graph**, and select **X-Y Plot.**
  2. Type **f(u.x),g(u.x)** in the left placeholder.
  3. Type **u.x** in the bottom placeholder.

The graph may be edited by, on the **Format** menu, clicking **Graph** or by right-clicking the graph and then clicking **Format**.

- Click the **Labels** tab.
  1. In the **Title** box, type **One-dimensional Velocity Distribution** as the title.
  2. For the x-axis, type **Velocity /(m/s)** as the label
  3. For the y-axis, type **Probability (dN/N)/du** as the label.
- Click the **Traces** tab.
  1. Change the **Trace 1** legend to **300K.**
     a. Set **Symbol** to **none.**
     b. Set **Line** to **Solid.**
     c. Set **Color** to **Black.**
     d. Set **Type** to **line.**
     e. Set **Weight** to **2.**
  2. Change the **Trace2** legend to **1500 K.**
     a. Set **Symbol, Color, Type,** and **Weight** as above.
     b. Set **Line** to **Dashed.**
- Select the **Hide Arguments** check box.

Notice the use of subscripts in both the text regions and the math regions of the Mathcad workbook. To write u_x in the text region, type **ux** and select the **x** by dragging the pointer over it. Then, on the **Format** menu, click **Text**, and select the subscript check box. A subscript used as part of the name of a variable is called a literal subscript. To write u_x in a variable name, type **u.x** and Mathcad changes the entry to u_x, where the $_x$ is a literal subscript.

Mathcad doesn't work out a reasonable range of the variable u over which to make the plot. You have to work that out in advance. A range of u from –1500 to +1500 is about right in the present example. Usually 25 to 100 points is sufficient, so an increment value of 50 was chosen. This choice gives 3000/50 equals 60 points. The value of the increment is done indirectly. In the range definition, the second term is 50 less that the first term: u:= –1500, –1450..1500. Mathcad assumes that you want to keep up this incrementation all the way to 1500 and does so. The default increment value is 1 if not otherwise specified. The value can be less than 1 as shown in the next example. You can experiment with the various **Format** menu commands as previously described.

Mathcad Worksheet 7
Plotting a Three-Dimensional Speed Distribution Function for N_2

The three-dimensional speed function, previously given by Equation 5-16, is:

$$\frac{dN/N}{du} = 4\pi\left(\frac{M}{2\pi RT}\right)^{3/2} u^2 e^{\frac{-Mu^2}{2RT}}$$

Plotting the three-dimensional speed distribution function is similar to plotting the one-dimensional velocity distribution function as seen by comparing Figure 19-11 and Figure 19-12.

Mathcad Worksheet 8
Multiple Plots: Three Atomic Wave Functions on One Graph

In Mathcad Worksheets 8 and 9, we'll plot some wave functions for the hydrogen atom. In Chapter 3, Excel Example 7 includes a short review of wave functions and the quantum mechanical treatment of the hydrogen atom.

In this Mathcad worksheet, we'll look at R, the radial part of the wave function for the ground state of the hydrogen atom, and prepare plots of R_{1s}, R_{1s}^2, and $r^2R_{1s}^2$ for the hydrogen atom. The radial wave function for a hydrogen atom in the $1s$ state (n = 0 and $\ell = 0$) is given by

$$R_{1s} = 2a_0^{-3/2}e^{-r/a_0} \tag{19-3}$$

If we let $x = r/a_0$ and plot in units of $a_0^{-3/2}$, then the function to be plotted becomes

$$R_{1s} = 2e^{-x} \tag{19-4}$$

A plot of this function gives us the shape of the $1s$ wave function or of the $1s$ orbital. It is also of interest to plot R_{1s}^2, the probability density. It is a measure of the probability per unit volume (unit length in our one-dimensional case) of finding the 1*s* electron at some particular *r*. Finally, $r^2 R_{1s}^2$ is the probability of finding the electron between *r* and *r* + *dr*, regardless of direction. With the same substitution ($x = r / a_0$), these functions become:

$$R_{1s}^2 = 4e^{-2x} \tag{19-5}$$

and

$$r^2 R_{1s}^2 = 4x^2 e^{-2x} \tag{19-6}$$

Plots of One-Dimensional Velocity Distribution Functions for Nitrogen at 298.15 K and 1500 K

$M := 0.028$ Molecular weight of Nitrogen (kg/mol)

$R := 8.314$ Gas constant (J/K/mol)

u_x is the velocity in the x direction (m/s)

$f(u_x)$ and $g(u_x)$ are the probability density dN/Ndu

$$u_x := -1500, -1450 .. 2000$$

$T_L := 298$ $\qquad$ $T_H := 1500$

$$f(u_x) := \sqrt{\frac{M}{2 \cdot \pi \cdot R \cdot T_L}} \cdot e^{\frac{-M \cdot u_x^2}{2 \cdot R \cdot T_L}} \qquad g(u_x) := \sqrt{\frac{M}{2 \cdot \pi \cdot R \cdot T_H}} \cdot e^{\frac{-M \cdot u_x^2}{2 \cdot R \cdot T_H}}$$

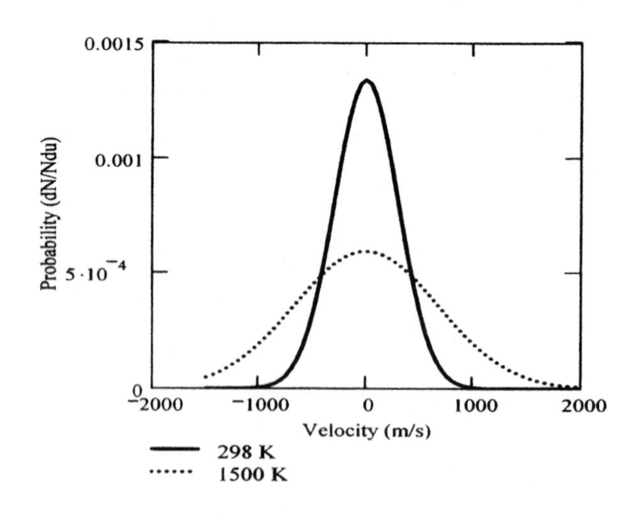

Figure 19-11. Mathcad Worksheet 6 Plotting a One-Dimensional Velocity Distribution Function for N_2

Plots of Three-Dimensional Speed Distribution Functions for Nitrogen at 298.15 K and 1500 K

$M := 0.028$ Molecular mass of Nitrogen (kg/mol)

$R := 8.314$ Gas constant ($J\ K^{-1} mol^{-1}$)

u is the velocity in the x direction (m/s)

f(u) and g(u) are the probability density dN/Ndu

$u := 0, 50 .. 2500$

$T_L := 298$ $T_H := 1500$

$$f(u) := 4\pi \left(\frac{M}{2\pi R \cdot T_L}\right)^{\frac{3}{2}} u^2 e^{\frac{-M \cdot u^2}{2R \cdot T_L}} \qquad g(u) := 4\pi \left(\frac{M}{2\pi R \cdot T_H}\right)^{\frac{3}{2}} u^2 e^{\frac{-M \cdot u^2}{2R \cdot T_H}}$$

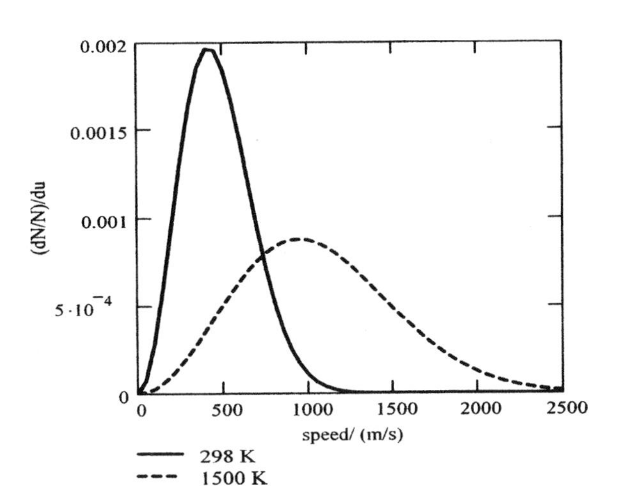

Figure 19-12. Mathcad Worksheet 7 for Plotting a Three-Dimensional Speed Distribution Function for N_2

Figure 19-13 shows three H atom radial wave functions plotted on one graph. The steps in plotting these three functions on one graph are as follows:

- For $x := 0, 0.01, 4$, type **x:0,0.01;4** to define the range variable.
- Define the function to be plotted:

1. For $R_{1s} = 2e^{-x}$, type **R(x):2*e[Shift]^-x.**
2. For $R_{1s}^2 = 4e^{-2x}$, type **R2(x): 4*e[Shift]^-2x.**
3. For $r^2 R_{1s}^2 = 4x^2 e^{-2x}$, type **r2R2(x): 4*x[Shift]^2*e[Shift]^-2x.**
4. Create the graph block by, on the **Insert** menu, clicking **Graph** and selecting **X-Y Plot**. In the left-middle placeholder, type **R(x),R2(x),r2R2(x).**

- In the bottom-middle placeholder, type **x** and click outside the plot area to see the plot.
- The graph may be edited by, on the **Format** menu, clicking **Graph** or by right-clicking the graph and then clicking **Format.** Add labels and format the traces as described for Mathcad Worksheet 6, above.

In this example, the range is zero to four, and the increment is 0.01, resulting in 40 plotted points. Regardless of quantum number, the units (vertical axis) are $a_0^{-3/2}$ for $R_{n,\ell}$, a_0^{-3} for $R_{n,\ell}^2$, and a_0^{-1} for $r^2R_{n,\ell}^2$. The unit for distance from the nucleus (horizontal axis) is r/a_0. It is easy to see that the maximum in the plot of $r^2R_{1s}^2$ versus r/a_0 lies at $r/a_0 = 1$, or $r = a_0$, the Bohr radius. That value is 52.92 pm.

If you feel that any of the plots created so far are too small, you can enlarge them. Just click outside the plot area to the left, hold down the mouse button, and drag the mouse across the plot until it's outside the plot area on the right. This action creates a rectangular dotted line encompassing the entire plot. In the bottom-middle and right-middle edges, you will see small black squares. If you click one of these, the pointer changes to a double-ended arrow, and if you hold down the mouse button and drag it down or to the right, the plot will increase in size.

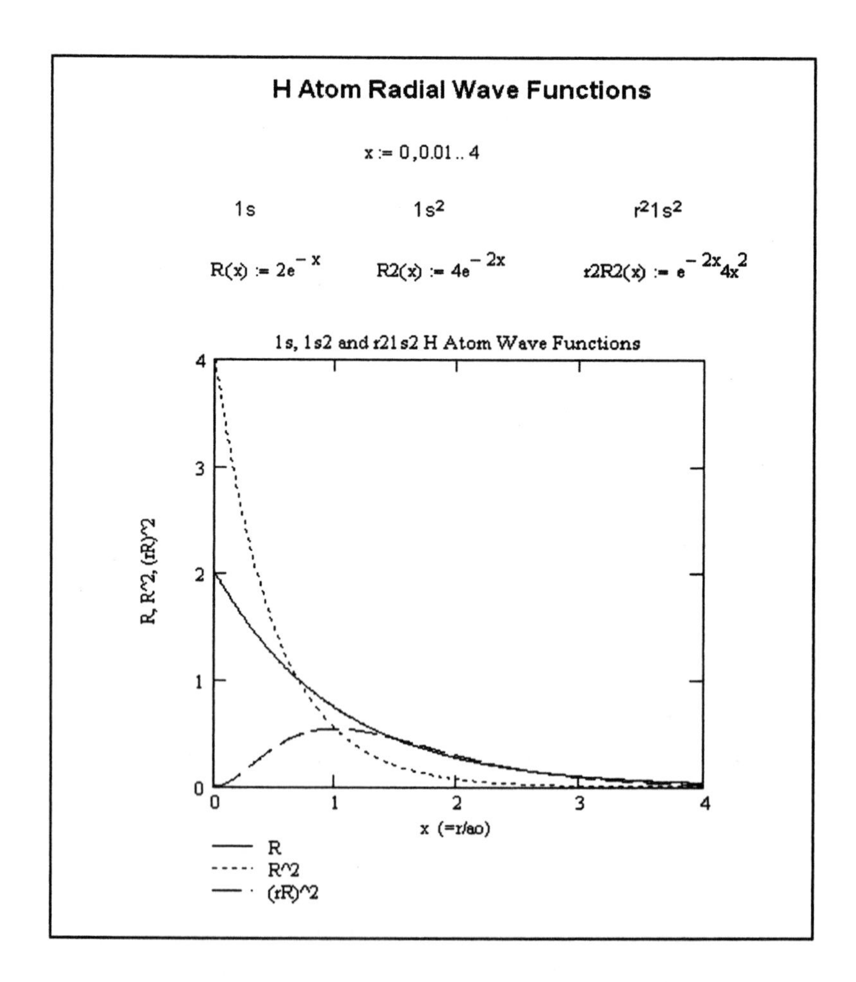

Figure 19-13. Mathcad Worksheet 8: Plotting H Atom Radial Wave Functions

Mathcad Worksheet 9
Plotting H Atom Angular Wave Functions in Two Dimensions

A summary of the quantum mechanics of the hydrogen atom was given in Example 7, Chapter 3. The first few angular wave functions for the hydrogen atom are

$$\Theta\Phi_{1s} = \left(\frac{1}{4\pi}\right)^{1/2} \tag{19-7}$$

$$\Theta\Phi_{2p_z} = \left(\frac{3}{4\pi}\right)^{1/2} \cos\theta \tag{19-8}$$

$$\Theta\Phi_{2p_x} = \left(\frac{3}{4\pi}\right)^{1/2} \sin\theta\cos\phi \tag{19-9}$$

$$\Theta\Phi_{2p_y} = \left(\frac{3}{4\pi}\right)^{1/2} \sin\theta\sin\phi \tag{19-10}$$

The first p orbital wave function (equation 19-8) is independent of ϕ, but dependent on θ, so we can set ϕ to anything. If we choose ϕ to be zero, then the plot lies in the xz plane according to Table 3-1. If we choose ϕ to be 90, then the plot lies in the yz plane. Let us choose ϕ to be zero. In either case, it turns out that the plot is symmetric about the z axis and is given the name p_z. The constant term in Equation 19-8, 19-9, and 19-10 does not affect the shape of the plot, and so it may be omitted. The plot of p_z. is just a plot of cos θ in the xz plane. The value of the p_z orbital depends only on θ according to

$$P_z = \cos\theta \tag{19-11}$$

The absolute value of the orbital at any θ is given by

$$r = |p_z| \tag{19-12}$$

This is the r shown in Figure 3-9 When θ varies from 0 to 360 degrees, the arrow on r sweeps out the coordinates of the Pz orbital in the xz plane (ϕ equals zero). For the purpose of plotting the position of r, we need its x and z coordinates. These coordinates are given by the transformation equations shown in Figure 3-5.

$$z = r\cos\theta \tag{19-13}$$

$$x = r\sin\theta\cos\phi = r\sin\theta \tag{19-14}$$

Each of the transformed coordinates x and z depend on r, so we must calculate r first and then x and y. In Mathcad, the steps in creating the x,y plot are:

- In a text region at the top of the worksheet, write some introductory remarks like those in Figure 19-14. Mathcad supports Microsoft Equation Editor 3.0. To open the Equation Editor, on the **Insert** menu, point to **Object**, and click **Microsoft Equation Editor 3.0**. (If **Microsoft Equation Editor 3.0** if not a listed object, go to Microsoft Word and add it by, on the **Tools** menu, clicking **Add-Ins** (Figure 1-16).
- Define the range n and the range variable i.
- Define the subscripted variable θ_i, which will vary from 0 to 2π in n steps.
- Define the wave function, $P_i := cos(\theta_i)$.
- Define the absolute value of the wave function, $r_i := |P_i|$.
- Define the Cartesian coordinates that locate the end of r (Figure 3-5). These are: $z_i := r_i * cos(\theta_I)$ and $x_i := r_i * sin(\theta_I)$.
- Optional: click r_i and z_i and x_i to see a table of their values.
- On the **Insert** menu, click **Graph**, and select **X-Y Plot**.
- In the left-middle placeholder, type $\mathbf{z_i}$.
- In the left-bottom placeholder, type $\mathbf{x_i}$.
- Click outside the graph block area, and the plot appears as two flattened tangent ellipses. The ellipses should be circles.
- Select the graph and stretch the bottom pull handle until x-axis and z-axis scales are equal. The graph will then appear as two tangent circles.
- Click the graph, and then, on the **Format** menu, click **Graph**, select **X-Y Plot**, and label as you wish.

The d_{xz} orbital is also independent of φ in the xz plane and may be graphed much the same as the P_z orbital. This graph is shown at the bottom of Figure 19-14.

Three-Dimensional Plots

Mathcad supports both surface plots and parametric plots in three dimensions. These plots are illustrated in the next two sections.

Surface Plots. A surface plot is a three-dimensional plot. Mathcad creates a surface plot with a matrix whose row and column numbers represent the x-axis and y-axis values. The matrix elements themselves are the z-values, and the matrix itself is the single argument of the Mathcad surface plot function.

A Plot of Two Angular Wave Functions: the P_z and the d_{xz} Orbitals

$$\Theta\Phi_{2p_z} = \left(\frac{3}{4\pi}\right)^{1/2} \cos\theta$$

To simplify nomenclature let $P = \Theta\Phi_{2p_z}$

$n := 100 \qquad i := 0..n \qquad \theta_i := \frac{2\pi i}{n}$

$P_i := \cos(\theta_i) \qquad r_i := |P_i|$

$x_i := r_i \cdot \sin(\theta_i) \qquad z_i := r_i \cdot \cos(\theta_i)$

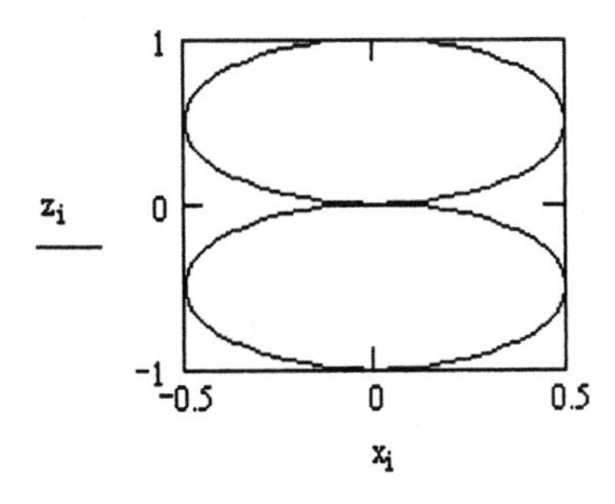

$$\Theta\Phi_{d_{xz}} = \frac{1}{2}\left(\frac{15}{\pi}\right)^{1/2} \sin\theta\cos\theta\cos\varphi$$

In the xz plane ϕ equals zero so $d_{xz} \sim \sin\theta\cos\theta$

To plot the dxz orbital:

$dxz_i := \sin(\theta_i)\cdot\cos(\theta_i) \qquad r_i := |dxz_i|$

$z_i := r_i \cdot \cos(\theta_i) \qquad x_i := r_i \cdot \sin(\theta_i)$

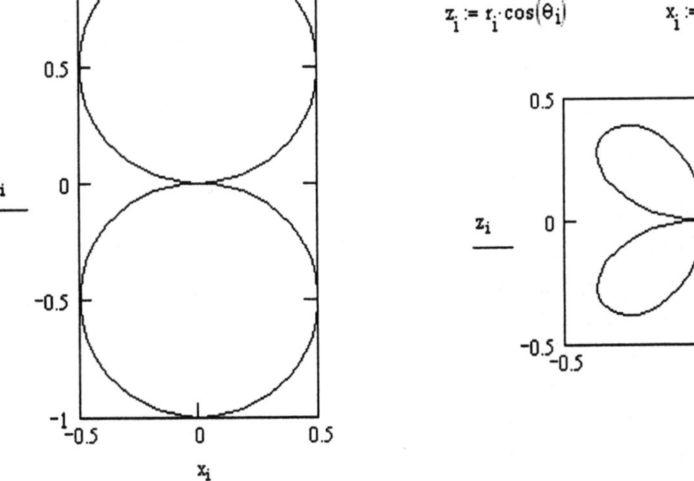

Figure 19-14. Mathcad Worksheet 9 for Two Angular Wave Functions of the Hydrogen Atom

Mathcad Worksheet 10
A Simple 3-D Surface Plot of a Simple Function of Two Variables

The values of a function of two variables typically lie on a surface. A simple example of such a function is

$$f(x,y) = x^2 + 3y^2 \tag{19-15}$$

The steps required to create a plot of a function like the preceding are as follows:

- Define the function $f(x,y)$.
- Define the number N points to be plotted in the x and in the y directions.
- Define the x_i and y_i points to be plotted.
- Fill a matrix P with the values of x_i and y_i.
- On the **Insert** menu, click **Graph**, and select **Surface Plot**.
- In the placeholder at the bottom of the plot area, type **P** as the name of the matrix.
- Double-click the graph to open the **3-D Plot Format** dialog box. Add labels.

A 3-D Surface Plot of f(x,y) = x² +3y²

$f(x,y) := \left(x^2 + 3 \cdot y^2\right)$

$N := 15$ $\quad M := 10$

$i := 0..N$ $\quad j := 0..M$

$x_i := 0.01 \cdot i$ $\quad y_j := 0.01 \cdot j$

$P_{i,j} := f\left(x_i, y_j\right)$

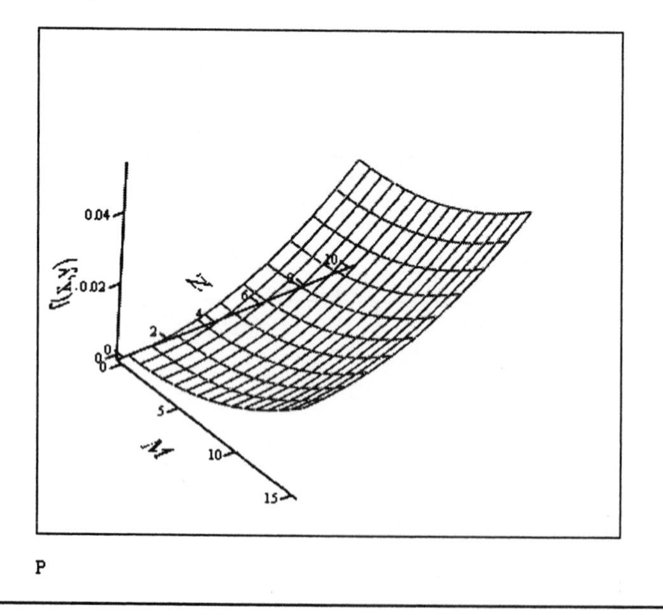

Figure 19-15 displays the Mathcad worksheet resulting from the steps for the sample function given above. Optional: If you would like to see the values of the matrix, place the pointer well below the plot and type **P=**. The matrix will be displayed. Click and drag anywhere on the graph to change its orientation with respect to the viewer.

Figure 19-15. Mathcad Worksheet 10: A 3-D Surface Plot of a Function of Two Variables

Mathcad Worksheet 11
A 3-D Surface Plot for a Particle in a Two-Dimensional Box

This exercise demonstrates a three-dimensional surface plot of the ψ_{11} and ψ_{22} wave functions of a particle in a two-dimensional box. The wave function for a particle in a two-dimensional box provides a typical example of a function of two variables. The quantum mechanics of the particle in a box problem are reviewed in Excel Example 26, Chapter 8.

Plotting the 3-D Surface Plot of a Particle in a Box

- Define the range variables *i* and *j*.
- Define the Cartesian coordinates over which to make the plot by entering the following sequence: **x_i:0+0.02*i** and **y_i:0+0.02*i**.
- Type **n := 2** to assign a value to the *n* quantum number.
- Define the wave function: *f(x,y):((sin(n*π*x)* sin(n*π*y))*.
- Copy, paste, rename, and add **^2** to the wave function to square it.
- Create a matrix that holds the values of the wave function. On the **Matrix** toolbar, click the $\mathbf{X_n}$ icon.
- Fill the first placeholder with the name of the matrix, say, *M*.
- Fill the next placeholder with **i**, then type a comma (**,**) to generate another placeholder, and fill it with **j**.
- Define the matrix $M_{i,j}$ to be equal to the function $f(x_i,y_j)$. Each *x,y* position in the matrix now has the value of the wave function *f(x,y)* at that coordinate.
- On the **Insert** menu, click **Graph**, and select **Surface Plot**, and a 3-D coordinate system appears in a graph block with a single placeholder at the bottom.
- In the placeholder (lower left), type **M** as the name of the matrix to be plotted.
- Click outside the graph block area, and the surface plot appears.
- Click the graph, then, on the **Format** menu, click **Graph**, and select **Surface Plot**; insert a title and label as you wish.
- In the same way, graph the square of the wave function. Define a matrix $N_{i,j}$ equal to *g(x,y)*, the square of *f(x,y)*.

The 3-D surface plots of Ψ and Ψ^2 for a particle in a two-dimensional box are shown in Figure 19-16. Change the values of *n* to other small integers to see how the wave function changes. When *n* equals one, the graph consists of a single peak in the center of the *x,y* plane, corresponding to the maximum probability. As *n* becomes larger, the peaks spread out over the *x,y* plane, gradually approaching classical behavior, where the probability of finding the particle is the same everywhere.

Three-Dimensional Parametric Surface Plot. In Mathcad, a parametric surface plot is created by passing to the surface plot three matrices. The three matrices (actually, one-dimensional arrays or lists) contain the *x, y,* and *z* coordinates of a point in space on the surface of the plot. The steps are as follows:

- Define the number of points to be plotted in the *x* direction and in the *y* direction.
- Define the range variables.
- Define the function.
- Define the three *x, y,* and *z* coordinate matrices.
- On the **Insert** menu, click **Graph**, and select **Surface Plot**.
- In the single placeholder at the bottom of the plot area, enter the name of the coordinate matrices in parentheses (**M** or **N**).
- Click and drag the graph to improve its orientation.

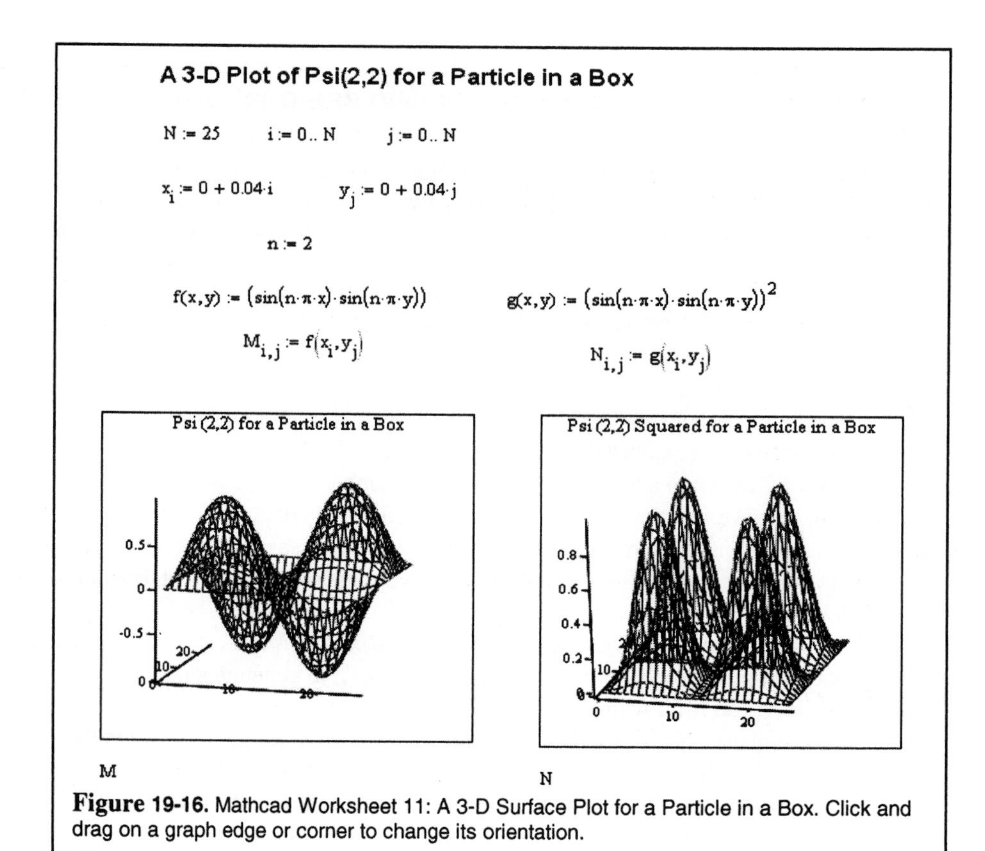

Figure 19-16. Mathcad Worksheet 11: A 3-D Surface Plot for a Particle in a Box. Click and drag on a graph edge or corner to change its orientation.

Mathcad Worksheet 12
A 3-D Parametric Surface Plot of a d_{z^2} Orbital

The three dimensional plot of a d_{z^2} orbital provides an attractive example of a three-dimensional parametric plot:

$$\Theta\Phi_{3d_{z^2}} = \left(\frac{5}{16\pi}\right)^{1/2}(3\cos^2\theta - 1) \qquad (19\text{-}16)$$

Examine Figure 19-17, which displays the final result. Think of a point on the surface of the plot as the end of a vector, the length of which is given by the value of $\Theta\Phi_{3d}$ for some particular value of θ. Relative to the Cartesian axes, a point at the end of that vector has some particular *(x,y,z)* coordinates. To create the Mathcad plot, we create three arrays, one containing the x_i, the second containing the y_i, and the third containing the z_i coordinates that lie on the surface to be plotted. These three arrays, expressed within parentheses and separated by commas *(x,y,z)* constitute the argument for the Mathcad **Surface Plot** command. The steps are as follows:

- Define the range. According to Figure 3-5 the θ ranges from 0 to π and ϕ ranges from 0 to 2π.
- Define the angular wave function, $P_{i,j} := |3*cos(\theta_i) - 1|$. On the **Matrix** toolbar, click the $\mathbf{X_n}$ icon; type **d** in the first placeholder, type **i** in the second, and type a comma (,) to get another placeholder; type **j**.

- Define $x_{i,j}$ similarly: $x_{i,j} := d_{i,j} * sin(\theta_i) * \cdot cos(\phi_i) - 1|$. The Cartesian x coordinate depends on the length of the vector (the magnitude of the wave function) and the θ, ϕ coordinates (in this case θ only).
- Repeat the definitions for the $y_{i,j}$ and $z_{i,j}$ coordinates.
- On the **Insert** menu, click **Graph**, and select **X-Y Plot**, and an empty graph block appears with a single placeholder at the bottom; in it type **(x, y, z)**.

Click outside the graph area, and the plot appears (Figure 19-17). Click inside the graph, and then, on the **Format** menu, click **Graph**, and select **Surface Plot** to format the graph according to your tastes.

Mathcad Worksheet 13
A 3-D Parametric Surface Plot of a $2p_x$ Orbital

The parametric surface plot of a p_x orbital shown in Figure 19-18. Mathcad Worksheet 13 is created in the same way. To show the axial direction, the x-axis was made bold. On the **Format** menu, click **Graph**, and select **3D Plot**; then click the **Axis** tab, and then the **X-Axis** tab. Change the axis weight to **2**.

3-D Parametric Plot of a $3d_{z2}$ Orbital

$N := 30$

$i := 0..N \qquad j := 0..N$

$\theta_i := \frac{\pi \cdot i}{N} \qquad \phi_j := \frac{2 \cdot \pi \cdot j}{N}$

$P_{i,j} := \left|3 \cdot \cos(\theta_i)^2 - 1\right|$

$x_{i,j} := P_{i,j} \cdot \sin(\theta_i) \cdot \cos(\phi_j)$

$y_{i,j} := P_{i,j} \cdot \sin(\theta_i) \cdot \sin(\phi_j)$

$z_{i,j} := P_{i,j} \cdot \cos(\theta_i)$

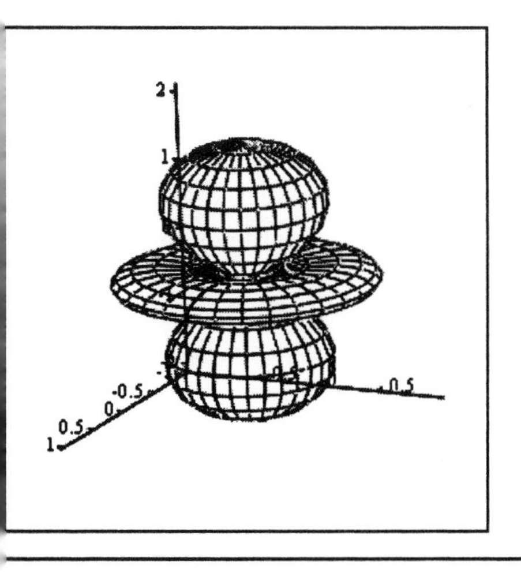

Figure 19-17. Mathcad Worksheet 12. Parametric 3-D Plot of a d_{z^2} Orbital

3-D Parametric Plot of a p_x Orbital

$N := 30$

$i := 0..N \qquad j := 0..N$

$\theta_i := \frac{\pi \cdot i}{N} \qquad \phi_j := \frac{2 \cdot \pi \cdot j}{N}$

$P_{i,j} := \left|\sin(\theta_i) \cdot \cos(\phi_j)\right|$

$x_{i,j} := P_{i,j} \cdot \sin(\theta_i) \cdot \cos(\phi_j)$

$y_{i,j} := P_{i,j} \cdot \sin(\theta_i) \cdot \sin(\phi_j)$

$z_{i,j} := P_{i,j} \cdot \cos(\theta_i)$

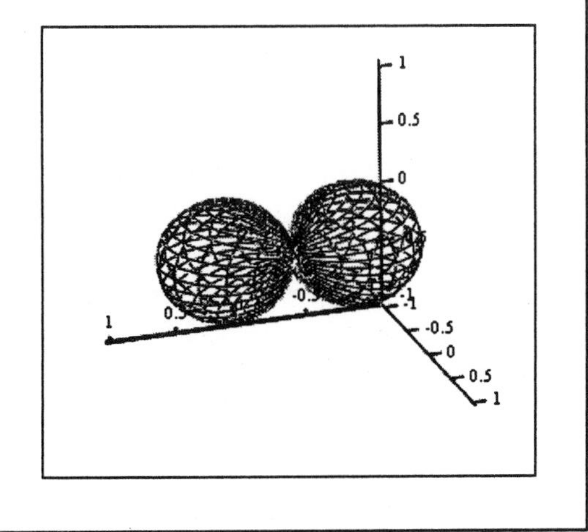

Figure 19-18. Mathcad Worksheet 13: Parametric 3-D Plot of a p_x Orbital

Symbolic Calculations

Up to this point we've looked at the way Mathcad handles numerical calculations, calculations that result in a number, and numbers or the graphical representation of numbers. For example:

$$a := 3;$$
$$b := 5;$$
$$c := 5;$$
$$VolBox := a \cdot b \cdot c;$$
$$VolBox = 75$$

The assignment operator (:=) assigns values. The evaluation equal sign (=) evaluates an expression.

Symbolic Operators and Keywords. Symbolic calculations involve various operations with expressions, but the result of the evaluation of an expression usually results in another expression. In a manner analogous to numerical calculations, the symbolic equal sign (→ or ■→ can return an expression. Other symbolic operators are listed in Table 19-4. You can access these signs by pressing CTRL+PERIOD or CTRL+SHIFT+PERIOD, respectively. Alternatively, click the **Symbolic** toolbar button (Figure 19-1), and then click the obvious icon. The ■ is a placeholder into which a *keyword* can be entered. The keyword is just a function or command that customizes the symbolic equal sign. Samples of Mathcad's keywords (Table 19-5) are demonstrated in the following sections.

Table 19-4

Some Symbolic Operators

Depressed Key	Operator Onscreen	Use
:	:=	Assignment
=	=	Computation equality
CTRL+=	=	Boolean equality
CTRL+PERIOD	→	Symbolic evaluation
CTRL+SHIFT+PERIOD	■→	Symbolic keyword evaluation

The Symbolic Equal Sign with Keywords: *Simplify, Expand, Factor, and Substitute*

Before using symbolic calculations, be sure that **Automatic Calculation** on the **Math** menu is checked. The steps in using the keywords for symbolic calculation are nearly the same for all of the keywords.

1. Enter your expression.
2. Press CTRL+SHIFT+PERIOD to get the symbolic equal sign, "■ →."
3. Type the desired the keyword **simplify** in the placeholder.
4. Click outside the expression, and Mathcad displays the resulting expression.

Table 19-5

Some Keywords and Their Functions

Keyword	Function
simplify	Simplifies an expression by performing arithmetic, canceling common factors, and so on
expand	Expands powers and products of sums in an expression
factor	Factors an expression into a product
substitute, var1=var2	Replaces all occurrences of a variable *var1* with an expression or variable *var2*
solve, var	Solves an equation for the variable, *var*

Simplify. To do symbolic calculations with the *simplify* keyword, carry out the following steps:

- Enter an expression, for example:

$$\frac{x^2 - 7x + 12}{x - 4}$$

  1. Type the numerator, and then press the SPACEBAR until the blue edit line lies under the entire numerator.
  2. Then press the slash (/) character for division. Again press the SPACEBAR until the entire expression is included by the blue edit line.
- Press CTRL+SHIFT+PERIOD to get the symbolic equal sign, "■ →."
- Enter the keyword **simplify** in the placeholder ■
- Click outside the expression, and Mathcad displays the resulting expression:

$$\frac{x^2 - 7x + 12}{x - 4} \text{ simplify} \rightarrow x - 3$$

The SPACEBAR and the left and right arrows permit you to control the position of the blue edit line, which looks like this: _⌋. The next operation, say division (/), will operate only on the part of the expression that lies over the horizontal portion of the edit line. To include more, or all, of the expression, just press the SPACEBAR, and the horizontal portion of the edit line grows to the left with each tap. When the numerator is entered, the blue edit line lies under the last term: $x^2 - 7 +$ 12⌋. Pressing the SPACEBAR repeatedly results with the entire numerator included within the edit line: $x^2 - 7 + 12$⌋. This is how the entire numerator $x^2 - 7 + 12$ was placed over the denominator $x - 4$. The SPACEBAR is also used to get the blue editor line down from an exponent. Type **x^2SPACEBAR.** The blue edit line lies under the 2 before the SPACEBAR is pressed. Pressing the SPACEBAR places the blue edit line under the entire *x*^2 expression. Now try some slightly more complex expressions.

$$\frac{\dfrac{3\cdot x - 6}{\left(x^2 - 4\cdot x + 4\right)}\cdot\dfrac{x^2 - 4}{3\cdot x - 3}}{\dfrac{5\cdot x+10}{x^3-1}} \text{ simplify } \rightarrow \frac{1}{5}\cdot x^2 + \frac{1}{5}\cdot x + \frac{1}{5}$$

$$\frac{1}{\tan(x) + \cot(x)} \text{ simplify } \rightarrow \cos(x)\cdot\sin(x)$$

$$\frac{\tan(x)\left(\csc(x)^2 - 1\right)}{\sin(x) + \cot(x)\cos(x)} \text{ simplify } \rightarrow \cos(x)$$

Notice that in Mathcad, $\csc^2(x)$ is written $\csc(x)^2$. In expressions, Mathcad sometimes hides the dot symbol for multiplication between terms.

Expand. Symbolic calculations with the *expand* keyword are carried out in the same way that *simplify* calculations are done, but with the keyword *expand.*

$$(x - 4)(2\cdot x + 3)(5\cdot x - 1)(3\cdot x - 7) \text{ expand } \rightarrow 30\cdot x^4 - 151\cdot x^3 + 24\cdot x^2 + 421\cdot x - 84$$

$$(3x + 2)^4 \text{ expand,x } \rightarrow 81\cdot x^4 + 216\cdot x^3 + 216\cdot x^2 + 96\cdot x + 16$$

Factor. Symbolic calculations with the *factor* keyword are carried out in a similar manner. If factors are there, Mathcad will find them.

$$x^2 + x - 6 \text{ factor } \rightarrow (x + 3)\cdot(x - 2)$$

$$x^6 + 7x^3 - 8 \text{ factor } \rightarrow (x - 1)\cdot(x + 2)\cdot\left(x^2 - 2\cdot x + 4\right)\cdot\left(x^2 + x + 1\right)$$

Substitute. Symbolic calculations with the *substitute* keyword are a little different from the previous examples. The results can often be refined with the keyword *simplify.* The keyword *simplify* can be entered immediately after the results obtained from using the keyword *substitute* as illustrated in the following example:

1. Enter your expression.
2. Click **substitute** in the **Symbolic** toolbar, which results in *substitute* ■ = ■ .
3. In the first placeholder, enter the name of the variable to be substituted for.
4. In the second variable, enter the expression you want to substitute:

$$x^2 + 3 \text{ substitute }, x = x^3 - 4 \rightarrow \left(x^3 - 4\right)^2 + 3 \text{ simplify } \rightarrow x^6 - 8\cdot x^3 + 19$$

Or, alternatively:

1. Enter your expression.
2. Press CTRL+SHIFT+PERIOD to get the symbolic equal sign, "■→."
3. Type **Substitute, x=x^3-4** (use CTRL= for the =).
4. Click outside the math region.
5. Place the pointer after the 3, and simplify if it appears necessary.

$$x^2 + 3 \text{ substitute }, x = x^2 - 4 \rightarrow \left(x^2 - 4\right)^2 + 3 \text{ simplify } \rightarrow x^4 - 8\cdot x^2 + 19$$

Your expression can have mixed variables, which is why you must specify which one you want to substitute.

$$x^2 - 2\cdot x\cdot y^2 + 4 \text{ substitute}, y = x^2 - 3 \rightarrow x^2 - 2\cdot x\cdot\left(x^2 - 3\right)^2 + 4 \text{ simplify } \rightarrow x^2 - 2\cdot x^5 + 12\cdot x^3 - 18\cdot x + 4$$

Occasionally, the use of symbolic operators results in an expression that you feel could be further simplified. In this case, just press CTRL+SHIFT+PERIOD immediately after the result appears from differentiation or integration. When you do this, the entire result disappears. Ignore this disturbing event; type **simplify** in the placeholder, click outside the expression, and everything reappears, including the simplified result.

Calculus

On the **View** menu, point to **Toolbars**, and click **Calculus** to open the **Calculus** toolbar, which provides easy access to icons for several operators such as operators for differentiation, integration, summation, and products.

Derivatives: Operating on a Function. The steps in taking the derivative of a function are as follows.

1. Define the function, for example: $f(x) := 4\cdot x^2 + 3\cdot x - 11$.
2. Click **d/dx** on the **Calculus** toolbar (Figure 19-1).
3. Type **x** in the lower placeholder.
4. Type **f(x)** in the upper placeholder.
5. Click the symbolic equal sign, "→" on the **Symbolic** toolbar.
6. Click outside the math area to see the result as it appears on a Mathcad worksheet.

Example 1. $f(x) := 4x^6 + 3x - 11$ $\quad \frac{d}{dx} f(x) \rightarrow 24 \cdot x^5 + 3$

Example 2. $g(x) := \frac{\ln(x)}{x}$ $\quad \frac{d}{dx} g(x) \rightarrow \frac{1}{x^2} - \frac{\ln(x)}{x^2}$ simplify $\rightarrow \frac{-(-1 + \ln(x))}{x^2}$

Derivatives: Operating on an Expression. Sometimes it is more convenient to operate directly on an expression instead of first defining a function and then operating on the function.

The van der Waals equation of state is

$$(P + \frac{a}{V^2})(V - b) = RT \tag{19-17}$$

To find *dP/dT*, solve this equation for *P*.

$$P = \left(\frac{RT}{V - b} - \frac{a}{V^2} \right) \tag{19-18}$$

1. Click **d/dx** on the **Calculus** toolbar, and enter **T** in the lower placeholder.
2. To write the preceding expression, type (**R*T SPACEBAR / V-b SPACEBAR SPACEBAR-a/V^2 SPACEBAR SPACEBAR**).
3. Click the → on the **Symbolic** toolbar.
4. Click outside the math area to get the result, shown below as it appears on a Mathcad worksheet.

Example 1. $\frac{d}{dT}\left(\frac{R \cdot T}{V - b} - \frac{a}{V^2} \right) \rightarrow \frac{R}{(V - b)}$

Example 2. $\frac{d}{dx}\left(\frac{1}{2} \right) \cdot \sec(2x) \rightarrow \sec(2 \cdot x) \cdot \tan(2 \cdot x)$

Integrals: Operating on a Function. First define the function you wish to integrate. Then:

1. Click the indefinite integral icon on the **Calculus** toolbar (Figure 19-1).
2. Type **f(x)** in the left placeholder.
3. Type **x** in the right placeholder.
4. Click the → arrow on the **Symbolic** toolbar.
5. Click outside the math area to see the Mathcad worksheet result.

Example 1. $f(x) := 4x^6 + 3x^2 - 11$ $\int f(x)\,dx \rightarrow$

Example 2. $g(x) := \frac{1 - \ln(x)}{x^2}$ $\int g(x)\,dx \rightarrow$

Example 3. $h(x) := \sin(3 \cdot x) \cdot \cos(x)$ $\int h(x)\,dx \rightarrow$

Integrals: Operating on an Expression. It is not necessary to define a function to integrate; integration can be carried out directly on the expression. To write the following expression, click the indirect integral on the **Calculus** toolbar, then type **1+uSPACEBAR/**. Next click the square root icon on the **Calculator** toolbar, and type **u**. Type **u** in the *du* placeholder. Click → on the **Symbolic** toolbar, and click outside the math area to get the following result:

Example 4. $\int \frac{1 + u}{\sqrt{u}}\,du \rightarrow$

Example 5. $\int \sin(5 \cdot x) \cdot \sin(2x)\,dx \rightarrow$

Definite Integrals. Similarly, definite integrals can be carried out with or without a defined function. The same function that was used above is integrated for *u* equals 0 to *u* equals 9, with the result shown in Example 6.

Example 6. $\int_0^9 \frac{1 + u}{\sqrt{u}}\,du \rightarrow$

The Keyword *Assume*. The keyword *assume* forces constraints on one or more variables. The syntax of *assume* consists of two parts: *assume, constraint.* Constraint is a Boolean expression; for example: $a > 0$ as in the preceding Example 10. Clicking the keyword *assume* on the **Symbolic** toolbar (Figure 19-1) results in *assume,* ■→. The placeholder contains the Boolean expression (the *constraint*). The **Boolean** toolbar (Figure 19-1) lists the permitted Boolean operators.

The preceding Example 9 displays the results of an integration without the *assume* keyword. This cumbersome result is correct. However the limit of the first expression is zero as t approaches infinity.

Integration with a Range Variable. In this example, the same function is integrated from 0 to 1, from 0 to 2, ... , up to from 0 to 10, and results are displayed in a vector (a list or one-dimensional table).

$$i := 1, 2 .. 10$$

$$f(u) := \frac{1+u}{\sqrt{u}}$$

$$g_i := \int_0^i f(u)\, du$$

$g_i =$

2.667
4.714
6.928
9.333
11.926
14.697
17.638
20.742
24
27.406

Summation and Product Operators. The symbolic sum operator (⊠) and product operator (✓) are used in a similar manner, but usually on a list. To create the Mathcad workspace reproduced below, follow these steps:

1. Type **i:1;5** to create the range variable.
2. Fill the elements of a list. Click the $\mathbf{X_n}$ icon in the **Matrix** toolbar.
    a. Type **x** to fill in the left placeholder.
    b. Type **i** to fill in the right placeholder.
    c. Type a colon (**:**), and Mathcad furnishes the assignment operator.
    d. Type **34,24,56,58,78** to create a table of x_i values.
3. Type **s:** and then click the $\sum_{n=1}^{n}$ icon on the **Calculus** toolbar.
    a. Type **i = 1** to fill in the bottom placeholder.
    b. Type **5** to fill in the top placeholder.
    c. In the right placeholder, click the $\mathbf{X_n}$ icon on the **Matrix** toolbar, and then type **x** and **i** to fill in the two placeholders.
    d. Outside this math area, type **s =** and the result **250** appears.
4. Repeat for the product of the elements in the list.

$i := 1..5$

$x_i :=$

34
24
56
58
78

$$s := \sum_{i=1}^{5} x_i \qquad p := \prod_{i=1}^{5} x_i$$

$$s = 250 \qquad p = 2.067 \times 10^{8}$$

Some Derivatives and Integrals from Statistical Thermodynamics

The goal of statistical thermodynamics is the calculation of macroscopic thermodynamic properties of a system from microscopic properties of individual molecules, especially their energies. Gaseous polyatomic molecules have energies arising from their translational, rotational, and vibrational degrees of freedom. These energies are quantized, and quantum mechanics provides a description of these energies. Mathcad Worksheets 14, 15, 16, and 17 give examples.

The statistical thermodynamic background for these worksheets is reviewed in Excel Examples 14 and 15 in Chapter 4. The translational, rotational, and vibrational partition functions were derived in Excel Examples 14 and 15.

Note: Mathcad does not always display the dot (·) that indicates the product of two terms.

Mathcad Worksheet 14
Partition Functions

This worksheet (Figure 19-19) shows how Mathcad can handle the integrals that arise in the derivation of Q_t, Q_r, and Q_v.

Mathcad Worksheet 15
Translational and Rotational Internal Energy

This worksheet (Figure 19-20) shows how Mathcad handles some of the derivatives that arise in the equations for the translational and rotational contributions to the internal energy.

Mathcad Worksheet 16
Vibrational Internal Energy

This worksheet (Figure 19-21) shows how Mathcad can take the derivatives that arise in the equations for the vibrational contribution to the internal energy of a diatomic molecule.

Mathcad Worksheet 17
Vibrational Heat Capacity

This worksheet (Figure 19-22) demonstrates how Mathcad takes the derivative of the vibrational contribution to the internal energy to obtain an expression for the vibrational contribution to the heat capacity.

The Translational Partition Function

$$\int_0^{\infty} e^{-b \cdot n^2}\, dn \text{ assume}, b > 0 \rightarrow \frac{1}{2} \cdot \frac{\pi^{\frac{1}{2}}}{b^{\frac{1}{2}}} \text{ substitute}, b = \frac{h^2}{8m \cdot l^2 \cdot k \cdot T} \rightarrow \frac{1}{2} \cdot \pi^{\frac{1}{2}} \cdot \frac{8^{\frac{1}{2}}}{\left(\frac{h^2}{m \cdot l^2 \cdot k \cdot T}\right)^{\frac{1}{2}}}$$

The keyword *simplify* after the last term doesn't simplify very much, but a pencil and paper shows that the last term does indeed simplify to

$$q_{trans} = \frac{(2\pi mkt)^{1/2}\, l}{h}$$

or in three dimensions $$Q_{trans} = \frac{(2\pi mkT)^{3/2}\, V}{h^3}$$

Rotational Partition Function

$$2 \cdot \int_0^{\infty} J \cdot e^{-a \cdot J^2}\, dJ \text{ assume}, a > 0 \rightarrow \frac{1}{a} \text{ substitute}, a = \frac{h^2}{8\pi^2 I \cdot k \cdot T} \rightarrow \frac{8}{h^2} \cdot \pi^2 \cdot I \cdot k \cdot T$$

The rotational partition function is $$Q_r = \frac{8\pi^2 IkT}{h^2}$$

Vibrational Partition Function

$$\sum_{v=0}^{\infty} e^{-v \cdot x} \rightarrow \frac{1}{(1 - \exp(-x))}$$

$$Q_v = (1 - e^{-x})^{-1}$$

Figure 19-19. Mathcad Worksheet 14 for Translational, Rotational, and Vibrational Partition Functions

Translational and Rotational Energy from the Partition Functions

The translational partition function is $Q_{tran} = (2\pi mkT)^{3/2}\dfrac{V}{h^3}$

The translational energy $(E^0 - E_0^0)_{tran}$ is given by

$$(E^0 - E_0^0)_{tran} = RT^2\left(\frac{\partial \ln Q_{tran}}{\partial T}\right)_V$$

Mathcad calculates RT² times the derivative at constant V

$$R\cdot T^2\left[\frac{d}{dT}\ln\left[\frac{(2\pi m\cdot k\cdot T)^{\frac{3}{2}}V}{h^3}\right]\right] \rightarrow \frac{3}{2}\cdot R\cdot T$$

Thus, the translational energy of a gas is given by

$$(E^0 - E_0^0)_{tran} = \frac{3RT}{2}$$

Rotational Energy from the Rotational Partition Function

The rotational partition function is $Q_{rot} = \dfrac{8\pi^2 IkT}{h^2}$

The rotational energy $(E^0 - E_0^0)_{rot}$ is given by

$$(E^0 - E_0^0)_{rot} = RT^2\left(\frac{\partial \ln Q_{rot}}{\partial T}\right)_V$$

Mathcad calculates RT² times the derivative at constant V

$$R\cdot T^2\left(\frac{d}{dT}\ln\left(\frac{8\pi^2 I\cdot k\cdot T\,V}{h^2}\right)\right) \rightarrow R\cdot T$$

Thus, rotational energy of a gas is given by

$$(E^0 - E_0^0)_{rot} = RT$$

Figure 19-20. Mathcad Worksheet 15: Translational and Rotational Energies

Statistical Thermodynamic Calculation of the Vibrational Contribution to the Internal Energy of a Gas

The vibrational partition function is $Q_V(a,T) := \left(1 - e^{\frac{-a}{T}}\right)^{-1}$

The translational energy $(E^0 - E_0^0)_{vib}$ is given by

$$(E^0 - E_0^0)vib = RT^2\left(\frac{\partial \ln Q_{vib}}{\partial T}\right)_V$$

$$LQ_V(a,T) := \ln(Q_V(a,T))$$

$$RT^2\left(\frac{d}{dT}LQ_V(a,T)\right) \rightarrow \frac{RT^2}{\left(1 - \exp\left(\frac{-a}{T}\right)\right)} \cdot \frac{a}{T^2} \cdot \exp\left(\frac{-a}{T}\right)$$

When x is subsituted wherever possible for a/T, the immediate result is

$$(E^0 - E_0^0)_{vib} = RT\frac{xe^{-x}}{(1-e^{-x})}$$

Further rearrangment leads to

$$(E^0 - E_0^0)_{vib} = RT\frac{x}{(e^x - 1)}$$

Figure 19-21. Mathcad Worksheet 16. Vibrational Contribution to the Internal Energy

Statistical Thermodynamic Calculation of the Vibrational Contribution to the Heat Capacity of a Gas

The vibrational contribution to the heat capacity of a gas may be determined from the derivative of the vibrational contribution to the energy of a gas

$$C_v = \frac{\partial (E^0 - E_0^0)_{trans}}{\partial T}$$

where

$$(E^0 - E_0^0)_{trans} = \frac{aR}{\left(e^{\frac{a}{T}} - 1\right)}$$

The right side of the above function may be defined as a Mathcad function

$$E(a,R,T) := \frac{a \cdot R}{e^{\frac{a}{T}} - 1}$$

from which the heat capacity may be obtained

$$\frac{d}{dT}E(a,R,T) \rightarrow a^2 \cdot \frac{R}{\left(\exp\left(\frac{a}{T}\right) - 1\right)^2 \cdot T^2} \cdot \exp\left(\frac{a}{T}\right) \text{ substitute}, \frac{a}{T} = x \rightarrow a^2 \cdot \frac{R}{(\exp(x) - 1)^2 \cdot T^2} \cdot \exp(x)$$

Since a^2/T^2 equals x^2, the vibrational contribution to the heat capacity is

$$C_v = \frac{Rx^2 e^x}{(e^x - 1)^2}$$

Figure 19-22. Mathcad Worksheet 17: Vibrational Contribution to the Heat Capacity

Mathcad Worksheet 18
The Population of Rotational Energy Levels

As the energy of a quantum level becomes high, it is less likely that molecules have sufficient energy to populate the level. The decrease in population is inverse exponentially proportional to the energy of a level. Thus, molecules tend to cluster in low-lying energy levels. However, when a quantum level is degenerate, the probability that the level will be populated increases in direct proportion to the degeneracy. The degeneracy of rotational energy levels increases as the rotational quantum number J increases. The Boltzmann distribution law permits the calculation of the relative populations of rotational energy levels. See Chapter 4, Excel Example 15.

Preparing the Worksheet. The *hc/k* factor appears so often in statistical thermodynamic calculation that it is convenient to group these constants together and give the result the symbol z. J is defined to be a range variable with values from 0 to 10, in increments of 1. The Mathcad worksheet for the rotational energy state of HCl(g) at 298 K is shown in Figure 19-23. The steps in preparing the Mathcad worksheet (Figure 19-23) are as follows:

- Define the constants h, c and k, all on one line.
- Define the constant z in terms of h, c, and k.
- Type **J:0,1;10** to define the quantum number J as a range variable.
- Define the values for T in Kelvin and B in m^{-1} (Table 3-2).
- Type **Qr:T/z*B** to define the partition function (Equation 4-41).

Define the Boltzmann distribution function for the population of the *Jth* rotational level relative to the population of the $J = 0$ level. Name the function *RelPop(J, B, T)* and note that three parameters are passed to it: J, B, and T. In the same way, define the function *FracPop(J,B,T)*.

- Type **RelPop(J,B,T):(2*J+1)*exp(-J*(J+1)*B*z [Sp][Sp]/T[Sp])**.
- Enter **FracPop(J,B,T): RelPop(J,B,T) [Sp]/Qr**.
- Type **RelPop(J,B,T)=** to evaluate a table of relative populations.
- Type **FraclPop(J,B,T)=** to evaluate a table of fractional populations.

Plotting the Population Distribution as a Bar Graph

- On the **Insert** menu, click **Graph**, and select **X-Y Plot**.
- Type **FracPop(J,B,T)** in the placeholder at the middle-left side of the blank square graph space.
- Type **J** in the placeholder at the middle-bottom side of the blank square graph space.
- On the **Format** menu, click **Graph**; select **X-Y Plot**, and then click the **Traces** tab (Figure 19-6).
- For **Trace 1**, change **Type** from **Lines** to **Bar**.
- Add titles and labels as usual.

The Total Population. The sum of all the fractional populations of the energy levels should equal unity. One can see that the population of levels with J greater than 10 is quite small, so a sum over the first eleven states should come out close to unity. You can check this sum with the Mathcad sum function for subscripted variables and a range function. Entering a dollar sign (**$**) on a Mathcad worksheet opens the sum function, a summation sign with two placeholders.

$$\sum_{\bullet} \bullet$$

In the placeholder on the right, type **FracPop(J,B,T) =** , and in the lower placeholder type **J**.

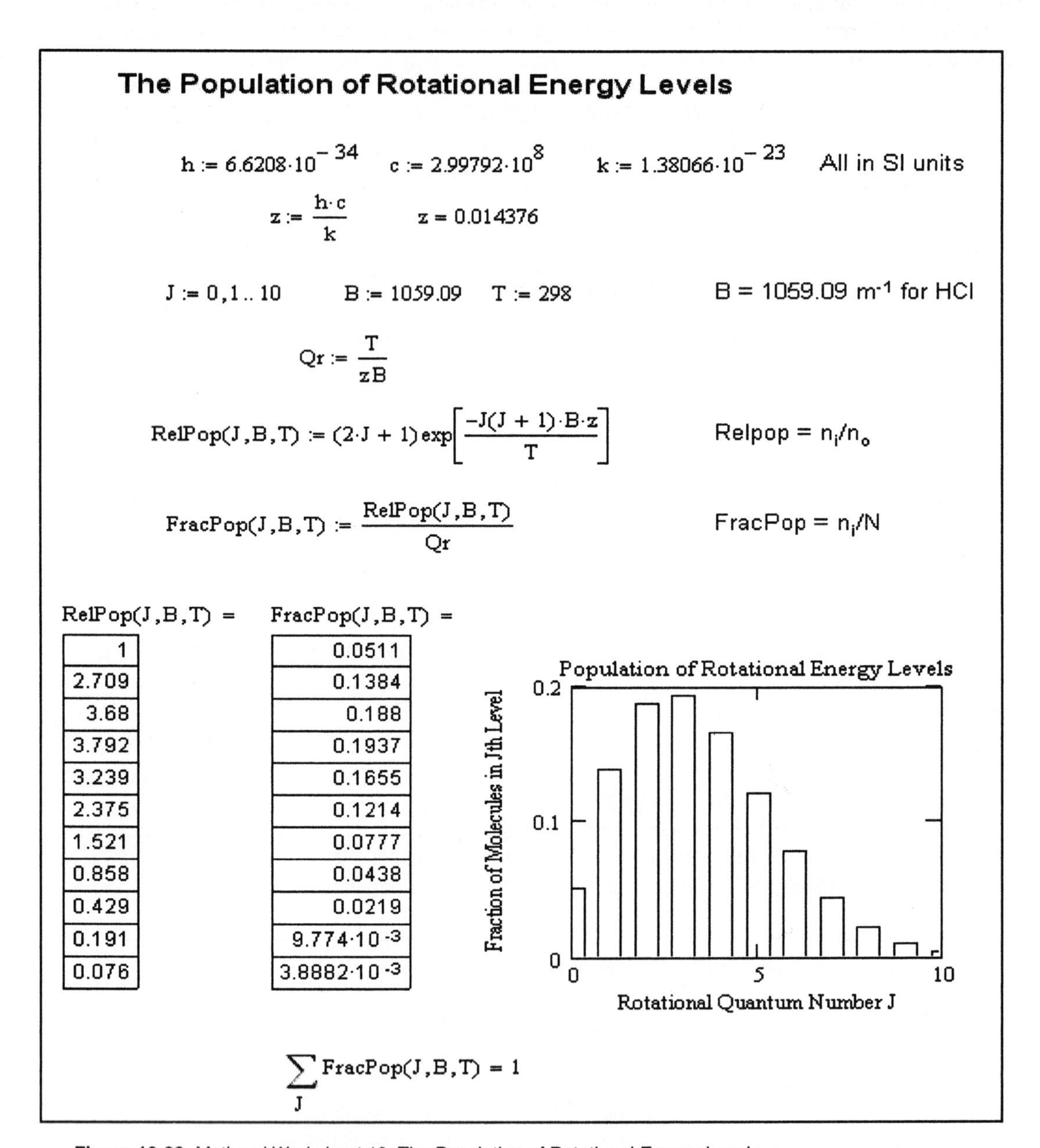

The Population of Rotational Energy Levels

$h := 6.6208 \cdot 10^{-34}$ $c := 2.99792 \cdot 10^{8}$ $k := 1.38066 \cdot 10^{-23}$ All in SI units

$z := \dfrac{h \cdot c}{k}$ $z = 0.014376$

$J := 0, 1 .. 10$ $B := 1059.09$ $T := 298$ $B = 1059.09\ m^{-1}$ for HCl

$Qr := \dfrac{T}{zB}$

$RelPop(J,B,T) := (2 \cdot J + 1)\exp\left[\dfrac{-J(J+1) \cdot B \cdot z}{T}\right]$ Relpop = n_i/n_o

$FracPop(J,B,T) := \dfrac{RelPop(J,B,T)}{Qr}$ FracPop = n_i/N

RelPop(J,B,T) =	FracPop(J,B,T) =
1	0.0511
2.709	0.1384
3.68	0.188
3.792	0.1937
3.239	0.1655
2.375	0.1214
1.521	0.0777
0.858	0.0438
0.429	0.0219
0.191	$9.774 \cdot 10^{-3}$
0.076	$3.8882 \cdot 10^{-3}$

$\sum_{J} FracPop(J,B,T) = 1$

Figure 19-23. Mathcad Worksheet 18: The Population of Rotational Energy Levels

The Effect of Temperature and Rotational Constant on the Distribution. To see how increasing the temperature increases the population of higher energy rotational state, try changing the temperature T to 500 and then to 100. To see how the rotational constant B (Table 3-2 affects the population of rotational energy states, change the temperature T back to 298 K, and then try B equals 608.9 m^{-1} for H_2 and 199.8 m^{-1} for N_2.

Mathcad Worksheet 19
The Heat Capacities of Iodine, Chlorine, and Nitrogen

The statistical thermodynamic background material for Mathcad Worksheet 19 is given in Chapter 4, "Statistical Thermodynamics," especially Excel Example 14. Compare Figure 4-1 with Mathcad Worksheet 19, shown in Figure 19-24.

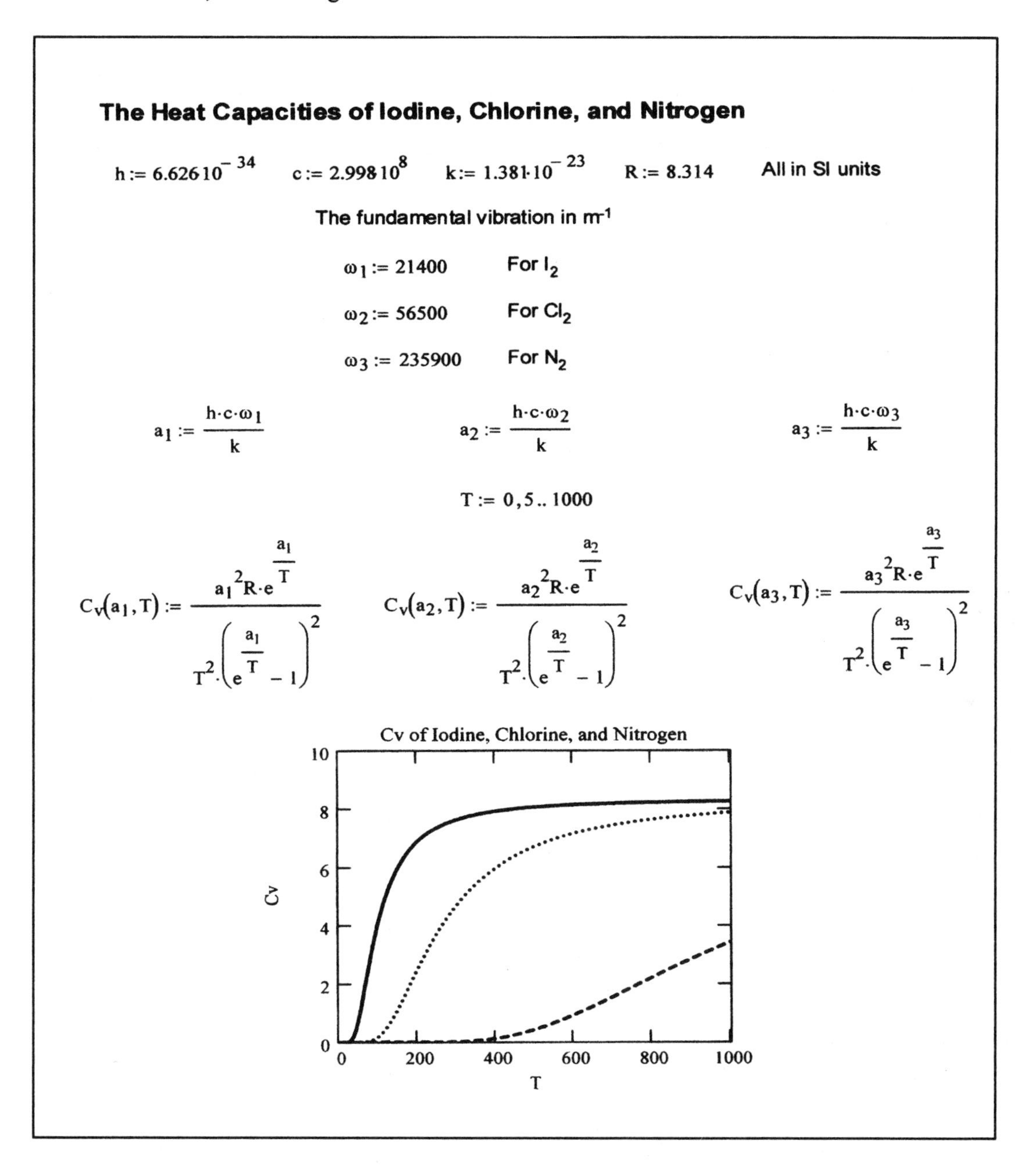

Figure 19-24. Mathcad Worksheet 19: Heat Capacity of Diatomic Molecules

Solutions of Algebraic Equations

To solve an equation, use the following three steps. Write the equation, and use the Boolean equal sign. The Boolean equal sign is a comparative operator that denotes logical equality. Access the Boolean equal sign either by typing CTRL= or by clicking its icon on the **Boolean** toolbar (Figure 19-1). The Boolean equal sign looks like an evaluation equal sign (=), except that it appears in bold (=).

1. from the keyboard type **$x^3 + x^2$ -34*x +56**
2. Click **=** on the Boolean Toolbar (not on the keyboard)
3. Type **0**
4. Click the symbolic equal sign (■→) in the **Symbolic** toolbar.
5..Type **solve,x** in the placeholder.
6. Click outside the math area to get the answer, which Mathcad presents as a column vector:

$$x^3 + x^2 - 34 \cdot x + 56 = 0 \text{ solve}, x \rightarrow \begin{pmatrix} 2 \\ -7 \\ 4 \end{pmatrix}$$

You do not actually need to set the equation equal to zero. Alternatively, you can enter just an expression and follow the same steps.

$$x^2 + x - 6 \text{ solve}, x \rightarrow \begin{pmatrix} -3 \\ 2 \end{pmatrix}$$

Another alternative to solve an equation is to define a function and solve it.

$$x^2 + x - 6 = 0$$

For this equation, you can define a function:

$$f(x) := x^2 + x - 6$$

Then, to solve the equation:

1. Type **f(x)**.
2. Click **solve** on the **Symbolic** toolbar.
3. Type **x** in the placeholder.
4. Click outside the math area.

$$f(x) \text{ solve}, x \rightarrow \begin{pmatrix} -3 \\ 2 \end{pmatrix}$$

Solving Equations That Arise in Physics and Chemistry

The keyword *solve* has a number of applications that can be used with some of the topics that have already been discussed. Several derivatives of functions lead to equations that must be solved, particularly those that involve a minimum or maximum in a curve.

Mathcad Worksheet 20
The Minimum and Maximum in a Curve

At the minimum or maximum in a curve, the first derivative equals zero. To find the minimum or maximum, take the first derivative, set it equal to zero, and solve the equation for the variable. The steps in Mathcad are as follows:

- Define the function, for example, *f(x).*
- Click the **d/dx** icon in the **Calculus** toolbar.
- Type **f(x)** in the first placeholder, and type **x** in the second placeholder.
- Click → on the **Symbolic** toolbar; click outside the math area.
- Click the resulting function, then click **substitute** in the **Symbolic** toolbar, which results in substitute, ■ →.
- Type **x** in the placeholder, and click outside the math area to see the result.

The results are shown in Mathcad Worksheet 20, shown in Figure 19-25, which displays two plots, one with a minimum and the other with a maximum.

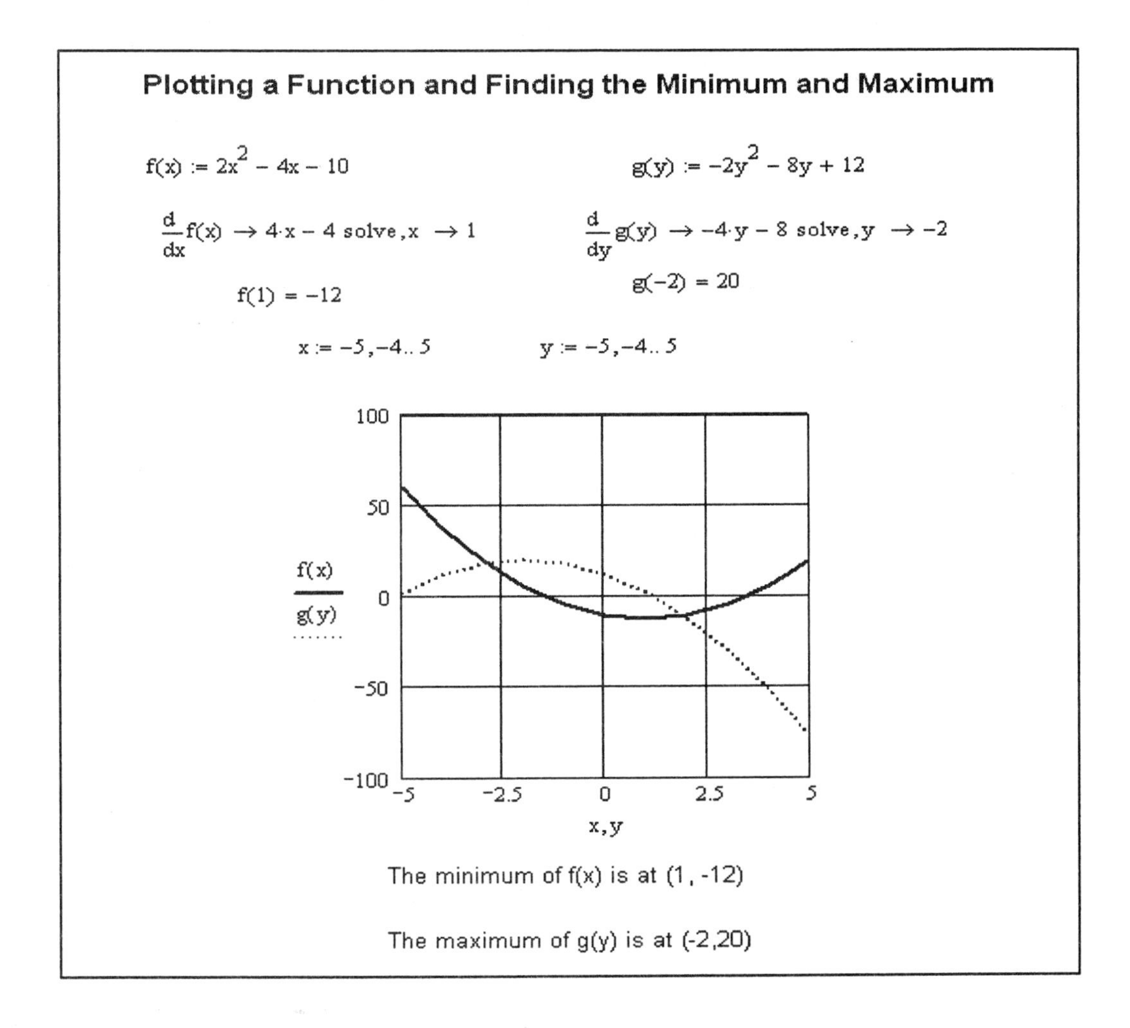

Figure 19-25. Mathcad Worksheet 20: Graphs with a Minimum and Maximum

Mathcad Worksheet 21
The Probability Distribution Function for a 1*s* Orbital: r_{mp} and r_{ave}

A previous worksheet (Figure 19-13) displayed a plot of the wave functions R_{1s}, R_{1s}^2, and $r^2R_{1s}^2$ versus *r*. The last plot is interpreted as the probability of finding an electron between *r* and *r* + *dr* regardless of direction. The location of the maximum in this curve (Figure 19-26) can be identified with the distance *r* from the nucleus where it is most probable to find the electron: r_{mp}. The derivative of $r^2R_{1s}^2$ with respect to *r* results in an equation that Mathcad can *solve* for r_{mp}. Figure 19-26 displays the plot of $r^2R_{1s}^2$ versus r/a_0, the derivative of the function, and its solution for r_{mp}. Because the probability function is not symmetrical about its maximum, the most probable *r* and the average *r* (r_{ave}) are not equal. The average value of a function x_{ave} is

$$x_{ave} = \int_0^{\infty} xf(x)dx = 1.5 \qquad (19\text{-}19)$$

where x_{ave} equals r_{ave}/a_0 so r_{ave} equals $1.5a_0$ as shown in Figure 19-26.

Mathcad Worksheet 22
The Most Probable Speed of a Gas

A previous graph (Figure 19-12) displayed a plot of a speed distribution function. The speed distribution function for a gas is

$$f(u) = 4\pi\left(\frac{M}{2\pi RT}\right)^{3/2} u^2 e^{\frac{-Mu^2}{2RT}} \qquad (19\text{-}20)$$

In Figure 19-27, the derivative of the function (Equation 19-20) was taken on one line. In order to keep the resulting equations on a single line, the resulting equation was copied, pasted, and solved on the next line for *u*, which equals u_{mp} at the maximum.

$$u_{mp} = \sqrt{\frac{2RT}{M}} \qquad (19\text{-}21)$$

Mathcad Worksheet 23
The Average Speed of a Gas

Determining u_{ave} for a gas (Figure 19-28) is similar to determining r_{ave} for a 1*s* orbital (Figure 19-28). The keyword *assume* applies three constraints to the function being integrated. You can try the keyword *simplify* on the results, but it is not very effective. With pencil and paper the result does simplify to

$$u_{ave} = \sqrt{\frac{8RT}{\pi M}} \qquad (19\text{-}22)$$

Plotting the Probability Distribution Function for a 1*s* Orbital
Calculating r_{mp} and r_{ave}

$f(x) := 4 \cdot x^2 e^{-2x}$

Note that $x = r/a_0$. The distance from the nucleus is r and a_0 is the Bohr radius (52.917 724 9 pm)

$$\frac{d}{dx}f(x) \rightarrow 8 \cdot x \cdot \exp(-2 \cdot x) - 8 \cdot x^2 \cdot \exp(-2 \cdot x) \text{ solve},x \rightarrow \begin{pmatrix} 0 \\ 1 \end{pmatrix}$$

$x := 0, 0.1 .. 5$

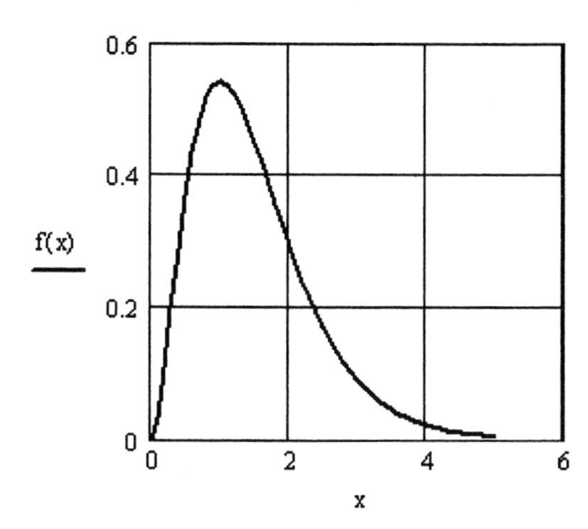

The value of x at the maximum is 1, so $\frac{r_{mp}}{a_0} = 1$ and $r_{mp} = a_0$

Thus, the most probable position for an electron in a hydrogen atom is a Bohr radius from the nucleus

The average r is obtained from $x_{ave} = \int_0^{\infty} g(x)dx$ where g(x) = x f(x)

$g(x) := x \cdot f(x)$

$$x_{ave} = \int_0^{\infty} g(x)\, dx \rightarrow x_{ave} = \frac{3}{2}$$

$$\frac{r_{ave}}{a_0} = \frac{3}{2} \quad \text{so} \quad r_{ave} = \frac{3}{2}a_0$$

r_{ave} is also the expectation value <*r*>

Figure 19-26. Mathcad Worksheet 21: Probability Distribution for a 1*s* Orbital

The Most Probable Speed of a Gas

$$f(u,M,R,T) := 4\pi\left(\frac{M}{2\pi R\cdot T}\right)^{\frac{3}{2}} u^2 e^{\frac{-M\cdot u^2}{2R\cdot T}}$$

$$\frac{d}{du}f(u,M,R,T) \rightarrow 4\cdot\frac{M}{R\cdot T}\cdot u\cdot\exp\left(\frac{-1}{2}\cdot\frac{M}{R\cdot T}\cdot u^2\right) - 2\cdot\frac{M^2}{R^2\cdot T^2}\cdot u^3\cdot\exp\left(\frac{-1}{2}\cdot\frac{M}{R\cdot T}\cdot u^2\right)$$

$$4\cdot M\cdot\frac{u}{(R\cdot T)}\cdot\exp\left[\frac{-1}{2}\cdot M\cdot\frac{u^2}{(R\cdot T)}\right] - 2\cdot M^2\cdot\frac{u^3}{\left(R^2\cdot T^2\right)}\cdot\exp\left[\frac{-1}{2}\cdot M\cdot\frac{u^2}{(R\cdot T)}\right] \text{ solve,u} \rightarrow \begin{bmatrix} 0 \\ \frac{1}{M}\cdot 2^{\frac{1}{2}}\cdot(M\cdot R\cdot T)^{\frac{1}{2}} \\ \frac{-1}{M}\cdot 2^{\frac{1}{2}}\cdot(M\cdot R\cdot T)^{\frac{1}{2}} \end{bmatrix}$$

The positive non-zero solution is the correct solution for the most probable velocity: U_{mp}

$$u_{mp} = \frac{1}{M}\sqrt{2}(MRT)^{\frac{1}{2}} = \sqrt{\frac{2RT}{M}}$$

Figure 19-27. Mathcad Worksheet 22: The Most Probable Speed of a Gas

Mathcad Worksheet 24
The Lennard-Jones Potential

The condensation of nonpolar gases into liquids suggests that even these molecules exert attractive forces between them. It's clear that polar molecules exert attractive forces between them, but even nonpolar atoms (rare gases) and molecules exert attractive forces. These forces, called London or dispersion forces, arise from the nonpolar molecule's instantaneous dipole. The instantaneous dipole, in turn, arises from the instantaneous asymmetry of the electrons in motion about a nucleus. In addition to these attractive forces between molecules, repulsive forces arise because in close proximity the electrons of one molecule repel the electrons of another molecule, and the nuclei of one molecule repel the nuclei of another molecule. J. E. Lennard-Jones studied the tug of war between intermolecular attractive and repulsive forces and proposed an empirical function for the potential energy between molecules. Table 5-1 lists a few value of ε/k and σ.

$$V = 4\varepsilon\left[\left(\frac{\sigma}{r}\right)^{12} - \left(\frac{\sigma}{r}\right)^{6}\right] \qquad (19\text{-}23)$$

The Average Speed of a Gas

$$u_{ave} = \int_0^{\infty} f(u,M,R,T)du$$

$$f(u,M,R,T) := 4\pi\left(\frac{M}{2\pi R\cdot T}\right)^{\frac{3}{2}} u^3 e^{\frac{-M\cdot u^2}{2R\cdot T}}$$

$$\int_0^{\infty} f(u,M,R,T)\,du \,\left|\begin{array}{l}\text{assume},M > 0\\ \text{assume},R > 0\\ \text{assume},T > 0\end{array}\right. \rightarrow \frac{2}{M^2}\cdot R^2\cdot T^2\cdot\pi\cdot 2^{\frac{1}{2}}\cdot\left(\frac{M}{\pi\cdot R\cdot T}\right)^{\frac{3}{2}}$$

Click three times on the keyword *assume* in the Symbolic toolbar to apply the three conditions

The keyword *simplify* does not simplify much, but the last term does simplify to give the average speed u_{ave}

$$u_{ave} = \left(\frac{8RT}{\pi M}\right)^{1/2}$$

Figure 19-28. Mathcad Worksheet 23: The Average Speed of a Gas

Figure 19-29 shows a display of the Lennard-Jones potential. It is customary to plot in units of ε/k where k is the Boltzmann constant. The steps in creating the plot are as follows:

- Define σ and ε.
- Define the function $f(r)$.
- Define the range variable r.
- On the **Insert** menu, click **Graph,** and select **X-Y Plot.**
  1. Type **f(r)** in the left placeholder, and type **r** in the bottom placeholder.
  2. Right-click the graph, and then click **Format.**
     a. Click the **X-Y Axes** tab, and select the **Gridlines** check box.
     b. Click the **Labels** tab; add a title, and type **r/pm** as the x-axis label and type **epsilon/k** as the y-axis label.
     c. Click the **Traces** tab, and select the **Hide Arguments** and **Hide Legend** check boxes.

Mathcad Worksheet 25
The Minimum in the Lennard-Jones Potential

Figure 19-30 shows the calculation of the minimum in the Lennard-Jones plot. Taking the derivative of the Lennard-Jones potential function results in an equation in the sixth power of r. Application of the keyword *solve* results in six roots. Four of the roots are imaginary, and one is negative. Only the first listed root is reasonable, so r_{min} equals $\sqrt[1/6]{2}\sigma$.

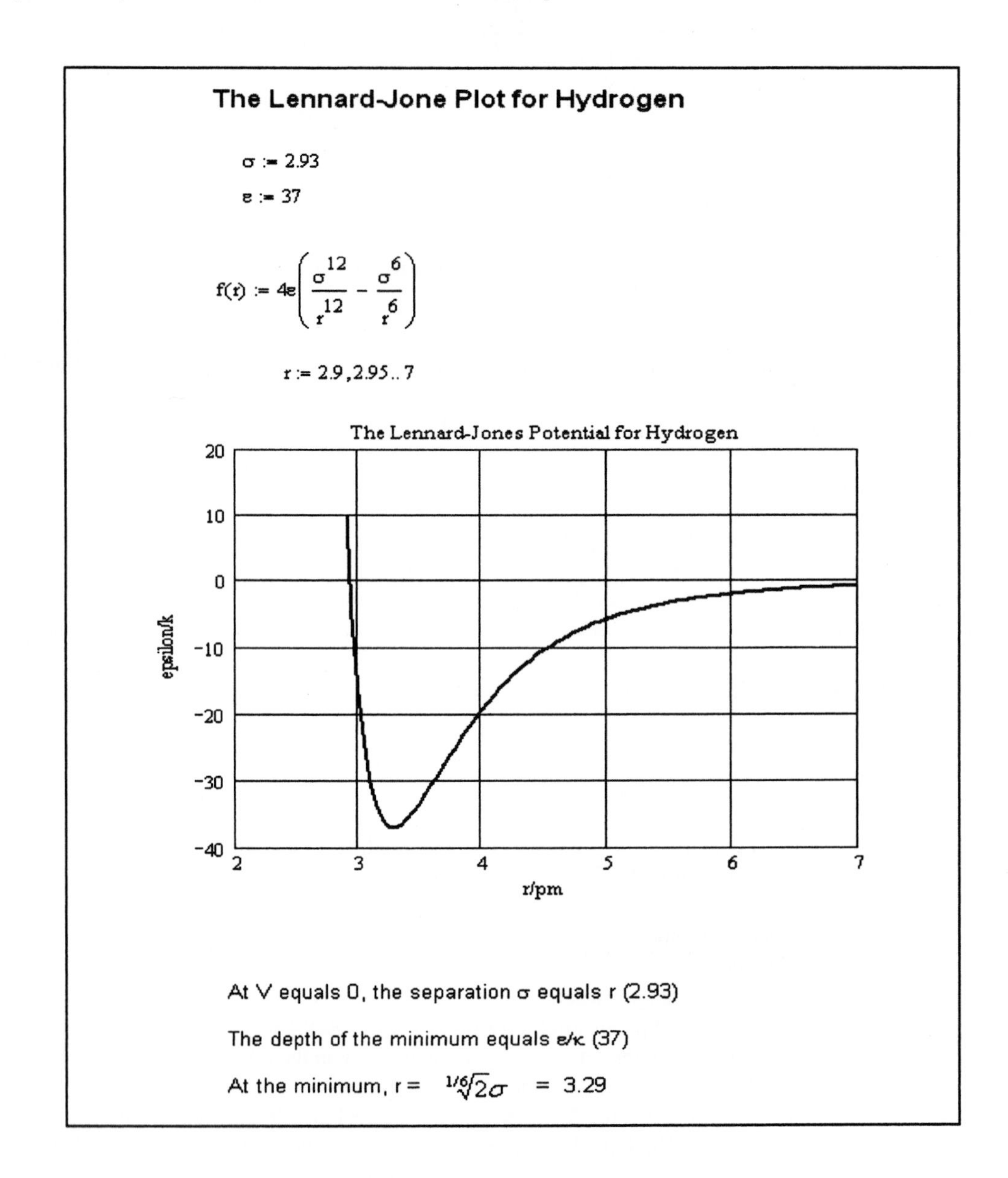

Figure 19-29. Mathcad Worksheet 24: The Lennard-Jones Potential

Minimum in the Lennard-Jones Potential

$$\frac{d}{dr}\left(\frac{\sigma^{12}}{r^{12}} - \frac{\sigma^{6}}{r^{6}}\right) \rightarrow -12\cdot\frac{\sigma^{12}}{r^{13}} + 6\cdot\frac{\sigma^{6}}{r^{7}}$$

$$\frac{-12\sigma^{12}}{r^{13}} + \frac{6\sigma^{6}}{r^{7}} \text{ solve,r} \rightarrow \begin{bmatrix} 2^{\frac{1}{6}}\cdot\sigma \\ \left(\frac{1}{2} + \frac{1}{2}\cdot i\cdot 3^{\frac{1}{2}}\right)\cdot 2^{\frac{1}{6}}\cdot\sigma \\ \left(\frac{-1}{2} + \frac{1}{2}\cdot i\cdot 3^{\frac{1}{2}}\right)\cdot 2^{\frac{1}{6}}\cdot\sigma \\ -2^{\frac{1}{6}}\cdot\sigma \\ \left(\frac{-1}{2} - \frac{1}{2}\cdot i\cdot 3^{\frac{1}{2}}\right)\cdot 2^{\frac{1}{6}}\cdot\sigma \\ \left(\frac{1}{2} - \frac{1}{2}\cdot i\cdot 3^{\frac{1}{2}}\right)\cdot 2^{\frac{1}{6}}\cdot\sigma \end{bmatrix}$$

$$r_{min} = \sqrt[6]{2}\sigma$$

Figure 19-30. Mathcad Worksheet 25: The Minimum in the Lennard-Jones Plot

Solutions of Simultaneous Equations

Mathcad offer several methods for solving systems of equations, linear and nonlinear. Three that are presented in this chapter are the Matrix Method, the *Solve* Keyword Method, and the Solve Block Method.

A typical set of three linear equations might look like the following:

$$3\cdot x - y + z = 11$$
$$x + 3\cdot y - z = 13$$
$$x + y - 3\cdot z = 11$$

A generalized set of linear equations can be represented by

$$\begin{aligned} &a_{11}x_1 + a_{12}x_2 + \ldots + a_{1j}x_j + \ldots + a_{1n}x_n = b_1 \\ &a_{21}x_1 + a_{22}x_2 + \ldots + a_{2j}x_j + \ldots + a_{2n}x_n = b_2 \\ &\ldots\ldots\ldots\ldots\ldots\ldots\ldots\ldots\ldots\ldots \\ &a_{i1}x_1 + a_{i2}x_2 + \ldots + a_{ij}x_j + \ldots + a_{in}x_n = b_i \\ &\ldots\ldots\ldots\ldots\ldots\ldots\ldots\ldots\ldots\ldots \\ &a_{n1}x_1 + a_{n2}x_2 + \ldots + a_{nj}x_j + \ldots + a_{nn}x_n = b_1 \end{aligned} \qquad (19\text{-}24)$$

Mathcad Worksheet 26
The Matrix Method for Solving Linear Equations

Equation 19-24 can be written in matrix form as

$$M \cdot s = v \qquad (19\text{-}25)$$

M is the matrix of the coefficients, and *v* is the right side written as a column vector, both of which are known. The solution to the equation is *s*, which is also a column vector. If M^{-1} is the inverse of *M,* then the solution *s* to the set of linear equations is given by

$$s = M^{-1}\cdot v \qquad (19\text{-}26)$$

With Mathcad it is easy to define *M* and *v* and to determine the inverse of *M.* The Matrix Method of solving linear equations is shown in a Mathcad worksheet (Figure 19-31), prepared as follows:

- To define the matrices:
  1. Type **M:**.
  2. Click the **Matrix** icon on the **Matrix** toolbar (Figure 19-1), and set columns and rows to **3**.
  3. Enter the coefficients, and press the TAB key between each entry.
  4. Repeat for the *v* vector with rows set to **3** and columns set to **1**.

- Define the solution vector s in terms of the inverse of M and v: $s = M^{-1} \cdot v$. The inverse, M^{-1}, is located on the **Matrix** toolbar.
- Display the solution by typing s =.

The solution then appears in a column vector.

Matrix Method for Solving Linear Equations

Here is a set of linear equations

3x - y + z = 11
x + 3y -z = 13
x + y -3z = 11

Define the matrix of the coefficients M.
Define the matrix of the right side v

$$M := \begin{pmatrix} 3 & -1 & 1 \\ 1 & 3 & -1 \\ 1 & 1 & -3 \end{pmatrix} \qquad v := \begin{pmatrix} 11 \\ 13 \\ 11 \end{pmatrix}$$

Define the solution in term of M^{-1} and v

$$s := M^{-1} \cdot v$$

Evaluate s to display the soluton

$$s = \begin{pmatrix} 4.857 \\ 2.286 \\ -1.286 \end{pmatrix}$$

Figure 19-31. Mathcad Worksheet 26: The Matrix Method

Mathcad Worksheet 27
The *Solve* Keyword Method for Solving Linear Equations

The *Solve* Keyword Method for Solving Linear Equations. This method is illustrated in Figure 19-32 and involves the following steps:

- Click the **Matrix** icon in the **Matrix** toolbar, and define a one-column matrix with as many rows as there are equations to solve.
- In the matrix's placeholders, enter each equation and write the equation with the Boolean equal sign, which is found on the **Boolean** toolbar.
  1. Type the keyword **solve**, and in the placeholder, enter the answer matrix as follows:
  2. Click the **Matrix** icon in the **Matrix** toolbar and define a one-column matrix with as many rows as there are equations to solve (just like the first matrix).
  3. In this column matrix, enter into the placeholder the names of the variables to be found (**x**, **y**, and **z**).

TheSolve Keyword Method for Solving Linear Equations

Here is a set of linear equations

3x - y + z = 11
x + 3y -z = 13
x + y -3z = 11

Place a matrix on the worksheet containing 1 column
and as many rows as linear equations (3 here).
Fill each placeholder with one of the equations

Click the keyword *solve*

In the *solve* placeholder, enter a one-column matrix.
Enter the variable names in its placeholders.

Click the arrow in the SymbolicToolbar.

$$\begin{pmatrix} 3x - y + z = 11 \\ x + 3y - z = 13 \\ x + y - 3z = 11 \end{pmatrix} \text{solve}, \begin{pmatrix} x \\ y \\ z \end{pmatrix} \rightarrow \begin{pmatrix} \frac{34}{7} & \frac{16}{7} & \frac{-9}{7} \end{pmatrix} \text{float}, 4 \rightarrow (4.857 \ \ 2.286 \ \ -1.286)$$

These are the answers ...here in decimal form.

Figure 19-32. Mathcad Worksheet 27: The *Solve* Keyword Method for Solving Linear Equations

- With the SPACEBAR, adjust the edit lines to hold the entire matrix, and then in the **Symbolic** toolbar, click → and then press ENTER. The answers appear in a row matrix.
- Optional: Type the function **float, n→** to get the answers in decimal form with *n* digits.

Mathcad Worksheet 28
The Solve Block Method for Solving Linear Equations

The Solve Block Method uses the keyword *Given* to establish a block containing a system of equations, and the **Find** function to obtain the solutions. The following steps illustrate the method in the Mathcad worksheet shown in Figure 19-33.

- To each variable, assign a value for the answer. This assignment is your guess and can be virtually anything.
- Type the word **Given** with one capital letter as shown. Enter this word in a math region, not in a text region.
- Write the *n* equations with a Boolean equal sign between the left and right sides of each equation. (The right side does not have to equal zero.)
- Choose any legal variable name (like *v*), and assign the solutions to it with the **Find** function, for example, **v := Find(x,y,z)**. The **Solve Block** consists of the lines from **Given** through **Find**.
- Use the evaluation equal sign (=) to display the answer. The answer is displayed as a vector, whose elements are the solutions to the system of equations.

The *Solve Block* Method for Solving a System of Equations

Three Simultaneous Equations in Three Unknowns

$x := 1$ Guess the answers. They don't have to be very close

$y := 1$

$z := 1$

Given — Before writing the equations, write the keyword word **Given**

$3 \cdot x - y + z = 11$

$x + 3 \cdot y - z = 13$

$x + y - 3 \cdot z = 11$

Then write the equation with a = Do this with a CTRL= or by clicking on the arithmetic icon.

$v := \text{Find}(x, y, z)$

To any variable (here "v,") assign the value of "Find(x,y,z)".

$$v = \begin{pmatrix} 4.857 \\ 2.286 \\ -1.286 \end{pmatrix}$$

Then type "v =" and the answers apppear

Four Simutaneous Equations in Four Unknowns

$x := 1 \quad y := 1 \quad w := 1 \quad z := 1$ Initial guesses (not very critical) are:

Given — Must be typed ahead of equations

$x + y + z - w = 1$
$x - y - z = 4$
$x + z - w = 4$
$x + y - z + w = -3$

Type the = with a CTRL=

$v := \text{Find}(x, y, z, w)$

$$v = \begin{pmatrix} 2 \\ -3 \\ 1 \\ -1 \end{pmatrix}$$

A vector is returned, the elements of which are the solutions. (type v= to display the vector elements)

Two Simultaneous Nonlinear Equations in Two Unknows

$p := 1 \quad q := 1$ Initial guesses

Given

$q^2 = 9p$

$q^2 - p^2 = 8$

$$\text{Find}(p, q) \rightarrow \begin{pmatrix} 1 & 1 & 8 & 8 \\ -3 & 3 & 6 \cdot 2^{\frac{1}{2}} & -6 \cdot 2^{\frac{1}{2}} \end{pmatrix} \text{float}, 4 \rightarrow \begin{pmatrix} 1. & 1. & 8. & 8. \\ -3. & 3. & 8.484 & -8.484 \end{pmatrix}$$

$p = 1, q = \pm 3$

$p = 8, q = \pm 8.484$

Figure 19-33. Mathcad Worksheet 28: The Solve Block Method for Solving a System of Equations

Chapter 20

Mathematica

Mathematica is an application for numerical calculations, for symbolic calculations, and for graphics. It can crunch numbers; it can factor polynomials, solve equations, and do differential and integral calculus. It can graph just about any function you can imagine, in two or three dimensions. And it knows a lot of mathematics and quantum mechanics. For example, it knows the wave functions for a harmonic oscillator and can list them, evaluate them, and plot them for you. And much more.

Introduction: The Mathematica Notebook

When you invoke Mathematica by clicking the **Mathematica** icon on the desktop of your computer, a screen like the one in Figure 20-1 appears. Unlike Figure 20-1, the white area under the title bar is entirely blank, except for the Mathematica pointer (a short horizontal line with split ends). Your mouse will move the Mathematica pointer around the screen, but regardless of where the pointer is initially located, the first character that you type appears in the upper left corner of the screen. The screen in Figure 20-1 is a Mathematica notebook. The notebook in Figure 20-1 is untitled. When you save a notebook, it is saved under the name you give it, but with the extension .nb.

Notebook Regions. Notebooks contain three different kinds of regions: text, numerical, and graphics. Several of these regions may alternate with each other. All of these regions will be illustrated in due course, but now let us begin with a numerical region. If you just start typing, Mathematica assumes that you are working in a numerical region. If the first character that you type is a space, then Mathematica assumes that you are going to type text and acts as a word processor that you can use to annotate the calculations going on in the numerical regions of the notebook.

The pair of text lines at the top of the notebook in Figure 20-1 lie in a *text region* (or cell) of the notebook. Notice the cell brackets at the far right that serve to delineate the text cell region. The pairs of lines that contain numbers and lie in the middle of the notebook are positioned in a *numerical region* of the notebook. Cell brackets at the far right also delineate this numerical region of the notebook. The last text line is in a text region, which is also delineated with cell brackets. Later on we'll look at a graphics region in Mathematica.

First Text Region. To reproduce Figure 20-1, open a new notebook (on the **File** menu, click **New**) and type text in the first text region:

- Type a single space. A vertical text entry pointer appears.
- Type the first line of text, and press the ENTER key at the end.

- Type the second line of text, and then move the pointer below the second line and click it. A horizontal line appears, indicating the end of this text region.

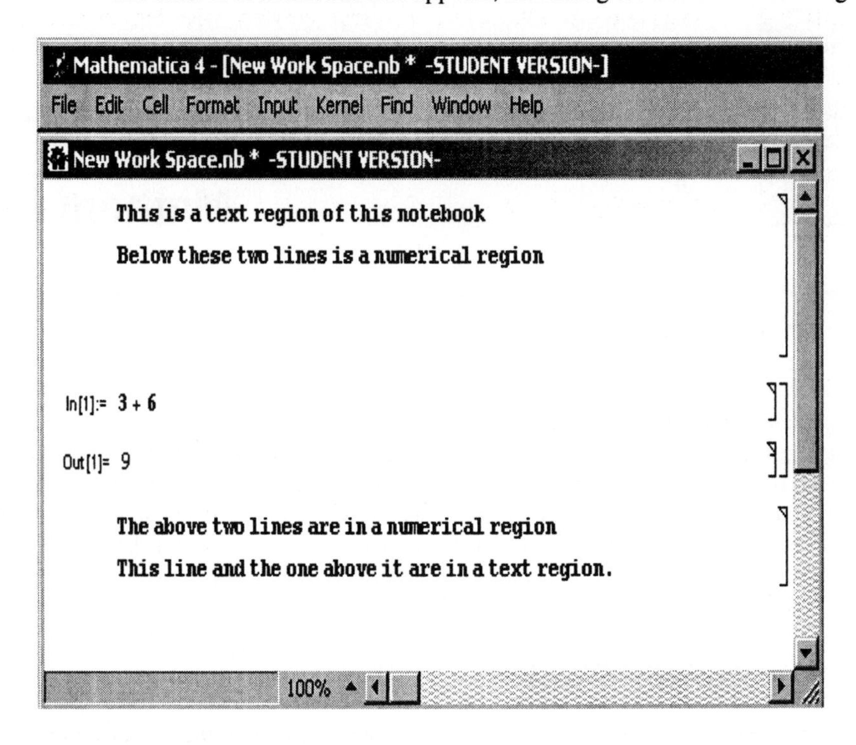

Figure 20-1. The Workspace of a New Mathematica Notebook

The Numerical Region. The [In]s and [Out]s of Mathematica. Mathematica assumes that the next region is numerical unless you type a space as the first character, so do not type a space if you want the next region to be a numerical region.

- Type **3+6** (do not type a space before the **3**).
- Press SHIFT+ENTER by holding the SHIFT key down while pressing the ENTER key (or press the SHIFT key located at the far lower right of your keyboard).
- Mathematica responds by inserting **In[1]:=** before the 3 + 6 and by inserting **Out[1]=** before the result (9) of adding 3 to 6.

Mathematica always labels the input expression as **In[n]:=** and the output value as **Out[*n*]=**. The values of *n* range from 1 to *n* for a given notebook. If you want to hide these labels, on the **Kernel** menu on the menu bar, uncheck **Show In/Out Names**. At the extreme lower right corner of your keyboard is another key labeled ENTER. Pressing this key has the same effect as pressing SHIFT+ENTER.

Second Text Region. Move the pointer below the last line (**Out[1] = 9**) on the screen, and press the ENTER key. Mathematica responds by inserting a horizontal black line, marking the end of the previous region.

- Type a single space, and then type the first of the last two lines shown on Figure 20-1.
- Press the ENTER key, and add the second line.

Within a text region, you can do all the things that you normally do with a word processor, such as click and drag to select, copy, cut, and paste.

Notice that on the right edge of the screen appears a group of what look like right square brackets, some long ones subtending two or three shorter square brackets. These cell brackets delineate the regions they bracket. If you click a little square bracket, it will blacken, indicating that it has been selected. Press the DELETE key, and the cell it defines disappears. If you click a bigger square bracket that subtends two or three little ones, they all blacken, and the DELETE key deletes all input and output to their immediate left.

Beside deleting a selected cell, you can copy, cut, and paste the selected cell, permitting you to rearrange the text regions of your notebook.

The File Menu

Save and Open. The File menu on the menu bar has several familiar commands: **New**, **Open**, **Close**, **Save**, and **Save As**. If you click **Save** on the **File** menu, you are prompted for a file name. Mathematica tacks on the file extension .nb to all of its files. If you print the file as shown if Figure 20-1, the **In[1]:**, **Out[1]**, and so on print along with your results of interest, but the cell brackets do not print. After you save a file and then click **Open** on the **File** menu to resume work on it, you will find that the **In[1]:** and **Out[1]** have disappeared. You can get them back if you wish by placing the pointer after each **In[ ]:** line (the first line of the **In-Out** pair) and pressing SHIFT+ENTER. The **In[*n*]:** and **Out[*n*]** groups will reappear in pairs.

Palettes. The **Palettes** command on the **File** menu provides a number of helpful symbols and syntaxes for various operations and functions. So far we've encountered only one arithmetic operation: addition. There are, of course, a great many more, and if you forget the syntax, one way to get the information quickly is with palettes.

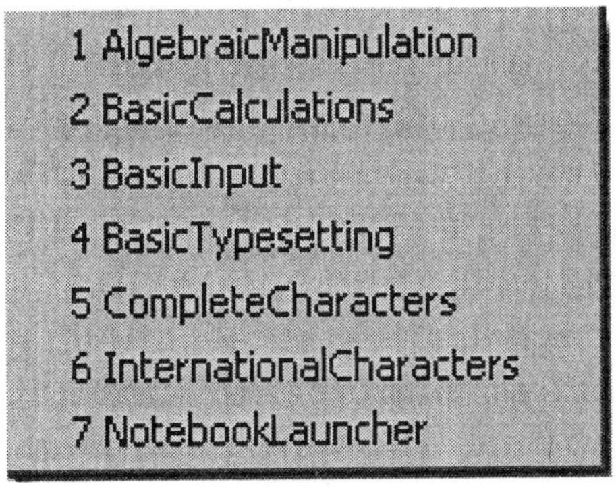

On the **File** menu, click **Palettes**, and you get a submenu that lists seven palettes (Figure 20-2). On the **Palettes** submenu, click **1 Algebraic Manipulation**, and this palette floats over your notebook in the upper left corner. Figure 20-3 displays all seven palettes at once, although you would normally select only one or two. In accordance with the numbers in Figure 20-2, the palettes in Figure 20-3 are arranged clockwise, with the **Algebraic Manipulation** palette in the upper left. The **Notebook Launcher** palette is at the bottom right, and in the figure, it is the selected palette.

1 AlgebraicManipulation
2 BasicCalculations
3 BasicInput
4 BasicTypesetting
5 CompleteCharacters
6 InternationalCharacters
7 NotebookLauncher

Figure 20-2. The **Palettes** Submenu

Using the Palettes. Before we go on to illustrate more arithmetic operations than addition, let us pause and look briefly at the palettes, especially the **Basic Calculations** palette (second from left on Figure 20-3). The operations listed for the **Basic Calculations** palette are:

- **Arithmetic and Numbers**
- **Algebra**
- **Lists and Matrices**
- **Trigonometric and Exponential Functions**
- **Calculus**
- **Other Functions**
- **Graphics**

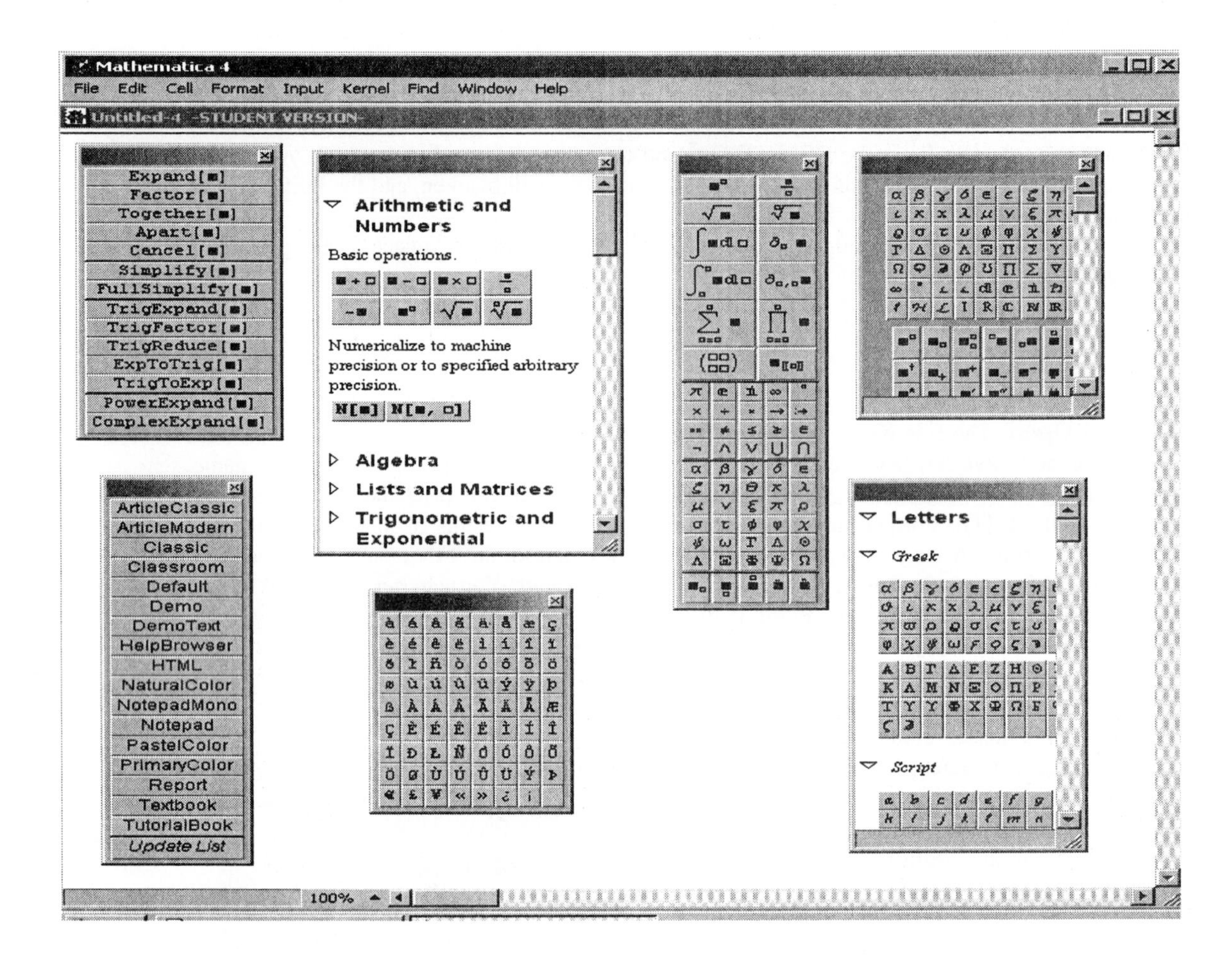

Figure 20-3. A Mathematica Notebook Nearly Completely Covered with the Seven Palettes

In Figure 20-3, the **Arithmetic and Numbers** option of palette 2 has been selected. This displays the symbols for the basic operations and one or two placeholders for the operands. Clicking one of these objects places it on your workspace. You fill in the placeholders by clicking each one and typing the desired numeric value. Then pressing SHIFT+ENTER results in the usual **In[*n*]:**, **Out[*n*]** pair with the input expression and the output value.

While palettes 1, 2, and 3 deal with activity in the numerical region of the notebook, palettes 4, 5, 6, and 7 deal with writing in the text region of the notebook. For example, you can use palette 4 (**Basic Typesetting**) to type into your text region strings such as:

$$\psi^2 \text{ and } {}^{238}_{92}U$$

Help! While the palettes provide brief, but readily accessible, help, the **Help** menu gives extensive help on all the issues in Mathematica. The **Help Browser** command is at the top of the **Help** menu. On the **Help** menu, click **Help Browser** to open the **Help Browser** window (Figure 20-4). The best way to get familiar with Mathematica **Help** is to click **Help** and explore the possibilities, especially the **Help Browser** window. Figure 20-4 shows the **Help Browser** opened for help on addition (**Plus**). The subdivisions for this selection are:

Below the selected operation (**Plus**), the **Help** page lists the syntax for the operation and several associated Mathematica nuances.

The definitive book on Mathematica is Stephen Wolfram's *Mathematica: A System for Doing Mathematics by Computer* (Addison Wesley, Reading, MA, 1991, 2nd ed.). This reference book covers all of Mathematica.

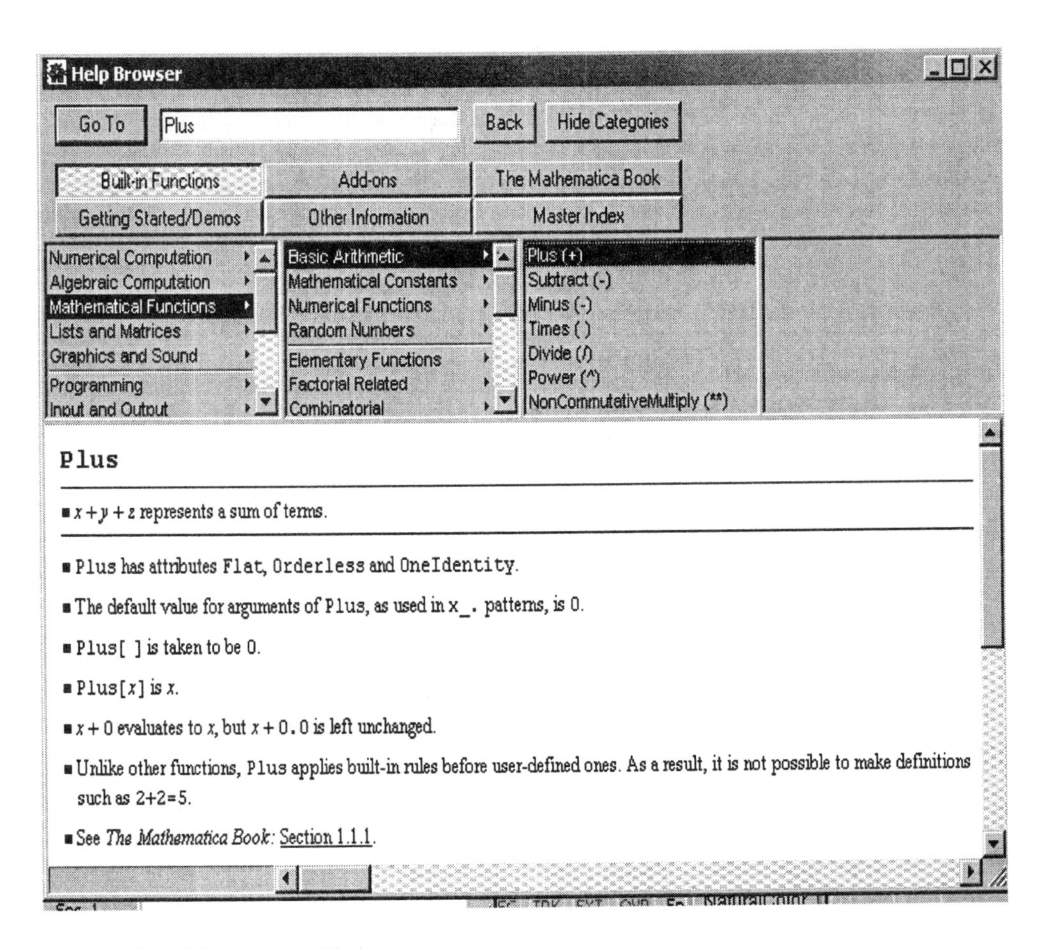

Figure 20-4. The **Help Browser** Window

Elementary Calculations

Arithmetic Operators. In this section, several of Mathematica's arithmetic operators are illustrated. Their symbols are +, -, *, /, **Mod**, **Div**, **&**, ^, and CTRL^.

Addition is pretty much as you would expect.

In[1]:= **3.123 + 12.765**

Out[1] = 15.888

Subtraction also produces no surprises.

In[2]:= **4.32 – 23.567**

Out[2] = -19.247

Multiplication usually uses a space between the multiplier and multiplicand.

In[3]:= **2.34 4.567**

Out[3] = 10.6868

However, you can multiply with * or a lowercase or uppercase x as the operator. In this case, the displayed product is followed by the operator x, which can be deleted to give the product alone, but as a text representation of the number.

Multiplication with x:

In[4]:= **2.34 x 4.567**

Out[4] = 10.6868 x
10.68678

Division produces no surprises if one or both of the numbers is a decimal number.

In[n]:= **3 / 7.12**

Out[n] = 0.421348

But if both of the numbers are integers

In[n]:= **3 / 7**

Out[n] = $\frac{3}{7}$

Mathematica tries to give an exact answer. It considers $\frac{3}{7}$ to be an exact answer, and it considers the decimal form to be an approximation.

You can get the decimal answer to 3/7 by following the input by **//N**.

In[n]:= **3 / 7 / / N**

Out[n] = 0.428571

Another way is to add a decimal point to one of the integers.

In[n]:= **3. / 7**

Out[n] = 0.428571

Finally, Mathematica has a built-in function for displaying decimal results directly. We will look into other functions later, but it is convenient to use the **N** function. Like any function, the N function takes an argument and returns a value.

The argument of the **N** function is any expression that has a numerical value.	In[1]:= **N[3/7]**
The **N** function returns a decimal value.	Out[1] = 0.4286

A Mathematica function always uses square brackets [ ] to contain the argument. The symbol % represents the result of the last calculation and often comes in handy, as the next example shows.

Input a fraction:	In[2]:= **5/6**
Output is the most precise value.	Out[2] = $\frac{5}{6}$
Use the result of the last **Out[*n*]** as the argument.	In[3]:= **N[%]**
The **N** function returns a decimal value.	Out[3] = 0.833333

Predefined Constants. Mathematica has a few useful predefined constants. The value of Pi (π) is a Mathematica constant. Either Pi or π may be used, but Pi is quicker, since you must get π from a palette.

Evaluating the predefined constant Pi:	In[n]:= **N[Pi]**
	Out[n] = 3.14159
The **N** function can take a second parameter that specifies the number of decimal places. The default number is six.	In[n]:= **N[Pi, 30]** Out[n] = 3.145926735897932384626433832 8
The names of all of Mathematica's predefined constants must begin with an uppercase letter.	In[n]:= **N[E]** Out[n] = 2.71828
Degree is really a conversion factor that has units of radian/degree (π radians/180 degrees equals 0.0174533 radian/degree).	In[n]:= **N[Degree]** Out[n] = 0.0174533
How many radians in 180 degrees?	In[n]:= **180 Degree / / N]**
	Out[n] = 3.14159
What is the cosine of 180 degrees?	In[n]:= **Cos [180 Degree / / N]**
	Out[n] = -1.

User-Defined Constants. You may define your own constant with an equal sign (=). Mathematica distinguishes between uppercase and lowercase, as the following examples demonstrate:

Define a lowercase constant.	In[1]:= **x = 6**
	Out[1] = 6
Define an uppercase constant.	In[2]:= **X = 7**
	Out[2] = 7
Square the lowercase constant.	In[3]:= **x ^ 2**
	Out[3] = 36
Square the uppercase constant.	In[4]:= **X^2**
	Out[4] = 49

The equal sign (=) is the usual assignment operator, although Mathematica also uses the symbol := for the assignment operator. In addition, Mathematica uses the double equal sign (= =) for logical equality as in equations. It is similar to the Boolean equal sign of Mathcad. We will explore these symbols later. In Mathematica's nomenclature, = is the **Set** operator, := is the **Delayed Set** operator, and = = is the **Equal** operator.

You can use scientific notation for your numbers and constants. Following, *h* (Planck's constant), *k* (Boltzmann's constant), and *c* (the speed of light) are defined.

Planck's constant:	In[5]:= **h = 6.626 !0^-34**
	Out[5] = 6.626 x 10^{-34}
The Boltzmann constant:	In[6]:= **k = 1.38 10^-23**
	Out[6] = 1.38 x 10^{-23}
The speed of light:	In[7]:= **c = 2.998 10^8**
	Out[7] = 2.998 x 10^{8}

These constants can now be used in calculations.	In[8]:= **h c/k**
	Out[8] = 0.0151639

Parentheses ( ) may be used freely to force the order of calculations. Brackets [ ] are reserved for containing the arguments of functions. Braces { } are used with lists. These usages are not interchangeable.

When *h, k,* and *c* were defined previously, the SHIFT+ENTER keys were pressed after each definition. In the next example, the ENTER key is pressed after each definition.

After typing the **5**, **3**, **2**, and **c**, press the ENTER key instead of SHIFT+ENTER.

```
In[n]:= a = 5
        b = 3
        c = 2
        a / b +c
Out[n] = 5
Out[n] = 3
Out[n] = 2
```

Out[n] = $\frac{11}{3}$

Use parentheses to force the order of calculation.

```
In[n]:= a / (b + c)
Out[n] = 1
```

The Semicolon as a Line Terminator. In the previous examples we terminated each line with the ENTER key. This action delays the **Out[*n*]** = for each value of *a, b,* and *c* until the SHIFT+ENTER key are pressed. Terminating each line with a semicolon (;) suppresses the **Out[*n*]** = even after SHIFT+ENTER are pressed.

Terminate the first three lines with a semicolon and SHIFT or SHIFT+ENTER. Place no semicolon after the **a*b*c** line. Press SHIFT+ENTER.

```
In[n]:= a = 3;
        b = 6;
        c = 7;
        a*b*c
Out[n] = 126
```

The semicolon has the effect of suppressing the output of these three lines, which is useful for making your page tidier and more compact.

The Clear Function. Several assignments were made in the preceding examples. If we wish to use the same variable names (*a*, *b*, or *c*) but wish to assign new values, the previous assignments can be cleared with the **Clear** function. Its arguments are the names of the variables to be cleared, enclosed (as always with Mathematica functions) within brackets.

```
In[n]:= Clear [a, b, c]
```

For the **Clear** function, Mathematica does not generate an **Out[*n*]:=**.

Functions

Mathematica has a large number of built-in functions, a few of which are demonstrated below. In addition, users can define their own functions.

Built-in Functions. Mathematica divides its built-in functions into two categories: elementary functions and numerical functions. Some of more common functions are shown in Table 20-1. The **Help Browser** gives detailed information on each function. If you know the name of the function, you can type it directly. Otherwise, use the **Help Browser** or the **Basic Calculations** palette.

On the **File** menu, point to **Palettes**, and click **Basic Calculations**. The **Basic Calculations** palette is shown in Figure 20-3 (second palette from the left). Click **Trigonometric and Exponential Functions** or **Other Functions** to get a list of functions. Clicking a member of the list places that function on your Mathematica notebook.

Table 20-1

Some Selected Built-in Functions for Elementary Calculations

Elementary Functions	Numerical Functions
Log[x]	Abs[x]
Exp[x]	Round[x]
Sqrt[x]	IntegerPart[x]
Sin[x]	FractionalPart[x]
Cos[x]	Max[x_1, ... ,]
Tan[x]	Min[x_1, x_n ...]
Csc[x]	Mod[x_1, x_n ...]
ArcSin[x]	Quotient[x, y]
ArcSinh[x]	x!
x^y	Random[x]

Mathematica includes **x!** as a numerical function even though it takes no argument as usual functions do and is written in lowercase.

In[1]:=**5!**

Out[1] =120

In[2]:=**10!**

Out[2] =628800

In[3]:=**100!**

Out[3]=93326215443944152681699238856266700490715968264381621468592963895217599993229915608941463976156518286253697920827223758251185210916864000000000000000000000000

If you save a Mathematica notebook and open it at a later time, Mathematica omits the **In[*n*]** and **Out[*n*]** that precede the lines as they are typed. To make them appear in the preceding section of a worksheet, place the pointer to the immediate right of an input line (boldface) and press the SHIFT+ENTER keys. Repeat for each input line. If **In[*n*]** and **Out[*n*]** do not appear, on the **Kernel** menu, click **Show In/Out Names**. It should be checked.

Elementary Functions. The next several examples include selections from Table 20-1.

The **Log[x]** function returns the log of x to the base e. Use the **N** function to force a decimal result.	In[n]:= **N[Log [100]]** Out[n] = 4.6051
The **Log[b,x]** function returns the log of x to the base b. This next example returns the log of 100 to the base 10.	In[n]:= **Log[10,100]** Out[n] = 2
This next example returns the log of 100 to the base 8.	In[n]:= **Log[8,100]** Out[n] = $\frac{Log[100]}{Log[8]}$ In[n]:= **N[%]** Out[n] = 2.21463
The **Sin[x]** function also returns a nondecimal number, but the result can be made into a decimal result with the **N** function.	In[n]:= **Sin[Pi / 13]** Out[n] = $\left[\frac{\pi}{13}\right]$ In[n]:= **N[%]** Out[n] = 0.239316
The **Degree** constant is especially useful with transcendental functions.	In[n]:= **Cos[30 Degree]** Out[n] = $\frac{\sqrt{3}}{2}$ In[n]:= **N[%]** Out[n] = 0.86602
Exponentiation can be done with ^ or **CTRL^**. First, ^ without pressing CTRL:	In[n]:= **2 ^ 8** Out[n] = 256
Second, with CTRL^:	In[n]:= $\mathbf{5^{3.5}}$ Out[n] = 279.508

Numerical Functions. The absolute value function, **Abs[x]**, like any function, can take as an argument any expression that evaluates to a number.

The absolute (**Abs[x]**) function:

In[n]:= **Abs[Cos[π]]**

Out[n] = 1

The **Round** function:

In[n]:= **Round [5.5]**

Out[n] = 6

The minimum (**Min**) function:

In[n]:= **Min [4, 3, 7, 9, 22]**

Out[n] = 3

The **Mod** function:

In[n]:= **Mod [13,5]**

Out[n] = 3

The **Quotient** (division) function:

In[n]:= **Quotient [13, 5**]
Out[n] = 2

The square root function:

In[n]:= **N[Sqrt [17]]**

Out[n] = 4.12311

The factorial function:

In[n]:= **17!**

Out[n] = 355687428096000

The exponential function:

In[n]:= **Exp[5]**

Out[n] = 148.413

User-Defined Functions. With Mathematica, the user defines a function with the following syntax:

NameOfFunction[var1_, var2,_ ...]:= polynomial expression

A few examples:

mySqrs[q_,r]:=q^2 - r^2

f[x_]:= 5x^4 -3x^2 + 11

g[x_]:= x^3 -4^2

conA[t_]:=Xo Exp[-k1 t]

y[t_]:= e^{-kt}

Notice that the base to natural logarithms is ***e***, not e. You can find ***e*** on the Mathematica **Basic Input** palette. Some characteristics of the function definition are:

- The last character in each variable name must be the underline character.
- The list of variables must be enclosed in brackets.
- The right bracket is followed by the := assignment operator.
- Usually user-defined names begin with a lowercase letter, reserving uppercase for Mathematica's predefined functions.

We have already seen the equal sign (=) used as an assignment operator. In the next section we will meet a third use of the equal sign with the double equal sign (= =), which denotes logical equality, rather than assignment.

Evaluating and Tabulating Functions. The next several examples illustrate defining a function, evaluating a function, and displaying the evaluation.

Define a function **f(x)**. No **Out[1]:=** appears.	In[1]:= **f[x_]:= x^2 – 7 x +12**
Type **f[7]** to evaluate at x = 7.	In[2]:= **f[7]**
	Out[2]:=12
Type **f[b-4]** to evaluate at x = b – 4.	In[3]:= **f[b – 4]**
	Out[3]:=12 – 7 (4 + b) + (4 + b)^2
Evaluate and tabulate from x = –2 to 7.	In[4]:= **Table [f [x] , { x, - 2, 7}]**
	Out[4]:={30, 20, 12, 6, 2, 0, 0, 2, 6, 12}

In line **In[4]**, typing **Table [f[x],{x,-2,7}]** creates a one-dimensional list (a row vector) of values of the function from x equals –2 to x equals 7.

The **Table** function, like any function, has its parameters enclosed in brackets [ ]. The first parameter is the name of the function, but without the underline. The second parameter is a list. A Mathematica list is always enclosed in braces { }. The elements of the **Table** function list are the name of the variable (x in this case) and the range of the variable (–2 to +7, in this case). Like any function, the **Table** function returns a value or vales. The **Table** function returns a list of values.

Applying the **Table** function to a function of two variables creates a list of lists (a table). When this action is followed by the **TableForm** function, Mathematica changes the list of lists to a conventional table.

Define a function **g(p,q)** and No **Out[1]:=** appears.

In[1] := **g[p_, q_] := 2p^2 – q^2**

Type **g[2, 7]** to evaluate at p = 2 and q = 7.

In[2] := **g[[2, 7]**

Out[2] = 41

Evaluate and tabulate from p = 1 to 4 and q = 2 to 4.

In[3] := **Table[g[p,q] ,{p, 1, 4}, {q, 2, 4}]**

Out[3] = {{-2,-7,-14 }, { 4, -1, -8}, {14, 9, 2 }, {28, 23, 16 }}

Change the list of lists to a table. Use the **TableForm** function on the most recent output [%], which is **Out[3]**.

In[4] := **TableForm[%]**

Out[4]=// TableForm-

-2	-7	-14
4	-1	-8
14	9	2
28	23	16

The right side of the function definition is an expression. In later sections on symbolic operations and calculus, we will see that expressions can also be expanded, factored, simplified, differentiated, and integrated (definite and indefinite).

Practical Calculations with Lists

Because lists are used so extensively in Mathematica, it is worthwhile to do some typical nongraphic calculations that involve lists. Mathematica uses lists for input, for output, and, as we have just seen, for containing the iteration range in the **Table** function.

HCl Infrared Spectral Lines. The theory for infrared (IR) spectra of diatomic molecules was given in Chapter 3, Excel Example 12. Chapter 14 also used this demonstration calculation to illustrate how Visual Basic accomplishes this calculation. Now for comparison we repeat this calculation for a third time with Mathematica. The lines in the infrared absorption spectrum of hydrogen chloride (HCl) are given by

$$v = 2885.90 + 20.577m - 0.3034m^2 - 0.00222m^3 \tag{20-1}$$

First calculate a list of transitions with the **Table** function.

```
In[1]:= Table[2885.90 + 20.557 m - 0.2024 m^2 - 0.00222 m^3, {m, -6, 6}]

Out[1]= {2755.75, 2778.33, 2800.58, 2822.47, 2843.99, 2865.14,
        2885.9, 2906.25, 2926.19, 2945.69, 2964.75, 2983.35, 3001.48}
```

With the output of the last calculation [%] as the argument of the **TableForm** function, put the list created by the **Table** function in table form. The horizontal list becomes vertical.

```
In[2]:= TableForm[%]

Out[2]//TableForm=
        2755.75
        2778.33
        2800.58
        2822.47
        2843.99
        2865.14
        2885.9
        2906.25
        2926.19
        2945.69
        2964.75
        2983.35
```

Mathematica Notebook 1
The DesLandres Table for the PN Molecule

In the HCl calculation, the **Table** function calculated a list. In this calculation, the **Table** function calculates a list of lists, that is, a table. In the electronic spectrum of the PN molecule, the energy E of the excited electronic state is 39699.01 cm^{-1} above the ground state. The vibrational quantum numbers of the upper state are given by *vp* and the vibrational quantum numbers for the lower state are given by *vpp*. The vibrational energies of the upper state (v') are given by

$$v' = 1094.8 * vp - 7.25 * vp \wedge 2 \qquad (20\text{-}2)$$

The vibrational energies of the ground state are given by

$$v'' = 1094.8 * vpp - 7.25 * vpp \wedge 2 \qquad (20\text{-}3)$$

The spectral lines are given by

$$v = E + v' - v'' \qquad (20\text{-}4)$$

Figure 20-5 displays the Mathematica notebook for calculating the DesLandres table. Notice that only two lines of input are required. Also, notice the use of (* ***comment*** *) to enclose a comment. Compare Figure 20-5 with Figure 3-17 and Table 14-15.

Mathematica Notebook 2
X-Ray Diffraction

Measurement of the diffraction angle θ from various families of planes permits calculation of the separation between planes $d_{hk\ell}$ and the size of the unit cell itself a_0. The length of the edge of a cubic cell is a_0. A brief discussion of x-ray diffraction of a powered crystal sample was given in Chapter 1, Excel Example 1. Summarized, the relevant equations are

$$d_{hkl} = \frac{\lambda}{2\sin\theta} \tag{20-5}$$

and

$$a_0 = d_{hk\ell}\sqrt{h^2 + k^2 + \ell^2} \tag{20-6}$$

DesLandres Table for the PN Molecule

```
In[1]:= Table[ 39699.0+ (1094.80 vp - 7.25 vp^2) - (1329.38 vpp - 6.98 vpp^2), {vp, 0, 6},
    {vpp, 0, 5}]

Out[1]= {{39699., 38376.6, 37068.2, 35773.7, 34493.2, 33226.6},
    {40786.6, 39464.2, 38155.7, 36861.2, 35580.7, 34314.2},
    {41859.6, 40537.2, 39228.8, 37934.3, 36653.8, 35387.2},
    {42918.2, 41595.8, 40287.3, 38992.8, 37712.3, 36445.8},
    {43962.2, 42639.8, 41331.4, 40036.9, 38756.4, 37489.8},
    {44991.8, 43669.4, 42360.9, 41066.4, 39785.9, 38519.4},
    {46006.8, 44684.4, 43376., 42081.5, 40801., 39534.4}}

In[2]:= TableForm[%]
    (*This is a DeLandres Table for the PN Bands calculated from the data
     given by HerzBerg, G.,"Spectra of Diatomic Molecules," Van Nostrand,
    New York, 2nd ed.,1950,
    page 40.*)

Out[2]//TableForm=
```

39699.	38376.6	37068.2	35773.7	34493.2	33226.6
40786.6	39464.2	38155.7	36861.2	35580.7	34314.2
41859.6	40537.2	39228.8	37934.3	36653.8	35387.2
42918.2	41595.8	40287.3	38992.8	37712.3	36445.8
43962.2	42639.8	41331.4	40036.9	38756.4	37489.8
44991.8	43669.4	42360.9	41066.4	39785.9	38519.4
46006.8	44684.4	43376.	42081.5	40801.	39534.4

Figure 20-5. Mathematica Notebook 1. The DesLandres Table.

The reflecting plane is characterized by its Miller indices h, k, and ℓ. The measurement of diffraction angles allows calculation of interplanar spacing $d_{hk\ell}$ and the length of the cubic unit cell a_0.The Miller indices h, k, and ℓ for each plane are determined analytically by a procedure described briefly in Chapter 1. These calculations provide a typical example of the way Mathematica makes extensive use of lists in numerical calculations. The steps in the x-ray diffraction calculation (Figure 20-6) are as follows:

- Define a **List** named **data** for the measured diffraction angles in degrees.
- Calculate a **List** named **radians** of diffraction angles in radians.
- Define a constant λ for the wavelength in angstroms.
- Calculate a **List d** of interplanar spacings $d_{hk\ell}$.
- Define **Lists** ***h*, *k*,** and ℓ of Miller indices (independently determined).
- In anticipation of calculating the mean of the cell dimension a_0, load the package **<<Statistics`DescriptiveStatistics`**.
- Calculate the **List a** of cell dimensions.
- Calculate the mean of the **List a**.

Mathematica Packages. Mathematica includes a large number of commands, but many more commands are contained in several packages. To use a package, you must know its name so that you can load it into Mathematica. To find the name, on the **Help** menu, click **Help Browser** to open the **Help Browser** window as it appears in Figure 20-4. Then click the **Add-ons** tab. Select **Standard Packages** in the list on the left. The middle column of the **Help Browser** window gives the directory names (for example, **Statistics**) as shown in Table 20-2. Click the directory name **Statistics** and the right column of the **Help Browser** lists several branches of statistics such as **Descriptive Statistics**. To load a package, type **<<Directory`*NameOfPackage*`**. Notice that the name of the package is included between back quotes, a character under the tilde in the upper left corner of your keyboard.

Go back to Mathematica Notebook 2: X-Ray Diffraction. After calculating the mean of *a*, you could try a couple of other **Descriptive Statistics** commands on the list a such as:

Median[*a*]
Mode[*a*]
GeometricMean[*a*]
SampleRange[*a*]
Variance[*a*]
StandardDeviation[*a*
DispersionReport[*a*]
LocationReport[*a*]

Table 20-2

Directory Names of Mathematica Packages

Introduction	LinearAlgebra
Algebra	Miscellaneous
Calculus	NumberTheory
DiscreteMath	NumericalMath
Geometry	Statistics
Graphics	Utilities

Here is a sample calculation of the mean and geometric mean of the list **MyList**.

<<Statistics`Descriptive Statistics`
In[n]: = **MyList = {1,2,5,4,33,5,66,77}**
Out[n] = {1,2,5,4,33,5,66,77}
In[n]: = **N[Mean[MyList]]**
Out[n] = 26.8571
In[n]: = **N[GeometricMean [MyList]]**
Out[n] = 9.44561

X – Ray Diffraction

The x - ray diffraction data are the diffraction angles in degrees.
These are placed in a list called *data*.

```
In[1]:=
        data = {21.811, 25.368, 37.218, 45.132, 47.754, 58.603};
```

The data in degrees are converted to radians by multiplying by Pi / 180

```
In[2]:= radians = data Pi/180

Out[2]= {0.380674, 0.442755, 0.649577, 0.787702, 0.833465, 1.02282}
```

This list of diffraction angles in radians has the name *radians*.
The d spacing between reflecting planes is calculated at each diffraction angles at a constant wavelength λ.

```
In[3]:= λ = 1.5418;
        d = λ/(2 Sin[radians])

Out[4]= {2.07484, 1.79936, 1.27453, 1.08771, 1.04138, 0.903139}
```

This list of interplanar spacings has the name *d*.
Examination of the reflections reveals the following set of Miller Indices h, k and ℓ. The unit cell dimension a is calculated for each reflection, characterized by the spacing between relecting planes and their Miller Indicies.
The following lines are *lists* of Miller indices.

```
In[5]:= h = {1, 2, 2, 3, 2, 4};
        k = {1, 0, 2, 1, 2, 0};
        ℓ = {1, 0, 0, 1, 2, 0};

In[8]:= << Statistics`DescriptiveStatistics`
        Note : In the following equation a, d, h, k and ℓ are lists.
```

$$a = d\sqrt{h^2 + k^2 + \ell^2}$$

```
In[9]:= a = d Sqrt[h^2 + k^2 + ℓ^2]

Out[9]= {3.59373, 3.59872, 3.60492, 3.60754, 3.60746, 3.61256}
```

The name of the above list of cell dimensions is a.

```
In[10]:= Mean[a]

Out[10]= 3.60415

In[11]:= DispersionReport[a]

Out[11]= {Variance → 0.0000463468, StandardDeviation → 0.00680785,
          SampleRange → 0.0188232, MeanDeviation → 0.00528632,
          MedianDeviation → 0.00385837, QuartileDeviation → 0.00441181}

In[12]:= For copper, the unit cell dimension is : a = 3.604 (7)
```

Figure 20-6. Mathematica Notebook 2. X-Ray Diffraction Calculations

More on Lists. Scientists and engineers frequently have lists of experimental data with which to work. Data reduction consists of operating on the lists of data. In this way, the list provides the input data that, after some operation, provide the output data in a new form. We have already seen how operations on a function over a range of the independent variable generate an output list of dependent variables. Let us assume that an engineer has a list of experimental pressures.

She defines a list of pressures named **Pmm**.

In[1]:= **Pmm = {10, 25,40, 60, 62.5}**

Out[1] = {10, 25,40, 60, 62.5}

Next, she converts all the elements in the list atmospheres (1 atm = 760 mm).

In[2]:= **Patm = N[Pmm/760]**

Out[2] = {0.0131579, 0.0328947, 0.0526326, 0.0789474, 0.0805263}

Finally, she takes the log of every element in the list.

In[3]:= **LogPatm = Log[Patm]**

Out[3] = {-4.33073, -3.41444, -2.94444, -2.53897, -2.51917}

Now suppose that she discovered a systematic error in her original pressure data: the pressures are too high by 1.3 mm. She corrects the data.

In[4]:= **Pcorrect = Pmm – 1.3**

Out[4] = {8.7, 23.7, 38.7, 58.7, 59.9}

Operating on Lists. Applying the **Plus** operator to a list is an efficient way to sum the elements of the list. The **Length** function returns the number of elements in the list. Let us define a list, name it *z,* and use it in some examples.

Define a list and name it *z*.

In[n]:= z = {1,4,3,7,5,9};

Calculate the sum of the elements in *z* and name the sum **MySum**.

In[n]: = **MySum = Apply[Plus,z]**
Out[n] = 29

Determine the length of *z* (the number of elements in *z*).

In[n]: = **Length [z]**
Out[n] = 6

Sort the elements in *z*.

In[n]: = **Sort[z]**
Out[n] = {1, 3, 4, 5, 7, 9}

Mathematica Notebook 3
The Heat Capacity of Iodine

The underlying theory for statistical thermodynamics and the calculation of the heat capacity of iodine is included in Chapter 4, Excel Example 14. The calculation is also shown in Chapter 14 with Visual Basic and again in Mathcad Worksheet 19.

This Mathematica notebook (Figure 20-7) illustrates the calculation of the heat capacity of iodine (g) from 300 K to 1000 K.

- In[1–4]: The physical constants *h, c, k,* and *R* are defined.
- In[5]: The fundamental vibration frequency of Iodine is defined to be *w.*
- In[6]: The five constants plus *w* are wrapped up in a single constant θ.

- In[7]: A list of temperatures from 300 to 1000 in steps of 100 is created and named *T*.
- In[8]: A list of vibrational contributions (C_{vib}) to the total heat capacities is calculated from 300 to 1000 K. Note that in this line, *T* is a list.
- Out[8]: A list of vibrational heat capacities from 300 to 1000 K.
- In[9]: The total heat capacity equals the vibration contribution plus the translational and rotational contribution (5*R*/2).
- In[10]: Put the list output of **Out[9]** into the form of a table (vertical list).
- Out[10] displays the calculated heat capacities.

The Heat Capacity of Iodine (g) from 300 K to 1000 K

```
In[1]:= h = 6.661 10^-34 (*Planck's constant*);
        c = 2.998 10^8  (* speed of light *);
        k = 1.381 10^-23 (* Boltzmann constant*);
        R = 8.3143      (* gas constant *);
        w = 21400       (*iodine fundamental in 1/meter *);
        θ = h*c*w/k (* The vibrational temperature of iodine *)

Out[6]= 309.45

In[7]:= T = Table[T, {T, 300, 1000, 100}]

Out[7]= {300, 400, 500, 600, 700, 800, 900, 1000}
```

In[8]:= $\text{Cvib} = R\,(\theta/T)^2 \dfrac{e^{\theta/T}}{(e^{\theta/T}-1)^2}$

```
Out[8]= {7.61472, 7.91175, 8.05392, 8.13243, 8.18021, 8.2114, 8.23287, 8.24827}
```

In[9]:= $\text{Ctotal} = \dfrac{5}{2} R + \text{Cvib}$

```
Out[9]= {28.4005, 28.6975, 28.8397, 28.9182, 28.966, 28.9972, 29.0186, 29.034}

In[10]:= TableForm[Ctotal]

Out[10]//TableForm=
        28.4005
        28.6975
        28.8397
        28.9182
        28.966
        28.9972
        29.0186
        29.034
```

Figure 20-7. Mathematica Notebook 3. The Heat Capacity of Iodine

Mathematica Notebook 4
Plotting Functions with Plot

The Mathematica function for plotting functions (built-in or user-defined) is the **Plot** function. The parameters for **Plot** consist of the name of the function followed by a list of the plot variables and limits.

Plot[*f*(*x*),{*x*, *xmin*, *xmax*}]

For multiple plots, the parameters of **Plot** consist of a list of the names of the functions followed by a list of the plot variable and limits.

Plot [{$f_1(x)$, $f_2(x)$ $f_3(x)$... },{*x*, *xmin*, *xmax*}]

This statement plots the functions named in the list from *x* equals *xmin* to *xmax*. Instead of *f*(*x*), an expression may be used.

Figure 20-8 displays the commands for creating single and multiple plots. The triple plot of the exponential functions represents the change in *A*, *B*, and *Z* for the decay sequence: *A*→*B*→ *Z*. Compare this Mathematica treatment with the Excel treatment shown in Chapter 6, Excel Example 19, where the science is also reviewed. Later sections provide more examples of plots.

Plot takes a number of options, a few of which are listed in Table 20-3. The plot in Figure 20-8 illustrates a few of these options. The upper plot of a quadratic utilizes no options for the **Plot** function, but in the lower plot, **In[9]** includes several option titles, labels, and their fonts. So far we've explored two kinds of plotting functions: **Plot** and **ListPlot**. They have many options in common. To get a complete list of their options, type the statements **Options[Plot]** or **Options[ListPlot]** followed by CTRL+ENTER. For a less compact, but easier to read list, type **TableForm[Options[Plot]]** or **TableForm[Options][ListPlot]]**.

Table 20-3

***Plot* Options**

Name of Option	Default Value	Description
Axes	True	Axes included if True
Axes Label*	None	Provides x- and y-axis labeling
AxesOrigin	Automatic	Sets the point where axes cross
DefaultFont*	$Default Font	Sets the font for text in plot labels
DisplayFunction	$DisplayFunction	$DisplayIdentity hides display of plot
Frame*	False	Frame included if True
FrameTicks	Automatic	None hides tick marks
GridLines*	None	True includes gridlines
PlotJoined	True	
PlotLabel*	None	Provides for a plot label
PlotRange	Automatic	All includes all plot points
PlotStyle	None	Provide for style of lines and points
Ticks	Automatic	None hides tick marks
xlabel	None	Provides for x-axis label only
ylabel	None	Provides for y-axis label only

*Illustrated in Figure 20-8.

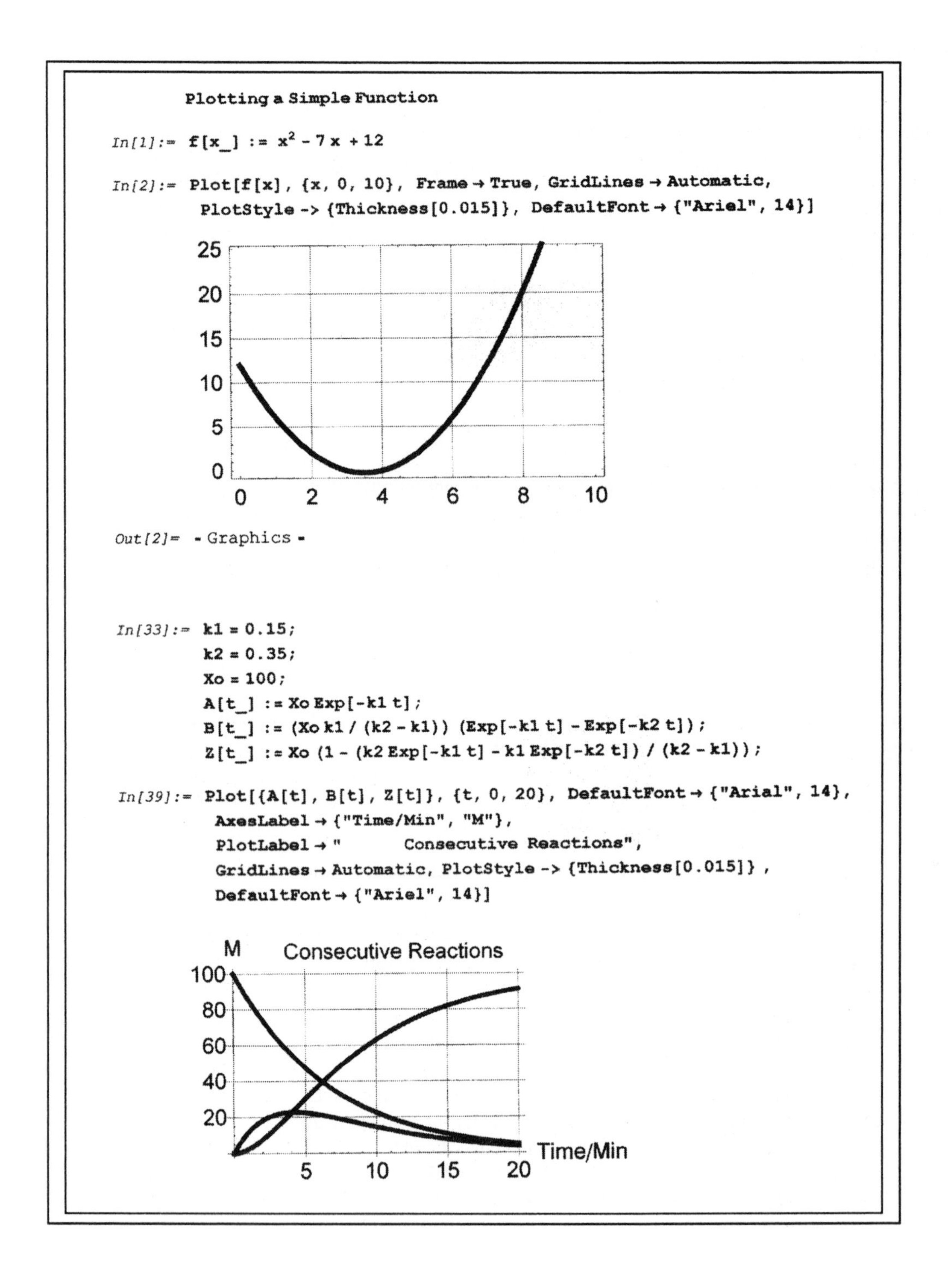

Plotting a Simple Function
In[1]:= f[x_] := x^2 - 7 x + 12
In[2]:= Plot[f[x], {x, 0, 10}, Frame → True, GridLines → Automatic,
PlotStyle -> {Thickness[0.015]}, DefaultFont → {"Ariel", 14}]
25
20
15
10
5
0
0 2 4 6 8 10
Out[2]= - Graphics -
In[33]:= k1 = 0.15;
k2 = 0.35;
Xo = 100;
A[t_] := Xo Exp[-k1 t];
B[t_] := (Xo k1 / (k2 - k1)) (Exp[-k1 t] - Exp[-k2 t]);
Z[t_] := Xo (1 - (k2 Exp[-k1 t] - k1 Exp[-k2 t]) / (k2 - k1));
In[39]:= Plot[{A[t], B[t], Z[t]}, {t, 0, 20}, DefaultFont → {"Arial", 14},
AxesLabel → {"Time/Min", "M"},
PlotLabel → "   Consecutive Reactions",
GridLines → Automatic, PlotStyle -> {Thickness[0.015]},
DefaultFont → {"Ariel", 14}]
M
Consecutive Reactions
100
80
60
40
20
5 10 15 20
Time/Min

Figure 20-8. Mathematica Notebook 4: Plotting Functions with Plot

Plotting Lists

Usually we think of a list as a one-dimensional array whose elements are of the same type. In Mathematica, a list is a one-dimensional array of objects separated by commas and enclosed in braces { }. The objects are usually of the same type, but they do not have to be. If you assign a list to a variable name, you can use the list as you would with other languages such as Visual Basic, C, or Pascal. You will encounter lists frequently in Mathematica, both as input and as output data structure. For example, we have already seen a list as the output of the **Table** function and as input to the **TableForm** function.

Mathematica Notebook 5
Plotting Lists of Numbers with ListPlot

An *x,y* pair of data points is just a short list (of two elements), so a list of *n* data pairs is a 2 x *n* table. Experimental scientists and engineers regularly collect sets of data pairs, consisting of a dependent and an independent variable. A table of vapor pressures is such a table. The pressure is the dependent variable and the temperature is the dependent variable. In this example we shall prepare two tables, one for benzene and one for chloroform, from the data in Table 2-3, Excel Example 5, Chapter 2. The **ListPlot** function plots one or more lists of data. The syntax is:

ListPlot[*ListName1*, *ListName2*, *options*]

In Mathematica Notebook 5 (Figure 20-9), the vapor pressures of benzene and chloroform are plotted on the same graph. **Benzene** and **Chloroform** are the names of our data pair lists, each of which is a list of lists. In order to create the graph with two plots shown at the bottom of the notebook, four separate plots are prepared: **Note:** the → symbol is formed by typing **->**. Mathematica changes this combination to →.

1. **benzPlot** is a **ListPlot** of the benzene data. The option **PlotJoined → True** produces a plot with lines joining the data points (which are not visible). The **PlotStyle** option sets the thickness of the line (0.01) and the gray level (0.3). A gray level of 0.1 is nearly black and a gray level of 0.9 is light gray. The option **DisplayFunction → Identity** suppresses the plot. This is done to make this notebook more compact. You should omit this so you can see the line plot that appears.
2. **benzDots** is a **ListPlot** of the benzene data that displays the data points, but not a line joining them. The option **PlotStyle → PointSize[0.03]** sets the size of the data points. The option **DisplayFunction → Identity** suppresses the plot.
3. **chlorPlot** is a **ListPlot** of the chloroform data. The default line thickness is used. The option **DisplayFunction → Identity** suppresses the plot.
4. **chlorDots** is a **ListPlot** of the chloroform data that shows only the data points. The option **PlotStyle → PointSize[0.25]** sets the size of the data points.

At this point, four graphic objects (**benzPlot**, **benzDots**, **chlorPlot**, and **chlorDots**) have been created but not displayed, because for each of them, the option **DisplayFunction → Identity** suppresses the plot. The **Show** function displays the four named graphic objects on the same graph. Since the **DisplayFunction** in the previous **ListPlots** command turned off the display of graphic objects, it is necessary to turn it back on by including in the **Show** function the option **DisplayFunction →$DisplayFunction**.

The Vapor Pressures of Benzene and Chloroform

In this Mathematica Notebook, the vapor pressures of benzene and chloroform are plotted on the same graph.

```
In[3]:= Benzene = {{3.0, 30}, {21.3, 80}, {35.3, 150}, {52.3, 300},
          {72.6, 600}, {78.4, 720}, {80.1, 760}};
        Chloroform = {{10, 100.5}, {20, 159.6}, {25, 199.1}, {30, 246.6},
          {35, 301.3}, {40, 366.4}, {45, 439.0}, {50, 526.0}, {55, 625.2},
          {60, 739.6}, {61.2, 760}};

In[5]:= benzPlot = ListPlot[Benzene, PlotJoined→ True,
          PlotStyle→ {Thickness[0.01], GrayLevel[0.3]},
          DisplayFunction → Identity];

In[6]:= benzDots = ListPlot[Benzene, PlotStyle→ PointSize[0.03],
          DisplayFunction → Identity];

In[7]:= chlorPlot = ListPlot[Chloroform, PlotJoined→ True,
          DisplayFunction → Identity];

In[8]:= chlorDots = ListPlot[Chloroform, PlotStyle→ PointSize[0.025],
          DisplayFunction → Identity];

In[9]:= Show[benzPlot, benzDots, chlorPlot, chlorDots,
          DisplayFunction → $DisplayFunction]
```

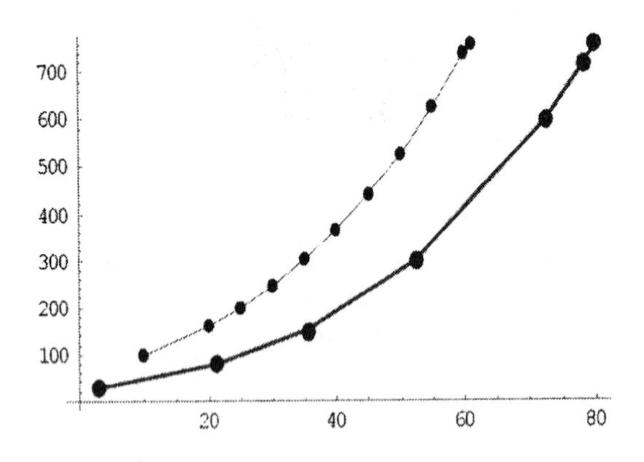


```
Out[9]= - Graphics -
```

Figure 20-9. Mathematica Notebook 5: Vapor Pressures of Benzene and Chloroform—Plotting a List of Numbers with **ListPlot**

Mathematica Notebook 6
The DisplayTogether Function

The syntax of the **DisplayTogether** function is:

DisplayTogether[*Plot1*, *Plot2*, *Plot3* ...]

The **DisplayTogether** function provides a compact and convenient method for creating several plots and displaying them together. In Figure 20-10, **Benzene** is the name of a list of vapor pressure, temperature pairs for benzene, and **Chloroform** is the name of a list of vapor pressure, temperature pairs for chloroform. **ListPlot** is a function that plots a list of points. The four arguments of **DisplayTogether** are the four **ListPlot** functions. Two of the **ListPlot** functions are line plots, and two are point plots. The **DisplayTogether** function returns all four plots combined, as displayed in Figure 20-10.

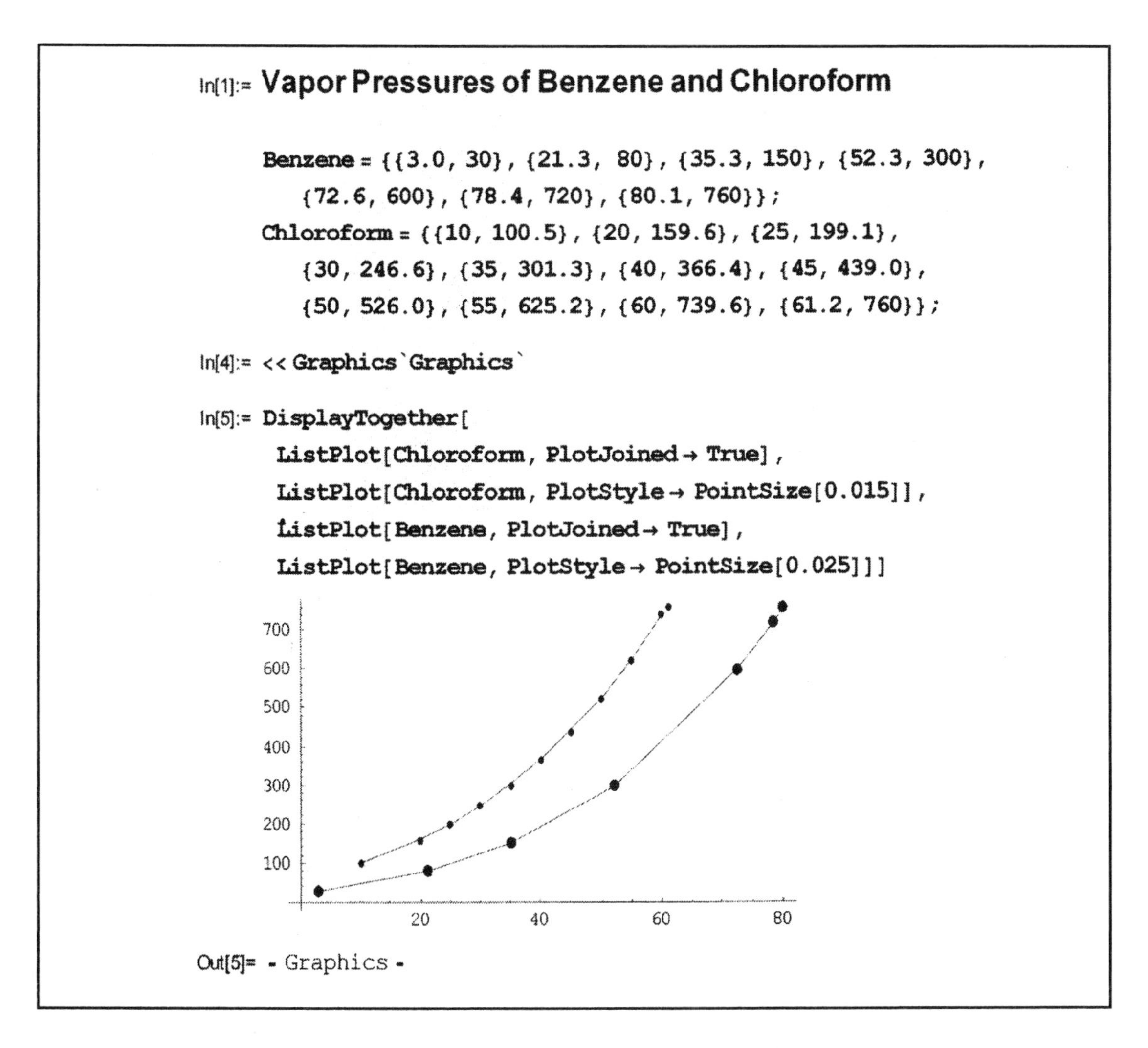

Figure 20-10. Mathematica Notebook 6: Two Vapor Pressure Plots with DisplayTogether

Linear Regression

Linear equations of the form $y = mx + b$ are common in science and engineering. Plots of the type ln*K* versus 1/*T* play an important role in many physics and chemistry phenomena. In this case, ln*K* serves as the dependent variable and 1/*T* serves at the independent variable. Let us use the benzene vapor pressure data from Table 2-3, Excel Example 5, in Chapter 2, to create a linear plot of ln*P* versus 1/*T*. Preparing the Mathematica worksheet shown in Figure 20-11 provides more practice in calculating and plotting with lists. The **Print** command serves to display intermediate results.

Preparing the Data for Linear Regression. In Figure 20-11 these data are plotted as ln*P*, 1/*T* pairs. As in the previous example with the plots of benzene and chloroform, separate plots are prepared for the line and for the points.

- Prepare a list *p* with benzene vapor pressures as the elements of the list.
- Prepare a list ln*P* of the natural logs of the vapor pressures.
- Prepare a list *t* of the temperatures in degrees Celsius.
- Prepare a list *T* of the temperatures in K.
- Prepare a list *R* of reciprocal temperatures (1/*T*).
- With the **Print** command, display the lists ln*P* and *R* for the next step.
- From the displayed lists ln*P* and *R*, type the corresponding *R,* ln*P* pairs, with each pair enclosed in braces { }. Enclose each list of pairs in braces and define the list of lists as **NewData**, as shown in **In[10]:,** Figure 20-11.
- Put the last list (%) in the form of a table with **TableForm(%)**, and proofread your data.

Mathematica Notebook 7
The Fit Function.

The **Fit** function has the form:

Fit[*data*, *funs*, *vars*]

- The parameter ***data*** refers to a list of *x,y* data pairs. The name of the list can be used (for example, **NewData**, as described above); or the list of data pairs that appears in braces can be used, for example, **{{1.2, 23.4}, {1.5, 27.8}, ...}**.
- The parameter ***funs*** refers to the function to which the data is to be fit. It has the form **{1, *x*}** if the function is first order in *x*, **{1, *x*, *x*^2}** if the function is second order in *x* and so on.
- The parameter ***vars*** refers to the name of the variable or variables, in this case, *x*.

With the data prepared as described above and shown in Figure 20-11, the **Fit** function is:

Fit[NewData, {1, *x*}, *x*]

This statement fits **NewData** to a linear equation in *x*.

Mathematica Notebook 8
Plotting the Linear Regression Line

Next we'll prepare the three plots shown in Figure 20-12, first, a plot of the linear regression line, second, a plot of the points (data pairs), and third a plot of the first two plots combined on the same graph. The steps for creating these three plots follow:

- Define **MyLine** as the plot of *f* from *x* equal 0.0028 to 0.0038.
- Define **MyDots** as the plot of the **NewData** pairs listed with **In[8]:** and **In[9]:**. Set the **PointSize** a little larger than the default size.
- Combine the **MyDots** and **MyLine** plots with **Show[MyDots,MyLine, ...]** with options for a frame, a label, and gridlines.

Notice that the **ListPlot** function specializes in plotting points or pairs of points. If a list of points is furnished, then the *x* values are 1, 2, 3, 4.... The linear regression line is defined as a function *f* and is plotted with **Plot**, the usual command for plotting functions and expressions. Try varying some of the option's parameters to see what the effect is. Also try labeling the axes with **AxesLabel→{"x-axis label","y-axis label"}**

Mathematica Notebook 9
The Linear Regression Line and DisplayTogether

The **DisplayTogether** function was discussed and illustrated in the previous section on the vapor pressure of benzene and chloroform. Figure 20-13 show how the **DisplayTogether** function is applied to plots of a linear regression line and the points from which it is calculated.

Linear Regression with the Fit Function

```
In[3]:= p := {30, 80, 150, 300, 600, 720, 760};
        lnP := N[Log[p]];

In[5]:= t := {3.0, 21.3, 35.3, 52.3, 72.6, 78.4, 80.1};

In[6]:= T := t + 273.16;

In[7]:= R := 1 / T;

In[8]:= Print[lnP]

        {3.4012, 4.38203, 5.01064, 5.70378, 6.39693, 6.57925, 6.63332}

In[9]:= Print[R]

        {0.00362109, 0.00339605, 0.00324191,
         0.00307257, 0.00289218, 0.00284446, 0.00283078}

In[10]:= NewData = {{0.00362109, 3.4012}, {0.00339605, 4.38203},
           {0.00324191, 5.01064}, {0.00307257, 5.70378}, {0.00289218, 6.39693},
           {0.00284446, 6.57925}, {0.0028308, 6.63332}};

In[11]:= TableForm[%]

Out[11]//TableForm=
        0.00362109      3.4012
        0.00339605      4.38203
        0.00324191      5.01064
        0.00307257      5.70378
        0.00289218      6.39693
        0.00284446      6.57925
        0.0028308       6.63332

In[12]:= f = Fit[NewData, {1, x}, x]

Out[12]= 18.1509 - 4061.77 x
```

Figure 20-11. Mathematica Notebook 7: Calculating the Linear Regression Line with the **Fit** Function

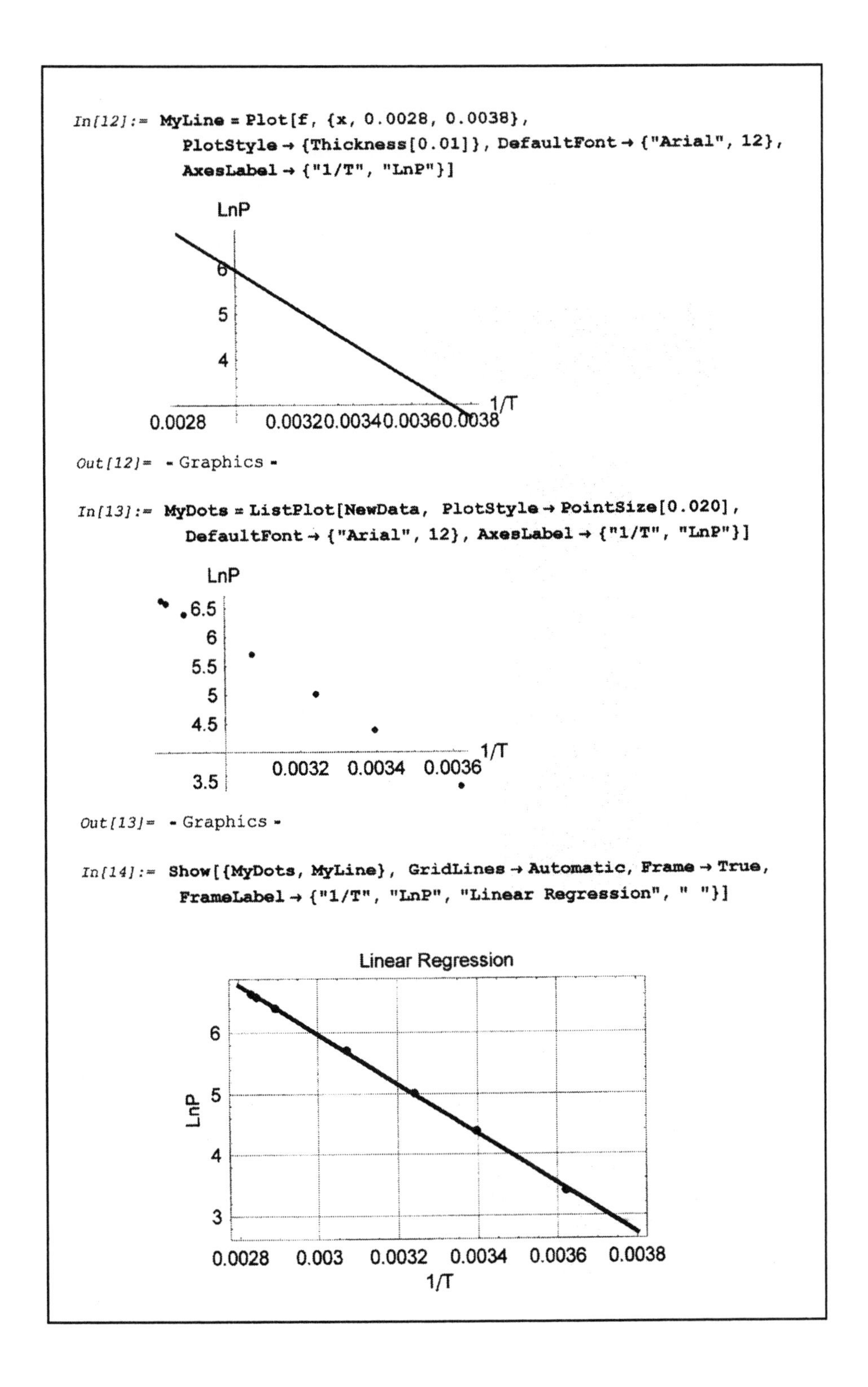

Figure 20-12. Mathematic Notebook 8: Plotting the Points and the Line, Which was Calculated from the Linear Regression Named f (See Figure 20-11, Mathematica Notebook 7).

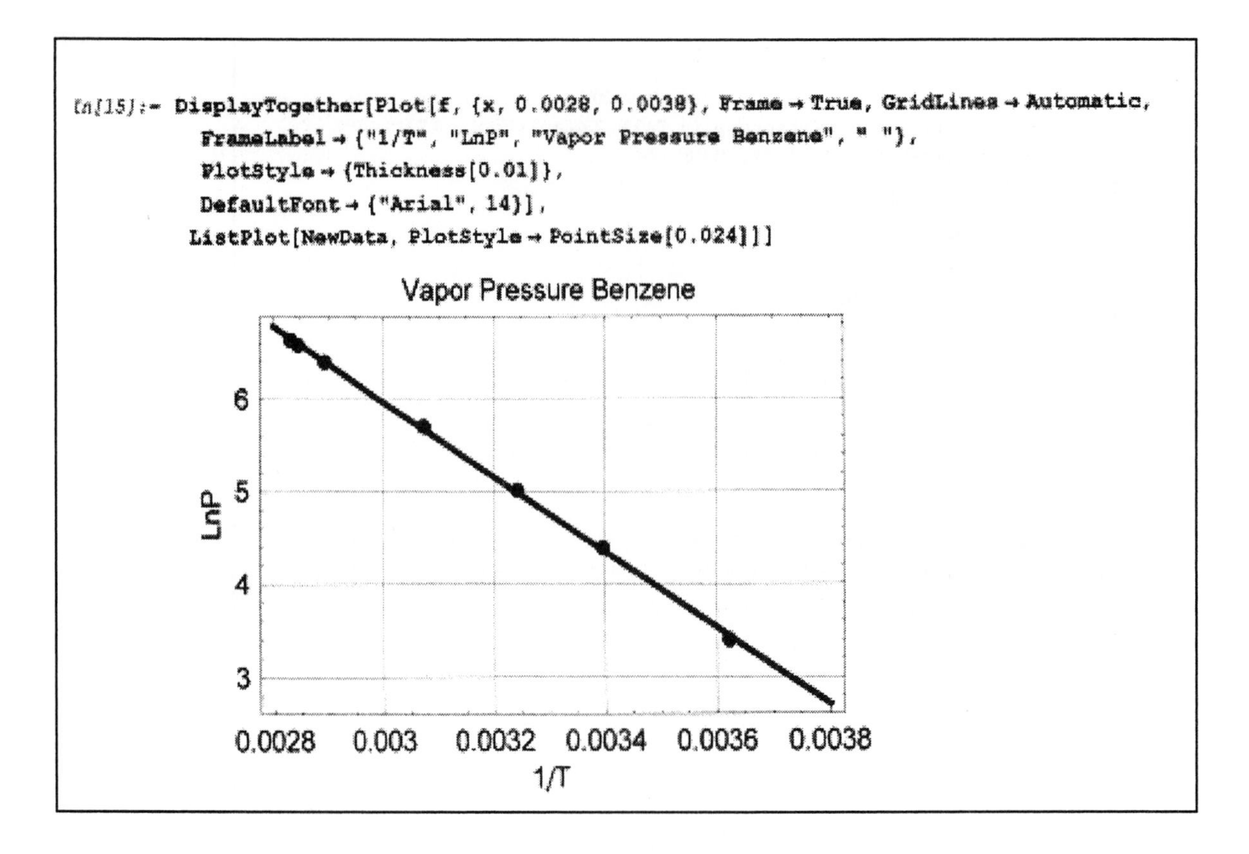

Figure 20-13. Mathematica Notebook 9: Using **DisplayTogether** to Plot the Experimental Points and the Linear Regression Line Named **f**. See Figure 20-11, Mathematica Notebook 7.

Plotting Functions from Quantum Mechanics

Chapter 3 gives a review of the underlying quantum mechanics for this section. This shows how to create plots of vibrational wave functions of a diatomic molecule and of the radial wave functions of the hydrogen Atom.

Vibrational Wave Functions. A vibrational wave function consists of three factors: a normalization factor N_v, a Hermite polynomial, and an exponential factor $e^{-q^2/2}$.

$$\psi_v(q) = N_v H_v(q) e^{\frac{-q^2}{2}} \tag{20-7}$$

The quantity q is a generalized coordinate given by

$$q = (2a)^{1/2}/2 \tag{20-8}$$

$$a = 2\pi^2 \nu\mu / h \tag{20-9}$$

where ν is the fundamental vibration frequency of the particular diatomic molecule and v is a vibrational quantum number with allowed values of 0, 1, 2, 3 .... The factors N_v are normalization factors and are given by

$$N_v = \left(\frac{(2a/\pi)^{1/2}}{2^v v!} \right)^{1/2} \tag{20-10}$$

The factors H_v are Hermite polynomials, well known to mathematicians and even to Mathematica, which lists the first seven below in **Out[1]** and **Out[2]**. The exponential factor $e^{\frac{-q^2}{2}}$ is the same for all quantum states.

Create a table named **hermpoly** of Hermite polynomials

```
In[1]:= hermpoly = Table [ HermiteH [ n, q],{n,0,6}]
Out [1]:= {1, 2 q, -2 + 4 q^2, -12 q + 8 q^3 , 12 – 48 q^2 + 16 q^4 ,
```

Change the table's vertical list.

```
In[2]:= TableForm[hermpoly]
Out[2]//TableForm=
        1
        2 q
        -2 + 4 q^2
        -12 q + 8 q^3
        12 – 48 q^2 + 16 q^4
        120 q -160 q^3 + 32 q^5
        -120 +720 q^2 - 480 q^4 +64 q^6
```

Mathematica Notebook 10
Plotting the Vibrational Wave Function

In Figure 20-14, ψ_1^2, ψ_6^2, and ψ_{30}^2 are graphed with the product of a Hermite polynomial and the exponential factor as the arguments of **Plot**. The plot range of q is –6 to 6 for v equals 6 and –9 to 9 for v equals 30. For simplicity, the normalization factor is omitted; its value is a constant for each state and does not affect the shape of the curve.

The probability density of finding the atom at a displacement q from the origin is proportional to the square of the wave function. Examination of the plots in Figure 20-14 demonstrates that the probability tends to pile up at the extremes of the oscillation, especially as the vibrational quantum number reaches high values. The system approaches classical behavior at very high quantum numbers.

To compare quantum mechanical behavior of the vibrating molecule with classical behavior, visualize the shadow on the floor of a pendulum illuminated from above. The shadow of the pendulum oscillates back and forth along a straight line. The shadow spends most of its time at the extremes of the oscillation, so the highest probability of locating the shadow is at the extremes of its oscillation. The lowest probability is at its equilibrium position where the shadow is traveling fastest. This is just the opposite behavior of a vibrating molecule in its ground state (v equals 0), the uppermost plot in Figure 20-14. The probability clearly piles up at the equilibrium position.

For the ground vibrational state, v equals 0, H equals 1, N is a constant, and the shape of the wave function (not shown here) is just the shape of $e^{-q^2/2}$, a bell-shaped Gaussian probability distribution (Figure 20-14, top). For the ground state, the probability reaches a maximum at the origin. Plot this simple curve for the sake of comparison with the higher-level quantum states.

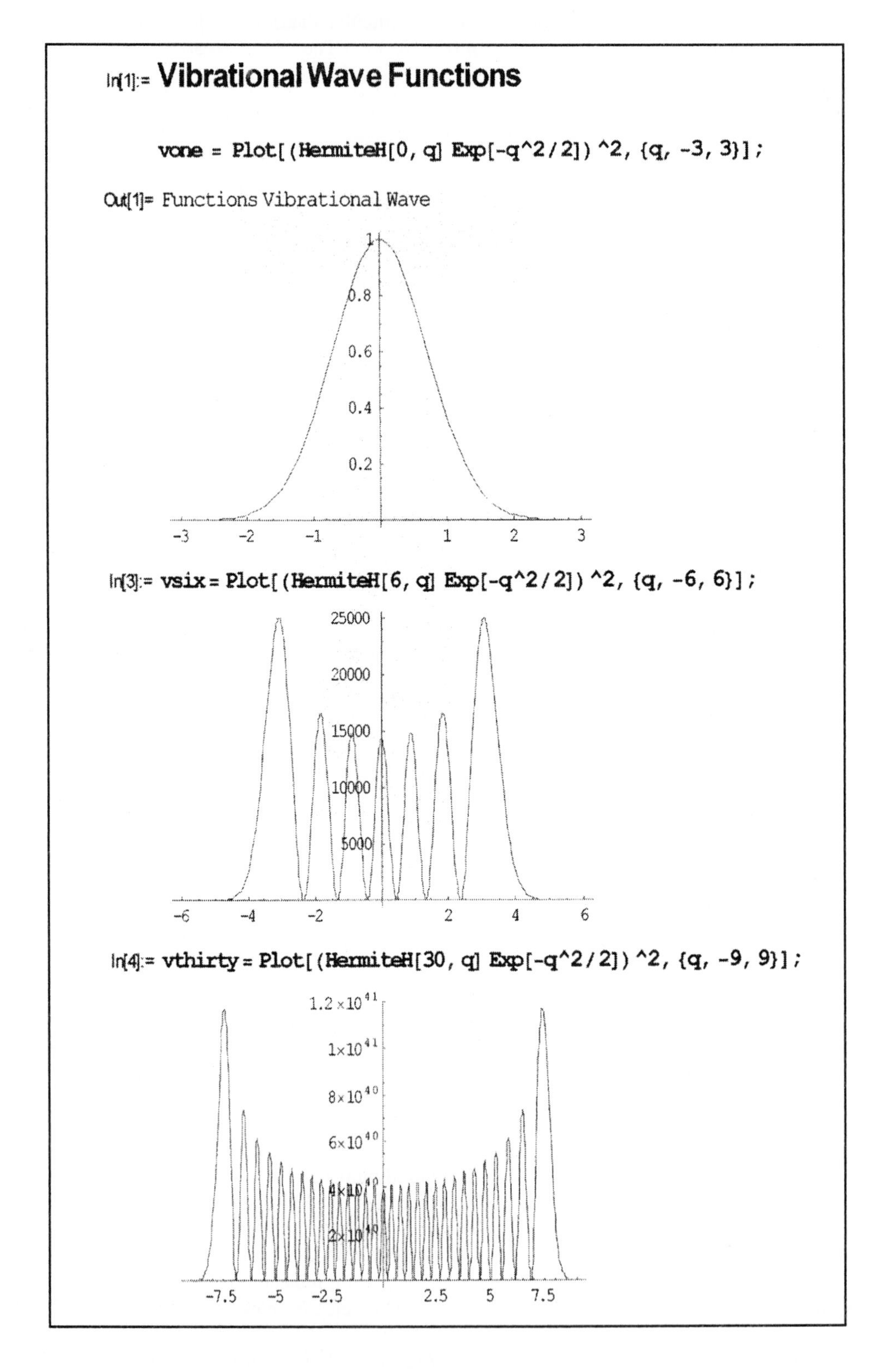

Figure 20-14. Mathematica Notebook 10: Vibrational Wave Functions

Hydrogen Atom radial Wave Functions. As we've already seen in Excel Example 7 in Chapter 3, a discussion of the hydrogen atom, the Schrödinger wave equation in polar coordinates

$$\frac{1}{r^2}\frac{\partial}{\partial r}\left(r^2\frac{\partial\psi}{\partial r}\right)+\frac{1}{r^2\sin\theta}\frac{\partial}{\partial\theta}\left(\sin\theta\frac{\partial\psi}{\partial\theta}\right)+\frac{1}{r^2\sin^2\theta}\frac{\partial^2\psi}{\partial\phi^2}+\frac{8\pi^2 m}{h^2}\left(E+\frac{e^2}{4\pi\varepsilon_0 r}\right)\psi=0 \quad (20\text{-}11)$$

can be separated into three much simpler equations, one a function of r only, a second equation a function of θ only, and a third equation a function of φ only. These three equations are relatively easily solved, and the complete solution is the product of the radial and angular parts:

$$\psi_{n,\ell,m}(r,\theta,\varphi)=R_{n,\ell}(r)\,\Theta_{\ell,m}(\theta)\,\Phi_m(\varphi) \quad (20\text{-}12)$$

In examining the spatial properties of $\psi_{n,\ell,m}(r,\theta,\varphi)$, it is convenient to look at the radial part and the angular parts separately. In shortened notation, $\psi = R\Theta\Phi$, where R is the radial part and is a function of r only and does not depend on direction and is the angular part that is independent of r, but depends on direction, that is, on θ and φ.

In this example, we'll look at R, the radial part of the wave function. More exactly we'll plot the 1*s* (R_{10}), the 2*s* (R_{20}), and the 3*s* (R_{30}) orbitals. And then, in the next section of this chapter, we'll look at $\Theta\Phi$, the angular part of the wave function. For the *s*-orbitals and n equal 1, 2, and 3, the hydrogen-like radial wave functions are written

n = 1, ℓ = 0, (1s)
$$R_{10}(r)=2\left(\frac{Z}{a_0}\right)^{3/2}e^{-\rho} \quad (20\text{-}13)$$

n = 2, ℓ = 0, (2s)
$$R_{20}(r)=\frac{1}{2\sqrt{2}}\left(\frac{Z}{a_0}\right)^{3/2}(2-\rho)e^{-\rho/2} \quad (20\text{-}14)$$

n = 3, ℓ = 0, (3s)
$$R_{30}(r)=\frac{2}{81\sqrt{3}}\left(\frac{Z}{a_0}\right)^{3/2}(27-18\rho+2\rho^2)e^{-\rho/3} \quad (20\text{-}15)$$

where

$$\rho=\frac{Zr}{a_0} \quad (20\text{-}16)$$

and a_0 is the Bohr radius, Z is the charge on the nucleus, n is the principal quantum number, and r is the distance from the nucleus. For most wave function plots, r/a_0 ranges from about zero to ten or so and is just the number of Bohr radii from the nuclei. With this change of variable and with Z set equal to one, the above wave equations become

$$\frac{R_{10}(r)}{a_0^{-3/2}} = 2e^{-x} \tag{20-17}$$

$$\frac{R_{20}(r)}{a_0^{-3/2}} = \frac{1}{2\sqrt{2}}(2-x)e^{-x/2} \tag{20-18}$$

$$\frac{R_{30}(r)}{a_0^{-3/2}} = \frac{2}{81\sqrt{3}}(27-18x+2x^2)e^{-x/3} \tag{20-19}$$

The units of these and all other radial wave functions are $a_0^{-3/2}$. The functions plotted are the above functions divided by $a_0^{-3/2}$, which means that the functions are plotted in units of $a_0^{-3/2}$. Some textbooks and monographs note this explicitly (Karplus and Porter 1970), most ignore the units by not labeling the axis, and others (for example, Herzberg), substitute for a_0 , 0.529 x 10^{-8} cm.

It should be noted that different sources use different definitions of ρ. The aforementioned ρ is used by Eyring, Walter, and Kimbal (1979); Berry, Rice, and Ross (1980); and Alberty (1992). On the other hand, Pauling and Wilson (1935) and Atkins (1990) use the following definition of ρ:

$$\rho = \frac{2Zr}{na_0} \tag{20-20}$$

The two definitions give the radial wave functions a slightly different appearance, but they are, of course, equivalent.

Mathematica Notebook 11
Plots of R_{10} (r), R_{20} (r), and R_{30} (r)

Figure 20-15 shows the Mathematica plots of R_{10} (r), R_{20} (r), and R_{30} (r). The argument of **Plot**, enclosed in brackets [ ], is the expression to be plotted, followed by a list enclosed in braces { }. The first element of the list is the name of the variable being plotted x, followed by the *xmin* and *xmax*. If Mathematica thinks that changes in the maximum abscissa or ordinate values would lead to a clearer presentation, it will do so. Change *xmax* in **In[1]:** to **10** and see what happens.

The plots of R_{20} (r) and R_{30} (r) shown in **In[2]** and **In[3]:**, Figure 20-15, include the option **PlotRange→All**. This option forces Mathematica to plot over all of the range that you specify without ad-libbing. Try these plots without ***PlotRange →All*** to see the results.

The **Show** command displays the previously named plots, namely R_{10}, R_{20}, and R_{30}. Seeking to display the position of these three graphs, whose values hover around zero, Mathematica succeeds well enough, but truncates the three plots at 0.3. The addition of the **PlotRange→All** option permits comparing the maximum values of the plots, although the resolution in the neighborhood of zero is diminished. The result you select may involve a compromise between various parameters that you wish to present.

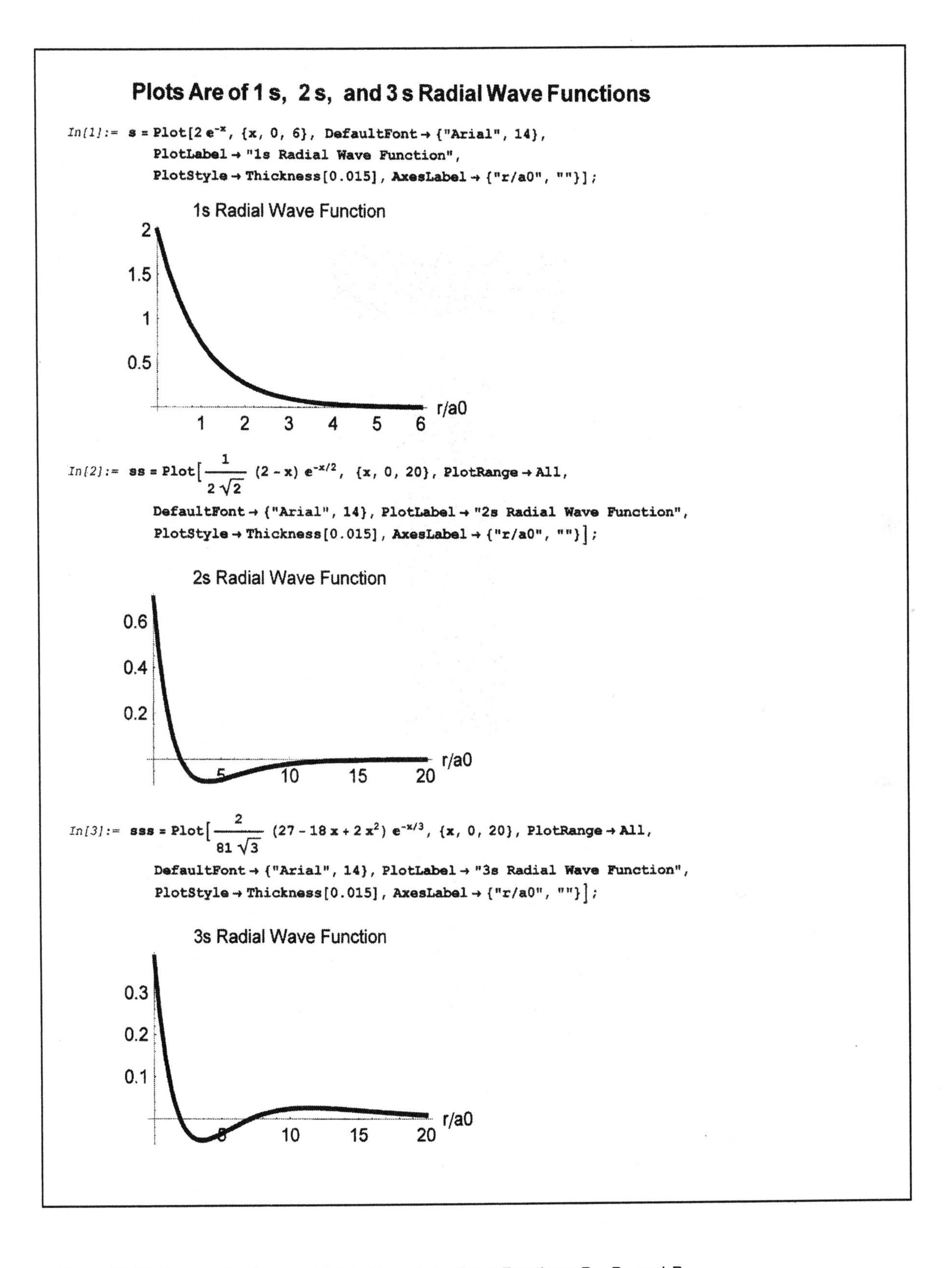

Figure 20-15. Mathematica Notebook 11: H Atom Radial Wave Functions: R_{1s}, R_{2s}, and R_{3s}

Three Dimensional Plots

Mathematica offers a large number of three-dimensional plots, each with many options. In the next sections we will demonstrate two of the more widely used functions: **Plot3D** and **SphericalPlot3D**.

Mathematica Notebook 12
Surface Graphics with Plot3D

To make a three-dimensional surface plot of a function f in Mathematica, load the graphics package:

<<graphics`graphics`

and then use the statement:

Plot3D[*f*, {*x*, *xmin*, *xmax*}, {*y*, *ymin*, *ymax*}]

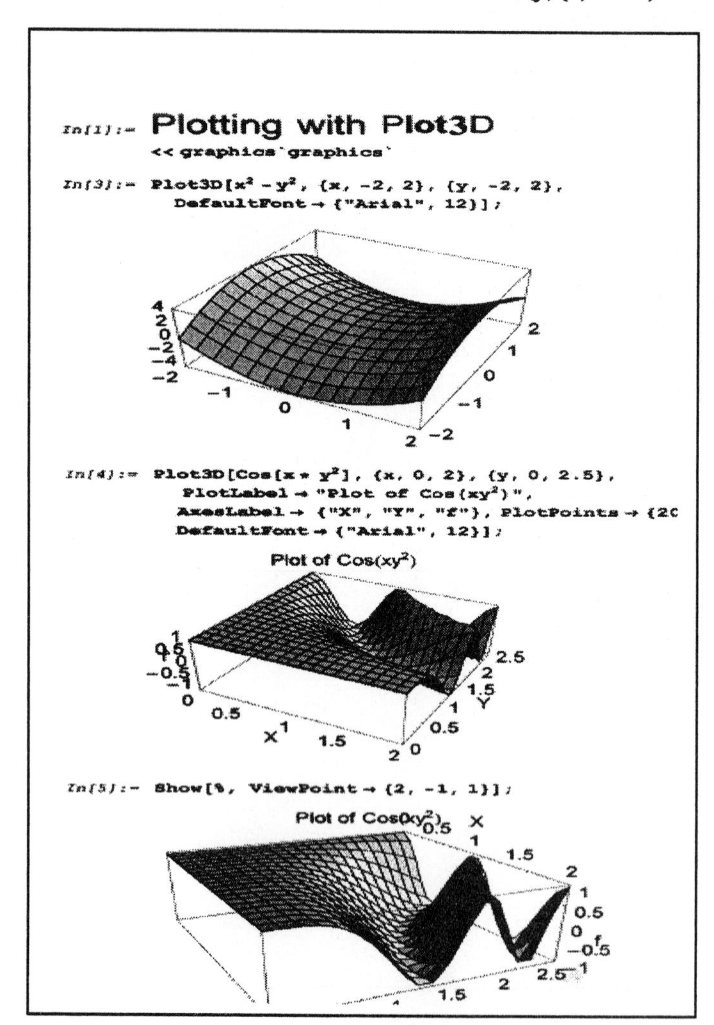

The first plot in Figure 20-16 shows a plot of the function $f = x^2 - y^2$ with default options. The **ViewPoint** is the default one:

{1.3, –2.4, 2}}

The second plot, of the function f = *Cos*[xy^2] shows **Plot3D** with several options. You can get a more complete list of options by typing **TableForm[Options[Plo3D]]**. Many of these options are the same as the options for the **Plot** and **ListPlot** functions described earlier. The **ViewPoint** is also the default one.

The **ViewPoint** option (the third plot in Figure 20-16) permits the user to change the viewpoint from which the plot is observed. The viewpoint in the bottom plot is {2, –1, 1}. Some other specific viewpoints are listed below.

{0, –2, 0} directly in front
{0, –2, 2} in front and up
{0, –2, –2} in front and down
{–2, –2, 0} left corner
{2, –2, 0} right corner
{0, 0, 2} directly above

Figure 20-16. Mathematica Notebook 12: Plots of Two Functions: $f = x^2 - y^2$ and f = Cos[xy^2]

Mathematica Notebook 13
Spherical Harmonics

The Mathematica **SphericalPlot3D** function provides a convenient method for plotting the angular wave function of a hydrogen atom. The total wave function for the solution of the Schrödinger wave equation, $H\psi = E\psi$, is of the form

$$\psi_{n,\ell,m}(r,\theta,\varphi) = R_{n,\ell}(r)\Theta_{\ell,m}(\theta)\Phi_m(\varphi) \tag{20-21}$$

The angular part of the wave function, $\Theta_{\ell,m}(\theta)\Phi_m(\varphi)$, corresponds to the spherical harmonics associated with the solution to the angular part of the wave equation. Mathematicians were familiar with spherical harmonics long before the development of quantum mechanics. Symbolized by $Y_{\ell,m}(\theta,\varphi)$, spherical harmonics have applications in the normal modes of vibration of an elastic sphere and in the classical physics of gravitation and electrostatics. It's not surprising then, that Mathematica knows spherical harmonics and will summon them up for us at the touch of a keyboard. The Mathematica **SphericalHarmonicY[ℓ, m, theta, phi]** command gives the spherical harmonic $Y_{\ell,m}(\theta,\varphi)$. Figure 20-17 shows a few spherical harmonics corresponding to the angular part of the total wave function.

```
In[1]:= Spherical Harmonics

        SphericalHarmonicY[0, 0, theta, phi]
```

Out[2]= $\frac{1}{2\sqrt{\pi}}$

```
In[3]:= SphericalHarmonicY[1, 0, theta, phi]
```

Out[3]= $\frac{1}{2}\sqrt{\frac{3}{\pi}}$ Cos[theta]

```
In[4]:= SphericalHarmonicY[2, 0, theta, phi]
```

Out[4]= $\frac{1}{4}\sqrt{\frac{5}{\pi}}$ (-1 + 3 Cos[theta]^2)

Figure 20-15. Mathematica Notebook 13. Some *Spherical Harmonics*

Figure 20-17. Mathematica Notebook 13: Some Spherical Harmonics

Mathematica Notebook 14

Plotting *s*, *p*, and *d* Angular Wave Functions with SphericalPlot3D

In order to plot the $\Theta_{\ell,m}(\theta)\Phi_m(\varphi)$ angular wave function, it is necessary to call up a Mathematica package with the command **<<Graphics`ParametricPlot3D`** as shown at the top of Figure 20-18. This package contains a number of plotting functions, but the one we will use is **SphericalPlot3D**, which has the syntax:

SphericalPlot3D[*r*,{theta, theta*min*, theta*max*},{phi, phi*min*, phi*max*}]

Plotting a Sphere. The coordinate system is described in Chapter 3, Excel Example 7. The parameter *r* is the distance of a point from the origin. For example, all the points lying on the surface of a sphere lie at a distance *r* from the origin. Let us let that distance *r* equal 2, a constant, and carry out the plot:

SphericalPlot3D[2,{theta, Pi, Pi/30},{phi, 0, Pi/15}]

Now think of that radius as a vector sweeping out over all thetas, every π/30, and over all phis, every π/15, jotting down a point and then connecting all the points with short lines (arcs, actually). The result is the sphere shown in the top plot of Figure 20-18.

Plotting an Angular Wave Function. When an angular wave function is plotted, *r* is not a constant (like 2), but is the absolute value of the wave function at a particular theta, phi pair. The plot is the absolute value of the wave function carried out over all space; that is, over theta ranging from zero to π (in increments of π/30) and phi ranging from zero to 2π (in increments of $\pi/15$). Leave out the increments, and Mathematica uses default values.

To plot a wave function, look at the parameter list for **SphericalPlot3D** in Figure 20-18. Instead of a constant 2, we use the absolute value of $Y_{\ell,m}(\theta,\varphi)$ for the orbital (here, the *p* orbital).

SphericalPlot3D[Abs[SphericalHarmonicY[1, 0, theta, phi]] ,{theta, Pi, Pi/30},{phi, 0, 2 Pi, Pi/15}]

The length of the vector sweeping out over all thetas and phis equals the value of the angular wave function at each value of theta and phi. A point is jotted down, the points are connected, and the result is the middle 3-D plot in Figure 20-18, a p_z orbital. The bottom plot in Figure 20-18 shows the shape of a d_{z^2} orbital.. The top plot of Figure 20-18 shows a sphere, which is the shape of an *s* orbital, although it was not plotted with a spherical harmonic. To plot an *s* orbital with a spherical harmonic, use **SphericalPlot3D[Abs[SphericalHarmonicY[0, 0, theta, phi]] ,{theta, Pi, Pi/30},{phi, 0, 2 Pi, Pi/15}]**.

SphericalPlot3D Plots of s, p, and d Orbitals

```
In[1]:= << Graphics`ParametricPlot3D`

In[2]:= SphericalPlot3D[2, {theta, 0, Pi, Pi / 30}, {phi, 0, 2 Pi, Pi / 15}];
```

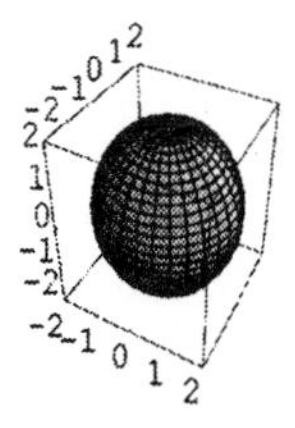

```
In[3]:= SphericalPlot3D[Abs[SphericalHarmonicY[1, 0, theta, phi]],
          {theta, 0, Pi, Pi / 30}, {phi, 0, 2 Pi, Pi / 15}];
```

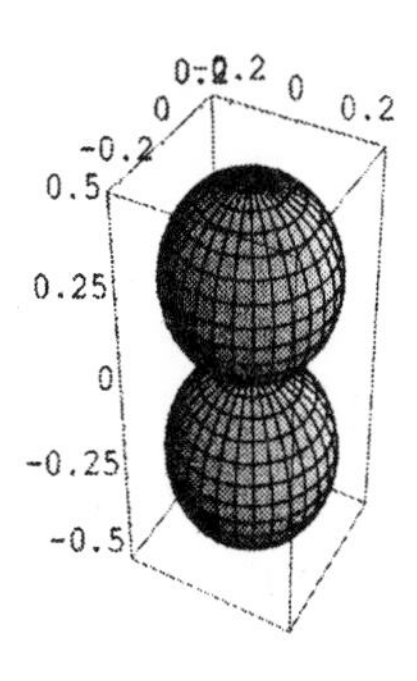

```
In[4]:= SphericalPlot3D[Abs[SphericalHarmonicY[2, 0, theta, phi]],
          {theta, 0, Pi, Pi / 30}, {phi, 0, 2 Pi, Pi / 15}];
```

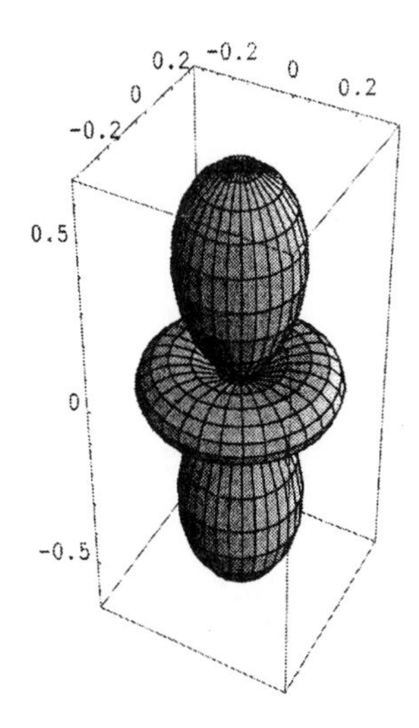

Figure 20-18. Mathematica Notebook 14: Plot of Angular Wave Functions s, p_z, and d_{z^2}

Mathematica Notebook 15
The Particle in a Box: ψ_{22} and ψ_{22}^2

The wave function for the particle in a box problem provides a good example of a three-dimensional surface plot. The quantum mechanics of the particle in a box problem are reviewed in Excel Example 26, Chapter 8.

Plotting a Particle in a Box Wave Function. A surface plot of a three-dimensional wave function requires a four-dimensional plot (*x, y, z,* and $\psi(x, y, z)$), so we will settle for two-dimensional plots of ψ_{22} and ψ_{22}^2, for which both n_x and $n_y = 2$. Figure 20-19 shows these two plots. The inclusion of a few optional directives improves the overall appearance of the plot. **PlotPoint → 30** increases the resolution, **BoxRatios → {1, 1, 1.1}** stretches the box in which the plot is enclosed in the *z* direction to improve clarity, and **ViewPoint → {1.3, -3.9, 1.4}** slightly rotates the whole box about the z-axis to improve visibility of the node at the bottom of the plot. It also tilts the whole box slight away from the viewer. The default parameter list for **ViewPoint** is **{1.3, -2.4, 2}**. The default parameter list for **BoxRatios** is **Automatic**.

Mathematica Notebook 16
3-D Hydrogen Atom Radial Wave Functions ψ_{2s} and ψ_{2s}^2

In the previous section, we prepared a series of plots of the 1*s* (R_{10}), 2*s* (R_{20}), and the 3*s* (R_{30}) orbitals. It is also possible to prepare a plot of the magnitude of the wave function in the *x,y* plane (Pilar 1990). For example R_{20}, is a simple function of *r*, the distance from the nucleus:

$$R_{20}(r) = \frac{1}{2\sqrt{2}} a_0^{-3/2} [2 - \left(\frac{r}{a_0}\right)] e^{\frac{-r}{a_0}} \tag{20-22}$$

At every point, the position of *r* can be expressed in terms of *x* and *y* (Figure 3-5):

$$r = (x^2 + y^2)^{1/2} \tag{20-23}$$

This last equation combined with R_{20} above results in an equation that gives magnitude of the R_{20} wave function as a function of *x* and *y*:

$$R_{20}(r) = \frac{1}{2\sqrt{2}} a_0^{-3/2} [2 - (x^2 + y^2)^{1/2}] e^{\frac{-(x^2+y^2)^{1/2}}{2}} \tag{20-24}$$

Plot3D applied to this equation in Figure 20-20, **In[1]:** gives the **SurfaceGraphics** plot shown in **Out[2]** coordinates.

A comparison of R_{20} here in Figure 20-20 and R_{20} in Figure 20- 15 shows that both plots reveal a node at which the sign of the wave function changes from positive to negative. In both plots the value of the wave function is negative at large distances from the nucleus and approaches zero asymptotically. Figure 20-20 also shows R_{2s}^2, which is positive everywhere, and the negative dip with R_{20} appears with R_{2s}^2 as a slight positive rise, circling the main peak. Again a few graphic optional directives were used to improve the presentation of the plots.

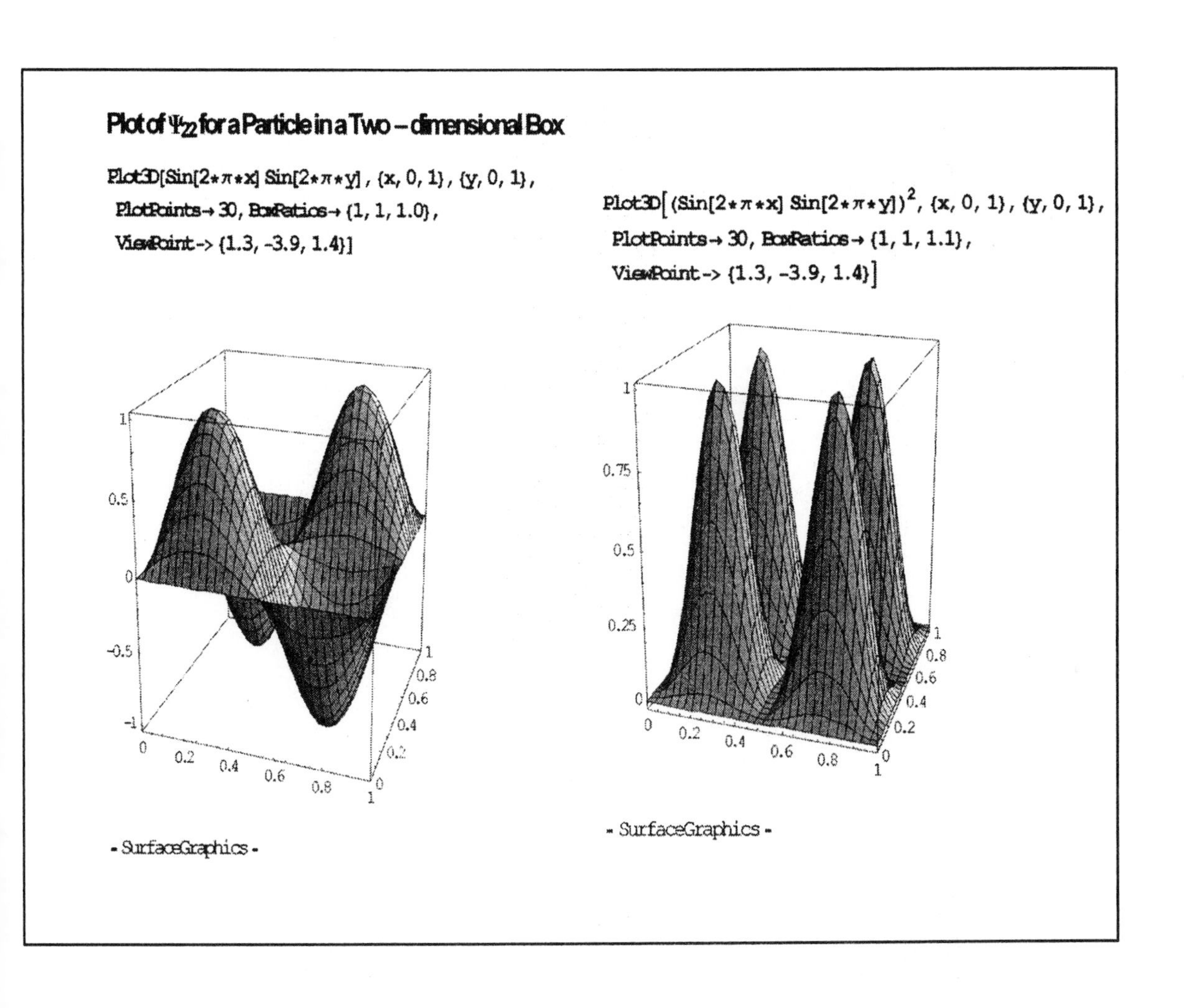

Figure 20-19. Mathematica Notebook 15: Wave Function for a Particle in a Two-Dimensional Box. To the left is $\psi_{2,2}$, and to the right is $\psi_{2,2}^2$.

Surface Plot of ψ_{2s} for a Hydrogen Atom

```
Plot3D[1/(2 Sqrt[2]) (2 - (x^2 + y^2)^(1/2)) E^(-(x^2+y^2)^(1/2)/2), {x, -8, 8}, {y, -8, 8}, PlotPoints -> 30,
  PlotRange -> All, BoxRatios -> {1, 1, 1.5}, ViewPoint -> {1.3, -2.4, 1.2}];
```

Out[1]= a Atom for Hydrogen of Plot Surface ψ2s

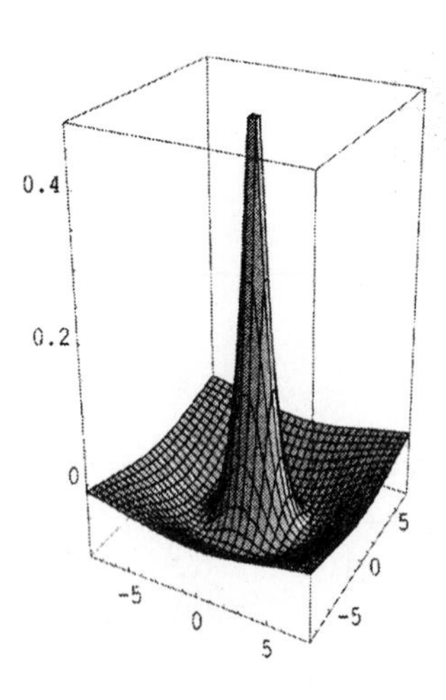

Surface Plot of ψ_{2s} Squared for a Hydrogen Atom

```
In[3]:= Plot3D[(1/(2 Sqrt[2]) (2 - (x^2 + y^2)^(1/2)) E^(-(x^2+y^2)^(1/2)/2))^2, {x, -8, 8}, {y, -7, 7}, PlotPoints -> 30,
  PlotRange -> All, BoxRatios -> {1, 1, 1.5}, ViewPoint -> {1.3, -2.4, 1.2}]
```

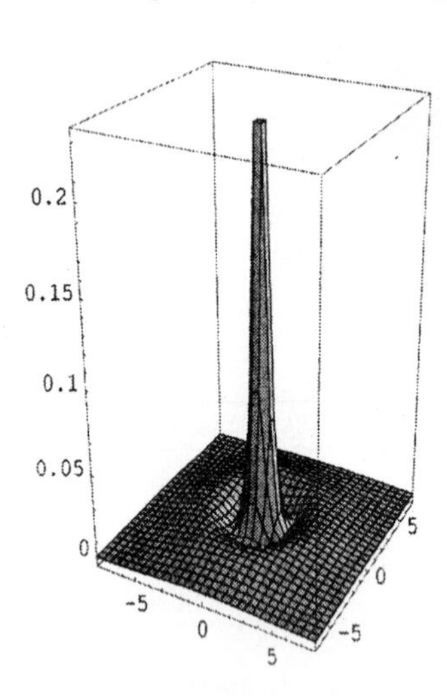

Out[3]= - SurfaceGraphics -

Figure 20-20. Mathematica Notebook 16: Plots of ψ_{2s} and ψ_{2s}^2 for a Hydrogen Atom

Bar Charts

The package **<<Graphics`Graphics`** contains many plotting tools including **BarChart** for creating bar charts and **PieChart** for creating pie charts.

Mathematica Notebook 17
Bar Chart of the Population of Rotational Energy Levels

Figure 19-23 shows a Mathcad worksheet for calculating and plotting a bar chart of the population of rotational energy levels according to the Boltzmann distribution law. The theory is reviewed in Chapters 3 and 4. Figure 20-21 shows the same calculation in Mathematica. The steps in preparing a similar Mathematic worksheet are as follows:

- In the first several steps the values of Planck's constant h, the speed of light c, the Boltzmann constant k, the rotational constant B, and the temperature T are defined. The constant z is defined and calculated from h, c, and k. The rotational partition function Q_r is defined and evaluated.
- A function for the population of the rotational energy stated is defined.
- A (list) table (**Tbl**) of these populations is calculated, but not displayed.
- The values of the populations are displayed with the **TableForm** function. These are the fractions of the total population in each rotational energy state.
- The **BarChart** function creates a bar chart of the populations of rotational energy levels. Setting the **BarStyle** option to **RGBColor[0,0,0]** makes the color of the bars black. In Mathematica the **RGBColor** parameters range from 0 to 1, not from 0 to 255, the usual case.
- The sum of the fractional populations of all the energy states equals unity, since all of the molecules populate all of the states. Figure 20-21 shows that the population of states higher than J equals 11 is very small. The Mathematica function **Apply[Plus, list name]** calculates the sum of all the elements of a list. The value of 1.0152 is in reasonable agreement with 1.

Population of Rotational Energy Levels

```
h = 6.6208 10^-34;
c = 2.99792 10^8;
k = 1.38066 10^-23;
B = 1059.09;
T = 298.15;
z = h c / k

Out[1]= Energy levels of Population Rotational

Out[7]= 0.0143762

In[8]:= Qr = T / (z B)

Out[8]= 19.5821

In[9]:= Population[J_] :=  ((2 J + 1) Exp[(-J (J + 1) B z) / T]) / Qr;
        Tbl = N[Table[Population[J], {J, 0, 10}]];
        TableForm[%]

Out[11]//TableForm=
        0.0510672
        0.138327
        0.18795
        0.193687
        0.165509
        0.121392
        0.0777322
        0.0438788
        0.0219665
        0.0097917
        0.00389727

In[14]:= << Graphics`graphics`;
         BarChart[Tbl, BarStyle → {RGBColor[0, 0, 0]},
           DefaultFont → {"Ariel,Bold", 14}, PlotLabel → "Population of Rotational Levels"];
```

Population of Rotational Levels

0.15
0.1
0.05

1 2 3 4 5 6 7 8 9 10 11

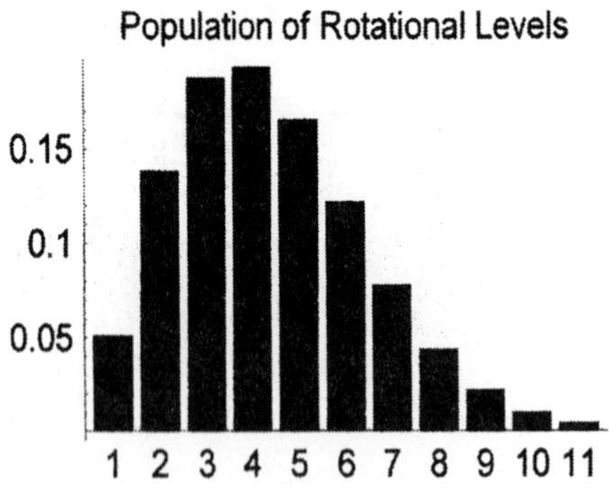

```
In[16]:= Apply[Plus, Tbl]

Out[16]= 1.0152
```

Figure 20-21. Mathematica Notebook 17: Population of Rotational Energy Levels

Symbolic Calculations

It is with symbolic calculation that Mathematica shows its muscle. Not that you cannot do numerical calculation (we have already done a few), but it is the scope of symbolic calculations that has made Mathematica a household word among scientists, mathematicians, and engineers. For the sake of clarity, the examples chosen here are simple enough for us to do without a computer. Try these examples and then use the same Mathematica syntax to operate on expressions that you would find difficult (if not impossible) to factor, expand, simplify, solve, integrate, or differentiate. *Difficult* is in our vocabulary, but not in Mathematica's.

Polynomial Expressions. Mathematica has functions for operating directly on a polynomial expression, for example, **Factor**, **Expand**, **Simplify**, **Integrate** (definite and indefinite), **Differentiate**, and **Plot**. Many **ListPlot** options, some of which we have already used, are the same for **Plot**.

With an expression, the basic format for **Plot** is **Plot[*Expression*,{*x*, *xmin*, *xmax*}]**, which generates a plot of the expression from ***xmin*** to ***xmax***.

Expand. Mathematica can expand complex expressions.

In[n]: = **Expand[**$(2x - y)^3(3x+y)^2$**]**
Out[n] = $72x^5 - 60x^4 y - 10x^3 y^2 + 15x^2 y^3 - y^5$

Factor. More difficult for us, but not for Mathematica, Mathematica can factor complex expressions.

In[n]: = **Factor[**$30x^4 - 151x^3 + 24x^2 + 421x - 84$**]**
Out[n] = $(-4 + x)(3 + 2x)(-7 + 3x)(-1 + 5x)$

Simplify. Mathematica can simplify complex expressions.

In[n]: = **Simplify[**$\frac{((x+y)^2 - x^2)}{y}$**]**

Out[n] = $2x + y$

The Replacement Operator (Substitution). The most obvious way to substitute a number in an expression is to define the number to be equal to a variable in the expression.

In[1]: = **z =3**
Out[1] = 3

In[2]: = $\mathbf{6x^3 - 7x^2 + 1}$
Out[2] = 100

Instead of a number, you can substitute one expression in another expression.

In[1]: = **x =y + 2**
Out[1] = 2 +y

In[2]: = $\mathbf{x^3}$ **-7x +6**
Out[2] = $6 - 7(2 + y) + (2 + y)^2$

In[3]: = **Simplify [%]**
Out[3] = $y(5 + 6y + y^2)$

The replacement operator accomplishes substitution in a more compact manner. The replacement operator is **/.** and it has no spaces before or after it. An arrow (→) separates a variable named *x* from the substituted expression *y* + 2.

In[1]: = $\mathbf{x^3}$ **-7x +6 /. x → y + 2**
Out[1] = $6 - 7(2 + y) + (2 + y)^2$

With the replacement operator, you can make multiple replacements.

In[1]: = $\mathbf{p^2 + q^2}$ **/. {p → 2 r, q → 2 s}**
Out[1] = $4r^2 - 27s^2$

The items to be replaced are the items in a list, and in Mathematica, a pair of braces { } delimits a list.

Mathematica Notebook 18
Multiple Plots with Plot

Mathematica can **Plot** polynomial expressions. Plots of expressions are similar to plots of functions, discussed earlier. As with functions, one or more expressions may be plotted on the same graph. Here, two expressions are plotted on the same graph.

In[n]: = **Plot[{**$\mathbf{x^2}$ **+4,** $\mathbf{x^3 - x^2}$ **– 5x + 5},{x, -4, 4}]**
Out[n] = -Graphics-

The **Plot** function plots a list of expressions. The plots of this operation are shown in Figure 20-22. See the section on **NSolve** in the later section on solving equations.

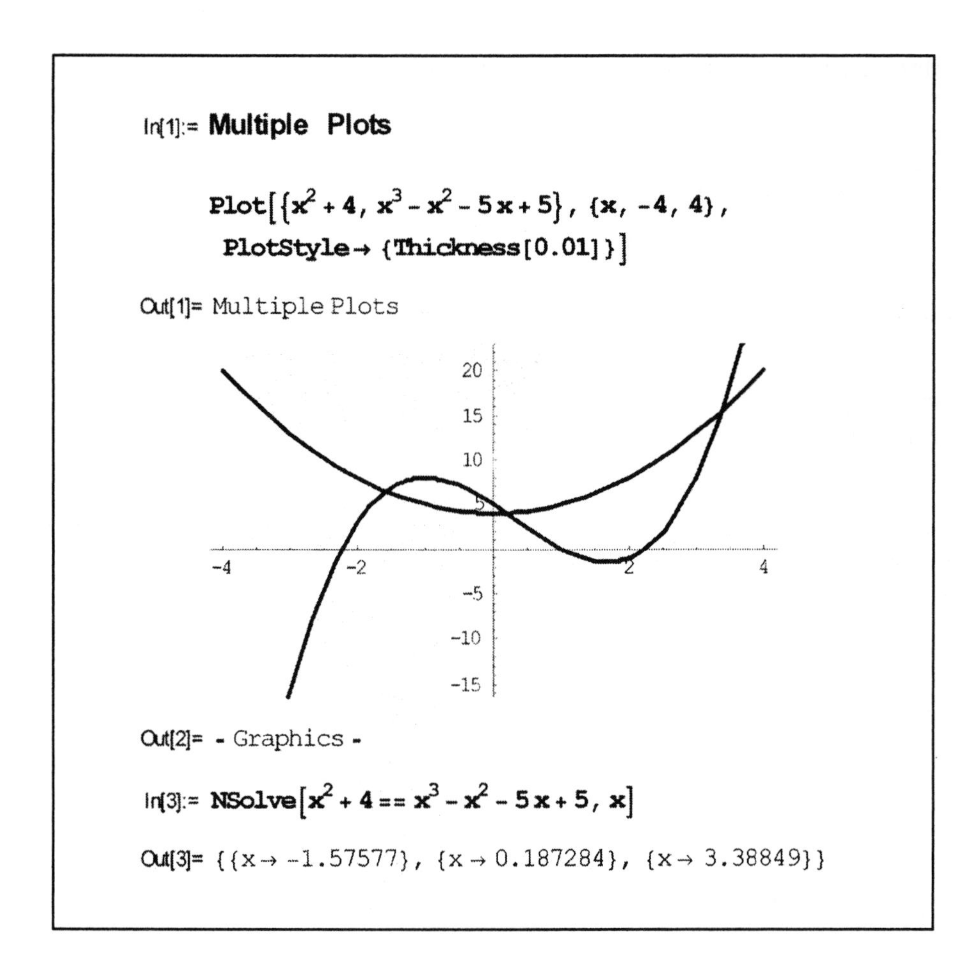

Figure 20-22. Mathematica Notebook 18: Plots of Two Polynomial Expressions

Calculus

Mathematica knows calculus and can integrate and differentiate expressions and functions.

Derivatives. Mathematica can differentiate both algebraic and transcendental expressions and functions.

In[n]: = **D[3x^2 -8x +5, x]**
Out[n] = - 8 + 6x

In[n]: = **D[**$\dfrac{Sin[x]}{Tan[x]}$**, x]**
Out[n] = - Sin[x]

Integral Calculus. Mathematica can handle both indefinite and definite integrals, and it can integrate both expressions and functions. First, integrate an expression.

In[1]: = **Integrate[3x^2 -8x +5, x]**
Out[2] = $5x - 4x^2 + x^3$

Alternatively, you can define a function and then integrate the function.

In[3]: = **f[x_]:= 3x^2 – 8x + 5**
In[4]: = **Integrate[f[x],x]**
Out[4] = $5x - 4x^2 + x^3$

Mathematica can also do definite integrals.

In[5]: = **Integrate[3x^2 -8x +5, { x, -2, 3}]**
Out[5] = 40

Int[6]: = **g[x_]:= 5*e*$^{-2x}$**
Int[6]: = **Integrate [g[x] , {x, 0, ∞ }]**
Out[6] = $\frac{5}{2}$

In[7]:= **f[x_]:= x^2 +2x +1**
In[7]:= **Integrate[f[x], {x, -3, 2}]**

Out[7] = $\frac{35}{3}$

In[8]:= **NIntegrate [f[x], {x, -3, 2}]**
Out[8] = 11.6667

Notice that the base to natural logarithms is *e*, not e. You can find *e* and ∞ on the Mathematica **Basic Input** palette.

In[9]:= $\int_0^\infty 2 * e^{-2*x^2} \, dx$

Out[9]= $\sqrt{\frac{\pi}{2}}$

Relational and Logical Operators

We have already used the equal sign (=) in Mathematica to make assignments, such as the assignment of a value to a constant. The double equal sign (= =) in Mathematica tests for equality. Mathematica supports the usual rational and logical operators beside equality as listed in Table 20-4.

In[n]: = **a = 4** (* this assigns 4 to a *)
Out[n] = 4

In[n]: = **b =4** (* this assigns 4 to b *)
Out[n] = 4

In[n]: = **c =5** (* this assigns 5 to c *)
Out[n*J* = 5

In[n]: = **a= =b** (* this tests whether a equals b *)
Out[n] = True

In[n]: = **a= =c** (* this tests whether a equals c *)
Out[n] = False

In[n]: = **a > c** (* this tests whether a is greater than c *)
Out[n] = False

Table 20-4

Relational and Logical Operators

Notation	Meaning
x = = y	Equal
x ! = y	Unequal
x > y	Greater than
x < y	Less than
x > = y	Greater than or equal to
x < = y	Less than or equal to

Equations in Mathematica

An expression like $x^2 - 5x + 6 = = 0$ is an equation in Mathematica. The = = (double equal sign) is just two equal signs. It may also be entered from the **Basic Calculations** palette. A few examples of entering and naming equations are shown below. You can type the equation directly.

In[n]: = $\mathbf{x^2 - 5x + 6 = = 0}$
Out[n] = $6 - 5x + x^2 = = 0$

You can name the equation when you type it.

In[n]: = **MyEq** = $\mathbf{x^2 - 5x + 6 = = 0}$
Out[n] = $6 - 5x + x^2 = = 0$

In[n]: = **MyEq**
Out[n] = $6 - 5x + x^2 = = 0$

Solving Equations. You can solve an equation for its variable, *x* in the preceding examples.

In[n]: = **Solve[x^2 - 5x +6 = = 0, x]**
Out[n] = {(x→ 2),(x→ 3)}

In[n]: = **Solve[MyEq,x]**
Out[n] = {(x→ 2),(x→ 3)}

To get a decimal result, use **NSolve** instead of **Solve**.

In[n]: = **NSolve[x^2 +3x -2 = = 0, x]**
Out[n*]* = {x → -3.56155},{x → 0}.

Figure 20-23 illustrates an example of plotting two expressions. With **NSolve** one can find the coordinates of their intersections. In Figure 20-23, the cubic equation (x^3 $-x^2$ **-5***x* **+5**) and the quadratic (x^2 **+ 4**) equation intersect at three points. At the points of intersection, the functions equal each other. **NSolve** finds the values of *x* at the three points. Substituting these three roots in either of the functions gives the *y* values of the three points of intersection.

In[n]: = **NSolve[x^2 + 4 = = x^3 –x^2 -5x +5, x]**
Out[n] = {{x → 1.57577},{x → 0.0187284},{x → 33.38849}}.

The keyword **Solve** can always solve algebraic equations if the highest power is less than five.
For higher powers, **Nsolve** can often find numerical solutions.
In difficult cases, but where it is possible to guess sensible roots, **FindRoot** can be useful. The argument for **FindRoot**, contained between square brackets is, first the equation to be solved, and second, a short list, enclosed in braces. The first element of the list is the variable name, and the second element is the numerical value of the estimated root. If more than one root is possible, then Mathematica tries to find the root closest to the estimated value.

In[n]: = **FindRoot[2 Sin[x] = = 0,{x, 3}]**
Out[n] = {x → 3.14159}

In[n]: = **FindRoot[2 Sin[x] = = 0,{x, 5}]**
Out[n] = {x → 9.42478}

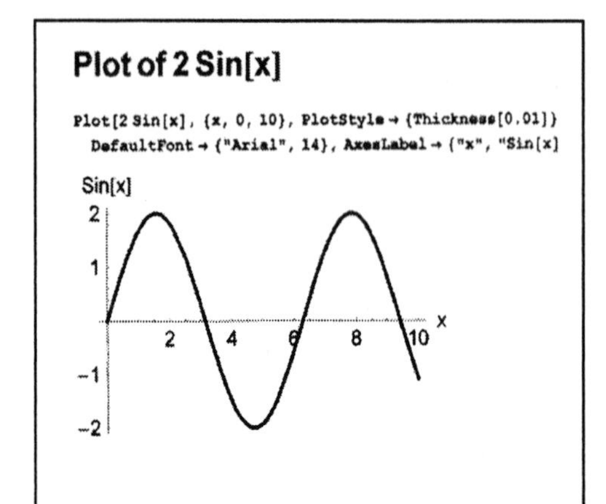

Actually, the function has an infinite number of roots. If we plot the expression (Figure 20-23), we see from the plot that we missed a root near *x* equals 6. So try **FindRoot** with 6 as an approximate solution.

In[1]: = **FindRoot[2 Sin[x] = = 0,{x, 6}]**
Out[1] = 6.23319

Figure 20-23. A Plot of 2 Sin[*x*] versus *x*

Let us try **Solve** with the next example: **Cos(x) – 2 log(x) = 0**.

In[2]: = **Solve[Cos[x]-2 Log[x]==0,x]**

This attempt results in the following message from Mathematica:

```
Solve :: eqf : Cos[x] – 2 Log [x]
Is not a well-formed equation.

Out[2] = Solve[Cos[x]-2 Log[x]==0,x]
```

That did not work, so let us try **FindRoot**.

In[3]: = **FindRoot[[Cos[x] - 2Log[x] = = 0, {x, 1.5}]**
Out[3] = {x → 1.19912}

FindRoot can also operate on a previously defined function, as the next example shows.

In[4]: = **g[x_]:= Cos[x]^2-Sin[x]**
In[5*]:* = **FindRoot [g[x] = = 0, {x, 0.5}]**
Out[3] = {x → 0.666239}

Mathematica Notebook 19
Solving Simultaneous Equations

You can solve a system of simultaneous equations with **Solve** or **NSolve**. The syntax is:

Solve[{*list of equations*},{*list of variables to solve for*}]

Commas delimit the members of the list. The equations may be linear or polynomial or a mixture, as illustrated in Figure 20-24.

```
In[1]:=
```

Solving Simultaneous Equations

```
Solve[{2 x + 3 y + z == 5, 2 x + y - z - 3 == 0, x - 3 y - 3 z == 1}, {x, y, z}]

Out[1]= Equations Simultaneous Solving

Out[2]= {{x → 4, y → -2, z → 3}}

In[3]:= NSolve[{3 x - y + z == 13, x + 3 y - z == 13, x + y - 3 z == 11}, {x, y, z}]

Out[3]= {{x → 5.42857, y → 2.14286, z → -1.14286}}

In[4]:= NSolve[{q^2 == 9 p, q^2 - p^2 == 8}, {q, p}]

Out[4]= {{p → 8., q → 8.48528}, {p → 8., q → -8.48528}, {p → 1., q → 3.}, {p → 1., q → -3.}}

In[5]:= TableForm[%]

Out[5]//TableForm=
        p → 8.      q → 8.48528
        p → 8.      q → -8.48528
        p → 1.      q → 3.
        p → 1.      q → -3.

In[6]:= Solve[{x^2 + y^2 == 12, x^2 - y == 4}, {x, y}]
```

Out[6]=

$$\left\{\left\{y \to -\frac{1}{2} - \frac{\sqrt{33}}{2},\ x \to -\sqrt{\frac{7}{2} - \frac{\sqrt{33}}{2}}\right\},\ \left\{y \to -\frac{1}{2} - \frac{\sqrt{33}}{2},\ x \to \sqrt{\frac{7}{2} - \frac{\sqrt{33}}{2}}\right\},\right.$$

$$\left.\left\{y \to -\frac{1}{2} + \frac{\sqrt{33}}{2},\ x \to -\sqrt{\frac{7}{2} + \frac{\sqrt{33}}{2}}\right\},\ \left\{y \to -\frac{1}{2} + \frac{\sqrt{33}}{2},\ x \to \sqrt{\frac{7}{2} + \frac{\sqrt{33}}{2}}\right\}\right\}$$

```
In[7]:= N[%]

Out[7]= {{y → -3.37228, x → -0.792287}, {y → -3.37228, x → 0.792287},
         {y → 2.37228, x → -2.52434}, {y → 2.37228, x → 2.52434}}

In[8]:= TableForm[%]

Out[8]//TableForm=
        y → -3.37228      x → -0.792287
        y → -3.37228      x → 0.792287
        y → 2.37228       x → -2.52434
        y → 2.37228       x → 2.52434
```

Figure 20-24. Mathematica Notebook 19: Solving Simultaneous Equations

Mathematica Notebook 20
The r_{mp} and r_{ave} for a 1*s* Orbital

The location of the maximum in this curve (Figure 20-25) can be identified with the distance r from the nucleus where it is most probable to find the electron: r_{mp}. Integration of the product of r and r^2R^2 gives the average value of r. The Mathcad treatment of these problems and more discussion is given in Chapter 19.

Mathematica Notebook 21
The Gas Speed Distribution Function at 298 K, 500 K, and 1000 K

The value of the speed u at the maximum in the speed distribution function for a gas (Figure 20-26) corresponds to the most probable speed u_{mp}. The fraction of molecules having speed between u and u + du rises rapidly because of the u^2 term, but eventually the inverses exponential term dominates and the plot falls to low values. It is apparent that the maxima (most probable speed) shifts to higher speeds as the temperature increases.

Minima and Maxima

The maximum or minimum of the independent variable in many plots from physics and chemistry leads to a more complete understanding of the function under study. Setting the first derivative of a function equal to zero and solving for the variable permits finding the value of the variable at the minimum or maximum. The following Mathematica notebooks illustrate a few maxima and minima of interest in physics and chemistry.

Mathematica Notebook 22
The u_{mp} and u_{ave} for a Gas

The value of the speed u at the maximum in the speed distribution function for a gas (Figure 20-27) corresponds to the most probable speed. The derivative of the function leads to the value of u at the maximum in the curve. Determining u_{ave} for a gas is similar to determining r_{ave} for a 1*s* orbital. The Mathcad treatment of u_{mp} and u_{ave} is given in Chapter 19.

Mathematica Notebook 23
The Lennard-Jones Plot

First the Lennard-Jones function is defined. Then variable names (*p*1, *p*2, *p*3, and *p*4) are given to the four plots for the Lennard-Jones potential for oxygen, methane, chlorine, and neon. The plots are distinguished by giving them different thicknesses. Plotting is repressed with the use of the statement **DisplayFunction→$DisplayIdentity**. In the final **Show** statement, the statement **DisplayFunction→$DisplayFunction** allows display of the plots. The theory underlying the Lennard-Jones potential (Figure 20-28) is given in Chapter 19.

Mathematica Notebook 24
The Lennard-Jones Minimum

In Figure 20-29 it is shown that at the minimum the value of r equals $\sqrt[6]{2}^{\,1}\sigma$, and the depth of the potential equals ε.

The 1 s Orbital

```
In[1]:= f[x_] := 4 x^2 e^(-2 x)

In[2]:= Plot[f[x], {x, 0, 6},
          PlotStyle→ {Thickness[0.015], GrayLevel[0.5]},
          PlotLabel→ "r^2R^2 vs. r", AxesLabel→ {"x/(r/a0)", "r^2R^2"}]
```

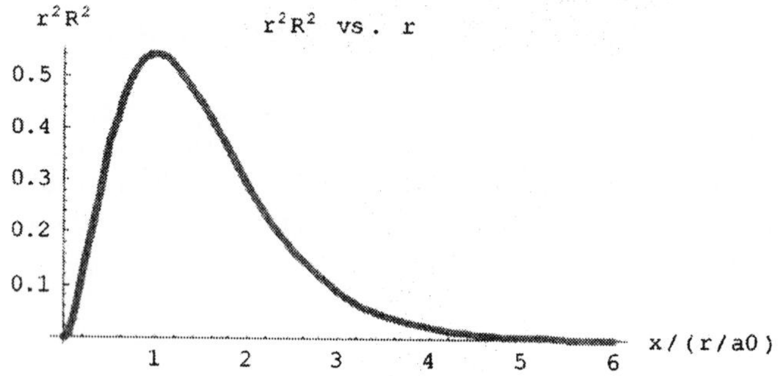

```
Out[2]= - Graphics -

In[3]:= D[f[x], x]

Out[3]= 8 e^(-2 x) x - 8 e^(-2 x) x^2

In[4]:= Solve[8 e^(-2 x) x == 8 e^(-2 x) x^2, x]

Out[4]= {{x → 0}, {x → 1}}
```

At the maximum probability, $x = 1 = \frac{r}{a0}$,

so at the most probable $r = a0$

In[5]:= $\int_0^\infty x\, f[x]\, dx$

Out[5]= $\frac{3}{2}$

On the average, $x = \frac{3}{2} = \frac{r}{a0}$

so the average $r = \frac{3}{2}\, a0$

Figure 20-25. Mathematica Notebook 20: The r_{ave} and r_{mp} for a 1*s* Orbital

In[1]:=

The Speed Distribution Function

```
R := 8.314;
```

In[3]:= `M := 0.028;`

In[4]:= $f[u\_, T\_] := 4\pi \left(\frac{M}{2\pi R T}\right)^{3/2} u^2 \, Exp\left[\frac{-M u^2}{2 R T}\right];$

In[5]:= `Plot[{f[u, 298.15], f[u, 500], f[u, 1000]}, {u, 0, 2500}, PlotStyle→{Thickness[0.01]}]`

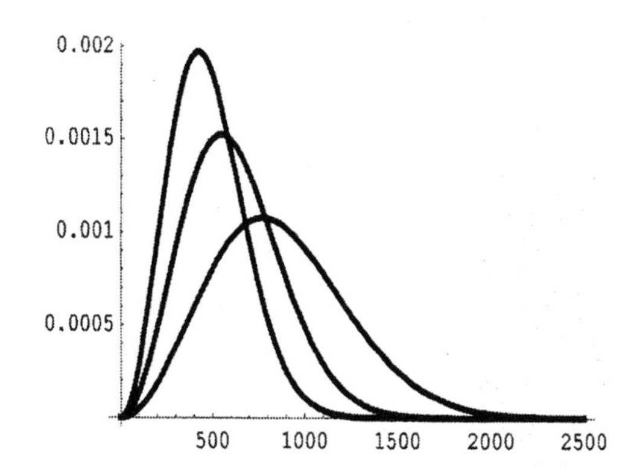

Out[5]= - Graphics -

$D[a u^2 Exp[-b u^2], u]$

Out[6]= $2 a e^{-b u^2} u - 2 a b e^{-b u^2} u^3$

In[8]:= $Solve[2 a e^{-b u^2} u == 2 a b e^{-b u^2} u^3, u]$

Out[8]= $\{\{u \to 0\}, \{u \to -\frac{1}{\sqrt{b}}\}, \{u \to \frac{1}{\sqrt{b}}\}\}$

Note : b = M / 2 RT so u most probable equals Sqrt (2 RT / M)

Figure 20-26. Mathematica Notebook 21: The Plots of the Speed Distribution Function, Its Derivative, and the Solution to the Equation

In[1]:=

Most Probable and Average Speed of a Gas

$$f[u\_] := 4\pi \left(\frac{M}{2\pi R T}\right)^{3/2} u^2 \,\mathrm{Exp}\left[\frac{-M u^2}{2 R T}\right]$$

In[3]:= D[f[u], u]

$$\text{Out[3]=}\ 2\, e^{-\frac{M u^2}{2RT}} \sqrt{\frac{2}{\pi}} \left(\frac{M}{RT}\right)^{3/2} u - \frac{e^{-\frac{M u^2}{2RT}} M \sqrt{\frac{2}{\pi}} \left(\frac{M}{RT}\right)^{3/2} u^3}{RT}$$

$$\text{In[4]:= Solve}\left[2\, e^{-\frac{M u^2}{2RT}} \sqrt{\frac{2}{\pi}} \left(\frac{M}{RT}\right)^{3/2} u == \frac{e^{-\frac{M u^2}{2RT}} M \sqrt{\frac{2}{\pi}} \left(\frac{M}{RT}\right)^{3/2} u^3}{RT}, u\right]$$

$$\text{Out[4]=}\ \left\{\{u \to 0\}, \left\{u \to -\frac{\sqrt{2}\sqrt{R}\sqrt{T}}{\sqrt{M}}\right\}, \left\{u \to \frac{\sqrt{2}\sqrt{R}\sqrt{T}}{\sqrt{M}}\right\}\right\}$$

$$\text{In[5]:= The average speed } u = \left(\frac{2RT}{M}\right)^{\frac{1}{2}}$$

$$\text{Out[5]=}\ \sqrt{2}\sqrt{\frac{RT}{M}}$$

$$\text{In[6]:=}\ \int_0^\infty u\, f[u]\, du$$

$$\text{Out[6]=}\ \sqrt{\frac{2}{\pi}} \left(\frac{M}{RT}\right)^{3/2} \mathrm{If}\left[\mathrm{Re}\left[\frac{M}{RT}\right] > 0, \frac{2 R^2 T^2}{M^2}, \int_0^\infty e^{-\frac{M u^2}{2RT}} u^3\, du\right]$$

$$\text{In[7]:= Simplify}\left[\sqrt{\frac{2}{\pi}} \left(\frac{M}{RT}\right)^{3/2} \frac{2 R^2 T^2}{M^2}\right]$$

$$\text{Out[7]=}\ \frac{2\sqrt{\frac{2}{\pi}}}{\sqrt{\frac{M}{RT}}}$$

$$\text{The most probable speed } u = \left(\frac{8RT}{\pi M}\right)^{\frac{1}{2}}$$

Figure 20-27. Mathematica Notebook 22: The Average and Most Probable Speeds of a Gas

In[1]:= **The Lennard – Jones Plot**

$$LJ[r\_, e\_, s\_] := 4\,e\left(\frac{s^{12}}{r^{12}} - \frac{s^{6}}{r^{6}}\right)$$

The Lennard-Jones parameters are e in Kelvin and s in picometer

e equals 196 and s equals 277 for neon
e equals 382 and s equals 148 for methane
e equals 358 and s equals 118 for oxygen
e equals 440 and s equals 258 for chlorine

```
In[4]:= p1 = Plot[LJ[r, 196, 277], {r, 270, 500}, PlotStyle→ {Thickness[0.015], GrayLevel[0.1]},
          GridLines→ Automatic, DefaultFont→ {"Arial,Bold", 14},
          PlotLabel→ "Lennard-Jones (L to R): Oxygen, Methane, Chlorine, Neon",
          AxesLabel→ {"r/pm", " "}, DisplayFunction→ Identity]

Out[4]= - Graphics -

In[5]:= p2 = Plot[LJ[r, 382, 148], {r, 145, 400}, PlotStyle→ {Thickness[0.009], GrayLevel[0.1]},
          GridLines→ Automatic, DefaultFont→ {"Arial,Bold", 14}, AxesLabel→ {"r/pm", " "},
          DisplayFunction→ Identity]

Out[5]= - Graphics -

In[6]:= p3 = Plot[LJ[r, 358, 118], {r, 115, 400}, PlotStyle→ {Thickness[0.007], GrayLevel[0.1]},
          GridLines→ Automatic, DefaultFont→ {"Arial,Bold", 14}, AxesLabel→ {"r/pm", " "},
          DisplayFunction→ Identity]

Out[6]= - Graphics -

In[7]:= p4 = Plot[LJ[r, 440, 256], {r, 253, 500}, PlotStyle→ {Thickness[0.011], GrayLevel[0.1]},
          GridLines→ Automatic, DefaultFont→ {"Arial,Bold", 14}, AxesLabel→ {"r/pm", " "},
          DisplayFunction→ Identity]

Out[7]= - Graphics -

In[8]:= Show[p1, p2, p3, p4, PlotRange→All, DisplayFunction→ $DisplayFunction];
```


Figure 20-28. Mathematica Notebook 23: The Lennard-Jones Potential

In[3]:=

The Minim um in the Lennard – Jones Potential

$$D[4\epsilon[(\sigma/r)^{12} - (\sigma/r)^6], r]$$

Out[3]= -Jones Potential + in Lennard Minimum the The

Out[4]= $4\left(\frac{6\sigma^6}{r^7} - \frac{12\sigma^{12}}{r^{13}}\right)\epsilon'\left[-\frac{\sigma^6}{r^6} + \frac{\sigma^{12}}{r^{12}}\right]$

In[5]:= $\mathrm{Solve}\left[\left(\frac{6\sigma^6}{r^7} - \frac{12\sigma^{12}}{r^{13}}\right) == 0, r\right]$

Out[5]= $\{\{r\to -2^{1/6}\sigma\}, \{r\to 2^{1/6}\sigma\}, \{r\to -(-1)^{1/3}2^{1/6}\sigma\}, \{r\to (-1)^{1/3}2^{1/6}\sigma\}, \{r\to -(-1)^{2/3}2^{1/6}\sigma\}, \{r\to (-1)^{2/3}2^{1/6}\sigma\}\}$

In[6]:= **TableForm[%]**

Out[6]//TableForm=

$r\to -2^{1/6}\sigma$
$r\to 2^{1/6}\sigma$
$r\to -(-1)^{1/3}2^{1/6}\sigma$
$r\to (-1)^{1/3}2^{1/6}\sigma$
$r\to -(-1)^{2/3}2^{1/6}\sigma$
$r\to (-1)^{2/3}2^{1/6}\sigma$

$r\to -2^{1/6}\sigma$
$r\to 2^{1/6}\sigma$
$r\to -(-1)^{1/3}2^{1/6}\sigma$
$r\to (-1)^{1/3}2^{1/6}\sigma$
$r\to -(-1)^{2/3}2^{1/6}\sigma$
$r\to (-1)^{2/3}2^{1/6}\sigma$

At the Minimum $r\to 2^{1/6}\sigma$

Next, find the value of the Lennard - Jones Potential at the Minimum

In[7]:= $r := 2^{\frac{1}{6}} s$

In[8]:= $4e\left[\left(\frac{s}{r}\right)^{12} - \left(\frac{s}{r}\right)^6\right]$

Out[8]= $4e\left[-\frac{1}{4}\right]$

At the minimum V = e

Figure 20-29. Mathematica Notebook 24: The Minimum in the Lennard-Jones Potential

Mathematica Notebook 25
Miscellaneous Information in Mathematica

`Audio`
`BlackBodyRadiation`
`Calendar`
`ChemicalElements
`CityData`
`Geodesy`
`Music`
`PhysicalConstants`
`RealOnly`
`ResonanceAbsorptionLines`
`StandardAtmosphere`
`Units`
`WorldData`
`WorldName`
`WorldPlot`

Selecting **Help Browser** on the **Help** menu, then selecting **Add-ons** in the **Help Browser** window (Figure 20-30), then **StandardPackages**, then **Miscellaneous** provides a drop-down menu listing several selections listed to the left. The Mathematica package **<<Miscellaneous`ChemicalElements`** provides the data for the bar chart in Figure 20-31. Mathematica contains much interesting information in the package **<<Miscellaneous**, listed to the left. Table 20-2 lists other standard Mathematica packages.

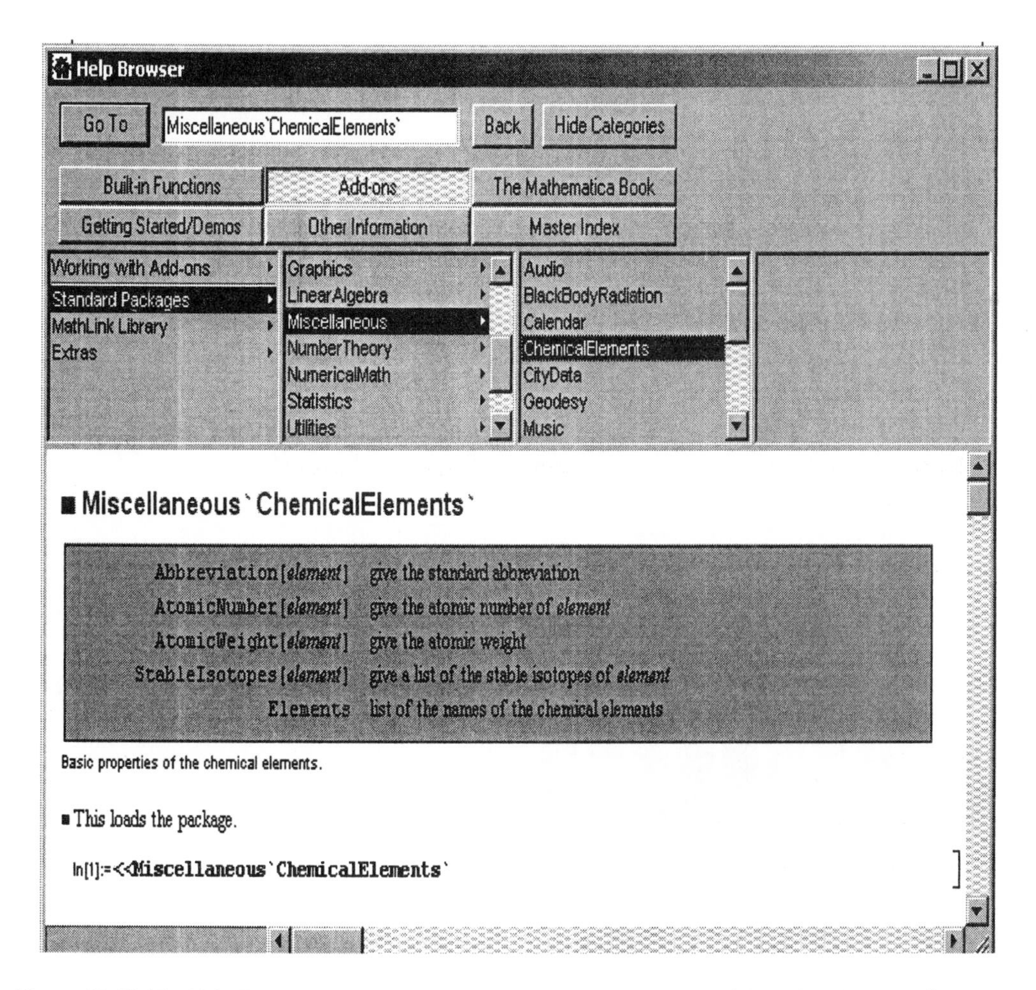

Figure 29-30. The Help Browser Displaying the Path to the **<<Miscellaneous `Chemical Elements`** Package

In[1]:= **Properties of Miscellaneous Elements**

```
<< Graphics`Graphics`
```

```
In[3]:= << Miscellaneous`ChemicalElements`

In[4]:= AtomicWeight[Copper]

Out[4]= 63.546

In[5]:= ElectronConfigurationFormat[U]
```

Out[5]= $1s^2\ 2s^2 2p^6\ 3s^2 3p^6 3d^{10}\ 4s^2 4p^6 4d^{10} 4f^{14}\ 5s^2 5p^6 5d^{10} 5f^3\ 6s^2 6p^6 6d^1\ 7s^2$

```
In[6]:= EarthCrustAbundance[U]
```

Out[6]= 2.7×10^{-6}

```
In[7]:= SolarSystemAbundance[H]

Out[7]= 0.91

In[8]:= Density[U]
```

Out[8]= $\frac{18950.\ \text{Kilogram}}{\text{Meter}^3}$

```
In[9]:= Density[{Sc, Ti, V, Cr, Mn, Fe, Co, Ni, Cu, Zn}] / (Kilogram / Meter^3)

Out[9]= {2989., 4540., 6110., 7190., 7440., 7874., 8900., 8902., 8960., 7133.}

In[10]:= BarChart[{2989.`, 4540.`, 6110.`, 7190.`, 7440.`, 7874.`, 8900.`, 8902.`,
    8960.`, 7133.`}, BarStyle -> {RGBColor[0, 0, 0]},
   PlotLabel -> "Densities of 3d Elements", AxesLabel -> "kg/m^2",
   BarLabels -> {"Sc", "Ti", "V", "Cr", "Mn", "Fe", "Co", "Ni", "Cu", "Zn"}]
```

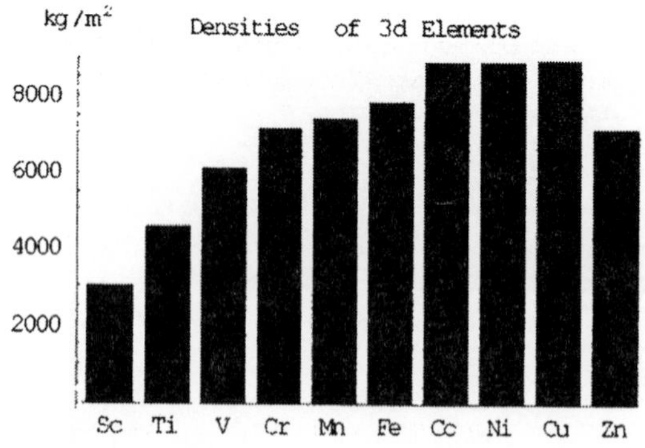

```
Out[10]= - Graphics -
```

Figure 29-31. Mathematica Notebook 25: Data and Plot from the Package **<<Miscellaneous `Chemical Elements`**

Bibliography

General

1. Alberty, R. A., and R. J. Silbey. *Physical Chemistry*. New York: Wiley, 2000.
2. Atkins, P. W. *Molecular Quantum Mechanics*. 3rd ed. New York: Oxford University Press, 1999.
3. Atkins, P. W. *Physical Chemistry*. 7th ed. New York: W. H. Freeman, 2002.
4. Berry, R. S., S. A. Rice, and J. Ross. *Physical Chemistry*. New York: Wiley, 2000.
5. Cornell, G. *Visual Basic 6*. New York: Osborn/McGraw-Hill, 1998.
6. Engel, T., and P. Reid. *Physical Chemistry*. San Francisco: Benjamin Cummings, 2005.
7. Harris, M. *Visual Basic for Applications*. Indianapolis: Sams Publishing, 1997.
8. *Mathcad User's Guide*, Cambridge, MA: MathSoft, 2000.
9. Sime, R. J. *Physical Chemistry: Methods, Techniques and Experiments*. Philadelphia: Saunders College Publishing, 1990.
10. Walkenbach, J. *Excel 2002 Power Programming with VBA*. New York: M & T Books, 2002.
11. Wolfram, S. *Mathematica: A System for Doing Mathematics by Computer*. 2nd ed. Reading, MA: Addison-Wesley, 1991.

Crystallography

1. Hammond, C. *The Basics of Crystallography and Diffraction*. New York: Oxford University Press, 2001.
2. Stout, G. H., and L. H. Jensen. *X-ray Structure Determination*. 2nd ed. London: Macmillan, 1989.
3. Woolfson, M. M. *An Introduction to X-ray Crystallography*. 2nd ed. Cambridge, England: Cambridge University Press, 1997.

Kinetics

1. Berry, R. S., S. A. Rice, and J. Ross. *Physical and Chemical Kinetics*, 2nd ed. New York: Oxford University Press, 2001.
2. Laidler, K. J. *Chemical Kinetics*. 2nd ed. New York: McGraw-Hill, 1965.
3. Tinoco, I., K. Sauer, and J. C. Wang. *Physical Chemistry: Principles and Applications in Biological Sciences*. Englewood Cliffs, NJ: Prentice-Hall, 1978.
4. Voet, D., and J. G. Voet. *Biochemistry*. 2nd ed. New York: John Wiley, 1995.

Quantum Mechanics

1. Barrow, G. M. *Introduction to Molecular Spectroscopy*. New York: McGraw-Hill, 1962.
2. Engel, T., and P. Reid. *Quantum Chemistry and Spectroscopy*. San Francisco: Benjamin Cummings, 2005.
3. Eyring, H., J. Walter, G. E. Kimball. *Quantum Chemistry*. New York: Wiley, 1944.
4. Greiner, W. *Quantum Mechanics: An Introduction*. 4th ed. New York: Springer, 2001.
5. Herzberg, G. *Atomic Spectra and Atomic Structure*. 2nd ed. New York: Dover, 1944.

6. Herzberg, G. *Molecular Spectra and Molecular Structure I. Spectra of Diatomic Molecules.* 2nd ed. Princeton, NJ: Van Nostrand, 1950.
7. Huber, K. P., and G. Herzberg. *Molecular Spectra and Molecular Structure.* Princeton, NJ: Van Nostrand, 1979.
8. Karplus, M., and R. N. Porter. *Atoms and Molecules.* Menlo Park, CA: Benjamin Cummings, 1970.
9. Lummer, O., and E. Pringsheim, "Radiation of Black Bodies and Platinum," *D. phys. Ges. Verhandlungen*, 1.12, 215-230 (1899)
10. McQuarrie, D. A. *Quantum Chemistry.* Mill Valley, CA: University Science Books, 1983.
11. Pauling, L., and E. B. Wilson. *Introduction to Quantum Mechanics.* New York: McGraw-Hill, 1935.
12. Pilar, F. L. *Elementary Quantum Chemistry.* 2nd ed. Dover: New York, 2001.
13. Planck, M. "Zur Theorie des Gesetzes der Energie Verteilung im Normalspekrum," *Ann. d. Physik* 4:553 (1901).

Statistics

1. Ambrose, H. W., and K. P. Ambrose. *A Handbook of Biological Investigations.* 5th ed. Winston-Salem, NC: Hunter Textbooks, 1995.
2. Barford, N. C. *Experiment Measurements: Precision, Error and Truth.* 2nd ed. New York: John Wiley & Sons, 1985.
3. Beers, Y. *Introduction to the Theory of Errors.* Reading: MA: Addison-Wesley, 1957.
4. Bethea, R. M., B. S. Duran, and T. L. Boullion. *Statistical Methods for Engineers and Scientists.* 3rd ed. 1995, New York: Marcel Dekker, 1995.
5. Dowdy, S., and S. Wearden. *Statistics for Research.* 2nd ed. New York: John Wiley & Sons, 1991.
6. Montgomery, D. C. *The Design and Analysis of Experiments.* 3rd ed. New York: John Wiley, 1991.
7. Triola, M. F. *Elementary Statistics.* New York: Addison-Wesley, 1998.
8. Young, H. D. *Statistical Treatment of Experimental Data.* New York: McGraw-Hill, 1962.

Thermodynamics

1. Boublík, T., V. Fried, and E. Hála. *The Vapor Pressures of Pure Substances.* Amsterdam: Elsevier Scientific Publishing Company, 1973.
2. Engel, T., and P. Reid. *Thermodynamics, Statistical Thermodynamics, and Kinetics.* San Francisco: Benjamin Cummings, 2005.
3. Gerke, R. H., "Temperature Coefficient of Electromotice Force of Galvanic Cells and the Entropy of Reaction," *J. Am Chem. Soc.* 44:1684 (1922).
4. Hála, E., J. Pick, V. Fried, and O. Vilim. *Vapor-Liquid Equilibrium.* 2nd ed. New York: Pergamon Press, 1967.
5. Ives, D. J. G., and J. J. Janz. *Reference Electrodes.* New York: Academic Press, 1961.
6. Klotz, I. M., and R. M. Rosenberg. *Chemical Thermodynamics: Basic Theory and Methods.* 6th Ed. New York: John Wiley, 2000.
7. Latimer, W. M. *The Oxidation States of the Elements and Their Potentials in Aqueous Solutions.* Englewood Cliffs, NJ: Prentice-Hall, 1952.
8. Meads, P. F., W. K. Forsythe, and W. F. Giauque. "The Heat Capacities and Entropies of Silver and Lead from 15° to 300° K," *J. Am. Chem. Soc.* 63:1902 (1941).
9. Pitzer, K. S., and L. Brewer. *Thermodynamics.* 3rd ed. New York: McGraw-Hill, 1995.
10. Timmermans, J. *Physico-Chemical Constants of Binary Systems.* New York: Interscience Publishers, 1968.

Index

A

.bmp, 427
.cur, 427
.gif, 427
.ico, 427
.jpg, 427
.wmf, 427
&, 241, 244, 277, 293
Absolute cell references, 9, 10
ActiveCell property, 488
ActiveChart parperty, 492
ActiveDocument property, 473
ActiveSheet property, 495
Add-ins, 22
AddItem method, 334, 338, 361
Add Module dialog box, 459
Add Procedure dialog box, 342
Alignment constants, 242
And logical operator
Analysis ToolPak, 22
Angular wave function, 55, 565, 637
Antoine constants, 33
Append mode, 392
Arcs, drawing, 444
Arithmetic operators, 293, 606
Array, control, 431
Arrays, 346, 376
ASCII character codes, 322, 326
Assignment operator, 247, 539
Assigning a value, 539
***Assume* keyword, 590**
Automatic completion, 229
Average
- function, 153
- quantities, 120
- *r* (1*s* orbital), 590, 653 - 654
- speed, 120, 590, 651, 653
- velocity, 112

AxesLabel, 635

B

Bar chart, 68–69, 585, 643, 660
Boltzmann distribution, 88–89, 98
Boolean type, 243
Boolean value, 263
BorderStyle property, 227
Bragg's law, 17
Bravais lattices, 13
Bubble sort, 383
Built-in functions, 5, 176, 541, 610

C

Calculus
- derivatives, 575, 647
- integral, 576, 648

Call statement, 278
Cell
- address, 2
- entries, 4
- references, 9, 10, 488

Cells property, 460
Chart
- bar chart, 68–69, 585, 660
- creating, 26–28, 179, 194, 548–549, 621–625
- editing, 21, 29, 196, 199
- Insert Chart dialog box,
- With, 3
- XY (scatter), 26, 27, 35, 38, 43

Chart Wizard, Excel, 26, 181
Circle Screen method, 441
Clear function, 609
Clear method, 288, 336, 366, 368, 374
Cls method, 249
Click event, 217
Code
- automatic completion, 229
- entering, 228, 470

Code window, 216, 470
- events list, 231
- procedure list, 230

Color
- constants, 234
- **QBColor** function, 438
- **RGB** function, 234

ColumnIndex property, 502
Column width, Excel, 8
Combo box, 359, 362–363
Command button, 217, 227, 233

Commenting, 228, 337
Common dialog boxes control, 410–425
Components dialog box, 413
Concatenation, 241, 244, 277, 293
Confidence interval, 155
Consecutive reaction, 122, 126, 618
Constants, 563, 586, 607–608
Constants, with absolute cell reference, 9
Control array, 381, 431
Controls
combo box, 359, 362
command button, 217
common dialog boxes, 410
image, 427
label, 237
line, 429
list box, 288, 359
menu, 296
naming conventions, 224
option button, 307
picture box, 429
scroll bar, 278, 523
shape, 218
styles, 364
text box, 236, 241, 359
timer, 329
Control structures
Do ... Until, 314
Do ... While, 314
For ... Next, 313
If ... Then, 268
Select Case, 288
Conversion function, 302
Coordinate system
quantum mechanics, 46
graphics, 433, 443, 456
properties, 434
Copy Down command, 186
Crystal planes, 15
Crystal systems, 13

D

Data analysis, 20, 137, 149, 203
Data types, 243, 245
Dates and time, 333
Define Name dialog box, 11
Derivatives, 575–576
Design window, 215, 235, 448, 509
DesLandres table, 386, 501, 615
Diffraction, 12–17
Dim reserved word, 246
DisplayFunction option, 623
DisplayTogether function, 625, 627
Distribution functions
Boltzmann, 88–89
hydrogen atom 1*s* orbital, 590
speed, 112, 561
velocity, 115, 561–562
Do ... Until control statement, 314
Do ... While control statement, 314
Docking windows, 217
Double type, 243
Drawing, 426
coordinate system, 433, 443, 456
properties, 433
DrawStyle property, 438
DrawWidth property, 438

E

Eadie-Hofstee plot, 130
Edit Copy Cell method, 207
Edit Fill Down method, 53
Edit Fill Series method, 52, 207
Edit menu Copy Down command, 186
Editing lines,
Einstein function, 354–355
Elementary functions, 611
***emf*, 144**
Enabled property, 332
EndDoc method, 258, 457
Energy
electronic, 82
internal, 88, 354
molecular, 86
rotational, 66, 77, 98, 580, 586, 643
vibrational, 70
vibrational-rotational, 78
Entering data
combo box, 359
cells, 4
files, reading from, 404
numeric regions, 538, 602
text box, 236
text regions, 538, 601
Enthalpy, 24, 134, 143
Entropy, 134, 142, 514
Enzymes, 126
EOF( ) statement, 406
Equations, solving
algebraic equations, 588, 648
FindRoot function, 650
Matrix Method, 596
NSolve function, 650
simultaneous, 596–599, 651
Solve Block Method, 599

Solve keyword, 650
Equilibrium constant, 134
Error handling, 303–304, 337–338
Errors of measurement, 153–157
Events
- **Activate**, 526
- **Change**, 278
- **Click**, 217, 248, 427
- **DblClick**, 248, 450
- **KeyPress**, 327
- **Load**, 259, 266, 277, 428
- **MouseMove**, 452
- **Resize**, 444
- **Scroll,** 278
- **Timer( )**, 332

Events list, 231
Executable VB file, 232
Expand function, 574

F

Factor function, 574
Faraday, 143
Files
- opening, closing, 392–394
- reading from, 404–410
- writing to, 394–404
- Visual Basic types, 220

File type, 220, 426
FillColor method, 444
Fill-Handle method, 10, 50, 182
Filling ranges
- Excel Edit Fill Down method, 51
- Excel Edit Fill Series method, 50
- Excel Fill-Handle method, 10, 50
- Lotus Copy Down command, 186
- Lotus Fill command, 185
- Lotus fill handle, 182
- Quattro Pro Edit Copy Cell method, 207
- Quattro Pro Edit Fill Series method, 206

FillStyle property, 226
FilterIndex property, 425
Find function, 599
FindRoot function, 650
First law of thermodynamics, 23, 515
Fit function, 626
Flags property, 414
Float function, 599
Form, 248
- events, 248
- methods, 248
- properties, 248
- removing, 458
- window, 215

Format
- axes, 30
- cells, 9, 25
- chart area, 30
- data series, 30
- dates and time, 257
- function, 255
- graph 546, 557
- numbers, 255
- plot area, 30
- plot options, 621
- plotted lines, 39, 104, 439, 551, 627

Formatting characters, 256
Formulas, entering, l, 10
For ... Next control statement, 313
Frame, 629
Frame label, 629
Free energy, 134, 143
Freezing point diagram, 533
Function Arguments dialog box, 6
Functions, 5, 42, 176, 306, 541, 610
- built-in, 5, 174, 541, 610
- **Clear**, 609
- conversion, 302
- **DisplayFunction**, 623
- **DisplayTogether**, 625, 627
- elementary functions, 611
- evaluating and tabulating, 613
- Excel statistical, 10
- **Find**, 599
- **Fit**, 626
- **Float**, 599
- **Format**, 255
- **Intercept**, 551
- **Line**, 551
- **ListPlot**, 623
- Lotus, 184
- **MsgBox**, 311
- numerical functions, 612
- **Plot**, 622, 646–647
- plotting, 32, 557, 559
- regression, 551, 556
- **RGB**, 234
- **Rnd**, 271
- **Show**, 623, 624
- **Slope**, 551
- statistical, 10
- **Stderr**, 551
- **Table**, 613–627
- **TableForm**, 613–627
- **Timer**, 272
- user-defined, 339, 542–543, 612
- **VarType**, 302

Fundamental vibration frequencies (cm^{-1}), 359

G

Gas
- Lennard-Jones potential, 102, 592, 653, 657, 658
- speed distribution function, 109
- van der Waals, 105
- velocity distribution function, 109

Given keyword, 598
Grade sheet, 17
Graphics
- coordinate system, 434–437
- formats, 427
- methods, 437
- printing, 456
- properties, 434, 436
- scaling, 433
- units, 433

Graphics controls
- image box, 427
- line, 429, 438
- picture box, 429
- properties, 433
- shape, 426

Graphics events
- **MouseMove**, 452
- **Resize**, 444

Graphics methods, 437, 440–441
- **Line**, 440–441
- **Pset**, 445
- **Scale**, 441

Graphing
- data, 23, 177, 192, 549, 623
- functions, 32, 184, 205, 446, 559, 621
- regression line, 126, 131–132
- three-dimensional, 161, 164
- with AutoGraph, 490

GrayLevel, 654, 657

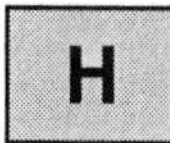

Hamiltonian operator, 46, 166
Heat capacity, 25, 85, 91, 354, 554, 587, 619
Heat of fusion, 529, 534
Heat of mixing, 534
Help browser, 605, 659
Histogram, 19–21
Hydrogen molecule ion, 59
Hydrogen sulfide equilibrium, 134

If ... Then statement, 268
Image box, 426, 427
Input mode, 392
Insert
- blank line, 254
- chart, 208
- Equation Editor, 566
- form, 476, 503
- function, 5, 176, 542
- graph, 552
- matrix, 554
- module, 477
- name,
- procedure, 478
- text region, 538

Integer type, 243
Integrals, 576–578
Intercept function, 138
Intercept(vx,vy) function, 551, 556
Internal energy, 24
Interval property, 331
Iteration, 312, 541

Keypress event, 327
Keywords, 572–575
Kinetics
- consecutive reactions, 122, 622
- enzyme, 126, 177–179

Label control, 241
Lennard-Jones potential, 102, 592–595
Linear regression, 626–627
- Excel, 126, 134
- Lotus, 177
- Mathcad, 550, 553
- Mathematica, 626
- Quattro Pro, 200–204
- comparison of Excel, Lotus, and trendline, 126, 130–132

Line control, 426
Line separators, 254
Line styles
- **DrawStyle**, 438
- editing, 39, 187, 199, 202, 209
- **PlotStyle**, 624
- trace line, 550

Line terminator, 609
LINEST function, 137
Lineweaver-Burk plot, 130
Line(vx, vy) function, 551
List box control, 288, 359, 363
ListCount property, 363

ListIndex property, 363
List(n) function, 363
ListPlot function, 623
Lists, 544–547, 614, 619
Lists, plotting with, 548–549
Literal subscript, 562
Load event, 259, 266, 277, 319, 427
Locked property, 363
Logical operators, 649
Loops, 313, 346, 535

M

Macros, 463
- dialog box, 466, 533
- enabling and disabling, 466–467
- global versus local, 470
- recording, 463–465, 485
- running Excel, 459, 486, 494, 504, 507, 512, 520, 532, 533
- running Word, 465, 469, 472, 477, 482, 483
- virus warning, 465

Make Project dialog box, 233
Math region, 539, 540
Matrix, 554
Matrix Method, 596–597
Max property, 280
Maximum and minimum, 589–596, 653–658
Mean, 10, 153–156, 546
Melting points, 528
Menu bar
- comparison for Excel, Lotus, and Quattro Pro, 191
- Excel, 3
- Lotus 1-2-3, 175
- Quattro Pro, 193

Menu control, 296
Method
- **AddItem**, 338, 360, 363
- **Box**, 440
- **Clear**, 288, 336, 366, 368, 374
- **Circle**, 441
- **Cls**, 356, 366, 368, 385
- **EndDoc**, 258, 457
- **Line**, 441
- **Print**, 249
- **Pset**, 445, 457
- **RemoveItem**, 363
- **Scale**, 441
- **Show**, 249

Michaelis-Menten equation, 126–128
Miller indices, 12–15
Min property, 281
Minima and maxima, 589–595, 653–658
Module, adding, 458
Moment of inertia, 66
Most probable *r* (H atom),
Most probable speed, 119, 590
Most probable velocity, 111
MouseMove event, 452
MsgBox constants, 273
MsgBox function, 311
MsgBox statement, 272, 304, 362
Multiline property, 363, 419

N

N function, 607
Name box, 4
Named date and time formats, 333
Name dialog box, 11
Name prefixes, 244
Name property, 213–218
Naming a range, 11
Naming controls, 224–225
Naming variables, 244–246
Nested loops, 318
Nonlinear regression, 149
Nonprinting ASCII character codes, 326
Normal error distribution, 153
Notebook, Mathematica, 604
NSolve function, 650
Number formats, 255–257
Numerical functions, 612

O

Object Browser, 219, 329, 475
Objects, 212, 489
On Error GoTo statement, 303–304, 337–338
Operator
- arithmetic, 293, 538–539, 606
- assignment, 247, 539, 572
- Boolean equality, 572
- computational equality, 572
- conditional, 264
- derivative, 576
- integral, 575
- Hamiltonian, 166
- logical, 649
- relational, 263, 649
- replacement, 648–649
- summation and product, 579
- symbolic, 572–575
- vectorize, 555

Option Explicit statement, 258
Options button control, 305

Options dialog box, 262
Orientation of 3-D Plots
ViewPoint, 641
View dialog box, 163
Output mode, 392

P

Packages, 617
Palettes, 603–604
Parametric surface plots, 569
Particle in a box, 166–170, 569, 640
Partition functions, 90–94, 99, 580–584, 643–644
pH, 280
Picture box control, 426, 429
Pie Chart, 643
Planck's radiation law, 40
Plot function, 622, 646
Plot options, 621
PlotLabel, 616
PlotPoints, 641
PlotStyle, 629
Plotting
functions, 32, 184, 205, 446, 559, 622, 630–635, 638–642
data, 23, 176, 192, 548
lists, 623
surface plots, 566
with AutoGraph, 490
PointSize, 629
Polynomial expressions, 645
Polynomial trendline, 150
Population of energy levels, 585, 644
Postulates of quantum mechanics, 165–166
Predefined shapes, 218
Prefixes for variable names, 244
Printer methods and properties, 258
Printing
numbers, dates, and time, 255–257, 333
separators, 254
to the form, 249
to the printer, 250
Probability distribution function
Boltzmann, 88–89
hydrogen atom 1*s* orbital, 590
speed, 112, 561
velocity, 116, 562–563
Procedure
structure, 217
list, 230
writing without a form, 492
Product, 579
Progression, 84
Project window, 214, 470
Properties
ActiveDocument, 473
Alignment, 241
BorderStyle, 227
Caption, 222
Cells, 497
CurrentX, 440
DrawStyle, 438
DrawWidth, 438
Enabled, 332
FillColor, 226, 444
FillStyle, 227, 444
FilterIndex, 425
Flags, 414
FullName, 473
Interval, 331
ListCount, 288
ListIndex, 288
of objects, 212, 223–226
Max, 280
Min, 280
Multiline, 237, 419
Name, 232
Picture, 426
PlotStyle, 623
ScaleBottom, 439
ScaleHeight, 439
ScaleMode, 434, 439
ScaleTop 439
ScaleWidth, 439
Selection, 473
setting, 224–225
Shape, 222
Sorted, 288, 361
Table, 224, 226
TabIndex, 242
Text, 240, 288
Visible, 357, 429
Properties of combo box, list box, and text box, 363
Properties window, 216
Pset method, 445, 457

Q

QBColor function, 438
Quantum mechanics
coordinate system, 46
DesLandres table, 82, 387, 615
electronic energies, 81–84
Hamiltonian operator, 166
hydrogen atom, 46, 54, 565
hydrogen molecule ion, 59–62
normalizing wave functions, 169
operator equations, 164

particle in a box, 166–170
Planck's radiation law, 40
postulates of, 165–166
rotational energies, 64–65, 643
Schrödinger wave equation, 46–48, 69, 167
units, 49, 61
vibrational energies, 69, 630
vibrational-rotational energies, 76, 348, 615
vibrational wave function, 70–72
Quantum statistics
Boltzmann distribution, 88–89
coordinate system, 46
molecular energies, 87–93
partition function and energy, 89
partition functions, 88–91, 580–585
population distribution, 100, 585, 643 - 644
rotational energy, 65, 76, 97, 579, 585, 643–644
vibrational energies, 69, 630
vibrational partition function, 581
vibrational-rotational energies, 76, 350, 614
vibrational temperature, 92, 94

R

Radial wave function, 45–59, 564, 633, 635
Randomize statement, 272
Range menu Fill command, 185
Range name, 11, 42, 498
Ranges, 495–496
Range variable, 541–542, 579
Reading text files, 404–410
Recording toolbar, 465
Record Macro dialog box, 466
Reduced mass, 65
Referencing objects, 489
Refractive indices, 148
Regression
dialog box, 137
functions, 140–41, 551
nonlinear, 148
other spreadsheets, 141
Relational Operators, 649
Relative cell references, 9, 10
RemoveItem method, 363
Resize event, 444
RGB Color function, 234
Rnd function, 271, 454
Rotational constants, 65
Rotational energy, 65, 78, 97, 580, 585, 644
Rotational partition function, 582, 644
Rotational spectrum, 64,
RowIndex property, 500
Running a Visual Basic program, 232

S

Scale method, 433, 441
ScaleMode property, 434
Schrödinger wave equation, 46–48, 69, 167
Scope of variables, 247
Scroll bar, 279, 522
Second law of thermodynamics, 515
Select Case statement, 288
Selection, 489
Separators, line, 254, 609
Sequences, 84
Set focus, 239 (see also Tab index)
Shape control, 218, 233, 426
Show function, 623, 624
Show method, 249
Simplify function, 573
Simultaneous equations, 596, 651
Single type, 243
Slope function, 138
Slope(vx, vy) function, 552, 557
***Solve* keyword, 573, 588, 590, 592, 595, 599, 652**
Solving equations
algebraic equations, 588
FindRoot function, 650–651
Matrix Method, 603
NSolve function, 650
simultaneous, 597–600, 651
Solve Block Method, 600
Sol*ve* keyword, 573, 650
Sorting, 18, 363
Spatial characteristics of angular wave functions, 55
Spectra
electronic, 81
rotational, 65, 76
vibrational, 69
vibrational-rotational, 77
Speed distribution function, 109, 560
Spherical harmonics, 637
Standard deviation, 154–156
Statistical functions, 10
Stderr (vx, vy) function, 551
String type, 243
Structure of a procedure, 218
Student grade sheet, 17

Student's *t*-distribution, 157
Subscript, literal, 561
Subscripted variables, 544, 551
Substitute, 575
Summation, 579
Surface graphics, 568, 636
Symbolic calculations
- **Assume** keyword, 578, 593
- derivative, 576,647
- **Expand** keyword, 573, 645
- **Factor** keyword, 573, 645
- **Integrate** keyword, 648
- **Simplify** keyword, 573, 645
- **Solve** function, 573, 650
- **Substitute** keyword, 573
- Substitution (Replacement operator), 645

Symbolic operators, 572, 645

T

***t*-values, 158**
Tab index, 239 (see also Set focus)
TableForm function, 613–627
Table function, 613–627
Text box control, 236, 241, 359, 363
Text region, 538
Thermodynamics
- chemical equilibrium, 32, 134, 279
- electrochemical cells, 143–144, 146–147
- energy, enthalpy, and heat capacity, 23, 93, 556
- First law, 23, 515
- Second law, 515
- Third Law, 515
- entropy, 134, 515–516
- phase equilibrium, 534–535
- work, 23

Thickness, 635, 657
Third law of thermodynamics, 515
Three-dimensional plots, 161, 170, 567, 638–642
Tiner( ) event, 332
Timer function, 272
Timer control, 329
Toolbars
- Excel, 3–4
- Mathcad, 537
- Mathematica (palettes), 602–603
- Quattro Pro, 191–192
- Lotus, 175–176
- Visual Basic, 214–217

Toolbox, 216
Tools menu, Data Analyis command, Regression dialog box, 137, 149
Traces tab, 549–550
Translational partition function, 601
Trapezoid integration, 370
Trace tab, 549
Trendline, 126, 138, 150
Two-dimensional plots, 23, 177, 192, 548
Twip, 434
Types of data, 243
Type prefixes, 246

U

Units, graphics, 435
User-defined functions, 339, 542
User, of a VBA program, 212

V

Van der Waals, 105
Vapor pressure, 32, 36, 548, 624–625
Variables
- declaring, 246
- names, 244
- **Option Explicit** statement, 258
- scope, 247
- types, 244

VarType function, 302
VB Programs, list
- Arithmetic, 294
- Array of Pressures, 347
- ASCII Print, 323
- Boolean Operator, 265
- Bubble Sort, 383
- CD Color, 414
- CD Fonts, 419
- CD Open File, 421
- CD Printer, 417
- Circle to Printer, 456
- ColorBox, 284
- Color Properties, 220
- Combo Box to Array, 363
- Coordinate Demo, 436
- DesLandres, 386, 402
- Functions, 302
- Graph Circle, 442
- Graph Vib, 450
- Heat Capacity, 354
- HKL, 320
- IfThenElse, 270
- Image, 427
- IR Spectrum, 350
- Justify, 237
- KeyPress, 327
- Line and Box, 439
- List Box, 289
- Loops, 315

Nested Loops, 318
NoForm, 458
OptExp1, OptExp2, OptImp1, and OptImp2, 258
pH, 279
Picture Box, 429
Print Demo, 250
PSet to Printer, 457
Random Guess, 271
Random Points, 454
Read a File of Names, 404
Read X_iY_i File, 407
Resize the Form, 445
Rotational Energy, 334
r2R2, 446
Show the Boxes, 412
Sketch, 452
Text Box to List Box, 359
Timer A to Z, 330
TotoLoto, 380
Trapezoid Integration, 370
2-D Array, 376
$2p_x$ Orbital, 448
Vapor Pressure, 340
Write to File, 395
Write to File 2, 397
Write to File 3, 398
Write to a File from a Table, 402
VBA Procedures, list
ABCScroll, 521
ABCUser, 516
Arial10B, 463
AutoGraph, 490
CheckNames, 492
ChessBoard, 498
ColorBlock, 496
CountStuff, 476
DateName, 479
DesLandres Table, 501
Eutectic, 528
First Form, 503
NumberStats, 507
SilverEntropy , 510
SwimHeader, 486
Table3_5, 471
Times12, 467
Vectorize, 555
Vectors, 553–556
Velocity distribution function, 109, 560–562
Vibrational energies, 69, 630
Vibrational partition function, 582
Vibrational-rotational energies, 76, 350, 614
Vibrational temperature, 93, 95
Vibrational wave function, 69–71
ViewPoint, 640
View dialog box, 163
Visible property, 357, 382

Wave equation, 46
Wave function
angular, 54
radial, 48
vibrational, 69
Window
code, 215
design, 214, 235
docking, 216
form, 214
project, 214, 524
properties, 215, 223
With ... End With statement, 288, 492
Workspace layout
Excel, 2, 4
Lotus 1-2-3, 175
Mathcad, 537
Mathematica, 601, 603
Quattro Pro, 192

X-ray diffraction, 618
XY (scatter) chart,
Excel, 26, 27, 34, 38, 43
Lotus, 176-181
Mathcad 548-549,
Mathematica, 623-624
Quattro Pro192-196